Foreword

Microbial Cultures are Still Essential in the Era of High-Throughput Sequencing

We live in an era when it is possible to sequence any sample, taken from any environment, and to know which organisms are present. It is an extraordinary achievement. In little over 40 years, from the very innovative scientific advances initiated by Frederick Sanger and others, and with the equally important development of technology, we have a present-day capability to generate metagenomes and metatranscriptome libraries, and to provide that information to every microbiologist. Why then, should microbiologists still be interested in isolating microorganisms from the natural environment, manipulating samples to form axenic cultures, and then working with those cultures in the very artificial environment of the laboratory? Surely, all the information that anyone could ever require can be gleaned from sequencing DNA and RNA. Are microbiologists too wedded to a culturing approach that, although very successful in the past, may now have outlived its usefulness? My view is that this is not the case and, more than ever, we need to bring important microbes into culture. But there is a caveat—we have to be selective and not attempt to culture everything.

In a sense, high-throughput sequencing has introduced new challenges for microbiologists. When marine scientists first turned their attention to the microbiology of the oceans, it seemed a relatively tractable problem. Microbiologists, such as Claude ZoBell, using approaches based on classical microbiology methods, estimated bacterial number from the colonies that developed on a nutrient agar plate. A core assumption of microbiologists is that a single bacterial cell will grow to form a single colony; early marine microbiologists had every expectation that this would be a very quantitative approach to estimate how many bacteria were in the sea. They found that there were only a few 100 bacteria per mL of seawater—so suggesting that bacteria were not at all a significant component of the pelagic ecosystem and were apparently much less abundant than phytoplankton. But in the 1970s with the introduction of epifluorescence microscopy, microbiologists discovered that culturing techniques were underestimating the number of bacteria in the euphotic zone by 4 or 5 orders of magnitude; i.e., there were not just a few hundred bacteria present, but millions of bacteria per mL. Most marine bacteria were not growing in the culture media and the numbers were grossly underestimated.

High-throughput sequencing techniques now allow us to describe that massive diversity. We can do many things with these data: infer the biogeochemical function of the bacterial assemblage, track changes in species richness with time, and describe novel sequences that are not presently in databases. There are the limits to what sequence data can reveal; but many of these challenges can easily be addressed if there is access to a laboratory culture. Unfortunately, many families of bacteria are known only from sequence data generated from natural assemblages and we do not have any cultured representatives of many of the phyla that are widely distributed and abundant in the ocean. We may know quite a lot about bacterial assemblages from sequence data, but we know very little of the basic biology of these organisms; phenotype cannot be adequately determined from sequence data. Access to laboratory cultures would be immensely beneficial in answering important questions about the role of microbes in the sea.

One very contentious issue is what constitutes a bacterial species. It is very common to approach the problem of phylogeny by using the 16S rRNA gene (or another genetic locus) to define an operational taxonomic unit (OTU). Although a pragmatic solution to the problem could describe what may be present in a natural population, 16S sequence does not encapsulate species information. It does not answer the basic question of what is this organism; what constitutes this organism; and how has evolution resulted in this particular biological entity that is the center of our interest? That entity is the result of many processes; mutation of the nuclear genome, acquisition and mutation of plasmids, what has been acquired among others by horizontal gene transfer, pathogenicity and/or genomic islands.

One vivid example of the difficulties faced in describing a bacterial species comes from medical microbiology and a study of pathogenic *Escherichia coli* using comparative genomic analysis (Rasko et al. (2008). *E. coli* is probably the best-studied bacterium ever and is the default organism for biochemical and genetic studies. Probably all microbiologists would recognize *E. coli* as a single species. As such, it might be expected that it could be defined precisely on the basis of genomic similarity. Rasko et al. compared the genomes of a total of 17 cultures, including a number of pathogenic strains, and found a great diversity. The mean genome size of the 17 isolates was 5,020 ± 446 genes, and of that total less than half the genes (2, 344 ± 43) could be considered as to comprise a "conserved core"—genes that were highly conserved in all 17 isolates. A significant number of genes (ca. 300) were unique to one isolate, i.e., were present in one, but none of the other 17 genomes. Given such great genetic diversity in our best-studied bacterial species, what is the likelihood that marine species can be described purely on the basis of metagenomic data? At present, it is probable that laboratory cultures are the only sure way to completely describe a bacterium. It is true that genomic analyses of single cells, as well as the technology of manipulating single cells from the natural environment, are making significant progress; but, for those whose research interests are phylogeny, evolution, and identification, the most cost-effective approach remains isolation and laboratory culture.

Cultures are also essential for commercially important research. Terrestrial microorganisms, particularly those derived from soils, have been a source of many secondary metabolites, enormously profitable for pharmaceutical companies, and responsible for huge advances in human health. Microbes from the oceans have been less important to date, not because marine microbes offer fewer possibilities for biodiscovery, but because they have not received the same attention. The case for focusing on marine microbes is strong; the seas offer a wide range of unusual environments in which microbes have evolved. Unusual environments have unusual microbes that do unusual things—many of which are likely to offer real opportunities for biodiscovery of commercially important products. The pathway into these new fields is through isolation and culturing of these novel microbes.

Clearly, with tens of thousands of different species of microorganisms, bacteria and archaea—as well as protists—in seas, we cannot isolate everything. The challenge is to prioritize, beginning with microbial families that are known to be abundant in the ocean but for which we have no cultured representative. Sequence data is clearly an essential tool in that prioritization process. Traditional microbiological approaches, such as dilution to extinction, will continue to be cost-effective but new culture media need to be devised. My personal belief is that we should move away from traditional media that are based on yeast extract and peptone, towards media that better mimic the composition of organic matter in the sea. Progress may depend on the marine chemists developing better ways to quantify and describe those organic compounds that are likely to be the major substrates used by natural assemblages (but which will be present at vanishing low concentrations). Evolving technologies, such as the manipulation of single cell, will clearly be an important aid to successful culturing.

There is no doubt that this is an exciting time for marine microbiology. DNA sequencing has shown which organisms are present in the seas—and the diversity is huge. The challenge now is to ascribe function to this diversity—to know who does what. Isolation and culturing will continue to be very important tools in meeting these challenges—and this book is an excellent description of the real progress that is being made.

Ian Joint
The Marine Biological Association
The Laboratory, Citadel Hill
Plymouth, UK

Reference

Rasko DA, Rosovitz MJ, Myers GSA et al. (2008). The pangenome structure of *Escherichia coli*: comparative genomic analysis of *E. coli* commensal and pathogenic isolates. J Bacteriol 190:6881–6893, doi:10.1128/JB.00619-08

Preface

When we accepted the invitation by Springer to edit a book on "marine microbiology," the next thing was to think about the contents of such book and about possible authors. There have been a fair number of books published on the topic and we did not want to duplicate any of them. The lucky coincidence was that we were leading a large European consortium "MaCuMBA," which stands for "Marine Microorganisms: Cultivation Methods for Improving their Biotechnological Applications," a 4-year one (2012–2016), with 22 partners from 11 European countries. MaCuMBA is a consortium of industrial and academic partners with great variety in expertise in marine microbiology. Hence, it seemed obvious to ask colleagues and principal investigators of the MaCuMBA consortium to contribute to this book, for which we gave the title: The marine microbiome—an untold resource of biodiversity and biotechnological potential.

The term "microbiome" is fashionable and is used to describe the whole microbial community, i.e., all microorganisms and their genetic information in a certain habitat or environment. The marine microbiome refers to the totality of microorganisms living in the ocean, its fringing seas, estuaries, and bays and fjords. This includes the intertidal areas of the coast but also the seafloor and the sub-seafloor, thousands of meters down in the bottom of the sea. It also includes the microorganisms living on and in marine animals, plants (seagrasses), and macroalgae, even though each of them forms its own microbiome.

Microorganisms are basically defined by their size and as a rule of thumb we consider any organism a microorganism when its size is too small to be observed in detail by the naked eye. This would mean everything smaller than 1 mm. In practice, most organisms that we consider as microorganisms are in the micrometer (μm) (one-thousandth of a millimeter) range, the smallest may be only 0.2 μm, but the biggest can be several hundreds of μm. Microorganisms comprise all three domains of life: Bacteria, Archaea, and Eukarya. While the former two domains are traditionally considered microorganisms (bacteria), all macroorganisms (plants, animals, macroalgae, and many fungi) are Eukarya. However, what is often forgotten that by far most Eukarya are in fact microorganisms (protists).

Bacteria and Archaea are often referred to as "Prokaryotes," (organisms without a nucleus "karyon" in their cells) to distinguish "bacteria" from the eukaryotes that do have a nucleus. We agree with Norman Pace ('Time for a change' Nature 441: 289, 2006, and 'It's time to retire the prokaryote'. Microbiology Today, May 2009, 85–87) who argues that a prokaryote is defined by what is does not have (a nucleus) and that this is not a good criterion. In this book we decided to avoid this term, even though some authors were not fully convinced. However, as with the knowledge we have today, there is no doubt that Bacteria and Archaea comprise very different domains of life, even though you cannot tell much from the morphology observed under the microscope. The morphology of many microorganisms is anyway devoid of much resolution. The diversity of microorganisms is in their genome and in what they do. Using "prokaryote" is often sloppy, because when one starts asking, often only one of the domains of Bacteria or Archaea is meant. It is therefore more accurate to name that domain. And in the more rare cases that both domains are meant, it is not a big deal to name both. That is what we consequently did in this book.

There is another biological entity that we have not named yet. Viruses are not representing exactly a domain of life (i.e., not belonging to any of the three domains of life), but they are surely a biological entity that needs to be considered when talking about any microbiome. Viruses are not "living" simply by the fact that they need a living cell to replicate. But viruses play an extremely important role in maintaining the biodiversity, maintaining the microbial foodweb, and transferring genetic information between organisms, perhaps even between domains. Recent discoveries also show that the border between bacteria and virus is vanishing. The number of viruses in the ocean is overwhelming with an order of magnitude larger than that of all microorganisms (10 and 1 million per milliliter of seawater, respectively).

It is exactly 70 years ago when Claude E. ZoBell's book "Marine Microbiology" appeared (C.E. ZoBell, Marine Microbiology. A monograph on hydrobacteriology, Waltham, Massachusetts, USA, 1946, 240 p.). ZoBell was at that time Associate Professor of Marine Microbiology at Scripps Institute of Oceanography, La Jolla, California. His monograph is still worth reading and presents us with a remarkably complete picture of marine microbial life and you could ask yourself how much more we know today. We hope nevertheless that the present edited volume does give a taste of what has been achieved in those 70 years and where we stand now.

The marine microbiome is not just interesting from a scientific point of view. Certainly, with 70 % of the Earth's surface covered by the ocean and the ocean probably being the largest continuous habitat, the marine microbiome plays a prominent role in the biogeochemical cycling of elements, is at the basis of the marine foodweb, critical for the ecology of the sea, and essential for climate regulation and counteracting the effects of global change. However, science has also made great discoveries of bioactive compounds made by marine microorganisms that have found applications in biotechnology, bioenergy, and pharmacy, and activities that are displayed by these organisms that find application in bioremediation. Being aware that we might not know most of the microbial diversity that is out there, marine microbiology holds a great promise for many more of such discoveries. Therefore, with this book we also wanted to give a podium for the application of marine microorganisms, for the legal issues using microbial resources, and to explain the methods for dissemination to a wider public.

We divided the book into three main sections. The first section contains chapters that describe the diversity and ecology of marine microorganisms. The second section contains the chapters about marine habitats, their inhabitants, and biogeochemical cycles and the third is about the marine resources (the hidden treasures).

We would like thank all the authors of the chapters in this book for their excellent work and for making this book what it is.

Finally, we would like to thank project coordinator of Springer, Dr. Andrea Schlitzberger, for giving us the opportunity to edit this book and for her help and patience.

March 2016

Lucas J. Stal
Mariana Silvia Cretoiu

Contents

Part I
Diversity and Ecology of Marine Microorganisms

Chapter 1
What is so Special About Marine Microorganisms? Introduction to the Marine Microbiome—From Diversity to Biotechnological Potential

Henk Bolhuis and Mariana Silvia Cretoiu

Abstract Marine microscopic life varies from single-celled organisms, simple multicellular, to symbiotic microorganisms encompassing all three domains of life: Bacteria, Archaea and Eukarya as well as biologically active entities such as viruses and viroids. Together they form the Ocean's "microbiome". Over billions of years of evolution this microbiome developed a plethora of adaptations and lifestyles and participates in the fluxes of virtually all chemical elements. The importance of the marine microbiome for human society and for the functioning of our living planet is not disputed. In this introductory chapter we bring to your attention some of the most important features of the marine microbiome and try to answer the question what distinguishes it from other microbial systems. Our main goal is to urge the reader to find more information about the taxonomic and functional diversity by exploring the specific chapters.

1.1 Introduction

Our blue planet is a water-dominated habitat with more than 70 % of its surface covered by the ocean and seas. With an estimated volume of about 1.332×10^9 km^3 and an average depth of 3688 m the ocean is the biggest continuous environment with a near unlimited diversity of ecological niches (Charette and Smith 2010). The enormous diversity of niches and extant microbial inhabitants makes a complete description of "The" marine microbiome an impossible task within the limits of this book chapter and therefore a certain focus is necessary. We provide here a basic description of the marine environment, the importance of microorganisms in the cycling of elements, the marine microbial diversity, and their

H. Bolhuis (✉) · M.S. Cretoiu
Department of Marine Microbiology and Biogeochemistry,
NIOZ Royal Netherlands Institute for Sea Research and Utrecht University,
PO Box 59, 1790 AB Den Burg, Texel, The Netherlands
e-mail: henk.bolhuis@nioz.nl

L.J. Stal and M.S. Cretoiu (eds.), *The Marine Microbiome*,
DOI 10.1007/978-3-319-33000-6_1

potential application for biotechnological purposes. In addition to some historical considerations we present some of the latest findings on the "microbial" inhabitants of the marine realm and if applicable refer to the specific topics covered by the dedicated chapters in this book.

The marine ecosystem includes the open waters of the ocean and of the seas, the estuaries and other tidal regions, the seafloor and the sub-seafloor, the polar sea ice masses, and brines. Five major marine water basins are identified and used in geological and hydrological classifications with no clear separation and boundaries: Atlantic Ocean, Pacific Ocean, Indian Ocean, Arctic Ocean, and Southern (Antarctic) Ocean. The Pacific is the deepest and largest ocean, covering more (28 %) than the total terrestrial area on Earth (Amante and Eakins 2009). The ocean is a continuous habitat and rather than distinguishing five oceans we consider just one world ocean. The marginal seas such as the Mediterranean Sea, Baltic Sea, South China Sea, and many others border this ocean and are more or less connected to it. Marginal seas influence the oceanic climate and water circulation through their input of freshwaters, heat, and nutrients.

Almost half of the ocean floor is continental shelf with a central basin interrupted by three submarine ridges. The continental shelf extends offshore for up to 100 km and to a depth of 200 m and descends further to the deep continental slope. The abyssal plain is a mostly flat surface interrupted by undersea mountains, volcanoes, ridges and trenches that covers the ocean floor (Thurman and Burton 2001). Many subdivisions have been made in names and country ownership in order to distinguish the limits of seas and gulfs that have economical, historical and ecological importance.

1.2 The Marine Ecosystem

1.2.1 Gyres, Tides and the Global Ocean Conveyor Belt

The marine and oceanic surfaces are not equally distributed with respect to the Equator. While the northern hemisphere is mostly covered by land, about 80 % of the southern hemisphere is ocean. This asymmetry in land–water and the solar radiation that is received by water influence the atmospheric circulation and global temperature. Wind belts created by differential heating of air masses induce the ocean surface currents. Moreover, the planet rotation deflects moving water (Coriolis Effect) and produces the boundary currents between water masses. Due to differences in deflection with the Earth velocity and therefore with the position on the Earth relative to its axe, the Coriolis effect is maximum at poles (closer to Earth axe) and minimum (considered null) along the Equator (farther from the Earth axe). Circular gyres moving in opposite direction (clockwise–anticlockwise in the northern and southern hemispheres, respectively) are generated by the combined physical effect of planet spinning motion and horizontal and vertical friction of air masses (the wind curl torque). The gyres mix water bodies with different temperature and influence the distribution of nutrients and organisms in the water column.

The water moves also due to the tidal currents generated by the interaction of several factors that form a system named "tidal constituents". Tidal constituents are the Earth's rotation, the position of the Sun and Moon relative to the Earth, the Moon elevation above the Earth's equator and the underwater depth (bathymetry). Water tides are not limited to the ocean and they can appear in estuaries and in the lower course of rivers. Tides affect also the solid Earth but the moving of the land is at lower scale (only few centimeters) compared to the moving of water bodies. Tides exhibit different variation of their period from half of day, days, months and years. The inertial forces that cause the Coriolis Effect together with all the gravitational forces of the tidal constituents conduct the formation of water waves with a moving pattern around a point with almost no vertical movement named "amphidromic point" (point of zero amplitude). Oceanic tides are influenced by the shape of coastline and near-shore depth and as well influence the geography of these zones (Cartwright 2000).

The oceanic waters move continuously due to the different currents generated by the combination of the gyres, winds and restrictions of the lateral movements of water cause by shorelines and shallow bottoms. In the open ocean and along coastlines the currents can flow up from beneath the surface and replace the water blown by the wind in a process known as "upwelling". The reverse process, "downwelling", occurs when the wind makes the coastal waters to build up as a mass along coastline and eventually to sink toward the bottom. Five types are described in oceanography: coastal, large-scale wind-driven, associated with physical obstacles or differences in pressure (eddies), topographically induced and broad-diffusive limited to some specific zones (such as Australian coasts). Geographically, five major coastal currents are associated with upwelling zones: Northwest Africa (Canary Current), Southern Africa (Benguela Current), California and Oregon (California Current), Peru and Chile (Humboldt Current), and Somalia and Oman (Somali Current). In the equatorial zone, the upwelling is associated with the intertropical convergence zone of winds with variation in weather patterns from stagnant to violent thunderstorms and leads to a remarkably enrichment with nutrients of the surface waters. Solid evidence was the detection from space of the equatorial region of Pacific Ocean as a broad line of high chlorophyll concentration indicating the presence of large phytoplankton populations.

The effect of mixing and wind-generated currents decreases below a depth of $\sim$200 m. Undersea "rivers" and currents that form a circulation system known as the "global ocean conveyor belt" or "thermohaline circulation" maintain the mixing of the deep ocean water. The whole circulation system is driven by density differences of the seawater, a feature dependent on small differences in temperature and salinity. The "global ocean conveyor belt" starts in the North Atlantic Ocean where the cold water becomes denser due to ice formation, sinks, and is replaced with warmer surface water. This process initiates a continuous mix of waters from the North Atlantic Ocean to the Indian Ocean and the Pacific Ocean. The conveyor belt moves at a slower speed than currents that are driven by wind or tides. Estimations reveal that 1 m^3 of seawater needs about 1000 years to complete its journey along the global ocean conveyor belt (Ross 1995). Moreover, the global conveyor belt and the upwelling and downwelling processes represent a vital component of the ocean sea life support system. The whole food chain depends on circulation of nutrients, organic matter and oxygen between surface waters and bottom sediments. The conveyor belt and upwelling maintain the ocean/marine productivity by flowing up to the surface from beneath the vital nutrients and organic matter. This enrichment of surface waters is coupled with the transfer of oxygen to the bottom sediments through downwelling. In this way the benthic deep-sea life can continue to exist not only as anaerobic microorganisms but also in its aerobic forms.

1.2.2 Oceanic Climate and Zones

Based on the temperature of surface waters, four bio-geographical zones are distinguished: tropical, warm-temperate (subtropical), cold-temperate and polar (cold). In the tropical ocean and seas the surface waters can reach 25–30 °C causing a thermocline at ~100 m, below which the temperature decreases to 10 °C or less. In the absence of wind little mixing of water layers occurs. Further away from the tropics the surface waters temperature drops to 20–15 °C in subtropical and cold-temperate zones. This is because of the continuous input of the cold water from polar regions through the global conveyor belt. Temperate waters have the largest seasonal variation in the thermocline depth, with extensive mixing of waters due to strong winds. In polar regions the water is permanently cold and the sea surface temperature varies from 3 to 0.5 °C. Ice is forming when the water temperature drops below −1.8 °C (salt lowers the freezing temperature) and the freezing point decreases with the increase of salt concentration ranging from −2.0 to −2.6 °C. In some regions of the Arctic Ocean the freezing process favors the accumulation of extreme cold (−1.56 to −1.98 °C) saline (34.04–35.29) waters in brines. Both poles of the Earth are covered by multilayered ice caps with different thickness, hardness and color based on salinity, temperature and age. Packed in a compact form in mid-Pleistocene epoch (0.85 Ma) the polar ice is continuously exposed to climatic changes. All physical processes that occur in the ocean are dependent on the chemical composition of the seawater and on solar radiation. These processes are locally influenced by various factors such as season, geographic location, cloud cover, polar caps or atmospheric dust (Brown and Mote 2009; Drinkwater 2006; Dyurgerov and Meier 2005).

The gigantic volume of water that is contained in the ocean has been produced during early Earth in the Hadean (ca. 4.5 Ga b.p.). Comparison of fossilized sea life and modern living organisms shows similarities between them that support the hypothesis according to which the composition of seawater ceased to change drastically at the end of the Proterozoic (ca. 600 Ma b.p.) (Hoffman et al. 1998; McCarthy and Rubisge 2005).

Seawater differs from river and lake water by its salt concentration (called salinity), which is on average 34–37 (salinity is often expressed as ‰ (weight of the total dissolved salts (g) per weight of seawater (kg), and therefore dimensionless). Salinity may be slightly lower or higher due to local freshwater discharge or evaporation. Brackish water bodies have considerably lower salinities. The Baltic Sea has for instance a salinity gradient from almost full salinity seawater where it connects to the North Sea and almost freshwater in the northern reaches. The whole Black Sea has a salinity of approximately half seawater concentration.

Seawater is a complex mixture of more than 80 elements, inorganic compounds, gases, and dissolved organic substances (Table 1.1). Edmond Halley initially addressed the origin of sea salt in 1715 and proposed that salts and other minerals were carried from the landmass by rivers and after rainfall a process he termed "continental weathering". More recent analyses show an important role for sodium leached out of the ocean floor when the ocean was formed and outgassing of chloride from Earth's interior via volcanoes and hydrothermal vents (Pinet 1996). It is conceived that the salt composition of the ocean has been stable for billions of years (Pinet 1996). An extensive list of constituents can be found in the pioneering work of Turekian (1968). The coastal areas are more prone to mixing with terrestrial freshwater discharge and from melting glaciers. Nearly enclosed seas, such as the Baltic Sea and the Black Sea, became brackish because the discharge of freshwater exceeded the mixing with full salinity seawater (Feistel et al. 2010). In the Red Sea, where evaporation exceeds water input, the salinity can increase up to 41. The high salt concentration of seawater is the main difference with freshwater and terrestrial environments and thus is probably the most important evolutionary driver of the marine microbiome. Moreover, the ocean is a continuous moving environment with vertical gradients of light and nutrients. Particularly, light plays an important role in the life of oceanic microorganisms because it drives photosynthesis and primary production that is the basis of the food web. Light is scattered or absorbed by suspended particles and organisms and is available only in the upper part of the water column of the ocean (150–200 m), which is termed the euphotic zone. In coastal areas were the turbidity due to suspended matter is high and phytoplankton blooms may occur, the euphotic zone may be as shallow as only a few meters.

Table 1.1 Seawater chemistry: contribution of the most important chemical elements (after Turekian 1968)

Element	g/l
Chloride	19.4
Sodium	10.8
Magnesium	1.29
Trace elements	~ 1.000
Sulfur	0.904
Calcium	0.411
Potassium	0.392
Bromine	0.067

1.3 Diversity in Life Style—Elements and Cycles

1.3.1 Ocean's Stoichiometry

The marine microbiome is involved in more than half of the Earth primary production and global nutrient cycling (Arrigo 2005). Historically, much attention has been paid to the study of the nutrient stoichiometry in water bodies and the ocean as well as in different life styles (free, attached, or symbiotic) (Arrigo et al. 1999; Arrigo 2005; Fanning 1989; Pahlow and Riebesell 2000). Starting with the year 1934 (Redfield 1934) the role of phytoplankton in the ocean life became under the permanent observation.

Based on measurements of the elemental composition of plankton and the concentration of dissolved NO_3^- and PO_4^- in seawater, Redfield concluded that the plankton is "uniform in a statistical sense" regarding its elemental composition and that the inorganic C, N and P variation in seawater is the "result of the synthesis or decomposition of organic matter" (Redfield 1958). The carbon–nitrate–phosphate (C:N:P) ratio of 106:16:1 of marine phytoplankton was considered to be constant and became known as the "Redfield ratio" (Redfield et al. 1963). Over time, variations from these values were reported by numerous studies of marine phytoplankton directly in the seawater and in laboratory experiments. A long-term discussion was generated around the question whether N and P can exhibit a simultaneous influence on phytoplankton growth. Also the role of ocean nutrient chemistry in controlling the physiology and the chemical composition of the phytoplankton was intensively and extensively discussed (Goldman et al. 1979, Sardans et al. 2012). The origins of the Redfield ratio were explored at cellular level and several discussed models (Falkowski 2000; Loladze and Elser 2011) concluded that roots of this stoichiometry were in the protein–rRNA ratio that every cell needs to maintain in order to maintain its homeostasis.

It was generally agreed that in the ocean all biological processes complement the geochemical ones and form together the foundation of Redfield ratios. Therefore the ratio is not a universal biochemical optimum but an average of ratios and community-specific due to the variation in C, N, and P of different organisms. The ratio has been extended to other elements (sulfur, potassium, iron) and helped in developing important ideas with regard to microbial nitrogen fixation, nitrification and denitrification and in exogenous nutrient changes in the ocean (Moore et al. 2013).

In phytoplankton ecology, three different growth strategies have been proposed based on N:P ratio measured in the seawater: the "generalist" (N:P ratio near the Redfield ratio; 16), the "bloomer" (low N:P ratio, <10) and the "survivalist" (high N:P ratio, >30) (Arrigo et al. 1999; Arrigo 2005; Klausmeier et al. 2004; Pahlow and Riebesell 2000; Sarmiento et al. 1998; Smetacek 2001).

The N:P ratio (basically the ratio of nitrate and phosphate in the water column) is often used to judge whether the system is nitrogen or phosphorus limited. A ratio above Redfield (16) is then considered to indicate phosphorus limited and below indicates a nitrogen limitation. In the latter case nitrogen-fixing cyanobacteria would have a selective advantage. The fixation of N_2 would increase the N:P ratio, both by increasing the fixed nitrogen as well as drawing down the phosphate. While, when nitrogen (N_2) fixation is not possible, the activity of non-diazotrophic phytoplankton might result in the opposite. It has been proposed that phosphate would be the ultimate limiting nutrient in the ocean (Tyrrel 1999). However, the nitrogen cycle is complex. Denitrification may lead to a depletion of bound nitrogen and whether or not nitrogen-fixing organisms proliferate and N_2 fixation occurs depends on several other factors (Stal and Zehr 2008). In large parts of the ocean the estimated losses of nitrate as the result of denitrification are consistently higher than the input of fixed nitrogen by N_2 fixation (Middelburg et al. 1996).

Current oceanographic research makes use of the Redfield ratio for estimation of carbon and nutrient fluxes in global circulation models. The ratio helps in determining nutrient limitations and to understand the formation of phytoplankton blooms. Gruber and Sarmiento (1997) introduced a tracer N^* that was used to derive whether different ocean basins were dominated by losses of nitrogen by denitrification or experienced net input from N_2 fixation. N^* basically describes the deviation of nitrate and phosphate concentrations in the seawater from the Redfield ratio. A low (negative) N^* indicates a dominance by denitrification and a high (positive) value indicates N_2 fixation. The advantage of this approach is that it makes use of extensive databases of nitrate and phosphate concentrations that are routinely measured during any oceanic research cruise.

1.3.2 Carbon

In the biosphere, the ocean is the largest reservoir of carbon, containing about 4×10^{13} tons of dissolved inorganic carbon (Post et al. 1990). CO_2 is highly soluble in seawater and reacts with water to carbonic acid (CO_3^{2-}), which dissociates rapidly to bicarbonate (HCO_3^-). Although the absorption of CO_2 at the interface between ocean and atmosphere and the transfer to deeper layers is mainly due to physical processes such as temperature gradients, surface winds and Earth rotation as well as the biological pump are major processes involved in the redistribution of the carbon in the ocean (Heinze et al. 2015; Ridgwell 2011). The development of the "microbial loop" concept (Azam et al. 1983) helped in better understanding the role of microbes in nutrient cycling and offered new perspectives in assessing species interactions in marine ecosystems. The marine carbon cycle is essential for interconversions of inorganic and organic carbon forms (primary production) that support life on Earth and depends on photosynthesis and carbon oxidation rates.

Photosynthesis in the euphotic layer drives primary production in the ocean and this forms the basis of the oceanic food web. Primary production depends on light and nutrients and their availability is different in different regions of the ocean and in coastal regions. Important contributors to primary production are the picocyanobacteria *Synechococcus* and *Prochlorococcus*, which are responsible for at least 20 % of the global carbon fixation (Li 1994). These organisms are common in the euphotic zones up to depths of 200 m but studies of the South China Sea (North Pacific) indicate that some *Prochlorococcus* populations are adapted to the deep ocean zone (Jiao et al. 2014; Zhang et al. 2014). To what extend these populations have an impact on ocean carbon cycle and carbon sequestration in the deep sediments remains to be investigated.

Unicellular eukaryotic organisms (protists) are also contributors to the carbon cycle by performing photosynthesis and exhibiting life styles that involve carbon processing (e.g., predation, parasitism, and symbioses). Different ecosystems biology models investigate how biotic and abiotic factors influence the interaction between protists and bacteria (Worden et al. 2015). The role of viruses in microbial communities and their contribution to the microbial loop is also acknowledged. Locally, primary production by photoautotrophs and secondary production by chemoautotrophs may be important. Such primary and secondary production can be found in symbiotic organisms, in hot and cold vents, microbial mats, and seafloor seeps (Worden et al. 2015).

1.4 Taxonomic Diversity

1.4.1 The Microbial Ocean

Microorganisms are omnipresent in the ocean. They exist as single organisms or as communities, planktonic or attached to substrates, living outside or inside of other organisms, and exhibiting different types of interactions among themselves and with their abiotic habitat. Studying marine microorganisms implies that one needs to identify their lifestyle as either benthic (attached to a substrate) or pelagic (free-living in the water column) and their localization in one of the oceanic zones. These zones have been defined and described based on light input, depth, pressure and temperature. The zones based on the depth of the water column and light penetration are the most used when pelagic microorganisms are considered. Below the water surface five zones are identified: epipelagic (from surface to 200 m), mesopelagic (between 200 and 1000 m), bathypelagic (between 1000 and 2000 m), abyssopelagic (between 2000 and 6000 m) and hadalpelagic (below 6000 m, the trenches) (Fig. 1.1). The average depth of the ocean is 3688 m and the deepest trenches (Mariana Trench) on Earth have maximum depth of 10,911 m.

Based on the amount of light received from the ocean surface the zones are: euphotic zone, twilight disphotic zone and aphotic (dark) zone. The maximum depth of light penetration is 200 m in epipelagic. All these zones are characterized by microscale heterogeneity in the distribution of nutrients and nearly an infinite number of life-suitable niches are formed. Microscopic life occupies and inhabits every thinkable niche. The photosynthetic algae and cyanobacteria are obviously confined to the euphotic zone. Chemotrophic bacteria and archaea could occupy even the most hostile environments such as the deep-sea with high hydrostatic

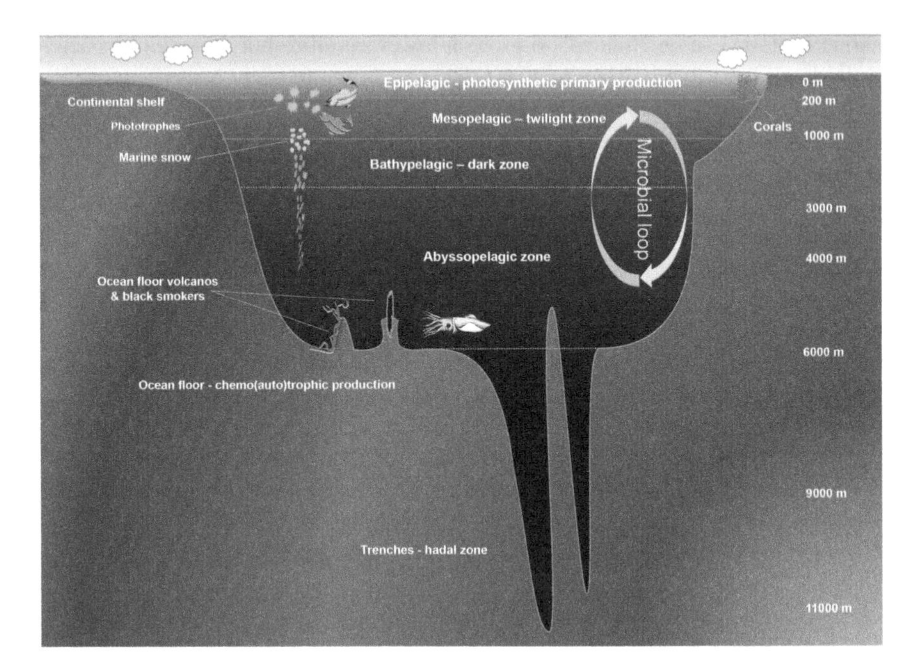

Fig. 1.1 Overview of the major oceanic zones

pressures, kilometers deep into the deep-seafloor, hydrothermal vents, and cold seeps, and sea ice in the polar waters; they all host a rich diversity of microbial life.

It has been estimated that the global biomass of bacteria exceeds the biomass of plants and animals and reaches in extremely high numbers (10^{30} cells, Whitman et al. 1998) and microorganisms account as much as 90 % of the total oceanic biomass. When we include the almost 4 billion years of history of Bacteria that thrived on Earth before multicellular organisms evolved we can only agree with the late Stephen Jay Gould that we are living on "The Planet of Bacteria" (Gould 1996), although "Planet of Microorganisms" might have been a more appropriate term given the large contribution of Archaea and unicellular eukaryotes. Moreover, phototrophic marine microorganisms are responsible for half of the primary production on the planet (Pedrós-Alió and Simó 2002). The marine microbiome as a whole serves as a catalyst of all biogeochemical cycles in the ocean.

In order to define and understand the concept of the marine microbiome it is important to know what makes a microorganism marine. Is it because an organism lives in a marine environment or that it requires elevated salt concentrations that we consider it as "marine"? Many species of marine fish and halophytic plants have evolved a life style that allow them to thrive at seawater salinity and with a few exceptions, most are not able to survive in freshwater due to osmotic imbalance. However, the existence of real marine species of bacteria has been argued for over 70 years (MacLeod 1965; Stanier 1941; ZoBell and Rittenberg 1938). In the late thirties and early forties several experiments were performed that showed that marine bacteria can be "trained" to grow at lower salinities but are often less active (ZoBell and Rittenberg 1938). Stanier (1941) also tried to adapt bacteria to grow at a lower salinity but found no evidence for a shift to a lower limit of salinity at which a bacterium could grow when compared to the original strain. MacLeod (1965) was the first to thoroughly review and questioning the existence of specific marine bacteria. He argued that "*If no differences exist between bacteria in the sea and their counterparts on land except superficial ones readily lost by training, there would be little purpose in studying the nutrition and metabolism of the same genera of bacteria in more than one habitat....and work on marine bacteria, apart from studies on gross transformations of matter, would have very little point*".

Although salinity is an obvious parameter to be considered when distinguishing marine from non-marine microorganisms, MacLeod suggested that salt requirement itself is not a good criterion. Many terrestrial microorganisms such as aerobic spore-formers and micrococci can tolerate much higher (up to 25 %) concentrations of salt than most marine bacteria. Nevertheless, most marine species require salts, especially the cations Na^+, K^+, and Mg^{2+} for maintaining cell integrity (MacLeod and Matula 1962) and for facilitation of growth and metabolism (Rhodes and Payne 1962). Moreover, MacLeod (1965) argues that temperature might be a better determinant for distinguishing marine from freshwater and terrestrial microorganism since most marine bacteria are moderately psychrophilic with optimal growth temperatures around 15 °C. Because the temperature of the ocean does not show large fluctuations, as is the case in many freshwater and terrestrial environments, the transition of marine species to a freshwater environment might be a bigger hurdle than vice versa. In contrast, freshwater microorganisms might experience a greater barrier in the higher salinity and in the increase of pressure with the depth (up to 1100 bar in the trenches). MacLeod (1965) further concludes that marine bacteria are unique and can be distinguished from their freshwater and terrestrial relatives by their requirements for inorganic ions.

Recent studies using molecular genetics approaches to identify the microbial community composition of different habitats give a better insight in the possible existence of "marine bacteria". Several authors concluded that saline and freshwater ecosystems generally show little overlap in abundant bacterial taxa (Comte et al. 2014; Logares et al. 2009; Lozupone and Knight 2007; Tamames et al. 2010). Lozupone and Knight (2007) clustered samples based on similarities in the phylogenetic lineages found in different types of habitats and concluded that the major environmental determinant of microbial community composition is salinity rather than extremes of temperature, pH, or other physical and chemical factors. Although most families of Bacteria are not confined to one ecosystem, some are significantly more abundant in marine environments than in freshwater or terrestrial environments (Table 1.2).

Based on the analysis of hydrothermal vents (Wächtershäuser 2006; Martin & Russell 2003) it is generally assumed that "cellular life" started with chemoautotrophic marine bacteria. Those first organisms migrated to freshwater and terrestrial habitats and an infrequent exchange between ocean and land was maintained (Hou et al. 2011; Hsia et al. 2013). Altogether we conclude that true

Table 1.2 Cosmopolitan bacterial taxa observed in the oceanic environment

Phylum	Family	%[a]	Oceanic environment
Alphaproteobacteria	Aurantimonadaceae	69.6	Open ocean
Alphaproteobacteria	Rhodobacteraceae	68.9	Symbiont
Alphaproteobacteria	Hyphomonadaceae	63.3	Symbiont
Alphaproteobacteria	Erythrobacteraceae	47.8	Symbiont
Bacteroidetes	Cryomorphaceae	72.8	Open ocean
Bacteroidetes	Flavobacteriaceae	59.5	Open ocean
Cyanobacteria	Prochlorococcaceae	96.5	Open ocean
Deltaproteobacteria	Nitrospinaceae	79.1	Open ocean
Deltaproteobacteria	Desulfuromonadaceae	74.1	Sediment
Deltaproteobacteria	Desulfohalobiaceae	72.8	Sediment
Deltaproteobacteria	Bacteriovoracaceae	63.8	Sediment
Deltaproteobacteria	Pelobacteraceae	63.2	Sediment
Deltaproteobacteria	Desulfobacteraceae	62.4	Sediment
Deltaproteobacteria	Desulfobulbaceae	57.1	Sediment
Epsilonproteobacteria	Campylobacteraceae	45.8	Sediment
Firmicutes	Halobacteroidaceae	78.6	Sediment
Gammaproteobacteria	Colwelliaceae	94.8	Sediment
Gammaproteobacteria	Oleiphilaceae	94.4	Open ocean
Gammaproteobacteria	Moritellaceae	88.0	Sediment
Gammaproteobacteria	Saccharospirillaceae	85.7	Sediment
Gammaproteobacteria	Pseudoalteromonadaceae	81.6	Open ocean
Gammaproteobacteria	Alteromonadaceae	80.7	Sediment
Gammaproteobacteria	Psychromonadaceae	78.1	Sediment
Gammaproteobacteria	Vibrionaceae	75.9	Symbiont
Gammaproteobacteria	Oceanospirillaceae	68.0	Open ocean
Gammaproteobacteria	Idiomarinaceae	65.0	Sediment
Gammaproteobacteria	Shewanellaceae	59.7	Sediment
Gammaproteobacteria	Ectothiorhodospiraceae	59.5	Sediment
Gammaproteobacteria	Halomonadaceae	58.8	Sediment
Gammaproteobacteria	Piscirickettsiaceae	56.9	Hydrothermal

[a]The % abundance represents the percentage of this family found in their most optimal environment based on distribution of 16S rRNA gene analysis reported by different studies summarizing 499 samples stored in the GenBank database with the environmental feature "saline-water" (adapted from Tamames et al. 2010). For example, 56.9 % of all identified Piscirickettsiaceae sequences were retrieved from hydrothermal oceanic waters

marine bacteria exist, and given the global impact of marine microbial activity, the study of the marine microbiome is crucial for understanding the biogeochemical cycles and climate change on Earth. They continue to evolve and diversify their genomes by interacting with non-organismal entities (viruses), mutations and epigenetic modifications. Bacteria–virus interactions are central to the ecology and evolution of marine microbial communities. Acquisition of immunity against viruses through the Clustered Regularly Interspaced Short Palindromic Repeats (CRISPR) occurs in pelagic and benthic bacteria and archaea irrespective of their preferred metabolism. CRISPRs are widespread in cyanobacteria with the exception of *Prochlorococcus* and *Synechococcus* (Cai et al. 2013), in deep-sea ammonia-oxidizing archaea (Park et al. 2014) and in actinomycetes of the genus *Salinispora* (Wietz et al. 2014), but hitherto they have not been found in symbiotic bacteria.

Using an imaginary transect from land to the sea, we identify the coastal zones and estuaries inundated by seawater. Together with the continental shelf these ecosystems are influenced by the tides and are driven by photosynthesis. In contrast to the deep sea they contain larger plants such as kelp (*Laminaria* sp.) and see grasses (e.g., *Posidonia oceanica* and *Zostera marina*). Mudflats and beaches are sometimes covered by microbial mats, which are among the richest microbial ecosystems on earth (Bolhuis et al. 2014). Microbial mats are small-scale self-sustaining ecosystems where photolithoautotrophic, chemolithoautotrophic, and chemoorganoheterotrophic communities form a complex food web in a layer of only a few millimeters (see Chap. 10). Cyanobacteria and diatoms are the primary producers in the top layer of the sediment while immediately below at less than 5 mm depth sulfate reducers prevail similar to the deep-sea sediments.

Amongst the most abundant marine photosynthetic organisms are the picocyanobacteria *Prochlorococcus* spp. and *Synechococus* spp., with an estimated total number of 10^{27} cells. These unicellular cyanobacteria (diameter 3 to $\sim$0.6 μm) fuel an important part of the ocean's food web via different trophic levels up to the major predators such as fish, whales and other marine mammals. The euphotic realm contributes through photosynthesis up to 50 % of Earth primary production (Field et al. 1998) and plays a critical role in global oxygen production and in carbon sequestration from the atmosphere (see Chaps. 3 and 7). The photosynthetically produced organic matter is transported from the surface to the deeper layers in the form of sinking particles such as fecal pellets, phytodetritus, and marine snow (Turner 2015). Marine snow in itself is a unique marine mini ecosystem (Alldredge et al. 1986; Silver 2015; Stocker 2012), an aggregate with its own dense community of bacteria, protists, and other microorganisms. The density of these marine snow communities is 2–5 orders of magnitude higher than in the surrounding seawater. However, because of the low density of the marine snow particles the bacteria in these aggregates represent less than 4 % of the total microbial biomass in the sea. The bacterial biomass in marine snow provides food for bacterivorous flagellates and other protozoans (Caron 1987) whereas dissolved organic matter released from marine snow particles is a substrate for free-living bacteria. Molecular analysis of marine snow from the North Atlantic Ocean revealed the accumulation of members of various phyla: Bacteroidetes, Cyanobacteria, Gammaproteobacteria (*Alteromonas* and *Pseudoalteromonas*), Alphaproteobacteria (*Roseobacter*) and Planctomycetes (Thiele et al. 2014). Surprisingly, the most abundant organism found by Thiele et al. (2014) was *Synechococcus* and this discovery provided evidence for the initial recruitment of these picocyanobacteria in marine snow at the surface but also leaving us with a puzzle because it is unlikely that these phototrophic organisms will be metabolically active or even survive in the aphotic deep sea.

In the water column the microbial communities change with depth (Wu et al. 2013). Crenarchaeota, Euryarchaeota, Thaumarchaeota together with Actinobacteria, Firmicutes, Beta-, Delta-, and Gammaproteobacteria emerge while Bacteroidetes and Alphaproteobacteria decrease in importance. It is not known how these shifts in the microbial community composition are linked to light, temperature, pressure and available food and energy sources.

In addition to absence of solar radiation, a constant low temperature and a high hydrostatic pressure are important characteristics of the abyssopelagic (Lauro and Bartlett 2008). This physically uniform environment has a low abundant but highly diverse community (see Chap. 9). Hot spots of activity and abundance are around hydrothermal vents (Prieur et al. 1995), whale falls (Smith and Baco 2003), anoxic hypersaline basins (Daffonchio et al. 2006; van der Wielen et al. 2005) and cold seeps (Elvert et al. 2000). The community composition of the warm (14 °C) deep Mediterranean (3000 m) was more similar with that of the mesopelagic (500–700 m) in the Pacific Ocean where the water temperature was in the same range than with the community composition at 3000 m where the water temperature is 2 ° C (Martín-Cuadrado et al. 2007). These authors concluded that after light, temperature is the second major factor that determines community composition, and that depth (hydrostatic pressure) is less important.

Although photosynthesis may still contribute indirectly to the food web as a result of sinking particles from the photic zone, metabolism in these special environments is dominated by chemoautotrophy and is fueled by the oxidation of reduced inorganic compounds such as hydrogen sulfide or methane. These processes are confined at or near the seafloor, which is another distinct habitat. In contrast to the various microbial hotspots and special environments mentioned above, most of the deep seafloor appears to be uniform and is low in diversity and abundance of microorganisms. At the interface between seawater and seafloor a sharp oxic/anoxic transition zone is present that allows a wide variety of microorganisms to co-exist and interact. The top part mainly consists of fine muddy sediment with a sometimes a high percentage of accumulated organic material from the photic zone.

Heterotrophic bacteria initially transform the organic matter and use up the available oxygen resulting in anoxic conditions at only a few millimeters depth in the sediment (Goffredi and Orphan 2010; Orcutt et al. 2011). In the anoxic sediment anaerobic metabolism will take over and can persist to more than hundred meters below the seafloor (Orcutt et al. 2011). In regions with low microbial activity, oxygen is not completely metabolized and can penetrate much deeper in the sediment floor where it can support aerobic respiration albeit at extremely low rates (D'Hondt et al. 2015; Røy et al. 2012) (see Chap. 8). Bacteria that are commonly found include members of the Gamma-, Delta- (especially families Desulfobacteraceae and Desulfobulbaceae), Epsilonproteobacteria, Firmicutes, and Bacteroidetes. These bacteria are able to perform simultaneous fermentation and sulfate reduction and create the conditions for methanogenesis.

1.5 Biotechnological Potential of Marine Microbes

1.5.1 Marine Microbes—Treasures of the Ocean

The short generation times and the nearly 4 billion years of evolution of (marine) microorganisms has resulted in an enormous biodiversity. Marine microorganisms evolved a plethora of metabolic pathways and are the exclusive drivers of biogeochemical cycling. A considerable part (estimated 80 %) of marine microorganisms so far remains uncultured and therefore their potential remains hidden in the abysses of marine microbial diversity. The progress made by marine bioprospecting is strongly linked to the development of "omics"-based methodologies. Numerous genomics and metagenomics-based studies investigated the marine microorganisms and habitats for their enzymatic potential (Kennedy et al. 2008; Trindade et al. 2015). A massive interest was directed also to produce optimized enzymes that are active under a broad range of harsh conditions so they can be used in wastewater treatment or as inhibitors of causative processes in different pathologies. Marine polysaccharides are being increasingly used in coatings, adhesives, feed stocks, substrates, foods, pharmaceuticals and biotechnological separations, as the gene pools of marine bacteria are tapped by recombinant DNA technology to increase polysaccharide yield. Rapid developments in the field of chemistry allowed the fast screening of many underexplored marine habitats such as cold seeps, deep-sea hypersaline basins, deep-sea hydrothermal vents, but also the endosymbionts of marine invertebrates (sponges, tunicates and corals). One of most cited successes of bioprospecting is the anti-cancer drug trabectedin obtained from *Candidatus* Endoecteinascidia frumentensis, a symbiotic gammaproteobacterium in the sea squirt *Ecteinascidia turbinate*. This drug is now produced in a semisynthetic process and is heavily used for the treatment of sarcomas and ovarian cancer. Marine microbiome derived molecules showed the high utility as industrial products or enhancers of industrial processes. New enzymes with high substrate specificity and stability in environments affected by anthropogenic pollution or new biopolymers and biomaterial with high thermo-stability are examples of high-value compounds targeted by marine biotechnology.

Many scientific articles have been written to address the biotechnological potential of marine microorganisms (Debnath et al. 2007; Kim 2015; Santos-Gandelman et al. 2014). Marine biotechnology is coined as "Blue Biotechnology" and as an efficient and sustainable contribution to the global economy. The European Union commissioned a study (European Commission-Marine Biotechnology ERA-NET) to assess the impact and growth potential of Blue Biotechnology. Blue Biotechnology is being defined as "the application of science and technology to transform marine bioresources (raw materials derived from living organisms) by biotechnological processes for use in the sectors health, cosmetics, food, aquaculture, energy and marine environmental services. It is estimated that the growth rate of Blue Biotechnology will be in the order of 6–8 % during the coming 5 years, which could lead to annual revenue up to one billion euros. The ultimate goal is a fast return of the investments to the society under the format of sound products and knowledge. Therefore, multidisciplinary initiatives and joint ventures of the public and private sectors, combining fundamental and applied research, are urgently required to develop blue biotechnology and to uncover the hidden treasures in the sea.

1.6 Conclusion

The ocean not only covers more than 70 % of our planet, it is also critical for the regulation of climate and the Earth system as a whole. The response of the ocean to climate change is far from understood, but it is certain that the marine microbiome plays a central role. The genetic diversity of the marine microbiome is untold. The microbiome is responsible for the biogeochemical cycling of virtually all elements and therefore critical for the marine ecosystem function.

The marine ecosystem is as immense as it is diverse with respect to potential niches, metabolic capacity, environmental fluctuations, geochemical composition, microbiome composition, and potential use for biotechnological applications. Many unidentified microorganisms, their genetic make-up, potential novel metabolic routes, their contribution to the major element cycles, and potential novel bioactive compounds for pharmaceutical and biotechnological purposes are yet to be uncovered and require an intricate collaboration between fundamental and applied sciences.

Acknowledgments The research leading to these results has received funding from the European Union Seventh Framework Programme (FP7/2007-2013) under grant agreement no. 311975. This publication reflects the views only of the authors, and the European Union cannot be held responsible for any use which may be made of the information contained therein.

References

Alldredge AL, Cole JJ, Caron DA (1986) Production of heterotrophic bacteria inhabiting macroscopic organic aggregates (marine snow) from surface waters. Limnol Oceanogr 31:68–78

Amante C, Eakins BW (2009) ETOPO1 1 arc-minute global relief model: procedures, data sources and analysis. National Oceanic and Atmospheric Administration (NOAA) Technical Memorandum NESDIS NGDC-24, p 19

Arrigo KR (2005) Marine microorganisms and global nutrient cycles. Nature 437:349–355

Arrigo KR, Robinson DH, Worthen DL, Robert B. Dunbar RB, DiTullio GR, van Woert M, Lizotte MP (1999) Phytoplankton community structure and the drawdown of nutrients and CO_2 in the Southern Ocean. Science 283:365–367

Azam F, Fenchel T, Field JG, Gray JS, Meyer-Reil LA, Thingstad F (1983) The ecological role of water-column microbes in the sea. Mar Ecol Prog Ser 10:257–263

Bolhuis H, Cretoiu MS, Stal LJ (2014) Molecular ecology of microbial mats. FEMS Microbiol Ecol 90:335–350

Brown RD, Mote W (2009) The response of northern hemisphere snow cover to a changing climate. J Clim 22:2124–2145
Cai F, Axen SD, Kerfeld CA (2013) Evidence for the widespread distribution of CRISPR-Cas system in the phylum cyanobacteria. RNA Biol 10(5):687–693
Caron DA (1987) Grazing of attached bacteria by heterotrophic microflagellates. Microb Ecol 13:203–218
Cartwright DE (2000) Tides: a scientific history. Cambridge University Press, p 243
Charette MA, Smith WHF (2010) The volume of Earth's ocean. Oceanography 23(2):104–106
Comte J, Lindstrom ES, Eiler A, Langenheder S (2014) Can marine bacteria be recruited from freshwater sources and the air? ISME J 8:2423–2430
D'Hondt S, Inagaki F, Zarikian CA, Abrams LJ, Dubois N, Engelhardt T, Evans H, Ferdelman T, Gribsholt B, Harris RN, Hoppie BW et al (2015) Presence of oxygen and aerobic communities from sea floor to basement in deep-sea sediments. Nat Geosci 8:299–304
Daffonchio D, Borin S, Brusa T, Brusetti L, van der Wielen PWJJ, Bolhuis H, Yakimov MM, D'Auria G, Giuliano L et al (2006) Stratified prokaryote network in the oxic–anoxic transition of a deep-sea halocline. Nature 440:203–207
Debnath M, Paul AK, Bisen PS (2007) Natural bioactive compounds and biotechnological potential of marine bacteria. Curr Pharm Biotechnol 8:253–260
Drinkwater KF (2006) The regime shift of the 1920s and 1930s in the North Atlantic. Prog Oceanogr 68:134–151
Dyurgerov MB, Meier MF (2005) Glaciers and the changing earth system: a 2004 snapshot. Occasional Paper No. 55, INSTAAR, University of Colorado
Elvert M, Greinert J, Suess E, Whiticar MJ (2000) Archaea mediating anaerobic methane oxidation in deep-sea sediments at cold seeps of the eastern aleutian subduction zone. Org Geochem 31:1175–1187
European Commission Seventh Framework Programme—Marine Biotechnology ERA-NET (2014) Marine study in support of impact assessment work on blue biotechnology
Falkowski PG (2000) Rationalizing elemental ratios in uni-cellular algae. J Phycol 36:3–6
Fanning KA (1989) Influence of atmospheric pollution on nutrient limitation in the ocean. Nature 339:460–463
Lauro FM & Bartlett DH (2008) Prokaryotic lifestyles in deep sea habitats. Extremophiles 12:15–25
Feistel R, Weinreben S, Wolf H, Seitz S, Spitzer P, Adel B, Nausch G, Schneider B, Wright DG (2010) Density and absolute salinity of the Baltic Sea 2006–2009. Ocean Sci 6:3–24
Field CB, Behrenfeld MJ, Randerson JT, Falkowski P (1998) Primary production of the biosphere: Integrating terrestrial and oceanic components. Science 281:237–240
Goffredi SK, Orphan VJ (2010) Bacterial community shifts in taxa and diversity in response to localized organic loading in the deep sea. Environ Microbiol 12:344–363
Goldman JG, McCarthy JJ, Peavey DG (1979) Growth rate influence on the chemical composition of phytoplankton in oceanic waters. Nature 279:210–215
Gould SJ (1996) Planet of the bacteria. Wash Post Horiz 119(344):H1
Gruber N, Sarmiento JL (1997) Global patterns of marine nitrogen fixation and denitrification. Global Biogeochem Cycles 11:235–266
Heinze C, Meyer S, Goris N, Anderson L, Steinfeldt R, Chang N, Le Quéré C, Bakker DCE (2015) The ocean carbon sink – impacts, vulnerabilities and challenges. Earth Syst Dynam 6:327–358
Hoffman PF, Kaufman AJ, Halverson GP, Schrag DP (1998) A Neoproterozoic snowball Earth. Science 281:1342–1346
Hou Z, Sket B, Fiser C, Li S (2011) Eocene habitat shift from saline to freshwater promoted Tethyan amphipod diversification. Proc Natl Acad Sci USA 108:14533–14538
Hsia CCW, Schmitz A, Lambertz M, Perry SF, Maina JN (2013) Evolution of air breathing: oxygen homeostasis and the transitions from water to land and sky. Compr Physiol 3:849–915
Jiao N, Luo T, Zhang R, Yan W, Lin Y, Johnson ZI, Tian J, Yuan D, Yang Q, Zheng Q, Sun J, Hu D, Wang P (2014) Presence of *Prochlorococcus* in the aphotic waters of the western Pacific Ocean. Biogeosciences 11:2391–2400

Kennedy J, Marchesi J, Dobson A (2008) Marine metagenomics: strategies for the discovery of novel enzymes with biotechnological applications from marine environments. Microb Cell Fact 7:27
Kim SK (2015) Handbook of marine biotechnology. Springer, Heidelberg, p 1512
Klausmeier CA, Litchman E, Levin SA (2004) Phytoplankton growth and stoichiometry under multiple nutrient limitation. Limnol Oceanogr 49:1463–1470
Li WKW (1994) Primary production of prochlorophytes, cyanobacteria, and eukaryotic ultraphytoplankton—measurements from flow cytometric sorting. Limnol Oceanogr 39:169–175
Loladzel I, Elser JJ (2011) The origins of the Redfield nitrogen-to-phosphorus ratio are in a homoeostatic protein-to-rRNA ratio. Ecol Lett 14:244–250
Logares R, Bråte J, Bertilsson S, Clasen JL, Shalchian-Tabrizi K, Rengefors K (2009) Infrequent marine–freshwater transitions in the microbial world. Trends Microbiol 17:414–422
Lozupone CA, Knight R (2007) Global patterns in bacterial diversity. Proc Natl Acad Sci USA 104:11436–11440
MacLeod RA (1965) The question of the existence of specific marine bacteria. Bacteriol Rev 29:9–23
MacLeod RA, Matula TI (1962) Nutrition and metabolism of marine bacteria: XI. Some characteristics of the lytic phenomenon. Can J Microbiol 8:883–896
Martin W, Russell MJ (2003) On the origins of cells: a hypothesis for the evolutionary transitions from abiotic geochemistry to chemoautotrophic prokaryotes, and from prokaryotes to nucleated cells. Philos Trans Roy Soc B 358:59–85
Martín-Cuadrado AB, Lopez-Garcia P, Alba JC, Moreira D, Monticelli L, Strittmatter A, Gottschalk G, Rodriguez-Valera F (2007) Metagenomics of the deep Mediterranean, a warm bathypelagic habitat. PLoS ONE 9:e914
McCarthy T, Rubisge B (2005) Story of Earth and life. University of the Witwatersrand, School of Geosciences (ed), p 70
Middelburg JJ, Soetaert K, Herman PMJ, Heip CHR (1996) Denitrification in marine sediments: a model study. Global Biogeochem Cycles 10:661–673
Moore CM, Mills MM, Arrigo KR, Berman-Frank I, Bopp L, Boyd PW, Galbraith ED, Geider RJ, Guieu C, Jaccard SL, Jickells TD, La Roche J, Lenton TM et al (2013) Processes and patterns of oceanic nutrient limitation. Nat Geosci 6:701–710
Orcutt BN, Sylvan JB, Knab NJ, Edwards KJ (2011) Microbial ecology of the dark ocean above, at, and below the seafloor. Microbiol Mol Biol Rev 75:361–422
Pahlow M, Riebesell U (2000) Temporal trends in deep ocean Redfield ratios. Science 287:831–833
Park SJ, Ghai R, Martín-Cuadrado AB, Rodríguez-Valera F, Chung WH, Kwon K et al (2014) Genomes of two new ammonia-oxidizing archaea enriched from deep marine sediments. PLoS ONE 9(5):e96449
Pedrós-Alió C, Simó R (2002) Studying marine microorganisms from space. Int Microbiol 5:195–200
Pinet PR (1996) Invitation to oceanography. West Publishing Company, St. Paul, pp 126, 134–135
Post WM, Peng T-H, Emanuel WR, King AW, Dale VH, DeAngelis DL (1990) The global carbon cycle. Am Sci 78:310–326
Prieur D, Erauso G, Jeanthon C (1995) Hyperthermophilic life at deep-sea hydrothermal vents. Planet Space Sci 43:115–122
Redfield AC (1934) On the proportions of organic derivations in seawater and their relation to the composition of plankton. In Daniel RJ (ed) James Johnstone memorial volume. University Press of Liverpool, Liverpool, pp 176–192
Redfield AC (1958) The biological control of chemical factors in the environment. Am Sci 46:205–221
Redfield AC, Ketchum BH, Richards FA (1963) The influence of organisms on the composition of sea water. In: Hill MN (ed) The sea. Interscience, New York, pp 26–77
Rhodes ME, Payne WJ (1962) Further observations on effects of cations on enzyme induction in marine bacteria. Antonie van Leeuwenhoek. J Microbiol Serol 28:302–314
Ridgwell A (2011) Evolution of the ocean's biological pump. In: Proceedings of the national academy of sciences of the United States of America, vol 108, pp 16485–16486
Ross D (1995) Introduction to oceanography. HarperCollins College Publishers, New York, pp 199–226, 339–343

Røy H, Kallmeyer J, Adhikar RR, Pockalny R, Jørgensen BB, D'Hondt S (2012) Aerobic microbial respiration in 86-million-year-old deep-sea red clay. Science 336:922–925
Santos-Gandelman JF, Giambiagi-deMarval M, Oelemann WM, Laport MS (2014) Biotechnological potential of sponge-associated bacteria. Curr Pharm Biotechnol 15:143–155
Sardans J, Rivas-Ubach A, Penuelas J (2012) The elemental stoichiometry of aquatic and terrestrial ecosystems and its relationships with organismic lifestyle and ecosystem structure and function: a review and perspectives. Biogeochemistry 111:1–39
Sarmiento JL, Hughes TMC, Stouffer RJ, Manabe S (1998) Simulated response of the ocean carbon cycle to anthropogenic climate warming. Nature 393:245–249
Silver M (2015) Marine snow: a brief historical sketch. Limnol Oceanogr Bull 24:5–10
Smetacek V (2001) A watery arms race. Nature 411:745
Smith CR, Baco AR (2003) Ecology of whale falls at the deep-sea floor. Oceanogr Mar Biol Annu Rev 41:311–354
Stal LJ, Zehr JP (2008) Cyanobacterial nitrogen fixation in the ocean: diversity, regulation, and ecology. In: Flores E, Herrero A (eds) The cyanobacteria: molecular biology, genomics and evolution. Caister Academic Publishers, Norfolk, pp 423–446
Stanier RY (1941) Studies on marine agar digesting bacteria. J Bacteriol 42:527–559
Stocker R (2012) Marine microbes see a sea of gradients. Science 338:628–633
Tamames J, Abellán JJ, Pignatelli M, Camacho A, Moya A (2010) Environmental distribution of prokaryotic taxa. BMC Microbiol 10:85
Thiele S, Fuchs BM, Amann R, Iversen MH (2014) Colonization in the photic zone and subsequent changes during sinking determines bacterial community composition in marine snow. Appl Environ Microbiol 81:1463–1471
Thurman H, Burton E (2001) Introductory oceanography, 9th edn. Prentice Hall, Upper Saddle River, New Jersey
Trindade M, Van Zyl L, Navarro-Fernández J, Abd Elrazak A (2015) Targeted metagenomics as a tool to tap into marine natural product diversity for the discovery and production of drug candidates. Front in Microbiol 6:890
Turekian KK (1968) Oceans. Prentice-Hall, Englewood Cliffs, New Jersey
Turner JT (2015) Zooplankton fecal pellets, marine snow, phytodetritus and the ocean's biological pump. Prog Oceanogr 130:205–248
Tyrrel T (1999) The relative influences of nitrogen and phosphorus on oceanic primary production. Nature 400:525–531
van der Wielen PWJJ, Bolhuis H, Borin S, Daffonchio D, Corselli C, Giuliano L et al (2005) The enigma of prokaryotic life in deep hypersaline anoxic basins. Science 307:121–123
Wächtershäuser G (2006) From volcanic origins of chemoautotrophic life to Bacteria, Archaea and Eukarya. Philos Trans Roy Soc B 361:1787–1808
Whitman WB, Coleman DC, Wiebe WJ (1998) Prokaryotes: the unseen majority. Proc Natl Acad Sci USA 95:6578–6583
Wietz M, Millán-Aguiñaga N, Jensen PR (2014) CRISPR-Cas systems in the marine actinomycete *Salinispora*: linkages with phage defense, microdiversity and biogeography. BMC Genom 25 (15):936
Worden AZ, Follows MJ, Giovannoni SJ, Wilken S, Zimmerman AE, Keeling PJ (2015) Rethinking the marine carbon cycle: factoring in the multifarious lifestyles of microbes. Science 347:1257594
Wu J, Gao W, Johnson RH, Zhang W, Meldrum DR (2013) Integrated metagenomic and metatranscriptomic analyses of microbial communities in the meso- and bathypelagic realm of North Pacific Ocean. Mar Drugs 11:3777–3801
Zhang Y, Zhao Z, Dai M, Jiao N, Herndl GJ (2014) Drivers shaping the diversity and biogeography of total and active bacterial communities in the South China Sea. Mol Ecol 23:2260–2274
Zobell CE, Rittenberg SC (1938) The occurrence and characteristics of chitinoclastic bacteria in the sea. J Bacteriol 35:275–287

Chapter 2
Marine Bacteria and Archaea: Diversity, Adaptations, and Culturability

Jörg Overmann and Cendrella Lepleux

Abstract With an estimated total number of 6.6×10^{29} cells, Bacteria and Archaea in marine waters and sediments constitute a major fraction of global microbial biomass. Most marine bacterial communities are highly diverse and individual samples can comprise over 20,000 species. Different marine habitats such as coastal surface waters, subsurface open ocean waters and sediments are colonized by distinct bacterial communities. Consequently, global marine bacterial diversity must be very high but has remained largely uncharted to date. One major obstacle that needs to be overcome is the persisting difficulty to culture most of the dominant marine bacterial and archaeal phylotypes. Typically, these difficulties relate to an insufficient appreciation of the specific physiological requirements and adaptations of marine Bacteria and Archaea. In many marine environments, concentrations of readily utilizable dissolved organic carbon (DOC) compounds or inorganic nutrients are present at submicromolar concentrations whereas suspended marine particles constitute spatially discrete hot spots of growth substrates. Known bacterial adaptations to oligotrophic growth conditions include high affinity uptake systems, low growth rates and cell sizes, streamlined genomes, little regulatory flexibility, physiological specialization and low loss rates due to grazing and viral lysis. On the opposite, lineages adapted to exploitation of nutrient hot spots are motile, chemotactically active, have large cells, adhere to particles, employ specialized uptake systems for high molecular weight substrates, excrete exoenzymes, and feature a broad substrate spectrum. Besides these canonical types of adaptations, interesting novel traits have been discovered over the past years, like the widely distributed proton-pumping bacteriorhodopsins, a multitude of carbohydrate-active enzymes, TonB-like receptors, thiosulfate oxidation, methylotrophic pathways, carbon monoxide oxidation,

J. Overmann (✉) · C. Lepleux
Leibniz-Institut DSMZ-Deutsche Sammlung von Mikroorganismen
und Zellkulturen GmbH, Inhoffenstraße 7B, 38124 Braunschweig, Germany
e-mail: joerg.overmann@dsmz.de

C. Lepleux
e-mail: cel13@dsmz.de

J. Overmann
Technische Universität Braunschweig, Braunschweig, Germany

L.J. Stal and M.S. Cretoiu (eds.), *The Marine Microbiome*,
DOI 10.1007/978-3-319-33000-6_2

metabolism of compatible solutes, and heavy-metal resistance. In order to retrieve and study representatives of not-yet-cultured bacterial lineages in the laboratory, future culture attempts need to be modified according to this improved knowledge of the specific adaptations of marine Bacteria and Archaea.

2.1 Introduction

Bacteria and Archaea represent major drivers of the biogeochemical cycles in the marine environment due to their large abundance, high cell-specific activity, and their unique metabolic capabilities that include, among others, nitrogen fixation, (anaerobic) ammonia oxidation, and methane oxidation. Laboratory analyses of isolated marine Bacteria and Archaea reveal their metabolic features and the evolutionary mechanisms underlying bacterial diversification. Vice versa, a better understanding of the activities and diversity of marine Bacteria and Archaea in their natural environment is the key to their successful cultivation. Finally, novel applications of microorganisms in bio-economy rely on an improved access to novel bacterial and archaeal isolates.

This chapter focuses on the extent, patterns and controls of bacterial and archaeal diversity in the marine environment, then reviews and discusses the current state of knowledge of their functions in situ and the underlying specific adaptations, and finally contrasts this information with available concepts and approaches for their cultivation.

2.2 Biomass, Diversity and Phylogenetic Composition of Marine Bacterial Communities

Total numbers of bacterial cells of 1.0×10^{29}, 1.7×10^{28} and 1.0×10^{26} have been calculated for the marine pelagic, surface sediments and coastal waters, respectively (Whitman et al. 1998). The global bacterial biomass in the sub-seafloor biosphere is estimated as 5.4×10^{29} cells (Parkes et al. 2014), representing $\sim$24 % of the global number and biomass of bacteria and archaea (recalculated based on the numbers of Whitman et al. 1998). These numbers demonstrate that the global bacterial biomass of the open ocean and the marine subsurface is similar to that of what is present in soils (2.6×10^{29} cells) and that marine Bacteria and Archaea constitute a considerable part of the microbial biomass on Earth.

Across their different habitats, marine bacteria are distributed rather unequally. Surface seawater harbors between 10^4 and 10^7 cells ml^{-1} with an average of 5×10^5 cells ml^{-1} for epipelagic and coastal shelf waters. Deeper water layers on average are colonized by 5×10^4 cells ml^{-1} (Whitman et al. 1998). In the sediment surface, however, cell densities range from 10^8 to 10^9 cells cm^{-3}. Coastal sediments contain higher cell numbers than those of the open ocean (Orcutt et al. 2011). In sub-seafloor sediments down to almost 2 km depths, numbers range between 10^{10} and 10^3 (the detection limit) cells cm^{-3} depending on the location (Parkes et al. 2014). The lowest numbers (10^3–10^4 cells cm^{-3}) were found in the first 10 m of sediments underlying the extremely low productivity South Pacific Gyre that receives extremely low organic carbon input through sedimentation (D'Hondt et al. 2009).

Next generation sequencing of 16S rRNA genes from 193 marine samples collected within the framework of the *Tara* Oceans project yielded a total species richness of 37,470 for marine Bacteria and Archaea (Sunagawa et al. 2015). These data corroborate earlier findings of highly diverse bacterial communities in individual samples that were estimated to harbor up to 23,315 species even in the deep-sea water column despite its supposedly energy-poor conditions (Sogin et al. 2006). In a worldwide comparison across all major habitat types, sediments contained the most diverse bacterial communities (Lozupone and Knight 2007). Accordingly, the species richness of marine environments is similar to those estimated for soils using comparable experimental approaches (Roesch et al. 2007). The total species diversity of marine Bacteria and Archaea most likely is underestimated by the above. This is suggested by the fact that a significantly higher number of 799,581 of 16S rRNA gene sequences in GenBank is of marine origin (Benson et al. 2013).

The differences in bacterial community composition are most pronounced between pelagic and sediment environments. Pelagic and benthic marine habitats share only 7.1 % of the bacterial species inventory (corresponding to 9900 species) based on a global comparison of 509 samples from the International Census of Marine Microbes (ICoMM) project (Zinger et al. 2011). Marine sediments harbor the phylogenetically most divergent bacterial communities, which maybe related to their highly stratified nature that provides very different ecological niches. Bacterial communities in the marine pelagic are much less even in species composition than the benthic environment (Zinger et al. 2011). Within the pelagic, bacterial communities in coastal surface waters, subsurface open ocean waters and anoxic habitats differ the most (Lozupone and Knight 2007). In a detailed comparison of marine bacterial communities across all latitudes, pelagic communities from the polar Southern and Arctic Oceans were distinct, but more similar to each other than to the bacterioplankton of lower latitudes. Mid-latitude open ocean communities resembled each other across the different oceans. Similarly, the deep ocean communities exhibited a composition that resembled that of other regions but was distinct from the respective surface communities (Ghiglione et al. 2012).

Open ocean surface marine bacterioplankton typically consists of a few dominant (sub)-phyla that decrease in abundance in the order Alphaproteobacteria > Gammaproteobacteria > Flavobacteria and Cyanobacteria. Additional groups that are usually detected at low abundances are the Betaproteobacteria, Firmicutes and Actinobacteria. At the level of genera/species, the diversity of marine bacterioplankton appears to follow a bimodal abundance distribution. Whereas few taxa almost invariantly occur at high abundance, the majority of the taxa typically are rare (Yooseph et al. 2010). Thus, members of the alphaproteobacterial SAR11 clade are ubiquitous and represent the most abundant marine organism in the oceans. They constitute about one third of all Bacteria and are the dominant bacterial heterotrophs (Morris et al. 2002; Sunagawa et al. 2015; Zinger et al. 2011). Another group of the Alphaproteobacteria, the *Roseobacter* clade-affiliated (RCA) cluster of the family Rhodobacteraceae, can constitute up to 35 % of the total bacterioplankton in temperate and polar oceans (Giebel et al. 2013). Additional abundant phylotypes belong to the gammaproteobacterial SAR86-clade and the Actinobacteria (Yooseph et al. 2010). Even across 41 different marine and coastal ecosystems in the Atlantic, Caribbean and Pacific that varied substantially with regard to salinity, nutrient content, temperature, and other environmental conditions, only 60 highly abundant ribotypes were identified by a metagenomic approach (Rusch et al. 2007). Based on the fragment recruitment of metagenomic sequences, *Pelagibacter*, *Prochlorococcus* and *Synechococcus* genomes constitute a large fraction of the bacterial community in the Sargasso Sea (Rusch et al. 2007; Venter et al. 2004). This typical composition has also been determined in the most ultra-oligotrophic waters of the South Pacific Gyre (Yin et al. 2013). In the latter, Archaea are two orders of magnitude less abundant than Bacteria and are dominated by Euryarchaeota of Marine Group II.

In oceanic deep waters, Gammaproteobacteria, Deltaproteobacteria and Actinobacteria occur in higher relative abundances than at the surface. In coastal waters, however, the relative proportion of Flavobacteria is significantly higher (17 %) than in all other marine pelagic habitats (Zinger et al. 2011). Other bacterial phyla such as the Marine Group A or Chloroflexi are less frequent, but still consistently present in pyrosequencing or Illumina rRNA sequence datasets from the marine pelagic. Marine members of these groups represent particularly interesting targets for future cultivation attempts since the physiology of cultivated relatives does not match the environmental conditions of the marine pelagic. It is therefore not surprising that attempts to isolate typical representatives of the marine pelagic Chloroflexi such as the SAR202 group have failed so far (Orcutt et al. 2011).

In oxic surface sediments, Gammaproteobacteria and Deltaproteobacteria are dominant groups; in addition Alphaproteobacteria, Acidobacteria, Actinobacteria and Planctomycetes are abundant. The diversity of Archaea at the sediment surface is low where marine group I Thaumarchaeota dominate (Orcutt et al. 2011). In the first meters below the sediment surface, the high phylum level diversity of bacterial communities persists, and also comprises Epsilonproteobacteria, Chloroflexi and Thaumarchaeota (Parkes et al. 2014).

In contrast, bacterial communities in the deep marine biosphere are dominated by Gammaproteobacteria, Chloroflexi, and members of the candidate phylum JS1, which together contribute an average of >50 % of all 16S rRNA gene sequences. Over half of the Archaea present in the deep marine biosphere belongs to the so-far uncharacterized groups the Miscellaneous Crenarchaeotal Group and the Marine Benthic Group B of Crenarchaeota. Thaumarchaeota, South African Gold Mine Euryarchaeotal Group and the Marine Benthic Group D of the Crenarchaeota represent the next most abundant groups (Parkes et al. 2014). Different types of subsurface sediments (seeps, hydrate-associated, shelf, ash, carbonates, turbidites) contain different percentages of the groups listed. The candidate phylum JS1 and members of the Miscellaneous Crenarchaeotal Group are the dominant microbial groups of the organic-rich subsurface sediments. Chloroflexi, but also Planctomycetes dominate the abyssal sediments of the oligotrophic oceanic provinces. In contrast, deltaproteobacterial sulfate reducers, Epsilonproteobacteria (including lithotrophic sulfur-oxidizers) and Archaea belonging to the anaerobic methanotrophic (ANME) group and methanogenic Methanosarcinales dominate the strongly reducing cold seeps that are characterized by cycling of sulfur and methane. Members of the candidate phylum JS1 and Crenarchaeota of Marine Benthic Group B are the predominant groups in gas hydrates (Orcutt et al. 2011; Parkes et al. 2014).

The estimates of the bacterial species richness even of individual marine samples are twice as high as the total number of species that has so far been validly described (<11,000; Overmann 2015). In line with this low coverage of marine bacterial diversity by culture collections, extensive metagenomic datasets from a large variety of water samples contain only few sequences of the genomes available for cultivated bacterial species (Rusch et al. 2007). As a result, the major fraction of marine Bacteria and Archaea is poorly characterized with respect to ecology, physiology, biochemistry, and possible applications. The probable cause of the low coverage of marine microbial diversity obtained by cultivation-based approaches is the labor-intensive character of enriching and isolating marine bacteria on multiple different media. However, cultivation trials usually do not yield the numerically dominant bacterial phylotypes (Eilers et al. 2001; Uphoff et al. 2001) and marine members of entire candidate phyla, or other high rank taxa continue to resist cultivation (e.g., Gammaproteobacteria cluster SAR86, Chloroflexi cluster SAR202, Deltaproteobacteria cluster SAR324, candidate phylum Marine Group A, and the marine Actinobacteria). Several explanations for these discrepancies have been offered, including dormancy, substrate-accelerated death at high substrate concentrations, a viable-but-not-culturable state of the cells, the presence of lysogenic phages, obligate oligotrophy, or the dependence on other microorganisms for growth (summarized in Overmann 2013). During the past two decades, some of these challenges have been successfully addressed and the culture success has been considerably improved in some instances (Bruns et al. 2002; Button et al. 1993; Cho and Giovannoni 2004). However, the currently available isolates of marine Bacteria and Archaea cover only a minute fraction of the existing diversity (see last section below). For a better representation of the marine microbial diversity, culture approaches need to be developed further based on a state-of-the-art analysis of the physicochemical and biological determinants of the marine microbial community composition and based on an improved knowledge of the specific adaptations of the not-yet-cultured lineages.

2.3 Patterns and Potential Drivers of Marine Bacterial Diversity

Correlating phylogenetic similarity of marine bacterial communities or the relative abundance of particular target phylotypes of Bacteria and Archaea with the physicochemical characteristics of their respective environment can yield initial information about potential drivers of bacterial diversity and the niches of individual phylotypes.

Based on analyses of similarities between metagenomic sequences of marine bacterioplankton communities, water temperature and salinity were identified as the major environmental factors that correlated with community composition. Water depth, primary productivity and proximity to land constituted additional relevant factors (Rusch et al. 2007).

Salinity was identified as the major environmental determinant of community composition on a global scale (Lozupone and Knight 2007). Salinity is also the major environmental variable that is associated with the phylogenetic differences of bacterial communities in freshwater, brackish and marine sediments (Wang et al. 2012) and constitutes an important driver of the composition of bacterioplankton in inland waters (Wu et al. 2006). Marked changes in the relative abundance of bacterial subphyla have been documented along salinity gradients in estuaries and in the Baltic Sea. With increasing salinity, the abundance of Alphaproteobacteria and Gammaproteobacteria increases whereas that of Betaproteobacteria, Actinobacteria and Verrucomicrobia decreases (Cottrell and Kirchman 2003; Herlemann et al. 2011; Zhang et al. 2006). The bacterioplankton in brackish waters of the Baltic Sea appears to be autochthonous and also harbors sequence types unique to this environment, e.g., an uncharacterized member of the Spartobacteriaceae (Verrucomicrobia). Unlike multicellular organisms, however, bacterial diversity in brackish waters is not decreased compared to marine or freshwater communities (Herlemann et al. 2011). The distinct community composition suggests that marine Bacteria and Archaea have specifically adapted to the environmental conditions of their habitat, such as salinity or compounds typically found in seawater like the osmolyte dimethylsulfoniopropionate (DSMP). In fact, bacteria of the typically marine *Roseobacter* clade have an absolute requirement for sodium ions due to their sodium-based transporters (Brinkhoff et al. 2008). 16S rRNA gene-based studies as well as genome comparisons show that marine members of the Chlorobiaceae (green sulfur bacteria) requiring NaCl concentrations >1 % (w/v) form separate clusters that are distinct from freshwater or brackish water lineages (Bryant et al. 2012; Imhoff 2003). Furthermore, various clades of marine bacteria, like the SAR116 group or the *Roseobacter* clade, have been shown to be capable of DMSP degradation (see next section).

Within the photic zone, temperature constitutes an important driver of microbial community composition (Sunagawa et al. 2015). While relatives of the alphaproteobacterial SAR11, the gammaproteobacterial SAR86, *Prochlorococcus*, *Synechococcus*, and Acidimicrobidae clades occur widespread across marine habitats, distinctly different SAR11 sub-clades are adapted to high latitude, low temperature (<10 °C) polar waters (see next section). *Prochlorococcus* spp. are absent from cold-water regions of the ocean and occur between latitudes of 40°N and 40°S where they reach cell abundances of $>10^5$ ml^{-1} and account for up to 43 % of the photosynthetic biomass (Johnson et al. 2006; Scanlan et al. 2009). On the opposite, members of the RCA clade are abundant in temperate but not in tropical waters (Rusch et al. 2007). Members of the gammaproteobacterial SAR86 occur in cold as well as warm oceanic waters (Yooseph et al. 2010). Sequence similarity determined between metagenomes from different marine environments showed consistent patterns with largely dissimilar bacterial metagenomes between tropical and temperate samples (Rusch et al. 2007). Similarly, single cell genomes were distinctly different between temperate regions and subtropical gyres, with temperature and latitude representing major determinants (Swan et al. 2013). While this differentiation can be detected at the genomic level, it is not apparent at the level of 16S rRNA gene sequence types that do not exhibit consistent latitudinal differences. This emphasizes the need to analyze the micro-diversity of individual phylogenetic lineages of marine bacteria (see below).

When analyzing the patterns of fragment recruitment of metagenomes against a suite of single cell genomes, it becomes evident that the effect of geographic latitude on the composition of bacterial communities is not the result of their biogeographic isolation (Swan et al. 2013). Even within the clades of numerically dominant planktonic bacteria like *Prochlorococcus marinus* or *P. ubique* individual genotypes do not follow a biogeographic pattern (Rusch et al. 2007). Obviously, dispersal rates in the different marine environments are high enough to prevent biogeographic isolation of bacterial species or even lineages within species. Since the time of overturn of the global ocean by surface currents and thermohaline circulation ranges between 1000 and 2000 years (Doos et al. 2012), geographic isolation is expected only when genotypes arise over shorter time periods, which are apparently too short for lasting genetic isolation of diverging lineages of marine bacteria. Consequently, speciation and geographic isolation are not major drivers of microbial diversity on a local scale.

Across a wide variety of sampling locations, water depth explains the major part of variance in community composition, indicating that vertical stratification determines most of the differences in the composition of bacterial communities (Sunagawa et al. 2015). Ecological theory predicts that competition for limiting resources decreases diversity and ultimately can lead to ecological instability, whereas strong predator-prey interactions enhance the number of coexisting species and dampen their fluctuations in abundance (Allesina and Tang 2012; Loreau and de Mazancourt 2013). Because of the severe energy limitation of deeper oceanic water layers, it was rather unexpected to find an increase of species richness with depth in the ocean (Sunagawa et al. 2015). This finding is suggestive of the presence of a considerable diversity of ecological niches that are created by the pronounced stratification of the mesopelagic zone. For example, anaerobic ammonia oxidizing (anammox) bacteria typically occur in oxygen minimum zones of coastal upwelling regions (Dalsgaard et al. 2012). The peak abundance of Group I and pSL12-like marine Crenarchaeota is below the photic zone at the position of the nitracline where nitrate concentrations are above several μM. The deltaproteobacterium *Nitrospina* shows a similar vertical distribution. In contrast, marine Euryarchaeota peak in oceanic surface waters (Mincer et al. 2007). Moreover, metagenomic analysis of bacterial communities revealed specific adaptations of the bacterial lineages that prevail in deeper water layers (see next section).

The high diversity of deep-sea bacterioplankton may also be related to specific strategies towards energy limitations. In multicellular organisms, higher energy availability is thought to enable a higher species richness, since it allows a larger number of individuals to coexists and hence more species to maintain viable populations (Gaston 2000). Following this concept for microorganisms, coastal sediments and surface waters in upwelling regions would thus be expected to harbor more diverse bacterial assemblages than deep-sea sediments or the bathypelagic. However, this concept will hold true for microorganisms only if different species have similar (high) energy requirements, if the growth yield for microbial growth substrates is more or less constant (given that productivity or energy availability is typically determined by bulk parameters such as organic carbon content), and if energy supply is more or less constant. Thus, vastly different growth yields of different physiological types of microorganisms, or particular adaptions to survival such as endospores or extremely low maintenance energy requirements (Marschall et al. 2010) would disturb this relationship. Also, highly productive marine systems typically are characterized by high temporal variability. This fact has been used to explain why metazoa diversity does not increase with productivity in this realm (Gaston 2000).

In addition to the mostly abiotic factors mentioned so far, the activity of predators like viruses and protists also modifies bacterial community structure. Protists can create a shift in marine bacterial communities, leading to the development of rare bacteria (Suzuki 1999) or the replacement of strains (Hahn and Höfle 1999). Larger celled and fast-growing copiotrophic bacteria are selectively removed and kept at low abundances (Pernthaler 2005; Rodriguez-Brito et al. 2010). This seems to be particularly the case for Gammaproteobacteria including *Alteromonas* spp. (Lebaron et al. 2001). As a result, the contribution of low abundance copiotrophic bacteria to bacterial biomass production and carbon flux through the microbial loop may be substantial despite their low abundance. The analysis of the effect of bottom-up factors (temperature, nutrient concentration, chlorophyll *a*) and of top down factors (viral abundance and mortality caused by bacteria and viruses) on the phylogenetic diversity of marine bacteria in coastal waters of the Mediterranean Sea showed that viral activity was more pronounced than protistan grazing (Boras et al. 2015). In this case of severe viral predation, viral lysis was a stronger factor in shaping the bacterial community than temperature or light.

In order to be culture-based approaches successful, the key biogeochemical parameters and processes that determine the composition of the natural bacterial communities and the abundance of target bacteria in their environment need to be taken into account and reproduced during laboratory incubations. During the past decade this strategy has been increasingly successful (compare sections below).

2.4 Life Strategies and Adaptation Mechanisms of Marine Bacteria and Archaea

2.4.1 Environmental Conditions and Canonical Life Strategies of Marine Bacteria and Archaea

Recent advances in molecular and analytical techniques have enabled a deeper elucidation of the metabolic potential and in situ physiological activity of marine Bacteria and Archaea (Fig. 2.1). This novel information provides the basis not only for a better understanding of their specific adaptations and ecological niches, but also for improving strategies for their cultivation.

The dominant primary producers in the nutrient limited euphotic zone of the open oceans are cyanobacteria of the genera *Synechococcus* and *Prochlorococcus* (see Chap. 3). *Prochlorococcus* is present between latitudes of 40°N and 40°S and absent in waters with temperatures <15 °C. Distribution patterns along environmental gradients and genome comparisons have been used to infer ecological

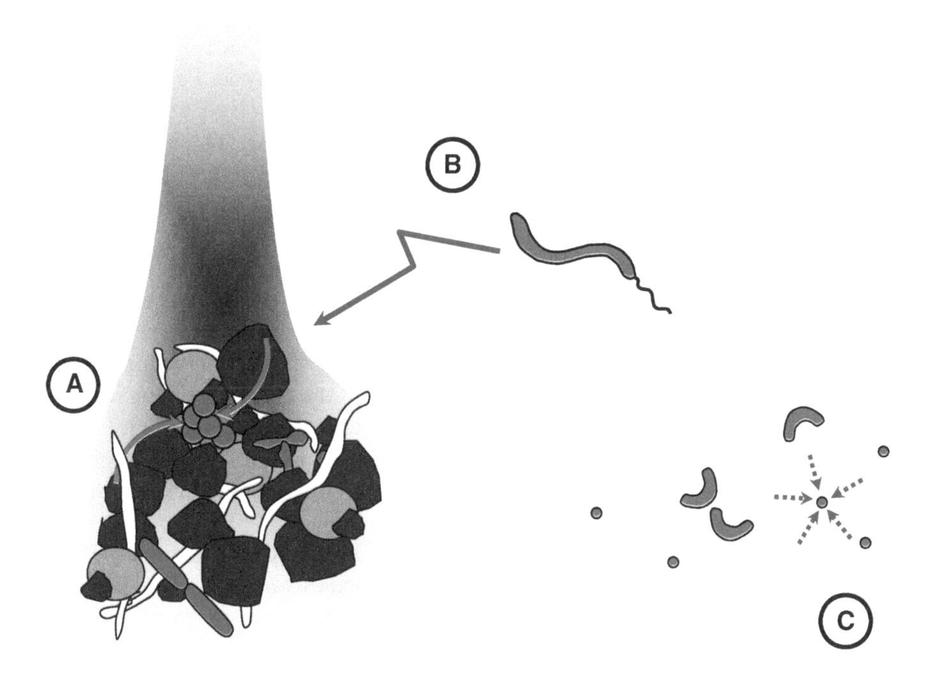

Fig. 2.1 Conceptual representation of the canonical adaptations of marine Bacteria and Archaea, including the copiotrophic lifestyle on marine aggregates (*A*), chemotaxis towards higher nutrient concentrations liberated from marine aggregates (*B*), and the adaptation to oligotrophic conditions in the free water (*C*)

niches of its different phylogenetically defined clades (Johnson et al. 2006; Rocap et al. 2003). The vertical distribution of different ecotypes of *Prochlorococcus* can be related to the specific adaptation to either high light and low nutrient or low light high nutrient conditions (Moore et al. 1998). Genomic streamlining (down to a size of 1.66–2.41 Mbp), *i.e.* selective genome reduction through loss of genes of central metabolic (e.g., nitrate utilization), signaling, and regulatory functions is one strategy for adaptation of *Prochlorococcus* to limiting nutrient concentrations. The discovery of a widespread, nitrogen-fixing cyanobacterium ('*Candidatus* Atelocyanobacterium thalassa'; UCYN-A) with a genome size of only 1.44 Mbp illustrates a second type of genome streamlining in marine cyanobacteria that resembles that known from obligate parasitic bacteria such as *Mycoplasma*: the small genome of UCYN-A encodes numerous regulatory and metabolic features but lacks many biosynthetic pathways (Tripp et al. 2010).

In the ocean, heterotrophic bacterioplankton plays the key role in the alteration and consumption of organic matter (Pomeroy et al. 2007) and consumes the major part of the daily primary production (Ducklow 2000). Primary production by microalgae and cyanobacteria constitutes the main source of organic carbon substrates of bacteria in the open ocean. However, concentrations of carbohydrate monomers and amino acids that can be taken up directly are only in the nanomolar range (Kirchman et al. 2001; McCarthy et al. 1996). Part of the photosynthetic products are already released by living cells and encompass up to 60 % of organic carbon >1 kDa. Up to one third of cellular organic carbon is released as dissolved organic carbon (DOC) compounds during grazing and viral lysis. Polysaccharides constitute the major fraction of high molecular weight dissolved organic matter since they represent about 70 % of the dry weight of algae. These compounds comprise anionic polysaccharides with sulfate and carboxyl groups, which distinguishes them from polysaccharides that occur in terrestrial environments (Mann et al. 2013). Marine bacteria rapidly utilize polysaccharides but proteins show lower decay constants and lipids are degraded even slower (Benner and Amon 2015).

Bacteria mineralize high molecular weight DOC more rapidly than the smaller size classes (Benner and Amon 2015). The DOC that is not degraded becomes increasingly refractive (Lou et al. 2010). The remaining small dissolved molecules represent the dominant organic carbon compounds in the marine environment and may persist either due to their inherent refractory nature, their inaccessibility due to embedding in gel-like structures, or because of their chemical diversity that lowers the encounter rates with bacterial transporter molecules or enzymes to the point that microbial growth is not possible (Azam and Malfatti 2007; Benner and Amon 2015). As a result, the majority of total organic carbon (TOC) in the ocean (77 %) occurs in the size fraction <1 nm, which corresponds to a molecular weight of 1 kDa. Only 22 % of the TOC is bigger than 1 nm and comprises macromolecules (up to 1000 kDa). Particulate organic carbon (POC) is operationally defined by retention on filters (typically with pore sizes 0.1–0.2 μm), accounts for <2 % of the TOC (Benner and Amon 2015) and ranges in size up to centimeters (Simon et al. 2002). Macro-aggregates or marine snow range between 5 to several tenths of mm and originate from coagulation of dead or senescent phytoplankton cells or fecal pellets of zooplankton. Macrogels and transparent exopolymer particles range between mm and μm in size, and microgels and colloidal nanogels range from the micro- to the nano-meter scale. This transparent matter contains polymers of microbial origin such as mucus (Azam and Malfatti 2007).

The direct uptake of DOC is limited to molecules below 600–800 Da when taken up through porins (Blanvillain et al. 2007). Alternatively, TonB-dependent receptors bind extracellular substrates before uptake and allow uptake also of molecules that are >800 Da and participate for instance in the uptake of glycans. Besides in Bacteroidetes, TonB-type receptors occur in marine Alphaproteobacteria (particularly the Sphingomonadales) and Gammaproteobacteria (e.g., Alteromonadales such as *Alteromonas macleodii*) (Neumann et al. 2015; Tang et al. 2012). For the utilization of degradable particulate and higher molecular dissolved organic matter, bacteria have to synthesize and exudate depolymerizing ectoenzymes before the bacterial cell can take up the degradation products. Extracellular enzyme activity is therefore highly relevant for the marine carbon cycle (Arnosti 2011) and bacterial exoenzymes are highly active on marine particles that contain large amounts of polymeric substrates (Smith et al. 1992) (Fig. 2.1). The hydrolysis of laminarin, chondroitin, xylan, pullulan, arabinogalactan and fucoidan was detected in coastal water, whereas the spectrum of hydrolyzed polysaccharides was narrow in offshore waters (D'Ambrosio et al. 2014). Exoenzymes can be exuded into the medium or stay bound to the surface of the cell. Modeling of diffusive losses of enzymes and the generated monomers has demonstrated that the production of free (non cell-bound) exoenzymes pays off only at concentrations of polymeric substrates >3.3 μM, whereas the expression of surface-attached exoenzymes becomes energetically profitable already at substrate concentrations of 0.4 nM (Traving et al. 2015). Production of free exoenzymes is also favorable if cells occur in denser populations and can degrade polymers cooperatively. In several bacterial species the expression of extracellular enzymes is regulated by quorum sensing (Chernin et al. 1998; Hmelo et al. 2011). Addition of inducers may therefore stimulate enzyme production at low cell densities and enable the utilization of polymers in laboratory enrichments which might explain the improved cultivation success upon addition of signaling molecules observed in some cases (Bruns et al. 2002).

Two canonical trophic strategies of bacteria are distinguished based on their very different adaptations towards substrate supply (Koch 2001). Copiotrophic bacteria are adapted to rapid growth upon encountering high substrate levels, e.g. during phytoplankton blooms or on detrital aggregates whereas they just persist in low abundances under severe substrate limitation. This so-called 'feast and famine strategy' involves rapid substrate uptake and high maximum growth rates (>1 h^{-1}), high cell volumes (>1 μm^3) in rich media, a high diversity of specific substrate transporters, reductive (*rpoS*-dependent) cell division upon starvation, cell motility and chemotaxis. At least some species maintain a high intracellular ribosome content and thereby can react quickly to sudden increases in growth substrates (Beardsley et al. 2003; Eilers et al. 2000). In contrast, oligotrophic bacteria are often highly abundant, possess few broad-spectrum uptake systems, have low growth rates (<0.2 h^{-1}) that are independent of high substrate concentrations, are small-celled (<0.1 μm^3), lack motility, and have streamlined genomes and little regulatory flexibility (Lauro et al. 2009). The transcriptomes also reflect these two life strategies and show that the fast-growing bacterial taxa in bacterioplankton represent metabolic generalists and are strongly controlled by top-down mechanisms. In contrast, slow-growing, abundant taxa are characterized by metabolic specialization (Gifford et al. 2013).

2.4.2 Adaptations to Temporal and Spatial Heterogeneity

Marine Bacteria and Archaea with a copiotrophic lifestyle proliferate at high substrate and nutrient concentrations. Examples of typical marine copiotrophs are found in the genera *Alteromonas*, *Pseudoalteromonas*, *Photobacterium*, and *Vibrio* (Lauro et al. 2009; Yooseph et al. 2010). In the ocean, higher substrate and nutrient concentrations are found on the continental shelves (<150 m water depth) that cover ~7 % of the surface area of the world ocean but due to their often higher productivity harbor a much larger proportion of the global number of bacterial cells (Kallmeyer et al. 2012). Under energy-replete conditions, e.g. during phytoplankton blooms, copiotrophic bacteria can form distinct blooms and reach cell numbers of 10^7 ml^{-1}. Most notably, copiotrophs show specific adaptations to the spatially heterogeneous distribution and transient availability of their substrates, and to the polymeric nature of carbon substrates that are produced by marine phytoplankton.

While satellite images as well as shotgun metagenomic analysis demonstrated that patchiness of microbial populations exist on scales of tens of kilometers (Venter et al. 2004), microscale spatial heterogeneity has been demonstrated in coastal seawater where the abundance of different bacterial cell types varies more than tenfold over horizontal distances of millimeters (Seymour et al. 2004). These patterns are due to the effects of (living or dead) organic particulates (Fig. 2.1).

Direct colonization of marine aggregates by bacteria occurs at high densities (10^8–10^9 ml^{-1}; Azam and Malfatti 2007) and can constitute up to 20 % of all bacterial biomass in ocean water (Teeling et al. 2012). Because of the high abundance of micro- and macro-gels that reaches 1000 ml^{-1} (as compared to 1–10 marine snow particles ml^{-1}), numerous immotile marine bacteria may adhere to them and thereby concentrated from the surrounding water. This might explain the high abundance of immotile or gliding bacteria (Bacteroidetes) on marine snow particles that form through aggregation of micron-sized gels (Azam and Malfatti 2007). Alternatively, encounter by chance and rapid multiplication stimulated by high substrate concentrations could lead to the colonization of the larger particles.

On the aggregates, bacteria highly express ectoenzymes such as proteases, lipases, chitinases and phosphatases (but less so glucosidases) (Smith et al. 1992). During degradation, members of the Flavobacteriales (*Zobellia*, *Polaribacter*, *Gramella*, *Formosa*, *Cellulophaga*), Planctomycetes, the genus *Roseobacter*, Alteromonadales (*Alteromonas*, *Marinobacter*, *Moritella*, *Pseudoalteromonas*), Oceanospirillales (*Thalassolituus*, *Kangiella*, *Alcanivorax*) *Methylophaga*, *Cycloclasticus*, *Rheinheimera*, and Deltaproteobacteria are enriched whereas Archaea constitute less than 10 % of the microbial assemblages on particles (Fontanez et al. 2015). The presence of presumptive chemoautotrophic sulfur-oxidizers (*Sulfurimonas*, *Sulfurovum*, *Sulfuricurvum*), sulfur- or nitrate-reducing heterotrophs (*Sulfurospirillum*), and chemolithotrophic hydrogen-oxidizing denitrifiers (*Nitratifractor*, *Nitratiruptor*) in marine particles maybe related to an active sulfur cycling and the presence of anoxic microniches such as intestinal tracts of zooplankton carcasses within the large particles (Fontanez et al. 2015).

Exudation by active phytoplankton cells, lysing cells, or degradation of marine snow particles create small patches of dissolved organic matter (Blackburn et al. 1998). Since the bacteria attached to aggregates do not incorporate or respire all liberated monomers, their ectoenzymatic activity produces a plume of readily utilizable carbon substrates behind sinking marine snow (Azam and Malfatti 2007). In a turbulent environment such patches persist at dimensions of up to 300 μm and hence can be exploited by motile Bacteria and Archaea at their typical swimming speeds of up to 40 μm. Motility thus confers a selective advantage over non-motile competitors as long as the energy gained through substrate uptake exceeds the energetic costs of locomotion (Fig. 2.1). The benefit of motility and chemotaxis for high swimming speeds of up to 170 $\mu m\ s^{-1}$, as they occur in marine bacterioplankton, is the higher substrate uptake (up to 133 %) compared to non-motile cells, but this advantage rapidly vanishes at lower frequency and size of nutrient patches (Taylor and Stocker 2012). The fact that up to 70 % of the bacterial cells in coastal waters are motile indicates the relevance of this adaptation for bacteria in this particular habitat (Grossart et al. 2001). At the typical concentrations of aggregates in the marine epipelagic, freely swimming bacteria will encounter an aggregate in less than one day, which will provide sufficient substrate to entirely support the growth of these bacteria (Kiørboe et al. 2002). Although marine aggregates occupy only a small fraction of the marine water volume they can support substantial growth of free-living bacteria in their solute plume at growth rates of 1–10 d^{-1}. Since mean growth rates of bacterioplankton range between 0.05 and 1.0 d^{-1} (Ducklow 2000) degradation of aggregates can promote a considerable part of the heterotrophic bacterial production (Kiørboe and Jackson 2001).

Typical copiotrophic marine bacteria appear to be involved in the transformation of high-molecular DOC in the sea. High-molecular-weight organic matter to concentrations of around 300 μM DOC elicits bacterial growth and specifically stimulates transcription in bacteria affiliated with the genera *Alteromonas* and *Idiomarina*, followed by *Methylophaga*. The first two groups show increased expression of TonB-associated transporters and nitrogen assimilation genes (of the glutamine synthase cycle) (McCarren et al. 2010). Besides genes involved in alginate and fucose utilization, *Alteromonas* cells associated with marine snow particles have also encode heavy-metal resistance genes (the *czsABCD* efflux pump for cobalt, zinc, cadmium, the *cusAB* and *cusRS* copper efflux and copper two-component sensor systems, and the mercury resistance and transport *merABR* and *merTP* systems) (Fontanez et al. 2015). Degradation of marine polysaccharides seems to be mediated also by other Gammaproteobacteria such as *Reinekea* and SAR92, as well as the Alphaproteobacteria of the RCA, which reach high abundances during late stages of phytoplankton blooms (Teeling et al. 2012). *Reinekea* features a broad substrate spectrum typical for a generalist type of bacterium. In contrast, *Methylophaga*-associated genes involved in methanol and formaldehyde oxidation/assimilation were also detected in association with marine particles and expression of genes for C1-compounds (methylotrophic pathways) was demonstrated (McCarren et al. 2010). Since 15 % of the carbohydrates present in DOM consist of methyl sugars, this temporal pattern of transcription suggests that the initial degradation through *Alteromonas* and *Idiomarina* generates methanol or formaldehyde that is subsequently utilized by *Methylophaga* at the end of this aerobic food chain (McCarren et al. 2010).

The Bacteroidetes constitute an abundant marine bacterial group and can reach up to 30 % or even 55 % of the total number of bacterial cells in coastal waters (Hahnke and Harder 2013). Their members have genomes with a particularly large percentage of genes for peptidases and glycoside hydrolases (Fernández-Gómez et al. 2013). Due to the gliding motility and the capacity to form biofilms, members of the Bacteroidetes accumulate on detritus such as fecal pellets and surfaces of living organisms such as phytoplanktonic algae, macroalgae or invertebrates. Many Bacteroidetes are capable of degrading algal cell wall and storage polysaccharides such as agarose, alginate, arabinan, fucosides, laminarins, mannan, porphyran, xylans and carrageenans. They take up and utilize mono-, di-, and trisaccharides that constitute typical algal polysaccharides (arabinose, xylose, galactose, glucose, fructose, mannitol, mannose, rhamnose, *N*-acetylglucosamine; trehalose, sucrose, maltose; raffinose) (Mann et al. 2013). The genes of many of the carbohydrate-active enzymes (CAZymes, that include carbohydrate-binding modules, glycoside hydrolases, polysaccharide lyases, carbohydrate esterases and glycosyltransferases) together with those of TonB-type receptors, sensor kinases, transporters and sulfatases form particular polysaccharide utilization loci (PULs) that constitute the characteristic genomic features of these bacteria.

The Flavobacteriales genera *Zobellia*, *Gramella*, *Formosa*, and *Cellulophaga* are specifically adapted to the degradation of algal-derived organic matter (Teeling et al. 2012). This trait appears to have emerged within this phylogenetic group (Thomas et al. 2012). Accordingly, these bacteria can be enriched on exudates and decay products of algae (Nelson and Carlson 2012; Sarmento and Gasol 2012) and many grow readily on solid media (Hahnke and Harder 2013). The facultative anaerobe *Formosa agariphila* features a 4.23 Mbp genome and is capable of mixed acid fermentation and denitrification, the degradation of algal polysaccharides and the utilization of the resulting mono-, di- and trisaccharides (Mann et al. 2013). The Flavobacterium lineage MS024-2A expresses genes for the oxidation of hydrogen, that might provide supplemental energy for growth, but also expresses genes for alkaline phosphatase, polysulfide reductase, and gliding (Gifford et al. 2013).

Polaribacter represents one of the most abundant Bacteroidetes in marine samples (Fernández-Gómez et al. 2013; Yooseph et al. 2010). While *Polaribacter* or *Dokdonia* also contain glycoside hydrolases and peptidases like other Bacteroidetes specialized in the degradation of polymeric algal matter in nutrient-rich environments, they possess much smaller genomes ($\sim$3 Mbp), are lacking certain genes for carbohydrate utilization, are not capable of fermentative growth, but contain proteorhodopsins as well as an increased fraction of genes involved in anaplerotic CO_2 fixation. Since the latter two features enable additional energy gain in the light (Gómez-Consarnau et al. 2007) they may represent a key adaptation of the two genera to life in illuminated but oligotrophic pelagic waters (Fernández-Gómez et al. 2013).

Since genes for CAZymes typically constitute 2 % and rarely more than 5 % of bacterial genomes, the specific enzyme composition determines the distinct types of polysaccharides that are used for growth of individual Bacteroidetes (Mann et al. 2013). As a result, specialized populations of different members of the Bacteroidetes participate during the decomposition of diatom blooms and hence are specialized to distinct ecological niches (Teeling et al. 2012). A fivefold increase in the abundance of Flavobacteria has been documented during a spring phytoplankton bloom and a succession of *Ulvibacter*, *Formosa* and *Polaribacter* was observed (Teeling et al. 2012). Because of the specific spectrum of polysaccharides utilized by different Bacteroidetes, this property could theoretically be used to enrich and isolate particular types of Bacteroidetes. It should be kept in mind, though, that marine Bacteroidetes from temperate waters encode many more peptidases than glycoside hydrolases and hence seem to be specialized even more to the degradation of proteins (Fernández-Gómez et al. 2013).

Members of the genera *Cycloclasticus*, *Thalassolituus*, *Alcanivorax* and *Marinobacter* are often obligate hydrocarbonoclastic, which might explain their occurrence on marine particles that contain adsorbed hydrocarbons and lipids as well as hydrocarbons of the alga present.

Vibrionales, most notably *Vibrio*, are specifically adapted to a lifestyle in association with eukaryotes. This involves genes for chitin binding and utilization, chemotaxis (towards *N*-acetylglucosamine), quorum sensing and anaerobic metabolism, in particular trimethyl ammonium oxide (TMAO, an abundant osmolyte of marine eukaryotes) utilization (Fontanez et al. 2015). The genes for hemolysins, RTX toxins, vibriolysin, and non-ribosomal peptide synthase detected in particular matter are affiliated with *Vibrio* and may reflect the adaptation of this and other species to the association with surfaces and intestinal tracts of zooplankton (Fontanez et al. 2015). Also, siderophore biosynthesis is typically found in taxa associated with eukaryotes and is required for colonization by the bacteria (Miethke and Marahiel 2007).

Members of the marine *Roseobacter* group have been classified as moderate copiotrophs to oligotrophs (Lauro et al. 2009). They typically utilize a wide array of organic carbon substrates and can oxidize reduced inorganic sulfur compounds such as sulfite or thiosulfate as well as carbon monoxide to supplement heterotrophy. Some members are capable of aerobic anoxygenic photosynthesis and DMSP demethylation. Genomes range between 3.5 and 5 Mbp and encode multiple genes for sensing and reacting to the environment, and for uptake and utilization of numerous substrates including peptides and amino acids. Members of the group also harbor a large number of conjugative plasmids and prophages (Brinkhoff et al. 2008). *Roseobacter* sp. and *Citreicella* sp. appear to utilize reduced inorganic sulfur compounds through the *sox*-pathway and take up amino acids and polyamines. *Citreicella* sp. may also use urea (Gifford et al. 2013).

POM that settles down constitutes the main source of organic carbon for dark, deep-water layers. The deep sea represents another habitat of copiotrophs as indicated by the dominance of genomic features of copiotrophs in deep-sea bacterial metagenomes (Konstantinidis et al. 2009). Occasionally, growth maybe possible for copiotrophs, e.g., in high nutrient patches surrounding sinking marine snow, but less so this is the case for oligotrophs (Yooseph et al. 2010).

2.4.3 Adaptations to Permanently Low Substrate and Nutrient Concentrations

The low productivity regions of oceanic gyres constitute 42 % of the world ocean (Kallmeyer et al. 2012) and hence represent a major habitat type to which oligotrophic marine bacteria have adapted. Nutrient concentrations in the most ultra-oligotrophic waters of the South Pacific Gyre are 0.5 μM for nitrate, 20 nM for ammonia and 50 nM for phosphate (Yin et al. 2013).

Culture-independent genome analysis suggests that oligotrophy represents a life strategy that is not only limited to the genera *Prochlorococcus* and *Pelagibacter* (SAR11) but also prevails in other groups such as some members of the *Roseobacter* group or the SAR116 clade (Swan et al. 2013) (Fig. 2.1). These highly abundant, often cosmopolitan taxa of marine bacterioplankton typically have smaller genomes and appear to be capable only of slow growth but cannot readily sense and adapt to substrate-rich environments. Many regulatory genes, particularly for transcriptional regulation, as well as numerous functional genes, including those encoding energy-linked sugar and amino acid uptake, siderophore synthesis or uptake, molybdate uptake and processing, efflux systems, extracellular enzymes, anaerobic- or micro-aerobic metabolism (including the cytochrome *bd* complex, genes for nitrate reduction or the regulators $^{54}\sigma$ (sigma factor) and *arcBA*), motility, sensory proteins, chemotaxis and quorum sensing are missing or greatly diminished. Genomes of oligotrophic representatives from various phylogenetic groups are characterized by a low GC content (average of 41 single cell genomes, 37.9 %; Swan et al. 2013), which may reflect adaptation to nitrogen limitation (Grzymski and Dussaq 2012) or the absence of effective DNA repair systems (Giovannoni et al. 2005a). Since the low GC content also affects the first two nucleotides of codons it leads to a strong bias towards utilization of tyrosine, phenylalanine, isoleucine, glutamate, asparagine, lysine and serine. Since photolyases are over-represented in the genomes, photodamage seems to exert a selective pressure on the surface-dwelling oligotrophic bacteria (Yooseph et al. 2010). The genomic and regulatory streamlining of oligotrophic bacteria results in a decreased demand for nutrients and energy and enables the cells to replicate under most limiting nutrient conditions. The fact that Clustered Regularly Interspaced Short Palindromic Repeat (CRISPR) arrays and Cas genes are under-represented in the genomes of oligotrophic bacteria suggests a low selection pressure of lytic bacteriophages. It has been suggested that the constitutively slow growing, small-celled bacterial taxa with small genomes cannot adapt to changing environmental conditions, but avoid grazing by protists or infection and lysis by bacteriophages and hence can maintain densities of 10^5 cells ml^{-1}.

Canonical marine oligotrophs such as members of the SAR11 clade lack the genes for several essential cell constituents such as reduced sulfur compounds (Carini et al. 2013; Tripp et al. 2008). SAR11 bacteria express light-dependent proteorhodopsin proton pumps and systems for the uptake and metabolism of the compatible solutes glycine betaine, ectoine and hydroxyectoine (Gifford et al. 2013). Growth of an isolate of the SAR11 clade is significantly improved by glycine betaine (Tripp et al. 2009). Osmolytes and amino acids may constitute important carbon substrates for these bacteria (Giovannoni et al. 2005b; Tripp et al. 2008, 2009). Single cell genomic studies of the alphaproteobacterial SAR116 group suggests that CO and reduced inorganic sulfur compounds might represent growth substrates for this marine bacterium (Swan et al. 2013). As another oligotrophic representative of the marine Alphaproteobacteria, *Sphingopyxis alaskensis* grows only slowly but has the ability of maintaining a constant specific growth rate also on substrate concentrations (Cavicchioli et al. 2003). Using thiosulfate as an additional electron donor besides glucose allowed the isolation of the first representative of the gammaproteobacterial sulfur oxidizer SUP05 group (Marshall and Morris 2013), supporting the view that ambient thiosulfate present in marine waters is utilized by ubiquitous sulfur-oxidizing marine planktonic bacteria. Other isolates of obligate oligotrophic marine Gammaproteobacteria cannot multiply at organic carbon concentrations above 30 μM and only part of the isolates tested grew on nutrient poor solid media such as 1/10 strength R2A agar after acclimatization to laboratory conditions. The temperature optima for growth were typically between 16 and 20 °C and most isolates could not grow above 25 °C (Cho and Giovannoni 2004).

2.4.4 Other Types of Adaptations

Other unique features suggested by high coverage transcriptional profiling of marine Bacteria and Archaea encompass the use of substrates such as methanesulfonate, taurine, tartrate and ectoine, as well as alternative energy conservation pathways such as by proteorhodopsin (Gifford et al. 2013).

Organosulfur compounds are relevant growth substrates for marine bacterioplankton and might serve as carbon and sulfur source or solely as sulfur source in the absence of an assimilatory sulfate reduction pathway. Whereas in coastal waters members of the *Roseobacter* clade cleave the osmolyte dimethylsulfoniopropionate (DMSP), bacteria of the SAR116 clade are responsible for the DMSP degradation in the euphotic zone of oligotrophic oceanic regions (Choi et al. 2015). In the marine euphotic zone, members of the SAR116 clade can constitute over 10 % of the bacterial community. "*Candidatus* Puniceispirillum marinum" was the first cultured representative of the alphaproteobacterial SAR116 clade. Besides genes for proteorhodopsin, the small (2.75 Mbp) genome of '*Candidatus* Puniceispirillum marinum' encodes aerobic-type carbon monoxide dehydrogenase, DMSP demethylase, DMSP lyase and dimethylsulfoxide reductase, genes for C1-metabolism (of methanol, formaldehyde, formate, formamide, methanesulfonate, and a gene for glycine/serine hydroxymethyltransferase), and a high-affinity ABC-type phosphate uptake system, polyphosphate kinase and exopolyphosphatase (Oh et al. 2010). This indicates the capability of chemoheterotrophic growth on DMSP and an adaptation to the low inorganic nutrient conditions in oligotrophic waters. Metatranscriptomic studies have shown that relatives of "*Candidatus* Puniceispirillum marinum" seem to be motile and express genes of the degradation pathway of protocatechuate, and for the transport and utilization of nitrate and polyphosphate (Gifford et al. 2013). Since protocatechuate is the central intermediate during the degradation of aromatic compounds such as lignin, its biochemical features may enable "*Candidatus* Puniceispirillum marinum" to participate in the mineralization of complex organic matter. Notably, "*Candidatus* Puniceispirillum marinum" is infected by a novel type of double stranded DNA bacteriophage that constitutes up to 25 % of the oceanic virome and maybe capable of altering the metabolism of its host through expression of a putative methanesulfonate monooxygenase (Kang et al. 2013). Another example for the adaptation to a specific substrate is the discovery of autotrophic, ammonia-oxidizing members of the planktonic marine Crenarchaeota that catalyze the major part of ammonia oxidation in many marine environments (*Nitrosopumilus maritimus*; Könneke et al. 2005).

Genomes of numerous but different bacterial phylotypes have been found to encode bacteriorhodopsin that allows a non-chlorophyll-based utilization of light energy. After its initial discovery in a marine gammaproteobacterium (Beja et al. 2000), hundreds of different and phylogenetically divergent homologs of bacteriorhodopsin were identified by using metagenomics (Venter et al. 2004). To date, bacteriorhodopsin has been found in the SAR11 clade, marine Bacteroidetes, and Euryarchaeota (Frigaard et al. 2006).

Mesopelagic blooms of bacteria of the SAR11 sub-clades Ib and II, Actinobacteria, and the SAR202 and OCS116 clades (Morris et al. 2005) maybe specifically adapted to degrade semi-labile DOM that is exported from upper water layers (Carlson et al. 2009). Based on metagenomic and metatranscriptomic data, several groups of Bacteria and Archaea in oceanic deep waters are adapted to suboxic zones of the deep ocean and capable of chemolithotrophic growth. Thus, representatives of the SAR324 group of the Deltaproteobacteria contain sulfur oxidation and carbon fixation genes. A population of these bacteria in and in the vicinity of a hydrothermal plume expressed genes for alkane oxidation such as hydrocarbon monooxygenase, enzymes involved in the degradation of the respective alcohols and fatty acids, but also formate dehydrogenase and nitrite reductase. The considerable metabolic versatility of these representatives of the SAR324 group may explain their ubiquity in the marine environment (Sheik et al. 2014). Bacteria of the candidate phylum Marine Group A (that encompasses the previous cluster SAR406, related to the genus *Fibrobacter* and the phylum Chlorobi) occur ubiquitously in the ocean. Members of this group are adapted to low oxygen concentrations (down to concentrations of 1–20 μM) and hence are associated with the oxyclines of oceanic oxygen minimum zones or stratified anoxic basins where they attain mean relative abundances of up to 11 % of total cell numbers. Besides low oxygen concentrations, these oxygen minimum zones are also characterized by lower water temperatures and elevated concentrations of nitrate and phosphate (Allers et al. 2013). In suboxic environments, bacteria of the Marine Group A might participate in a cryptic marine sulfur cycle since metagenomics has revealed the presence of genes involved in sulfur-based energy metabolism, in particular polysulfide reductase that may catalyze either dissimilatory polysulfide reduction to hydrogen sulfide or dissimilatory sulfur oxidation (Wright et al. 2014).

Finally, deep-sea sediments constitute a highly specific habitat to which different lineages of Bacteria and Archaea have been adapted. The Planctomycetes and Thaumarchaeota that inhabit deep-sea surface sediments down to 2 m below the seafloor maybe involved in the anaerobic or aerobic oxidation of ammonium, respectively (Parkes et al. 2014). The sub-seafloor Chloroflexi that dominate below the surface of many deep-sea sediments, belong to different subgroups. Members of the *Anaerolinea*, like their cultured relatives, maybe organoheterotrophs, whereas other deep-sea lineages are related to the Dehalococcoides that encompass obligate anaerobic reductive dehalogenating bacteria. Yet, at least one representative of this group from Baltic Sea sediments, based on single cell genomics, seems to be metabolically more versatile and may utilize fatty acids and aromatic compounds but does not contain genes known to be involved in reductive dehalogenation (Wasmund et al. 2013). Archaea of the Miscellaneous Crenarchaeotal Group may degrade detrital proteins (Lloyd et al. 2013). These examples stress that the physiology of so-far-uncultured bacterial phylotypes cannot directly be inferred from the known phenotype of phylogenetically close relatives.

The deeper, sub-seafloor sediments are characterized by mostly recalcitrant organic matter, low temperatures and elevated pressure and hence constitute a particular type of habitat with bacterial cells so far detected down to 1922 m of sediment depth. Bacteria and Archaea might even exist in deeper layers, only limited by temperatures higher than 122 °C (Parkes et al. 2014). Although the energy available in this habitat is 1000 times lower than the energy requirements determined in laboratory cultures (Hoehler and Jørgensen 2013), the majority of Bacteria and Archaea present appear to be active (Schippers et al. 2005). Evidently, such organisms must have adapted to the subsurface habitats with respect to mechanisms of substrate utilization, maintenance energy requirements, and survival. Higher bacterial cell numbers in layers rich in diatomaceous organic matter, higher hydrolytic exoenzyme activities and anaerobic glucose degradation rates suggest that some supposedly recalcitrant organic matter may actually provide substrates to the bacterial populations present (Coolen et al. 2002).

2.5 Microdiversity, Population Genetics and Ecotypes of Marine Bacteria

It is well established that phylogenetically closely related strains of the same species may occupy different ecological niches and hence represent different so-called ecotypes. Even strains with identical 16S rRNA gene sequence can exhibit considerable genetic and phenotypic differences (Jaspers and Overmann 2004). Large-scale metagenomic analysis revealed a small population size of clonal bacteria (below 1 % relative abundance in the bacterial communities) in different marine environments (Rusch et al. 2007). Across different genomic regions, nucleotide sequences differ by 3–5 %. While many of these differences are silent and hence neutral with respect to selective processes, *Prochlorococcus*, the SAR11 clade, as well as other marine bacteria, represent physiologically highly diverse populations of phylogenetically related bacteria (Rusch et al. 2007; Venter et al. 2004). These findings are corroborated by analyses of single cell genomes of highly similar bacterial ribotypes that revealed substantial differences between their entire genomes (Swan et al. 2013). The micro-diversity within the clades of dominant bacterioplankton influences the productivity and nutrient cycling in the ocean. Therefore, isolation and characterization of the adaptations of sub-clades is of relevance for the modeling the biogeochemical cycling in the ocean (Follows et al. 2007).

The alphaproteobacterial SAR11 clade (*Pelagibacter ubique*) has served as a model for the study of ecotypes (Giovannoni et al. 2005a) and comprises at least 4 distinct sub-clades with sequence divergence between 3 and 11 %. Genomic variation in the clade is not unstructured but consists of distinct subtypes in terms of sequence similarity, gene content and distribution; the individual genotypes may have diverged already millions of years ago (Rusch et al. 2007). The loss of the mutLS mismatch repair system may have caused this extensive sequence divergence of the clade (Viklund et al. 2012). Co-occurring members of three of the sub-clades diverge by >4 % in their 16S rRNA gene sequences, exhibit distinct vertical and seasonal patterns, and different adaptations to nutrient limitation and UV stress (Carlson et al. 2009). These patterns indicate the existence of (non-geographical) barriers of genetic exchange (Rusch et al. 2007). Surface-dwelling members of the SAR11 clade express high-affinity uptake systems for phosphate, amino acids and carbohydrates and outcompete other bacteria for amino acids, sugars, aromatic monomers and DMSP (Malmstrom et al. 2004; Mou et al. 2007). Members of SAR11 that are adapted to high latitude, low temperature (<10 °C) polar waters contain specific genes for cell wall/membrane/envelope biogenesis, for inorganic ion transport and metabolism, and for energy production and conversion (Brown et al. 2012). Numerous genes like those encoding glycosyl transferases and glycosyl synthetases, Lpx, or FtsZ show indications for positive selection and have been related to the adaption of the cells to lower ambient temperatures.

The evolutionary mechanisms that underlie the segregation of ecotypes are different. For example, the sequences of the two spectral variants of proteorhodopsin in SAR11 form coherent, but multiple clusters. This phylogeny indicates several independent (convergent) adaptation events to the specific light environment in blue open ocean and green coastal waters with subsequent evolution within the separate sub-clades. Conversely, the presence of the high-affinity inorganic phosphate transporter PstS is entirely independent of the phylogeny of SAR11. Thus, complex physiological properties like light-dependent generation of a proton motive force by proteorhodopsin seems to be less rapidly acquired or lost by horizontal gene transfer than single genes/gene clusters. As a result, certain phenotypes (e.g. of phosphate acquisition) may occur across different ecotypes with otherwise distinct physiological properties (Rusch et al. 2007).

Niche separation of phylogenetically closely related lineages is also well documented for the genus *Prochlorococcus*. This genus represents a genetically diverse group that consists of several distinct ecotypes. Phylogenetic clades with a divergence of <3 % of the 16S rRNA gene are adapted to different temperatures, light intensities, and show distinct pigmentation, maximum growth rates, nutrient preferences, and metal tolerances (Coleman et al. 2006; Johnson et al. 2006; Mann et al. 2002; Moore and Chisholm 1999; Moore et al. 2002) and these different ecotypes show distinct differences in abundance patterns along environmental gradients. Distinct patterns were even observed for two clades that differed in their 16S rRNA gene sequence by less than 1 % (Johnson et al. 2006).

Several other bacterial species such as *Vibrio splendidus* (Hunt et al. 2008) may contain a considerable number of up to 15 of micro-diverse clusters that are phylogenetically closely related (>99 % sequence similarity of the 16S rRNA genes) but nevertheless adapted to distinct ecological niches (Hunt et al. 2008). In these cases, ecological differentiation could even be detected for sequence types that differed by as little as one selectively neutral base substitution in a protein-coding housekeeping gene. Similarly, ecotypes within the marine *Roseobacter* group appear to correspond to sequence clusters with ≤ 1 % divergence (Buchan et al. 2005). These micro-diverse sequence clusters may represent evolutionary units that actually correspond to a species.

Based on this cumulative information, single isolates of bacterial species will only insufficiently represent the full ‘micro’-diversity present in the marine environment. A more extensive isolation effort is highly likely to yield strains differing in physiology, ecology, virulence, or biotechnological relevance. This emphasizes the necessity to improve and extend culture-based approaches for the study of marine microbial diversity. Because of the considerable diversity within similar or identical 16S rRNA-phylotypes, alternative, high resolution, but still generally applicable detection techniques are required. This will help to detect the different sub-clades, to deduce their potential ecological niches, and to follow their enrichment in laboratory media. Internally transcribed spacer (ITS) sequences have been used to distinguish sub-clades and to determine potential ecological niches because of their high sequence variability that maybe similar to that of protein-coding genes (Brown et al. 2012). However, since databases for ITS sequences are currently much smaller than for 16S rRNA genes, ITS sequences will have to be determined for the natural subpopulations of most bacterial target groups.

2.6 The Cultivation-Based Approach Towards Marine Microbial Diversity

2.6.1 The Current Public Archive of Cultivated Marine Bacteria and Archaea

Despite ongoing efforts to isolate marine Bacteria and Archaea, only a minor fraction of the currently estimated ~40,000 marine species and ~800,000 marine phylotypes could be isolated to date. The Global Catalogue of Microorganisms (GCM; Wu et al. 2013) currently lists 1604 of the 350,758 strains in 77 culture collections as originating from "water"; hence an even smaller fraction is expected to be of marine origin. Based on the strain-associated data of the collections hosted by the Leibniz-Institute DSMZ, a total of 1356 strains of Bacteria and 74 strains of Archaea deposited at DSMZ were isolated from marine environments. Only 870 of these strains have a larger set of meta-data associated with them as documented in the bacterial meta-database Bac*Dive* (Söhngen et al. 2015). The small fraction of marine Bacteria and Archaea in culture collections, their low absolute numbers, and the even smaller number of strains that are well documented, emphasize the need for innovative, more extensive, and much better documented, culture efforts in marine research. An improved coverage of marine microbial diversity by well-characterized isolates is a precondition for a profound understanding of microbial processes in the largest ecosystems on Earth.

2.6.2 Current Improvements in Cultivation Technology

Previous reviews (e.g., Overmann 2013) provide an overview of the challenges of culturing-based approaches and possible improvements in culturing technology. The following discussion therefore focuses on the specific aspects of the culturing of marine Bacteria and Archaea.

A pronounced discrepancy between phylotypes recovered by culturing and those decisive for the oceans biogeochemical cycles has been recognized early on. Despite recent advancements (see below), the challenges in obtaining relevant Bacteria and Archaea continue to persist and are particularly relevant for the nutrient-limited surface layer of the open ocean as also demonstrated by the results of metagenomics studies (Rusch et al. 2007). Most of the marine isolates obtained so far represent copiotrophs. If suboptimal growth conditions are established in culturing attempts, only rare, but potentially rapid growing bacteria overgrow the dominant species. Thus, even for the most ultra-oligotrophic open ocean samples, standard approaches involving plating on Marine Agar or Marine R2A media yield only typical copiotrophic species, mostly Gammaproteobacteria (Yin et al. 2013).

While it is evident that the media composition and incubation conditions should mimic the physicochemical conditions of the natural habitat of previously uncultured Bacteria and Archaea, the life strategies and adaptations deduced from metagenomic data can be used as guidelines for improving culturing approaches further. Although many of the specific adaptations of dominant planktonic bacteria remain unresolved to date, the consequent application of low-nutrient media together with high-throughput culturing technology and specific substrate amendments have successfully been applied over the past decade to isolate obligate oligotrophic marine bacteria.

By scaling up the dilution to extinction method (Button et al. 1993), Connon and Giovannoni (2002) were able to culture two strains of the abundant SAR11 clade. Miniaturized culturing plates and liquid handling robots were combined with using natural seawater as low substrate concentration medium. Subsequently, using the dilution to extinction method in natural seawater amended with phosphate (0.1 μM) and ammonium (1 μM), 11 strains of the SAR11 clade were isolated (Rappé et al. 2002). Notably, adding natural DOC in or from seawater provides only very little readily utilizable organic carbon substrates in a pM to nM concentration range for bacteria (see Sect. 2.4.2). Subsequent improvements of the culture media recipe included the addition of a source of reduced sulfur such as methionine or DMSP (Tripp et al. 2009) as well as glycine (Carini et al. 2013; Tripp et al. 2009). Some of the SAR11 strains required glucose as carbon source but the majority required pyruvate or a precursor of pyruvate such as oxaloacetic acid (Carini et al. 2013). Culturing approaches can exploit the fact that SAR11 members are able to produce energy from the oxidation of C1-moieties of methylated compounds (Sun et al. 2011) and can use sunlight (Giovannoni et al. 2005b).

Planktomarina temperata, the first representative of the RCA clade, was isolated from high dilution steps of seawater inoculums in liquid media consisting of autoclaved seawater supplemented with 10 mM thiosulfate, vitamin- and trace element solutions, or alternatively on nutrient poor R2A agar (Giebel et al. 2013). Long incubation times (1–2 months) and incubation temperatures between 15 and 20 °C also were a key for growth of the target bacteria. In artificial media, so-far unidentified growth factors other than vitamins but present in yeast extract or peptone were required for growth and similar requirements have been detected for two other members of the *Roseobacter* clade (Giebel et al. 2013). Growth of these, as well as some other fastidious marine bacteria continues to be unpredictable even on nutrient-poor agar plates. As a consequence, enrichment and subsequent culturing has to be performed in liquid media (Giebel et al. 2013).

A similar application of high-throughput methodology that combined dilution to extinction approaches with the use of very low nutrient growth media yielded numerous novel types of marine Gammaproteobacteria that were previously only known from culture-independent surveys (Cho and Giovannoni 2004).

Other high-throughput methods like the encapsulation method coupled with flow cytometry (Zengler et al. 2002) or the filtration-acclimatization method (Hahn et al. 2004) successfully permitted the isolation of novel types of marine bacteria. The encapsulation of single cells in agar beads enables their growth separated from the other cells and at environmental nutrient concentrations when the agar beads are suspended in a flow-through column. Subsequently, gel droplets containing micro-colonies are detected by flow cytometry and sorted into a micro-plate for further culturing. Bacteria of the SAR11 and SAR116 clades, and Bacteroidetes were cultured by this method (Zengler et al. 2002). Similarly, diffusion chambers constructed from sandwiches of membrane filters permit an exchange of low molecular compounds with the environment while preventing the cells to escape. This allows the enrichment of bacteria in their quasi-natural environment. When inoculated with intertidal sediments and incubated in an aquarium with natural seawater, new species of Bacteroidetes were enriched (Kaeberlein et al. 2002).

While some representatives of dominant phylogenetic clades were isolated from nutrient-poor oceanic waters using improved culturing approaches, single cell genomic studies revealed that many dominant representatives of the alphaproteobacterial *Roseobacter* or SAR116 clades differ from the isolates by their streamlined genomes, high AT-content, and lower numbers of genes for transcription, signal transduction and non-cytoplasmatic proteins (Swan et al. 2013). These features have been taken as evidence for specific adaptations of the different marine planktonic bacterial species to an oligotrophic life style (Giovannoni et al. 2005a) and emphasize that culturing approaches towards marine oligotrophs require substantial improvement. For instance, abundant oligotrophic marine bacteria seem to use non-chelated ferrous or ferric iron but cannot take up siderophores (Yooseph et al. 2010). Therefore, strongly chelated iron, as present in some trace element solutions, should not be used when enrichments or isolation of these types of bacteria is attempted. In addition, the lack of extracellular enzymes suggests that these bacteria are not adapted to the utilization of polymeric extracellular substrates or to biofilm formation (Yooseph et al. 2010).

Bacteria and Archaea that were retrieved from sub-seafloor sediments are affiliated with Proteobacteria, Firmicutes, Actinobacteria and Bacteroidetes as well as with *Methanoculleus* and *Methanococcus*. Although some of the Proteobacteria (e.g., *Pseudomonas*, *Halomonas*, *Marinobacter* or *Rhizobium*) are also detectable by culture-independent techniques in the marine subsurface sediments, typical inhabitants of surface sediments are usually not recovered by the existing enrichment and isolation approaches, despite some adjustments of substrate concentrations (Parkes et al. 2014). Another discrepancy between the results of culture-independent and -dependent studies of sediments is the frequent isolation of Actinobacteria and Firmicutes that maybe the result of insufficient detection of endospores by molecular techniques and/or the rapid growth of endospores of Firmicutes when encountering the favorable growth conditions in laboratory cultivation trials (Parkes et al. 2014).

Culturing success for samples from sub-surface sediments have often been extremely low (0.0003 % of total cell numbers; Parkes et al. 2014), which correlates with the unusual environmental conditions of these habitats. The pronounced energy limitation of the sub-seafloor is probably the reason for the very low cultivation success of resident Bacteria and Archaea in standard laboratory media. Accordingly, decreasing the concentrations of substrates below mM concentrations can increase the culturing success by up to 10,000 fold (Sub et al. 2004). It has been suggested that new approaches such as continuous-flow-type bioreactors or culturing at sub-nanomolar H_2 concentrations are required for the growth of bacteria from the deep sub-surface (Parkes et al. 2014). A culture apparatus for maintaining H_2 at sub-nanomolar concentrations was designed (Valentine et al. 2000) and the maintenance of H_2 concentrations at ~ 10 pM permitted culturing of the ethanol-oxidizing *Pelobacter acetylenicus*. New isolates of the Chloroflexi phylum and the genus *Spirochaete* were isolated using a continuous-flow-type bioreactor filled with polyurethane sponges mimicking the porous layers of the sub-seafloor sediments (Imachi et al. 2011).

Similar to the lifestyle of free-living oligotrophs, the specific adaptations to a spatially heterogeneous distribution of substrates (see Sect. 2.4.2) can be exploited for the selective culturing of Bacteria and Archaea.

The cultivation success of marine Flavobacteria was significantly improved by using solid media based on artificial seawater, complex carbon sources (yeast extract, peptone or casamino acids) or defined carbohydrates, washed agar, low incubation temperature (11 °C) and extended incubation times (up to 110 days). Similar to oligotrophic bacteria, high throughput dilution culturing (here by using a 96-pin replicator; Hahnke and Harder 2013) proved highly successful. Furthermore, addition of kanamycin at concentrations of 50 $\mu g\ ml^{-1}$ resulted in the selective growth of Flavobacteria.

The selective enrichment of bacteria capable of adhesion in cultured biofilms has been proven to yield novel, previously unknown phylotypes (Gich et al. 2012). The marine heterotrophic bacterial communities colonizing glass slides under limiting concentrations of organic carbon are distinctly different, and also exhibit a considerable higher diversity than the accompanying planktonic communities. The higher diversity of biofilm communities is due to the enrichment of phylotypes with low abundance in the planktonic community and may be caused by a higher diversity of the biofilm habitat with respect to oxygen, nutrient and pH gradients (Pepe-Ranney and Hall 2015). The biofilm enrichment approach thus should yield novel types of bacteria especially under limiting supply of organic carbon substrates (e.g. at a molar ratio of C:P $\leq$ 100).

Naturally forming particles can be exploited for the selective culturing of target bacteria. Transparent exopolymeric particles (TEP) forming after an algal bloom (Engel 2000) are enriched in carbon and appear to be a key element in the fluxes of organic matter in oceanic environment (Engel et al. 2004). Mesocosms filled with post-bloom seawater, amended with nutrients (N, P and Si) and submitted to turbulence showed an increased bacterial colonization of TEP. *Cytophaga*-Flavobacterium members were the most abundant on TEP amounting up to 59 % of the total cells (Pedrotti et al. 2009).

Theoretically, an enrichment of bacteria adapted to the hydrolysis of exopolymers should be possible on polymeric substances that are kept under continuous flow to eliminate other, free-living bacteria that exploit monomers liberated. This could also be done by using gel beads inoculated with single cells and kept in a continuous flow and in a monolayer.

Alternatively, motile, chemotactically active bacteria can be enriched from marine samples using a chemotactic assay (Tout et al. 2015). This has been shown to yield previously unknown bacteria (Tout et al. 2015). This strategy is particularly powerful to rapidly screen the chemotactic response to numerous substrates in order to determine potential substrates for subsequent cultivation trials (Overmann 2005). Tout et al. (2015) demonstrated that some taxa were enriched using only one chemo-attractant. These specialists included taxa belonging to Endozoicomonaceae, Alteromonadaceae and Vibrionaceae families.

In their natural environment, many marine bacteria interact with accompanying (micro)organisms. Since the obligate syntrophic co-culture of an ethanol-oxidizing organism (S organism) and a hydrogen-utilizing methanogenic microorganism (*Methanobacillus omelianskii*) was documented for *Methanobacterium omelianskii* (Bryant et al. 1967), it has been suggested to consider biotic interactions also in other culture attempts. The co-culturing of the unicellular cyanobacterium *Synechococcus* with the heterotrophic bacterium *Ruegeria pomeroyi* allowed the latter to produce a set of secreted proteins for shaping the possible interaction between these organisms (Christie-Oleza et al. 2015). The alphaproteobacterium *Phaeobacter inhibens* in co-culture with the marine microalga *Emiliana huxleyi* secretes hormones and antibiotics for the algae as long as the latter provides it with food. Upon senescence of *E. huxleyi*, the bacterium produces roseobactericides to kill the alga (Seyedsayamdost et al. 2011, 2014). The culturing of the alphaproteobacterium *Dinoroseobacter shibae* was strongly improved by the presence of its host, the dinoflagellate *Prorocentrum minimum* (Wang et al. 2014). Another example is the effect of the addition of signaling compounds such as cAMP to enrichment media that could improve the culturability of microorganisms from natural samples (Bruns et al. 2002).

Based on the distinct composition of bacterial communities in different marine habitats, it could theoretically be concluded that the origin of a marine inoculum chosen for culturing attempts should determine the success of growing particular target bacteria. Yet, given the rapid dispersal of bacteria and the capability of many bacteria to survive unfavorable environmental conditions, even bacterial phylotypes with very low abundance and inactive phylotypes can be obtained from natural samples when employing highly selective enrichment conditions. An example for the recovery of unexpected types of bacteria is the successful enrichment and isolation of thermophilic sulfate-reducing bacteria from permanently cold surface sediments of the Arctic (Hubert et al. 2010). In order to physically separate and pre-enrich rare bacteria prior to cultivation attempts, sucrose density centrifugation may be employed as successfully demonstrated for different taxa of actinomycetes (Xiong et al. 2013). As an alternative, isoelectric focusing of entire bacterial cells provides another means for separating different types of bacterial cells according to their surface charge (Jaspers and Overmann 1997).

2.6.3 Ex Situ Preservation of Marine Bacteria and Archaea

Bacterial strains that are kept as continuously growing cultures will change due to the accumulation of genetic changes such as spontaneous point mutation, loss of plasmids or transposons, or selection of mutants by laboratory growth and storage conditions. The feasibility of cryo-preservation (down to −190 °C) without loss in viability was already demonstrated at the end of the nineteenth century (Macfadyen 1899). The detrimental effects of repeated freeze-and-thaw cycles (Rivers 1927) and the method of freeze-drying (Swift 1921) were described two decades later.

In most research laboratories today, bacterial isolates are stored at −80 °C for convenience and financial reasons. The concentration of the inoculum and the type of cryoprotectant employed are important for the survival and regrowth of the preserved strains when using this approach. After 12–18 months, the viability of common pathogenic bacterial species stored without cryoprotectant decreases to <20 %, while adding blood serum raised the viability to 80–90 % (Hubàlek 2003; Moore and Shaw 2001). Similarly, the survival of a cryo-preserved culture of *Haemophilus influenzae* after 1 year at −20 °C and in a mixture of cryoprotectant (skim milk, glucose, glycerol, yeast extract) was >80 % (De Saab 2001). Additional cryoprotectants that are currently in use include dimethyl sulfoxide (DMSO), serum albumin, peptone, saccharose, methanol, polyvinylpyrrolidone (PVP), sorbitol and malt extract (Hubálek 2003). Most commonly now, 10–25 % (v/v, final concentration) glycerol or 5–10 % (v/v, final concentration) DMSO are used for cryopreservation (Prakash et al. 2013). Because frequent freezing and thawing may decrease the viability of a strain (Harrison 1955; Squires and Hartsell 1955; Walker et al. 2006) glass or plastic beads are added to bacterial cultures before freezing. The bacterial cells will coat the beads and individual beads may be removed after freezing, avoiding thawing of the remainder of the sample. Typically, strains can be then stored at −20 °C for one to three years, at −80 °C for one to ten years, and in liquid nitrogen at least for up to 30 years (De Paoli 2005).

Freeze-drying (lyophilization) techniques are used by public collections to preserve microorganisms and quickly deliver strains on demand. This method comprises freezing of the bacterial cell pellet and then the removal of the ice and then residual water by vaporization under vacuum. The freeze-dried sample can then be stored at temperatures between 5 and 10 °C in ampoules sealed under vacuum. Although this technique is efficient for many bacteria, some strains cannot be freeze-dried. Finally, the storage of bacterial cultures in liquid nitrogen and in the presence of cryoprotectant (usually DMSO) within hermetically sealed, thin glass capillaries remains the method of choice to preserve and keep the vast majority of bacterial isolates over extended periods of time, *i.e.* decades (Tindall 2007).

As outlined in this chapter, the enrichment and isolation of phenotypically and genetically novel types of marine bacteria require a major effort and need the input of a considerable number of person hours. A detailed compilation of all actual costs that are associated with the isolation and characterization of a fastidious bacterial strain yielded a monetary value of 9836 Euro per isolate (Overmann 2015). This estimate includes costs for personnel, consumables, and the depreciation of the equipment, but not the costs for sampling. In countries with an emerging economy and lower wages (e.g., India), this monetary value would be somewhat lower (5042 Euro; Overmann 2015). In contrast, the costs for depositing and curating of microbial resources are modest. The deposit of a bacterial strain in a culture collection comprises the culturing and initial preservation (liquid nitrogen and freeze-drying), identification and quality control (e.g., by 16S rRNA gene sequencing, MALDI-TOF, biochemical testing, fatty acid analysis and microscopy), which together incur costs of 918 Euros. Costs for the long-term preservation and maintenance of live cultures, are much lower and amount to 3.60 Euro per strain and year, which represents less than 0.1 % of their overall monetary value. A deposit of available isolates at public culture collections therefore represents a cost-effective way of maintaining the monetary value of isolated strains (Overmann 2015) and hence is essential not only from a scientific, but also from an economical perspective.

Acknowledgments The research leading to these results has received funding from the European Union Seventh Framework Programme (FP7/2007-2013) under grant agreement no 311975. This publication reflects the views only of the author, and the European Union cannot be held responsible for any use which may be made of the information contained therein.

References

Allers E, Wright JJ, Konwar KM et al (2013) Diversity and population structure of Marine Group A bacteria in the Northeast subarctic Pacific Ocean. ISME J 7:256–268

Allesina S, Tang S (2012) Stability criteria for complex ecosystems. Nature 483:205–208

Arnosti C (2011) Microbial extracellular enzymes and the marine carbon cycle. Ann Rev Mar Sci 3:401–425

Azam F, Malfatti F (2007) Microbial structuring of marine ecosystems. Nat Rev Microbiol 5:782–791

Beardsley C, Pernthaler J, Wosniok W et al (2003) Are readily culturable bacteria in coastal North Sea waters suppressed by selective grazing mortality? Appl Environ Microbiol 69:2624–2630

Beja O, Aravind L, Koonin EV et al (2000) Bacterial rhodopsin: evidence for a new type of phototrophy in the sea. Science 289:1902–1906

Benner R, Amon RMW (2015) The size-reactivity continuum of major bioelements in the ocean. Ann Rev Mar Sci 7:185–205

Benson DA, Cavanaugh M, Clark K et al (2013) GenBank. Nucleic Acids Res 41(Database issue): D36–D42

Blackburn N, Fenchel T, Mitchell J (1998) Microscale nutrient patches in planktonic habitats shown by chemotactic bacteria. Science 282:2254–2256

Blanvillain S, Meyer D, Boulanger A et al (2007) Plant carbohydrate scavenging through TonB-dependent receptors: a feature shared by phytopathogenic and aquatic bacteria. PLoS ONE 2:e224

Boras JA, Vaqué D, Maynou F et al (2015) Factors shaping bacterial phylogenetic and functional diversity in coastal waters of the NW Mediterranean Sea. Estuar Coast Shelf Sci 154:102–110

Brinkhoff T, Giebel HA, Simon M (2008) Diversity, ecology, and genomics of the *Roseobacter* clade: a short overview. Arch Microbiol 189:531–539

Brown MV, Lauro FM, DeMaere MZ et al (2012) Global biogeography of SAR11 marine bacteria. Mol Syst Biol 8:595

Bruns A, Cypionka H, Overmann J (2002) Cyclic AMP and acyl homoserine lactones increase the cultivation efficiency of heterotrophic bacteria from the central Baltic Sea. Appl Environ Microbiol 68:3978–3987

Bryant MP, Wolin EA, Wolin MJ et al (1967) *Methanobacillus omelianskii*, a symbiotic association of two species of bacteria. Arch Microbiol 59:20–31

Bryant DA, Liu Z, Li T et al (2012) Comparative and functional genomics of anoxygenic green bacteria from the taxa *Chlorobi*, *Chloroflexi*, and *Acidobacteria*. In: Burnap RL, Vermaas WFJ (eds) Advances in photosynthesis and respiration. Functional genomics and evolution of photosynthetic systems, vol 33. Springer, Dordrecht, pp 47–102

Buchan A, Gonzalez JM, Moran MA (2005) Overview of the marine *Roseobacter* lineage. Appl Environ Microbiol 71:5665–5677

Button DK, Schut F, Quang P et al (1993) Viability and isolation of marine bacteria by dilution culture: theory, procedures, and initial results. Appl Environ Microbiol 59:881–891

Carini P, Steindler L, Beszteri S et al (2013) Nutrient requirements for growth of the extreme oligotroph "Candidatus *Pelagibacter ubique*" HTCC1062 on a defined medium. ISME J 7:592–602

Carlson CA, Morris R, Parsons R et al (2009) Seasonal dynamics of SAR11 populations in the euphotic and mesopelagic zones of the northwestern Sargasso Sea. ISME J 3:283–295

Cavicchioli R, Ostrowski M, Fegatella F (2003) Life under nutrient limitation in oligotrophic marine environments: an eco/physiological perspective of *Sphingopyxis alaskensis* (formerly *Sphingomonas alaskensis*). Microb Ecol 45(3):203–217

Chernin LS, Winson MK, Thompson JM et al (1998) Chitinolytic activity in *Chromobacterium violaceum*: substrate analysis and regulation by quorum sensing. J Bacteriol 180:4435–4441

Cho JC, Giovannoni SJ (2004) Cultivation and growth characteristics of a diverse group of oligotrophic marine *Gammaproteobacteria*. Appl Environ Microbiol 70:432–440

Choi DH, Park KT, An SM et al (2015) Pyrosequencing revealed SAR116 clade as dominant dddP-containing bacteria in oligotrophic NW Pacific Ocean. PLoS ONE 10:e0116271

Christie-Oleza JA, Scanlan DJ, Armengaud J (2015) "You produce while I clean up", a strategy revealed by exoproteomics during *Synechococcus-Roseobacter* interactions. Proteomics 15:3454–3462

Coleman ML, Sullibvan MB, Martiny AC et al (2006) Genomic islands and the ecology and evolution of *Prochlorococcus*. Science 311:1768–1770

Connon SA, Giovannoni SJ (2002) High-throughput methods for culturing microorganisms in very low nutrient media yield diverse new marine isolates. Appl Environ Microbiol 68 (8):3878–3885

Coolen MJL, Cypionka H, Smock A et al (2002) Ongoing modification of Mediterranean Pleistocene sapropels mediated by prokaryotes. Science 296:2407–2410

Cottrell MT, Kirchman DL (2003) Contribution of major bacterial groups to bacterial biomass production (thymidine and leucine incorporation) in the Delaware estuary. Limnol Oceanogr 48:168–178

D'Ambrosio L, Ziervogel K, MacGregor B et al (2014) Composition and enzymatic function of particle-associated and free-living bacteria: a coastal/offshore comparison. ISME J 8:2167–2179

D'Hondt S, Spivack A, Pockalny R et al (2009) Subseafloor sedimentary life in the South Pacific Gyre. Proc Natl Acad Sci USA 106:11651–11656

Dalsgaard T, Thamdrup B, Farías L, Revsbech NP (2012) Anammox and denitrification in the oxygen minimum zone of the eastern South Pacific. Limnol Oceanogr 57:1331–1346

De Paoli P (2005) Bio-banking in microbiology: from sample collection to epidemiology, diagnosis and research. FEMS Microbiol Rev 29:897–910

De Saab O (2001) A comparative study of preservation and storage of Haemophilus influenzae. Mem Inst Oswaldo Cruz 96:583–586

Doos K, Nilsson J, Nycander J et al (2012) The world ocean thermohaline circulation. J Phys Oceanogr 42:1445–1460

Ducklow H (2000) Bacterial production and biomass in the ocean. In: Kirchman DL (ed) Microbial ecology of the oceans. Wiley-Liss Inc, New York, pp 85–120

Eilers H, Pernthaler J, Amann R (2000) Succession of pelagic marine bacteria during enrichment: a close look at cultivation-induced shifts. Appl Environ Microbiol 66:4634–4640

Eilers H, Pernthaler J, Peplies J et al (2001) Isolation of novel pelagic bacteria from the German Bight and their seasonal contribution to surface picoplankton. Appl Environ Microbiol 67:5134–5142
Engel A (2000) The role of transparent exopolymer particles (TEP) in the increase in apparent particle stickiness (alpha) during the decline of a diatom bloom. J Plankton Res 22:485–497
Engel A, Thoms S, Riebesell U et al (2004) Polysaccharide aggregation as a potential sink of marine dissolved organic carbon. Nature 428(6986):929–932
Fernández-Gómez B, Richter M, Schüler M et al (2013) Ecology of marine Bacteroidetes: a comparative genomics approach. ISME J 7:1026–1037
Follows MJ, Dutiewicz S, Grant S et al (2007) Emergent biogeography of microbial communities in a model ocean. Science 30:1843–1846
Fontanez KM, Eppley JM, Samo TJ et al (2015) Microbial community structure and function on sinking particles in the North Pacific subtropical gyre. Front Microbiol 6:469
Frigaard NU, Martinez A, Mincer TJ et al (2006) Proteorhodopsin lateral gene transfer between marine planktonic Bacteria and Archaea. Nature 439:847–850
Gaston KJ (2000) Global patterns in biodiversity. Nature 405:220–227
Ghiglione JF, Galand PE, Pommier T et al (2012) Pole-to-pole biogeography of surface and deep marine bacterial communities. Proc Natl Acad Sci 109:17633–17638
Gich F, Janys MA, König M et al (2012) Enrichment of previously uncultured bacteria from natural complex communities by adhesion to solid surfaces. Environ Microbiol 14:2984–2997
Giebel HA, Kalhoefer D, Gahl-Janssen R et al (2013) *Planktomarina temperata* gen. nov., sp. nov., belonging to the globally distributed RCA cluster of the marine *Roseobacter* clade, isolated from the German Wadden Sea. Int J Syst Evol Mircobiol 63:4207–4217
Gifford SM, Sharma S, Booth M et al (2013) Expression patterns reveal niche diversification in a marine microbial assemblage. ISME J 7:281–298
Giovannoni SJ, Tripp JH, Givan S et al (2005a) Genome streamlining in a cosmopolitan oceanic bacterium. Science 309(5738):1242–1245
Giovannoni SJ, Bibbs L, Cho JC et al (2005b) Proteorhodopsin in the ubiquitous marine bacterium SAR11. Nature 438:82–85
Gómez-Consarnau L, González JM, Coll-Lladó M et al (2007) Light stimulates growth of proteorhodopsin-containing marine *Flavobacteria*. Nature 445:210–213
Grossart H, Riemann L, Azam F (2001) Bacterial motility in the sea and its ecological implications. Aquat Microb Ecol 25:247–258
Grzymski JJ, Dussaq AM (2012) The significance of nitrogen cost minimization in proteomes of marine microorganisms. ISME J 6(1):71–80
Hahn M, Höfle M (1999) Flagellate predation on a bacterial model community: interplay of size-selective grazing, specific bacterial cell size, and bacterial community composition. Appl Environ Microbiol 65(11):4863–4872
Hahn MW, Stadler P, Wu QL et al (2004) The filtration-acclimatization method for isolation of an important fraction of the not readily cultivable bacteria. J Microbiol Methods 57:379–390
Hahnke RL, Harder J (2013) Phylogenetic diversity of *Flavobacteria* isolated from the North Sea on solid media. Syst Appl Microbiol 36:497–504
Harrison JA (1955) Survival of bacteria upon repeated freezing and thawing. J Bacteriol 711–715
Herlemann DPR, Labrenz M, Jürgens K et al (2011) Transitions in bacterial communities along the 2000 km salinity gradient of the Baltic Sea. ISME J 5:1571–1579
Hmelo LR, Mincer TJ, Van Mooy BAS (2011) Possible influence of bacterial quorum sensing on the hydrolysis of sinking particulate organic carbon in marine environments. Environ Microbiol Rep 3:682–688
Hoehler TM, Jørgensen BB (2013) Microbial life under extreme energy limitation. Nat Rev Microbiol 11:83–94
Hubálek Z (2003) Protectants used in the cryopreservation of microorganisms. Cryobiology 46:205–229

Hubert C, Arnosti C, Brüchert V et al (2010) Thermophilic anaerobes in Arctic marine sediments induced to mineralize complex organic matter at high temperature. Environ Microbiol 12:1089–1104
Hunt DE, David LA, Gevers D et al (2008) Resource partitioning and sympatric differentiation among closely related bacterioplankton. Science 320:1081–1085
Imachi H, Aoi K, Tasumi E et al (2011) Cultivation of methanogenic community from subseafloor sediments using a continuous-flow bioreactor. ISME J 5:1913–1925
Imhoff JF (2003) Phylogenetic taxonomy of the family *Chlorobiaceae* on the basis of 16S rRNA and *fmo* (Fenna–Matthews–Olson protein) gene sequences. Int J Syst Evol Microbiol 53:941–951
Jaspers E, Overmann J (1997) Separation of bacterial cells by isoelectric focusing, a new method for analysis of complex microbial communities. Appl Environ Microbiol 63:3176–3181
Jaspers E, Overmann J (2004) The ecological significance of "microdiversity": identical 16S rRNA gene sequences represent bacteria with highly divergent genomes and physiology. Appl Environ Microbiol 70:4831–4839
Johnson ZI, Zinser ER, Coe A et al (2006) Niche partitioning among *Prochlorococcus* ecotypes along ocean-scale environmental gradients. Science 311:1737–1741
Kaeberlein T, Lewis K, Epstein SS (2002) Isolating "uncultivable" microorganisms in pure culture in a simulated natural environment. Science 296:1127–1129
Kallmeyer J, Pockalny R, Adhikari RR et al (2012) Global distribution of microbial abundance and biomass in subseafloor sediment. Proc Natl Acad Sci USA 109:16213–16216
Kang I, Oh HM, Kang D et al (2013) Genome of a SAR116 bacteriophage shows the prevalence of this phage type in the oceans. Proc Natl Acad Sci USA 110:12343–12348
Kiørboe T, Jackson G (2001) Marine snow, organic solute plumes, and optimal chemosensory behavior of bacteria. Limnol Oceanogr 46:1309–1318
Kiørboe T, Grossart HP, Ploug H et al (2002) Mechanisms and rates of bacterial colonization of sinking aggregates. Appl Environ Microbiol 68:3996–4006
Kirchman DL, Meon B, Ducklow HW et al (2001) Glucose fluxes and concentrations of dissolved combined neutral sugars (polysaccharides) in the Ross Sea and Polar Front Zone, Antarctica. Deep Sea Res II Topical Stud Oceanogr 48:4179–4197
Koch AL (2001) Oligotrophs versus copiotrophs. BioEssays 23:657–661
Könneke M, Bernhard AE, de la Torre JR et al (2005) Isolation of an autotrophic ammonia-oxidizing marine archaeon. Nature 437:543–546
Konstantinidis KT, Braff J, Karl DM et al (2009) Comparative metagenomic analysis of a microbial community residing at a depth of 4000 m at station ALOHA in the North Pacific subtropical gyre. Appl Environ Microbiol 75:5345–5355
Lauro FM, McDougald D, Thomas T et al (2009) The genomic basis of trophic strategy in marine bacteria. Proc Natl Acad Sci USA 106:15527–15533
Lebaron P, Servais P, Troussellier M et al (2001) Microbial community dynamics in Mediterranean nutrient-enriched seawater mesocosms: changes in abundances, activity and composition. FEMS Microbiol Ecol 34:255–266
Lloyd KG, Schreiber L, Petersen DG et al (2013) Predominant Archaea in marine sediments degrade detrital proteins. Nature 496:215–218
Loreau M, de Mazancourt C (2013) Biodiversity and ecosystem stability: a synthesis of underlying mechanisms. Ecol Lett 16:106–115
Lou YW, Friedrichs MAM, Doney SC et al (2010) Oceanic heterotrophic bacterial nutrition by semilabile DOM as revealed by data assimilative modeling. Aquat Microb Ecol 60:273–287
Lozupone C, Knight R (2007) Global patterns in bacterial diversity. Proc Natl Acad Sci USA 104 (27):11436–11440
Macfadyen A (1899) On the influence of the temperature of liquid air on bacteria. Proc R Soc Lond 66:180–182
Malmstrom RR, Kiene RP, Cottrell MT et al (2004) Contribution of SAR11 bacteria to dissolved dimethylsulfoniopropionate and amino acid uptake in the North Atlantic ocean. Appl Environ Microbiol 70:4129–4135

Mann EL, Ahlgren N, Moffett JW et al (2002) Copper toxicity and cyanobacteria ecology in the Sargasso Sea. Limnol Oceanogr 47:976–988
Mann AJ, Hahnke RL, Huang S et al (2013) The genome of the alga-associated marine flavobacterium *Formosa agariphila* KMM 3901T reveals a broad potential for degradation of algal polysaccharides. Appl Environ Microbiol 79:6813–6822
Marschall E, Jogler M, Henssge U et al (2010) Large scale distribution and activity patterns of an extremely low-light adapted population of green sulfur bacteria in the Black Sea. Environ Microbiol 12:1348–1362
Marshall KT, Morris RM (2013) Isolation of an aerobic sulfur oxidizer from the SUP05/Arctic96BD-19 clade. ISME J 7(2):452–455
McCarren J, Becker JW, Repeta DJ et al (2010) Microbial community transcriptomes reveal microbes and metabolic pathways associated with dissolved organic matter turnover in the sea. Proc Natl Acad Sci USA 107:16420–16427
McCarthy M, Hedges J, Benner R (1996) Major biochemical composition of dissolved high molecular weight organic matter in seawater. Mar Chem 55:281–297
Miethke M, Marahiel MA (2007) Siderophore-based iron acquisition and pathogen control. Microbiol Mol Biol Rev 71:413–451
Mincer TJ, Church MJ, Taylor LT et al (2007) Quantitative distribution of presumptive archaeal and bacterial nitrifiers in Monterey Bay and the North Pacific subtropical gyre. Environ Microbiol 9:1162–1175
Moore LR, Chisholm SW (1999) Photophysiology of the marine cyanobacterium *Prochlorococcus*: ecotypic differences among cultured isolates. Limnol Oceanogr 44:628–638
Moore J, Shaw A (2001) Long-term preservation of strains of *Burkholderia cepacia*, *Pseudomonas* spp. and *Stenotrophomonas maltophilia* isolated from patients with cystic fibrosis. Lett Appl Microbiol 33:82–83
Moore L, Rocap G, Chisholm SW (1998) Physiology and molecular phylogeny of coexisting *Prochlorococcus* ecotypes. Nature 393:464–467
Moore LR, Post AF, Rocap G et al (2002) Utilization of different nitrogen sources by the marine cyanobacteria *Prochlorococcus* and *Synechococcus*. Limnol Oceanogr 47:989
Morris RM, Rappé MS, Connon SA et al (2002) SAR11 clade dominates ocean surface bacterioplankton communities. Nature 420:806–810
Morris RM, Vergin KL, Cho JC et al (2005) Temporal and spatial response of bacterioplankton lineages to annual convective overturn at the Bermuda Atlantic Time-series Study site. Limnol Oceanogr 50:1687–1696
Mou XZ, Hodson RE, Moran MA (2007) Bacterioplankton assemblages transforming dissolved organic compounds in coastal seawater. Environ Microbiol 9:2025–2037
Nelson CE, Carlson CA (2012) Tracking differential incorporation of dissolved organic carbon types among diverse lineages of Sargasso Sea bacterioplankton. Environ Microbiol 14:1500–1516
Neumann AM, Balmonte JP, Berger M et al (2015) Different utilization of alginate and other algal polysaccharides by marine *Alteromonas macleodii* ecotypes. Environ Microbiol 17:3857–3868
Oh HM, Kwon KK, Kang I et al (2010) Complete genome sequence of "*Candidatus* Puniceispirillum marinum" IMCC1322, a representative of the SAR116 clade in the *Alphaproteobacteria*. J Bacteriol 192:3240–3241
Orcutt BN, Sylvan JB, Knab NJ et al (2011) Microbial ecology of the dark ocean above, at, and below the seafloor. Microbiol Mol Biol Rev 75:361–422
Overmann J (2005) Chemotaxis and behavioral physiology of not-yet-cultivated microbes. Methods Enzymol 397:133–147
Overmann J (2013) Principles of enrichment, isolation, cultivation, and preservation of bacteria. In: Rosenberg E, DeLong EF, Lory S, Stackebrandt E, Thompson F (eds) The prokaryotes, 4th edn, Prokaryotic biology and symbiotic associations. Springer, New York, pp 149–207
Overmann J (2015) Significance and future role of microbial resource centers. Syst Appl Microbiol 38:258–265

Parkes RJ, Cragg B, Roussel E (2014) A review of prokaryotic populations and processes in sub-seafloor sediments, including biosphere: geosphere interactions. Mar Geol 352:409–425
Pedrotti ML, Beauvais S, Kerros ME et al (2009) Bacterial colonization of transparent exopolymeric particles in mesocosms under different turbulence intensities and nutrient conditions. Aquat Microb Ecol 55:301–312
Pepe-Ranney C, Hall EK (2015) The effect of carbon subsidies on marine planktonic niche partitioning and recruitment during biofilm assembly. Front Microbiol 6:703
Pernthaler J (2005) Predation on prokaryotes in the water column and its ecological implications. Nat Rev Microbiol 3:537–546
Pomeroy LR, Williams PJ, Azam F et al (2007) The microbial loop. Oceanography 20(2):28–33
Prakash O et al (2013) Practice and prospects of microbial preservation. FEMS Microbiol Lett 339:1–9
Rappé M, Connon S, Vergin K et al (2002) Cultivation of the ubiquitous SAR11 marine bacterioplankton clade. Nature 418:630–633
Rivers T (1927) Effect of repeated freezing (−185 °C) and thawing on colon bacilli, virus III, vaccine virus, herpes virus, bacteriophage; complement and trypsin. J Exp Med 11–21
Rocap G, Larimer FW, Lamerdin J et al (2003) Genome divergence in two *Prochlorococcus* ecotypes reflects oceanic niche differentiation. Nature 424:1042
Rodriguez-Brito B, Li LL, Wegley L et al (2010) Viral and microbial community dynamics in four aquatic environments. ISME J 4:739–751
Roesch LFW, Fulthorpe RR, Riva A (2007) Pyrosequencing enumerates and contrasts soil microbial diversity. ISME J 1:283–290
Rusch DB, Halpern AL, Sutton G (2007) The Sorcerer II global ocean sampling expedition: Northwest Atlantic through eastern tropical Pacific. PLoS Biol 5:e77
Sarmento H, Gasol JM (2012) Use of phytoplankton-derived dissolved organic carbon by different types of bacterioplankton. Environ Microbiol 14:2348–2360
Scanlan DJ, Ostrowski M, Mazard S et al (2009) Ecological genomics of marine picocyanobacteria. Microbiol Mol Biol Rev 73:249–299
Schippers A, Neretin LN, Kallmeyer J et al (2005) Prokaryotic cells of the deep sub-seafloor biosphere identified as living bacteria. Nature 433:861–864
Seyedsayamdost MR, Carr G, Kolter R et al (2011) Roseobacticides: small molecule modulators of an algal-bacterial symbiosis. J Am Chem Soc 133:18343–18349
Seyedsayamdost MR, Wang R, Kolter R et al (2014) Hybrid biosynthesis of roseobacticides from algal and bacterial precursor molecules. J Am Chem Soc 136:15150–15153
Seymour JR, MItchell JG, Seuront L (2004) Microscale heterogeneity in the activity of coastal bacterioplankton communities. Aquat Microb Ecol 35:1–16
Sheik CS, Jain S, Dick GJ (2014) Metabolic flexibility of enigmatic SAR324 revealed through metagenomics and metatranscriptomics. Environ Microbiol 16:304–317
Simon M, Grossart HP, Schweitzer B et al (2002) Microbial ecology of organic aggregates in aquatic ecosystems. Aquat Microb Ecol 28:175–211
Smith DC, Simon M, Alldredge AL et al (1992) Intense hydrolytic enzyme activity on marine aggregates and implications for rapid particle dissolution. Nature 359:139–142
Sogin ML, Morrison HG, Huber JA et al (2006) Microbial diversity in the deep sea and the underexplored “rare biosphere”. Proc Natl Acad Sci USA 103:12115–12120
Söhngen C, Podstawka A, Bunk B et al (2015) BacDive—the bacterial diversity metadatabase in 2016. Nucleic Acids Res. doi:10.1093/nar/gkv983
Squires R, Hartsell S (1955) Survival and growth initiation of defrosted *Escherichia coli* as affected by frozen storage menstrua. Appl Microbiol 40–45
Sub J, Engelen B, Cypionka H et al (2004) Quantitative analysis of bacterial communities from Mediterranean sapropels based on cultivation-dependent methods. FEMS Microbiol Ecol 51:109–121
Sun J, Steindler L, Thrash JC et al (2011) One carbon metabolism in SAR11 pelagic marine bacteria. PLoS ONE 6(8):e23973

Sunagawa S, Coelho L, Chaffron S et al (2015) Structure and function of the global ocean microbiome. Science 348(6237):1261359-1–1261359-10
Suzuki M (1999) Effect of protistan bacterivory on coastal bacterioplankton diversity. Aquat Microb Ecol 20:261–272
Swan BK, Tupper B, Sczyrba A et al (2013) Prevalent genome streamlining and latitudinal divergence of planktonic bacteria in the surface ocean. Proc Natl Acad Sci USA 110:11463–11468
Swift H (1921) Preservation of stock cultures of bacteria by freezing and drying. J Exp Med 69–75
Tang K, Jiao N, Liu K (2012) Distribution and functions of TonB-dependent transporters in marine bacteria and environments: implications for dissolved organic matter utilization. PLoS ONE 7: e41204
Taylor JR, Stocker R (2012) Trade-offs of chemotactic foraging in turbulent water. Science 338:675–679
Teeling H, Fuchs BM, Becher D et al (2012) Substrate-controlled succession of marine bacterioplankton populations induced by a phytoplankton bloom. Science 336:608–611
Thomas F, Barbeyron T, Tonon T (2012) Characterization of the first alginolytic operons in a marine bacterium: from their emergence in marine Flavobacteriia to their independent transfers to marine Proteobacteria and human gut Bacteroides. Environ Microbiol 14:2379–2394
Tindall BJ (2007) Vacuum-drying and cryopreservation of prokaryotes. Methods Mol Biol 368:73–97
Tout J, Jeffries TC, Petrou K et al (2015) Chemotaxis by natural populations of coral reef bacteria. ISME J 9:1–14
Traving SJ, Thygesen UH, Riemann L et al (2015) A model of extracellular enzymes in free-living microbes: which strategy pays off. Appl Environ Microbiol 81:7385–7393
Tripp HJ, Kitner JB, Schwalbach MS et al (2008) SAR11 marine bacteria require exogenous reduced sulphur for growth. Nature 452(7188):741–744
Tripp HJ, Schwalbach MS, Meyer MM et al (2009) Unique glycine-activated riboswitch linked to glycine-serine auxotrophy in SAR11. Environ Microbiol 11:230–238
Tripp HJ, Bench SR, Turk KA et al (2010) Metabolic streamlining in an open-ocean nitrogen-fixing cyanobacterium. Nature 464:90–94
Uphoff HU, Felske A, Fehr W et al (2001) The microbial diversity in picoplankton enrichment cultures: a molecular screening of marine isolates. FEMS Microbiol Ecol 35:249–258
Valentine DL, Reeburgh WS, Blanton DC (2000) A culture apparatus for maintaining H_2 at sub-nanomolar concentrations. J Microbiol Methods 39:243–251
Venter JC, Remington K, Heidelberg JF et al (2004) Environmental genome shotgun sequencing of the Sargasso Sea. Science 304:66–74
Viklund J, Ettema TJ, Andersson SG (2012) Independent genome reduction and phylogenetic reclassification of the oceanic SAR11 clade. Mol Biol Evol 29:599–615
Walker V, Palmer G, Voordouw G (2006) Freeze-thaw tolerance and clues to the winter survival of a soil community. Appl Environ Microbiol 72:1784–1792
Wang Y, Sheng HF, He Y et al (2012) Comparison of the levels of bacterial diversity in freshwater, intertidal wetland, and marine sediments by using millions if Illumina tags. Appl Environ Microbiol 78:8264–8271
Wang H, Tomasch J, Jarek M (2014) A dual-species co-cultivation system to study the interactions between *Roseobacters* and dinoflagellates. Front Microbiol 5:1–11
Wasmund K, Schreiber L, Lloyd KG et al (2013) Genome sequencing of a single cell of the widely distributed marine subsurface *Dehalococccoidea*, phylum *Chloroflexi*. ISME J 8:383–397
Whitman WB, Coleman DC, Wiebe WJ (1998) Prokaryotes: the unseen majority. Proc Natl Acad Sci USA 95:6578–6583
Wright JJ, Mewis K, Hanson NW et al (2014) Genomic properties of Marine Group A bacteria indicate a role in the marine sulfur cycle. ISME J 8:455–468
Wu QL, Zwart G, Schauer M et al (2006) Bacterioplankton community composition along a salinity gradient of sixteen high-mountain lakes located on the Tibetan Plateau, China. Appl Environ Microbiol 72:5478–5485

Wu L, Sun Q, Sugawara H et al (2013) Global catalogue of microorganisms (gcm): a comprehensive database and information retrieval, analysis, and visualization system for microbial resources. BioMed Cent Genomics 14:933
Xiong ZQ, Wang JF, Hao YY (2013) Recent advances in the discovery and development of marine microbial natural products. Mar Drugs 11:700–717
Yin Q, Fu B, Li B et al (2013) Spatial variations in microbial community composition in surface seawater from the ultra-oligotrophic center to rim of the South Pacific Gyre. PLoS ONE 8: e55148
Yooseph S, Nealson KH, Rusch DB et al (2010) Genomic and functional adaptation in surface ocean planktonic prokaryotes. Nature 468:60–66
Zengler K, Toledo G, Rappe M et al (2002) Cultivating the uncultured. Proc Natl Acad Sci 99:15681–15686
Zhang Y, Jiao N, Cottrell MT et al (2006) Contribution of major bacterial groups to bacterial biomass production along a salinity gradient in the South China Sea. Aquat Microb Ecol 43:233–241
Zinger L, Amaral-Zettler LA, Fuhrman JA et al (2011) Global patterns of bacterial beta-diversity in seafloor and seawater ecosystems. PLoS ONE 6:e24570

Chapter 3
Phototrophic Microorganisms: The Basis of the Marine Food Web

Wolfgang R. Hess, Laurence Garczarek, Ulrike Pfreundt and Frédéric Partensky

Abstract Although numerous marine microorganisms can exploit solar energy for photosynthesis or photoheterotrophy, cyanobacteria and microalgae are the only ones able to perform oxygenic photosynthesis and to produce organic carbon, an essential brick of life that sustains the whole marine trophic web. Here we review recent advances in the investigation of marine oxygenic microorganisms, with a special focus on cyanobacteria. We discuss novel insights into the ecology, evolution and diversity of *Synechococcus* and *Prochlorococcus*, the two most abundant and certainly the best known oxyphototrophs at all scales of organization from the gene to the global ocean. A particular emphasis is also made on diazotrophic cyanobacteria, which constitute an important source of bioavailable nitrogen to oceanic surface waters, possibly the most important external nitrogen source, before atmospheric and riverine inputs. Diazotrophic cyanobacteria are polyphyletic and display a remarkably large range of physiologies and morphologies. These include both multicellular cyanobacteria, such as the colonial *Trichodesmium* or the heterocyst-forming *Calothrix*, *Richelia* and *Nodularia*, and unicellular cyanobacteria belonging to three major groups: the symbiotic *Candidatus* Atelocyanobacterium thalassa (UCYN-A), the free-living *Crocosphaera* sp. (UCYN-B) and the UCYN-C cluster that notably encompasses *Cyanothece*. Whereas some of these species can form immense blooms (*Nodularia*, *Trichodesmium*), others can also have a major ecological impact even

W.R. Hess (✉) · U. Pfreundt
Faculty of Biology, Institute of Biology 3, Genetics and Experimental Bioinformatics, University of Freiburg, Schänzlestr. 1, 79104 Freiburg, Germany
e-mail: wolfgang.hess@biologie.uni-freiburg.de

U. Pfreundt
e-mail: upfreundt@gmail.com

L. Garczarek · F. Partensky
CNRS UMR 7144, Department of Adaptation and Diversity in the Marine Environment, Marine Photosynthetic Prokaryotes Team, Station Biologique, Sorbonne Universités-Université Paris 06, CS 90074, 29688 Roscoff Cedex, France
e-mail: laurence.garczarek@sb-roscoff.fr

F. Partensky
e-mail: frederic.partensky@sb-roscoff.fr

L.J. Stal and M.S. Cretoiu (eds.), *The Marine Microbiome*,
DOI 10.1007/978-3-319-33000-6_3

though they represent only a minor fraction of the bacterioplankton (UCYN-C). After about one billion years of evolution, which led them to colonize any single marine niche reached by solar light, cyanobacteria appear as truly fascinating organisms that constitute a major component of the marine microbial communities and are the matter of an ebullient research area. The considerable amount of omics information recently becoming available on both isolates and natural populations of marine oxyphototrophs provide a solid basis for investigating their molecular ecology, their contribution to biogeochemical cycles, as well as their possible utilization in biotechnology, data mining, or biomimetics.

3.1 Introduction

Phototrophs have been defined by Eiler as organisms capable of using electromagnetic energy (photons) as a source of energy to produce chemical energy (ATP and organic compounds) (Eiler 2006). Marine phytoplankton accounts for a net primary production of about 48.5 Pg (10^{15}) C year^{-1}, which is close to the 56.4 Pg C year^{-1} calculated for terrestrial plants (Field et al. 1998). Whilst the ability to use light energy was long thought to be a functional trait restricted to phytoplankton in marine systems, there is now compelling evidence that a large proportion of bacteria inhabiting the upper lit layer of oceans is able to exploit to some extent solar energy for photosynthesis and/or (facultative) photoheterotrophy (Béjà and Suzuki 2008; Béjà et al. 2000; Ferrera et al. 2015). There are indeed three main types of phototrophs in marine microbial communities:

(i) Cyanobacteria and microalgae perform 'oxygenic photosynthesis' using two photosystems (PSI and PSII) connected via an electron transfer chain, like in plants (Falkowski and Raven 2013). Photons are collected by antenna complexes coupled to photosystems and transferred to special chlorophyll (Chl) molecules (generally Chl *a*, except in the atypical cyanobacteria *Prochlorococcus* and *Acaryochloris*; see below) located in the photosystem core. Photon energy is used to break water molecules and produce electrons and protons as well as reducing power in the form of nicotinamide adenine dinucleotide phosphate (NADPH). In this process, oxygen is only a by-product. NADPH is ultimately used to synthesize organic carbon molecules from carbon dioxide via the Calvin-Benson-Bassham cycle. It is important to note that the cyanobacteria *Prochlorococcus* and *Synechococcus* are not strict photoautotrophs since they are also capable of photoheterotrophic uptake and assimilation of considerable amounts of organic molecules, such as amino acids or dissolved organic matter (Björkman et al. 2015; Gómez-Pereira et al. 2013; Sharma et al. 2014; Zubkov 2009).

(ii) Aerobic anoxygenic photosynthetic bacteria (AAnPB) perform photoheterotrophy thanks to a single (type II) reaction center containing bacteriochlorophyll *a*. These bacteria do not produce oxygen since they use compounds other than water (e.g., hydrogen, hydrogen sulfide, thiosulfate) as electron donors (Eiler 2006). AAnPB are polyphyletic, with members belonging to either Alpha-, Beta- or Gammaproteobacteria, and are widespread in the upper layer of the ocean, constituting up to 15 % of the total bacterial community (Béjà et al. 2002; Boeuf et al. 2013; Cottrell et al. 2010; Koblížek 2015). Their anaerobic counterparts, so-called AnAnPB, are restricted to oxygen-free habitats, such as sediments and mats (see e.g., Hubas et al. 2011).

(iii) Proteorhodopsin-containing bacteria (PRB) possess a light-driven proton pump that generates membrane potential used for ATP synthesis and notably confers cells enhanced survival during starvation (Akram et al. 2013; Fuhrman et al. 2008; Gómez-Consarnau et al. 2010). Proteorhodopsin absorption properties are tuned to match the surrounding environment, i.e., green light in surface and blue light at greater depths (Béjà et al. 2001; Man et al. 2003). PRB are abundant and phylogenetically diverse, encompassing members of the abundant and ubiquitous marine groups SAR11 (Alphaproteobacteria) and SAR86 (Gammaproteobacteria) as well as Flavobacteria (Béjà and Suzuki 2008; Béjà et al. 2001).

Even though AAnPB and PRB are important components of marine photo (hetero)trophic communities (for recent reviews, see, e.g., DeLong and Béjà 2010; Evans et al. 2015; Koblížek 2015), this chapter will focus on marine microorganisms performing oxygenic photosynthesis and make an overview of some recent advances on their biology, ecology, evolution, and exploitability. Cyanobacteria and microalgae are well known for their ability to assimilate CO_2 and synthesize organic carbon, a key function that place them at the basis of the marine food web, but they also perform several other important functions within the marine ecosystem. For instance, many marine cyanobacteria are diazotrophs, i.e., capable to uptake atmospheric dinitrogen (N_2) and transform it into ammonium that is assimilated and can be transferred to other members of microbial communities. The latter can be either free-living planktonic microbes, benthic organisms co-occurring with cyanobacteria in microbial mats or even eukaryotic partners in symbioses, these mutually beneficial associations being frequently encountered in nutrient-depleted areas of the ocean (see Chap. 11 for details). Marine oxygenic phototrophs also constitute a unique resource for biotechnological applications. Among bacteria this is illustrated by the identification of *Cyanothece* strains (cyanobacteria) with the natural potential for high yield hydrogen production (Bandyopadhyay et al. 2010; Melnicki et al. 2012), the discovery a biosynthetic pathway for short-to-medium chain alkanes in *Prochlorococcus* (Lea-Smith et al. 2015; Schirmer et al. 2010), or marine cyanobacteria producing beneficial metabolites (Jones et al. 2011; Mevers et al. 2014; Salvador-Reyes and Luesch 2015; Shao et al. 2015). Eukaryotic marine microalgae are frequently considered as promising feedstock for developing functional foods, bioactive pharmaceuticals, and cosmetics (Desbois et al. 2009; Fu et al. 2015), as a rich source of beneficial metabolites such as natural antioxidants (Xia et al. 2014), for biomass (Stengel and

Connan 2015) or for biofuel production (Daboussi et al. 2014; Matsumoto et al. 2010). However, some microalgae can develop harmful blooms, impacting on other forms of marine life as well as on fisheries, aquaculture and tourism (Anderson et al. 2012). Last but not least, marine diatoms have been providing inspiration for biomimetic approaches (Nichols 2015; You et al. 2014).

3.2 Phytoplankton Biodiversity and the Next Generation Sequencing Revolution

The availability of complete genome sequences from representatives of the major groups of marine phototrophic microorganisms is of uttermost importance to understand the physiological and ecological potential of these organisms at the most fundamental level. The first published complete marine cyanobacterial genome sequences were those of three different strains of *Prochlorococcus* (Dufresne et al. 2003; Rocap et al. 2003) and one *Synechococcus* (Palenik et al. 2003), while the first marine eukaryotic genome, *Thalassiosira pseudonana* CCMP 1335 appeared only one year later, allowing unprecedented insights into the metabolic capabilities of this cosmopolitan centric diatom (Armbrust et al. 2004). In fact, the very first published sequence from a marine phototroph was that of the tiny nucleomorph of the cryptophyte alga *Guillardia theta* (Douglas et al. 2001). Nucleomorphs are remnants of the secondary endosymbiont's nucleus in cryptophyte and chlorarachniophyte algae. The complete draft nuclear genome of *G. theta* and of the chlorarachniophyte *Bigelowiella natans* became available in 2012, shedding further light on nucleomorph functions and fate (Curtis et al. 2012).

Since these pioneer works, there is a steady increase in the number of available genome sequences (Fig. 3.1), with a clear acceleration since 2012, in part due to the sequencing of many different isolates of *Prochlorococcus* and *Synechococcus*. Still, the true biodiversity of these picocyanobacteria might exceed vastly the information obtained from the genomic analysis of laboratory strains (Biller et al. 2015; Kashtan et al. 2014). At the time of writing, 186 genome sequences of marine oxyphototrophic microorganisms were publicly available, among which 167 were of cyanobacteria and 19 of eukaryotic microalgae, but these numbers are moving targets as sequencing efforts continue. This intense sequencing effort focused on laboratory cultures is nowadays often complemented by the genomic analysis of single amplified genomes (SAGs) from the environment (see, e.g., Kashtan et al. 2014; Malmstrom et al. 2013) and genome assemblies from whole metagenomes (see, e.g., Rusch et al. 2010; Shi et al. 2011; Vaulot et al. 2012). Single cell genome sequencing highlighted the impressive heterogeneity within *Prochlorococcus* populations, revealing hundreds of coexisting subpopulations (Kashtan et al. 2014). This work was a major step towards understanding the global success of these small cyanobacteria by showing how population-level adaptability can be maintained even for organisms with such tiny genomes.

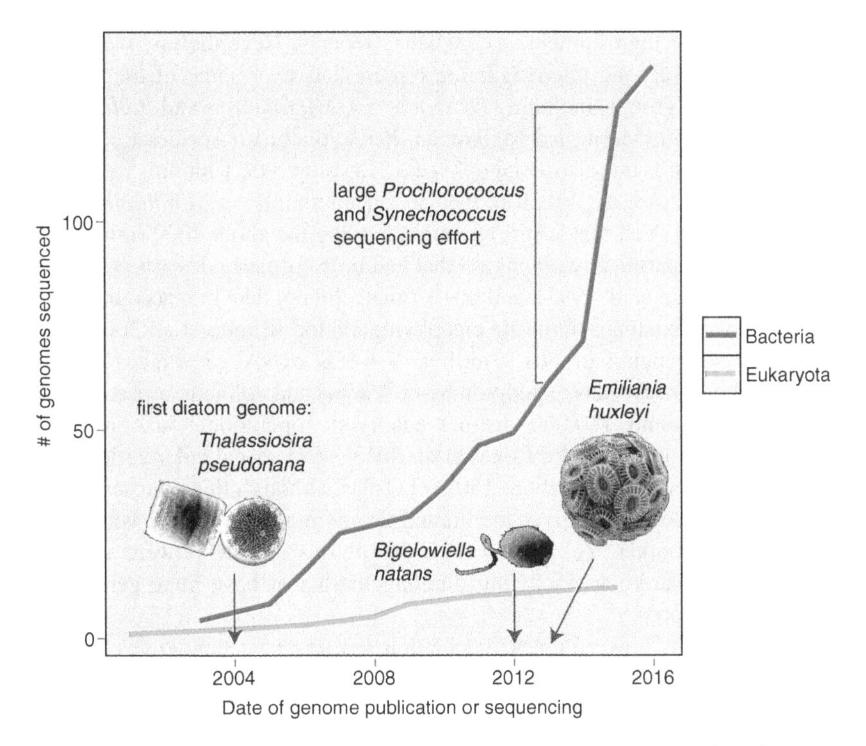

Fig. 3.1 Sequencing effort for marine phototrophic microorganisms. The number of sequenced genomes is shown separately for cyanobacteria and photosynthetic eukaryotes. For bacterial genomes, numbers are based on data available at the Integrated Microbial Genomes (IMG, Joint Genome Institute) database in January 2016 with manual selection of marine species where no metadata was published, plus 32 yet unpublished *Synechococcus* genomes sequenced by the Genoscope (Evry, France) and the Center for Genome Research (Liverpool, UK) at the initiative of the Roscoff Biological Station (France) and the Warwick University (David J. Scanlan, Coventry, UK). For eukaryotes, numbers are based on manual inspection of the Phytozome database (Joint Genome Institute) and the NCBI database of eukaryotic genomes. We are aware that there are probably more unpublished genomes that thus are not part of this analysis. Please note that the >90 partial single cell genomes (SAGs) of wild *Prochlorococcus* (Kashtan et al. 2014) are not present in IMG yet and thus are not included here. Parts of this figure were provided by Geoff McFadden (*Bigelowiella natans*), the Alfred-Wegener-Institut Bremerhaven (*Thalassiosira pseudonana*) and the International Nanoplankton Association (*Emiliania huxleyi*), with the courtesy of the authors, institutions and photographers.

Whereas cyanobacterial genomic information is steeply increasing, genome sequencing of marine eukaryotic phototrophs is progressing rather slowly, despite their global importance to CO_2 sequestration and nutrient cycling, especially in environments drastically impacted by climate change like the polar regions (Smetacek and Nicol 2005). Part of this is due to the fact that, compared to Bacteria and Archaea, eukaryotic nuclear genomes are larger (sometimes considerably, e.g., in the case of dinoflagellates), contain substantially longer non-coding regions, including introns of variable number and size within gene sequences, and frequently include high numbers of sequence repeats. Nevertheless, the available sequences for eukaryotic microalgae are representative of some of the most ecologically relevant groups, including *Ostreococcus, Micromonas* and *Bathycoccus* (Mamiellophyceae) (Marin and Melkonian 2010), diatoms (Armbrust et al. 2004; Bowler et al. 2008), but also Pelagophyceae, Cryptophyta, Chlorarachniophyceae (*B. natans*) (Curtis et al. 2012) as well as coccolithophores (*Emiliania huxleyi*) (Read et al. 2013). Yet, this is still far from covering the about 4000 (range: 3444–4375) species of marine phytoplankton that had been formerly described by the end of the 80's (Sournia et al. 1991), and this estimate did not take into account the wide cryptic diversity existing within the picophytoplankton (Simon et al. 2009). In this context, the sequencing of ~1.7 million V9 18S rRNA gene reads from 334 samples of the *Tara* Oceans expedition by de Vargas and collaborators revealed the occurrence of around 110,000 distinct eukaryotic operational taxonomic units (OTUs) along the transect (de Vargas et al. 2015). The global eukaryotic plankton richness was estimated to be about 150,000 OTUs, although this represented likely a lower-boundary estimate, given the limited taxonomic resolution power of the V9 18S rRNA gene marker. Yet, among these OTUs, less than 20 % were assigned to photosynthetic eukaryotes, including dinoflagellates that have huge genomes and are often heterotrophic.

The tremendous power of next generation sequencing technologies associated with rapid progresses in bioinformatics and data processing is currently revolutionizing our view of the genetic and functional diversity of marine oxyphototrophs (and marine plankton at large) and there is little doubt that this field will evolve considerably in the forthcoming years. Another relevant aspect is that transcriptomic approaches provide the expressed share of a genome, leaving out large chunks of non-expressed or repetitive genomic DNA. Therefore, it is highly productive to complement the ongoing efforts in analyzing the genomes by sequencing the transcriptomes (Keeling et al. 2014).

3.3 Eukaryotic Phytoplankton

Through their physiology, which is based on oxygenic photosynthesis and carbon fixation, eukaryotic phytoplankton plays a key role in the marine food web and within the global biogeochemical processes (Falkowski et al. 2004; Jardillier et al. 2010; Worden et al. 2004, 2015) and in the export and the sequestration of organic carbon to the deep ocean (Richardson and Jackson 2007). Most marine algae are unicellular (then also referred to as protists), but there are also multicellular forms (macroalgae), e.g., Florideophyceae (Rhodophyta) and Ulvophyceae (Chlorophyta). Small eukaryotic unicellular phytoplankton is of bewildering morphological and phylogenetic diversity (Fig. 3.2). It includes species within most of the eukaryotic super-groups (Archibald 2012; Not et al. 2012). There is a substantial number of lineages without cultured representatives and there have been new groups being discovered also in recent years, such as the Rappemonads (Kim et al. 2011), or

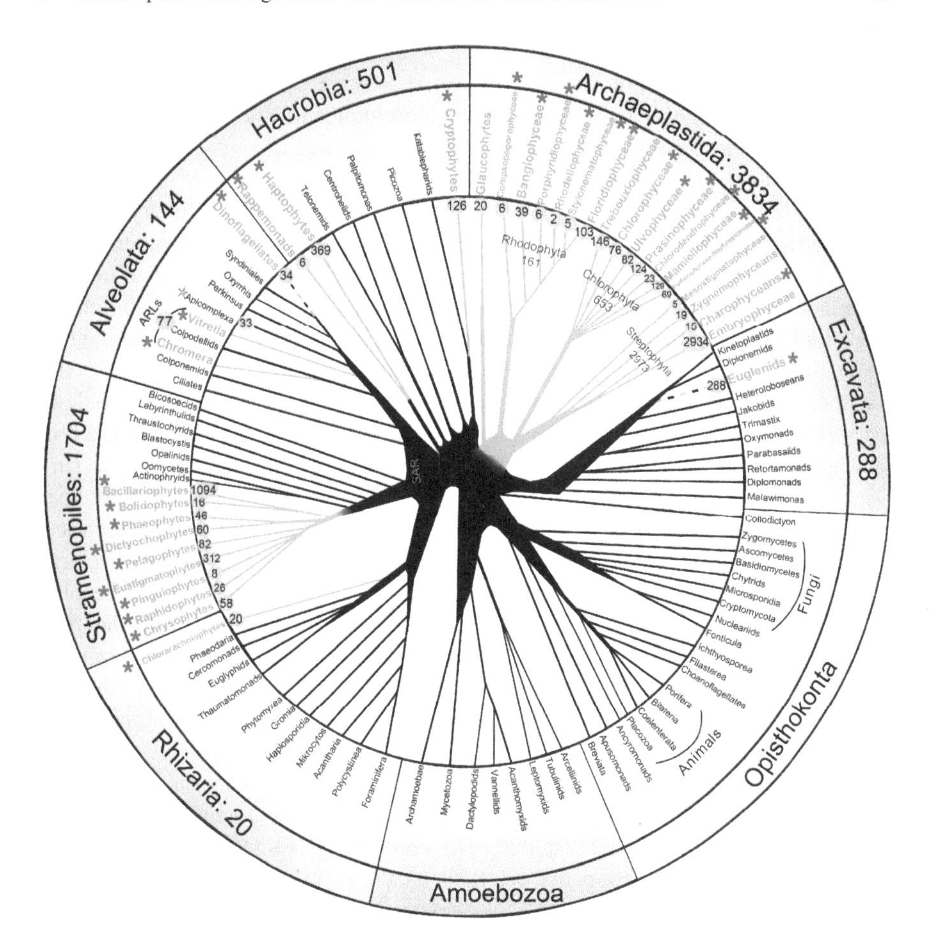

Fig. 3.2 Marine algae within the tree of eukaryotic life. All plastid-containing eukaryotic lineages are highlighted in *green*, and those containing marine algae are additionally labeled by a *blue star*. The numbers in *small grey circles* refer to the plastidial 16S rRNA gene sequences available from the PhytoREF database (Decelle et al. 2015). Note that the shown taxa are of different taxonomic ranks. SAR, the group encompassing Stramenopiles, Alveolata and Rhizaria (Archibald 2009). The phylogenetic tree was drawn according to recent phylogenomics and morphological evidence (Burki and Keeling 2014) and the figure was provided by Johan Decelle (CNRS Station Biologique de Roscoff, France, now at the Helmholtz Center for Environmental Research Leipzig, Germany) and is modified from Fig. 2 in Decelle et al. (2015), with the courtesy of the author and the publisher.

Chromera veolia (Moore et al. 2008) and *Vitrella brassicaformis* (Oborník et al. 2012), two apicomplexan marine algae whose discovery raised substantial interest because they are the phototrophic relatives of apicomplexan parasites (Oborník and Lukeš 2015).

Eukaryotic microalgae are paraphyletic. Green and red algae are the only two classes with primary endosymbionts, i.e., their plastids and mitochondria are surrounded by double membranes and can phylogenetically be directly traced back to the primary endosymbiosis with a cyanobacterium or alphaproteobacterium. Typical green algae widely represented throughout marine environments are Mamiellophyceae such as *Bathycoccus*, *Ostreococcus* and *Micromonas*. In contrast, most microalgal lineages originated from secondary endosymbiosis, i.e., different heterotrophic eukaryotic host cells took up another photosynthetic eukaryote. As a consequence, plastids resulting from secondary endosymbiosis are frequently surrounded by four, sometimes by three (e.g., Dinoflagellates) membranes. Examples for important marine primary producers include diatoms, pelagophytes, haptophytes and dinoflagellates. Many different additional groups exist, among them chlorarachniophytes, cryptophytes, stramenopiles, dinoflagellates, and apicomplexans. As a consequence of the multiple and paraphyletic origin of eukaryotic microalgae, a bewildering diversity of morphologies, lifestyles, genome content and incredible metabolic versatility has evolved. Moreover, the results obtained during the *Tara* Oceans expedition suggest the existence of several more lineages that have not been accounted for (de Vargas et al. 2015). To add to this further, an impressive variety of symbiotic to parasitic interactions between different eukaryotic and bacterial unicellular microorganisms has evolved that led to cell consortia that can be photosynthetic under some circumstances and heterotrophic when without the phototrophic partner. On example of high ecological importance is the symbiosis between the diazotrophic cyanobacterium *candidatus* Atelocyanobacterium thalassa and the unicellular prymnesiophyte *Braarudosphaera bigelowii* (Hagino et al. 2013; Krupke et al. 2015; Thompson et al. 2014; Zehr 2015).

To be able to track microalgae in their respective environment and to interpret lineages only known through environmental sequences, or potentially novel lineages, molecular markers are required. The chloroplast 16S rRNA gene is such a marker. The recently developed PhytoREF database (http://phytoref.fr) organizes chloroplast 16S rRNA gene reference sequences from various sources representing all known major photosynthetic lineages (Decelle et al. 2015). The information is provided in the context of a curated and normalized taxonomy and allows exploring the total diversity of photosynthetic eukaryotes in any given ecosystem.

Marine microalgae are of extreme heterogeneity and therefore it is not surprising that they have evolved also specific mechanisms to cope with different environmental cues. Phototrophic microorganisms can be limited by different factors. In the contemporary ocean, the amount of bioavailable iron is frequently a limiting factor. Therefore, eukaryotic microalgae have special mechanisms developed to specifically scavenge this mineral. Metatranscriptomic studies revealed the genes and molecular responses fundamental for the physiological responses to varying iron availability in the diatom community (Chappell et al. 2015; Marchetti et al. 2012; Morrissey et al. 2015). Laboratory studies with the diatom *Phaeodactylum tricornutum* then uncovered the function of a previously uncharacterized surface-associated protein, ISIP2a, which concentrates and facilitates uptake of Fe (III), which is then reduced by the activity of ferrireductase to Fe(II), the biologically active form (Morrissey et al. 2015). The comparison of the *Phaeodactylum* ISIP2a sequence against available genome sequences of other marine algae reveals that similar proteins exist also in diverse other marine phytoplankton taxa, including *E. huxleyi*, the haptophyte *Chrysochromulina*, and also in the brown alga *Ectocarpus siliculosus*. Studying metatranscriptomic datasets from Antarctica and Monterey Bay further showed that homologs of ISIP2a are expressed by algae from diverse marine lineages (including diatoms, haptophytes, and dinoflagellates) in situ (Morrissey et al. 2015). Furthermore, the authors found a similar gene expressed in marine green algae, illustrating the ecological significance of ISIP2a further (Morrissey et al. 2015).

3.4 Cyanobacteria

3.4.1 Cyanobacterial Origin and Evolution

Cyanobacteria are the most ancient microorganisms capable of oxygenic photosynthesis and were undoubtedly responsible for the 'Great Oxidation Event' (GOE), i.e. the first sharp rise of atmospheric O_2 concentration in Earth history, which occurred ca. 2.4–2.1 billion years (Gyr) before present (Lyons et al. 2014). Yet, there is still a vivid controversy about when exactly oxygenic photosynthesis started, with estimates spanning from 3.7 (Rosing and Frei 2004) to 2.3 Gyr ago (Kirschvink and Kopp 2008), based on geological or geochemical evidence. For instance, a study of the distribution of chromium isotopes and redox-sensitive metals in paleosols indicated that there were already appreciable levels of atmospheric oxygen (i.e. 3×10^{-4} times present levels) about 3.0 Gyr ago, suggesting that ancestral cyanobacteria (hereafter 'procyanobacteria') might have evolved by this time (Crowe et al. 2013).

One may also wonder how these procyanobacteria looked like and where they first occurred: in the ocean or on land? Among the most reliable early cyanobacterial microfossils are colonial ellipsoids that have been assigned to the modern genus *Gloeobacter* (Golubic and Seong-Joo 1999), a coccoid, rock-dwelling organism that has the unique 'primitive' property to lack thylakoids (Mareš et al. 2013; Rippka et al. 1974). Thus, it is reasonable to assume that procyanobacteria were terrestrial *Gloeobacter*-like coccoids and many phylogenetic analyses indeed use *Gloeobacter* to root their trees (see, e.g., Blank and Sánchez-Baracaldo 2010; Larsson et al. 2011; Shih et al. 2013). Yet, filamentous forms of cyanobacteria must have evolved soon after (possibly during or just after the GOE) since they were prominent components of microbial mats during most of the Proterozoic Eon (2.5–0.54 Gyr; Knoll and Semikhatov 1998). Comparative genome analyses of present-day cyanobacteria tend to confirm hypotheses about their terrestrial origin. Indeed, the facts that marine lineages (i) are not monophyletic within the cyanobacterial radiation but dispersed among terrestrial species (see Fig. 3.3) and (ii) display distinct morphological complexity and habitats, strongly suggest that cyanobacteria appeared first on land (either on rock or in freshwater) and that their descendants colonized the marine environment later on, most likely through several independent colonization events

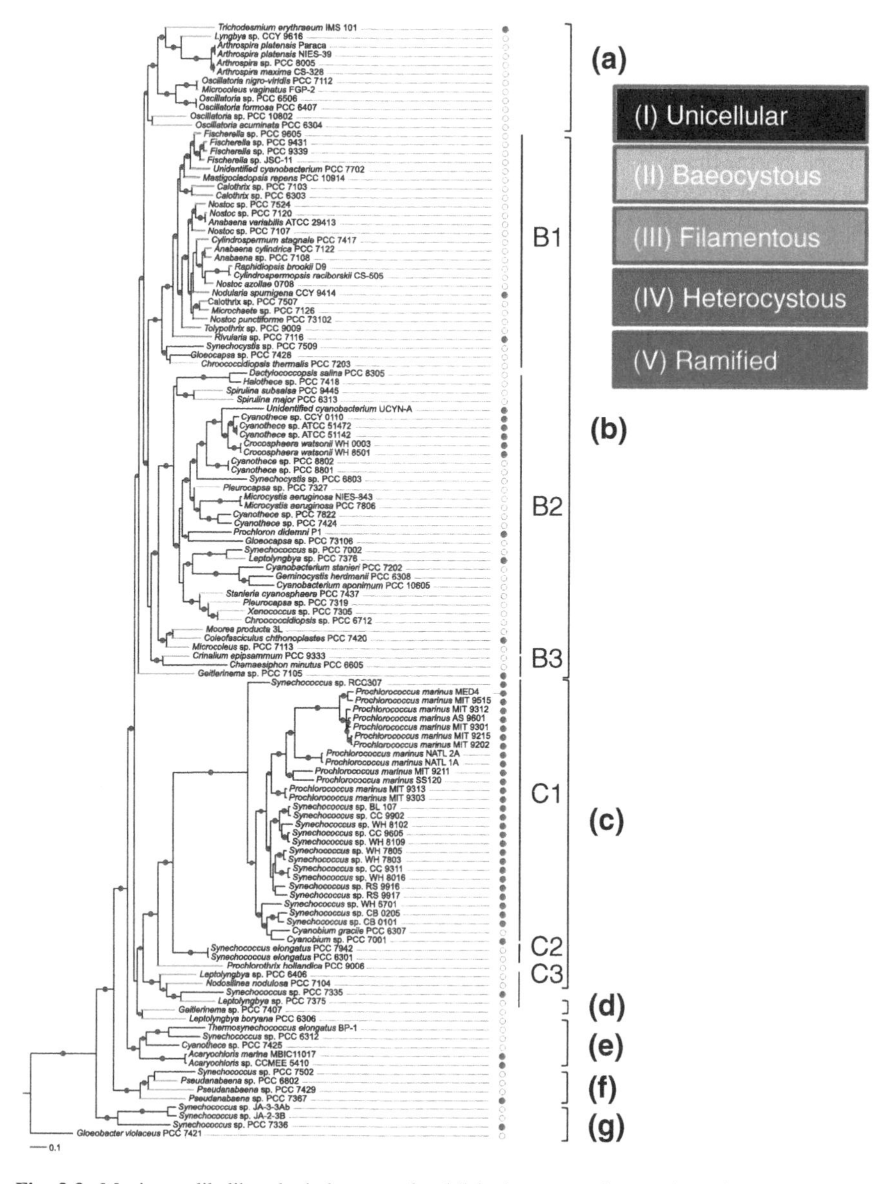

Fig. 3.3 Maximum-likelihood phylogeny of published sequenced cyanobacteria. Branches are color-coded according to morphological subsection. Taxa names marked with *blue dots* correspond to marine species. Phylogenetic subclades are grouped into seven major subclades (*a–g*), some of which are split into smaller subgroups. Nodes supported with a bootstrap of ≥70 % are indicated by *grey dots*. Adapted with permission from Fig. 1a in Shih et al. (2013)

(Blank and Sánchez-Baracaldo 2010; Coelho et al. 2013; Larsson et al. 2011). A recent phylogenomic study (Sánchez-Baracaldo 2015) proposed that, while marine benthic cyanobacteria arose early during the evolution of the phylum, the first marine planktonic cyanobacterial lineages were diazotrophic and evolved only during the Neoproterozoic (0.54–1.0 Gyr), possibly explaining why the open ocean remained anoxic during most of the Proterozoic Eon (Reinhard et al. 2013).

Representative strains of all major marine planktonic genera known to date have been sequenced in recent years, including the free-living *Prochlorococcus*, *Synechococcus*, *Crocosphaera*, *Trichodesmium*, and *Nodularia* as well as the uncultured symbionts UCYN-A (provisionally called *Candidatus* Atelocyanobacterium thalassa; (Thompson et al. 2012), *Richelia* and *Calothrix* (the latter two are missing in Fig. 3.3, but see (Hilton et al. 2013). Comparative genomic analyses brought several insights into how cyanobacteria have adapted to the marine habitat. In particular, it seems that adaption to salinity has been made possible by the development of specific machineries allowing cells to actively export ions (the "salt-out-strategy") and to increase the cellular osmolarity by accumulating small organic molecules, called compatible solutes (Klähn et al. 2010; Pade and Hagemann 2014; Scanlan et al. 2009). In agreement with the polyphyly of marine genera, the nature of these compatible solutes strongly varies depending on lineages, with glucosylglycerol in halotolerant species, glycine betaine in halophiles and stromatolite-forming cyanobacteria, sucrose and glucosylglycerate in *Prochlorococcus* and trehalose in *Crocosphaera watsonii*, while no known compatible solute biosynthesis genes have been identified so far in the *Trichodesmium* genome (Pade and Hagemann 2014).

The second important parameter that has strongly influenced the evolution of marine cyanobacteria is the availability of nutrients. Nitrogen (N) is the main limiting element in the global ocean, although limitation by phosphorus or iron also occurs (Canfield et al. 2010; Falkowski 1997; Karl et al. 1997; Moore et al. 2001). Cyanobacteria have developed two main strategies to deal with N limitation: either fixing atmospheric N_2 or decreasing their cell and genome size in order to limit their cellular requirement for N. The first one has led to many different physiological and morphological adaptations that aim at circumventing the main drawback of N_2 fixation, i.e. the oxygen sensitivity of nitrogenase, the key enzyme of this process. In *Trichodesmium*, N_2 fixation is separated spatially and temporally from oxygenic photosynthesis (Berman-Frank et al. 2001). Early results pointed to a split of functions inside colonies between peripheral nitrogenase-lacking cells that solely perform oxygenic photosynthesis and nitrogenase-containing cells in the central part of colonies that fix dinitrogen but not CO_2 (Carpenter and Price 1976).

The unicellular *Crocosphaera* has opted for a much more stringent temporal separation between photosynthesis and N_2 fixation, restricted to day and night, respectively (Compaoré and Stal 2010; Shi et al. 2010). The endosymbiotic filamentous *Richelia* and *Calothrix* both form heterocysts, thick-walled cells specialized in N_2 fixation without the oxygen-evolving photosystem II. These heterocysts occur at the basis of the trichomes of *Richelia* and *Calothrix* and differentiate from typical, CO_2-fixing cells (Hilton et al. 2013). At last, the symbiotic UCYN-A literally lost its photosystem II, making it the sole cyanobacterium known so far that is incapable of oxygenic photosynthesis, while it is an efficient N_2-fixer (Tripp et al. 2010; Zehr et al. 2008). The second strategy, consisting in drastic decrease in cell size, a process conferring an advantage for nutrient uptake by increasing the surface/volume ratio, is observed in the marine coccoid picocyanobacteria *Synechococcus* (average cell diameter ~1 μm) and is pushed to its limits in all but one *Prochlorococcus* lineages that have an average small cell size cell diameter of only about 0.6 μm. This decrease in cell size goes hand in hand with genomic streamlining. *Prochlorococcus* genomes decreased in size by about 30 % compared to their ancestral size via an extensive streamlining process (Dufresne et al. 2005; Kettler et al. 2007; Partensky and Garczarek 2010; Scanlan et al. 2009).

Another factor that primitive cyanobacteria had to deal with during their adaption to the marine habitat is the wide range of photosynthetically available radiation (PAR) in the water column, spanning from red wavelengths in turbid subsurface coastal waters to blue wavelengths in deep open ocean waters (Kirk 1994). Most cyanobacteria possess large, extrinsic light-harvesting complexes, called phycobilisomes, constituted of a core of allophycocyanin surrounded by rods with variable phycobiliprotein composition, i.e. either phycocyanin (PC) alone or in combination with phycoerythrocyanin (PEC), phycoerythrin-I (PEI) and/or phycoerythrin-II [PEII; (Sidler 1994; Six et al. 2007)]. These phycobiliproteins exhibit distinct absorption properties that directly depend on their respective assemblages of covalently bound chromophores (phycobilins): phycourobilin (PUB; A_{max}: 495 nm), phycoerythrobilin (PEB; A_{max}: 550 nm), phycoviolobilin (PVB; A_{max}: 590 nm) and/or phycocyanobilin (PCB; A_{max}: 620 nm). While all cyanobacterial phycobilisomes contain PCB, PEB is found only in PE-containing cells and is the major phycobilin in habitats where green light is predominant. PVB is specific of PEC-containing cells, a pigment type found in soils, freshwater, hot springs, and some marine benthic habitats. In contrast, PUB, which is the phycobilin best suited to collect blue photons, is strictly marine; it is often bound to PE-I and is the major phycobilin in PE-II, a *Synechococcus*-specific phycobiliprotein located at the extremity of rods (Ong et al. 1984; Six et al. 2007). While both PCB and PEB derive from biliverdin IXa and exist as free pigments, PVB and PUB necessarily result from the binding and isomerization of PCB and PEB, respectively (Blot et al. 2009; Shukla et al. 2012; Zhao et al. 2000). Interestingly, although PVB and PUB occur in different organisms, the lyase-isomerases PecE-F and RpcG that catalyze these binding and isomerization reactions at the equivalent binding site (α-84) of PEC and PC, respectively, are chemically similar and phylogenetically related. This is a nice example of adaption to blue light resulting from lateral gene transfer followed by a change in gene function (Blot et al. 2009). All genera of free-living cyanobacteria that thrive in the open ocean possess phycobilisomes with high (*Crocosphaera*, *Synechococcus*) or intermediate PUB to PEB ratios (*Trichodesmium*), while symbionts such as *Richelia* have a low PUB:PEB, that possibly better complement the absorption properties of their microalgal host (Neveux et al. 1999, 2006; Ong and Glazer 1991; Six et al. 2007). In this context, *Prochlorococcus* constitutes an exception among open ocean cyanobacteria in that it has no phycobilisomes but instead possesses membrane-intrinsic antenna complexes binding divinyl derivatives of Chl *a* and *b*, a unique pigment complement that allows this microorganism to collect with even more efficacy than PUB the blue wavelengths prevailing at the bottom of the lit layer in clear oceanic waters (Goericke and Repeta 1992; Morel et al. 1993; Ting et al. 2002). It remains enigmatic that, in addition, *Prochlorococcus* has retained a small set of phycobiliprotein genes that are expressed and from which chromophorylated phycoerythrin is produced (Hess et al. 1996, 1999, 2001; Steglich et al. 2003, 2005).

After about one billion years of evolution, which led them to colonize any single marine niche reached by solar light—and even below since live *Prochlorococcus* cells have been observed in the aphotic zone (Jiao et al. 2014)—, cyanobacteria appear as truly fascinating organisms that are the matter of an ebullient research area, and the following paragraphs will detail some of the recent advances on these major components of microbial communities.

3.4.2 Marine Picocyanobacteria

Despite their recent discovery by the end of the twentieth century (Chisholm et al. 1988; Waterbury et al. 1979), the non-diazotrophic, marine unicellular cyanobacteria *Prochlorococcus* and *Synechococcus* are nowadays most certainly the best known marine cyanobacteria at all scales of organization from the gene to the global ocean (Biller et al. 2015; Coleman and Chisholm 2007; Flombaum et al. 2013; Scanlan et al. 2009). Indeed, they have a number of advantages that make them particularly relevant models for ecological, physiological and evolutionary studies, including their abundance and ubiquity and thence strong contribution to global marine primary productivity (Flombaum et al. 2013; Partensky et al. 1999a, b), their culturability (Moore et al. 2007; Morris et al. 2008), allowing refined physiological characterization of representative isolates (see e.g., Berube et al. 2015; Blot et al. 2011; Humily et al. 2013; Krumhardt et al. 2013; Mella-Flores et al. 2012), as well as their small genome size with little gene redundancy, which favored the acquisition of a large number of genomes, SAGs, and metagenomes within the last decade (see e.g., Biller et al. 2014a; Kashtan et al. 2014; Malmstrom et al. 2013; Scanlan et al. 2009).

With cell densities reaching up to 2–3 $\times$ 10^5 cells mL^{-1} in the upper layer of warm, nutrient-poor central gyres, and a distribution area extending between 40°S and 45°N, *Prochlorococcus* is undoubtedly the most abundant photosynthetic organism on Earth (Flombaum et al. 2013; Partensky et al. 1999b). It generally co-occurs with *Synechococcus*, but the latter is even more widespread, since its distribution extends to sub polar areas and brackish waters such as the Baltic Sea (Cottrell and Kirchman 2009; Haverkamp et al. 2008; Larsson et al. 2014; Partensky et al. 1999a). *Synechococcus* abundance is typically two orders of magnitude lower than *Prochlorococcus* in central gyres, but it often outcompetes *Prochlorococcus* in nutrient-rich regions and can even reach cell densities above 10^6 cells mL^{-1} in the Costa Rica Dome, likely due to high concentrations of cobalt

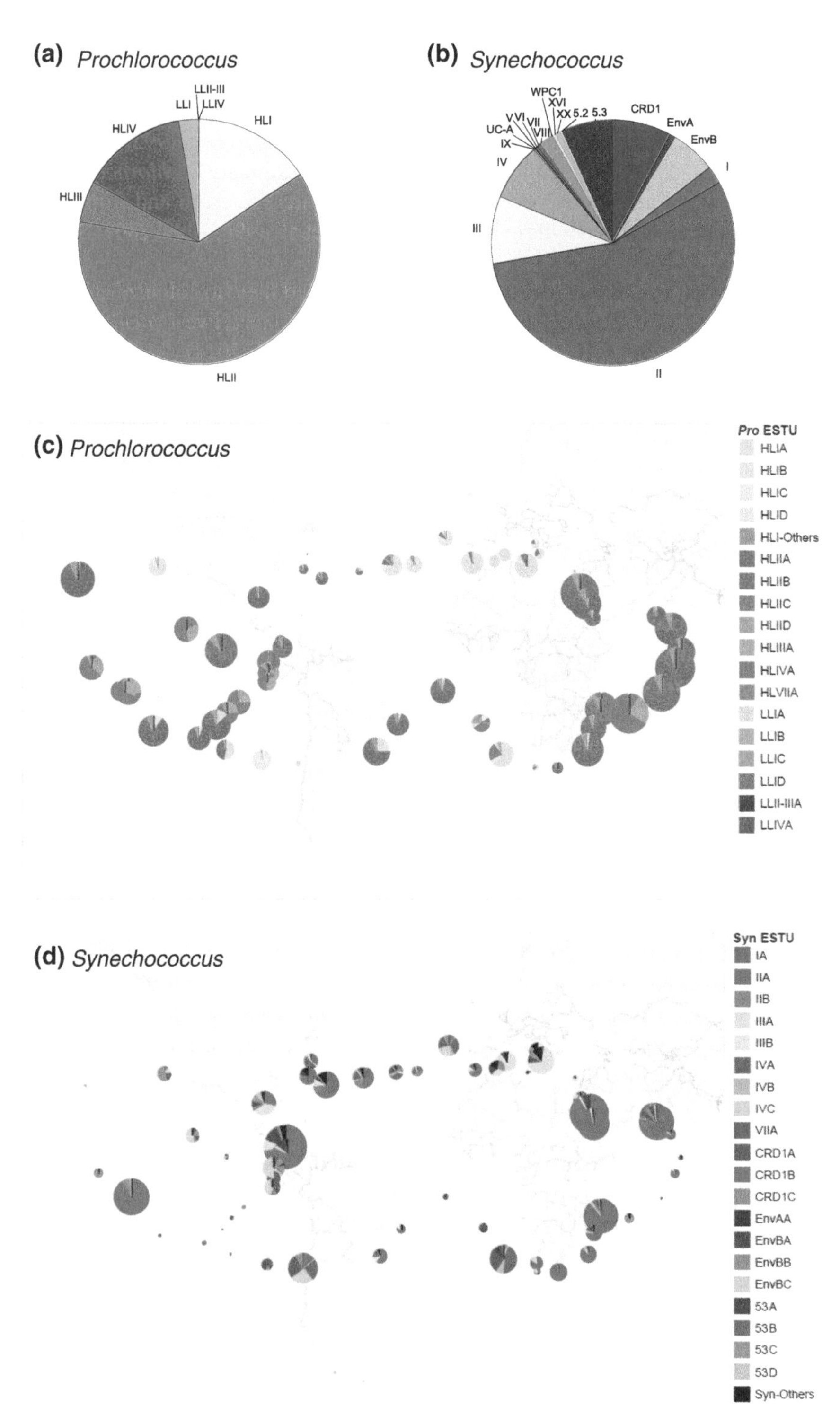
(a) Prochlorococcus
LLII-III
LLI
LLIV
HLI
HLIV
HLIII
HLII
(b) Synechococcus
WPC1
XVI
V VI
VII
XX
5.2
5.3
CRD1
EnvA
EnvB
UC-A
VIII
IX
IV
I
III
II
(c) Prochlorococcus
Pro ESTU
HLIA
HLIB
HLIC
HLID
HLI-Others
HLIIA
HLIIB
HLIIC
HLIID
HLIIIA
HLIVA
HLVIIA
LLIA
LLIB
LLIC
LLID
LLII-IIIA
LLIVA
(d) Synechococcus
Syn ESTU
IA
IIA
IIB
IIIA
IIIB
IVA
IVB
IVC
VIIA
CRD1A
CRD1B
CRD1C
EnvAA
EnvBA
EnvBB
EnvBC
53A
53B
53C
53D
Syn-Others

◀ **Fig. 3.4** Relative abundance of *Prochlorococcus* and *Synechococcus* clades in surface (**a–b**) and global distribution along the *Tara* Oceans expedition transect of ecologically significant taxonomic units (ESTUs, defined as within-clade 94 % OTUs sharing a similar distribution pattern; **c–d**). The size of *circles* is proportional to the number of *Prochlorococcus* or *Synechococcus* reads at the corresponding station. Reproduced with permission from Farrant et al. (2016) (**a–b**) or drawn using data from this study (**c–d**)

and iron (Ahlgren et al. 2014; Saito et al. 2005). Using a regression-based modeling approach to analyze the *Tara* Ocean metagenomes and their associated metadata, Guidi et al. (2016) suggested that *Synechococcus* and their phages, but surprisingly not *Prochlorococcus,* would be key players of the plankton networks driving carbon export and thence the oceanic biological pump.

Marine *Synechococcus* and *Prochlorococcus* are phylogenetically closely related and together with *Cyanobium* form a monophyletic branch that long diverged from all other cyanobacteria, including several freshwater *Synechococcus* (Fig. 3.3; Coutinho et al. 2016; Shih et al. 2013). Members of this branch, which so far account a single freshwater representative (*Synechococcus* sp. PCC 6307), all share the property to have alpha-carboxysomes encapsulating a Form-IA ribulose-bisphosphate carboxylase oxygenase (RuBisCO; Badger and Price 2003). Genes encoding these different components are phylogenetically more related to those coding for the microcompartments of chemoautotrophic proteobacteria, such as thiobacilli, than to those encoding beta-carboxysomes and Form-IB RuBisCO found in all other cyanobacteria. Yet, a recent comparative study has revealed no obvious functional differences between these two types of carboxysomes (Whitehead et al. 2014).

The marine *Synechococcus* radiation is divided into three main groups called subclusters 5.1–5.3 (Dufresne et al. 2008). The first one is much more widespread and diversified than the other two, with subcluster 5.2 encompassing halotolerant strains thriving in estuaries and near coastal areas (Chen et al. 2006), while subcluster 5.3 members are strictly oceanic and can constitute locally an abundant component of *Synechococcus* communities, e.g., in the Mediterranean Sea or in the Gulf of Mexico (Fig. 3.4c; see also Chen et al. 2006; Farrant et al. 2016; Huang et al. 2012; Sohm et al. 2015). A major diversification event seemingly occurred within subcluster 5.1 soon after its divergence from 5.2 and 5.3, producing the large genetic diversity known nowadays within this group, with 12–19 distinct clades, depending on the resolution power of different genetic markers (Ahlgren and Rocap 2012), which can be further split into about 30 subclades (Mazard et al. 2012). This event seemingly occurred near simultaneously with the advent of the *Prochlorococcus* genus and possibly generated it (Urbach et al. 1998).

The evolutionary history of marine picocyanobacteria has been strongly influenced by environmental factors (Biller et al. 2015; Scanlan 2012; Scanlan et al. 2009). For *Prochlorococcus*, Moore and co-workers were the first to demonstrate the key role played by light in the vertical niche partitioning of high-light (HL) adapted ecotypes in the upper lit layer and low-light (LL) ecotypes further down (Moore et al. 1998). Yet, more recent studies have considerably refined this simplistic picture, by showing that while some LL clades (LLII-VI) are confined to the bottom of the euphotic layer, with LLV-VI being specific of oxygen minimum zones (Lavin et al. 2010), the LLI clade is more photo-tolerant (Johnson et al. 2006; Malmstrom et al. 2010; Partensky and Garczarek 2010). In warm, stratified oligotrophic areas, LLI thrives at the base of the upper mixed layer, whereas in temperate mixed waters it can stand temporary exposure to high irradiance (Johnson et al. 2006; Malmstrom et al. 2010). Other physicochemical parameters have been critical for the diversification of clades within the HL branch (Martiny et al. 2015). Iron availability has likely conditioned the differentiation between the HLI-II clades that thrive in iron-replete waters, and HLIII-IV clades that co-occur in warm, iron-depleted areas, notably in high-nutrient low-Chl (HNLC) areas of the Pacific Ocean (Farrant et al. 2016; Rusch et al. 2010; West et al. 2011). This area has been largely explored during the *Tara* Oceans expedition, explaining the large proportion of HLIII-IV clades in this global metagenome dataset, even though HLII remains globally the most abundant clade (Fig. 3.4a). Thereafter, temperature would have favored the separation between HLI, which predominates at high latitudes and HLII that dominates in warm, oligotrophic waters, but it is worth noting that the minor ecotype is never outcompeted to extinction (Chandler et al. 2016; Farrant et al. 2016). Such shifts between HLI- and HLII-dominated communities can also occur over short geographical distances, along sharp temperature gradients, e.g., between the core and outside of Agulhas rings, huge anticyclonic eddies formed in the southern Indian Ocean and that then drift across the South Atlantic Ocean (Villar et al. 2015). Lastly, the availability of nitrogen and phosphorus, which are limiting factors in wide expanses of the ocean, has also played a role but only in the recent evolutionary history of the *Prochlorococcus* genus. A number of studies have indeed shown that some P-depleted HLII populations can acquire genes involved in the metabolism or regulation of P uptake by lateral gene transfer (Martiny et al. 2006, 2009a) and a similar phenomenon has also been observed for nitrite and/or nitrate assimilation (Berube et al. 2015; Kent et al. 2016; Martiny et al. 2009b; Villar et al. 2015). Despite this apparent solid relationship between *Prochlorococcus* phylogeny and community structure, several recent studies have shown the occurrence of a large genetic microdiversity, and identified different sets of within-clade phylogenetic groups exhibiting distinct distribution and/or seasonal patterns. Indeed, Kashtan et al. (2014) showed that *Prochlorococcus* HLII populations are actually composed of hundreds of closely related subpopulations possessing distinct genomic ‘backbones’, each consisting of a different set of core genes associated with a defined set of accessory genes. These discrepancies are sufficient to allow differentiated responses of these subpopulations to seasonal changes in the environment. Interestingly, the occurrence of minor within-clade subpopulations exhibiting drastically distinct ecological niches from that of the dominant population occurs in other clades and are even more pronounced for HLI and LLI than for HLII (Farrant et al. 2016; Larkin et al. 2016). This includes some LLI populations, which are adapted to iron-limited surface waters, much like HLIIIA-IVA, as well as HLI sub-populations that thrive not only in cold temperate waters, as do the more typical HLI, but also in warm subtropical waters, thus extending the global niche occupied by these clades. Even though this deviant behavior concerns only minor components of the *Prochlorococcus* population, these observations constitute a piece of evidence for the ongoing diversification within *Prochlorococcus* clades.

For *Synechococcus*, the temporal succession of physicochemical parameters that have driven the evolution of this genus is difficult to establish because of the rapid diversification of lineages (Urbach et al. 1998) that has resulted in a much fuzzier relationship between phylogeny and adaptation to specific environmental factors than for *Prochlorococcus*. Temperature has seemingly played a major selective role, since clades I and IV mainly occur in cold, nutrient-rich waters, while clades II and III preferentially thrive in warm, oligotrophic waters (Zwirglmaier et al. 2008). Like for *Prochlorococcus* HLI and HLII (Johnson et al. 2006; Zinser et al. 2007), these differences in latitudinal distributions can be explained by the different temperature growth ranges of these phylotypes, as demonstrated on representative isolates (Mackey et al. 2013; Pittera et al. 2014; Varkey et al. 2015). Yet, phylogenies made from 16S rRNA or other core genes do not split apart cold and warm-adapted clades, since clade I and IV strains generally fall in distantly related branches (see e.g., Dufresne et al. 2008; Scanlan et al. 2009). Furthermore, the study of the global distribution of *Synechococcus* at high taxonomic resolution (Fig. 3.4d) showed that a minor ecologically significant taxonomic unit (ESTU) within clade II, ESTU IIB, was able to colonize cold niches and that CRD1 and EnvB clades encompassed three distinct ESTUs exhibiting distinct thermal preferences (Farrant et al. 2016). As for *Prochlorococcus*, several recent studies have indicated that iron availability could also be an important driving factor of the composition of *Synechococcus* communities, since CRD1, CRD2 and/or EnvB clades—the latter two have been defined with ITS and *petB* markers, respectively, and might correspond to the same lineage—were shown to co-dominate in iron-depleted areas (Ahlgren et al. 2014; Farrant et al. 2016; Sohm et al. 2015). At last, phosphorus availability also seem to have influenced *Synechococcus* genetic diversification because ESTUs IIIA and 5.3A appear to be adapted to P-limited conditions, notably occurring in the Mediterranean Sea and Gulf of Mexico (Fig. 3.4d). Thus, unlike *Prochlorococcus*, the co-occurrence in similar niches of *Synechococcus* genotypes belonging to phylogenetically distant clades indicates that adaptation to specific environmental parameters (temperature, iron and/or phosphorus availability) has likely happened several times independently during the evolution of the *Synechococcus* radiation through convergent evolution.

The availability of numerous genomes and metagenomes of marine picocyanobacteria brings further important insights into their biology and the evolution of genome structure in this radiation (Fig. 3.3). A striking peculiarity of this group is that even though the 16S rRNA gene of all marine picocyanobacterial strains sequenced so far share >95 % identity, there is a remarkable genome divergence between and within *Prochlorococcus* and marine *Synechococcus*, as measured by average nucleotide identity (ANI) (Coutinho et al. 2016; Dufresne et al. 2008; Zhaxybayeva et al. 2009). Indeed, ANI values, observed between and within clades, are often significantly lower than the value of 94 % ANI thought to be equivalent to the bacterial species threshold of 70 % DNA-DNA hybridization (Konstantinidis and Tiedje 2005), with e.g., only 91.3 % ANI between the two clade IV *Synechococcus* strains BL107 and CC9902 that are identical at the 16S rRNA level (Dufresne et al. 2008).

The evolutionary history of the *Prochlorococcus* genus is characterized by a major genome streamlining event that has affected all but one lineages, while no comparable decrease has seemingly occurred during the evolution of marine *Synechococcus* (Dufresne et al. 2005; Kettler et al. 2007; Partensky and Garczarek 2010). This process has led to an overall decrease of about one third of the *Prochlorococcus* genome size—corresponding to about 930 genes—with regard to its most recent common ancestor with *Synechococcus*, with a concomitant drop of the G + C content down to ca. 30 % in HL clades (Dufresne et al. 2005; Kettler et al. 2007; Partensky and Garczarek 2010). While many gene gains and losses have taken place in all *Prochlorococcus* lineages, this gene flow was balanced in the basal LLIV lineage, whose members have a genome size around 2.45 Mb—i.e. in the range of *Synechococcus* genome sizes: 2.1 to >3.0 Mbp— but strongly unbalanced toward losses in all other lineages that have genome sizes between 1.64 and 1.86 Mbp (Biller et al. 2014a; Kettler et al. 2007). Only $\sim$130 genes are absent from all streamlined *Prochlorococcus* genomes compared to LLIV and *Synechococcus* genomes, indicating that the different *Prochlorococcus* lineages have lost different sets of genes. In both picocyanobacteria, genomes are constituted by a core genome shared by all strains, which encompasses 1050 $\pm$ 50 and 1350 $\pm$ 50 genes in *Prochlorococcus* and *Synechococcus*, respectively, and an accessory genome of variable size, mainly due to the presence of unique (i.e. strain-specific) genes that can constitute from 1.6 to 33.2 % of the gene repertoire (Kettler et al. 2007; Dufresne et al. 2008; Scanlan et al. 2009; Garczarek and Partensky, unpublished data). A large fraction of these unique genes, most of which being uncharacterized, is localized in hypervariable regions called genomic islands (Coleman et al. 2006; Dufresne et al. 2008), and might be involved in specific adaptations to the local environment, such as resistance to grazers or phage attacks (Avrani et al. 2011; Dufresne et al. 2008; Palenik et al. 2006). In *Prochlorococcus*, some genomic islands are specialized in nitrogen and phosphate uptake (see above), while in marine *Synechococcus* strains, some island genes code for ‘giant’ proteins (>1 MDa) that owe their huge size to a high number of repeats, such as the cell-surface located protein, SwmB that is required for a special form of motility, found only in clade III members (Dufresne et al. 2008; McCarren and Brahamsha 2007).

While the core genome of picocyanobacteria is fairly small, their pangenome is huge, since each new sequenced genome contains, on average, 277 new genes (Baumdicker et al. 2010). Using a quantitative, evolutionary model for the distributed genome, Baumdicker et al. (2012) predicted that the pangenome is finite and would contain 57,792 genes, although in a more recent study based on many more genomes, this number climbed to 84,872 genes (Biller et al. 2015). Yet, this vast ‘collective genome’ might still miss some important gene functions, since *Prochlorococcus* cells were shown to depend upon free-living ‘helper’ bacteria from their immediate environment, e.g., to fight against oxidative stress using catalases released by such helpers (Morris et al. 2008). This exchange might not be one-way since *Prochlorococcus* can itself release lipid vesicles containing proteins and nucleic acids that might be used for transferring material to their helper bacteria but also possibly be used as lures for phages (Biller et al. 2014b).

Another advantage of this wealth of genomes is to inform us about the ability of marine picocyanobacteria to synthesize products of potential biotechnological interest. Due to their decreased genome size, picocyanobacteria possess few genes involved in the biosynthesis of secondary metabolites compared to their larger size marine, freshwater or extremophile counterparts (Mandal and Rath 2015). Yet, picocyanobacteria not only possess and express alkane biosynthesis genes (Klähn et al. 2014), but also produce and accumulate hydrocarbons, which may constitute between 0.02 and 0.37 % of their dry cell weight (Lea-Smith et al. 2015; Schirmer et al. 2010). Thus, at the global ocean scale, picocyanobacteria produce about 540 million tons of hydrocarbons annually. A number of genes that are involved in cell defense mechanisms are also worth mentioning. These include antibiotic resistance genes (Hatosy and Martiny 2015), a possible toxin–antitoxin system that might be beneficial for cell survival under unfavorable growth conditions (Steglich et al. 2008) and genes involved in allelopathy (Paz-Yepes et al. 2013). Other potential exploitable products include phycobiliproteins that are used in a number of fluorescence applications. Phycoerythrin-II, which is specific of marine *Synechococcus* (Ong et al. 1984; Six et al. 2007) might constitute interesting molecules to exploit for these technologies, especially forms binding large amounts of PUB (see above), a chromophore whose absorption properties fits well with blue lasers.

3.4.3 Nitrogen-Fixing Cyanobacteria

Nitrogen is the predominant limiting nutrient for primary productivity within the euphotic zone of much of the surface low-latitude oceans (Canfield et al. 2010; Falkowski 1997; Moore et al. 2013). By assimilating dinitrogen (N_2), the simplest and most abundant form in the atmosphere and in seawater, diazotrophs constitute an important source of bioavailable nitrogen to oceanic surface waters, possibly the most important external nitrogen source, before atmospheric and riverine inputs (Deutsch et al. 2007). In the North Pacific Subtropical Gyre, arguably the largest biome on Earth, N_2 contributed $\sim$50 % of 'new nitrogen' (Dugdale and Goering 1967) to the euphotic zone, substantially increasing export production (Karl et al. 1997). Accordingly, microbial diazotrophy is a major and crucial biogeochemical process in the ocean that compensates bioavailable nitrogen lost due to processes such as denitrification and anaerobic ammonium oxidation and allows new primary production (Zehr and Kudela 2011).

Even though cyanobacteria are not the sole N_2-fixers in marine ecosystems, the contribution of heterotrophic diazotrophs such as some Gammaproteobacteria and Deltaproteobacteria is poorly understood (Turk-Kubo et al. 2014; Zehr and Kudela 2011) and will not be addressed further in this chapter. Comparative genomic analyses provided evidence that the genes encoding the N_2 fixation machinery are frequently clustered and in some cases may have been transferred laterally, causing diazotrophy to occur in taxa that are otherwise not known to fix nitrogen, such as in *Microcoleus chthonoplastes* (Bolhuis et al. 2010) or in an isolate of the Chl *d* containing genus *Acaryochloris* (Pfreundt et al. 2012).

3.4.3.1 Filamentous Marine Diazotrophs

Multicellular, filamentous marine diazotrophs are cyanobacteria belonging to the genera *Trichodesmium* (non-heterocystous, free-living) (Capone 1997; Ehrenberg 1830; Gomont 1892; Karl et al. 2002), heterocystous *Richelia* living symbiotic with diatoms of the genera *Rhizosolenia* (Het-1) or *Hemiaulus* (also referred to as Het-2) (Foster and Zehr 2006; Foster et al. 2009; Goebel et al. 2010; Janson et al. 1999; Villareal 1991, 1992, 1994; Zeev et al. 2008), and the heterocyst-forming *Calothrix* (Het-3) associated with the diatom *Chaetoceros* (Carpenter and Foster 2002; Foster and Zehr 2006; Goebel et al. 2010). Additional filamentous diazotrophs are found in coastal and reef environments, among them the non-heterocystous *Lyngbya* (Omoregie et al. 2004; Woebken et al. 2015) and in brackish water environments such as the Baltic Sea, the heterocystous *Nodularia* (Ploug et al. 2011).

Trichodesmium, a Diazotroph of Tropical and Subtropical Marine Waters

Diazotrophic cyanobacteria of the genus *Trichodesmium* can form huge surface blooms of tens of thousands of km^2 (Dupouy et al. 1988) in the tropical and subtropical ocean and constitute an important source of new nitrogen to these oligotrophic environments (Capone 1997; Davis and McGillicuddy 2006; Mahaffey et al. 2005).

Trichodesmium is taxonomically close to filamentous cyanobacteria of the genus *Oscillatoria* (Larsson et al. 2011). Unique for diazotrophic filamentous cyanobacteria is that *Trichodesmium* expresses the oxygen-sensitive nitrogenase and fixes N_2 during the day (Dugdale et al. 1961) at the same time when also photosynthetic oxygen evolution takes place. Whereas other diazotrophs such as *Calothrix*, *Nodularia* or *Anabaena* develop heterocysts, differentiated cells specialized for nitrogen fixation (Muro-Pastor and Hess 2012), *Trichodesmium* uses a different, not terminally differentiated and not fully understood cell type for this purpose, called diazocyte (Berman-Frank et al. 2001; El-Shehawy et al. 2003; Sandh et al. 2009, 2012). In addition, *Trichodesmium* can form colonies or multicellular aggregates of surprisingly varying morphologies, including threads (trichomes), radial puffs, vertically aligned fusiform tufts, and bowties (Hynes et al. 2012; Olson et al. 2015;

Post et al. 2002; Webb et al. 2007). Although the physiological relevance of these morphologies is not well understood, there is clear evidence for true multicellular behavior of *Trichodesmium* colonies, e.g., in the acquisition of mineral-rich dust particles (Rubin et al. 2011).

Different factors have been discussed to influence termination of *Trichodesmium* blooms, including bacteriophage-induced lysis (Hewson et al. 2004; Ohki 1999) or grazing by copepods (O'Neil 1998; O'Neil and Roman 1994). An important factor seems to be the activation of an autocatalytic mechanism leading to a programmed cell death, resembling apoptosis in metazoans. Programmed cell death is a documented process in marine phototrophs (Bidle and Falkowski 2004; Franklin et al. 2006) but many of the involved details have remained enigmatic. In *Trichodesmium*, programmed cell death may be triggered by several environmental factors, such as P or Fe starvation, high irradiance, or oxidative stress (Berman-Frank et al. 2004, 2007) and involves the activation of caspase-like activity (Berman-Frank et al. 2004). The analysis of a *Trichodesmium* bloom in oligotrophic lagoon waters of New Caledonia demonstrated an increased caspase-like activity during bloom demise, including the specific activation of 11 out of 12 predicted metacaspase genes (Spungin et al. 2016) that previously were defined in *Trichodesmium* (Asplund-Samuelsson 2015; Asplund-Samuelsson et al. 2012; Berman-Frank et al. 2004).

The full genome sequence of the reference strain *Trichodesmium erythraeum* IMS101 with a length of 7.75 Mb and 4451 annotated genes belongs to the larger cyanobacterial genomes (Shih et al. 2013). Very different to the genome sequences of other free-living cyanobacteria with an average coding capacity of $\sim$85 %, only 64 % of the *T. erythraeum* genome encodes proteins (Larsson et al. 2011). This unusual high non-coding genome share appears to be conserved in the genus *Trichodesmium*, supported by several metagenomic datasets (Walworth et al. 2015). Transcription from such non-coding genome space can produce non-coding RNAs (sRNAs), which frequently have a regulatory function (Storz et al. 2011), also in cyanobacteria (Georg et al. 2014; Klähn et al. 2015; Kopf and Hess 2015). Indeed, the analysis of the primary transcriptome (the sequencing of enriched nascent transcript starts) of *T. erythraeum* IMS101 revealed that at least 40 % of all promoters active under standard laboratory conditions produce non-protein-coding transcripts and that these accumulate in much larger amounts than mRNA (Pfreundt et al. 2014).

Amongst those non-coding transcripts is an unusually high number of retroelement-like genetic elements resembling group II introns, an actively splicing twintron (Pfreundt and Hess 2015), and also a Diversity Generating Retroelement (DGR) for the targeted mutation of specific genes. DGRs were first described in a phage of *Bordetella* where modification of a single target gene ensures inducible receptor diversity for host recognition and have the potential to create protein diversity in the order of the human immune system (Liu et al. 2002). These elements consist of at least two components, the first being a non-coding RNA called template repeat (TR) RNA that serves as a template for the second element, an error-prone reverse transcriptase that converts this template into cDNA for recombination into the protein-coding region of the target gene(s) (Doulatov et al. 2004; Guo et al. 2008). Thus, the identification of the TR RNA is essential to identify the target genes of this mechanism. Although there was previous evidence for the existence of this mechanism in *Trichodesmium* (Doulatov et al. 2004), the target gene(s) remained unknown. The primary transcriptome enabled the exact definition of the DGR components in *Trichodesmium* and revealed 12 putative target genes (Pfreundt et al. 2014), an unprecedented potential for in vivo protein diversification in bacteria. Although none of these genes possesses a clear functional assignment, some appear connected to putative signaling proteins (kinases), possibly constituting their receptor component. Two target genes are in close proximity to the rest of the DGR elements, but the remaining 10 genes are distributed all over the genome. It can be speculated that at least some of the targeted proteins are involved in the defense against bacteriophages by systematic variation of a surface receptor. It is further possible that this system is involved in generating phenotypic variability by diversifying a few key genes. This could play a role, for instance, in the cell-cell recognition required for multicellular behavior and colony formation.

Nodularia, a Bloom-Forming Cyanobacterium Specifically Adapted to Salinity Gradients

Almost every summer, massive blooms of toxic cyanobacteria occur in the central regions of the Baltic Sea. These cyanobacteria cope well with the salinity gradient and brackish conditions that characterize the Baltic Sea. The dominating organism within these blooms is *Nodularia spumigena*, a filamentous cyanobacterium that produces several bioactive metabolites. Among these compounds is the hepatotoxin nodularin (Fewer et al. 2013; Mazur-Marzec et al. 2012) and protease inhibitors of the pseudoaeruginosin family (Liu et al. 2015). *Nodularia* is diazotrophic using heterocysts for this process (Ploug et al. 2011; Sivonen et al. 2007). Excess phosphorus combined with low nitrogen concentrations are thought to favor the growth and bloom formation of diazotrophic cyanobacteria in summer (Sellner 1997), most pronounced under stably stratified warm water conditions. Then, gas vesicles that provide buoyancy to *Nodularia* and related cyanobacteria lead to the formation of large surface scums in the absence of mixing.

Thus, *Nodularia* appears to have a selective advantage under the conditions of the Baltic Sea (Möke et al. 2013). The question what these advantages could be were followed by transcriptomic analyses of laboratory cultures after exposure to high light and oxidative stress, mimicking the extreme environmental conditions in the surface layer of the Baltic Sea in summer (Kopf et al. 2015). As the basis for this work served the draft genome sequence for *N. spumigena* strain CCY9414 (Voss et al. 2013). Genes encoding enzymes for the synthesis of toxins were among the up-regulated genes, implying that these compounds may play an important role in the cellular acclimation to conditions of bloom formation, consistent with similar observations for the freshwater cyanobacterium *Microcystis* (Zilliges et al. 2011).

The measured photosynthetic activity in the tested *N. spumigena* CCY9414 trichomes remained high also at the highest light intensities of 1200 μmol photons m^{-2} s^{-1}, and showed signs of an increase in photorespiratory flux (Kopf et al. 2015). This observation is of interest for understanding the acclimation of cyanobacterial trichomes to the combination of high light, high oxygen partial pressure and low nutrients, including low iron and CO_2 in the surface layer. Photorespiration cooperates with Mehler-like reactions catalyzed by flavodiiron proteins to dissipate excess absorbed energy in cyanobacteria (Allahverdiyeva et al. 2011, 2013; Hackenberg et al. 2009). The observed activation of photorespiratory flux to a physiologically meaningful level was furthermore consistent with the observation of many up-regulated genes encoding photorespiratory enzymes and flavodiiron proteins. Thus, specialized phototrophs such as *Nodularia* may have evolved a mechanism involving activation of photorespiratory flux to be able to cope with the high light and oxidative stress conditions during the Baltic Sea summer. The work by Kopf et al. (2015) identified many additional, previously unknown, stress-regulated genes, some of which encoding proteins of known functions in metabolism, transport, DNA stability and structure, but the majority were of entirely unknown function. These results suggested the existence of stress-related mechanisms in the surface layer cyanobacterial blooms that remain to be identified and which operate in addition to the observed activation of photorespiratory flux (Kopf et al. 2015).

Richelia and Calothrix

Whereas for *Richelia* symbiosis is obligatory, for *Calothrix* (Het-3) an epiphytic lifestyle with *Chaetoceros* was reported (Carpenter and Foster 2002) and that it can also grow free-living (Foster et al. 2010). These differences between obligatory or facultative interaction with their respective diatom host match distinct differences in the genetic capacity of these two cyanobacteria in a dramatic way. Whereas the genome of a marine *Calothrix rhizosoleniae* SC01 isolated from outside the frustule of *Chaetoceros* was at least 6.0 Mb and similar to those of free-living heterocyst-forming cyanobacteria, *Richelia intracellularis* HH01 isolated from inside the siliceous frustule of *Hemiaulus* was decreased to a size of only 3.2 Mb. Moreover, the *Richelia* genome sequence lacked genes for ammonium transporters, nitrate/nitrite reductases and glutamine:2-oxoglutarate aminotransferase (Hilton et al. 2013). With these features, the decreased genome size of *R. intracellularis* HH01 shows clear signs of adaptation as an obligate symbiont. Despite some evidence of genome size decrease, it is currently unknown whether *Richelia* that occurs in association with the diatom *Rhizosolenia* (then also called Het-1) is an obligate symbiont as well (Hilton et al. 2013; Villareal 1992).

3.4.3.2 Unicellular Marine Diazotrophs

Marine unicellular cyanobacterial diazotrophs are phylogenetically divided into three different groups. The uncultivated UCYN-A (Tripp et al. 2010; Zehr et al. 2001, 2008) which live in association with the prymnesiophyte alga *B. bigelowii* (Hagino et al. 2013; Thompson et al. 2012, 2014); the free-living *Crocosphaera* sp. (also referred to as unicellular group B or UCYN-B); and the unicellular group C (UCYN-C), to which several free-living cyanobacteria belong, including group C TW3 (Taniuchi et al. 2012) and *Cyanothece* sp. ATCC 51142 (Reddy et al. 1993). The latter strain has become a major model organism in biotechnology and synthetic biology (Aryal et al. 2013; Bandyopadhyay et al. 2013; Bernstein et al. 2015; Červený et al. 2013; Chou et al. 2015; Melnicki et al. 2012).

Symbiotic interactions are important for major sources of fixed nitrogen in the world's ocean. Particular insight on the relevance of symbiotic interactions has been gained from the analysis of *Candidatus* Atelocyanobacterium thalassa (UCYN-A). UCYN-A and its prymnesiophyte (haptophyte) host are abundant and widely distributed (Krupke et al. 2014). Therefore, UCYN-A are ecologically relevant players in the marine biogeochemical cycles (Cabello et al. 2015; Goebel et al. 2010; Jardillier et al. 2010; Montoya et al. 2004; Zehr and Kudela 2011). The UCYN-A group consists of at least three distinct clades (Thompson et al. 2014), called UCYN-A1, UCYN-A2 and UCYN-A3 and the whole group is monophyletic within the clade that includes also the UCYN-B and UCYN-C species *Crocosphaera* sp. and *Cyanothece* sp. (Bombar et al. 2014).

In a pioneering work, the streamlined genome of UCYN-A was determined after flow cytometry based cell sorting. The most striking result was the lack of all genes encoding the photosystem II complex, the Calvin-Benson-Bassham cycle for carbon fixation, as well as other pathways that are normally essential (Tripp et al. 2010; Zehr et al. 2008). Thus, it appeared logical that UCYN-A lives in a symbiotic interaction, an assumption that now has been confirmed (Hagino et al. 2013; Thompson et al. 2012, 2014). In this mutualistic relationship, UCYN-A provides fixed nitrogen to its host in exchange for fixed carbon. The details of this interaction are intricate. Before any carbon is transported from the haptophyte to UCYN-A, transfer of nitrogen is required from UCYN-A to the haptophyte (Krupke et al. 2015).

UCYN-B, *C. watsonii* and related cyanobacteria are unicellular and free-living diazotrophs, but also capable of colonial aggregation (Foster et al. 2013) and to form symbioses with the diatom *Climacodium frauenfeldianum* (Carpenter and Janson 2000; Foster et al. 2011). Otherwise, it is a typical unicellular N_2-fixing cyanobacterium in the sense that it is photosynthetic during the day and fixes N_2 during the night.

3.4.4 *Abundance and Contribution of the Different Diazotroph Taxa to the Biogeochemical Cycle*

A small share of diazotrophic cells in the total microbial population can have a substantial impact. During the VAHINE mesocosm experiment performed in 2013 in the shallow water of the New Caledonia lagoon (Bonnet et al. 2016b), N_2 fixation rates reached >60 nmol N L^{-1} d^{-1}, which are among the highest rates reported for marine waters (Bonnet et al. 2016a; Luo et al. 2012). Evidence from ^{15}N isotope labeling analyses indicated that the dominant source of nitrogen fueling export production shifted from subsurface nitrate assimilated prior to the start of the 23 day experiment to N_2 fixation by the end (Knapp et al. 2015). During days 15–23 of the VAHINE experiment, N_2 fixation rates increased dramatically (Bonnet et al. 2016a) and N_2-fixing cyanobacteria of the UCYN-C type dominated the diazotroph community in the mesocosms (Turk-Kubo et al. 2015). Based on relative 16S rRNA gene copy numbers that were normalized by comparison against the flow cytometry counts of abundant marine picocyanobacteria, a maximum of only 500 cells mL^{-1} was calculated for *Cyanothece*-like diazotrophs (Pfreundt et al. 2016). This matched reasonably well to the maximum of 100 UCYN-C *nifH* copies mL^{-1} determined for the same population (Turk-Kubo et al. 2015), especially when taking into account that *Cyanothece* spp. usually have two to three 16S rRNA gene copies per cell compared to *nifH*, which is a single-copy gene. The total number of bacteria (heterotrophs and picocyanobacteria) was between 5 and 7×10^6 cells mL^{-1}, hence *Cyanothece*-like diazotrophs had a share of <0.01 % in the total microbial population but impacted its biogeochemical properties in a profound way.

3.5 Photoheterotrophy and Phototroph-Heterotroph Interactions

Interactions within marine microbial populations and especially between phototrophic and heterotrophic organisms are of considerable complexity and may involve metabolic factors as well as specific regulatory signals. These interactions are likely to impact the physiology of both partners, the chemistry of their environment, and shape ecosystem diversity, but details are only slowly emerging.

Sulfitobacter-related bacteria impact the cell division of marine diatoms via secretion of indole-3-acetic acid (Amin et al. 2015), an otherwise well studied plant hormone. Its synthesis by these *Sulfitobacter*-related bacteria depends on endogenously produced tryptophan as well as on tryptophan of diatom origin, adding an additional facet to this interaction.

Auxotrophy for the soluble B vitamins (B1, B7, and B12) is well-known for marine autotrophs as well as for heterotrophs, both in oligotrophic and eutrophic environments (Sañudo-Wilhelmy et al. 2014). Especially vitamin B12 is an essential factor that has been implicated in several examples of phototroph-heterotroph interactions (Kazamia et al. 2012). Remarkably, dissolved vitamin B12 can be measured in surface waters at concentrations useful for possibly auxotrophic species (Bonnet et al. 2013).

Experimental exoproteome analysis of marine *Synechococcus* showed transport systems for inorganic nutrients and an interesting array of strain-specific exoproteins likely involved in mutualistic or hostile interactions (i.e. hemolysins, pilins, adhesins), and exoenzymes with a potential mixotrophic goal (i.e. exoproteases and chitinases) (Christie-Oleza et al. 2015a). Exoproteomic analysis of especially designed synthetic communities in which specific *Roseobacter* strains (*Ruegeria pomeroyi* DSS-3, *Roseobacter denitrificans* OCh114, and *Dinoroseobacter shibae* DFL-12) were combined with two different cyanobacteria (*Synechococcus* spp. WH7803 and WH8102) revealed the set of hydrolytic enzymes secreted by *Roseobacter*, that is thought to degrade the biopolymers of cyanobacterial origin outside the cell (Christie-Oleza et al. 2015b). Evidence is growing that many eukaryotic microalga but also marine cyanobacteria are photoheterotrophs or mixotrophs, i.e., they can photosynthesize but also feed on various sources of organic carbon (Gómez-Baena et al. 2008; Gómez-Pereira et al. 2013; McKie-Krisberg and Sanders 2014; Muñoz-Marín et al. 2013; Unrein et al. 2014). These examples illustrate the complexity of lifestyles and organismic interactions between marine phototrophic microorganisms and within the greater microbial community, but it is very likely that the majority of these interrelationships still await their discovery.

Acknowledgments The research leading to these results has received funding from the European Union Seventh Framework Programme MaCuMBA (FP7/2007-2013) under grant agreement n° 311975. This publication reflects the views only of the authors, and the European Union cannot be held responsible for any use which may be made of the information contained therein.

References

Ahlgren NA, Rocap G (2012) Diversity and distribution of marine *Synechococcus*: multiple gene phylogenies for consensus classification and development of qPCR assays for sensitive measurement of clades in the ocean. Front Microbiol 3:213

Ahlgren NA, Noble A, Patton AP, Roache-Johnson K, Jackson L, Robinson D, McKay C, Moore LR, Saito MA, Rocap G (2014) The unique trace metal and mixed layer conditions of the Costa Rica upwelling dome support a distinct and dense community of *Synechococcus*. Limnol Ocean 59:2166–2184

Akram N, Palovaara J, Forsberg J, Lindh MV, Milton DL, Luo H, González JM, Pinhassi J (2013) Regulation of proteorhodopsin gene expression by nutrient limitation in the marine bacterium *Vibrio* sp. AND4. Environ Microbiol 15:1400–1415

Allahverdiyeva Y, Ermakova M, Eisenhut M, Zhang P, Richaud P, Hagemann M, Cournac L, Aro E-M (2011) Interplay between flavodiiron proteins and photorespiration in *Synechocystis* sp. PCC 6803. J Biol Chem 286:24007–24014

Allahverdiyeva Y, Mustila H, Ermakova M, Bersanini L, Richaud P, Ajlani G, Battchikova N, Cournac L, Aro E-M (2013) Flavodiiron proteins Flv1 and Flv3 enable cyanobacterial growth and photosynthesis under fluctuating light. Proc Natl Acad Sci USA 110:4111–4116

Amin SA, Hmelo LR, van Tol HM, Durham BP, Carlson LT, Heal KR, Morales RL, Berthiaume CT, Parker MS, Djunaedi B et al (2015) Interaction and signalling between a cosmopolitan phytoplankton and associated bacteria. Nature 522:98–101

Anderson DM, Cembella AD, Hallegraeff GM (2012) Progress in understanding harmful algal blooms: paradigm shifts and new technologies for research, monitoring, and management. Annu Rev Mar Sci 4:143–176

Archibald JM (2009) The puzzle of plastid evolution. Curr Biol CB 19:R81–R88

Archibald JM (2012) The evolution of algae by secondary and tertiary endosymbiosis. Adv Bot Res 64:87–118

Armbrust EV, Berges JA, Bowler C, Green BR, Martinez D, Putnam NH, Zhou S, Allen AE, Apt KE, Bechner M et al (2004) The genome of the diatom *Thalassiosira pseudonana*: ecology, evolution, and metabolism. Science 306:79–86

Aryal UK, Callister SJ, Mishra S, Zhang X, Shutthanandan JI, Angel TE, Shukla AK, Monroe ME, Moore RJ, Koppenaal DW et al (2013) Proteome analyses of strains ATCC 51142 and PCC 7822 of the diazotrophic cyanobacterium *Cyanothece* sp. under culture conditions resulting in enhanced H_2 production. Appl Environ Microbiol 79:1070–1077

Asplund-Samuelsson J (2015) The art of destruction: revealing the proteolytic capacity of bacterial caspase homologs. Mol Microbiol 98:1–6

Asplund-Samuelsson J, Bergman B, Larsson J (2012) Prokaryotic caspase homologs: phylogenetic patterns and functional characteristics reveal considerable diversity. PLoS ONE 7:e49888

Avrani S, Wurtzel O, Sharon I, Sorek R, Lindell D (2011) Genomic island variability facilitates *Prochlorococcus*-virus coexistence. Nature 474:604–608

Badger MR, Price GD (2003) CO_2 concentrating mechanisms in cyanobacteria: molecular components, their diversity and evolution. J Exp Bot 54:609–622

Bandyopadhyay A, Stöckel J, Min H, Sherman LA, Pakrasi HB (2010) High rates of photobiological H_2 production by a cyanobacterium under aerobic conditions. Nat Commun 1:139

Bandyopadhyay A, Elvitigala T, Liberton M, Pakrasi HB (2013) Variations in the rhythms of respiration and nitrogen fixation in members of the unicellular diazotrophic cyanobacterial genus *Cyanothece*. Plant Physiol 161:1334–1346

Baumdicker F, Hess WR, Pfaffelhuber P et al (2010) The diversity of a distributed genome in bacterial populations. Ann Appl Probab 20:1567–1606

Baumdicker F, Hess WR, Pfaffelhuber P (2012) The infinitely many genes model for the distributed genome of bacteria. Genome Biol Evol 4:443–456

Béjà O, Suzuki MT (2008) Photoheterotrophic marine prokaryotes. In: Kirchman DL (ed) Microbial ecology of the oceans. Wiley, New York, pp 131–157

Béjà O, Aravind L, Koonin EV, Suzuki MT, Hadd A, Nguyen LP, Jovanovich SB, Gates CM, Feldman RA, Spudich JL et al (2000) Bacterial rhodopsin: evidence for a new type of phototrophy in the sea. Science 289:1902–1906

Béjà O, Spudich EN, Spudich JL, Leclerc M, DeLong EF (2001) Proteorhodopsin phototrophy in the ocean. Nature 411:786–789

Béjà O, Suzuki MT, Heidelberg JF, Nelson WC, Preston CM, Hamada T, Eisen JA, Fraser CM, DeLong EF (2002) Unsuspected diversity among marine aerobic anoxygenic phototrophs. Nature 415:630–633

Berman-Frank I, Lundgren P, Chen Y-B, Küpper H, Kolber Z, Bergman B, Falkowski P (2001) Segregation of nitrogen fixation and oxygenic photosynthesis in the marine cyanobacterium *Trichodesmium*. Science 294:1534–1537

Berman-Frank I, Bidle KD, Haramaty L, Falkowski PG (2004) The demise of the marine cyanobacterium, *Trichodesmium* spp., via an autocatalyzed cell death pathway. Limnol Oceanogr 49:997–1005

Berman-Frank I, Rosenberg G, Levitan O, Haramaty L, Mari X (2007) Coupling between autocatalytic cell death and transparent exopolymeric particle production in the marine cyanobacterium *Trichodesmium*. Environ Microbiol 9:1415–1422

Bernstein HC, Charania MA, McClure RS, Sadler NC, Melnicki MR, Hill EA, Markillie LM, Nicora CD, Wright AT, Romine MF et al (2015) Multi-omic dynamics associate oxygenic photosynthesis with nitrogenase-mediated H_2 production in *Cyanothece* sp. ATCC 51142. Sci Rep 5:16004

Berube PM, Biller SJ, Kent AG, Berta-Thompson JW, Roggensack SE, Roache-Johnson KH, Ackerman M, Moore LR, Meisel JD, Sher D et al (2015) Physiology and evolution of nitrate acquisition in *Prochlorococcus*. ISME J 9:1195–1207

Bidle KD, Falkowski PG (2004) Cell death in planktonic, photosynthetic microorganisms. Nat Rev Microbiol 2:643–655

Biller SJ, Berube PM, Berta-Thompson JW, Kelly L, Roggensack SE, Awad L, Roache-Johnson KH, Ding H, Giovannoni SJ, Rocap G et al (2014a) Genomes of diverse isolates of the marine cyanobacterium *Prochlorococcus*. Sci Data 1

Biller SJ, Schubotz F, Roggensack SE, Thompson AW, Summons RE, Chisholm SW (2014b) Bacterial vesicles in marine ecosystems. Science 343:183–186

Biller SJ, Berube PM, Lindell D, Chisholm SW (2015) *Prochlorococcus*: the structure and function of collective diversity. Nat Rev Microbiol 13:13–27

Björkman KM, Church MJ, Doggett JK, Karl DM (2015) Differential assimilation of inorganic carbon and leucine by *Prochlorococcus* in the oligotrophic North Pacific Subtropical Gyre. Front Microbiol 6

Blank CE, Sánchez-Baracaldo P (2010) Timing of morphological and ecological innovations in the cyanobacteria—a key to understanding the rise in atmospheric oxygen. Geobiology 8:1–23

Blot N, Wu X-J, Thomas J-C, Zhang J, Garczarek L, Böhm S, Tu J-M, Zhou M, Plöscher M, Eichacker L et al (2009) Phycourobilin in trichromatic phycocyanin from oceanic cyanobacteria is formed post-translationally by a phycoerythrobilin lyase-isomerase. J Biol Chem 284:9290–9298

Blot N, Mella-Flores D, Six C, Le Corguillé G, Boutte C, Peyrat A, Monnier A, Ratin M, Gourvil P, Campbell DA et al (2011) Light history influences the response of the marine cyanobacterium *Synechococcus* sp. WH7803 to oxidative stress. Plant Physiol 156:1934–1954

Boeuf D, Cottrell MT, Kirchman DL, Lebaron P, Jeanthon C (2013) Summer community structure of aerobic anoxygenic phototrophic bacteria in the western Arctic Ocean. FEMS Microbiol Ecol 85:417–432

Bolhuis H, Severin I, Confurius-Guns V, Wollenzien UIA, Stal LJ (2010) Horizontal transfer of the nitrogen fixation gene cluster in the cyanobacterium *Microcoleus chthonoplastes*. ISME J 4:121–130

Bombar D, Heller P, Sanchez-Baracaldo P, Carter BJ, Zehr JP (2014) Comparative genomics reveals surprising divergence of two closely related strains of uncultivated UCYN-A cyanobacteria. ISME J 8:2530–2542

Bonnet S, Tovar-Sánchez A, Panzeca C, Duarte CM, Ortega-Retuerta E, Sañudo-Wilhelmy SA (2013) Geographical gradients of dissolved vitamin B12 in the Mediterranean sea. Front Microbiol 4:126

Bonnet S, Berthelot H, Turk-Kubo K, Fawcett S, Rahav E, l'Helguen S, Berman-Frank I (2016a) Dynamics of N_2 fixation and fate of diazotroph-derived nitrogen in a low nutrient low chlorophyll ecosystem: results from the VAHINE mesocosm experiment (New Caledonia). Biogeosciences 13:2653–2673

Bonnet S, Moutin T, Rodier M, Grisoni JM, Louis F, Folcher E, Bourgeois B, Boré JM, Renaud M (2016b) Introduction to the project VAHINE: VAriability of vertical and tropHIc transfer of fixed N_2 in the south wEst Pacific. Biogeosci Discuss. doi:10.5194/bg-2015-615

Bowler C, Allen AE, Badger JH, Grimwood J, Jabbari K, Kuo A, Maheswari U, Martens C, Maumus F, Otillar RP et al (2008) The *Phaeodactylum* genome reveals the evolutionary history of diatom genomes. Nature 456:239–244

Burki F, Keeling PJ (2014) Rhizaria. Curr Biol CB 24:R103–R107

Cabello AM, Cornejo-Castillo FM, Raho N, Blasco D, Vidal M, Audic S, de Vargas C, Latasa M, Acinas SG, Massana R (2015) Global distribution and vertical patterns of a prymnesiophyte-cyanobacteria obligate symbiosis. ISME J. doi:10.1038/ismej.2015.147

Canfield DE, Glazer AN, Falkowski PG (2010) The evolution and future of Earth's nitrogen cycle. Science 330:192–196

Capone DG (1997) *Trichodesmium*, a globally significant marine cyanobacterium. Science 276:1221–1229

Carpenter EJ, Foster RA (2002) Marine cyanobacterial symbioses. In: Cyanobacteria in symbiosis. Springer, Berlin, pp 11–17

Carpenter EJ, Janson S (2000) Intracellular cyanobacterial symbionts in the marine diatom *Climacodium frauenfeldianum* (Bacillariophyceae). J Phycol 36:540–544

Carpenter EJ, Price CC (1976) Marine *Oscillatoria* (*Trichodesmium*): explanation for aerobic nitrogen fixation without heterocysts. Science 191:1278–1280

Červený J, Sinetova MA, Valledor L, Sherman LA, Nedbal L (2013) Ultradian metabolic rhythm in the diazotrophic cyanobacterium *Cyanothece* sp. ATCC 51142. Proc Natl Acad Sci USA 110:13210–13215

Chandler JW, Lin Y, Gainer PJ, Post AF, Johnson ZI, Zinser ER (2016) Variable but persistent coexistence of *Prochlorococcus* ecotypes along temperature gradients in the ocean's surface mixed layer. Microbiol Rep, Environ. doi:10.1111/1758-2229.12378

Chappell PD, Whitney LP, Wallace JR, Darer AI, Jean-Charles S, Jenkins BD (2015) Genetic indicators of iron limitation in wild populations of *Thalassiosira oceanica* from the northeast Pacific ocean. ISME J 9:592–602

Chen F, Wang K, Kan J, Suzuki MT, Wommack KE (2006) Diverse and unique picocyanobacteria in Chesapeake Bay, revealed by 16S-23S rRNA internal transcribed spacer sequences. Appl Environ Microbiol 72:2239–2243

Chisholm SW, Olson RJ, Zettler ER, Goericke R, Waterbury JB, Welschmeyer NA (1988) A novel free-living prochlorophyte abundant in the oceanic euphotic zone. Nature 334:340–343

Chou YL, Lee YL, Yen CC, Chen LFO, Lee LC, Shaw JF (2015) A novel recombinant chlorophyllase from cyanobacterium *Cyanothece* sp. ATCC 51142 for the production of bacteriochlorophyllide a. Biotechnol Appl Biochem. doi:10.1002/bab.1380

Christie-Oleza JA, Armengaud J, Guerin P, Scanlan DJ (2015a) Functional distinctness in the exoproteomes of marine *Synechococcus*. Environ Microbiol 17:3781–3794

Christie-Oleza JA, Scanlan DJ, Armengaud J (2015b) "You produce while I clean up", a strategy revealed by exoproteomics during *Synechococcus*-Roseobacter interactions. Proteomics 15:3454–3462

Coelho SM, Simon N, Ahmed S, Cock JM, Partensky F (2013) Ecological and evolutionary genomics of marine photosynthetic organisms. Mol Ecol 22:867–907

Coleman ML, Chisholm SW (2007) Code and context: *Prochlorococcus* as a model for cross-scale biology. Trends Microbiol 15:398–407

Coleman ML, Sullivan MB, Martiny AC, Steglich C, Barry K, DeLong EF, Chisholm SW (2006) Genomic islands and the ecology and evolution of *Prochlorococcus*. Science 311:1768–1770

Compaoré J, Stal LJ (2010) Oxygen and the light-dark cycle of nitrogenase activity in two unicellular cyanobacteria. Environ Microbiol 12:54–62

Cottrell MT, Kirchman DL (2009) Photoheterotrophic microbes in the Arctic ocean in summer and winter. Appl Environ Microbiol 75:4958–4966

Cottrell MT, Ras J, Kirchman DL (2010) Bacteriochlorophyll and community structure of aerobic anoxygenic phototrophic bacteria in a particle-rich estuary. ISME J 4:945–954

Coutinho F, Tschoeke DA, Thompson F, Thompson C (2016) Comparative genomics of *Synechococcus* and proposal of the new genus *Parasynechococcus*. PeerJ 4:e1522

Crowe SA, Døssing LN, Beukes NJ, Bau M, Kruger SJ, Frei R, Canfield DE (2013) Atmospheric oxygenation three billion years ago. Nature 501:535–538

Curtis BA, Tanifuji G, Burki F, Gruber A, Irimia M, Maruyama S, Arias MC, Ball SG, Gile GH, Hirakawa Y et al (2012) Algal genomes reveal evolutionary mosaicism and the fate of nucleomorphs. Nature 492:59–65

Daboussi F, Leduc S, Maréchal A, Dubois G, Guyot V, Perez-Michaut C, Amato A, Falciatore A, Juillerat A, Beurdeley M et al (2014) Genome engineering empowers the diatom *Phaeodactylum tricornutum* for biotechnology. Nat Commun 5:3831

Davis CS, McGillicuddy DJ (2006) Transatlantic abundance of the N_2-fixing colonial cyanobacterium *Trichodesmium*. Science 312:1517–1520
de Vargas C, Audic S, Henry N, Decelle J, Mahé F, Logares R, Lara E, Berney C, Le Bescot N, Probert I et al (2015) Eukaryotic plankton diversity in the sunlit ocean. Science 348:1261605
Decelle J, Romac S, Stern RF, Bendif EM, Zingone A, Audic S, Guiry MD, Guillou L, Tessier D, Le Gall F et al (2015) PhytoREF: a reference database of the plastidial 16S rRNA gene of photosynthetic eukaryotes with curated taxonomy. Mol Ecol Resour 15:1435–1445
DeLong EF, Béjà O (2010) The light-driven proton pump proteorhodopsin enhances bacterial survival during tough times. PLoS Biol 8:e1000359
Desbois AP, Mearns-Spragg A, Smith VJ (2009) A fatty acid from the diatom *Phaeodactylum tricornutum* is antibacterial against diverse bacteria including multi-resistant *Staphylococcus aureus* (MRSA). Mar Biotechnol 11:45–52
Deutsch C, Sarmiento JL, Sigman DM, Gruber N, Dunne JP (2007) Spatial coupling of nitrogen inputs and losses in the ocean. Nature 445:163–167
Douglas S, Zauner S, Fraunholz M, Beaton M, Penny S, Deng LT, Wu X, Reith M, Cavalier-Smith T, Maier UG (2001) The highly reduced genome of an enslaved algal nucleus. Nature 410:1091–1096
Doulatov S, Hodes A, Dai L, Mandhana N, Liu M, Deora R, Simons RW, Zimmerly S, Miller JF (2004) Tropism switching in *Bordetella* bacteriophage defines a family of diversity-generating retroelements. Nature 431:476–481
Dufresne A, Salanoubat M, Partensky F, Artiguenave F, Axmann IM, Barbe V, Duprat S, Galperin MY, Koonin EV, Le Gall F et al (2003) Genome sequence of the cyanobacterium *Prochlorococcus marinus* SS120, a nearly minimal oxyphototrophic genome. Proc Natl Acad Sci USA 100:10020–10025
Dufresne A, Garczarek L, Partensky F (2005) Accelerated evolution associated with genome reduction in a free-living prokaryote. Genome Biol 6:R14
Dufresne A, Ostrowski M, Scanlan DJ, Garczarek L, Mazard S, Palenik BP, Paulsen IT, de Marsac NT, Wincker P, Dossat C et al (2008) Unraveling the genomic mosaic of a ubiquitous genus of marine cyanobacteria. Genome Biol 9:R90
Dugdale RC, Goering JJ (1967) Uptake of new and regenerated forms of nitrogen in primary productivity. Limnol Oceanogr 12:196–206
Dugdale RC, Menzel DW, Ryther JH (1961) Nitrogen fixation in the Sargasso sea. Deep Sea Res 1953(7):297–300
Dupouy C, Petit M, Dandonneau Y (1988) Satellite detected cyanobacteria bloom in the southwestern tropical Pacific. Int J Remote Sens 9:389–396
Ehrenberg CG (1830) Neue Beobachlungen über blutartige Erscheinungen in Aegypten, Arabien und Sibirien, nebst einer Uebersicht und Kritik der früher bekannnten. Ann Phys 94:477–514
Eiler A (2006) Evidence for the ubiquity of mixotrophic bacteria in the upper ocean: implications and consequences. Appl Environ Microbiol 72:7431–7437
El-Shehawy R, Lugomela C, Ernst A, Bergman B (2003) Diurnal expression of hetR and diazocyte development in the filamentous non-heterocystous cyanobacterium *Trichodesmium erythraeum*. Microbiol Read Engl 149:1139–1146
Evans C, Gómez-Pereira PR, Martin AP, Scanlan DJ, Zubkov MV (2015) Photoheterotrophy of bacterioplankton is ubiquitous in the surface oligotrophic ocean. Prog Oceanogr 135:139–145
Falkowski PG (1997) Evolution of the nitrogen cycle and its influence on the biological sequestration of CO_2 in the ocean. Nature 387:272–275
Falkowski PG, Raven JA (2013) Aquatic photosynthesis, 2nd edn. Princeton University Press, Princeton
Falkowski PG, Katz ME, Knoll AH, Quigg A, Raven JA, Schofield O, Taylor FJR (2004) The evolution of modern eukaryotic phytoplankton. Science 305:354–360
Farrant GK, Doré D, Cornejo-Castillo FM, Partensky F, Ratin M, Ostrowski M, Pitt F, Wincker P, Scanlan DJ, Iudicone D, Acinas SG, Garczarek L (2016) Delineating ecologically significant taxonomic units from global patterns of marine picocyanobacteria. Proc Natl Acad Sci USA (in press)

Ferrera I, Sebastian M, Acinas SG, Gasol JM (2015) Prokaryotic functional gene diversity in the sunlit ocean: stumbling in the dark. Curr Opin Microbiol 25:33–39

Fewer DP, Jokela J, Paukku E, Österholm J, Wahlsten M, Permi P, Aitio O, Rouhiainen L, Gomez-Saez GV, Sivonen K (2013) New structural variants of aeruginosin produced by the toxic bloom forming cyanobacterium *Nodularia spumigena*. PLoS ONE 8:e73618

Field CB, Behrenfeld MJ, Randerson JT, Falkowski P (1998) Primary production of the biosphere: integrating terrestrial and oceanic components. Science 281:237–240

Flombaum P, Gallegos JL, Gordillo RA, Rincón J, Zabala LL, Jiao N, Karl DM, Li WKW, Lomas MW, Veneziano D et al (2013) Present and future global distributions of the marine Cyanobacteria *Prochlorococcus* and *Synechococcus*. Proc Natl Acad Sci USA 110:9824–9829

Foster RA, Zehr JP (2006) Characterization of diatom-cyanobacteria symbioses on the basis of *nifH, hetR* and 16S rRNA sequences. Environ Microbiol 8:1913–1925

Foster RA, Subramaniam A, Zehr JP (2009) Distribution and activity of diazotrophs in the Eastern Equatorial Atlantic. Environ Microbiol 11:741–750

Foster RA, Goebel NL, Zehr JP (2010) Isolation of *Calothrix rhizosoleniae* (cyanobacteria) strain sc01 from *Chaetoceros* (bacillariophyta) spp. diatoms of the subtropical North Pacific Ocean. J Phycol 46:1028–1037

Foster RA, Kuypers MM, Vagner T, Paerl RW, Musat N, Zehr JP (2011) Nitrogen fixation and transfer in open ocean diatom—cyanobacterial symbioses. ISME J 5:1484–1493

Foster RA, Sztejrenszus S, Kuypers MM (2013) Measuring carbon and N_2 fixation in field populations of colonial and free-living unicellular cyanobacteria using nanometer-scale secondary ion mass spectrometry1. J Phycol 49:502–516

Franklin DJ, Brussaard CP, Berges JA (2006) What is the role and nature of programmed cell death in phytoplankton ecology? Eur J Phycol 41:1–14

Fu W, Wichuk K, Brynjólfsson S (2015) Developing diatoms for value-added products: challenges and opportunities. New Biotechnol 32:547–551

Fuhrman JA, Schwalbach MS, Stingl U (2008) Proteorhodopsins: an array of physiological roles? Nat Rev Microbiol 6:488–494

Georg J, Dienst D, Schürgers N, Wallner T, Kopp D, Stazic D, Kuchmina E, Klähn S, Lokstein H, Hess WR et al (2014) The small regulatory RNA SyR1/PsrR1 controls photosynthetic functions in cyanobacteria. Plant Cell 26:3661–3679

Goebel NL, Turk KA, Achilles KM, Paerl R, Hewson I, Morrison AE, Montoya JP, Edwards CA, Zehr JP (2010) Abundance and distribution of major groups of diazotrophic cyanobacteria and their potential contribution to N_2 fixation in the tropical Atlantic Ocean. Environ Microbiol 12:3272–3289

Goericke R, Repeta DJ (1992) The pigments of *Prochlorococcus marinus*: the presence of divinylchlorophyll *a* and *b* in a marine procaryote. Limnol Oceanogr 37:425–433

Golubic S, Seong-Joo L (1999) Early cyanobacterial fossil record: preservation, palaeoenvironments and identification. Eur J Phycol 34:339–348

Gómez-Baena G, López-Lozano A, Gil-Martínez J, Lucena JM, Diez J, Candau P, García-Fernández JM (2008) Glucose uptake and its effect on gene expression in *Prochlorococcus*. PLoS ONE 3:e3416

Gómez-Consarnau L, Akram N, Lindell K, Pedersen A, Neutze R, Milton DL, González JM, Pinhassi J (2010) Proteorhodopsin phototrophy promotes survival of marine bacteria during starvation. PLoS Biol 8:e1000358

Gómez-Pereira PR, Hartmann M, Grob C, Tarran GA, Martin AP, Fuchs BM, Scanlan DJ, Zubkov MV (2013) Comparable light stimulation of organic nutrient uptake by SAR11 and *Prochlorococcus* in the North Atlantic subtropical gyre. ISME J 7:603–614

Gomont M (1892) Monographie des Oscillariées (Nostocacées Homocystées). Deuxième partie. - Lyngbyées. Ann Sci Nat Bot Sér 7(16):91–264

Guidi L, Chaffron S, Bittner L, Eveillard D, Larhlimi A, Roux S et al (2016) Plankton networks driving carbon export in the global ocean. Nature 532:465–4701

Guo H, Tse LV, Barbalat R, Sivaamnuaiphorn S, Xu M, Doulatov S, Miller JF (2008) Diversity-generating retroelement homing regenerates target sequences for repeated rounds of codon rewriting and protein diversification. Mol Cell 31:813–823

Hackenberg C, Engelhardt A, Matthijs HCP, Wittink F, Bauwe H, Kaplan A, Hagemann M (2009) Photorespiratory 2-phosphoglycolate metabolism and photoreduction of O_2 cooperate in high-light acclimation of *Synechocystis* sp. strain PCC 6803. Planta 230:625–637

Hagino K, Onuma R, Kawachi M, Horiguchi T (2013) Discovery of an endosymbiotic nitrogen-fixing cyanobacterium UCYN-A in *Braarudosphaera bigelowii* (Prymnesiophyceae). PLoS ONE 8:e81749

Hatosy SM, Martiny AC (2015) The ocean as a global reservoir of antibiotic resistance genes. Appl Environ Microbiol 81:7593–7599

Haverkamp T, Acinas SG, Doeleman M, Stomp M, Huisman J, Stal LJ (2008) Diversity and phylogeny of Baltic sea picocyanobacteria inferred from their ITS and phycobiliprotein operons. Environ Microbiol 10:174–188

Hess WR, Partensky F, van der Staay GW, Garcia-Fernandez JM, Börner T, Vaulot D (1996) Coexistence of phycoerythrin and a chlorophyll a/b antenna in a marine prokaryote. Proc Natl Acad Sci USA 93:11126–11130

Hess WR, Steglich C, Lichtlé C, Partensky F (1999) Phycoerythrins of the oxyphotobacterium *Prochlorococcus marinus* are associated to the thylakoid membrane and are encoded by a single large gene cluster. Plant Mol Biol 40:507–521

Hess WR, Rocap G, Ting CS, Larimer F, Stilwagen S, Lamerdin J, Chisholm SW (2001) The photosynthetic apparatus of *Prochlorococcus*: insights through comparative genomics. Photosynth Res 70:53–71

Hewson I, Govil SR, Capone DG, Carpenter EJ, Fuhrman JA (2004) Evidence of *Trichodesmium* viral lysis and potential significance for biogeochemical cycling in the oligotrophic ocean. Aquat Microb Ecol 36:1–8

Hilton JA, Foster RA, Tripp HJ, Carter BJ, Zehr JP, Villareal TA (2013) Genomic deletions disrupt nitrogen metabolism pathways of a cyanobacterial diatom symbiont. Nat Commun 4:1767

Huang S, Wilhelm SW, Harvey HR, Taylor K, Jiao N, Chen F (2012) Novel lineages of *Prochlorococcus* and *Synechococcus* in the global oceans. ISME J 6:285–297

Hubas C, Jesus B, Passarelli C, Jeanthon C (2011) Tools providing new insight into coastal anoxygenic purple bacterial mats: review and perspectives. Res Microbiol 162:858–868

Humily F, Partensky F, Six C, Farrant GK, Ratin M, Marie D, Garczarek L (2013) A gene island with two possible configurations is involved in chromatic acclimation in marine *Synechococcus*. PLoS ONE 8:e84459

Hynes AM, Webb EA, Doney SC, Waterbury JB (2012) Comparison of cultured *Trichodesmium* (cyanophyceae) with species characterized from the field. J Phycol 48:196–210

Janson S, Wouters J, Bergman B, Carpenter EJ (1999) Host specificity in the Richelia-diatom symbiosis revealed by *hetR* gene sequence analysis. Environ Microbiol 1:431–438

Jardillier L, Zubkov MV, Pearman J, Scanlan DJ (2010) Significant CO_2 fixation by small prymnesiophytes in the subtropical and tropical northeast Atlantic ocean. ISME J 4:1180–1192

Jiao N, Luo T, Zhang R, Yan W, Lin Y, Johnson ZI, Tian J, Yuan D, Yang Q, Zheng Q et al (2014) Presence of *Prochlorococcus* in the aphotic waters of the western Pacific ocean. Biogeosciences 11:2391–2400

Johnson ZI, Zinser ER, Coe A, McNulty NP, Woodward EMS, Chisholm SW (2006) Niche partitioning among *Prochlorococcus* ecotypes along ocean-scale environmental gradients. Science 311:1737–1740

Jones AC, Monroe EA, Podell S, Hess WR, Klages S, Esquenazi E, Niessen S, Hoover H, Rothmann M, Lasken RS et al (2011) Genomic insights into the physiology and ecology of the marine filamentous cyanobacterium *Lyngbya majuscula*. Proc Natl Acad Sci USA 108:8815–8820

Karl D, Letelier R, Tupas L, Dore J, Christian J, Hebel D (1997) The role of nitrogen fixation in biogeochemical cycling in the subtropical North Pacific ocean. Nature 388:533–538

Karl D, Michaels A, Bergman B, Capone D, Carpenter E, Letelier R, Lipschultz F, Paerl H, Sigman D, Stal L (2002) Dinitrogen fixation in the world's oceans. Biogeochemistry 57–58:47–98

Kashtan N, Roggensack SE, Rodrigue S, Thompson JW, Biller SJ, Coe A, Ding H, Marttinen P, Malmstrom RR, Stocker R et al (2014) Single-cell genomics reveals hundreds of coexisting subpopulations in wild *Prochlorococcus*. Science 344:416–420

Kazamia E, Czesnick H, Nguyen TTV, Croft MT, Sherwood E, Sasso S, Hodson SJ, Warren MJ, Smith AG (2012) Mutualistic interactions between vitamin B12-dependent algae and heterotrophic bacteria exhibit regulation. Environ Microbiol 14:1466–1476

Keeling PJ, Burki F, Wilcox HM, Allam B, Allen EE, Amaral-Zettler LA, Armbrust EV, Archibald JM, Bharti AK, Bell CJ et al (2014) The marine microbial eukaryote transcriptome sequencing project (MMETSP): illuminating the functional diversity of eukaryotic life in the oceans through transcriptome sequencing. PLoS Biol 12(6):e1001889

Kent AG, Dupont CL, Yooseph S, Martiny AC (2016) Global biogeography of *Prochlorococcus* genome diversity in the surface ocean. ISME J. doi:10.1038/ismej.2015.265

Kettler GC, Martiny AC, Huang K, Zucker J, Coleman ML, Rodrigue S, Chen F, Lapidus A, Ferriera S, Johnson J et al (2007) Patterns and implications of gene gain and loss in the evolution of *Prochlorococcus*. PLoS Genet 3:e231

Kim E, Harrison JW, Sudek S, Jones MDM, Wilcox HM, Richards TA, Worden AZ, Archibald JM (2011) Newly identified and diverse plastid-bearing branch on the eukaryotic tree of life. Proc Natl Acad Sci USA 108:1496–1500

Kirk JTO (1994) Light and photosynthesis in aquatic ecosystems. Cambridge University Press, Cambridge

Kirschvink JL, Kopp RE (2008) Palaeoproterozoic ice houses and the evolution of oxygen-mediating enzymes: the case for a late origin of photosystem II. Philos Trans R Soc B Biol Sci 363:2755–2765

Klähn S, Steglich C, Hess WR, Hagemann M (2010) Glucosylglycerate: a secondary compatible solute common to marine cyanobacteria from nitrogen-poor environments. Environ Microbiol 12:83–94

Klähn S, Baumgartner D, Pfreundt U, Voigt K, Schön V, Steglich C, Hess WR (2014) Alkane biosynthesis genes in cyanobacteria and their transcriptional organization. Front Bioeng Biotechnol 2:24

Klähn S, Schaal C, Georg J, Baumgartner D, Knippen G, Hagemann M, Muro-Pastor AM, Hess WR (2015) The sRNA NsiR4 is involved in nitrogen assimilation control in cyanobacteria by targeting glutamine synthetase inactivating factor IF7. Proc Natl Acad Sci USA 112:E6243–E6252

Knapp AN, Fawcett SE, Martínez-Garcia A, Leblond N, Moutin T, Bonnet S (2015) Nitrogen isotopic evidence for a shift from nitrate-to diazotroph-fueled export production in VAHINE mesocosm experiments. Biogeosci Discuss 12:19901–19939

Knoll AH, Semikhatov MA (1998) The genesis and time distribution of two distinctive proterozoic stromatolite microstructures. Palaios 13:408–422

Koblížek M (2015) Ecology of aerobic anoxygenic phototrophs in aquatic environments. FEMS Microbiol Rev 39:854–870

Konstantinidis KT, Tiedje JM (2005) Genomic insights that advance the species definition for prokaryotes. Proc Natl Acad Sci USA 102:2567–2572

Kopf M, Hess WR (2015) Regulatory RNAs in photosynthetic cyanobacteria. FEMS Microbiol Rev 39:301–315

Kopf M, Möke F, Bauwe H, Hess WR, Hagemann M (2015) Expression profiling of the bloom-forming cyanobacterium *Nodularia spumigena* CCY9414 under high light and oxidative stress conditions. ISME J 9:2139–2152

Krumhardt KM, Callnan K, Roache-Johnson K, Swett T, Robinson D, Reistetter EN, Saunders JK, Rocap G, Moore LR (2013) Effects of phosphorus starvation versus limitation on the marine cyanobacterium *Prochlorococcus* MED4 I: uptake physiology. Environ Microbiol 15:2114–2128

Krupke A, Lavik G, Halm H, Fuchs BM, Amann RI, Kuypers MMM (2014) Distribution of a consortium between unicellular algae and the N_2 fixing cyanobacterium UCYN-A in the North Atlantic ocean. Environ Microbiol 16:3153–3167

Krupke A, Mohr W, LaRoche J, Fuchs BM, Amann RI, Kuypers MMM (2015) The effect of nutrients on carbon and nitrogen fixation by the UCYN-A-haptophyte symbiosis. ISME J 9:1635–1647

Larkin AA, Blinebry SK, Howes C, Lin Y, Loftus SE, Schmaus CA, Zinser ER, Johnson ZI (2016) Niche partitioning and biogeography of high light adapted *Prochlorococcus* across taxonomic ranks in the North Pacific. ISME J. doi:10.1038/ismej.2015.244

Larsson J, Nylander JA, Bergman B (2011) Genome fluctuations in cyanobacteria reflect evolutionary, developmental and adaptive traits. BMC Evol Biol 11:187

Larsson J, Celepli N, Ininbergs K, Dupont CL, Yooseph S, Bergman B, Ekman M (2014) Picocyanobacteria containing a novel pigment gene cluster dominate the brackish water Baltic sea. ISME J 8:1892–1903

Lavin P, González B, Santibáñez JF, Scanlan DJ, Ulloa O (2010) Novel lineages of *Prochlorococcus* thrive within the oxygen minimum zone of the eastern tropical South Pacific. Environ Microbiol Rep 2:728–738

Lea-Smith DJ, Biller SJ, Davey MP, Cotton CAR, Perez Sepulveda BM, Turchyn AV, Scanlan DJ, Smith AG, Chisholm SW, Howe CJ (2015) Contribution of cyanobacterial alkane production to the ocean hydrocarbon cycle. Proc Natl Acad Sci USA 112:13591–13596

Liu M, Deora R, Doulatov SR, Gingery M, Eiserling FA, Preston A, Maskell DJ, Simons RW, Cotter PA, Parkhill J et al (2002) Reverse transcriptase-mediated tropism switching in *Bordetella* bacteriophage. Science 295:2091–2094

Liu L, Budnjo A, Jokela J, Haug BE, Fewer DP, Wahlsten M, Rouhiainen L, Permi P, Fossen T, Sivonen K (2015) Pseudoaeruginosins, nonribosomal peptides in *Nodularia spumigena*. ACS Chem Biol 10:725–733

Luo Y-W, Doney SC, Anderson LA, Benavides M, Berman-Frank I, Bode A, Bonnet S, Boström KH, Böttjer D, Capone DG et al (2012) Database of diazotrophs in global ocean: abundance, biomass and nitrogen fixation rates. Earth Syst Sci Data 4:47–73

Lyons TW, Reinhard CT, Planavsky NJ (2014) The rise of oxygen in Earth/'s early ocean and atmosphere. Nature 506:307–315

Mackey KR, Paytan A, Caldeira K, Grossman AR, Moran D, McIlvin M, Saito MA (2013) Effect of temperature on photosynthesis and growth in marine *Synechococcus* spp. Plant Physiol 163:815–829

Mahaffey C, Michaels AF, Capone DG (2005) The conundrum of marine N_2 fixation. Am J Sci 305:546–595

Malmstrom RR, Coe A, Kettler GC, Martiny AC, Frias-Lopez J, Zinser ER, Chisholm SW (2010) Temporal dynamics of *Prochlorococcus* ecotypes in the Atlantic and Pacific oceans. ISME J 4:1252–1264

Malmstrom RR, Rodrigue S, Huang KH, Kelly L, Kern SE, Thompson A, Roggensack S, Berube PM, Henn MR, Chisholm SW (2013) Ecology of uncultured *Prochlorococcus* clades revealed through single-cell genomics and biogeographic analysis. ISME J 7:184–198

Man D, Wang W, Sabehi G, Aravind L, Post AF, Massana R, Spudich EN, Spudich JL, Béjà O (2003) Diversification and spectral tuning in marine proteorhodopsins. EMBO J 22:1725–1731

Mandal S, Rath J (2015) Secondary metabolites of cyanobacteria and drug development. In: Extremophilic cyanobacteria for novel drug development. Springer, Berlin, pp 23–43

Marchetti A, Schruth DM, Durkin CA, Parker MS, Kodner RB, Berthiaume CT, Morales R, Allen AE, Armbrust EV (2012) Comparative metatranscriptomics identifies molecular bases for the physiological responses of phytoplankton to varying iron availability. Proc Natl Acad Sci USA 109:E317–E325

Mareš J, Hrouzek P, Kaňa R, Ventura S, Strunecký O, Komárek J (2013) The primitive thylakoid-less cyanobacterium *Gloeobacter* Is a common rock-dwelling organism. PLoS ONE 8:e66323

Marin B, Melkonian M (2010) Molecular phylogeny and classification of the Mamiellophyceae class. nov. (Chlorophyta) based on sequence comparisons of the nuclear- and plastid-encoded rRNA operons. Protist 161:304–336

Martiny AC, Coleman ML, Chisholm SW (2006) Phosphate acquisition genes in *Prochlorococcus* ecotypes: evidence for genome-wide adaptation. Proc Natl Acad Sci USA 103:12552–12557

Martiny AC, Huang Y, Li W (2009a) Occurrence of phosphate acquisition genes in *Prochlorococcus* cells from different ocean regions. Environ Microbiol 11:1340–1347

Martiny AC, Kathuria S, Berube PM (2009b) Widespread metabolic potential for nitrite and nitrate assimilation among *Prochlorococcus* ecotypes. Proc Natl Acad Sci USA 106:10787–10792

Martiny JB, Jones SE, Lennon JT, Martiny AC (2015) Microbiomes in light of traits: a phylogenetic perspective. Science 350:aac9323

Matsumoto M, Sugiyama H, Maeda Y, Sato R, Tanaka T, Matsunaga T (2010) Marine diatom, *Navicula* sp. strain JPCC DA0580 and marine green alga, *Chlorella* sp. strain NKG400014 as potential sources for biodiesel production. Appl Biochem Biotechnol 161:483–490

Mazard S, Ostrowski M, Partensky F, Scanlan DJ (2012) Multi-locus sequence analysis, taxonomic resolution and biogeography of marine *Synechococcus*. Environ Microbiol 14:372–386

Mazur-Marzec H, Kaczkowska MJ, Blaszczyk A, Akcaalan R, Spoof L, Meriluoto J (2012) Diversity of peptides produced by *Nodularia spumigena* from various geographical regions. Mar Drugs 11:1–19

McCarren J, Brahamsha B (2007) SwmB, a 1.12-megadalton protein that is required for nonflagellar swimming motility in *Synechococcus*. J Bacteriol 189:1158–1162

McKie-Krisberg ZM, Sanders RW (2014) Phagotrophy by the picoeukaryotic green alga *Micromonas*: implications for Arctic oceans. ISME J 8:1953–1961

Mella-Flores D, Six C, Ratin M, Partensky F, Boutte C, Le Corguillé G, Marie D, Blot N, Gourvil P, Kolowrat C et al (2012) *Prochlorococcus* and *Synechococcus* have evolved different adaptive mechanisms to cope with light and UV stress. Front Microbiol

Melnicki MR, Pinchuk GE, Hill EA, Kucek LA, Fredrickson JK, Konopka A, Beliaev AS (2012) Sustained H(2) production driven by photosynthetic water splitting in a unicellular cyanobacterium. mBio 3:e00197–e00112

Mevers E, Matainaho T, Allara' M, Di Marzo V, Gerwick WH (2014) Mooreamide A: a cannabinomimetic lipid from the marine cyanobacterium *Moorea bouillonii*. Lipids 49:1127–1132

Möke F, Wasmund N, Bauwe H, Hagemann M (2013) Salt acclimation of *Nodularia spumigena* CCY9414–a cyanobacterium adapted to brackish water. Aquat Microb Ecol 70:207–214

Montoya JP, Holl CM, Zehr JP, Hansen A, Villareal TA, Capone DG (2004) High rates of N_2 fixation by unicellular diazotrophs in the oligotrophic Pacific ocean. Nature 430:1027–1032

Moore LR, Rocap G, Chisholm SW (1998) Physiology and molecular phylogeny of coexisting *Prochlorococcus* ecotypes. Nature 393:464–467

Moore JK, Doney SC, Glover DM, Fung IY (2001) Iron cycling and nutrient-limitation patterns in surface waters of the World ocean. Deep Sea Res Part II Top Stud Oceanogr 49:463–507

Moore LR, Coe A, Zinser ER, Saito MA, Sullivan MB, Lindell D, Frois-Moniz K, Waterbury J, Chisholm SW (2007) Culturing the marine cyanobacterium *Prochlorococcus*. Limnol Oceanogr Methods 5:353–362

Moore RB, Oborník M, Janouskovec J, Chrudimský T, Vancová M, Green DH, Wright SW, Davies NW, Bolch CJS, Heimann K et al (2008) A photosynthetic alveolate closely related to apicomplexan parasites. Nature 451:959–963

Moore CM, Mills MM, Arrigo KR, Berman-Frank I, Bopp L, Boyd PW, Galbraith ED, Geider RJ, Guieu C, Jaccard SL et al (2013) Processes and patterns of oceanic nutrient limitation. Nat Geosci 6:701–710

Morel A, Ahn Y-H, Partensky F, Vaulot D, Claustre H (1993) *Prochlorococcus* and *Synechococcus*: a comparative study of their optical properties in relation to their size and pigmentation. J Mar Res 51:617–649

Morris JJ, Kirkegaard R, Szul MJ, Johnson ZI, Zinser ER (2008) Facilitation of robust growth of *Prochlorococcus* colonies and dilute liquid cultures by "helper" heterotrophic bacteria. Appl Environ Microbiol 74:4530–4534

Morrissey J, Sutak R, Paz-Yepes J, Tanaka A, Moustafa A, Veluchamy A, Thomas Y, Botebol H, Bouget F-Y, McQuaid JB et al (2015) A novel protein, ubiquitous in marine phytoplankton, concentrates iron at the cell surface and facilitates uptake. Curr Biol 25:364–371

Muñoz-Marín M del C, Luque I, Zubkov MV, Hill PG, Diez J, García-Fernández JM (2013) *Prochlorococcus* can use the Pro1404 transporter to take up glucose at nanomolar concentrations in the Atlantic Ocean. Proc Natl Acad Sci USA 110:8597–8602

Muro-Pastor AM, Hess WR (2012) Heterocyst differentiation: from single mutants to global approaches. Trends Microbiol 20:548–557

Neveux J, Lantoine F, Vaulot D, Marie D, Blanchot J (1999) Phycoerythrins in the southern tropical and equatorial Pacific ocean: evidence for new cyanobacterial types. J Geophys Res-Oceans 10:3311–3321

Neveux J, Tenírio MMB, Cé Dupouy, Villareal TA (2006) Spectral diversity of phycoerythrins and diazotroph abundance in tropical waters. Limnol Oceanogr 51:1689–1698

Nichols WT (2015) Designing biomimetic materials from marine organisms. J Nanosci Nanotechnol 15:189–191

Not F, Siano R, Kooistra WH, Simon N, Vaulot D, Probert I (2012) 1 Diversity and ecology of eukaryotic marine phytoplankton. Adv Bot Res 64:1–53

O'Neil JM (1998) The colonial cyanobacterium *Trichodesmium* as a physical and nutritional substrate for the harpacticoid copepod *Macrosetella gracilis*. J Plankton Res 20:43–59

O'Neil JM, Roman MR (1994) Ingestion of the cyanobacterium *Trichodesmium* spp. by pelagic harpacticoid copepods *Macrosetella*, Miracia and Oculosetella. Hydrobiologia 292:235–240

Oborník M, Lukeš J (2015) The organellar genomes of Chromera and Vitrella, the phototrophic relatives of apicomplexan parasites. Annu Rev Microbiol 69:129–144

Oborník M, Modrý D, Lukeš M, Cernotíková-Stříbrná E, Cihlář J, Tesařová M, Kotabová E, Vancová M, Prášil O, Lukeš J (2012) Morphology, ultrastructure and life cycle of *Vitrella brassicaformis* n. sp., n. gen., a novel chromerid from the Great Barrier Reef. Protist 163:306–323

Ohki K (1999) A possible role of temperate phage in the regulation of *Trichodesmium* biomass. Bull Inst Océan 287–291

Olson EM, McGillicuddy DJ Jr, Dyhrman ST, Waterbury JB, Davis CS, Solow AR (2015) The depth-distribution of nitrogen fixation by *Trichodesmium* spp. colonies in the tropical–subtropical North Atlantic. Deep Sea Res Part Oceanogr Res Pap 104:72–91

Omoregie EO, Crumbliss LL, Bebout BM, Zehr JP (2004) Determination of nitrogen-fixing phylotypes in *Lyngbya* sp. and *Microcoleus chthonoplastes* cyanobacterial mats from Guerrero Negro, Baja California, Mexico. Appl Environ Microbiol 70:2119–2128

Ong LJ, Glazer AN (1991) Phycoerythrins of marine unicellular cyanobacteria. I. Bilin types and locations and energy transfer pathways in *Synechococcus* spp. phycoerythrins. J Biol Chem 266:9515–9527

Ong LJ, Glazer AN, Waterbury JB (1984) An unusual phycoerythrin from a marine cyanobacterium. Science 224:80–83

Pade N, Hagemann M (2014) Salt acclimation of cyanobacteria and their application in biotechnology. Life Basel Switz 5:25–49

Palenik B, Brahamsha B, Larimer FW, Land M, Hauser L, Chain P, Lamerdin J, Regala W, Allen EE, McCarren J et al (2003) The genome of a motile marine *Synechococcus*. Nature 424:1037–1042

Palenik B, Ren Q, Dupont CL, Myers GS, Heidelberg JF, Badger JH, Madupu R, Nelson WC, Brinkac LM, Dodson RJ et al (2006) Genome sequence of *Synechococcus* CC9311: insights into adaptation to a coastal environment. Proc Natl Acad Sci USA 103:13555–13559

Partensky F, Garczarek L (2010) *Prochlorococcus*: advantages and limits of minimalism. Annu Rev Mar Sci 2:305–331

Partensky F, Blanchot J, Vaulot D (1999a) Differential distribution and ecology of *Prochlorococcus* and *Synechococcus* in oceanic waters : a review. Bull Inst Océan 457–475

Partensky F, Hess WR, Vaulot D (1999b) *Prochlorococcus*, a marine photosynthetic prokaryote of global significance. Microbiol Mol Biol Rev 63:106–127

Paz-Yepes J, Brahamsha B, Palenik B (2013) Role of a Microcin-C–like biosynthetic gene cluster in allelopathic interactions in marine *Synechococcus*. Proc Natl Acad Sci 110:12030–12035

Pfreundt U, Hess WR (2015) Sequential splicing of a group II twintron in the marine cyanobacterium *Trichodesmium*. Sci Rep 5:16829

Pfreundt U, Stal LJ, Voß B, Hess WR (2012) Dinitrogen fixation in a unicellular chlorophyll d-containing cyanobacterium. ISME J 6:1367–1377

Pfreundt U, Kopf M, Belkin N, Berman-Frank I, Hess WR (2014) The primary transcriptome of the marine diazotroph *Trichodesmium erythraeum* IMS101. Sci Rep 4:6187

Pfreundt U, Wambeke FV, Caffin M, Bonnet S, Hess WR (2016) Succession within the prokaryotic communities during the VAHINE mesocosms experiment in the New Caledonia lagoon. Biogeosciences 13:2319–2337

Pittera J, Humily F, Thorel M, Grulois D, Garczarek L, Six C (2014) Connecting thermal physiology and latitudinal niche partitioning in marine *Synechococcus*. ISME J 8:1221–1236

Ploug H, Adam B, Musat N, Kalvelage T, Lavik G, Wolf-Gladrow D, Kuypers MMM (2011) Carbon, nitrogen and O(2) fluxes associated with the cyanobacterium *Nodularia spumigena* in the Baltic sea. ISME J 5:1549–1558

Post AF, Dedej Z, Gottlieb R, Li H, Thomas DN, El-Absawi M, El-Naggar A, El-Gharabawi M, Sommer U (2002) Spatial and temporal distribution of *Trichodesmium* spp. in the stratified Gulf of Aqaba, Red Sea. Mar Ecol Prog Ser 239:241–250

Read BA, Kegel J, Klute MJ, Kuo A, Lefebvre SC, Maumus F, Mayer C, Miller J, Monier A, Salamov A et al (2013) Pan genome of the phytoplankton *Emiliania* underpins its global distribution. Nature 499:209–213

Reddy KJ, Haskell JB, Sherman DM, Sherman LA (1993) Unicellular, aerobic nitrogen-fixing cyanobacteria of the genus *Cyanothece*. J Bacteriol 175:1284–1292

Reinhard CT, Planavsky NJ, Robbins LJ, Partin CA, Gill BC, Lalonde SV, Bekker A, Konhauser KO, Lyons TW (2013) Proterozoic ocean redox and biogeochemical stasis. Proc Natl Acad Sci USA 110:5357–5362

Richardson TL, Jackson GA (2007) Small phytoplankton and carbon export from the surface ocean. Science 315:838–840

Rippka R, Waterbury J, Cohen-Bazire G (1974) A cyanobacterium which lacks thylakoids. Arch Microbiol 100:419–436

Rocap G, Larimer FW, Lamerdin J, Malfatti S, Chain P, Ahlgren NA, Arellano A, Coleman M, Hauser L, Hess WR et al (2003) Genome divergence in two *Prochlorococcus* ecotypes reflects oceanic niche differentiation. Nature 424:1042–1047

Rosing MT, Frei R (2004) U-rich Archaean sea-floor sediments from Greenland—indications of > 3700 Ma oxygenic photosynthesis. Earth Planet Sci Lett 217:237–244

Rubin M, Berman-Frank I, Shaked Y (2011) Dust-and mineral-iron utilization by the marine dinitrogen-fixer *Trichodesmium*. Nat Geosci 4:529–534

Rusch DB, Martiny AC, Dupont CL, Halpern AL, Venter JC (2010) Characterization of *Prochlorococcus* clades from iron-depleted oceanic regions. Proc Natl Acad Sci USA 107:16184–16189

Saito MA, Rocap G, Moffett JW (2005) Production of cobalt binding ligands in a *Synechococcus* feature at the Costa Rica upwelling dome. Limnol Oceanogr 50:279–290

Salvador-Reyes LA, Luesch H (2015) Biological targets and mechanisms of action of natural products from marine cyanobacteria. Nat Prod Rep 32:478–503

Sánchez-Baracaldo P (2015) Origin of marine planktonic cyanobacteria. Sci Rep 5:17418

Sandh G, El-Shehawy R, Di-ez B, Bergman B (2009) Temporal separation of cell division and diazotrophy in the marine diazotrophic cyanobacterium *Trichodesmium erythraeum* IMS101. FEMS Microbiol Lett 295:281–288

Sandh G, Xu L, Bergman B (2012) Diazocyte development in the marine diazotrophic cyanobacterium *Trichodesmium*. Microbiol Read Engl 158:345–352
Sañudo-Wilhelmy SA, Gómez-Consarnau L, Suffridge C, Webb EA (2014) The role of B vitamins in marine biogeochemistry. Annu Rev Mar Sci 6:339–367
Scanlan DJ (2012) Marine picocyanobacteria. In: Ecology of cyanobacteria II. Springer, Berlin, pp 503–533
Scanlan DJ, Ostrowski M, Mazard S, Dufresne A, Garczarek L, Hess WR, Post AF, Hagemann M, Paulsen I, Partensky F (2009) Ecological genomics of marine picocyanobacteria. Microbiol Mol Biol Rev 73:249–299
Schirmer A, Rude MA, Li X, Popova E, del Cardayre SB (2010) Microbial biosynthesis of alkanes. Science 329:559–562
Sellner KG (1997) Physiology, ecology, and toxic properties of marine cyanobacteria blooms. Limnol Oceanogr 42:1089–1104
Shao CL, Linington RG, Balunas MJ, Centeno A, Boudreau P, Zhang C, Engene N, Spadafora C, Mutka TS, Kyle DE et al (2015) Bastimolide A, a potent antimalarial polyhydroxy macrolide from the marine cyanobacterium *Okeania hirsuta*. J Org Chem 80:7849–7855
Sharma AK, Becker JW, Ottesen EA, Bryant JA, Duhamel S, Karl DM, Cordero OX, Repeta DJ, DeLong EF (2014) Distinct dissolved organic matter sources induce rapid transcriptional responses in coexisting populations of *Prochlorococcus*, *Pelagibacter* and the OM60 clade. Environ Microbiol 16:2815–2830
Shi T, Ilikchyan I, Rabouille S, Zehr JP (2010) Genome-wide analysis of diel gene expression in the unicellular N(2)-fixing cyanobacterium *Crocosphaera watsonii* WH 8501. ISME J 4:621–632
Shi Y, Tyson GW, Eppley JM, DeLong EF (2011) Integrated metatranscriptomic and metagenomic analyses of stratified microbial assemblages in the open ocean. ISME J 5:999–1013
Shih PM, Wu D, Latifi A, Axen SD, Fewer DP, Talla E, Calteau A, Cai F, Tandeau de Marsac N, Rippka R et al (2013) Improving the coverage of the cyanobacterial phylum using diversity-driven genome sequencing. Proc Natl Acad Sci USA 110:1053–1058
Shukla A, Biswas A, Blot N, Partensky F, Karty JA, Hammad LA, Garczarek L, Gutu A, Schluchter WM, Kehoe DM (2012) Phycoerythrin-specific bilin lyase-isomerase controls blue-green chromatic acclimation in marine *Synechococcus*. Proc Natl Acad Sci USA 109:20136–20141
Sidler WA (1994) Phycobilisome and phycobiliprotein structures. In: Bryant DA (ed) The molecular biology of cyanobacteria. Springer, Netherlands, pp 139–216
Simon N, Cras AL, Foulon E, Lemée R (2009) Diversity and evolution of marine phytoplankton. C R Biol 332:159–170
Sivonen K, Halinen K, Sihvonen LM, Koskenniemi K, Sinkko H, Rantasärkkä K, Moisander PH, Lyra C (2007) Bacterial diversity and function in the Baltic sea with an emphasis on cyanobacteria. AMBIO 36:180–185
Six C, Thomas JC, Garczarek L, Ostrowski M, Dufresne A, Blot N, Scanlan DJ, Partensky F (2007) Diversity and evolution of phycobilisomes in marine *Synechococcus* spp.: a comparative genomics study. Genome Biol 8:R259
Smetacek V, Nicol S (2005) Polar ocean ecosystems in a changing world. Nature 437:362–368
Sohm JA, Ahlgren NA, Thomson ZJ, Williams C, Moffett JW, Saito MA, Webb EA, Rocap G (2015) Co-occurring *Synechococcus* ecotypes occupy four major oceanic regimes defined by temperature, macronutrients and iron. ISME J 10(2):333–345
Sournia A, Chrdtiennot-Dinet MJ, Ricard M (1991) Marine phytoplankton: how many species in the world ocean? J Plankton Res 13:1093–1099
Spungin D, Pfreundt U, Berthelot H, Bonnet S, AlRoumi D, Natale F, Hess WR, Bidle KD, Berman-Frank I (2016) Mechanisms of Trichodesmium bloom demise within the New Caledonia Lagoon during the VAHINE mesocosm experiment. Biogeosci Discuss. doi:10.5194/bg-2015-613

Steglich C, Post AF, Hess WR (2003) Analysis of natural populations of *Prochlorococcus* spp. in the northern Red sea using phycoerythrin gene sequences. Environ Microbiol 5:681–690

Steglich C, Frankenberg-Dinkel N, Penno S, Hess WR (2005) A green light-absorbing phycoerythrin is present in the high-light-adapted marine cyanobacterium *Prochlorococcus* sp. MED4. Environ Microbiol 7:1611–1618

Steglich C, Futschik ME, Lindell D, Voss B, Chisholm SW, Hess WR (2008) The challenge of regulation in a minimal photoautotroph: non-coding RNAs in *Prochlorococcus*. PLoS Genet 4: e1000173

Stengel DB, Connan S (2015) Marine Algae: a source of biomass for biotechnological applications. Methods Mol Biol Clifton NJ 1308:1–37

Storz G, Vogel J, Wassarman KM (2011) Regulation by small RNAs in bacteria: expanding frontiers. Mol Cell 43:880–891

Taniuchi Y, Chen YL, Chen HY, Tsai ML, Ohki K (2012) Isolation and characterization of the unicellular diazotrophic cyanobacterium Group C TW3 from the tropical western Pacific ocean. Environ Microbiol 14:641–654

Thompson AW, Foster RA, Krupke A, Carter BJ, Musat N, Vaulot D, Kuypers MMM, Zehr JP (2012) Unicellular cyanobacterium symbiotic with a single-celled eukaryotic alga. Science 337:1546–1550

Thompson A, Carter BJ, Turk-Kubo K, Malfatti F, Azam F, Zehr JP (2014) Genetic diversity of the unicellular nitrogen-fixing cyanobacteria UCYN-A and its prymnesiophyte host. Environ Microbiol 16:3238–3249

Ting CS, Rocap G, King J, Chisholm SW (2002) Cyanobacterial photosynthesis in the oceans: the origins and significance of divergent light-harvesting strategies. Trends Microbiol 10:134–142

Tripp HJ, Bench SR, Turk KA, Foster RA, Desany BA, Niazi F, Affourtit JP, Zehr JP (2010) Metabolic streamlining in an open-ocean nitrogen-fixing cyanobacterium. Nature 464:90–94

Turk-Kubo KA, Karamchandani M, Capone DG, Zehr JP (2014) The paradox of marine heterotrophic nitrogen fixation: abundances of heterotrophic diazotrophs do not account for nitrogen fixation rates in the Eastern Tropical South Pacific. Environ Microbiol 16:3095–3114

Turk-Kubo KA, Frank IE, Hogan ME, Desnues A, Bonnet S, Zehr JP (2015) Diazotroph community succession during the VAHINE mesocosm experiment (New Caledonia lagoon). Biogeosciences 12:7435–7452

Unrein F, Gasol JM, Not F, Forn I, Massana R (2014) Mixotrophic haptophytes are key bacterial grazers in oligotrophic coastal waters. ISME J 8:164–176

Urbach E, Scanlan DJ, Distel DL, Waterbury JB, Chisholm SW (1998) Rapid diversification of marine picophytoplankton with dissimilar light-harvesting structures inferred from sequences of *Prochlorococcus* and *Synechococcus* (Cyanobacteria). J Mol Evol 46:188–201

Varkey D, Mazard S, Ostrowski M, Tetu SG, Haynes P, Paulsen IT (2015) Effects of low temperature on tropical and temperate isolates of marine *Synechococcus*. ISME J. doi:10.1038/ismej.2015.179

Vaulot D, Lepère C, Toulza E, De la Iglesia R, Poulain J, Gaboyer F, Moreau H, Vandepoele K, Ulloa O, Gavory F et al (2012) Metagenomes of the picoalga *Bathycoccus* from the Chile coastal upwelling. PLoS ONE 7:e39648

Villar E, Farrant GK, Follows M, Garczarek L, Speich S, Audic S, Bittner L, Blanke B, Brum JR, Brunet C et al (2015) Environmental characteristics of Agulhas rings affect interocean plankton transport. Science 348:1261447

Villareal TA (1991) Nitrogen-fixation by the cyanobacterial symbiont of the diatom genus *Hemiaulus*. Mar Ecol Prog Ser Oldendorf 76:201–204

Villareal TA (1992) Marine nitrogen-fixing diatom-cyanobacteria symbioses. In: Marine pelagic cyanobacteria: *Trichodesmium* and other diazotrophs. Springer, Berlin, pp 163–175

Villareal TA (1994) Widespread occurrence of the *Hemiaulus*-cyanobacterial symbiosis in the southwest North Atlantic ocean. Bull Mar Sci 54:1–7

Voss B, Bolhuis H, Fewer DP, Kopf M, Möke F, Haas F, El-Shehawy R, Hayes P, Bergman B, Sivonen K et al (2013) Insights into the physiology and ecology of the brackish-water-adapted

Cyanobacterium *Nodularia spumigena* CCY9414 based on a genome-transcriptome analysis. PLoS ONE 8:e60224
Walworth N, Pfreundt U, Nelson WC, Mincer T, Heidelberg JF, Fu F, Waterbury JB, Glavina del Rio T, Goodwin L, Kyrpides NC et al (2015) *Trichodesmium* genome maintains abundant, widespread noncoding DNA in situ, despite oligotrophic lifestyle. Proc Natl Acad Sci USA 112:4251–4256
Waterbury JB, Watson SW, Guillard RRL, Brand LE (1979) Widespread occurrence of a unicellular, marine, planktonic, cyanobacterium. Nature 277:293–294
Webb EA, Jakuba RW, Moffett JW, Dyhrman ST (2007) Molecular assessment of phosphorus and iron physiology in *Trichodesmium* populations from the western Central and western South Atlantic. Limnol Oceanogr 52:2221–2232
West NJ, Lebaron P, Strutton PG, Suzuki MT (2011) A novel clade of *Prochlorococcus* found in high nutrient low chlorophyll waters in the South and Equatorial Pacific ocean. ISME J 5:933–944
Whitehead L, Long BM, Price GD, Badger MR (2014) Comparing the in vivo function of α-carboxysomes and β-carboxysomes in two model cyanobacteria. Plant Physiol 165:398–411
Woebken D, Burow LC, Behnam F, Mayali X, Schintlmeister A, Fleming ED, Prufert-Bebout L, Singer SW, Cortés AL, Hoehler TM et al (2015) Revisiting N_2 fixation in Guerrero Negro intertidal microbial mats with a functional single-cell approach. ISME J 9:485–496
Worden AZ, Nolan JK, Palenik B et al (2004) Assessing the dynamics and ecology of marine picophytoplankton: the importance of the eukaryotic component. Limnol Oceanogr 49:168–179
Worden AZ, Follows MJ, Giovannoni SJ, Wilken S, Zimmerman AE, Keeling PJ (2015) Environmental science. Rethinking the marine carbon cycle: factoring in the multifarious lifestyles of microbes. Science 347:1257594
Xia S, Gao B, Li A, Xiong J, Ao Z, Zhang C (2014) Preliminary characterization, antioxidant properties and production of chrysolaminarin from marine diatom *Odontella aurita*. Mar Drugs 12:4883–4897
You I, Lee TG, Nam YS, Lee H (2014) Fabrication of a micro-omnifluidic device by omniphilic/omniphobic patterning on nanostructured surfaces. ACS Nano 8:9016–9024
Zeev EB, Yogev T, Man-Aharonovich D, Kress N, Herut B, Béjà O, Berman-Frank I (2008) Seasonal dynamics of the endosymbiotic, nitrogen-fixing cyanobacterium *Richelia intracellularis* in the eastern Mediterranean sea. ISME J 2:911–923
Zehr JP (2015) Evolution. How single cells work together. Science 349:1163–1164
Zehr JP, Kudela RM (2011) Nitrogen cycle of the open ocean: from genes to ecosystems. Annu Rev Mar Sci 3:197–225
Zehr JP, Waterbury JB, Turner PJ, Montoya JP, Omoregie E, Steward GF, Hansen A, Karl DM (2001) Unicellular cyanobacteria fix N_2 in the subtropical North Pacific ocean. Nature 412:635–638
Zehr JP, Bench SR, Carter BJ, Hewson I, Niazi F, Shi T, Tripp HJ, Affourtit JP (2008) Globally distributed uncultivated oceanic N_2-fixing cyanobacteria lack oxygenic photosystem II. Science 322:1110–1112
Zhao KH, Deng MG, Zheng M, Zhou M, Parbel A, Storf M, Meyer M, Strohmann B, Scheer H (2000) Novel activity of a phycobiliprotein lyase: both the attachment of phycocyanobilin and the isomerization to phycoviolobilin are catalyzed by the proteins PecE and PecF encoded by the phycoerythrocyanin operon. FEBS Lett 469:9–13
Zhaxybayeva O, Doolittle WF, Papke RT, Gogarten JP (2009) Intertwined evolutionary histories of marine *Synechococcus* and *Prochlorococcus marinus*. Genome Biol Evol 1:325–339
Zilliges Y, Kehr JC, Meissner S, Ishida K, Mikkat S, Hagemann M, Kaplan A, Börner T, Dittmann E (2011) The cyanobacterial hepatotoxin microcystin binds to proteins and increases the fitness of microcystis under oxidative stress conditions. PLoS ONE 6:e17615

Zinser ER, Johnson ZI, Coe A, Karaca E, Veneziano D, Chisholm SW (2007) Influence of light and temperature on *Prochlorococcus* ecotype distributions in the Atlantic ocean. Limnol Oceanogr 52:2205–2220

Zubkov MV (2009) Photoheterotrophy in marine prokaryotes. J Plankton Res 31:933–938

Zwirglmaier K, Jardillier L, Ostrowski M, Mazard S, Garczarek L, Vaulot D, Not F, Massana R, Ulloa O, Scanlan DJ (2008) Global phylogeography of marine *Synechococcus* and *Prochlorococcus* reveals a distinct partitioning of lineages among oceanic biomes. Environ Microbiol 10:147–161

Chapter 4
Marine Fungi

Vanessa Rédou, Marine Vallet, Laurence Meslet-Cladière, Abhishek Kumar, Ka-Lai Pang, Yves-François Pouchus, Georges Barbier, Olivier Grovel, Samuel Bertrand, Soizic Prado, Catherine Roullier and Gaëtan Burgaud

Abstract Marine fungi have long been considered as exotic microorganisms fascinating only a few scientists. However, during the last two decades there has been an increasing interest in marine fungal communities resulting in a considerable advance in our knowledge of marine fungi. Marine fungi have been retrieved from various marine habitats, ranging from coastal waters to the deep subseafloor, and their ecologically important roles have been demonstrated. The purpose of this chapter is to review the increasing amount of culture-based and molecular-based data along with metabolomics and to summarize our current knowledge of the diversity, adaptive capabilities, functions, ecological roles, and biotechnological potential of marine fungi. The availability of this amount of complementary data

V. Rédou · L. Meslet-Cladière · G. Barbier · G. Burgaud (✉)
EA 3882 Laboratoire Universitaire de Biodiversité et Ecologie Microbienne, Technopôle Brest-Iroise, Université de Brest, 29280 Plouzané, France
e-mail: Gaetan.Burgaud@univ-brest.fr

V. Rédou
e-mail: vanessa.redou@univ-brest.fr

L. Meslet-Cladière
e-mail: Laurence.Meslet@univ-brest.fr

G. Barbier
e-mail: Georges.Barbier@univ-brest.fr

M. Vallet · S. Prado
Molécules de Communication et Adaptation des Micro-organismes, UMR 7245 CNRS/MNHN, Muséum National d'Histoire Naturelle, 57 rue Cuvier (CP54), 75231 Paris Cedex 05, France
e-mail: mvallet@edu.mnhn.fr

S. Prado
e-mail: sprado@mnhn.fr

A. Kumar
Department of Genetics and Molecular Biology in Botany, Institute of Botany, Christian-Albrechts-University, 24118 Kiel, Germany
e-mail: abhishek.abhishekkumar@gmail.com

L.J. Stal and M.S. Cretoiu (eds.), *The Marine Microbiome*,
DOI 10.1007/978-3-319-33000-6_4

allows a revision of the consensual but likely out-of-date definition of marine fungi. Since the field of marine fungal natural products continues to expand rapidly, another aim of this chapter is to provide some innovative approaches to optimize the search for novel bioactive compounds using genomics and metabolomics.

4.1 Introduction

The ocean harbors a tremendous diversity of habitats ranging from coastal waters to deep-sea chemosynthetic-based environments where microorganisms are the major actors of the biogeochemical cycling of elements. Bacteria, Archaea, and some microbial Eukarya (mostly micro-algae) are the most commonly studied microorganisms in the marine environment. However, recent studies strongly support the idea that marine microbial communities comprise also fungi as an important component. We will not propose another exhaustive summary of marine fungal diversity and classification (for that, see Jones et al. 2015; Nagano et al. 2010; Richards et al. 2012), but rather to consider marine fungi from a different angle. Ideas here are (i) to take advantage of the huge advances in omic-based methods to better understand the function, role, and biotechnological potential of marine fungi and (ii) to give clues about how to access to the untapped marine fungal metabolome, in particular the secondary metabolites (SM).

A. Kumar
Division of Molecular Genetic Epidemiology, German Cancer Research Center, Heidelberg, Germany

K.-L. Pang
Institute of Marine Biology and Center of Excellence for the Oceans, National Taiwan Ocean University, 2 Pei-Ning Road, Keelung 20224, Taiwan R.O.C.
e-mail: klpang@ntou.edu.tw

Y.-F. Pouchus · O. Grovel · S. Bertrand · C. Roullier
Faculty of Pharmacy, University of Nantes, EA2160 Nantes, France
e-mail: yves-francois.pouchus@univ-nantes.fr

O. Grovel
e-mail: olivier.grovel@univ-nantes.fr

S. Bertrand
e-mail: samuel.bertrand@univ-nantes.fr

C. Roullier
e-mail: catherine.roullier@univ-nantes.fr

4.2 Toward a New Consensual Definition of Marine Fungi? Let's Think Outside the Box!

(Micro)biologists need definitions to precisely describe their model organisms, usually proposing a concept and debating ideas in order to try glimpsing a consensual rationale, a process that may take many years. Marine fungi have not escaped such a long debate. Barghoorn and Linder (1944) initiated the first exhaustive study of marine fungi, demonstrating the occurrence of an '*indigenous marine mycota'* on submerged wood. However, it took marine mycologists almost two decades to suggest a definition of marine fungi based on their physiology (Johnson and Sparrow 1961), i.e., their ability to grow at certain seawater concentrations. This postulate was later criticized by Kohlmeyer and Kohlmeyer (1979), who argued that terrestrial fungi are able to cope with low water activity, including salinity tolerance up to concentrations that equal those in the marine environment. Based on this strong adaptive tolerance, many authors debated on the physiology-based definition of marine fungi and suggested a broad double box ecological definition dividing marine fungi into obligate and facultative marine organisms. Obligate marine fungi are those that grow and sporulate exclusively in a marine or estuarine habitat while facultative marine fungi are those from freshwater or terrestrial habitats able to grow and possibly to sporulate in the marine environment (Kohlmeyer and Kohlmeyer 1979).

This consensual definition, widely accepted by the scientific community, has however generated a split between obligate marine fungi and the "ugly duckling" group of facultative marine fungi. Indeed, research on marine mycology has primarily been focused on obligate marine fungi revealed by the direct microscopic observations of fruiting structures (ascomata, basidiomata) and subsequent microscopic identification of spores with prominent oil globules and appendages which keep them suspended and increase their chance of attaching to a substrate (Kohlmeyer and Kohlmeyer 1979). Advocates of this method always tended to consider facultative marine fungi (i) as common dust and windborne forms that seem to occur as dormant spores or (ii) as terrestrial fungi occurring on the vertical portion of salt marsh plants that occasionally receive seawater splash (Kohlmeyer and Volkmann-Kohlmeyer 2003a). The term "marine-derived" has also been proposed to define fungi isolated from marine samples that could not be classified as obligate or facultative marine fungi (Osterhage et al. 2000). However, this "marine-derived" term seems to be used only by scientists interested in bioactive compounds and who are not interested to know whether their fungal strains are adapted or not to the marine environment. Such an artificial "catchall" group does not help defining marine fungi, but has quite the opposite effect because it enhances complexity of the definition.

Numerous obligate marine fungi have been described and studied in the twentieth century, but it appears that studies dealing with facultative marine fungi are now totally outshined. A recent count reports only 1112 species of marine fungi retrieved exclusively from the marine environment (Jones et al. 2015). Taking the actual global culture collections of fungi that contain 74,000–120,000 species (Hawksworth 2001; Kis-Papo 2005), marine fungal diversity may account for 0.93–1.50 %. Richards et al. (2012) reported a close value of 0.6 % [with 467 species of obligate marine fungi as reported by Kis-Papo (2005)], and discussed about the meaning of such small percentage knowing that the ocean covers more than 70 % of the Earth's surface and that fungal communities have been reported from air–water surface interface up to thousands of meters below the seafloor (Ciobanu et al. 2014; Rédou et al. 2014; Kis-Papo 2005). The question whether we are overlooking a large proportion of marine fungi should be answered with yes if we only consider the obligate marine organisms.

Numerous studies targeting microeukaryotes in different marine habitats highlighted unexpected fungal communities, ranging from new species to ones close to terrestrial representatives using both molecular and culture-based approaches (see Burgaud et al. 2014; Mahé et al. 2013; Richards et al. 2012). Modern high-throughput omic methods now allow targeting and sequencing rRNA and mRNA including that obtained from deep marine sediments (Orsi et al. 2013a; Rédou et al. 2014). Such an approach has been used to uncover fungi as the dominant organisms among microeukaryotes (Edgcomb et al. 2011). Most of these fungi were close to terrestrial representatives and could be defined as facultative marine fungi, according to Kohlmeyer's definition. However, these fungi seem to show high activity and biotic interactions with other microbial communities. For example these fungi synthesize antimicrobial compounds and degrade complex organic matter (Orsi et al. 2013a). Culture-based approaches coupled with ecophysiological analyses reveal the ability of many fungal isolates to cope with various abiotic factors that shape different marine habitats such as temperature, salinity, and high hydrostatic pressure. For instance, fungal isolates have been obtained from an almost 2000 m deep sediment core (Rédou et al. 2015). Different fungal species that were close relatives of terrestrial organisms were obtained from different sediment depths. Surprisingly, the strains obtained from the greatest depths were much better adapted to their habitat conditions compared to the same species isolated from shallower depths. As elaborated by Rédou et al. (2015), the shift from terrestrial adapted to marine-adapted lifestyles along the sediment depth may indicate a transition from shallow layers, where fungi are not specifically adapted but are able to survive, to deeper layers, where fungi are better adapted to higher temperature and higher salinity. Such ecophysiological analyses provide concrete and relevant data that highlight specific fungal adaptations and that help to understand the true activities and roles of fungi in different marine habitats. Compiling these approaches may provide the current missing pieces of the puzzle of the nature of marine fungi. Facultative marine fungi seem to be much more important than expected with respect to their structural diversity as well as their functional implications.

So, let us think outside the box regarding this consensual but potentially out-of-date definition of marine fungi of Kohlmeyer. The fact that some terrestrial fungi seem to be able to adapt to the marine environment emphasizes the complexity of using a box-based definition. Mahé et al. (2013) proposed a new definition of marine fungi based on 3 levels of occurrence with (i) strict endemic active marine fungi, (ii) ubiquitous fungi, metabolically active in marine habitats, and (iii) ubiquitous fungi, metabolically inactive in marine habitats. However, a simple "box less" definition of marine fungi integrating the notions of activity and time would be much more user-friendly. Here follows an attempt for a new definition in an omics context: "Marine fungi display long-term presence and metabolic activities in a marine habitat, as revealed by their adaptations (ecophysiological profile), active metabolism (rRNA), gene expression (mRNA), catalytic functions (proteome), or specific metabolites (metabolome) resulting from their biotic and abiotic interactions." Such a simple definition should be completed with more detailed analyses such as for instance comparative omics. The idea is to process comparative genomics, transcriptomics, proteomics, and metabolomics from different representatives of the same species isolated from marine and terrestrial habitats in order to identify specific markers of fungal adaptation to marine habitats. Such omics approach would also help to demonstrate that obligate marine fungi are never encountered in any terrestrial habitat.

4.3 From Broad-Scale to Habitat Specific Distribution Patterns of Marine Fungal Communities

4.3.1 From Culture-Based to Next-Generation Sequencing Methods to Barcode Marine Fungal Life

Fungi are known from different marine environments ranging from shallow waters to the deep-sea, using various approaches such as (i) morphology-based observations, (ii) culture-based plating techniques, and (iii) molecular methods, including next-generation sequencing. In the following part, only information obtained from culture- and molecular-based methods is summarized.

Culture-based approach. The first exhaustive investigation of culturable fungal communities in seawater was from the northwestern subtropical Atlantic Ocean (Roth et al. 1964). Since then, it was estimated that a milliliter of average seawater contains over one thousand fungal cells (Gao et al. 2008). Marine fungi have been reported from mangroves, marine algae, and salt marshes, mostly in association with marine invertebrates. Sponges are the most studied habitat for marine fungi. Consequently, the largest number of marine fungi was obtained from sponges (Bugni and Ireland 2004), e.g., sponges from the Atlantic Ocean (Baker et al. 2009; Menezes et al. 2010), from the Mediterranean Sea (Paz et al. 2010; Wiese et al. 2011), from the Indian Ocean (Thirunavukkarasu et al. 2012), from the South China Sea (Ding et al. 2011; Liu et al. 2010; Zhou et al. 2011), from the Pacific Ocean (Li and Wang 2009; Wang et al. 2008), and from the Antarctic Sea (Henriquez et al. 2014). Sponge inner tissue was usually homogenized with artificial seawater and spread onto nutrient-rich culture media at different dilutions. These sponge-inhabiting fungi belong mostly to the Ascomycota (e.g. Leotiomycetes, Dothideomycetes, Eurotiomycetes and Sordiaromycetes) and few Basidiomycota (e.g. Agaricomycetes and Wallemiomycetes), and Zygomycetes (e.g. early diverging fungal lineages affiliated to the order Mucorales). However, it should be noted that such studies were dedicated to biotechnology and authored by chemists who did not investigate such habitats exhaustively and thus a conclusion regarding fungal diversity on sponges is challenging.

Going deeper into the ocean from the epipelagic to the abyss, numerous studies reported fungal communities in the deep-sea. Some fungal communities were found in extreme habitats such as deep-sea hydrothermal vents, where fungi were retrieved in association with endemic animals at Mid-Atlantic Ridge and East Pacific Rise (Burgaud et al. 2009, 2010; Gadanho and Sampaio 2005; Le Calvez et al. 2009). Conventional culture media enriched with glucose, maltose, peptone, and supplemented with sea salts and sometimes sulfur were used. Indeed, elemental sulfur is abundant in hydrothermal vents and is usually incorporated in culture media employed for the isolation of bacteria and archaea in order to mimic their natural environment (Gadanho and Sampaio 2005). Numerous filamentous fungal strains have been isolated and were mostly indigenous species (Burgaud et al. 2009). Only Ascomycota and Basidiomycota belonging to the Dothideomycetes, Eurotiomycetes, Sordariomycetes, and Exobasidiomycetes classes were retrieved. A large fraction of the yeast taxa found in the Mid-Atlantic Ridge represents new phylotypes and seems to be composed of autochthonous species (Gadanho and Sampaio 2005). Only *Candida oceani* and *Rhodotorula pacifica*, endemic yeasts occurring in hydrothermal vents, have been analyzed (Burgaud et al. 2011; Nagahama et al. 2006) but some others remain to be described (Gadanho and Sampaio 2005). Yeast diversity in deep-sea hydrothermal vents is represented by the *Candida*, *Debaryomyces*, *Exophiala*, *Hortaea*, *Phaeotheca,* and *Pichia* genera among the Ascomycota and the *Cryptococcus*, *Rhodotorula*, *Rhodosporidium*, *Sporobolomyces,* and *Trichosporon* ones among the Basidiomycota.

Again, going deeper into the oceans, from the abyss to the subseafloor, some recent studies discovered some "buried but not dead" culturable fungal communities in sediments. Fungi retrieved belong exclusively to the Dikarya and mostly to the Ascomycota phylum. No representative of basal fungal lineages was reported although molecular signatures suggested their presence (Nagahama et al. 2011; Nagano et al. 2010). The method that is conventionally used for the isolation of marine fungi is the particle plating technique (Cathrine and Raghukumar 2009; Damare et al. 2006; Jebaraj et al. 2010; Singh et al. 2011, 2012; Zhang et al. 2013), which consists of spreading sediment slurry onto Petri dishes. This strategy usually selects for ubiquitous fast-growing organisms and needs to be combined with dilution plating (Damare et al. 2006; Mouton et al. 2012; Raghukumar et al. 2004; Rédou et al. 2015; Singh et al. 2011, 2012; Zhang et al. 2014) that allows the isolation of slow-growing microorganisms and results in higher sensitivity and cultivation efficiency. Because the habitability in deep subseafloor sediments is set by a variety of physical and chemical characteristics, including elevated hydrostatic pressure, culture methods involving incubation under pressure has been performed.

The recovery of fungal isolates obtained using enrichments under elevated hydrostatic pressure was better compared to conventional methods, suggesting that hydrostatic pressure is a key physical parameter for deep subseafloor fungal growth (Damare et al. 2006; Rédou et al. 2015; Singh et al. 2011, 2012). Numerous filamentous fungal and yeast strains closely related to terrestrial species have been isolated. Ascomycota and Basidiomycota retrieved belong to the classes Dothideomycetes, Eurotiomycetes, Saccharomycetes, Sordariomycetes, and Agaricomycetes, Exobasidiomycetes, Microbotryomycetes, Tremellomycetes, respectively.

Several attempts to decipher the culturable fungal communities occurring in deep-sea ecosystems have been performed, but only a minor fraction of the microbial community has been isolated using conventional selective media (Alain and Querellou 2009). Most of the marine microorganisms are indeed refractory to culturing and the estimated culturing efficiency of endemic microorganisms using the standard plating techniques is between 0.1 and 0.25 % (Amann et al. 1995; D'Hondt et al. 2004). Thus, culture collections of microorganisms isolated by using conventional methods do not reflect the functional and phylogenetic diversity of marine habitats. Most of the culture-based methods used are biased due to the elimination of cell-to-cell communication, dormant spores germination induced by starvation or other stress, or too short incubation time. Recent culture strategies, including the refinement of culture media based on cell-to-cell communication and the development of high-throughput culture methods, aim at limiting the biases inherent to traditional culture methods and increase the isolation efficiency to bring into culture the uncultured. Traditionally, the concentration of nutrients in synthetic culture media is much higher than that typically found in marine environments and this represents an important issue for the culturing of marine oligotrophic microorganisms that are adapted to low concentrations of substrates. To decrease the stress related to the exposure to massive amounts of substrate, dilution of culture medium has been used for the isolation of fungal strains (Damare et al. 2006; Rédou et al. 2015; Singh et al. 2011). Connon and Giovannoni (2002) have developed a high-throughput culturing method based on the concept of dilution to extinction culturing to isolate cultures in small volumes of low-nutrient media. This technique allowed the isolation of four unique cell lineages that belong to previously uncultured and not described marine clades of proteobacteria. This high-throughput bacterial culturing method was adapted to isolate terrestrial fungi and to generate large numbers of fungal dilution to extinction cultures from forest litter (Collado et al. 2007), leaves (Unterseher and Schnittler 2009), and sub-Antarctic soil (Ferrari et al. 2011). These novel culturing approaches have hitherto not yet been applied for the isolation of fungi from marine environmental samples and they could therefore become a promising tool for the isolation and culturing of not yet cultured marine fungi. Another culturing issue is cell-to-cell communication. Microorganisms are normally living associated with other microorganisms or macroorganisms, establishing complex relationships that can be allegorized as a complex puzzle of molecules. However, the paradox of the microbiologist is that he aims at obtaining pure cultures (Alain and Querellou 2009). To isolate cells and preserve the endogenous cell-to-cell communication mechanisms, Zengler et al. (2002) developed an encapsulation technique of single cells in gel microdroplets for massive parallel microbial cultivation under low nutrient flux conditions, followed by flow cytometric sorting of microdroplets containing microcolonies of microorganisms. This high-throughput cultivation method can provide more than 10,000 bacterial and fungal isolates per environmental sample (Zengler et al. 2005). Ingham et al. (2007) developed a Petri dish divided into millions compartments, called the MicroDish Culture Chip (MDCC), in which nutrients are supplied through a porous membrane. Theoretically, this technique would allow the growth and screening of millions of strains and is an efficient method to overcome the effects of intercellular competition and allows growth of microorganisms often inhibited by opportunistic fast-growing organisms (Alain & Querellou 2009). This innovative method is used for the screening and isolation of marine bacteria and could be adapted for the isolation of marine fungi.

Culture-independent approach. Analysis of environmental samples using culture-independent methods, such as cloning-sequencing or next-generation sequencing paired with taxonomic and phylogenetic analysis, brings another view to the microbial diversity of marine ecosystems. Recent advances in marine molecular ecology have allowed to revealing an astonishing diversity of microorganisms. Analysis of environmental clone libraries in order to assess fungal diversity usually targets the small subunit (SSU) rRNA. Diez et al. (2001) analyzed eukaryotic clone libraries from five surface samples taken from three distant marine regions, i.e., the Mediterranean Sea, the Southern Ocean, and the North Atlantic Ocean. Among a total of 225 clones, only 5 belonged to fungi representing 2.2 % of the total eukaryotic diversity. This low fungal diversity in surface waters was confirmed by the study of Massana and Pedrós-Alió (2008) who sampled 23 coastal water libraries and 12 open ocean libraries and revealed a total of only 16 fungal clones, representing 0.8 % of the eukaryotic SSU rRNA gene sequences. Using shallow and deep-sea water samples from different parts of the world ranging depths from 250 to 4000 m, Bass et al. (2007) reported a fungal diversity largely represented by ascomycetes and basidiomycetes but also suggested that fungal diversity in deep-sea ecosystems is low. The dominance of Dikarya in marine habitats is well established (Richards et al. 2012), and molecular marine fungal signatures belong to classes such as the Dothideomycetes, Eurotiomycetes, Leotiomycetes, Saccharomycetes, Sordariomycetes of the Ascomycota, and the classes Agaricomycetes, Agaricostilbomycetes, Atractiellomycetes, Cystobasidiomycetes, Exobasidiomycetes, Microbotryomycetes, Tremellomycetes, Ustilaginomycetes, Wallemiomycetes of the Basidiomycota. Few sequences of Glomeromycota, Chytridiomycota, and other early fungal lineages have been found.

The occurrence of highly divergent fungal sequences in deep-sea marine environments is evidenced from the analysis of SSU rRNA-based clone libraries. Indeed, the identification by molecular methods of new phylotypes representing mainly basal fungal lineages was done in samples obtained from deep oceans, hydrothermal vents, oxygen-depleted regions, and deep-sea sediments (Bass et al.

2007; Jebaraj et al. 2010; Le Calvez et al. 2009; Nagahama et al. 2011; Nagano et al. 2010). These new phylotypes may have unique ecophysiological features, which would explain why no representatives have been already obtained by cultivation methods (Nagahama and Nagano 2012). The internal transcribed spacer (ITS) region was recently validated as DNA barcode marker for the identification of fungal species for its variability and thus its high resolution for taxonomic assignment (Schoch et al. 2012). However, the ITS of the rRNA operon has a low phylogenetic resolution and it seems that the combination with the SSU RNA gene could improve the phylogenetic placement of many orphan environmental ITS sequences (Richards et al. 2012). Sequencing of the clone libraries of the ribosomal ITS and SSU rRNA gene, has revealed the fungal diversity in deep-sea sediments of the Central Indian Basin (Singh et al. 2012). Even though fungal diversity may appear low in the water column, some deep-sea marine environments have been identified as habitats for filamentous fungi and yeasts. Indeed, a total of 42 fungal OTUs was recovered out of 192 clones and represented distinct phylotypes, belonging to Ascomycota and Basidiomycota. This indicated the presence of a high diversity of fungi at the sampling site (Singh et al. 2012). Several studies have shown the dominance of fungal communities among deep-sea microeukaryotes (Ciobanu et al. 2014; Edgcomb et al. 2011; Orsi et al. 2013a, b; Rédou et al. 2014). Sequencing of both the ribosomal ITS and the 18S rRNA gene V1-V3 diversity tags from the subseafloor sediments of the Canterbury basin using 454 pyrosequencing hinted to fungal communities that belong exclusively to the Dikarya (Rédou et al. 2014). These fungal communities occurred at record depths in deep-sea sediments albeit at a low phylogenetic diversity.

Most of the fungi that were recovered in diversity studies are closely related to known terrestrial groups, raising interesting ecological questions regarding their origin and abilities to adapt to deep subseafloor conditions. Several studies emphasized that fungi are metabolically active in deep groundwater of crystalline rock fractures and up to hundreds of meters below the seafloor (Edgcomb et al. 2011; Orsi et al. 2013b; Sohlberg et al. 2015). The application of metatranscriptomics in samples from the deep biosphere demonstrated the occurrence of microbial activity, including fungal activity, in the subsurface, (Orsi et al. 2013b). This activity was inferred from transcripts that were estimated to represent 20 and 5 % of the whole metatranscriptome at 5 and 159 m below the seafloor, respectively. Fungi in deep sediments appeared to be involved in carbohydrate, amino acid, and lipid metabolism, indicating a potentially important role in organic carbon turnover in sub-seafloor sediments. Eventually, the analysis of fungal transcripts from various geographical areas will provide a more accurate view of the involvement of fungal communities in the active part of the deep biosphere. However, it should be noted that none of the existing molecular tools are free from limitations such as PCR- and cloning biases, sequencing errors, low phylogenetic resolution of marker genes, lack of taxonomic annotation, and errors in taxonomic assignments of databases. The current era of metagenomics and metatranscriptomics coupled with high-throughput culture-based methods, as outstanding tools, will allow integrated analyses to obtain a more precise picture of marine fungal life.

4.3.2 Habitat Specific Community Structure or Over-Dispersion?

The picture of marine fungal diversity remains largely pixelated even when in the last decades ecological studies shed new light on this group of organisms. This new data allows us to catch a glimpse of distribution patterns. Answering the question whether marine fungi are overdispersed or endemic would help to better understand the broad-scale habitat and geographic differences among marine fungal communities.

The first exhaustive analysis of marine fungal distribution patterns was based on temperature-determined biogeographical zones: temperate, subtropical, tropical, arctic, and antarctic (Hughes 1974). Different fungal species were only retrieved in a specific zone, e.g., *Spathulospora antarcticum* was only retrieved from the red algae *Ballia callitricha* in Antarctic seawater. Some other species were retrieved in different zones as *Corollospora maritima*, *Halosphaeria appendiculata,* or *Lignicola laevis* each occurring in both temperate and tropical waters (Jones 2000). These examples strongly support the idea that marine fungal communities can be divided into habitat/substrate-specific species and overdispersed species. Panzer et al. (2015) generated an 18S rRNA gene sequence reference dataset using all publicly available marine fungal 18S rRNA gene sequences. Fungal taxonomic composition of different marine habitats varied considerably suggesting the existence of habitat-specific biomes for marine fungal communities, with some specific habitats differing more than others, e.g., hydrothermal vents, sediment, and seawater. Phylogenetic differences between habitats indicate that the type of habitat strongly affects fungal community structure and thus that environmental factors are the main drivers, rather than competition between species. In order to better understand marine fungal distribution patterns at habitat-scale, the following sections focus on three distinct marine habitats, namely plant-based habitats, algae, and the deep-sea.

Plant-based habitats. Saprophytic fungi occurring on plant substrates are the most well-studied group of marine fungi. Since the monumental study of wood-inhabiting fungi by Barghoorn and Linder (1944), many new species of marine fungi have been described from wood from diverse habitats, including wood buried in sandy beaches, decaying wood in mangroves, and drift or trapped wood on rocky shores. Wood-inhabiting marine fungi form fruiting bodies on wood and cause soft-rot decay mainly by producing cellulases and laccases (Bucher et al. 2004). Marine Dothideomycetes and Sordariomycetes belonging to the Ascomycota are dominant with few Basidiomycota. These observations were based on the identification of fruiting bodies and also on culture-independent techniques using tag-encoded 454 pyrosequencing of the ribosomal ITS and of the 18S rRNA gene (Arfi et al. 2012; Jones et al. 2015). Marine lignicolous basidiomycetes are mainly intertidal species and belong to the Agaricomycetes with reduced and enclosed fruiting bodies, loss of ballistospory, and evolution of spore appendages (Hibbett and Binder 2001). Marine lignicolous ascomycetes are phylogenetically diverse, but

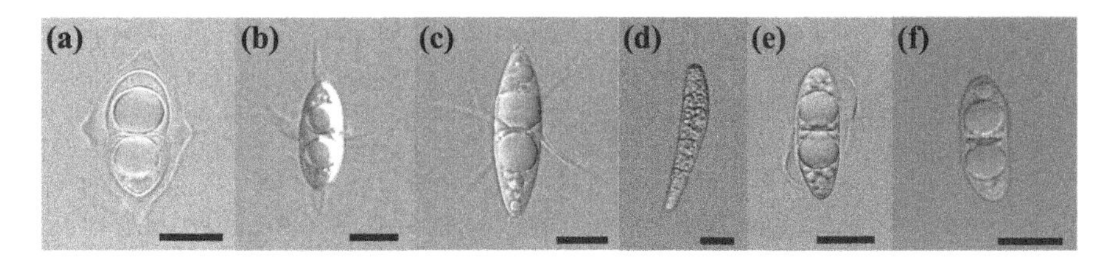

Fig. 4.1 Plant-inhabiting marine fungi from diverse habitats. **a** Ascospore of *Ebullia octonae* with a sheath. **b** Ascospore of *H. appendiculata* with polar and equatorial spoon-shaped appendages. **c** Ascospore of *Halosphaeriopsis mediosetigera*. **d** Clavate-shaped ascospore of *Buergenerula spartinae*. **e** Ascospore of *Natantispora retorquens* with bipolar unfurling appendages. **f** Ascospore of *Lignincola laevis* (*Scale bar* 10 μm)

mainly belonging to the Pleosporales in the Dothideomycetes and Microascales (Halosphaeriaceae) in the Sordariomycetes class, forming exposed or immersed perithecia (Jones et al. 2015). As revealed by Rämä et al. (2014), Hypocreales and Helotiales also represent important taxonomic groups in which fungal communities from Arctic intertidal and seafloor logs were obtained using culture techniques. Apothecium-type of ascomata is uncommon with only 10 described species (Baral and Rämä 2015), possibly due to the inability to withstand wave action (Suetrong and Jones 2006). These ascomycetes have also evolved diverse morphologies to adapt to a marine lifestyle, i.e., deliquescing asci and ascospore appendages of different morphology and ontogeny (Fig. 4.1a–c) (Pang 2002). Two of the largest lineages of marine lignicolous ascomycetes, the Halosphaeriaceae and the Lulworthiales, were inferred to have derived from terrestrial ancestors (Spatafora et al. 1998). Recent phylogenetic studies of the ribosomal RNA genes revealed further independent lineages into the marine environment in the Ascomycota: Dyfrolomycetales (Hyde et al. 2013), Tirisporellales, Torpedosporales (Jones et al. 2015), Savoryellales (Boonyuen et al. 2011), confirming their high diversity on wood substrates.

Marine fungi also grow on the decaying intertidal part of saltmarsh plants, such as *Spartina* spp., *Juncus roemerianus* and *Phragmites australis*, and the palm *Nypa fruticans*; many species being host-/substrate-specific (Calado and Barata 2012; Pilantanapak et al. 2005). In particular, fungi associated with *Spartina* spp. have been well studied in US and Portuguese salt marshes, where they are involved in nutrient cycling (Newell and Wasowski 1995). Diverse laccase genes were detected from the fungal community associated with *Spartina alterniflora*, which may suggest their involvement in lignin mineralization (Lyons et al. 2003). A total of 132 species of marine fungi are documented living saprophytically on *Spartina* spp.; the dominant groups belonging to the Dothideomycetes and Sordariomycetes (Calado and Barata 2012). *Phaeosphaeria halima*, *Phaeosphaeria spartinicola*, *Mycosphaerella* sp., *Byssothecium obiones*, and *Buergenerula spartinae* are common taxa living on *Spartina* spp. in US saltmarshes (Buchan et al. 2002; Newell et al. 1996; Walker and Campbell 2010). Many of these species have fully functional asci for forcible expulsion of spores (Newell 2001). These species are also common in Portuguese saltmarshes along with *Natantispora retorquens* (Fig. 4.1e)

(Barata 2002; Calado et al. 2015). Based on an automated ribosomal intergenic spacer analysis (ARISA), the fungal community composition of *J. roemerianus* appears to be different from the one of *Spartina alternifora* in the US saltmarsh. This suggests a host/substrate specificity of these plants (Torzilli et al. 2006). A total of 136 taxa have been recorded on *J. roemerianus* (Calado and Barata 2012); many of them are not marine but grow on the terrestrial, exposed parts of the plant. Common taxa on *J. roemerianus* include *Loratospora aestuarii*, *Papulosa amerospora*, *Aropsiclus junci*, *Anthostomella poecila*, *Physalospora citogerminans*, *Scirrhia annulata*, *Massarina ricifera*, and *Tremateia halophile* (Newell and Porter 2000). Intertidal *P. australis* support 109 species of marine fungi, although over 300 fungal species have been documented on this plant (Calado and Barata 2012). Common fungi are *Cladosporium* spp., *Collectotrichum* sp., *Didymella glacialis*, *Halosarpheia phragmiticola*, *L. laevis* (Fig. 4.1f), *Phaeosphaeria* sp., *Phoma* sp., *Phomatospora berkeleyi*, *Phomopsis* sp., *Septoriella* spp., and *Trichoderma* sp. (Luo and Pang 2014), most of which are asexual fungi.

N. fruticans is a palm tree that can be found at low salinity areas of estuaries. Hyde (1992) discovered 43 species on such substrates in Brunei, while a recent figure report 135 species with 90 Ascomycota, 3 Basidiomycota, and 42 asexual fungi (Loilong et al. 2012). A higher fungal diversity occurred on leaf base compared to the other tissues of this intertidal plant, including inflorescence, leaf, leaf midrib, rachis, and aerial parts (Hyde and Alias 2000). Host specificity is pronounced with an estimated 40 endemic species (Hyde and Alias 2000). For example, *Aniptodera intermedia* and *Linocarpon appendiculatum* was found only on *N. fruticans* although this palm grows alongside other mangrove tree species (Besitulo et al. 2010; Loilong et al. 2012).

Sea grasses have also been examined for endophytic fungi albeit colonization frequency and diversity were relatively low compared to terrestrial plants (Alva et al. 2002; Devarajan et al. 2002). This can be attributed to the unfavorable physical (low oxygen) or chemical (high salinity) conditions of seagrass beds curbing fungal infection, or interference competition with other marine microorganisms such as diatoms and bacteria (Venkatachalam et al. 2015). Ascomycetes mostly belonging to the Eurotiales, Hypocreales, and Capnodiales appear dominant on seagrasses, (Sakayaroj et al. 2012; Venkatachalam et al. 2015). Frequent genera are *Aspergillus*, *Cladosporium*, *Paecilomyces*, and *Penicillium*, which are common asexual fungi of seawater and sediment (Sakayaroj et al. 2012; Venkatachalam et al. 2015). The few basidiomycetes found on *Enhalus acoroides* may represent mycorrhizal relationships (Sakayaroj et al. 2010), suggesting that fungal communities occurring on plant-based habitats are diverse and play different ecological roles.

Algae. Marine macroalgae form a diverse and ubiquitous group of photosynthetic organisms that contribute importantly to global primary production. They have essential functions in nutrient cycling and represent a specific habitat strongly impacting marine life from the ocean surface to the seafloor (Dayton 1985). Indeed, as other eukaryotic organisms, macroalgae harbor associated microorganisms and evidence is accumulating that macroalgae interact with microbial communities for their growth, defense, development, and nutrient supply (Dittami et al. 2014).

Based on these findings emerged the concept of holobionte, defined as a "super-organism," which includes the entity of the macroalga and its associated microbial communities (Egan et al. 2013). While the vast majority of studies related to the microbiome of macroalgae are focusing on bacteria, other host-associated microbes have also been described, including fungi. Indeed, after sponges algae represent the second most important source of fungi that are responsible for the production of a huge diversity of natural products (Bugni and Ireland 2004; Schulz et al. 2008).

The biotic interactions between fungi and their algae host are highly diverse, ranging from parasitism to symbiosis (Jones and Pang 2012; Potin et al. 2002). Up to now, pathogenic and saprophytic fungi associated with seaweed have been described (Kohlmeyer and Demoulin 1981; Kohlmeyer and Volkmann-Kohlmeyer 2003b) and many diseases and infections in the marine environment have been assigned to fungi (Zuccaro and Mitchell 2005). Thus, the well-known fungal pathogens *Lindra thalassiae* induce the softening and collapsing of the air vesicles in brown algae *Sargassum* sp. (Kohlmeyer 1971) while *Guignardia gliopeltidis* is responsible for the well-known black dot disease (Andrews 1976). The fungus *Haloguignardia irritans* associated with the inner tissue of the *Cystoseira* sp. and *Halidrys* sp. brown algae induce galls formation in its hosts (Harvey and Goff 2010), whereas *Phycomelaina laminariae* cause the stipe blotch in kelps (Schatz 1983). Finally, marine protists belonging to the Labyrinthulomycetes class are among the most described fungal-like pathogens of algae, infecting the green algae *Cladophora* sp. and *Rhizoclonium* sp. (Raghukumar 1986).

Fungi also maintain symbiotic relationships with algae as in the mycophycobiosis of *Stigmidium ascophylli* (formerly *Mycophycias ascophylli*, Aproot 2006) and the brown algae *Pelvetia caniculata* and *Ascophyllum nodosum* (Fig. 4.2) (Jones and Pang 2012). In such interactions, host and fungus appear to depend on each other for their survival, as they never occur separately in nature (Kingham and Evans 1986). It is hypothesized that the fungus protects the algae host from desiccation (Garbary and MacDonald 1995). In the same vein, the ascomycete *Turgidosculum ulvae* was specifically found associated with the inner algal tissue of the green alga *Blidingia minima*, causing black discolorations, which are never eaten by grazers (Kohlmeyer and Volkmann-Kohlmeyer 2003c). Historically, the asymptomatic fungi associated with algae were mainly described through microscopy study and in situ observations (Kohlmeyer and Demoulin 1981). The

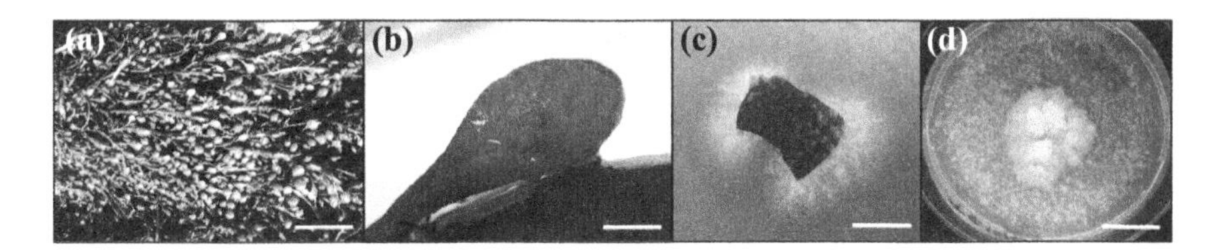

Fig. 4.2 Isolation of a marine fungus from *Ascophyllum nodosum*. **a** *A. nodosum*. **b** Receptacle of *A. nodosum*. **c** Marine fungus growing on the *A. nodosum* receptacle. **d** Culture of a marine fungal isolate. *Scale bars* **a** 10 cm, **b**, **c** 1 cm, **d** 1.5 cm

development of molecular identification has become a powerful tool for the identification of fungi. This has led to the identification of the Ascomycetes and anamorphic fungi as the dominant fungal endophytes of seaweeds. However, description of the biodiversity of asymptomatic marine fungi associated to algae remains an underexplored topic (Flewelling et al. 2013; Fries 1979; Harvey and Goff 2010; Loque et al. 2010; Zuccaro et al. 2003, 2008) and still relies mainly on culture-based approaches. Molecular analyses based on the large subunit rRNA PCR-DGGE of healthy and the decaying *Fucus serratus* allowed the identification of representatives of the Halosphaeriales, Lulworthiales, Hypocreales, and Pleosporales (Zuccaro et al. 2003, 2008) and this started a new era in the study of the fungal community associated to algae. Thus, our current understanding of algal-fungus relationships is still limited and the ecological role of fungi associated to algae in their health and function is speculative and is largely unknown. This lack of knowledge opens attractive new perspectives for deciphering the chemical ecology of fungi associated with algae using multidisciplinary approaches.

Deep-sea. Deep-sea environments harbor a large diversity of fungal species, some endemic such as the recently described *C. oceani* (Burgaud et al. 2011) or *R. pacifica* (Nagahama et al. 2006) isolated from deep-sea hydrothermal vents or the deep sunken wood-associated *Alisea longicolla* (Fig. 4.3a, b) and *Oceanitis scuticella* (Fig. 4.3c, d) (Dupont et al. 2009), but a large fraction of the fungal species or sequences are ubiquitous organisms. DNA signatures of the genus *Malassezia* have been predominantly retrieved from deep-sea samples (Bass et al. 2007; Jebaraj et al. 2010; Lai et al. 2007; Le Calvez et al. 2009; Lopez-Garcia et al. 2007; Nagahama et al. 2011; Singh et al. 2011; Sohlberg et al. 2015; Stock et al. 2012; Xu et al. 2014). Ribosomal RNA-based approaches also revealed metabolic activities of the genus *Malassezia* in deep-sea sediments (Edgcomb et al. 2011; Rédou et al. 2014). Phylogenetic analysis has demonstrated the occurrence and diversity of *Malassezia*-like sequences from marine environments (Richards et al. 2012). However, to date no isolates of *Malassezia* have been cultured using marine environment samples. *Malassezia* is a well-known genus from terrestrial environments and known as causative agent of skin diseases (Amend 2014). An emerging hypothesis is that

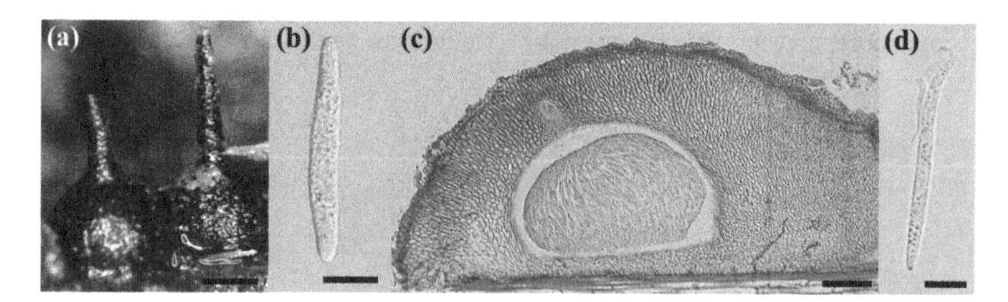

Fig. 4.3 Deep-sea marine fungi on sunken wood. **a** Superficial ascomata of the deep-sea fungus *Alisea longicolla*. **b** Cylindrical ascospore of *A. longicolla*. **c** A parafilm section showing thick-walled ascoma of the deep-sea fungus *Oceanitis scuticella*. **d** Cylindrical ascospore of *O. scuticella* with a unipolar unfurling appendage. *Scale bars* **a** 500 μm, **b–d** 10 μm

deep-sea *Malassezia* species may be opportunistic pathogens of deep-sea marine invertebrates. Indeed, Gao et al. (2008) revealed a high diversity of *Malassezia* lineages associated with marine sponges. Future studies are required to reveal knowledge about the ecological role of *Malassezia* in the marine environment. Other examples of ubiquitous fungi in deep-sea habitats include species belonging to the well-known *Cryptococcus* and *Rhodotorula* genera. *Cryptococcus* has been reported from deep-sea environments including polar regions (Connell et al. 2008; Turchetti et al. 2008), cold seeps (Takishita et al. 2006, 2007), hydrothermal vents (Burgaud et al. 2010; Le Calvez et al. 2009), sediments (Burgaud et al. 2013; Edgcomb et al. 2011), or the deep biosphere (Orsi et al. 2013a; Rédou et al. 2014). Almost half of the sequences recovered from RNA-based libraries from deep-sea sediment cores were closely related to *Cryptococcus* and that suggests that this phylotype is highly abundant but also metabolically active in these sedimentary ecosystems (Edgcomb et al. 2011). This is consistent with another molecular analysis of microeukaryotic diversity in deep-sea methane-rich sediments that showed that the yeast *Cryptococcus curvatus* with a closely related genotype to those recovered from deep sediments (Edgcomb et al. 2011) can dominate deep-sea microbial eukaryotic communities (Takishita et al. 2006). Positive correlation of representative sequences of the *Cryptococcus* genus with methane/ethane has been found in the deep biosphere. This supports the idea that such basidiomycetes might be indirectly involved in the deep subseafloor methane cycle (Rédou et al. 2014) and interact with deep-sea methanogenic/ethanogenic archaea/bacteria (Hinrichs et al. 2006). *Rhodotorula mucilaginosa* appears widely distributed and common in deep-sea environments (Bass et al. 2007; Burgaud et al. 2010; Gadanho and Sampaio 2005; Nagahama et al. 2001; Singh et al. 2011). This (red) yeast, known as halotolerant (Lahay et al. 2002), was the most abundant one in this environment (Rédou et al. 2015). Cultivation of *Rhodotorula rubra* under simulated deep-sea conditions has shown growth at 40 MPa, corresponding to the pressure present at depths of 4000 m (Lorenz and Molitoris 1997). Although representative yeasts belonging to *Rhodotorula* are ubiquitous, they are able to withstand high hydrostatic pressures and thus seem to be adapted to deep-sea conditions (Burgaud et al. 2015).

4.4 Adaptation of Marine Fungi

Fungi have been found in different marine habitats, from coastal waters to the deep subseafloor. The detection of numerous fungi related to known terrestrial groups is striking and suggests effective adaptive capabilities to many biotic and abiotic stresses. Adaptation is here defined as any adjustment of an organism that makes it better suited to live in the environment to which it adapted. As the deep-sea is characterized by many environmental constraints, we summarize the adaptation strategies of fungi that allow them to cope with deep-sea environmental conditions and thrive in this model habitat.

Deep-sea habitats are extreme environments where harsh in situ conditions may be summarized by (i) high hydrostatic pressure, (ii) extreme temperature gradient, and (iii) variable sea salt concentrations (from 3 % to close to saturation).

(i) Many factors influence the biodiversity in the marine environment, but hydrostatic pressure appears as a key physical parameter in deep-sea ecosystems (deep ocean, cold seeps, hydrothermal vents, and deep sediments). Pressure increases gradually with 10 MPa for every kilometer of depth and affects all biological reactions that involve volume modifications. Although piezophilic fungi have not yet been isolated from deep-sea environments, some deep-sea fungal isolates need enrichment at elevated hydrostatic pressure in order to initiate growth (Rédou et al. 2015). Few studies have examined spore germination and mycelial growth of ubiquitous filamentous fungi and yeasts under elevated hydrostatic pressures (Damare and Raghukumar 2008; Damare et al. 2006; Lorenz and Molitoris 1997; Raghukumar and Raghukumar 1998; Singh et al. 2011). High-pressure effects on biological processes are diverse such as for instance pressure-sensitive lipids that modify fluidity, and permeability and cell membrane functions. Some fungi adapt to elevated hydrostatic pressure by actively setting up defense mechanisms that counteract the pressure effects. Membrane composition modifications have been observed in *Saccharomyces cerevisiae*, a model yeast that increased the proportion of unsaturated fatty acids and ergosterol (Simonato et al. 2006). After 30 min of high (200 MPa) hydrostatic pressure, *S. cerevisiae* upregulated *ole1* and *erg25* gene expression in order to increase the proportion of unsaturated fatty acids and ergosterol biosynthesis, respectively (Fernandes et al. 2004). This strategy aims to increase membrane fluidity and to maintain the functionality of cell membranes, suggesting that unsaturated fatty acids and ergosterol could be important for this process.

(ii) In the ocean, temperature decreases with depth until 2–4 °C, whereas in the subseafloor sediments the average thermal gradient is 25–30 °C/km (Oger and Jebbar 2010). Most fungi are mesophilic (growing at moderate temperatures). However, few fungal species can grow at low or high temperature (Deacon 2005). Many fungi are able to adapt to cold environments such as polar regions (Robinson 2001) or the ocean (Kohlmeyer and Kohlmeyer 1979) with a stable temperature of a few degrees Celsius. High concentrations of low-molecular solutes such as trehalose often accumulate in psychrophilic or psychrotrophic fungi in response to low temperatures. Trehalose is thought to act as a general stress protectant in the cytosol and is known to stabilize membranes during dehydration. Cryoprotectants, polyols such as glycerol and mannitol also tend to accumulate in response to stress conditions (Robinson 2001). Thermal stress also involves modifications in the fatty acid composition of the membrane lipids to ensure optimal membrane fluidity for the functional integrity of membrane transporters and enzymes. In response to high temperatures, heat-shock proteins can be synthesized in high concentrations. These proteins act as chaperones ensuring that the cell proteins are correctly folded to prevent from protein damages (Deacon 2005).

(iii) Some of the most extreme marine ecosystems known are the Deep Hypersaline Anoxic Basins (DHABs) (Stoeck et al. 2014). Indeed, the combination of extreme physicochemical parameters, including high-hydrostatic pressure, anoxia, high-sulfide concentration, and nearly saturated salt concentration that results in low water activity, were thought to be hostile for life. However, over the last decade, these DHABs have been shown to harbor communities of bacteria and archaea (See Daffonchio et al. 2006; La Cono et al. 2011; van der Wielen et al. 2005; Yakimov et al. 2007a, b) and microbial eukarya (Edgcomb and Orsi 2013). Small subunit rRNA sequence analyses based on amplification of environmental cDNA identified fungi as the most abundant and diverse taxonomic group of Eukarya in the lower haloclines of DHAB water columns (Pachiadaki et al. 2014; Stock et al. 2012). Such studies revealed only sequences belonging to Ascomycota and Basidiomycota phyla with basidiomycetes as the dominant group. To date, no fungal cultures have been isolated from DHABs. However, many species such as *Debaryomyces hansenii*, *Hortaea werneckii*, and *Wallemia ichthyophaga* have been isolated from natural hypersaline environments (Gunde-Cimerman et al. 2009), suggesting that it may be possible to isolate halophilic/halotolerant fungal strains from DHABs. Such species have developed different strategies to cope with ion concentration toxicity and low water availability. Ecophysiological and molecular methods allowed the identification of specific adaptations. Indeed, isolated fungi were able to synthesize and accumulate glycerol or different polyols as compatible solutes for maintaining their intracellular concentrations of Na^{+} below the toxic levels for the cells (Hohmann 2002). Changes in the membrane composition and properties also have important functions in the adaptation to osmotic stress. Indeed, an increase in sterol-to-phospholipid ratio and fatty acid unsaturation increases the membrane fluidity (Turk et al. 2004, 2007).

Fungi are highly adaptive microorganisms able to withstand many physicochemical stressors that shape various marine habitats. The many ways of coping with pressure, temperature, and salinity seem to explain why so many ubiquitous fungal species have been retrieved from numerous marine habitats, including the deep-sea. Such adaptations and the fact that marine habitats are governed by complex biotic interactions that can be allegorized as a chemical war with a huge diversity of bioactive compounds, clearly position marine fungi as an interesting untapped resource of biodiversity and biotechnological potential of secondary metabolites.

4.5 Inferring Ecological Roles and Dynamics of Marine Fungal Communities Using Omics

4.5.1 Secondary Metabolites, a Definition

Living organisms produce small molecules called natural products that are generally biosynthesized by enzymes from simple building blocks available in the cell. Such natural products can be classified into primary and secondary metabolites

(SMs). Primary metabolites are essential to drive the activities of the cells and involve metabolic pathways more or less common to all living organisms. Secondary metabolites fulfill other functions and are thought to be beneficial to the producing organism and exhibit an astonishing chemical diversity. These compounds have a much more limited distribution in nature compared to primary metabolites and are only found in specific organisms or taxonomic groups. Although the function of most of these secondary metabolites is unknown, it is commonly accepted that some of them play a vital role in the well being of the producing organism.

Specific enzymes encoded by biosynthetic gene clusters (BGCs) produce in most cases fungal secondary metabolites. These enzymes assemble simple building blocks derived from the primary metabolism to complex and reactive chemical compounds. Fungi derived from marine sources are considered to represent a huge reservoir of biologically active secondary metabolites and are often produced by multifunctional enzyme complexes such as PolyKetide Synthases (PKSs) and Non-Ribosomal Peptide Synthetases (NRPSs). Rapid progress in metabolomics and genome mining stimulates the full characterization of BGCs and rationalizes the search for effective producers of secondary metabolites. Indeed, the advent of metabolomics and genomics dedicated to marine fungi has already demonstrated that their genome encode for the biosynthesis of many unknown metabolites and thus represent an untapped resource of new natural compounds.

4.5.2 New Methods to Access the Marine Fungal Metabolome

It is generally admitted that a small part of the marine fungal metabolic potential is observed with classical experiments consisting of cultivating a fungus as axenic culture in a common medium over a certain period of time. Under laboratory conditions, only subsets of fungal biosynthetic pathways encoding for secondary metabolite production are ever expressed. This is greatly limiting the potential of drug discovery from fungi (Brakhage 2013). Since marine fungal DNA features many secondary metabolites clusters coding for the biosynthesis of yet unidentified products, antisilencing approaches could be essential to trigger their production by fungi in the laboratory and thus help to identify their nature and biological effects.

Many research teams have investigated ways to unravel cryptic biosynthetic pathways in order to access a wider chemodiversity (Gram 2015; Pettit 2011; Reen et al. 2015). Different methods, either directly or indirectly targeting transcription and translation processes in the cells, have been applied with interesting results as reported below (and summarized in Fig. 4.4). Such approaches may generate large amounts of data, which need considerable computer power to analyze the metabolome. This corresponds to the advent of the "metabolomics" and "genomics" studies for which bioinformatics is essential for data mining.

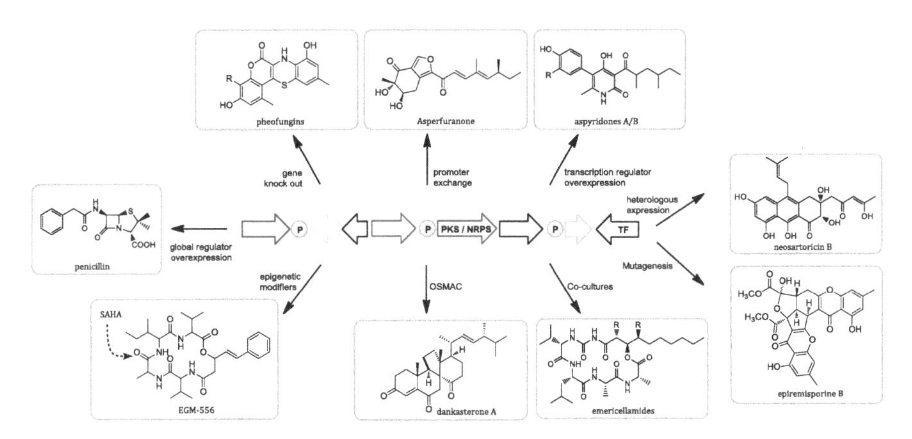

Fig. 4.4 Example of the recent strategies used to discover novel fungal secondary metabolites by the activation of cryptic gene clusters. *P* promoter; *TF* transcriptional factor. All the molecules presented here correspond to molecules previously cited in the text

4.5.2.1 Methods, not Based on Culture, to Access the Marine Fungal Metabolome

Recent advances in DNA sequencing technologies have enabled multiple genomes to be sequenced rapidly and inexpensively. First focusing on Bacteria and Archaea, these technologies now have been extended to microbial Eukarya such as fungi including those of well-known producers of secondary metabolites, such as *Aspergillus oryzae* and *Penicillium chrysogenum* (Ma and Fedorova 2010). This has shined new light on the synthetic capabilities of these organisms, revealing metabolic pathways that are silent under standard culture conditions. Discovering this hidden metabolism is one of the major goals to reach an exhaustive description of the marine fungal chemodiversity and the non-culture-based approaches to achieve this are promising (Brakhage and Schroeckh 2011; Brakhage et al. 2008).

– Genome mining: silent biosynthetic gene clusters screening

Between 2006 and 2014, hundreds of new chemical structures have been described from marine fungi (Ebada and Proksch 2015; Rateb and Ebel 2011). While the major compounds in fungi were polyketides, peptides, and terpenoids, now novel alkaloids, lipids, and shikimates were newly discovered. The pathways responsible for the biosynthesis of these major metabolites have been identified and characterized, and key biosynthetic genes encoding enzymes such as PKSs, NRPSs, or TerPene Synthases (TPSs) (Medema et al. 2011; Yamada et al. 2012; Ziemert and Jensen 2012) are now well described.

BGCs can be screened using PCR-based approaches. Different primers have been designed to target specific genes involved in the synthesis of secondary metabolites such as type I PKS (Amnuaykanjanasin et al. 2005; Bingle et al. 1999; Proctor et al. 1999), TPS (Kawaide et al. 1997), NRPS (Amnuaykanjanasin et al.

Table 4.1 Primers used to amplify targeted genes

Primers	Targeted gene	Sequence (5′–3′)	References
KAF1	PKS I	GAR KSI CAY GGI ACI GGI AC	Amnuaykanjanasin et al. (2005)
KAR1		CCA YTG IGC ICC RTG ICC IGA RAA	
XKS1	PKS/NRPS	TTY GAY GCI BCI TTY TTY RA	Nicholson et al. (2001)
XKS2		CRT TIG YIC CIC YDA AIC CAA A	
CHS1	PKS III	GAY-TGG-GCI-VTN-CAY-CCB-GGI-GGD	Rédou et al. (2015)
CHS2		YTC-NAY-NKT-RAK-VCC-IGG-VCC-RAA	
AUG003	NRPS	CCGGCACCACCggnaarcchaa	Slightom et al. (2009)
AUG007		GCTGCATGGCGGTGATGswrtsnccbcc	
TPS1	TPS	GCI TAY GAY ACI GCI TGG GT	Kawaide et al. (1997)
TPS2		RAA IGC ATI GCI GTR TCR TC	
LC1	PKS I	GAY CCI MGI TTY TTY AAY ATG	Bingle et al. (1999)
LC2c		GT ICC IGT ICC GTG CAT TTC	

Degenerated bases and inosine are described: *R* = A/G, *Y* = C/T, *B* = no A (C/G/T), *K* = G/T, *N* = A/T/G/C, *W* = A/T, *S* = C/G, *I* = Inosine, *M* = A/C, *H* = no G (A/T/C), *V* = no T (A/C/G), *D* = no C (G/T/A)

2009; Slightom et al. 2009), and hybrids PKS/NRPS (Nicholson et al. 2001) as described in Table 4.1. A screening for the presence/absence of such genes in a sample/microbial community/strain allows estimating its putative biotechnological potential, i.e., the synthesis potential of putatively interesting secondary metabolites (Rédou et al. 2015). As a complementary approach, amplification and sequencing of such genes coupled with a phylogenetic analysis allow determining the functions and/or novelty of BGCs (Bushley and Turgeon 2010; Cacho et al. 2014). However, because the presence of genes does not give evidence of their expression, this approach needs to be complemented by semiquantitative RT-PCR or transcriptomic analyses in order to ensure the expression of the targeted genes under one or several specific culture conditions (Fisch et al. 2009; Umemura et al. 2013). Another approach called “The Natural Product Proteomining” has been proposed (Gubbens et al. 2014) with the aim to combine metabolomics and quantitative proteomics in order to identify relevant BGCs and to better associate gene clusters with natural products.

– Genetic engineering

Following genome mining resulting in the identification and location of a silent biosynthetic gene cluster, the key issue for the success of such approaches is to find ways to induce or enhance their expression. Different methods have been reported for microorganisms and will be discussed here, with regard to their potential application to fungi.

Heterologous expression

One method that has been successfully applied to bacteria and yeast is heterologous expression of the targeted gene clusters in organisms such as *Escherichia coli* or *S. cerevisiae*. A 67-kb NRPS biosynthetic gene cluster from the marine actinomycete *Saccharomonospora* sp. CNQ-490 was expressed in this way and allowed the discovery of the lipopeptide antibiotic taromycin A (Yamanaka et al. 2014). Another study focused on silent biosynthetic polyketide gene clusters in fungi belonging to *Trichophyton* and *Arthroderma* genera and led to the isolation of neosartoricin B after heterologous expression in *Aspergillus nidulans* (Yin et al. 2013). However, the application of this strategy shows some limitations such as difficulties in handling large gene clusters or finding appropriate expression hosts. Especially in fungi, the heterologous production of biosynthetic enzymes can be cumbersome, emphasizing the need to find alternative or complementary techniques (Schümann and Hertweck 2006). Although, it may take a while until fungal BGCs heterologous expression techniques expand as much as those for bacterial, it is an emerging field which promises to be highly rewarding.

Homologous expression

Silent gene clusters can be awaken in a given organism by activating gene transcription using inducible promoters or promoter exchange in homologous recombination experiments, as performed by Chiang et al. (2009) leading to the novel polyketide asperfuranone in *A. nidulans*. Bypassing the limitations of homologous recombination, overexpression of a transcription regulator by ectopic integration is also an interesting way to express silent gene clusters as successfully applied by Bergmann et al. (2007). The latter approach allowed the identification of novel PKS-NRPS metabolites, Aspyridone A and B, from *A. nidulans*.

Gene knockout and epigenetics

The manipulation of global regulators, including enzymes involved in post-translational modifications, can also be used to unravel the secondary metabolome. Deletion of N-acetyltransferase gene in *A. nidulans* led to the formation of novel pheofungins (Scherlach et al. 2011). Another prominent example of manipulation of a global regulator is the overexpression of *laeA*, which led to increased production of various secondary metabolites in several fungi, such as penicillin in *A. nidulans* and *P. chrysogenum* (Bok and Keller 2004; Kosalková et al. 2009), aflatoxin in *A. flavus* (Kale et al. 2008), and T-toxin in *Cochliobolus heterostrophus* (Wu et al. 2012). So far, most of these concepts and molecular biology approaches have been applied and proved successful on model organisms such as *A. nidulans*. They may apply in the future to many other fungi including marine fungi. However, the requirements for complex gene manipulations in those approaches have restricted their applications in general microbial product research.

- Mutagenesis/Ribosome engineering

The concept of bacterial ribosome engineering was introduced by Ochi (2007) and arose from observations of bacteria in stress conditions. The experiments suggested that bacterial gene expression can be altered by modulating ribosomal proteins or rRNA, which can lead to the activation of dormant genes. One way to modulate the ribosome is the use of drug-resistance mutation techniques, such as the streptomycin resistance conferred by an altered ribosomal S12 protein. This has been well studied in bacteria and has led to the discovery of different new compounds such as piperidamycins (Hosaka et al. 2009; Ochi and Hosaka 2012). This strategy has been applied to fungi and more specifically also to marine fungi. Introduction of gentamicin resistance in *Penicillium purpurogenum* G59 altered its secondary metabolome leading to compounds that had never been reported from this organism before (Chai et al. 2012). The deep-sea *A. versicolor* ZBY-3 with acquired resistance to neomycin was also reported to produce new metabolites including the new phenethyl 5-oxo-L-prolinate (Dong et al. 2014). Finally, diethyl sulfate mutagenesis of the marine fungal strain *Penicillium purpurogenum* G59 led to three new chromone derivatives, epiremisporine B, epiremisporine B1, and isoconiochaetone C (Xia et al. 2015). With less technical constraints, this modulation of ribosomal function by mutagenesis may be applicable to a variety of other fungi and elicit their potential secondary metabolism.

4.5.2.2 Culture-Based Methods to Access the Marine Fungal Metabolome

- Epigenetic modifiers

Modulation of the structure of the chromatin organization is critical for the regulation of gene expression, since it determines the accessibility and the sequential recruitment of regulatory factors to the underlying DNA. The less compacted euchromatin regions are thus most accessible for transcription, whereas highly compacted heterochromatin regions are refractory to transcription. Accordingly, the same gene sequence can be either well expressed or transcriptionally silent depending on whether it lies in euchromatin or heterochromatin.

Depending on the transcriptional states, the structure of the chromatin may be altered, among others, by covalent modifications of its constituents such as DNA methylation at cytosine residues, and post-translational modifications of histone tails. These stable alterations are named "epigenetic" because they are heritable in the short term but do not involve mutations of the DNA itself.

Fungi are biosynthetically talented organisms capable of producing a wide range of chemically diverse and biologically intriguing small molecules. In addition, recent genome sequencing of fungi showed that many putative natural product BGCs exist in a heterochromatin state whose constitutive genes are often transcriptionally controlled by administering small-molecule epigenetic modifiers

(Williams et al. 2008). Accordingly, the uses of compounds able to inhibit DNA and histone, post-translational modifying proteins have thus emerged as a seductive approach to manipulate the fungal metabolome. Indeed, many small molecules acting as epigenetic modifiers have been reported and their addition to growth media leads to the production of new natural products and/or to enhanced level expression of secondary metabolites from fungi (Cichewicz 2010).

The field of marine fungal metabolomics has not escaped such approach even though few examples have been reported. For instance, addition of 5-azacytidine (a DNA methyltransferase inhibitor) to the culture medium to grow an Atlantic-Forest-Soil-Derived *Penicillium citreonigrum* fungus, resulted in a broth that was highly enriched in meroterpenoids and azaphilones compounds (Wang et al. 2010). In another case, the addition of the histone deacetylase inhibitors suberoylanilide hydroxamic acid (SAHA) turned on the biosynthesis of EGM-556, a new cyclodepsipeptide of hybrid biosynthetic origin produced by fungus *Microascus* sp., isolated from marine sediment from Florida (Vervoort et al. 2011). Finally, sodium butyrate (another histone deacetylase inhibitor) has also been successfully used as an epigenetic modifying agent for the modulation of the metabolome of marine fungi (Beau et al. 2012).

The addition of different types of elicitors or chemical agents in culture media, for which the mechanism of action is not well described, could fall into this category of epigenetic modifiers as they may also interfere with this type of regulation. However, classification of these agents is not straightforward as they could also, depending on their concentration, just impose stress on the organisms. When added in huge amounts they lead to dramatic changes in the growth medium composition and act in the same way as nutrients changes, which can be used in the OSMAC (One Strain MAny Compounds) approach (described below). Nevertheless, the use of elicitors or chemical agent addition in culture media is also an interesting field to investigate to access cryptic gene clusters of fungal secondary metabolites (Pimentel-Elardo et al. 2015).

- OSMAC approach

The regulation of expression of fungal biosynthetic enzymes by nutrients is a well-known phenomenon. It has been shown that ammonium ions greatly affect the production of β-lactam antibiotics, whereas L-asparagine and L-arginine enhance their yields (Shen et al. 1984). A number of polyketide biosyntheses is repressed by high contents of nitrogenous compounds in the culture medium (Rollins and Gaucher 1994). On the contrary, manganese ions are mandatory for the expression of the isoepoxydon dehydrogenase gene implicated in the synthesis of patuline intermediates (Puel et al. 2010; Scott et al. 1986).

Manipulating the growth conditions of marine fungi in order to stimulate the production of secondary metabolites encoded by silent BGCs and/or to enhance the production of constitutively produced secondary metabolites has been the initial postulate of the group of Axel Zeeck (University of Oldenburg, Germany). His group developed the concept of what is now a paradigm in fungal natural products research: the One Strain MAny Compounds (OSMAC) approach (Bode et al. 2002). This methodology resulted from the observation that even small changes in the cultivation conditions may completely shift the metabolic profile of microorganisms. This general statement depends on the species of microorganism but is particularly relevant for actinomycetes and fungi. In a screening program, the systematic alteration of culture parameters is now routinely used as an effective way to expand the structural diversity of metabolites in a single fungal strain. These culture parameters may include a change from static to agitated fermentation, the use of liquid or solid media, variations of pH, temperature, light quality (UV) and quantity, oxygen supply, nitrogen and carbon source, use of more or less mineral or metallic salts, or addition of precursors. The use of both seawater and distilled water alternatively is also of interest when revealing the chemical machinery of marine fungi. In this way, it has been shown that employing 8–12 culture conditions for each strain statistically enhances the probability of triggering the production of unusual compounds (Bills et al. 2008).

Since the initial publication by Bode et al. (2002), numerous results obtained using the OSMAC approach have demonstrated its effectiveness, not only in terms of discovery of new compounds, but also of novel chemical skeletons. In this way, Guo et al. (2013) showed that, when changing from static to agitated fermentation condition, the production of meleagrin and roquefortines by a deep-sea strain of *Penicillium* sp. was stopped, whereas various sorbicillinoids alkaloids appeared in the HPLC chromatograms. Investigating a single strain of a sponge fungus *Stachylidium* sp. and with only small changes in the duration of fermentation (40 vs. 60 days) and the concentration of biomalt (10 g or 20 g/L) added to the agar, the group of Gabriele König isolated either prenilated phtalides [marilones A-C (Almeida et al. 2011a)], O-prenilated tyrosine-derived compounds [stachylines A-D (Almeida et al. 2011b)], or a series of unusual phtalide-related polyketides [cyclomarinone, maristachones and marilactone (Almeida et al. 2013)]. The use of another culture medium yielded a fully different series of phtalimidine derivatives (Almeida et al. 2012). Lin et al. (2009) described for the first time the production of two types of cytochalasins, macrocyclic polyketides bearing a cyclized amino acid feature, by a single marine *Spicaria elegans* strain. By manipulating the culture conditions (a panel of 10 media was screened), these authors observed a selective regulation of the use of the amino acids involved in the biosynthesis of cytochalasans, leading to the inclusion of a phenylalanine residue for one type, and a leucine one for the other (Lin et al. 2009). In this way, they showed that one of the media abolished the production of cytochalasins but induced the biosynthesis of spicochalasin A and aspochalasins. A strain of *Penicillium* sp. isolated from a green alga and cultivated on a yeast extract-based seawater medium produced two different classes of novel alkaloids, communesins (Numata et al. 1993) and penochalasins (Numata et al. 1996). The replacement of yeast extract by malt extract and seawater by distilled water led to the discovery of completely different major compounds produced by the same *Penicillium* sp., the polyketide penostatins (Iwamoto et al. 1998; Takahashi et al. 1996). Applying a similar strategy, the same research group shifted the steroid metabolism of a *Gymnascella dankaliensis* isolated from a *Halichondria* sponge by using soluble starch instead of glucose in a malt-yeast extract medium (Amagata et al. 2007). These authors isolated dankasterones, intriguing novel steroids based on a rare four-ring skeleton, whereas the also unusual steroid alkaloid gymnasterone A and steroids gymnasterones B-D were initially obtained. Several studies pointed out the importance of using host-derived or sample-derived media (meaning adding to culture growth media a sample or an extract from the sampling site of the fungal strain) to mimic the environmental conditions and then reveal its influence on the expressed metabolome (Geiger et al. 2013; Overy et al. 2005).

- Cocultivation

The marine environment, which is characterized by its salinity, temperature, hydrostatic pressure, and which hosts a multitude of living macro- and microorganisms can be considered as harsh and competitive. It can be assumed that marine fungi coming from such environment invest a lot of efforts in secondary metabolites production in order to be able to survive. Indeed, they may have to produce compounds for their defense such as antibacterial, or they may use metabolites to communicate and establish beneficial relationships with other organisms. However, under standard laboratory growth conditions these metabolites are often not expressed and the corresponding BGCs may remain silent. One strategy to mimic the natural environment of these fungi is to grow them in the presence of other microbes. These mixed fermentations, while used for centuries in food and beverage production, have only been applied to the field of natural product discovery, since the last decade (Pettit 2009). One notable exception is the unintended co-culture of *Penicillium notatum* (now recognized as *Penicillium chrysogenum*) and *Staphylococcus aureus* in the same Petri dish that marked the discovery of penicillin, the world's first antibiotic (Fleming 1929). Recent reviews on co-cultures have discussed this approach in detail (Bertrand et al. 2014b; Goers et al. 2014; Marmann et al. 2014; Netzker et al. 2015). Most experiments co-culture two fungi or a fungus and a bacterium using liquid or solid media. In some cases, no contact between the colonies seems to be required to stimulate the production of new compounds (Bertrand et al. 2013a) while in other cases, intimate physical interaction between the organisms is required to elicit a response (Schroeckh et al. 2009). Different reports of new compounds produced through co-culturing suggest a specific response to the competing organism rather than a general reaction (Ola et al. 2013). Some successful examples from marine fungi include pestalone produced by a marine fungus in response to bacterial challenge (Cueto et al. 2001), emericellamides A and B (Oh et al. 2007), libertellenones A-D (Oh et al. 2005), marinamide and methyl marinamide (Zhu et al. 2013; Zhu and Lin 2006), aspergicin (Zhu et al. 2011), (−)-Byssochlamic acid bisdiimide (Li et al. 2010), a xanthone derivative (Li et al. 2011), and new cyclopeptides (Li et al. 2014).

- Time-scale studies

The above-mentioned approaches to stimulate the fungal metabolism through culture modulation usually compare the metabolic profiles generated after a selected period of time. However, in every organism biochemical processes vary with time. Most organisms acclimate their metabolism to the prevailing environmental fluctuations such as diurnal changes. Most biosynthetic processes are time-dependent because the enzymes needed for the various reactions involved have to be synthesized, modified, or degraded depending on the prevailing condition. Consequently, since many metabolites are the products of enzyme-catalyzed reactions and have a finite half-life, it is safe to assume that these metabolites will naturally vary in a dynamic way. The metabolome expressed by cultured fungi may therefore also vary considerably just depending on the factor time. Time-course metabolomics studies would therefore be an interesting approach to browse the vast array of metabolites, an organism which is able to produce and then to access a wider chemodiversity. Moreover, elucidating temporal relationships among genes and metabolites in relation to the growth of the fungus and its culture conditions could also lead to a better understanding of the way biosynthetic processes occur.

Time-series metabolomics studies then appear to be an interesting field to investigate with regard to the potential insights into the metabolism and biosynthetic pathways and complement the information obtained from proteomics and transcriptomics (Alam et al. 2010; Soanes et al. 2011). Many examples are given in the literature for diverse organisms and in different contexts. Most of the time, these examples consist of targeted analyses (on selected compounds), which purpose is to follow the behavior of an organism after a treatment or after imposing stress (Jones et al. 2010; Kim et al. 2007; Mahdavi et al. 2015; Sato et al. 2008; Zulak et al. 2008). So far, few studies investigated metabolic profiles of cultured fungi overtime (Bertrand et al. 2014a; Choi et al. 2010). When co-culturing strains belonging to *Fusarium* and *Aspergillus*, Bertrand et al. (2014a) observed two features that were only detected after 4 and 7 days of culturing and not detected after 9 days. When growing *Cordyceps militaris* on germinated soybeans, Choi et al. (2010) also showed differences in the metabolite composition overtime and this led to the identification of novel compounds. These studies confirm the dynamic nature of fungal metabolism and the potential novelty hidden in the time frame.

However, because time-course measurements are not independent, meaning the same compounds can be observed for multiple times in several time points, they are complex to analyze and multivariate modeling of dynamic metabolomics data remains challenging as discussed hereafter (Berk et al. 2011; Jansen et al. 2005; Peters et al. 2010; Xia et al. 2011).

4.5.2.3 Data Mining and Metabolomics

Various strategies can be applied in order to induce secondary metabolite production. This may lead to the discovery of new chemical compounds. In some cases, the metabolite overproduction is striking and can be easily monitored by traditional targeted approaches (quantification of particular compounds) such as thin layer chromatography (TLC) (Peiris et al. 2008), high-performance liquid chromatography with UV detector (HPLC-UV) (Dashti et al. 2014; Ola et al. 2013; Van der Molen et al. 2014; Wu et al. 2015), or gas chromatography (GC) (Do Nascimento et al. 2013; Dos Santos Dias et al. 2015). However, this overproduction is rare and therefore various analytical strategies must be applied, such as differential metabolite profiling and metabolomics or mass spectrometry (MS) imaging (Wolfender et al. 2015).

Metabolomics approaches are based on an untargeted metabolite profiling of fungal extract replicates using mainly direct ionization-MS (DI-MS), GC-MS, LC-MS, or nuclear magnetic resonance (NMR) (Bertrand et al. 2014b; Wolfender et al. 2015). These advanced analytical methods use generic approaches, which are devised to be simple (such as linear gradient in HPLC) and non-specific to any type of compounds. They aim at being as comprehensive as possible to be able to detect any changes in the chemical composition of the fungal extract in relation to applied induction strategies. The resulting data have to be carefully examined in order to detect secondary metabolites that are either produced de novo (induced compound which was not present/detected in "normal" conditions) or are up- or down-regulated as the result of the induction strategies. De novo induction can generally be easily detected by qualitative comparison of the recorded profiles (Ola et al. 2013), such as the apparition of a new peak in an HPLC-UV chromatogram. Similarly, strong up- or down-regulation is most of the time easy to reveal by qualitative exploration of the data. However, weak modification can only be highlighted by a careful analysis of the data based on simple differential analyses or advances chemometric analysis (Nguyen et al. 2012; Wolfender et al. 2013). The differential analysis generally used to detect metabolic changes in relation to direct or indirect induction strategies generally consist of univariate or multivariate data analysis applied to the selection of metabolic features (detected ions or chemical shifts) related to the "normal" or "induced" state. For that purpose, traditional statistical approaches are used, such as principal component analysis (PCA), partial least squares regression coupled with discriminant analysis (PLS), or orthogonal projections to latent structure (OPLS) coupled with discriminant analysis (DA) (Gromski et al. 2015; Vinaixa et al. 2012). In addition, specific statistical approaches may be needed for the analysis of time-dependent series (Bertrand et al. 2014a; Boccard et al. 2011; Soanes et al. 2011). A particular case is represented by fungal co-culture where the resulting induced extract consists of the mixed fungal metabolome. Therefore, the analysis could be either analyzed by traditional approach (PCA, (O)PLS-DA) (Bertrand et al. 2013b; Combès et al. 2012; Peiris et al. 2008) or by more innovative strategies taking into account the particular experimental design, such as the Projected Orthogonalized CHemical Encounter MONitoring (POChEMon) approach (Jansen et al. 2015). Both strategies lead to complementary results. Alternatively, such as in the case of co-culture, the induction of metabolites is usually highly localized to the confrontation zone.

Therefore, molecular imaging approaches are well suited to visualize their induction (Moree et al. 2013).

Once evidence of a modified metabolome is obtained, the identification of the induced compounds from their observed features is still challenging (Creek et al. 2014), even though large databases with more than 250,000 natural compounds exist (Chapman & Hall 2014). In the case of NMR spectral features, the structural information provided (chemical shifts, spin–spin coupling, and peak intensity) is sometimes sufficient for direct identification of compounds from an extract (Bingol and Brüschweiler 2014; Breton and Reynolds 2013; Pauli et al. 2014) or for identification by comparison with databases (Alm et al. 2012). In the case of GC-MS features, the identification is easier due to robustness and reproducibility of electronic impact (EI) ionization. Therefore, the comparison of the EI-MS spectra with reference compounds and data provided by database is usually sufficient (Peiris et al. 2008). For LC-MS features, identification remains much more challenging unless a large library of already identified compounds is available (Klitgaard et al. 2014). However, dereplication strategies based on mass and spectral accuracy combined with MS/MS interpretation in regards to database search remain an efficient but rather time-consuming strategy to identify compounds or at least find previously unreported structures (Wolfender et al. 2015). Ultimately, when compound identification is in doubt, the only solution that remains is the isolation and structural identification by NMR. Recently, software-driven protocol optimization combined with LC-MS-targeted isolation drastically accelerated the purification procedures and speeded up the identification of unknown compounds and valorization process (Bertrand et al. 2013a).

4.5.2.4 Valorization of Marine Fungal Natural Products

The search for new bioactive metabolites in marine fungi began in the late 1990s (Fig. 4.5), with a unique exception, i.e., the discovery of cephalosporin C by Giuseppe Brotzu in 1945 from a marine *Acremonium chrysogenum* (Bugni and Ireland 2004). Since, that time an increasing number of novel bioactive molecules has been isolated. This was mainly attributed to the development of the new techniques and methods. These bioactives were subsequently evaluated for pharmacological applications such as anticancer, antiviral, antidiabetic, antiinflammatory, antiplasmodial, antioxidant, or neuritogenic products (Gomes et al. 2015; Moghadamtousi et al. 2015). One of the major interests of marine fungal natural metabolites is the diversity, the originality, and sometimes the complexity of their chemical skeletons, suggesting that they could support new mechanism of actions or new types of interaction with pharmacological targets.

The diketopiperazine halimide was the first—and still the only one—that allowed the development of a candidate drug, plinabulin (NPI-2358). It was isolated in the late 1990s from cultures of an *Aspergillus* strain collected from the green alga *Halimeda lacrimosa*. This compound acts as microtubule-disrupting and vascular disrupting agent (Nicholson et al. 2006; Yamazaki et al. 2010). Plinabulin has

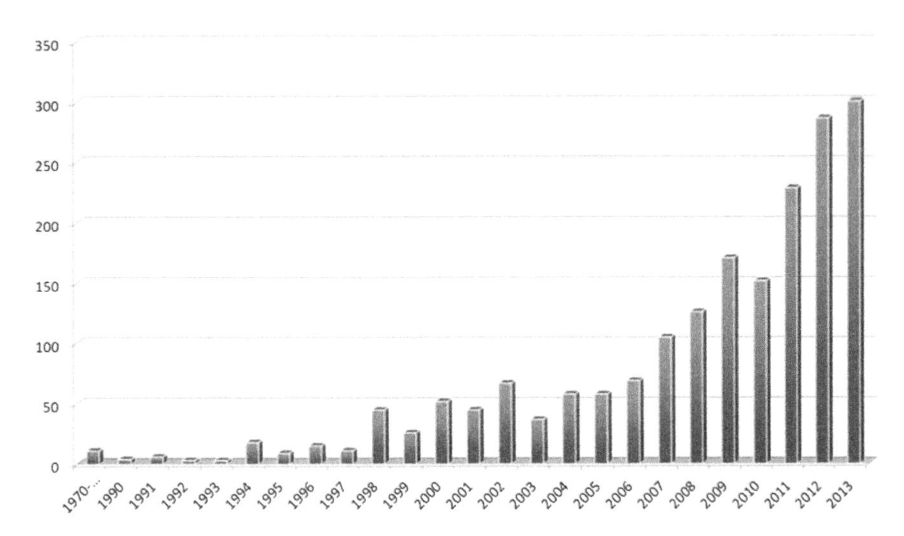

Fig. 4.5 Marine fungal natural products chemistry as a young but fast-growing science, as revealed by the annual amount of new compounds isolated from marine-sourced fungi. Adapted from Blunt et al. (2013)

reached phase II clinical trials for the treatment of advanced non-small cell lung cancer (Newman and Cragg 2014). Although it appears that clinical trials are suspended for the moment, studies on this molecule continues since some new plinabulin analogs exhibit increased activity to cancer cells (Hayashi et al. 2014). One can also mention other promising compounds, such as herqueidiketal, an original difuranonaphtoquinone, which was isolated in 2013 from a *Penicillium* strain collected in marine sediments in Korea. It exhibits activity against sortase A, an enzyme essential for the virulence of *S. aureus* and could therefore inspire the development of new antibiotics (Julianti et al. 2013). A strain of *Scopulariopsis brevicaulis* isolated from a marine sponge collected in Croatia, produces some unusual lipocyclodepsipeptides, scopularides A and B, patented in 2008 for their ability to decrease the proliferation of various cancer cell lines, (Imhoff et al. 2009; Yu et al. 2008). The nitrone-containing prenylated indole alkaloid avrainvillamide, isolated from a marine *Aspergillus* sp. strain, is also an example of intriguing skeletons rare in nature. The dimeric analog stephacidin B is especially active against testosterone-dependent prostate LNCaP cancer cells, possibly via a new mode of action, and has shown in vivo activity in preclinical models (Fenical et al. 2000; von Nussbaum 2003). Besides the valorization of marine fungal natural products for pharmacology, there are other kinds of industrial relevant compounds from marine fungi, including the development of alternative biocides for antifouling paints (Fusetani 2011) as well as ingredients for the cosmetic industry, such as fungal carotenoids used as natural pigments, or kojic acid employed as a skin-whitening component (Hong et al. 2015).

4.5.3 Hints for Ecological Roles Inferred from Secondary Metabolites

Fungi are responsible for many diseases and infections in the marine environment, many of them being induced by well-know terrestrial fungi (Richards et al. 2012). For instance, the Caribbean sea-fan (*Gorgonia* spp.) mortality has been attributed to an infection by pathogenic strains of *Aspergillus sydowii*, a well-known species in terrestrial habitats (Alker et al. 2001; Smith et al. 1996). To date, the chemical factors that mediate such host-pathogen relationships have not been elucidated. In contrast with the well-characterized mycotoxins involved in plant diseases, the secondary metabolites produced by marine fungal pathogens have not yet been investigated. Moreover, whether or not phylogenetically related strains from terrestrial and marine habitats produce the same chemicals remains an open question. We can assume that marine fungi are physiologically adapted to their environment and thus produce secondary metabolites characteristic of their habitat. Accordingly, biosynthetic pathways, silent in their terrestrial counterpart, could be activated as the result of the drastically different conditions in marine habitats, such as salinity and presence of halide. An additional unresolved question is whether or not fungi are living inside their host-tissue and, if so, whether the marine environment affects the metabolic production of these fungi. Further studies focusing on the detection of secondary metabolites or the fungal gene expression in situ will provide further clues to solve these issues.

Fungi are able to maintain mutualistic relationships with their host such as algae or sponges (Höller et al. 2000). Thus, fungi belonging to the *Koralionastes* genus are always found in close association with incrusting sponges (Kohlmeyer and Volkmann-Kohlmeyer 1990). Unfortunately, no chemical mediation in this interaction has been described. Even when few data suggest the involvement of chemical mediators in the relationship between marine fungi and their hosts, it is likely that such mediations represent an untapped resource of chemical diversity with a supposed wide range of biological activities, e.g., antimicrobial, cytotoxic, and antioxidant activities (for a review, see Ebada and Proksch 2015). Linking biological activities detected in the laboratory to any ecological role remains a difficult task. However, the amazing range of bioactivities of marine fungi suggests that these secondary metabolites could be defensive compounds and therefore is likely to play important roles in competition with other organisms in their habitats.

Aside from such host-fungi interactions in the mutualism–parasitism continuum, the concept of the holobiont, defined as an entity restricted to the host and its associated mutualistic symbionts, has emerged. It is now recognized that many other interactions within the microbiome exist and involve chemical mediators. In this context, it has been demonstrated that crude extracts of a collection of fungal strains isolated from aquatic habitats in Mexico displayed a potent effect on bacterial quorum sensing, a phenomenon that involves cell-to-cell communication and growth control processes (Martín-Rodríguez et al. 2014). Thus, cocultivation of marine fungi has been used to mimic a natural ecological situation, where microbes coexist within a complex microbial community and this approach has led to the production and isolation of numerous new natural compounds (Marmann et al. 2014). Furthermore, competition for limited resources and antagonism are characteristics of these microhabitats and thus responsible for the activation of various defense mechanisms, leading to the production of compounds displaying potent cytotoxic and antimicrobial activities.

The study of chemical mediation between host and fungus in the marine microbiome and the comprehension of the ecological role of fungal secondary metabolites remains one of the most challenges to overcome in the future. The advent of increasingly powerful and sensitive technologies (e.g. high field NMR and mass spectrometry) along with the use of more comprehensive approaches such as metabolomics and metagenomics provided a considerable improvement in the progress made in this subject. Further efforts aiming at mimicking in vivo processes in laboratory conditions and deciphering the mechanisms of what is happening in situ should provide more clues about the chemical communication mediated by fungal secondary metabolites.

4.5.4 From Genomes to Bioactive Molecules

Rapid progresses in DNA-sequencing methods in the last decade from Sanger sequencing to various types of high-throughput short-read sequencing have allowed modern biology and biotechnology reaching a point where exploring genetic information of any species of interest has become quest of a few months. This era of genome sequencing is often called "Desktop Genomics." These short-read sequencing methods are also known as next-generation sequencing (NGS) methods and are based on different platforms such as Roche GS-FLX 454 pyrosequencer, MiSeq, HiSeq, and Genome Analyzer II platforms (Illumina), SOLiD system (Life Technologies/Applied Biosystems), Ion Torrent and Ion Proton (Life Technologies), and the PacBio RS II (Pacific Biosciences) (Culligan et al. 2013; Metzker 2010).

Fungal genome sequencing and analyses have started with yeasts (Goffeau 2000; Goffeau et al. 1996). *Neurospora crassa* was sequenced as a model filamentous fungus (Galagan et al. 2003) and within the next 14–15 years, about 50 fungus genomes were sequenced using Sanger sequencing. In 2010, *Sordaria macrospora* genome sequencing was performed using more than one NGS method (Nowrousian et al. 2010). This accelerated other fungal genome sequencing projects using NGS methods. To date, hundreds of fungal genomes have been sequenced, covering all major groups of fungi (1000 Fungal Genome Consortium 2014) and now allowing a detailed search for SM-encoding genes. Fungi are generally producing four different types of secondary metabolites, namely PolyKetides (PKs), Non-Ribosomal Peptides (NRPs), terpenoids, or hybrid molecules such as PKs-NRPs or PKs-terpenes (Keller et al. 2005). The majority of the fungal secondary metabolites are derived from either NRPSs or PKSs, while only few fungal secondary metabolites are hybrid NRPS-PKS (Brakhage 2013). These include toxins like aflatoxins or fumonisins, and drugs like penicillin, cephalosporin, or cyclosporine.

Generally, SM-encoding genes are organized in discrete clusters in Dikarya (Brakhage 2013; Keller et al. 2005), which are called BGCs. A typical genome of a filamentous fungus possesses up to 70 BGCs encompassing a region from 30 to 80 kb on the fungal chromosome or scaffolds (Bok et al. 2015). Production of SMs is regulated by cluster-specific transcription factors and global regulators within the BGC (Brakhage 2013; Keller et al. 2005). SMs are highly relevant for drug discovery, particularly for new antibiotics and pharmaceuticals (Newman and Cragg 2007). Such a huge potential in producing bioactive compounds in fungi serves the future of drug discovery (Montaser and Luesch 2011). With genome sequencing becoming cost-effective, genome-based mining of BGCs has become attractive although only few examples of marine fungi exist. Three marine fungal genomes (*S. brevicaulis, Pestalotiopsis* sp., and *Calcariosporium* sp.) were sequenced and analyzed. Their BGCs were mined for further genetic engineering and production of selected bioactive compounds (Kramer et al. 2015; Kumar et al. 2015; Lukassen et al. 2015). We propose a general workflow of marine fungal genome mining for the characterization of the BGCs (Fig. 4.6).

The first step is to establish a collection of fungal strains from the marine environment. A metagenomic approach can also be performed when the aim is to study not yet cultured fungi. The second step consists of DNA extraction, which can be performed with known methods with little modifications (see as an example Kumar et al. 2015). The third step is genome sequencing using NGS methods, such as Illumina (HiSeq 2000), which is currently the predominant player in the market and also a budget choice. However, a hybrid assembly of two or more sequencing methods is recommended. As an example, the marine fungus *S. brevicaulis* LF580 has recently been sequenced using three methods (Kumar et al. 2015). Step four is genome assembly, which consists in assembling small reads into genomic fragments (contigs or larger scaffolds). Genome assembly can be achieved by broadly two ways: (i) de novo genome assembly, when a closely related fungal genome is not known, or (ii) reads assembled by genome assemblers without prior knowledge of chromosomes. Alternatively, when the genome of a closely related fungus is known, smaller reads can be mapped to the chromosomes of the known genome. This method is called genome mapping. When more than one type of genome sequencing reads are collected, a hybrid genome assembly can be performed. There are several genome assemblers known, and should be chosen depending on the specific requirements and type of genomic reads. A good example is documented in Kumar et al. (2015) with an initial genome assembly performed on Roche 454 reads using Newbler assembler (Margulies et al. 2005). Additionally, several genome assemblies can be processed using de Bruijn graph-based method by de novo assembly tool such as the CLCBio Genomic workbench using different types of reads from Illumina and Ion-Torrent. The genome assembly normally gives contigs that can be joined to form scaffolds, a method known as genome scaffolding. As an example, Kumar et al. (2015) performed genome scaffolding of *S. brevicaulis* LF580 using the genome-finishing module of the CLCBio Genomic workbench.

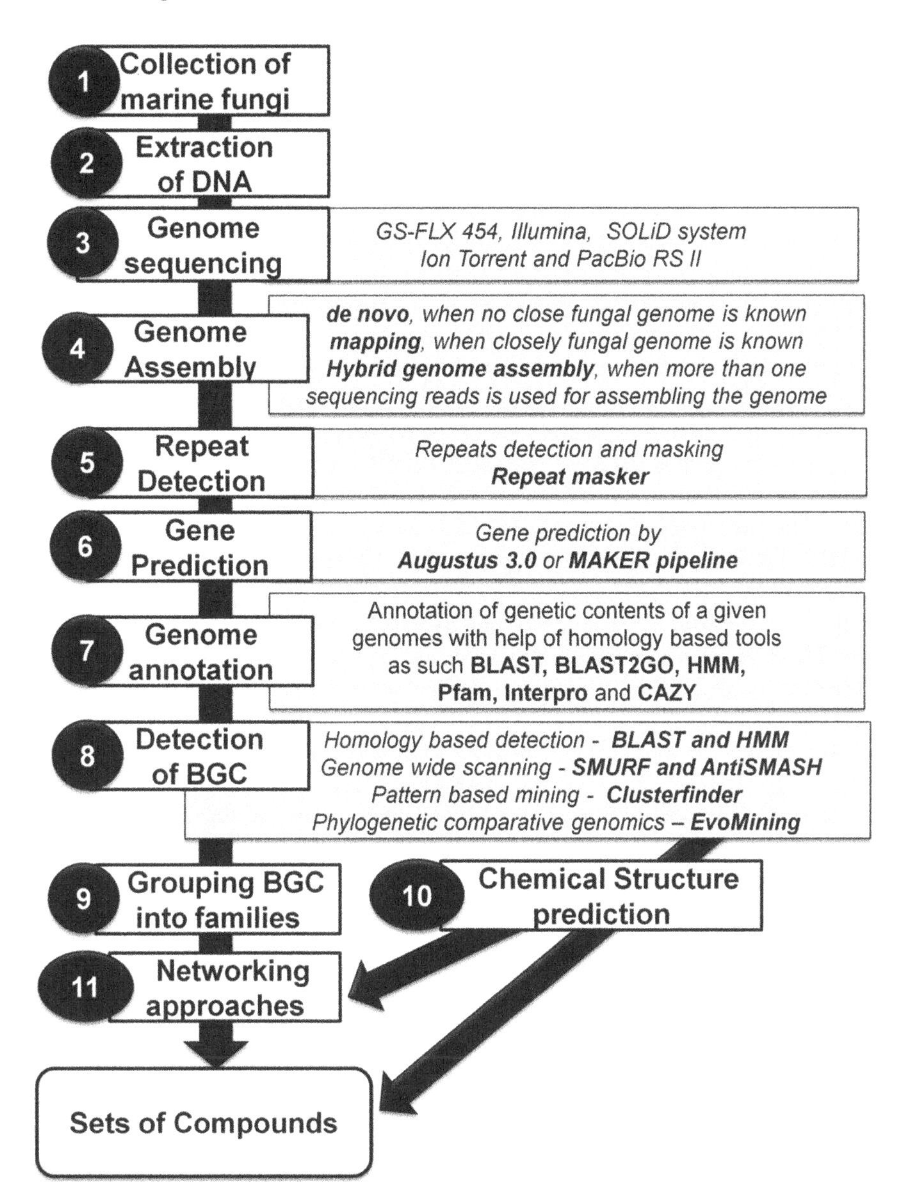

Fig. 4.6 Overview of next-generation sequencing strategies for the genome mining of bioactive compounds from marine fungi. This strategy can be applied to whole genome sequencing as well as metagenome sequencing

The detection of repeats and masking repetitive elements is an essential 5th step, because these elements may hamper gene predictions in the next step. RepeatMasker and RepeatProteinMasker software programs (Smit, Hubley & Green RepeatMasker

Open-4.0.0 1996–2013 http://www.repeatmasker.org) were used to detect and mask repeats using the fungal transposon species library as input. Gene prediction can be performed as a 6th step using either Augustus (Stanke and Morgenstern 2005) or within MAKER 2.0 pipeline (Cantarel et al. 2008). Gene annotation can be performed as a 7th step with the help of homology detection tools such as BLAST homology searches (Altschul et al. 1997), BLAST2GO tool (Götz et al. 2008), and the Kyoto Encyclopedia of Genes and Genomes (KEGG) (Kanehisa et al. 2010). Predicted proteins of a given genome can be scanned for protein domains using Pfam (Finn et al. 2014) and Interpro (Hunter et al. 2012) protein domains collections using HMMER 3.0 (Finn et al. 2011). Additionally, gene annotation can be performed when the objective of the analyses is clear.

After general annotation, detection of BGCs is performed as a 8th step using four different ways. Initially, putative biosynthetic genes are identified using either BLAST (Altschul et al. 1997) or HMMER 3.0 (Finn et al. 2011). This scanning step is performed using a collection of known biosynthetic genes. Second, two state-of-the-art tools namely Secondary Metabolite Unique Regions Finder (SMURF) (Khaldi et al. 2010) and antibiotics and Secondary Metabolite Analysis SHell (antiSMASH) (Blin et al. 2013; Weber et al. 2015) are recommended to be used for whole genome scanning of putative BGC. Alternatively, the MIDDAS-M method, allowing a motif-independent de novo detection of BGCs, can be used (Umemura et al. 2013). Furthermore, the functional domains of PKSs and NRPSs can be identified as previously described (Hansen et al. 2014), using a combination of tools namely antiSMASH, NCBI Conserved Domain Database (Marchler-Bauer et al. 2015), InterPro (Hunter et al. 2012), and the PKS/NRPS Analysis Web-site (Bachmann and Ravel 2009). Pattern-based approaches can be employed using cluster finder modules within antiSMASH. Phylogenetic and comparative genome methods can also be used for BGC detection and analysis with the help of genome browsers and specialized tools such as EvoMining (Cruz-Morales et al. 2015).

Based on sequence and domain similarities, the BGCs can be grouped into families of similar BGCs in a 9th step. Mass-spectroscopy based networking approaches will the help in the direction of scanning molecules from genomes. Predictions of small molecule structure from knowledge of BGCs are notorious tasks, although antiSMASH makes an attempt to predict. It is challenging as our knowledge of fungal BGC is poor and a rigorous classification and indexing system of BGC does not exist. The scientific community decided for a standard for careful cataloging of BGCs and potentially important enzyme and structural records, which are dispersed into the literature. This will be part of the genome standard as the Minimum Information about a Biosynthetic Gene cluster (MIBiG) data standard (Medema et al. 2015). This will improve to obtain an accurate picture of the genome to molecule approach. Further details for genome mining of BGC can be found in Boddy (2014) and (Medema and Fischbach 2015).

As an example, the genome sequence of the marine *S. brevicaulis* LF580 revealed 18 BGC clusters including one hybrid NRPS-PKS cluster (Fig. 4.7), which produces anticancerous scopularides (Kumar et al. 2015). During this study, this cluster and remaining clusters were characterized using UV-mediated mutagenesis for

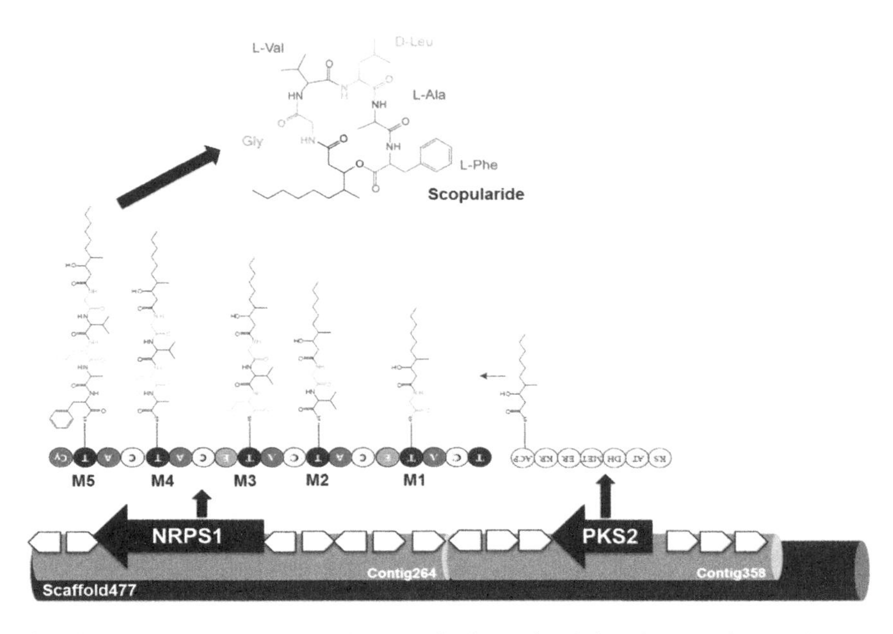

Fig. 4.7 An example of genome mining strategies for marine-derived fungal strain *Scopulariopsis brevicaulis* LF580 illustrates production of anticancerous scopularides by the hybrid BGC of NRPS1 (five modular M1-M5) and PKS2. Modified from Kumar et al. (2015) (Lukassen et al. 2015)

production of SM compounds using RNA-sequencing, proteomics, and comparative genomic approaches (Kramer et al. 2015; Kumar et al. 2015; Lukassen et al. 2015). Two other marine fungi, *Pestalotiopsis* sp. and *Calcariosporium* sp. have genomes with a total of 67 and 60 BGC clusters, respectively, producing several SM-encoding genes for putative bioactive production.

Computational genome mining approaches alone cannot solve all problems associated with the fungal BGCs. Several BGC are silent under standard laboratory conditions, leaving out a plethora of uncharacterized bioactive compounds (Brakhage 2013). Hence, testing different salinity conditions and other stressors to marine fungi is essential to activate the silent clusters.

The "genome mining to BGC mining" approach holds great potential. Initial genome mining reveals different types of BGCs in marine fungal genomes. Also, heterologous expression systems have already been developed for the entire set of intact fungal BGC using fungal artificial chromosomes (Bok et al. 2015). Thus, based on this proof of concept it is a matter of time when scientists will start characterizing marine fungal BGCs in heterologous systems. This holds the key for the future of drugs where genome mining of BGCs will be instrumental in fast and cost-effective roles in the field of marine fungal natural products.

The fungi harbor, a huge diversity of organisms occurring in numerous habitats and having different kind of lifestyles, e.g., they may be saprophytes, symbionts, or parasites/pathogens. Fungal lifestyles can be allegorized as the legendary movie

"The Good, The Bad and The Ugly." Indeed, some fungi are involved in nutrient recycling, food fermentation, synthesis of some of the gold standard drugs used in human medicine such as antibiotics as penicillins, immunosuppressive agents such as cyclosporin and mycophenolic acid, or cholesterol-lowering medications such as statins (The Good). Other fungi are involved in plant/crop diseases, food spoiling, and parasitism (The Bad). For the Ugly, just type "mycosis" in Google Images. While terrestrial fungi are well described, marine fungi still constitute an under-explored topic when compared to other marine microorganisms. This is mostly due to scientists who consider fungi terrestrial microorganisms. Now, we have become aware of the diversity, activity, function, and ecological role of marine fungi in almost all marine habitats, from coastal waters to the deep subseafloor. We now start to realize that these organisms are capable of synthesizing an astonishing array of bioactive compounds. The aim of this chapter was to establish a state-of-the-art of marine fungi, from the diversity and biotechnological points of view. This chapter specifically emphasized omic-based approaches that can be used to modulate and stimulate the marine fungal metabolome. This could increase the chance to reveal an original chemodiversity made of putative novel bioactive compounds.

Acknowledgments The research leading to these results has received funding from the European Union Seventh Framework Program (FP7/2007-2013) under grant agreement no. 311975. This publication reflects the views only of the author, and the European Union cannot be held responsible for any use which may be made of the information contained therein.

References

Alain K, Querellou J (2009) Cultivating the uncultured: limits, advances and future challenges. Extremophiles 13(4):583–594. doi:10.1007/s00792-009-0261-3

Alam MT, Merlo ME, The SC, Hodgson DA, Wellington EMH, Takano E, Breitling R (2010) Metabolic modeling and analysis of the metabolic switch in *Streptomyces coelicolor*. BMC Genom 11(1):202. doi:10.1186/1471-2164-11-202

Alker A, Smith G, Kim K (2001) Characterization of *Aspergillus sydowii* (Thom et Church), a fungal pathogen of Caribbean sea fan corals. Hydrobiologia 460(1–3):105–111. doi:10.1023/a:1013145524136

Alm E, Slagbrand T, Åberg K, Wahlström E, Gustafsson I, Lindberg J (2012) Automated annotation and quantification of metabolites in 1H NMR data of biological origin. Anal Bioanal Chem 403(2):443–455. doi:10.1007/s00216-012-5789-x

Almeida C, Kehraus S, Prudêncio M, König GM (2011a) Marilones A-C, phthalides from the sponge-derived fungus *Stachylidium* sp. Beilstein J Org Chem 7:1636–1642. doi:10.3762/bjoc.7.192

Almeida C, Part N, Bouhired S, Kehraus S, König GM (2011b) Stachylines A-D from the sponge-derived fungus *Stachylidium* sp. J Nat Prod 74(1):21–25. doi:10.1021/np1005345

Almeida C, Hemberger Y, Schmitt SM, Bouhired S, Natesan L, Kehraus S, Dimas K, Gütschow M, Bringmann G, König GM (2012) Marilines A-C: novel Phthalimidines from the sponge-derived fungus *Stachylidium* sp. Chem Eur J 18(28):8827–8834. doi:10.1002/chem.201103278

Almeida C, Eguereva E, Kehraus S, König GM (2013) Unprecedented polyketides from a marine sponge-associated *Stachylidium* sp. J Nat Prod 76(3):322–326. doi:10.1021/np300668j

Altschul SF, Madden TL, Schäffer AA, Zhang J, Zhang Z, Miller W, Lipman DJ (1997) Gapped BLAST and PSI-BLAST: a new generation of protein database search programs. Nucleic Acids Res 25(17):3389–3402. doi:10.1093/nar/25.17.3389

Alva PME, Pointing SB, Pena-Muralla R, Hyde KD (2002) Do sea grasses harbour endophytes? In: Hyde K (ed) Fungi in marine environments. Fungal Diversity Press, Hong Kong, pp 167–178

Amagata T, Tanaka M, Yamada T, Doi M, Minoura K, Ohishi H, Yamori T, Numata A (2007) Variation in cytostatic constituents of a sponge-derived *Gymnascella dankaliensis* by manipulating the carbon source. J Nat Prod 70(11):1731–1740. doi:10.1021/np070165m

Amann RI, Ludwig W, Schleifer KH (1995) Phylogenetic identification and in situ detection of individual microbial cells without cultivation. Microbiol Rev 59(1):143–169

Amend A (2014) From dandruff to deep-sea vents: malassezia-like fungi are ecologically hyper-diverse. PLoS Pathog 10(8):e1004277. doi:10.1371/journal.ppat.1004277 (PPATHOGENS-D-14-00979 [pii])

Amnuaykanjanasin A, Punya J, Paungmoung P, Rungrod A, Tachaleat A, Pongpattanakitshote S, Cheevadhanarak S, Tanticharoen M (2005) Diversity of type I polyketide synthase genes in the wood-decay fungus *Xylaria* sp. BCC 1067. FEMS Microbiol Lett 251(1):125–136. doi:10.1016/j.femsle.2005.07.038 (S0378-1097(05)00527-6 [pii])

Amnuaykanjanasin A, Phonghanpot S, Sengpanich N, Cheevadhanarak S, Tanticharoen M (2009) Insect-specific polyketide synthases (PKSs), potential PKS-nonribosomal peptide synthetase hybrids, and novel PKS clades in tropical fungi. Appl Environ Microbiol 75(11):3721–3732. doi:10.1128/AEM.02744-08 (AEM.02744-08 [pii])

Andrews JH (1976) The pathology of marine algae. Biol Rev 51(2):211–252. doi:10.1111/j.1469-185X.1976.tb01125.x

Arfi Y, Buee M, Marchand C, Levasseur A, Record E (2012) Multiple markers pyrosequencing reveals highly diverse and host-specific fungal communities on the mangrove trees *Avicennia marina* and *Rhizophora stylosa*. FEMS Microbiol Ecol 79(2):433–444. doi:10.1111/j.1574-6941.2011.01236.x

Bachmann BO, Ravel J (2009) Methods for in silico prediction of microbial polyketide and nonribosomal peptide biosynthetic pathways from DNA sequence data. Methods Enzymol 458:181–217. doi:10.1016/S0076-6879(09)04808-3

Baker PW, Kennedy J, Dobson AD, Marchesi JR (2009) Phylogenetic diversity and antimicrobial activities of fungi associated with *Haliclona simulans* isolated from Irish coastal waters. Mar Biotechnol (NY) 11(4):540–547. doi:10.1007/s10126-008-9169-7

Baral HO, Rämä T (2015) Morphological update on *Calycina marina* (Pezizellaceae, Helotiales, Leotiomycetes), a new combination for *Laetinaevia marina*. Bot Mar

Barata M (2002) Fungi on the halophyte *Spartina maritima* in salt marshes. In: Hyde K (ed) Fungi in marine environments. Fungal Diversity Press, Hong Kong, pp 179–193

Barghoorn E, Linder D (1944) Marine fungi: their taxonomy and biology. Farlowia 1:395–467

Bass D, Howe A, Brown N, Barton H, Demidova M, Michelle H, Li L, Sanders H, Watkinson SC, Willcock S, Richards TA (2007) Yeast forms dominate fungal diversity in the deep oceans. Proc Biol Sci 274(1629):3069–3077. doi:10.1098/rspb.2007.1067 (3600443776631874 [pii])

Beau J, Mahid N, Burda WN, Harrington L, Shaw LN, Mutka T, Kyle DE, Barisic B, van Olphen A, Baker BJ (2012) Epigenetic tailoring for the production of anti-infective cytosporones from the marine fungus *Leucostoma persoonii*. Marine Drugs 10(4):762

Bergmann S, Schümann J, Scherlach K, Lange C, Brakhage AA, Hertweck C (2007) Genomics-driven discovery of PKS-NRPS hybrid metabolites from *Aspergillus nidulans*. Nat Chem Biol 3(4):213–217. doi:10.1038/nchembio869

Berk M, Ebbels T, Montana G (2011) A statistical framework for biomarker discovery in metabolomic time course data. Bioinformatics 27(14):1979–1985. doi:10.1093/bioinformatics/btr289

Bertrand S, Schumpp O, Bohni N, Bujard A, Azzollini A, Monod M, Gindro K, Wolfender J-L (2013a) Detection of metabolite induction in fungal co-cultures on solid media by high-throughput differential ultra-high pressure liquid chromatography-time-of-flight mass spectrometry fingerprinting. J Chromatogr A 1292:219–228. doi:10.1016/j.chroma.2013.01.098

Bertrand S, Schumpp O, Bohni N, Monod M, Gindro K, Wolfender J-L (2013b) De novo production of metabolites by fungal co-culture of *Trichophyton rubrum* and *Bionectria ochroleuca*. J Nat Prod 76(6):1157–1165. doi:10.1021/np400258f Epub 2013 Jun 4

Bertrand S, Azzollini A, Schumpp O, Bohni N, Schrenzel J, Monod M, Gindro K, Wolfender J-L (2014a) Multi-well fungal co-culture for de novo metabolite-induction in time series studies based on untargeted metabolomics. Mol BioSyst 10(9):2289–2298. doi:10.1039/c4mb00223g

Bertrand S, Bohni N, Schnee S, Schumpp O, Gindro K, Wolfender J-L (2014b) Metabolite induction via microorganism co-culture: a potential way to enhance chemical diversity for drug discovery. Biotechnol Adv 32(6):1180–1204. doi:10.1016/j.biotechadv.2014.03.001

Besitulo A, Moslem MA, Hyde KD (2010) Occurrence and distribution of fungi in a mangrove forest on Siargao Island, Philippines. Botanica Marina 53. doi:10.1515/bot.2010.065

Bills G, Platas G, Fillola A, Mr Jiménez, Collado J, Vicente F, Martín J, González A, Bur-Zimmermann J, Jr Tormo, Peláez F (2008) Enhancement of antibiotic and secondary metabolite detection from filamentous fungi by growth on nutritional arrays. J Appl Microbiol 104(6):1644–1658. doi:10.1111/j.1365-2672.2008.03735.x

Bingle LE, Simpson TJ, Lazarus CM (1999) Ketosynthase domain probes identify two subclasses of fungal polyketide synthase genes. Fungal Genet Biol 26(3):209–223. doi:10.1006/fgbi.1999.1115 (S1087-1845(99)91115-5 [pii])

Bingol K, Brüschweiler R (2014) Multidimensional approaches to NMR-Based metabolomics. Anal Chem 86(1):47–57. doi:10.1021/ac403520j

Blin K, Medema MH, Kazempour D, Fischbach MA, Breitling R, Takano E, Weber T (2013) AntiSMASH 2.0—a versatile platform for genome mining of secondary metabolite producers. Nucleic Acids Res 41 (Web Server issue):W204–W212. doi:10.1093/nar/gkt449

Blunt JW, Copp BR, Keyzers RA, Munro MHG, Prinsep MR (2013) Marine natural products. Natural product reports 30(2):237–323. doi:10.1039/c2np20112g

Boccard J, Badoud F, Grata E, Ouertani S, Hanafi M, Mazerolles G, Lantéri P, Veuthey J-L, Saugy M, Rudaz S (2011) A steroidomic approach for biomarkers discovery in doping control. Forensic Sci Int 213(1–3):85–94. doi:10.1016/j.forsciint.2011.07.02310.1016/j.forsciint.2011.07.023 Epub 2011 Aug 9

Boddy CN (2014) Bioinformatics tools for genome mining of polyketide and non-ribosomal peptides. J Ind Microbiol Biotechnol 41(2):443–450

Bode HB, Bethe B, Höfs R, Zeeck A (2002) Big effects from small changes: possible ways to explore nature's chemical diversity. ChemBioChem 3(7):619–627

Bok JW, Keller NP (2004) LaeA, a regulator of secondary metabolism in *Aspergillus* spp. Eukaryot Cell 3(2):527–535. doi:10.1128/ec.3.2.527-535.2004

Bok JW, Ye R, Clevenger KD, Mead D, Wagner M, Krerowicz A, Albright JC, Goering AW, Thomas PM, Kelleher NL, Keller NP, Wu CC (2015) Fungal artificial chromosomes for mining of the fungal secondary metabolome. BMC Genom 16(1):343. doi:10.1186/s12864-015-1561-x

Boonyuen N, Chuaseeharonnachai C, Suetrong S, Sri-Indrasutdhi V, Sivichai S, Jones EB, Pang KL (2011) Savoryellales (Hypocreomycetidae, Sordariomycetes): a novel lineage of aquatic ascomycetes inferred from multiple-gene phylogenies of the genera Ascotaiwania, Ascothailandia, and Savoryella. Mycologia 103(6):1351–1371. doi:10.3852/11-102 (11-102 [pii])

Brakhage AA (2013) Regulation of fungal secondary metabolism. Nat Rev Microbiol 11(1):21–32. doi:10.1038/nrmicro2916

Brakhage AA, Schroeckh V (2011) Fungal (SM)—strategies to activate silent gene clusters. Fungal Genet Biol 48(1):15–22. doi:10.1016/j.fgb.2010.04.004

Brakhage AA, Schuemann J, Bergmann S, Scherlach K, Schroeckh V, Hertweck C (2008) Activation of fungal silent gene clusters: a new avenue to drug discovery. Prog Drug Res 66:1:3–12

Breton RC, Reynolds WF (2013) Using NMR to identify and characterize natural products. Natural product reports 30(4):501–524. doi:10.1039/c2np20104f

Buchan ANS, Lyons Moreta JI, Moran MA (2002) Analysis of internal transcribed spacer (ITS) regions of rRNA genes in fungal communities in a southeastern U.S. salt marsh. Microb Ecol 43:329–340

Bucher VVC, Pointing SB, Hyde KD, Reddy CA (2004) Production of wood decay enzymes, loss of mass, and lignin solubilization in wood by diverse tropical freshwater fungi. Microb Ecol 48 (3):331–337. doi:10.1007/s00248-003-0132-x

Bugni TS, Ireland CM (2004) Marine-derived fungi: a chemically and biologically diverse group of microorganisms. Nat Prod Rep 21(1):143–163. doi:10.1039/b301926h

Burgaud G, Le Calvez T, Arzur D, Vandenkoornhuyse P, Barbier G (2009) Diversity of culturable marine filamentous fungi from deep-sea hydrothermal vents. Environ Microbiol 11(6):1588–1600. doi:10.1111/j.1462-2920.2009.01886.x (EMI1886 [pii])

Burgaud G, Arzur D, Durand L, Cambon-Bonavita MA, Barbier G (2010) Marine culturable yeasts in deep-sea hydrothermal vents: species richness and association with fauna. FEMS Microbiol Ecol 73(1):121–133. doi:10.1111/j.1574-6941.2010.00881.x (FEM881 [pii])

Burgaud G, Arzur D, Sampaio JP, Barbier G (2011) *Candida oceani* sp. nov., a novel yeast isolated from a Mid-Atlantic Ridge hydrothermal vent (-2300 meters). Antonie Van Leeuwenhoek 100(1):75–82. doi:10.1007/s10482-011-9566-1

Burgaud G, Woehlke S, Rédou V, Orsi W, Beaudoin D, Barbier G, Biddle JF, Edgcomb VP (2013) Deciphering the presence and activity of fungal communities in marine sediments using a model estuarine system. Aquat Microb Ecol 70(1):45–62

Burgaud G, Meslet-Cladière L, Barbier G, Edgcomb VP (2014) Astonishing fungal diversity in deep-sea hydrothermal ecosystems: an untapped resource of biotechnological potential? In: Outstanding marine molecules: chemistry, biology, analysis, pp 85–98. doi:10.1002/9783527681501.ch04

Burgaud G, Hue NT, Arzur D, Coton M, Perrier-Cornet JM, Jebbar M, Barbier G (2015) Effects of hydrostatic pressure on yeasts isolated from deep-sea hydrothermal vents. Res Microbiol. doi:10.1016/j.resmic.2015.07.005

Bushley KE, Turgeon BG (2010) Phylogenomics reveals subfamilies of fungal nonribosomal peptide synthetases and their evolutionary relationships. BMC Evol Biol 10:26. doi:10.1186/1471-2148-10-26 (1471-2148-10-26 [pii])

Cacho RA, Tang Y, Chooi Y-H (2014) Next-generation sequencing approach for connecting secondary metabolites to biosynthetic gene clusters in fungi. Front Microbiol 5:774. doi:10.3389/fmicb.2014.00774

Calado MDL, Barata M (2012). Salt marsh fungi. In: Jones EBG, Jones PK (ed) Marine fungi and fungal-like organisms. De Gruyter, Berlin, pp 345–381

Calado MD, Carvalho L, Pang KL, Barata M (2015) Diversity and ecological characterization of sporulating higher filamentous marine fungi associated with *Spartina maritima* (Curtis) fernald in two Portuguese salt marshes. Microb Ecol. doi:10.1007/s00248-015-0600-0

Cantarel BL, Korf I, Robb SM, Parra G, Ross E, Moore B, Holt C, Sanchez Alvarado A, Yandell M (2008) MAKER: an easy-to-use annotation pipeline designed for emerging model organism genomes. Genome Res 18(1):188–196. doi:10.1101/gr.6743907

Cathrine SJ, Raghukumar C (2009) Anaerobic denitrification in fungi from the coastal marine sediments off Goa. India. Mycol Res 113(1):100–109. doi:10.1016/j.mycres.2008.08.009

Chai Y-J, Cui C-B, Li C-W, Wu C-J, Tian C-K, Hua W (2012) Activation of the dormant secondary metabolite production by introducing gentamicin-resistance in a marine-derived *Penicillium purpurogenum* G59. Marine Drugs 10(12):559–582. doi:10.3390/md10030559

Chapman & Hall (2014) Dictionary of natural products on DVD (23:1), vol 23(1). CRC Press, Taylor & Francis Group. URL http://dnp.chemnetbase.com/

Chiang Y-M, Szewczyk E, Davidson AD, Keller N, Oakley BR, Wang CCC (2009) A gene cluster containing two fungal polyketide synthases encodes the biosynthetic pathway for a polyketide, asperfuranone, in *Aspergillus nidulans*. J Am Chem Soc 131(8):2965–2970. doi:10.1021/ja8088185

Choi JN, Kim J, Lee MY, Park DK, Hong Y-S, Lee CH (2010) Metabolomics revealed novel isoflavones and optimal cultivation time of *Cordyceps militaris* Fermentation. J Agric Food Chem 58(7):4258–4267. doi:10.1021/jf903822e

Cichewicz RH (2010) Epigenome manipulation as a pathway to new natural product scaffolds and their congeners. Nat Prod Rep 27(1):11–22. doi:10.1039/b920860g

Ciobanu MC, Burgaud G, Dufresne A, Breuker A, Rédou V, Ben Maamar S, Gaboyer F, Vandenabeele-Trambouze O, Lipp JS, Schippers A, Vandenkoornhuyse P, Barbier G, Jebbar M, Godfroy A, Alain K (2014) Microorganisms persist at record depths in the subseafloor of the Canterbury Basin. ISME J 8(7):1370–1380. doi:10.1038/ismej.2013.250 (ismej2013250 [pii])

Collado J, Platas G, Paulus B, Bills GF (2007) High-throughput culturing of fungi from plant litter by a dilution-to-extinction technique. FEMS Microbiol Ecol 60(3):521–533. doi:10.1111/j.1574-6941.2007.00294.x (FEM294 [pii])

Combès A, Ndoye I, Bance C, Bruzaud J, Djediat C, Dupont J, Nay B, Prado S (2012) Chemical communication between the endophytic fungus *Paraconiothyrium variabile* and the phytopathogen *Fusarium oxysporum*. PLoS ONE 7(10):e47313. doi:10.1371/journal.pone.0047313

Connell L, Redman R, Craig S, Scorzetti G, Iszard M, Rodriguez R (2008) Diversity of soil yeasts isolated from South Victoria Land. Antarctica. Microb Ecol 56(3):448–459. doi:10.1007/s00248-008-9363-1

Connon SA, Giovannoni SJ (2002) High-throughput methods for culturing microorganisms in very-low-nutrient media yield diverse new marine isolates. Appl Environ Microbiol 68 (8):3878–3885

Creek DJ, Dunn WB, Fiehn O, Griffin JL, Hall RD, Lei Z, Mistrik R, Neumann S, Schymanski EL, Sumner LW, Trengove R, Wolfender J-L (2014) Metabolite identification: are you sure? And how do your peers gauge your confidence? Metabolomics 10(3):350–353. doi:10.1007/s11306-014-0656-8

Cruz-Morales P, Martínez-Guerrero CE, Morales-Escalante MA, Yáñez-Guerra LA, Kopp JF, Feldmann J, Ramos-Aboites HE & Barona-Gómez F (2015). Recapitulation of the evolution of biosynthetic gene clusters reveals hidden chemical diversity on bacterial genomes. doi:10.1101/020503

Cueto M, Jensen PR, Kauffman C, Fenical W, Lobkovsky E, Clardy J (2001) Pestalone, a new antibiotic produced by a marine fungus in response to bacterial challenge. J Nat Prod 64 (11):1444–1446. doi:10.1021/np0102713

Culligan EP, Sleator RD, Marchesi JR, Hill C (2013) Functional environmental screening of a metagenomic library identifies stlA; a unique salt tolerance locus from the human gut microbiome. PLoS ONE 8(12):e82985. doi:10.1371/journal.pone.0082985

Daffonchio D, Borin S, Brusa T, Brusetti L, van der Wielen PW, Bolhuis H, Yakimov MM, D'Auria G, Giuliano L, Marty D, Tamburini C, McGenity TJ, Hallsworth JE, Sass AM, Timmis KN, Tselepides A, de Lange GJ, Hubner A, Thomson J, Varnavas SP, Gasparoni F, Gerber HW, Malinverno E, Corselli C, Garcin J, McKew B, Golyshin PN, Lampadariou N, Polymenakou P, Calore D, Cenedese S, Zanon F, Hoog S (2006) Stratified prokaryote network in the oxic-anoxic transition of a deep-sea halocline. Nature 440(7081):203–207. doi:10.1038/nature04418 (nature04418 [pii])

Damare S, Raghukumar C (2008) Fungi and macroaggregation in deep-sea sediments. Microb Ecol 56(1):168–177. doi:10.1007/s00248-007-9334-y

Damare S, Raghukumar C, Raghukumar S (2006) Fungi in deep-sea sediments of the Central Indian Basin. Deep Sea Res Part I 53(1):14–27. doi:10.1016/j.dsr.2005.09.005

Dashti Y, Grkovic T, Abdelmohsen U, Hentschel U, Quinn R (2014) Production of induced secondary metabolites by a co-culture of sponge-associated actinomycetes, *Actinokineospora* sp. EG49 and *Nocardiopsis* sp. RV163. Marine Drugs 12(5):3046–3059. doi:10.3390/md12053046

Dayton PK (1985) Ecology of kelp communities. Annu Rev Ecol Syst 16(1):215–245. doi:10.1146/annurev.es.16.110185.001243

Deacon J (2005). Fungal growth. In: Fungal biology. Blackwell Publishing Ltd., London, pp 67–84. doi:10.1002/9781118685068.ch4

Devarajan P, Suryanarayanan T, Geetha V (2002) Endophytic fungi associated with the tropical seagrass *Halophila ovalis* (Hydrocharitaceae). Ind J Mar Sci 31(1):73–74

D'Hondt S, Jorgensen BB, Miller DJ, Batzke A, Blake R, Cragg BA, Cypionka H, Dickens GR, Ferdelman T, Hinrichs KU, Holm NG, Mitterer R, Spivack A, Wang G, Bekins B, Engelen B, Ford K, Gettemy G, Rutherford SD, Sass H, Skilbeck CG, Aiello IW, Guerin G, House CH, Inagaki F, Meister P, Naehr T, Niitsuma S, Parkes RJ, Schippers A, Smith DC, Teske A, Wiegel J, Padilla CN, Acosta JL (2004) Distributions of microbial activities in deep subseafloor sediments. Science 306(5705):2216–2221. doi:10.1126/science.1101155 (306/5705/2216 [pii])

Diez B, Pedrós-Alió C, Massana R (2001) Study of genetic diversity of eukaryotic picoplankton in different oceanic regions by small-subunit rRNA gene cloning and sequencing. Appl Environ Microbiol 67(7):2932–2941. doi:10.1128/AEM.67.7.2932-2941.2001

Ding B, Yin Y, Zhang F, Li Z (2011) Recovery and phylogenetic diversity of culturable fungi associated with marine sponges *Clathrina luteoculcitella* and *Holoxea* sp. in the South China Sea. Mar Biotechnol (NY) 13(4):713–721. doi:10.1007/s10126-010-9333-8

Dittami SM, Eveillard D, Tonon T (2014) A metabolic approach to study algal-bacterial interactions in changing environments. Mol Ecol 23(7):1656–1660. doi:10.1111/mec.12670

Do Nascimento M, MdlA Dublan, Ortiz-Marquez JCF, Curatti L (2013) High lipid productivity of an Ankistrodesmus-Rhizobium artificial consortium. Bioresour Technol 146:400–407. doi:10.1016/j.biortech.2013.07.085

Dong Y, Cui C-B, Li C-W, Hua W, Wu C-J, Zhu T-J, Gu Q-Q (2014) Activation of dormant secondary metabolite production by introducing neomycin resistance into the deep-sea fungus, *Aspergillus versicolor* ZBY-3. Mar Drugs 12(8):4326–4352. doi:10.3390/md12084326

Dos Santos Dias AC, Ruiz N, Couzinet-Mossion A, Bertrand S, Duflos M, Pouchus Y-F, Barnathan G, Nazih H, Wielgosz-Collin G (2015) The marine-derived fungus *Clonostachys rosea*, source of a rare conjugated 4-Me-6E,8E-hexadecadienoic acid reducing viability of MCF-7 breast cancer cells and gene expression of lipogenic enzymes. Mar Drugs 13(8):4934

Dupont J, Magnin S, Rousseau F, Zbinden M, Frebourg G, Samadi S, Richer de Forges B, Gareth-Jones EB (2009) Molecular and ultrastructural characterization of two ascomycetes found on sunken wood off Vanuatu Islands in the deep Pacific Ocean. Mycol Res 112:1351–1364. doi:10.1016/j.mycres.2009.08.015

Ebada S, Proksch P (2015) Marine-derived fungal metabolites. In: Kim S-K (ed) Springer handbook of marine biotechnology. Springer, Berlin, pp 759–788. doi:10.1007/978-3-642-53971-8_32

Edgcomb V, Orsi W (2013) Microbial eukaryotes in hypersaline anoxic deep-sea basins. In: Seckbach J, Oren A, Stan-Lotter H (eds) Polyextremophiles, vol 27. Cellular origin, life in extreme habitats and astrobiology. Springer, Netherlands, pp 517–530. doi:10.1007/978-94-007-6488-0_23

Edgcomb VP, Beaudoin D, Gast R, Biddle JF, Teske A (2011) Marine subsurface eukaryotes: the fungal majority. Environ Microbiol 13(1):172–183. doi:10.1111/j.1462-2920.2010.02318.x

Egan S, Harder T, Burke C, Steinberg P, Kjelleberg S, Thomas T (2013) The seaweed holobiont: understanding seaweed-bacteria interactions. FEMS Microbiol Rev 37(3):462–476. doi:10.1111/1574-6976.12011

Fenical W, Jensen PR, Cheng XC (2000) Avraincillamide, a cytotoxic marine natural product, and derivatives thereof. U.S. Patent

Fernandes PM, Domitrovic T, Kao CM, Kurtenbach E (2004) Genomic expression pattern in *Saccharomyces cerevisiae* cells in response to high hydrostatic pressure. FEBS Lett 556(1–3):153–160 (S0014579303013966 [pii])

Ferrari BC, Zhang C, van Dorst J (2011) Recovering greater fungal diversity from pristine and diesel fuel contaminated sub-antarctic soil through cultivation using both a high and a low nutrient media approach. Front Microbiol 2:217. doi:10.3389/fmicb.2011.00217

Finn RD, Clements J, Eddy SR (2011) HMMER web server: interactive sequence similarity searching. Nucleic Acids Res 39(Web Server issue):W29–W37. doi:10.1093/nar/gkr367

Finn RD, Bateman A, Clements J, Coggill P, Eberhardt RY, Eddy SR, Heger A, Hetherington K, Holm L, Mistry J, Sonnhammer EL, Tate J, Punta M (2014) Pfam: the protein families database. Nucleic Acids Res 42(Database issue):D222–230. doi:10.1093/nar/gkt1223

Fisch KM, Gillaspy AF, Gipson M, Henrikson JC, Hoover AR, Jackson L, Najar FZ, Wagele H, Cichewicz RH (2009) Chemical induction of silent biosynthetic pathway transcription in *Aspergillus niger*. J Ind Microbiol Biotechnol 36(9):1199–1213. doi:10.1007/s10295-009-0601-4

Fleming A (1929) On the antibacterial action of cultures of a penicillium, with special reference to their use in the isolation of B. influenzae. Br J Exp Pathol 10(3):226

Flewelling AJ, Johnson JA, Gray CA (2013) Isolation and bioassay screening of fungal endophytes from North Atlantic marine macroalgae. Botanica Marina 56. doi:10.1515/bot-2012-0224

Fries N (1979) Physiological characteristics of *Mycosphaerella ascophylli*, a fungal endophyte of the marine brown alga *Ascophyllum nodosum*. Physiol Plant 45(1):117–121. doi:10.1111/j.1399-3054.1979.tb01674.x

Fusetani N (2011) Antifouling marine natural products. Nat Prod Rep 28(2):400–410. doi:10.1039/c0np00034e

Gadanho M, Sampaio JP (2005) Occurrence and diversity of yeasts in the mid-atlantic ridge hydrothermal fields near the Azores Archipelago. Microb Ecol 50(3):408–417. doi:10.1007/s00248-005-0195-y

Galagan JE, Calvo SE, Borkovich KA, Selker EU, Read ND, Jaffe D, FitzHugh W, Ma LJ, Smirnov S, Purcell S, Rehman B, Elkins T, Engels R, Wang S, Nielsen CB, Butler J, Endrizzi M, Qui D, Ianakiev P, Bell-Pedersen D, Nelson MA, Werner-Washburne M, Selitrennikoff CP, Kinsey JA, Braun EL, Zelter A, Schulte U, Kothe GO, Jedd G, Mewes W, Staben C, Marcotte E, Greenberg D, Roy A, Foley K, Naylor J, Stange-Thomann N, Barrett R, Gnerre S, Kamal M, Kamvysselis M, Mauceli E, Bielke C, Rudd S, Frishman D, Krystofova S, Rasmussen C, Metzenberg RL, Perkins DD, Kroken S, Cogoni C, Macino G, Catcheside D, Li W, Pratt RJ, Osmani SA, DeSouza CP, Glass L, Orbach MJ, Berglund JA, Voelker R, Yarden O, Plamann M, Seiler S, Dunlap J, Radford A, Aramayo R, Natvig DO, Alex LA, Mannhaupt G, Ebbole DJ, Freitag M, Paulsen I, Sachs MS, Lander ES, Nusbaum C, Birren B (2003) The genome sequence of the filamentous fungus *Neurospora crassa*. Nature 422 (6934):859–868. doi:10.1038/nature01554

Gao Z, Li B, Zheng C, Wang G (2008) Molecular detection of fungal communities in the Hawaiian marine sponges *Suberites zeteki* and *Mycale armata*. Appl Environ Microbiol 74(19):6091–6101. doi:10.1128/AEM.01315-08 (AEM.01315-08 [pii])

Garbary DJ, MacDonald KA (1995) The ascophyllum/polysiphonia/mycosphaerella symbiosis. IV. Mutualism in the ascophyllum/mycosphaerella interaction. Botanica Marina 38. doi:10.1515/botm.1995.38.1-6.221

Geiger M, Guitton Y, Vansteelandt M, Kerzaon I, Blanchet E, Robiou du Pont T, Frisvad J, Hess P, Pouchus Y, Grovel O (2013). Cytotoxicity and mycotoxin production of shellfish-derived *Penicillium* spp., a risk for shellfish consumers. Lett Appl Microbiol 57 (5):385–392. doi:10.1111/lam.12143

Goers L, Freemont P, Polizzi KM (2014) Co-culture systems and technologies: taking synthetic biology to the next level. J R Soc Inter 11(96). doi:10.1098/rsif.2014.0065

Goffeau A (2000) Four years of post-genomic life with 6,000 yeast genes. FEBS Lett 480(1): 37–41

Goffeau A, Barrell B, Bussey H, Davis R, Dujon B, Feldmann H, Galibert F, Hoheisel J, Jacq C, Johnston M (1996) Life with 6000 genes. Science 274(5287):546–567

Gomes N, Lefranc F, Kijjoa A, Kiss R (2015) Can some marine-derived fungal metabolites become actual anticancer agents? Marine Drugs 13(6):3950

Götz S, García-Gómez JM, Terol J, Williams TD, Nagaraj SH, Nueda MJ, Robles M, Talón M, Dopazo J, Conesa A (2008) High-throughput functional annotation and data mining with the Blast2GO suite. Nucleic Acids Res 36(10):3420–3435. doi:10.1093/nar/gkn176

Gram L (2015) Silent clusters - speak up! Silent clusters—speak up! Microb Biotechnol. doi:10.1111/1751-7915.12181

Gromski PS, Muhamadali H, Ellis DI, Xu Y, Correa E, Turner ML, Goodacre R (2015) A tutorial review: Metabolomics and partial least squares-discriminant analysis—a marriage of convenience or a shotgun wedding. Analytica Chimica Acta 879:doi:10.1016/j.aca.2015.1002.1012

Gubbens J, Zhu H, Girard G, Song L, Florea BI, Aston P, Ichinose K, Filippov DV, Choi YH, Overkleeft HS, Challis GL, van Wezel GP (2014) Natural product proteomining, a quantitative proteomics platform, allows rapid discovery of biosynthetic gene clusters for different classes of natural products. Chem Biol 21(6):707–718. doi:10.1016/j.chembiol.2014.03.011 (S1074-5521(14)00123-9 [pii])

Gunde-Cimerman N, Ramos J, Plemenitas A (2009) Halotolerant and halophilic fungi. Mycol Res 113(Pt 11):1231–1241. doi:10.1016/j.mycres.2009.09.002 (S0953-7562(09)00161-0 [pii])

Guo W, Peng J, Zhu T, Gu Q, Keyzers RA, Li D (2013) Sorbicillamines A-E, nitrogen-containing sorbicillinoids from the deep-sea-derived fungus *Penicillium* sp. F23–2. J Nat Prod 76 (11):2106–2112. doi:10.1021/np4006647

Hansen FT, Gardiner DM, Lysoe E, Romans Fuertes P, Tudzynski B, Wiemann P, Sondergaard TE, Giese H, Brodersen DE & Sorensen JL (2014). An update to polyketide synthase and non-ribosomal synthetase genes and nomenclature in Fusarium. Fungal Genetics and Biology (In Press)

Harvey JB, Goff LJ (2010) Genetic covariation of the marine fungal symbiont *Haloguignardia irritans* (Ascomycota, Pezizomycotina) with its algal hosts Cystoseira and Halidrys (Phaeophyceae, Fucales) along the west coast of North America. Fungal biology 114(1):82–95. doi:10.1016/j.mycres.2009.10.009

Hawksworth DL (2001) The magnitude of fungal diversity: the 1.5 million species estimate revisited. Mycol Res 105(12):1422–1432

Hayashi Y, Takeno H, Chinen T, Muguruma K, Okuyama K, Taguchi A, Takayama K, Yakushiji F, Miura M, Usui T, Hayashi Y (2014) Development of a new benzophenone–diketopiperazine-type potent antimicrotubule agent possessing a 2-pyridine structure. ACS Med Chem Lett 5(10):1094–1098. doi:10.1021/ml5001883

Henriquez M, Vergara K, Norambuena J, Beiza A, Maza F, Ubilla P, Araya I, Chavez R, San-Martin A, Darias J, Darias MJ, Vaca I (2014) Diversity of cultivable fungi associated with Antarctic marine sponges and screening for their antimicrobial, antitumoral and antioxidant potential. World J Microbiol Biotechnol 30(1):65–76. doi:10.1007/s11274-013-1418-x

Hibbett DS, Binder M (2001) Evolution of marine mushrooms. Biol Bull 201(3):319–322

Hinrichs KU, Hayes JM, Bach W, Spivack AJ, Hmelo LR, Holm NG, Johnson CG, Sylva SP (2006) Biological formation of ethane and propane in the deep marine subsurface. Proc Natl Acad Sci USA 103(40):14684–14689. doi:10.1073/pnas.0606535103 (0606535103 [pii])

Hohmann S (2002) Osmotic stress signaling and osmoadaptation in yeasts. Microbiol Mol Biol Rev 66(2):300–372

Höller U, Wright AD, Matthee GF, Konig GM, Draeger S, Aust H-J, Schulz B (2000) Fungi from marine sponges: diversity, biological activity and secondary metabolites. Mycol Res 104 (11):1354–1365. doi:10.1017/S0953756200003117

Hong J-H, Jang S, Heo Y, Min M, Lee H, Lee Y, Lee H, Kim J-J (2015) Investigation of marine-derived fungal diversity and their exploitable biological activities. Marine Drugs 13 (7):4137

Hosaka T, Ohnishi-Kameyama M, Muramatsu H, Murakami K, Tsurumi Y, Kodani S, Yoshida M, Fujie A, Ochi K (2009) Antibacterial discovery in actinomycetes strains with mutations in RNA polymerase or ribosomal protein S12. Nat Biotechnol 27(5):462–464. doi:10.1038/nbt.1538

Hughes GC (1974) Geographical distribution of the higher marine fungi. Veröff Inst Meeresforsch Bremerh 5:419–441

Hunter S, Jones P, Mitchell A, Apweiler R, Attwood TK, Bateman A, Bernard T, Binns D, Bork P, Burge S, de Castro E, Coggill P, Corbett M, Das U, Daugherty L, Duquenne L, Finn RD, Fraser M, Gough J, Haft D, Hulo N, Kahn D, Kelly E, Letunic I, Lonsdale D, Lopez R, Madera M, Maslen J, McAnulla C, McDowall J, McMenamin C, Mi H, Mutowo-Muellenet P, Mulder N, Natale D, Orengo C, Pesseat S, Punta M, Quinn AF, Rivoire C, Sangrador-Vegas A, Selengut JD, Sigrist CJ, Scheremetjew M, Tate J, Thimmajanarthanan M, Thomas PD, Wu CH, Yeats C, Yong SY (2012). InterPro in 2011: new developments in the family and domain prediction database. Nucleic Acids Res 40(Database issue):D306–D312. doi:10.1093/nar/gkr948

Hyde KD (1992) Fungi from decaying intertidal fronds of Nypa fruticans, including three new genera and four new species. Bot J Linn Soc 110(2):95–110. doi:10.1111/j.1095-8339.1992.tb00284.x

Hyde KD, Alias SA (2000) Biodiversity and distribution of fungi associated with decomposing Nypa palm. Biodivers Conserv 9:393–402

Hyde KDJE, Liu JK, Ariyawansha H, Boehm E, Boonmee S, Braun U, Chomnunti P, Crous PW, Dai D, Diederich P, Dissanayake A, Doilom M, Doveri F, Hongsanan S, Jayawardena R, Lawrey JD, Li YM, Liu YX, Lücking R, Monkai J, Muggia L, Nelsen MP, Pang KL, Phookamsak R, Senanayake I, Shearer CA, Suetrong S, Tanaka K, Thambugala KM, Wijayawardene N, Wikee S, Wu HX, Zhang Y, Aguirre-Hudson B, Alias SA, Aptroot A, Bahkali AH, Bezerra JL, Bhat JD, Camporesi E, Chukeatirote E, Gueidan C, Hawksworth DL, Hirayama K, De Hoog S, Kang JC, Knudsen K, Li WJ, Li XH, Liu ZY, Mapook A, McKenzie EHC, Miller AN, Mortimer PE, Phillips AJL, Raja HA, Scheuer C, Schumm F, Taylor J, Tian Q, Tibpromma S, Wanasinghe DN, Wang Y, Xu JC, Yacharoen S, Yan JY, Zhang M (2013) Families of dothideomycetes. Fungal Divers 63:1–313. doi:10.1007/s13225-013-0263-4

Imhoff J, Yu Z, Lang G, Wiese J, Kalthoff HSK (2009) Production and use of antitumoral cyclodepsipeptides

Ingham CJ, Sprenkels A, Bomer J, Molenaar D, van den Berg A, van Hylckama Vlieg JE, de Vos WM (2007) The micro-Petri dish, a million-well growth chip for the culture and high-throughput screening of microorganisms. Proc Natl Acad Sci USA 104(46):18217–18222. doi:10.1073/pnas.0701693104 (0701693104 [pii])

Iwamoto C, Minoura K, Hagishita S, Nomoto K, Numata A (1998) Penostatins F-I, novel cytotoxic metabolites from a *Penicillium* species separated from an Enteromorpha marine alga. J Chem Soc Perkin Trans 1(3):449–456

Jansen JJ, Hoefsloot HCJ, van der Greef J, Timmerman ME, Smilde AK (2005) Multilevel component analysis of time-resolved metabolic fingerprinting data. Anal Chim Acta 530 (2):173–183. doi:10.1016/j.aca.2004.09.074

Jansen JJ, Blanchet L, Buydens LMC, Bertrand S, Wolfender J-L (2015) Projected orthogonalized chemical encounter MONitoring (POCHEMON) for microbial interactions in co-culture. Metabolomics 11(4):908–919. doi:10.1007/s11306-014-0748-5

Jebaraj CS, Raghukumar C, Behnke A, Stoeck T (2010) Fungal diversity in oxygen-depleted regions of the Arabian Sea revealed by targeted environmental sequencing combined with cultivation. FEMS Microbiol Ecol 71(3):399–412. doi:10.1111/j.1574-6941.2009.00804.x (FEM804 [pii])

Johnson TW, Sparrow FK (1961) Fungi in oceans and estuaries

Jones E (2000) Marine fungi: some factors influencing biodiversity. Fungal Divers 4(193):53–73

Jones EBG, Pang K-L (2012). Marine fungi and fungal-like organisms. Marine and freshwater part of: marine and freshwater botany. Walter De Gruyter, Berlin

Jones OAH, Maguire ML, Griffin JL, Jung Y-H, Shibato J, Rakwal R, Agrawal GK, Jwa N-S (2010) Using metabolic profiling to assess plant-pathogen interactions: an example using rice (*Oryza sativa*) and the blast pathogen *Magnaporthe grisea*. Eur J Plant Pathol 129(4):539–554. doi:10.1007/s10658-010-9718-6

Jones EBG, Suetrong S, Sakayaroj J, Bahkali AH, Abdel-Wahab MA, Boekhout KLP (2015) Classification of marine Ascomycota, Basidiomycota, Blastocladiomycota and Chytridiomycota. Fungal Divers 3:1–72. doi:10.1007/s13225-015-0339-4

Julianti E, Lee J-H, Liao L, Park W, Park S, Oh D-C, Oh K-B, Shin J (2013) New polyaromatic metabolites from a marine-derived fungus *Penicillium* sp. Org Lett 15(6):1286–1289. doi:10.1021/ol4002174

Kale SP, Milde L, Trapp MK, Frisvad JC, Keller NP, Bok JW (2008) Requirement of LaeA for secondary metabolism and sclerotial production in *Aspergillus flavus*. Fungal genetics and biology: FG & B 45(10):1422–1429. doi:10.1016/j.fgb.2008.06.009

Kanehisa M, Goto S, Furumichi M, Tanabe M, Hirakawa M (2010). KEGG for representation and analysis of molecular networks involving diseases and drugs. Nucleic Acids Res 38(Database issue):D355–D360. doi:10.1093/nar/gkp896

Kawaide H, Imai R, Sassa T, Kamiya Y (1997) Ent-kaurene synthase from the fungus *Phaeosphaeria* sp. L487. cDNA isolation, characterization, and bacterial expression of a bifunctional diterpene cyclase in fungal gibberellin biosynthesis. J Biol Chem 272(35):21706–21712

Keller NP, Turner G, Bennett JW (2005) Fungal secondary metabolism—from biochemistry to genomics. Nat Rev Microbiol 3(12):937–947

Khaldi N, Seifuddin FT, Turner G, Haft D, Nierman WC, Wolfe KH, Fedorova ND (2010) SMURF: genomic mapping of fungal secondary metabolite clusters. Fungal Genet Biol 47 (9):736–741. doi:10.1016/j.fgb.2010.06.003

Kim JK, Bamba T, Harada K, Fukusaki E, Kobayashi A (2007) Time-course metabolic profiling in *Arabidopsis thaliana* cell cultures after salt stress treatment. J Exp Bot 58(3):415–424. doi:10.1093/jxb/erl216

Kingham DL, Evans LV (1986) The Pelvetia-Mycosphaerella interrelationship. In: Moss ST (ed) The biology of marine fungi, vol 17. Cambridge University Press, Cambridge, pp 177–187

Kis-Papo T (2005). Marine fungal communities. In: The fungal community. mycology. CRC Press, New York, pp 61–92. doi:10.1201/9781420027891.ch4

Klitgaard A, Iversen A, Andersen MR, Larsen TO, Frisvad JC, Nielsen KF (2014) Aggressive dereplication using UHPLC–DAD–QTOF: screening extracts for up to 3000 fungal secondary metabolites. Anal Bioanal Chem 406(7):1933–1943. doi:10.1007/s00216-013-7582-x Epub 2014 Jan 18

Kohlmeyer J (1971) Fungi from the Sargasso Sea. Mar Biol 8(4):344–350. doi:10.1007/bf00348012

Kohlmeyer J, Demoulin V (1981) Parasitic and symbiotic fungi on marine algae. Botanica Marina 24. doi:10.1515/botm.1981.24.1.9

Kohlmeyer J, Kohlmeyer E (1979) Marine mycology: the higher fungi, p 690

Kohlmeyer J, Volkmann-Kohlmeyer B (1990) New species of Koralionastes (Ascomycotina) from the Caribbean and Australia. Can J Bot 68(7):1554–1559. doi:10.1139/b90-199

Kohlmeyer J, Volkmann-Kohlmeyer B (2003a) Fungi from coral reefs: a commentary. Mycol Res 107(04):386–387

Kohlmeyer J, Volkmann-Kohlmeyer B (2003b) *Octopodotus stupendus* gen. & sp. nov. and *Phyllachora paludicola* sp. nov., two marine fungi from *Spartina alterniflora*. Mycologia 95 (1):117–123

Kohlmeyer J, Volkmann-Kohlmeyer B (2003c) Marine ascomycetes from algae and animal hosts. Botanica Marina 46. doi:10.1515/bot.2003.026

Kosalková K, García-Estrada C, Ullán RV, Godio RP, Feltrer R, Teijeira F, Mauriz E, Martín JF (2009) The global regulator LaeA controls penicillin biosynthesis, pigmentation and sporulation, but not roquefortine C synthesis in Penicillium chrysogenum. Biochimie 91 (2):214–225. doi:10.1016/j.biochi.2008.09.004

Kramer A, Beck HC, Kumar A, Kristensen LP, Imhoff JF, Labes A (2015) Proteomic analysis of anti-cancerous scopularide production by a marine *Microascus brevicaulis* strain and Its UV mutant. PLoS ONE 10(10):e0140047. doi:10.1371/journal.pone.0140047

Kumar A et al (2015) De novo assembly and genome analyses of the marine sponge derived *Scopulariopsis brevicaulis* strain LF580 unravels life-style traits and anticancerous scopularide producing hybrid NRPS-PKS cluster. PLoS ONE (In Press)
La Cono V, Smedile F, Bortoluzzi G, Arcadi E, Maimone G, Messina E, Borghini M, Oliveri E, Mazzola S, L'Haridon S, Toffin L, Genovese L, Ferrer M, Giuliano L, Golyshin PN, Yakimov MM (2011) Unveiling microbial life in new deep-sea hypersaline Lake Thetis. Part I: prokaryotes and environmental settings. Environ Microbiol 13(8):2250–2268. doi:10.1111/j.1462-2920.2011.02478.x
Lahav R, Fareleira P, Nejidat A, Abeliovich A (2002) The identification and characterization of osmotolerant yeast isolates from chemical wastewater evaporation ponds. Microb Ecol 43 (3):388–396. doi:10.1007/s00248-002-2001-4
Lai X, Cao L, Tan H, Fang S, Huang Y, Zhou S (2007) Fungal communities from methane hydrate-bearing deep-sea marine sediments in South China Sea. ISME J 1(8):756–762. doi:10.1038/ismej.2007.51 (ismej200751 [pii])
Le Calvez T, Burgaud G, Mahe S, Barbier G, Vandenkoornhuyse P (2009) Fungal diversity in deep-sea hydrothermal ecosystems. Appl Environ Microbiol 75(20):6415–6421. doi:10.1128/AEM.00653-09 (AEM.00653-09 [pii])
Li Q, Wang G (2009) Diversity of fungal isolates from three Hawaiian marine sponges. Microbiol Res 164(2):233–241. doi:10.1016/j.micres.2007.07.002 (S0944-5013(07)00088-2 [pii])
Li C-Y, Ding W-J, Shao C-L, She Z-G, Lin Y-C (2010) A new diimide derivative from the co-culture broth of two mangrove fungi (strain no. E33 and K38). J Asian Nat Prod Res 12 (9):809–813. doi:10.1080/10286020.2010.497757
Li C, Zhang J, Shao C, Ding W, She Z, Lin Y (2011) A new xanthone derivative from the co-culture broth of two marine fungi (strain No. E33 and K38). Chem Nat Compd 47(3):382–384. doi:10.1007/s10600-011-9939-8
Li C, Cox D, Huang S, Ding W (2014) Two new cyclopeptides from the co-culture broth of two marine mangrove fungi and their antifungal activity. Pharmacognosy Magazine 10(40):410. doi:10.4103/0973-1296.141781
Lin Z, Zhu T, Wei H, Zhang G, Wang H, Gu Q (2009) Spicochalasin A and new aspochalasins from the marine-derived fungus *Spicaria elegans*. Eur J Org Chem 2009(18):3045–3051. doi:10.1002/ejoc.200801085
Liu W, Li C, Zhu P, Yang J, Cheng K (2010) Phylogenetic diversity of culturable fungi associated with two marine sponges: *Haliclona simulans* and *Gelliodes carnosa*, collected from the Hainan Island coastal waters of the South China Sea. Fungal Divers 42(1):1–15
Loilong A, Sakayaroj J, Rungjindamai N, Choeyklin R, Jones EBG (2012) Biodiversity of fungi on the palm *Nypa fruticans*. Marine Fungi and Fungal-like Organisms. doi:10.1515/9783110264067.273
Lopez-Garcia P, Vereshchaka A, Moreira D (2007) Eukaryotic diversity associated with carbonates and fluid-seawater interface in Lost City hydrothermal field. Environ Microbiol 9 (2):546–554. doi:10.1111/j.1462-2920.2006.01158.x (EMI1158 [pii])
Loque C, Medeiros A, Pellizzari F, Oliveira E, Rosa C, Rosa L (2010) Fungal community associated with marine macroalgae from Antarctica. Polar Biol 33(5):641–648. doi:10.1007/s00300-009-0740-0
Lorenz R, Molitoris HP (1997) Cultivation of fungi under simulated deep sea conditions. Mycol Res 101(11):1355–1365. doi:10.1017/S095375629700405X
Lukassen M, Saei W, Sondergaard T, Tamminen A, Kumar A, Kempken F, Wiebe M, Sørensen J (2015) Identification of the scopularide biosynthetic gene cluster in *Scopulariopsis brevicaulis*. Marine Drugs 13(7):4331
Luo ZH, Pang KL (2014) Fungi on substrates in marine environment. In: Misra JK, Deshmukh SK, Vágvölgyi C (eds) (ed) Progress in mycological research, vol III. Fungi in/on various substrates. CRC Press, Baca Raton, pp 97–114
Lyons JI, Newell SY, Buchan A, Moran MA (2003) Diversity of ascomycete laccase gene sequences in a southeastern US salt marsh. Microb Ecol 45(3):270–281. doi:10.1007/s00248-002-1055-7

Ma L-J, Fedorova ND (2010) A practical guide to fungal genome projects: strategy, technology, cost and completion. Mycology 1(1):9–24. doi:10.1080/21501201003680943

Mahdavi V, Farimani MM, Fathi F, Ghassempour A (2015) A targeted metabolomics approach toward understanding metabolic variations in rice under pesticide stress. Anal Biochem 478:65–72. doi:10.1016/j.ab.2015.02.021

Mahé S, Rédou V, Le Calvez T, Vandenkoornhuyse P, Burgaud G (2013) Fungi in deep-sea environments and metagenomics. In: Martin F (ed) The ecological genomics of the fungi. Wiley, New York

Marchler-Bauer A, Derbyshire MK, Gonzales NR, Lu S, Chitsaz F, Geer LY, Geer RC, He J, Gwadz M, Hurwitz DI, Lanczycki CJ, Lu F, Marchler GH, Song JS, Thanki N, Wang Z, Yamashita RA, Zhang D, Zheng C, Bryant SH (2015). CDD: NCBI's conserved domain database. Nucleic Acids Res 43(Database issue):D222–D226. doi:10.1093/nar/gku1221

Margulies M, Egholm M, Altman WE, Attiya S, Bader JS, Bemben LA, Berka J, Braverman MS, Chen Y-J, Chen Z, Dewell SB, Du L, Fierro JM, Gomes XV, Goodwin BC, He W, Helgesen S, Ho CH, Irzyk GP, Jando SC, Alenquer MLI, Jarvie TP, Jirage KB, Kim J-B, Knight JR, Lanza JR, Leamon JH, Lefkowitz SM, Lei M, Li J, Lohman KL, Lu H, Makhijani VB, McDade KE, McKenna MP, Myers EW, Nickerson E, Nobile JR, Plant R, Puc BP, Ronan MT, Roth GT, Sarkis GJ, Simons JF, Simpson JW, Srinivasan M, Tartaro KR, Tomasz A, Vogt KA, Volkmer GA, Wang SH, Wang Y, Weiner MP, Yu P, Begley RF, Rothberg JM (2005) Genome sequencing in open microfabricated high density picoliter reactors. Nature 437 (7057):376–380. doi:10.1038/nature03959

Marmann A, Aly A, Lin W, Wang B, Proksch P (2014) Co-cultivation—a powerful emerging tool for enhancing the chemical diversity of microorganisms. Mar Drugs 12(2):1043

Martín-Rodríguez A, Reyes F, Martín J, Pérez-Yépez J, León-Barrios M, Couttolenc A, Espinoza C, Trigos Á, Martín V, Norte M, Fernández J (2014) Inhibition of bacterial quorum sensing by extracts from aquatic fungi: first report from marine endophytes. Mar Drugs 12 (11):5503

Massana R, Pedrós-Alió C (2008) Unveiling new microbial eukaryotes in the surface ocean. Curr Opin Microbiol 11(3):213–218. doi:10.1016/j.mib.2008.04.004 (S1369-5274(08)00056-8 [pii])

Medema MH, Fischbach MA (2015) Computational approaches to natural product discovery. Nat Chem Biol 11(9):639–648. doi:10.1038/nchembio.1884

Medema MH, Blin K, Cimermancic P, de Jager V, Zakrzewski P, Fischbach MA, Weber T, Takano E, Breitling R (2011) AntiSMASH: rapid identification, annotation and analysis of secondary metabolite biosynthesis gene clusters in bacterial and fungal genome sequences. Nucleic Acids Res 39(Web Server issue):W339–W346. doi:10.1093/nar/gkr466 (gkr466 [pii])

Medema MH, Kottmann R, Yilmaz P, Cummings M, Biggins JB, Blin K, de Bruijn I, Chooi YH, Claesen J, Coates RC, Cruz-Morales P, Duddela S, Dusterhus S, Edwards DJ, Fewer DP, Garg N, Geiger C, Gomez-Escribano JP, Greule A, Hadjithomas M, Haines AS, Helfrich EJN, Hillwig ML, Ishida K, Jones AC, Jones CS, Jungmann K, Kegler C, Kim HU, Kotter P, Krug D, Masschelein J, Melnik AV, Mantovani SM, Monroe EA, Moore M, Moss N, Nutzmann H-W, Pan G, Pati A, Petras D, Reen FJ, Rosconi F, Rui Z, Tian Z, Tobias NJ, Tsunematsu Y, Wiemann P, Wyckoff E, Yan X, Yim G, Yu F, Xie Y, Aigle B, Apel AK, Balibar CJ, Balskus EP, Barona-Gomez F, Bechthold A, Bode HB, Borriss R, Brady SF, Brakhage AA, Caffrey P, Cheng Y-Q, Clardy J, Cox RJ, De Mot R, Donadio S, Donia MS, van der Donk WA, Dorrestein PC, Doyle S, Driessen AJM, Ehling-Schulz M, Entian K-D, Fischbach MA, Gerwick L, Gerwick WH, Gross H, Gust B, Hertweck C, Hofte M, Jensen SE, Ju J, Katz L, Kaysser L, Klassen JL, Keller NP, Kormanec J, Kuipers OP, Kuzuyama T, Kyrpides NC, Kwon H-J, Lautru S, Lavigne R, Lee CY, Linquan B, Liu X, Liu W, Luzhetskyy A, Mahmud T, Mast Y, Mendez C, Metsa-Ketela M, Micklefield J, Mitchell DA, Moore BS, Moreira LM, Muller R, Neilan BA, Nett M, Nielsen J, O'Gara F, Oikawa H, Osbourn A, Osburne MS, Ostash B, Payne SM, Pernodet J-L, Petricek M, Piel J, Ploux O, Raaijmakers JM, Salas JA, Schmitt EK, Scott B, Seipke RF, Shen B, Sherman DH, Sivonen K, Smanski MJ, Sosio M, Stegmann E, Sussmuth RD, Tahlan K, Thomas CM, Tang Y,

Truman AW, Viaud M, Walton JD, Walsh CT, Weber T, van Wezel GP, Wilkinson B, Willey JM, Wohlleben W, Wright GD, Ziemert N, Zhang C, Zotchev SB, Breitling R, Takano E, Glockner FO (2015) Minimum Information about a biosynthetic gene cluster. Nat Chem Biol 11(9):625–631. doi:10.1038/nchembio.1890. http://www.nature.com/nchembio/journal/v11/n9/abs/nchembio.1890.html#supplementary-information

Menezes CB, Bonugli-Santos RC, Miqueletto PB, Passarini MR, Silva CH, Justo MR, Leal RR, Fantinatti-Garboggini F, Oliveira VM, Berlinck RG, Sette LD (2010) Microbial diversity associated with algae, ascidians and sponges from the north coast of Sao Paulo state, Brazil. Microbiol Res 165(6):466–482. doi:10.1016/j.micres.2009.09.005 (S0944-5013(09)00077-9 [pii])

Metzker ML (2010) Sequencing technologies—the next generation. Nat Rev Genet 11(1):31–46. doi:10.1038/nrg2626

Moghadamtousi S, Nikzad S, Kadir H, Abubakar S, Zandi K (2015) Potential antiviral agents from marine fungi: an overview. Mar Drugs 13(7):4520

Montaser R, Luesch H (2011) Marine natural products: a new wave of drugs? Futu Med Chem 3 (12):1475–1489. doi:10.4155/fmc.11.118

Moree WJ, Yang JY, Zhao X, Liu W-T, Aparicio M, Atencio L, Ballesteros J, Sánchez J, Gavilán RG, Gutiérrez M, Dorrestein PC (2013) Imaging mass spectrometry of a coral microbe interaction with fungi. J Chem Ecol 39(7):1045–1054. doi:10.1007/s10886-013-0320-1

Mouton M, Postma F, Wilsenach J, Botha A (2012) Diversity and characterization of culturable fungi from marine sediment collected from St. Helena Bay, South Africa. Microb Ecol 64 (2):311–319. doi:10.1007/s00248-012-0035-9

Nagahama T, Nagano Y (2012) Cultured and uncultured fungal diversity in deep-sea environments. Prog Mol Subcell Biol 53:173–187. doi:10.1007/978-3-642-23342-5_9

Nagahama T, Hamamoto M, Nakase T, Takami H, Horikoshi K (2001) Distribution and identification of red yeasts in deep-sea environments around the northwest Pacific Ocean. Antonie Van Leeuwenhoek 80(2):101–110

Nagahama T, Hamamoto M, Horikoshi K (2006) *Rhodotorula pacifica* sp. nov., a novel yeast species from sediment collected on the deep-sea floor of the north-west Pacific Ocean. Int J Syst Evol Microbiol 56(Pt 1):295–299. doi:10.1099/ijs.0.63584-0

Nagahama T, Takahashi E, Nagano Y, Abdel-Wahab MA, Miyazaki M (2011) Molecular evidence that deep-branching fungi are major fungal components in deep-sea methane cold-seep sediments. Environ Microbiol 13(8):2359–2370. doi:10.1111/j.1462-2920.2011.02507.x

Nagano Y, Nagahama T, Hatada Y, Nunoura T, Takami H, Miyazaki J, Takai K, Horikoshi K (2010) Fungal diversity in deep-sea sediments—the presence of novel fungal groups. Fungal Ecology 3(4):316–325. doi:10.1016/j.funeco.2010.01.002

Netzker T, Fischer J, Weber J, Mattern DJ, König CC, Valiante V, Schroeckh V, Brakhage AA (2015) Microbial communication leading to the activation of silent fungal secondary metabolite gene clusters. Front Microbiol 6. doi:10.3389/fmicb.2015.00299

Newell SY (2001) Spore-expulsion rates and extents of blade occupation by ascomycetes of the smooth-cordgrass standing-decay system. Botanica Marina, 44. doi:10.1515/bot.2001.036

Newell S, Porter D (2000) Microbial secondary production from salt marsh-grass shoots, and its known and potential fates. In: Weinstein M, Kreeger D (eds) Concepts and controversies in tidal marsh ecology. Springer, Netherlands, pp 159–185. doi:10.1007/0-306-47534-0_9

Newell S, Wasowski J (1995) Sexual productivity and spring intramarsh distribution of a key salt-marsh microbial secondary producer. Estuaries 18(1):241–249. doi:10.2307/1352634

Newell SY, Porter D, Lingle WL (1996) Lignocellulolysis by ascomycetes (fungi) of a saltmarsh grass (smooth cordgrass). Microsc Res Tech 33(1):32–46 doi:10.1002/(SICI)1097-0029 (199601)33:1<32:AID-JEMT5>3.0.CO;2-2 [pii]

Newman DJ, Cragg GM (2007) Natural products as sources of new drugs over the last 25 years. J Nat Prod 70(3):461–477. doi:10.1021/np068054v

Newman D, Cragg G (2014) Marine-sourced anti-cancer and cancer pain control agents in clinical and late preclinical development. Mar Drugs 12(1):255

Nguyen Q-T, Merlo ME, Medema MH, Jankevics A, Breitling R, Takano E (2012) Metabolomics methods for the synthetic biology of secondary metabolism. FEBS Lett 586(15):2177–2183. doi:10.1016/j.febslet.2012.02.008

Nicholson TP, Rudd BA, Dawson M, Lazarus CM, Simpson TJ, Cox RJ (2001) Design and utility of oligonucleotide gene probes for fungal polyketide synthases. Chem Biol 8(2):157–178 (S1074-5521(00)90064-4 [pii])

Nicholson B, Lloyd GK, Miller BR, Palladino MA, Kiso Y, Hayashi Y, Neuteboom STC (2006) NPI-2358 is a tubulin-depolymerizing agent: in-vitro evidence for activity as a tumor vascular-disrupting agent. Anticancer Drugs 17(1):25–31

Nowrousian M, Stajich JE, Chu M, Engh I, Espagne E, Halliday K, Kamerewerd J, Kempken F, Knab B, Kuo HC, Osiewacz HD, Poggeler S, Read ND, Seiler S, Smith KM, Zickler D, Kuck U, Freitag M (2010) De novo assembly of a 40 Mb eukaryotic genome from short sequence reads: *Sordaria macrospora*, a model organism for fungal morphogenesis. PLoS Genet 6(4):e1000891. doi:10.1371/journal.pgen.1000891

Numata A, Takahashi C, Ito Y, Takada T, Kawai K, Usami Y, Matsumura E, Imachi M, Ito T, Hasegawa T (1993) Communesins, cytotoxic metabolites of a fungus isolated from a marine alga. Tetrahedron Lett 34(14):2355–2358

Numata A, Takahashi C, Ito Y, Minoura K, Yamada T, Matsuda C, Nomoto K (1996) Penochalasins, a novel class of cytotoxic cytochalasans from a *Penicillium* species separated from a marine alga: structure determination and solution conformation. J Chem Soc Perkin Trans 1(3):239–245. doi:10.1039/p19960000239

Ochi K (2007) From microbial differentiation to ribosome engineering. Biosci Biotechnol Biochem 71(6):1373–1386. doi:10.1271/bbb.70007

Ochi K, Hosaka T (2012) New strategies for drug discovery: activation of silent or weakly expressed microbial gene clusters. Appl Microbiol Biotechnol 97(1):87–98. doi:10.1007/s00253-012-4551-9

Oger PM, Jebbar M (2010) The many ways of coping with pressure. Res Microbiol 161(10):799–809. doi:10.1016/j.resmic.2010.09.017 (S0923-2508(10)00219-6 [pii])

Oh D-C, Jensen PR, Kauffman CA, Fenical W (2005) Libertellenones A-D: Induction of cytotoxic diterpenoid biosynthesis by marine microbial competition. Bioorg Med Chem 13(17):5267–5273. doi:10.1016/j.bmc.2005.05.068

Oh D-C, Kauffman CA, Jensen PR, Fenical W (2007) Induced production of emericellamides A and B from the marine-derived fungus *Emericella* sp. in competing co-culture. J Nat Prod 70 (4):515–520. doi:10.1021/np060381f

Ola ARB, Thomy D, Lai D, Brötz-Oesterhelt H, Proksch P (2013) Inducing secondary metabolite production by the endophytic fungus *Fusarium tricinctum* through coculture with *Bacillus subtilis*. J Nat Prod 76(11):2094–2099. doi:10.1021/np400589h

Orsi W, Biddle JF, Edgcomb V (2013a) Deep sequencing of subseafloor eukaryotic rRNA reveals active fungi across marine subsurface provinces. PLoS ONE 8(2):e56335. doi:10.1371/journal.pone.0056335 (PONE-D-12-30276 [pii])

Orsi WD, Edgcomb VP, Christman GD, Biddle JF (2013b). Gene expression in the deep biosphere. Nature (nature12230)

Osterhage C, Kaminsky R, König GM, Wright AD (2000) Ascosalipyrrolidinone A, an antimicrobial alkaloid, from the obligate marine fungus *Ascochyta salicorniae*. J Org Chem 65 (20):6412–6417. doi:10.1021/jo000307g

Overy DP, Zidorn C, Petersen BO, Duus JØ, Dalsgaard PW, Larsen TO, Phipps RK (2005) Medium dependant production of corymbiferone a novel product from *Penicillium hordei* cultured on plant tissue agar. Tetrahedron Lett 46(18):3225–3228. doi:10.1016/j.tetlet.2005.03.043

Pachiadaki MG, Yakimov MM, LaCono V, Leadbetter E, Edgcomb V (2014) Unveiling microbial activities along the halocline of Thetis, a deep-sea hypersaline anoxic basin. ISME J 8 (12):2478–2489. doi:10.1038/ismej.2014.100 (ismej2014100 [pii])

Pang KL (2002) Systematics of the halosphaeriales: which morphological characters are important? In: Hyde K (ed) Fungi in marine environments. Fungal Diversity Press, Hong Kong, pp 35–57
Panzer K, Yilmaz P, Weiß M, Reich L, Richter M, Wiese J, Schmaljohann R, Labes A, Imhoff JF, Glöckner FO, Reich M (2015) Identification of habitat-specific biomes of aquatic fungal communities using a comprehensive nearly full-length 18S rRNA dataset enriched with contextual data. PLoS ONE 10(7):e0134377. doi:10.1371/journal.pone.0134377
Pauli GF, Chen S-N, Lankin DC, Bisson J, Case RJ, Chadwick LR, Gödecke T, Inui T, Krunic A, Jaki BU, McAlpine JB, Mo S, Napolitano JG, Orjala J, Lehtivarjo J, Korhonen S-P, Niemitz M (2014) Essential parameters for structural analysis and dereplication by 1H NMR spectroscopy. J Nat Prod 77(6):1473–1487. doi:10.1021/np5002384
Paz Z, Komon-Zelazowska M, Druzhinina IS, Aveskamp MM, Shnaiderman A, Aluma Y, Carmeli S, Ilan M, Yarden O (2010) Diversity and potential antifungal properties of fungi associated with a Mediterranean sponge. Fungal Divers 42(1):17–26. doi:10.1007/s13225-010-0020-x
Peiris D, Dunn W, Brown M, Kell D, Roy I, Hedger J (2008) Metabolite profiles of interacting mycelial fronts differ for pairings of the wood decay basidiomycete fungus, *Stereum hirsutum* with its competitors *Coprinus micaceus* and *Coprinus disseminatus*. Metabolomics 4(1):52–62. doi:10.1007/s11306-007-0100-4
Peters S, Janssen H-G, Vivó-Truyols G (2010) Trend analysis of time-series data: a novel method for untargeted metabolite discovery. Anal Chim Acta 663(1):98–104. doi:10.1016/j.aca.2010.01.038
Pettit RK (2009) Mixed fermentation for natural product drug discovery. Appl Microbiol Biotechnol 83(1):19–25. doi:10.1007/s00253-009-1916-9
Pettit RK (2011) Small-molecule elicitation of microbial secondary metabolites. Microb Biotechnol 4(4):471–478. doi:10.1111/j.1751-7915.2010.00196.x
Pilantanapak A, Jones EBG, Eaton Rod A (2005) Marine fungi on *Nypa fruticans* in Thailand. Botanica Marina 48. doi:10.1515/bot.2005.049
Pimentel-Elardo S, Sørensen D, Ho L, Ziko M, Bueler SA, Lu S, Tao J, Moser A, Lee R, Agard DA, Fairn G, Rubinstein JL, Shoichet BK, Nodwell J (2015) Activity-independent discovery of secondary metabolites using chemical elicitation and cheminformatic inference. ACS Chem Biol. doi:10.1021/acschembio.5b00612
Potin P, Bouarab K, Salaun JP, Pohnert G, Kloareg B (2002) Biotic interactions of marine algae. Curr Opin Plant Biol 5(4):308–317
Proctor RH, Desjardins AE, Plattner RD, Hohn TM (1999) A polyketide synthase gene required for biosynthesis of fumonisin mycotoxins in *Gibberella fujikuroi* mating population A. Fungal Genet Biol 27(1):100–112. doi:10.1006/fgbi.1999.1141 (S1087-1845(99)91141-6 [pii])
Puel O, Galtier P, Oswald IP (2010) Biosynthesis and toxicological effects of patulin. Toxins (Basel) 2(4):613–631. doi:10.3390/toxins2040613 (toxins-02-00613 [pii])
Raghukumar C (1986). Fungal parasites of the marine green algae, Cladophora and Rhizoclonium. Botanica Marina 29. doi:10.1515/botm.1986.29.4.289
Raghukumar C, Raghukumar S (1998) Barotolerance of fungi isolated from deep-sea sediments of the Indian Ocean. Aquat Microb Ecol 15(2):153–163. doi:10.3354/ame015153
Raghukumar C, Raghukumar S, Sheelu G, Gupta SM, Nagender Nath B, Rao BR (2004) Buried in time: culturable fungi in a deep-sea sediment core from the Chagos Trench, Indian Ocean. Deep Sea Res Part I 51(11):1759–1768. doi:10.1016/j.dsr.2004.08.002
Rämä T, Nordén J, Davey ML, Mathiassen GH, Spatafora JW, Kauserud H (2014) Fungi ahoy! Diversity on marine wooden substrata in the high North. Fungal Ecology 8:46–58. doi:10.1016/j.funeco.2013.12.002
Rateb ME, Ebel R (2011) Secondary metabolites of fungi from marine habitats. Nat Prod Rep 28 (2):290–344. doi:10.1039/c0np00061b
Rédou V, Ciobanu MC, Pachiadaki MG, Edgcomb V, Alain K, Barbier G, Burgaud G (2014) In-depth analyses of deep subsurface sediments using 454-pyrosequencing reveals a reservoir

of buried fungal communities at record-breaking depths. FEMS Microbiol Ecol 90(3):908–921. doi:10.1111/1574-6941.12447

Rédou V, Navarri M, Meslet-Cladiere L, Barbier G, Burgaud G (2015) Species richness and adaptation of marine fungi from deep-subseafloor sediments. Appl Environ Microbiol 81 (10):3571–3583. doi:10.1128/AEM.04064-14 (AEM.04064-14 [pii])

Reen F, Romano S, Dobson A, O'Gara F (2015) The sound of silence: activating silent biosynthetic gene clusters in marine microorganisms. Mar Drugs 13(8):4754–4783. doi:10. 3390/md13084754

Richards TA, Jones MDM, Leonard G, Bass D (2012) Marine fungi: their ecology and molecular diversity. Ann Rev Mar Sci 4(1):495–522. doi:10.1146/annurev-marine-120710-100802

Robinson CH (2001) Cold adaptation in Arctic and Antarctic fungi. New Phytol 151(2):341–353. doi:10.1046/j.1469-8137.2001.00177.x

Rollins MJ, Gaucher GM (1994) Ammonium repression of antibiotic and intracellular proteinase production in *Penicillium urticae*. Appl Microbiol Biotechnol 41(4):447–455

Roth F Jr, Orpurt P, Ahearn DG (1964) Occurrence and distribution of fungi in a subtropical marine environment. Can J Bot 42(4):375–383

Sakayaroj JPS, Supaphon O, Jones EBG, Phongpaichit S (2010) Phylogenetic diversity of endophyte assemblages associated with the tropical seagrass *Enhalus acoroides* in Thailand. Fungal Divers 42:27–45. doi:10.1007/s13225-009-0013-9

Sakayaroj J, Preedanon S, Phongpaichit S, Buatong J, Chaowalit P, Rukachaisirikul V (2012) Diversity of endophytic and marine-derived fungi associated with marine plants and animals. Mar Fungi Fungal-Like Org. doi:10.1515/9783110264067.291

Sato S, Arita M, Soga T, Nishioka T, Tomita M (2008) Time-resolved metabolomics reveals metabolic modulation in rice foliage. BMC Syst Biol 2(1):51. doi:10.1186/1752-0509-2-51

Schatz S (1983) The developmental morphology and life history of *Phycomelaina laminariae*. Mycologia 75(5):762–772. doi:10.2307/3792768

Scherlach K, Nützmann H-W, Schroeckh V, Dahse H-M, Brakhage AA, Hertweck C (2011) Cytotoxic pheofungins from an engineered fungus impaired in posttranslational protein modification. Angew Chem Int Ed 50(42):9843–9847. doi:10.1002/anie.201104488

Schoch CL, Seifert KA, Huhndorf S, Robert V, Spouge JL, Levesque CA, Chen W (2012) Nuclear ribosomal internal transcribed spacer (ITS) region as a universal DNA barcode marker for Fungi. Proc Natl Acad Sci USA 109(16):6241–6246. doi:10.1073/pnas.1117018109 (1117018109 [pii])

Schroeckh V, Scherlach K, Nützmann H-W, Shelest E, Schmidt-Heck W, Schuemann J, Martin K, Hertweck C, Brakhage AA (2009) Intimate bacterial–fungal interaction triggers biosynthesis of archetypal polyketides in *Aspergillus nidulans*. Proc Natl Acad Sci USA 106(34):14558–14563. doi:10.1073/pnas.0901870106

Schulz B, Draeger S, dela Cruz Thomas E, Rheinheimer J, Siems K, Loesgen S, Bitzer J, Schloerke O, Zeeck A, Kock I, Hussain H, Dai J, Krohn K (2008) Screening strategies for obtaining novel, biologically active, fungal secondary metabolites from marine habitats. Botanica Marina 51. doi:10.1515/bot.2008.029

Schümann J, Hertweck C (2006) Advances in cloning, functional analysis and heterologous expression of fungal polyketide synthase genes. J Biotechnol 124(4):690–703. doi:10.1016/j. jbiotec.2006.03.046

Scott RE, Jones A, Lam KS, Gaucher GM (1986) Manganese and antibiotic biosynthesis. I. A specific manganese requirement for patulin production in *Penicillium urticae*. Can J Microbiol 32(3):259–267

Shen YQ, Heim J, Solomon NA, Wolfe S, Demain AL (1984) Repression of beta-lactam production in *Cephalosporium acremonium* by nitrogen sources. J Antibiot 37(5):503–511

Simonato F, Campanaro S, Lauro FM, Vezzi A, D'Angelo M, Vitulo N, Valle G, Bartlett DH (2006) Piezophilic adaptation: a genomic point of view. J Biotechnol 126(1):11–25. doi:10. 1016/j.jbiotec.2006.03.038 (S0168-1656(06)00316-6 [pii])

Singh P, Raghukumar C, Verma P, Shouche Y (2011) Fungal community analysis in the deep-sea sediments of the Central Indian Basin by culture-independent approach. Microb Ecol 61 (3):507–517. doi:10.1007/s00248-010-9765-8
Singh P, Raghukumar C, Meena RM, Verma P, Shouche Y (2012) Fungal diversity in deep-sea sediments revealed by culture-dependent and culture-independent approaches. Fungal Ecol 5 (5):543–553. doi:10.1016/j.funeco.2012.01.001
Slightom JL, Metzger BP, Luu HT, Elhammer AP (2009) Cloning and molecular characterization of the gene encoding the Aureobasidin A biosynthesis complex in *Aureobasidium pullulans* BP-1938. Gene 431(1–2):67–79. doi:10.1016/j.gene.2008.11.011 (S0378-1119(08)00594-5 [pii])
Smith GW, Ives LD, Nagelkerken IA, Ritchie KB (1996) Caribbean sea-fan mortalities. Nature 383(6600):487. doi:10.1038/383487a0
Soanes KH, Achenbach JC, Burton IW, Hui JPM, Penny SL, Karakach TK (2011) Molecular characterization of Zebrafish embryogenesis via DNA microarrays and multiplatform time course metabolomics studies. J Proteome Res 10(11):5102–5117. doi:10.1021/pr2005549
Sohlberg E, Bomberg M, Miettinen H, Nyyssonen M, Salavirta H, Vikman M, Itavaara M (2015) Revealing the unexplored fungal communities in deep groundwater of crystalline bedrock fracture zones in Olkiluoto. Finland. Front Microbiol 6:573. doi:10.3389/fmicb.2015.00573
Spatafora JW, Volkmann-Kohlmeyer B, Kohlmeyer J (1998) Independent terrestrial origins of the Halosphaeriales (marine Ascomycota). Am J Bot 85(11):1569–1580 (85/11/1569 [pii])
Stanke M, Morgenstern B (2005) AUGUSTUS: a web server for gene prediction in eukaryotes that allows user-defined constraints. Nucleic Acids Res 33(Web Server issue):W465–W467. doi:10.1093/nar/gki458
Stock A, Breiner HW, Pachiadaki M, Edgcomb V, Filker S, La Cono V, Yakimov MM, Stoeck T (2012) Microbial eukaryote life in the new hypersaline deep-sea basin Thetis. Extremophiles 16(1):21–34. doi:10.1007/s00792-011-0401-4
Stoeck T, Filker S, Edgcomb V, Orsi W, Yakimov MM, Pachiadaki M, Breiner H-W, LaCono V, Stock A (2014) Living at the limits: evidence for microbial eukaryotes thriving under pressure in deep anoxic, hypersaline habitats. Adv Ecol 2014:9. doi:10.1155/2014/532687
Suetrong S, Jones EBG (2006) Marine discomycetes: a review. Ind J Mar Sci 35:291–296
Takahashi C, Numata A, Yamada T, Minoura K, Enomoto S, Konishi K, Nakai M, Matsuda C, Nomoto K (1996) Penostatins, novel cytotoxic metabolites from a *Penicillium* species separated from a green alga. Tetrahedron Lett 37(5):655–658. doi:10.1016/0040-4039(95)02225-2
Takishita K, Tsuchiya M, Reimer JD, Maruyama T (2006) Molecular evidence demonstrating the basidiomycetous fungus *Cryptococcus curvatus* is the dominant microbial eukaryote in sediment at the Kuroshima Knoll methane seep. Extremophiles 10(2):165–169. doi:10.1007/s00792-005-0495-7
Takishita K, Yubuki N, Kakizoe N, Inagaki Y, Maruyama T (2007) Diversity of microbial eukaryotes in sediment at a deep-sea methane cold seep: surveys of ribosomal DNA libraries from raw sediment samples and two enrichment cultures. Extremophiles 11(4):563–576. doi:10.1007/s00792-007-0068-z
Thirunavukkarasu N, Suryanarayanan T, Girivasan K, Venkatachalam A, Geetha V, Ravishankar J, Doble M (2012) Fungal symbionts of marine sponges from Rameswaram, southern India: species composition and bioactive metabolites. Fungal Divers 55(1):37–46. doi:10.1007/s13225-011-0137-6
Torzilli AP, Sikaroodi M, Chalkley D, Gillevet PM (2006) A comparison of fungal communities from four salt marsh plants using automated ribosomal intergenic spacer analysis (ARISA). Mycologia 98(5):690–698
Turchetti B, Buzzini P, Goretti M, Branda E, Diolaiuti G, D'Agata C, Smiraglia C, Vaughan-Martini A (2008) Psychrophilic yeasts in glacial environments of Alpine glaciers. FEMS Microbiol Ecol 63(1):73–83. doi:10.1111/j.1574-6941.2007.00409.x (FEM409 [pii])

Turk M, Mejanelle L, Sentjurc M, Grimalt JO, Gunde-Cimerman N, Plemenitas A (2004) Salt-induced changes in lipid composition and membrane fluidity of halophilic yeast-like melanized fungi. Extremophiles 8(1):53–61. doi:10.1007/s00792-003-0360-5
Turk M, Montiel V, Zigon D, Plemenitas A, Ramos J (2007) Plasma membrane composition of *Debaryomyces hansenii* adapts to changes in pH and external salinity. Microbiology 153(Pt 10):3586–3592. doi:10.1099/mic.0.2007/009563-0 (153/10/3586 [pii])
Umemura M, Koike H, Nagano N, Ishii T, Kawano J, Yamane N, Kozone I, Horimoto K, Shin-ya K, Asai K, Yu J, Bennett JW, Machida M (2013) MIDDAS-M: motif-independent de novo detection of secondary metabolite gene clusters through the integration of genome sequencing and transcriptome data. PLoS ONE 8(12):e84028. doi:10.1371/journal.pone.0084028 (PONE-D-13-32652 [pii])
Unterseher M, Schnittler M (2009) Dilution-to-extinction cultivation of leaf-inhabiting endophytic fungi in beech (*Fagus sylvatica* L.)—different cultivation techniques influence fungal biodiversity assessment. Mycol Res 113(5):645–654. doi:10.1016/j.mycres.2009.02.002
Van der Molen KM, Darveaux BA, Chen W-L, Swanson SM, Pearce CJ, Oberlies NH (2014). Epigenetic manipulation of a filamentous fungus by the proteasome-inhibitor bortezomib induces the production of an additional secondary metabolite. RSC Adv 4(35):18329–18335. doi:10.1039/c4ra00274a
van der Wielen PW, Bolhuis H, Borin S, Daffonchio D, Corselli C, Giuliano L, D'Auria G, de Lange GJ, Huebner A, Varnavas SP, Thomson J, Tamburini C, Marty D, McGenity TJ, Timmis KN (2005) The enigma of prokaryotic life in deep hypersaline anoxic basins. Science 307(5706):121–123. doi:10.1126/science.1103569 (307/5706/121 [pii])
Venkatachalam A, Thirunavukkarasu N, Suryanarayanan TS (2015) Distribution and diversity of endophytes in seagrasses. Fungal Ecol 13:60–65. doi:10.1016/j.funeco.2014.07.003
Vervoort HC, Drašković M, Crews P (2011) Histone deacetylase inhibitors as a tool to up-regulate new fungal biosynthetic products: isolation of EGM-556, a cyclodepsipeptide, from *Microascus* sp. Org Lett 13(3):410–413. doi:10.1021/ol1027199
Vinaixa M, Samino S, Saez I, Duran J, Guinovart JJ, Yanes O (2012) A guideline to univariate statistical analysis for LC/MS-based untargeted metabolomics-derived data. Metabolites 2 (4):775–795. doi:10.3390/metabo2040775
von Nussbaum F (2003) Stephacidin B—a new stage of complexity within prenylated indole alkaloids from fungi. Angew Chem Int Ed 42(27):3068–3071. doi:10.1002/anie.200301646
Walker AK, Campbell J (2010) Marine fungal diversity: a comparison of natural and created salt marshes of the north-central Gulf of Mexico. Mycologia 102(3):513–521
Wang G, Li Q, Zhu P (2008) Phylogenetic diversity of culturable fungi associated with the Hawaiian Sponges *Suberites zeteki* and *Gelliodes fibrosa*. Antonie Van Leeuwenhoek 93(1–2):163–174. doi:10.1007/s10482-007-9190-2
Wang X, Sena Filho JG, Hoover AR, King JB, Ellis TK, Powell DR, Cichewicz RH (2010) Chemical epigenetics alters the secondary metabolite composition of guttate excreted by an Atlantic-forest-soil-derived *Penicillium citreonigrum*. J Nat Prod 73(5):942–948. doi:10.1021/np100142h
Weber T, Blin K, Duddela S, Krug D, Kim HU, Bruccoleri R, Lee SY, Fischbach MA, Müller R, Wohlleben W, Breitling R, Takano E, Medema MH (2015) AntiSMASH 3.0—a comprehensive resource for the genome mining of biosynthetic gene clusters. Nucleic Acids Research 43 (Web Server issue):W237–W243. doi:10.1093/nar/gkv437
Wiese J, Ohlendorf B, Blumel M, Schmaljohann R, Imhoff JF (2011) Phylogenetic identification of fungi isolated from the marine sponge *Tethya aurantium* and identification of their secondary metabolites. Mar Drugs 9(4):561–585. doi:10.3390/md9040561 (marinedrugs-09-00561 [pii])
Williams RB, Henrikson JC, Hoover AR, Lee AE, Cichewicz RH (2008) Epigenetic remodeling of the fungal secondary metabolome. Org Biomol Chem 6(11):1895–1897. doi:10.1039/b804701d

Wolfender J-L, Rudaz S, Hae Choi Y, Kyong Kim H (2013) Plant metabolomics: from holistic data to relevant biomarkers. Curr Med Chem 20(8):1056–1090. doi:10.2174/092986713805288932

Wolfender J-L, Marti G, Thomas A, Bertrand S (2015) Current approaches and challenges for the metabolite profiling of complex natural extracts. J Chromatogr A 1382:136–164. doi:10.1016/j.chroma.2014.10.091

Wu D, Oide S, Zhang N, Choi MY, Turgeon BG (2012) ChLae1 and ChVel1 regulate T-toxin production, virulence, oxidative stress response, and development of the maize pathogen *Cochliobolus heterostrophus*. PLoS Pathog 8(2):e1002542. doi:10.1371/journal.ppat.1002542

Wu CJ, Yi L, Cui CB, Li CW, Wang N, Han X (2015) Activation of the silent secondary metabolite production by introducing neomycin-resistance in a marine-derived *Penicillium purpurogenum* G59. Marine Drugs 13(4):2465–2487. doi:10.3390/md13042465

Xia J, Sinelnikov IV, Wishart DS (2011) MetATT: a web-based metabolomics tool for analyzing time-series and two-factor datasets. Bioinformatics 27(17):2455–2456. doi:10.1093/bioinformatics/btr392

Xia M-W, Cui C-B, Li C-W, Wu C-J, Peng J-X, Li D-H (2015) Rare chromones from a fungal mutant of the marine-derived *Penicillium purpurogenum* G59. Marine Drugs 13(8):5219–5236. doi:10.3390/md13085219

Xu W, Pang KL, Luo ZH (2014) High fungal diversity and abundance recovered in the deep-sea sediments of the Pacific Ocean. Microb Ecol 68(4):688–698. doi:10.1007/s00248-014-0448-8

Yakimov M, Giuliano L, Cappello S, Denaro R, Golyshin P (2007a) Microbial community of a hydrothermal mud vent underneath the deep-sea anoxic Brine Lake Urania (Eastern Mediterranean). Orig Life Evol Biosph 37(2):177–188. doi:10.1007/s11084-006-9021-x

Yakimov MM, La Cono V, Denaro R, D'Auria G, Decembrini F, Timmis KN, Golyshin PN, Giuliano L (2007b) Primary producing prokaryotic communities of brine, interface and seawater above the halocline of deep anoxic lake L'Atalante, Eastern Mediterranean Sea. ISMEJ 1(8):743–755

Yamada Y, Cane DE, Ikeda H (2012) Diversity and analysis of bacterial terpene synthases. In: David AH (ed) Methods in enzymology, vol 515. Academic Press, New York, pp 123–162. doi:10.1016/B978-0-12-394290-6.00007-0

Yamanaka K, Reynolds KA, Kersten RD, Ryan KS, Gonzalez DJ, Nizet V, Dorrestein PC, Moore BS (2014) Direct cloning and refactoring of a silent lipopeptide biosynthetic gene cluster yields the antibiotic taromycin A. Proc Natl Acad Sci USA 111(5):1957–1962. doi:10.1073/pnas.1319584111

Yamazaki Y, Sumikura M, Hidaka K, Yasui H, Kiso Y, Yakushiji F, Hayashi Y (2010) Anti-microtubule 'plinabulin' chemical probe KPU-244-B3 labeled both α- and β-tubulin. Bioorg Med Chem 18(9):3169–3174. doi:10.1016/j.bmc.2010.03.037

Yin W-B, Chooi YH, Smith AR, Cacho RA, Hu Y, White TC, Tang Y (2013) Discovery of cryptic polyketide metabolites from dermatophytes using heterologous expression in *Aspergillus nidulans*. ACS Synth Biol 2(11):629–634. doi:10.1021/sb400048b

Yu Z, Lang G, Kajahn I, Schmaljohann R, Imhoff JF (2008) Scopularides A and B, cyclodepsipeptides from a marine sponge-derived fungus, *Scopulariopsis brevicaulis*. J Nat Prod 71(6):1052–1054. doi:10.1021/np070580e

Zengler K, Toledo G, Rappe M, Elkins J, Mathur EJ, Short JM, Keller M (2002) Cultivating the uncultured. Proc Natl Acad Sci USA 99(24):15681–15686. doi:10.1073/pnas.252630999

Zengler K, Walcher M, Clark G, Haller I, Toledo G, Holland T, Mathur EJ, Woodnutt G, Short JM, Keller M (2005) High-throughput cultivation of microorganisms using microcapsules. Methods Enzymol 397:124–130. doi:10.1016/S0076-6879(05)97007-9 (S0076-6879(05)97007-9 [pii])

Zhang XY, Zhang Y, Xu XY, Qi SH (2013) Diverse deep-sea fungi from the South China Sea and their antimicrobial activity. Curr Microbiol 67(5):525–530. doi:10.1007/s00284-013-0394-6

Zhang XY, Tang GL, Xu XY, Nong XH, Qi SH (2014) Insights into deep-sea sediment fungal communities from the East Indian Ocean using targeted environmental sequencing combined

with traditional cultivation. PLoS ONE 9(10):e109118. doi:10.1371/journal.pone.0109118 (PONE-D-14-25039 [pii])
Zhou K, Zhang X, Zhang F, Li Z (2011) Phylogenetically diverse cultivable fungal community and polyketide synthase (PKS), non-ribosomal peptide synthase (NRPS) genes associated with the South China Sea sponges. Microb Ecol 62(3):644–654. doi:10.1007/s00248-011-9859-y
Zhu F, Lin Y (2006) Marinamide, a novel alkaloid and its methyl ester produced by the application of mixed fermentation technique to two mangrove endophytic fungi from the South China Sea. CHINESE SCI BULL 51(12):1426–1430. doi:10.1007/s11434-006-1426-4
Zhu F, Chen G, Chen X, Huang M, Wan X (2011) Aspergicin, a new antibacterial alkaloid produced by mixed fermentation of two marine-derived mangrove epiphytic fungi. Chem Nat Compd 47(5):767–769. doi:10.1007/s10600-011-0053-8
Zhu F, Chen G, Wu J, Pan J (2013) Structure revision and cytotoxic activity of marinamide and its methyl ester, novel alkaloids produced by co-cultures of two marine-derived mangrove endophytic fungi. Nat Prod Res 27(21):1960–1964. doi:10.1080/14786419.2013.800980
Ziemert N, Jensen PR (2012) Phylogenetic approaches to natural product structure prediction. Methods Enzymol 517:161–182 (B978-0-12-404634-4.00008-5 [pii])
Zuccaro A, Mitchell JI (2005) Fungal communities of seaweed In: Dighton J, White J (eds) The fungal community: its organization and role in the ecosystem, 3rd edn. CRC Press, New York, pp 533–580
Zuccaro A, Schulz B, Mitchell JI (2003) Molecular detection of ascomycetes associated with *Fucus serratus*. Mycol Res 107(Pt 12):1451–1466
Zuccaro A, Schoch CL, Spatafora JW, Kohlmeyer J, Draeger S, Mitchell JI (2008) Detection and identification of fungi intimately associated with the brown seaweed *Fucus serratus*. Appl Environ Microbiol 74(4):931–941. doi:10.1128/AEM.01158-07
Zulak KG, Weljie AM, Vogel HJ, Facchini PJ (2008) Quantitative 1H NMR metabolomics reveals extensive metabolic reprogramming of primary and secondary metabolism in elicitor-treated opium poppy cell cultures. BMC Plant Biol 8:5. doi:10.1186/1471-2229-8-5

Chapter 5
Marine Viruses

Corina P.D. Brussaard, Anne-Claire Baudoux and Francisco Rodríguez-Valera

Abstract With an estimated global abundance of 10^{30}, viruses represent the most abundant biological entities in the ocean. There is emergent awareness that viruses represent a driving force not only for the genetic evolution of the microbial world but also the functioning marine ecosystems. Culture studies advance our understanding how viruses regulate host population dynamics, but retrieving virus and host in pure culture can be difficult. Recent developments in high-throughput sequencing provide insights into the diversity and complexity of viral populations. This chapter describes current milestones in the burgeoning field of marine viral ecology, including the different aspects of marine viral action, viral diversity, ecological and biogeochemical implications of marine viruses, the cultivation of virus-host systems, and biotechnological applications of these astonishing microorganisms.

5.1 Introduction

Marine viruses are like any other viruses defined as small infectious agents, consisting of a core of nucleic acids (RNA or DNA) in a protein coat, that replicate only in the living cells of a host. A lipid envelope may be found outside or inside of

C.P.D. Brussaard (✉)
Department of Microbiology and Biogeochemistry, NIOZ Royal Netherlands Institute for Sea Research, PO Box 59, Den Burg, Texel, The Netherlands
e-mail: corina.brussaard@nioz.nl

C.P.D. Brussaard
Department of Aquatic Microbiology, Institute for Biodiversity and Ecosystem Dynamics (IBED), University of Amsterdam, Amsterdam, The Netherlands

A.-C. Baudoux
Sorbonne Universités, UPMC Université de Paris 06, CNRS, Adaptation et Diversité en Milieu Marin (AD2M UMR7144), Station Biologique de Roscoff, 29680 Roscoff, France

F. Rodríguez-Valera
Evolutionary Genomics Group, Universidad Miguel Hernandez, Campus de San Juan, Alicante, Spain

L.J. Stal and M.S. Cretoiu (eds.), *The Marine Microbiome*,
DOI 10.1007/978-3-319-33000-6_5

the capsid. In the strict sense, complete viral particles found outside host cells and constituting the infective form are called virions, however, throughout this chapter the term viruses will be used for both the intracellular and extracellular forms. Viruses are classified by their genome type (DNA or RNA, single or double stranded, segmented or not, circular or linear) and size (from few Kbp to 2.5 Mbp), particle structure (e.g., icosahedral or helical) and size (from $\sim$20 to 500 nm), capsid coat protein composition, whether it is enveloped or not, the replication strategy, the latent period (i.e., the time until progeny viruses are released from the host cell), sensitivity to physicochemical factors (e.g., temperature, pH, UV), and last but not least the host they infect. There are no universal oligonucleotide primers for marine viruses due to their huge genetic diversity, but the presence of specific conservative genes aids to the detection and classification of marine viruses, e.g., primers for fragments of the conserved cyanophage structural gene g20, or for the DNA polymerase gene (*pol*B) are used to detect large dsDNA algal viruses (Chen and Suttle 1995; Short and Suttle 2005).

Unable to replicate without a host cell, viruses are generally not considered living organisms or alive (Forterre 2010). The finding of large genome-sized viruses displaying a localized viral factory in the host's cytoplasm where the viral genome is replicated and virions are produced (La Scola et al. 2003; Santini et al. 2013), together with the discovery of viruses infecting the viral factory (c.q. virophages meaning virus eaters; La Scola et al. 2008; Fischer and Suttle 2011), renewed the discussion as to whether viruses are alive. Both ways, viruses have the ability to pass on genetic information to upcoming generations and as such they play an important role in biodiversity and evolution. Due to their mode of replication, viruses can take some genetic material from their host in their progeny but can also add their own genetic material to the host, thereby increasing genetic diversity and biodiversity as a whole. Actually, viruses can in a way be considered part of the genetic reservoir of their host, i.e., the genomes of viruses are part of the larger pan-genome (e.g., Bacteria or Archaea) like other extra-chromosomal elements. Given their rapid reproduction rates, and often high mutation rate, viruses represent an important source of genetic innovation. Besides the enormous reservoir of uncharacterized genetic diversity (Suttle 2007), viruses have the potential to be interesting material for biotechnological use (Sánchez-Paz et al. 2014).

The research on marine viruses, their population dynamics, diversity and ecological relevance has expanded exponentially over the last years. Marine viruses are the most abundant biological entities in the seas and oceans, ranging between 1 and $100{,}000 \times 10^6\ L^{-1}$, whereby the higher abundances are generally found in the more coastal, eutrophic surface waters and the lowest numbers in the deep ocean. Per unit volume, viruses in marine sediments (benthic viruses) with the higher load of bacterial hosts exceed pelagic viruses by one or more orders of magnitude (Danovaro et al. 2008; Glud and Middelboe 2004; Hewson et al. 2001). The numerically dominant hosts are the marine microorganisms belonging to all three domains of life, i.e., Bacteria, Archaea, and Eukarya. With the development of sensitive nucleic acid-specific dyes in combination with epifluorescence microscopy (Noble and Fuhrman 1998), pelagic viruses could be more easily counted

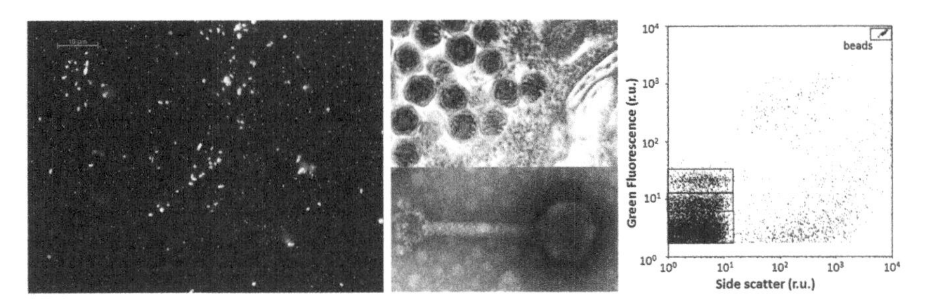

Fig. 5.1 Different techniques to detect marine viruses, i.e., epifluorescence microscopy upon staining the viral nucleic acid (*left panel*), transmission electron microscopy (TEM) thin section of host cell showing virions PgV infecting the phytoplankton *Phaeocystis globosa* (virus 150 nm diameter; *mid upper panel*), TEM of free bacteriophage (virus head diameter 100 nm, tail length 180 nm; *mid lower panel*), and flow cytometry of natural seawater sample showing several virus clusters (*right panel*)

than by transmission electron microscopy (Fig. 5.1). Moreover, the availability of fluorescent dyes allowed counting of the total viral community and not just the infectious viruses for specific hosts in culture as determined traditionally by plaque assay (using agar plates) and most probable number (end-point dilution of liquid cultures). More recently, staining of the viral nucleic acids combined with flow cytometry allowed a more objective and faster analysis (Brussaard et al. 2000; Brussaard 2004a) opening further the research field of marine viruses to a better spatial and temporal sample coverage, as well as higher degree of replication. Besides these methods for direct detection of virus particles, there are molecular approaches helping to identify and quantify (by qPCR) marine viruses (Sandaa and Larsen 2006; Tomaru et al. 2011a, b, c). Combining techniques such as, for example, flow cytometry sorting of virus and/or infected host with molecular sequencing will further advance prospective research aiming at answering questions on viral diversity and adaptation (Labonté et al. 2015; Martínez-Martínez et al. 2014; Zeigler et al. 2011). Metagenomic analysis has even revealed new virus families, e.g., ssDNA viruses infecting a wide range of marine hosts (Labonté and Suttle 2013a, b).

Although essential, quantification and identification by itself is not enough to answer the major research questions in marine viral ecology, e.g., who infects whom, what is the impact of viral infection on host population dynamics and subsequent species succession, and how do viruses affect ecosystem efficiency and consequently biogeochemical fluxes? One needs to be able to determine actual viral infection and lysis rates in the field. At the same time, there is need to isolate and bring into culture more (and new) virus-host model systems in order to allow optimal understanding of the mechanisms underlying successful infection and alterations in virus–host interactions due to change in environmental factors.

Table 5.1 Overview of some of the milestones obtained in marine virus ecology during the last 1–2 decades

Milestone	Impact
Counting viruses using sensitive dyes	Opened up the field as it allowed for more replicates and better spatial and temporal coverage
Viral lysis rates in the field	Demonstrate that virally induced mortality at least as important as traditional loss by grazing
New dsRNA virus family	Unique dsRNA virus infecting marine protist
Use of PCR primers	Specific virus groups can be detected
Full genome sequencing	Discovery of new metabolic pathways
Metagenomics	Deep coverage of viral diversity and discovery of new virus families (e.g., ssDNA viruses)
Large genome-sized virus discovery	With their huge size (1–2.5 Mb) and cellular trademark genes, these viruses are evolutionary challenging
Virophages	Viruses infecting viruses, thereby decreasing mortality of the host organism
Viral shunt	Realization that viral activity promotes flux of organic carbon and nutrients, making remineralized nutrients directly available for phytoplankton

The outcome of viral infection has implications for ocean function and it is an exciting time of discovery (some milestones are listed in Table 5.1). This chapter describes some of the milestones in marine viral ecology (Table 5.1) and outlines the different aspects of marine viral action, i.e., the different viral infection strategies, horizontal gene transfer, viral diversity, ecological influence on host mortality, microbial biodiversity, food web structure, organic carbon and nutrient cycling, the culturing and storage of virus-host systems, and concludes with several biotechnological applications.

5.2 Viral Infection Strategies

Microbial host cells and their viral predators have been evolving together for billions of years. At first glance, one might think that their interactions are simpler than those of other predator or parasite-host pairs but they are actually extremely complex. The main predators of microbes typically discriminate for prey size, along rough lines of taxonomy and morphology and can eat many individual prey of diverse genomic make up throughout its individual cell life span (Sintes and Del Giorgio 2014). In contrast, viruses are single opportunity killers and highly host-specific. One virus particle has only one chance to reproduce and requires a refined adjustment to the cell machinery on which it depends for ultimate replication and virion component. Therefore, viruses are specialized parasites that are fine-tuned to the host structure and metabolism (Leggett et al. 2013). Actually, the delicate discriminatory power of viruses has been used classically for typing, for example, bacteria. Precise host cell recognition is an essential step in the life cycle of any virus. Accordingly, viruses have refined molecular recognition modules that allow discrimination of encountered cells before the infection process in triggered (Garcia-Doval and van Raaij 2013). These molecular recognition devices are among the most sensitive in nature and match in complexity, and probably in diversity as well, the vertebrate immune system (Fraser et al. 2007). Mirroring this diversity, host cells have an enormous diversity of virus potential receptors. They can be any outermost exposed structure of the cell, frequently proteins such as pillins, porins or transporters but, above all, polysaccharides or the glycosidic moiety of glycoproteins. Actually, the interaction with viruses might be behind the commonality of polysaccharides and glycoproteins in exposed structures of Bacteria, Archaea, and Eukarya.

Through the lytic cycle, viruses can reproduce inside the host cell directly upon penetration, causing destruction of the host cell to release the progeny viruses into the surrounding water. Alternatively, viruses may integrate into the host genome as a provirus and propagate together with the host. Thus, in contrast to lytic viruses, these temperate (or latent) phages establish a stable, but reversible, relationship with their host, which is termed lysogeny. For the prophage being inside the host provides protection against virion decay by environmental factors (Mojica and Brussaard 2014). Although at first glance it does not seem beneficial for the host to carry a prophage, it generally protects from infection by similar and related phages. Furthermore, prophage may provide useful genes and improve host fitness (Paul 2008). This transfer of genes via prophages is a form of horizontal gene transfer (HGT) that aids to the genetic diversity within the host species. The prophage can be released again by an inducing event triggering a lytic cycle. The environmental factors responsible for the transition from a lysogenic to a lytic cycle are not well identified. UV and pollution have been reported to induce the lytic stage, potentially to produce progeny viruses before potential cell death of the host (Paul et al. 1999). Alternatively, the triggering factor may represent more optimal growth conditions for the host (e.g., release of P-limitation; Wilson et al. 1996), whereby the shift from prophage to lytic phage is a favorable one as many more progeny viruses can be produced than under the lysogenic life style (one progeny every time the host divides). In agreement with this potential mechanism, the percentage lysogeny of total mortality of bacteria was found highest under oligotrophic conditions with low host abundance, both geographically as well as seasonally (McDaniel et al. 2002; Weinbauer et al. 2003).

Besides lytic and lysogenic cycles, the host cell can also release viruses by continuous or intermittent budding or extrusion without its immediate lysis. This replication strategy, known as chronic cycle or carrier state, is far less investigated in marine environments, yet it has been described for viruses that infect marine eukaryotic photoautotrophs (Mackinder et al. 2009; Thomas et al. 2011), but also for archaeal viruses (Geslin et al. 2003; Gorlas et al. 2012).

5.3 Virus Characteristics

The characterization of virus isolates typically includes a phenotypic analysis (particle morphology, symmetry and size; genome type and size; composition of structural proteins) in order to assign a taxonomical status to these isolates. Additionally, parameters related to virus life cycle, specificity pattern, environmental stability or resistance, are other important viral features. The information generated by one-step growth experiments, also referred to as "single burst experiments," have proven essential in this respect. This classical procedure developed by Ellis and Delbrück (1939) is designed to quantitate and monitor the growth of lytic viruses and it provides two fundamental viral properties: the latent period (that is the duration between virus adsorption and extracellular release from the infected host cell) and the burst size (that is the number of progeny viruses released per infected host cell). These viral parameters vary depending on the virus isolate and the host they infect. Collectively these variables constitute important components to model virus–host interactions in the ocean. For example, the infection strategy of a given virus largely determines the ecological impact of virus interactions. Gaining fundamental knowledge on the infection strategy of relevant virus-host model systems, the conditions that induce the transition between these different modes of infection, and the potential influence environmental factors have on the intensity and the kinetics of infection are thus a prerequisite for understanding the impact of viral infection on microbial assemblages.

The infection range (the range of action of a phage measured in terms of the varieties of bacteria in which it can grow) has been a major conundrum of virus studies. The host specificity of bacteriophages (i.e., viruses that infect bacteria; there are also referred to as phage or bacteriovirus) was recognized very soon after their discovery in the early twentieth century (Abedon 2000). The idea of phage therapy was based on this early discovery and was the main driver of the original interest on bacteriophages, later displaced by the enormous role played by phages in the development of genetics and molecular biology. The success or failure of adsorption and subsequent lysis of the host typically, but not always, determines the host infection range. There is a wide diversity of mechanisms within the cell to prevent phage replication, including receptor variation, restriction–modification, abortive infection, lysogenic immunity, or innate immunity conferred by CRISPRs (see for example Stern and Sorek 2011). In the laboratory, host specificities are determined by pairwise infection, in which a clonal virus lineage is added to a collection of host cultures, infection is scored positive if lysis occurs. These infection assays suggest that marine viruses exist along a continuum of specialists to generalists. The specialist viruses can only infect a restricted range of strains, sometimes even only the one used for their initial isolation. The generalist viruses can infect many different strains and occasionally can infect also isolates from different species and even different genera (e.g., Baudoux et al. 2012; Johannessen et al. 2015; Matsuzaki et al. 1992; Sullivan et al. 2003; Suttle et al. 1995). However, in general the genus is the taxonomic unit that encompasses the most sensitive strains for the vast majority of phages. It is sensible to say that the boundary of natural host range of most phages is the genus.

Cross-infections between virus and hosts are challenging to interpret. In spite of 25 years of investigation, there is as yet no consensus on either the global pattern of viral infection or the environmental drivers that shape these patterns of interactions. For example, to what extent does virus–host infection vary in space and/or time, are there factors that favor generalists upon specialists, and are virus–host interactions structured at all (Avrani et al. 2012)? It is presently inconceivable to bring into culture and test the specificity pattern of all individual virus-host model systems dispersed in marine systems. As an alternative, theoretical approaches that rely on empirical data have been proposed to study of virus–host interactions as networks (Flores et al. 2011, 2013; Weitz et al. 2013). Network-based analysis tests whether virus–host interactions are structured (i.e., they possess a structured pattern) or random. The most common patterns in ecological networks are being nested or modular. Nested virus–host interaction networks are characterized by a hierarchy of resistance among hosts and infection ability among viruses (Flores et al. 2011; Weitz et al. 2013). This pattern is thought to result from coevolutionary arms race (Red Queen hypothesis) where viruses evolve to broaden host ranges and hosts evolve to increase the number of viruses to which they are resistant (Flores et al. 2011). In contrast, modularity is characterized by groups of viruses and hosts (referred to as modules) that preferentially interact with another (Flores et al. 2011; Weitz et al. 2013). This ecological pattern should emerge if virus–host interactions result from evolutionary processes that lead to specialization.

A network-based analysis of cross-infections of 215 phages against 286 bacteria collected across the Atlantic Ocean (Moebus and Nattkemper 1981) revealed that infection network possessed a multiscale structure (Flores et al. 2013). At a global network scale, virus–host interactions network displayed a modular structure. This modularity was explained, at least in part, by geography. In other words, viruses isolated from a given sampling site were more likely to infect cooccurring hosts, supporting the need to account for biogeography in the analysis of viral diversity. However, individual modules were either further modular or nested, which suggests that different coevolutionary processes drive virus–host interaction at different scale (Flores et al. 2013). Network-based analyses of virus–host interactions have emerged quite recently. Yet, these analyses represent a promising tool to quantify the functional complexity of viral infections in natural systems and to identify the drivers of microbial species interactions.

5.4 Virus and Host Diversity

The role of viruses in shaping the evolution of their hosts is acquiring new dimensions both as selective pressure that keeps populations diverse (Rodriguez-Valera et al. 2009) and as sources of genetic innovation (Suttle 2007). Viruses may carry metabolic genes of the host to enhance infection performance, but can also introduce genetic diversity to the cognate pathways. Microbiology has failed to provide a sensible representation of the diversity of microbes using only the pure culture approach. Although great advances are being made, until recently the vast majority of microbes in nature has been inaccessible to microbiologists. If this is the situation of cellular microbes, the status of the viruses preying on them is even direr. Most of what we know about viruses derives from the studies of human, cattle and plant crop viral pathogens. Lately large efforts to sequence more virus genomes of different hosts have been made (Pope et al. 2015), and the approaches based on high-throughput sequencing metagenomics and single cell genomics are also applied to marine viruses (Angly et al. 2006; Rodriguez-Brito et al. 2010).

The advent of high-throughput sequencing technologies permitted to determine the complete genome sequence of more than one hundred cultured marine viruses over the past 15 years and many more are currently being analyzed. Virus genomics provides invaluable insights into the understanding of how viral infection and virus evolutionary relationships function. One notable feature is that marine viruses encode remarkable panoplies of biological functionalities. They display core genes, found in most viruses, involved into the replication of viral DNA and the structure and the assembly of virions. Genome analyses also indicate that viruses encode and transfer a variety of auxiliary metabolic genes (AMG) derived from their hosts (Rohwer et al. 2000; Sullivan et al. 2005; Wilson et al. 2005). Although the ability to shuffle genetic information through HGT is commonly reported in viruses, the repertoire of host-derived genes encode by marine viruses differ from those of non-marine origins. The most emblematic AMG is probably the *psb*A photosynthesis gene reported in cyanophage genomes (Mann et al. 2003; Sullivan et al. 2005). This gene encodes the protein D1, which forms part of the photosystem II reaction center in the host cell. During the course of viral infection, the D1 proteins derived from the host decline dramatically while those encoded by the cyanophages increase (Clockie et al. 2006). This led to the hypothesis that the expression of virus-derived *psb*A helps maintaining photosynthesis throughout the lytic cycle, which, in turn, 'boosts' the host metabolism to support virus replication. The analysis of AMG in virus genomes suggests that marine viruses have evolved a wide range of strategies to hijack host metabolism towards their own replication (Hurwitz et al. 2013). AMG are widespread and they appear to be involved into phosphate scavenging, carbon, or nitrogen metabolism (Brum and Sullivan 2015; Hurwitz et al. 2013; Rohwer et al. 2000; Sullivan et al. 2005). The genome analysis of viruses that infect the bloom-forming alga *Emiliania huxleyi* (EhV) also revealed a unique cluster of sphingolipid biosynthetic genes (Wilson et al. 2005) that facilitate virus replication and assembly (Rosenwasser et al. 2014). Genomics of marine virus isolates provides unprecedented knowledge about complex metabolic pathways that play a pivotal in role in host–virus interactions.

In parallel to genome analysis of cultured isolates, the development of culture-independent tools was a breakthrough in virus ecology in the ocean. Metagenomics largely contributed to unveil the extent of viral diversity and the dynamics of viral communities (Angly et al. 2006; Breitbart et al. 2004; Brum et al. 2015). The genome sequence of relevant virus models is thus essential to build a reference database in order to identify the members of the viral communities and, importantly, whom they infect (Kang et al. 2013; Zhao et al. 2013). Despite its success, a general problem in the application of metagenomics techniques is that although the number of viral particles is high, the amount of viral DNA in natural environments is much smaller than cellular DNA. Therefore, since the beginning, DNA amplification approaches were applied to retrieve the required amount for Next Generation Sequencing (NGS) (Angly et al. 2006). In addition, the short NGS reads complicate enormously the already shaky grounds of viral genome annotation. Finally, one classical major limitation of viral metagenomics or metaviromics was assembly. By assembly of high coverage metagenomes large genome contigs can be reconstructed that are much more reliably annotated and classified. However, for unknown reasons, assembly works much more poorly with phage genomes in metaviromes. This problem can be partially by-passed by using cellular metagenomes. Cellular DNA is also rich in viral DNA derived from cells that are undergoing the lytic cycle. In metabolically active communities such as the deep chlorophyll maximum (DCM) 10–15 % of the bacterial and archaeal DNA derives from such phages (Ghai et al. 2010). For example, by using advanced binning approaches, a bacteroidetes phage that seems to be extremely widespread and conserved in the human microbiome could be assembled (Dutilh et al. 2014). By using metagenomic fosmid libraries Mizuno et al. (2013) described large genomic fragments from more than a 1000 viral genomes obtained from a single sample from the Mediterranean DCM. Fosmids have similar size to typical *Caudovirales* genomes and many of these could be retrieved. They could be gathered by genome comparison into 21 major sequence groups and a vast genomic diversity was present within each group. Host could be assigned to some of the 10 totally novel groups by comparison to putative host genomes what allowed assigning the first phages detected to marine Verrucomicrobia and the recently described marine actinobacteria Actinomarinales (Ghai et al. 2013). This work also revealed high microdiversity among the highly related sequence groups (Rodriguez-Valera et al. 2014), particularly in the host recognition modules. Specific single stranded (ss) DNA amplification has allowed the reconstruction of ssDNA viruses that are prevalent in marine habitats and have small genomes that simplify their reconstruction from metaviromes (Labonté and Suttle 2013a, b). The newly developed technology of single cell genomics is also providing data about viruses taking advantage of the same phenomenon, i.e., sorted cells are often undergoing lysis and eventually the active virus genome is abundantly represented in the resulted amplified genome (Labonté et al. 2015). This technique allowed the description of the first genome of a virus preying on the ubiquitous group I marine archaea or Thaumarchaeota.

5.5 Ecological Importance

Under natural conditions the richness of microbial species is high, while laboratory research indicates that competition for the same resources ultimately will lead to the dominance of the best competitor and the extinction of the other competitor(s). The most renowned example is the Plankton Paradox (Hutchinson 1961), whereby according to the competitive exclusion principle only a small number of plankton species should be able to coexist on the growth-limiting resources while in reality large numbers of plankton species are found to coexist within small regions of open sea. Because ecological factors stay hardly ever constant, spatio-temporal heterogeneity, but also selective grazing, symbiosis and commensalism have been suggested as factors responsible for resolving the Plankton Paradox. Another top-down controlling factor more recently acknowledged is viral infection. Marine viruses have typically narrow host ranges, which makes them highly relevant as underlying cause of coexistence of microbial species (Suttle 2007; Thingstad 2000). As viral infection is dependent on contact with the host, higher abundances of susceptible host species are expected to favor successful infection. This Killing-the-Winner (KtW) model describes how the abundance of the competition strategist is top-down controlled by viruses, thereby ensuring coexistence of competing species since the growth-limiting resources can be used by others (Thingstad 2000). Then again, a high abundance of host is not necessarily the result of rapid growth as host species might be resistant to top-down control (by zooplankton predators and/or viral lysis), or there might be substantial loss of viral infectivity or virus particles that prevent further infection of the developing host population. The traditional rank-abundance curves (illustrating the relative abundance of species) have to be read differently then, i.e., the least abundant taxa are not rare (or dormant) but are instead actively growing but at the same time also predated and/or lost by viral lysis (Suttle 2007). They illustrate r-selection, with high reproductive output, while the few most abundant taxa are K-selected with low maximum growth rates, better competitors for resources and less sensitive to cell loss (Suttle 2007; Våge et al. 2013). At the same time, the marine viruses rank-abundance curve shows a contrasting pattern with most viruses r-selected, being virulent with small genome sizes and high decay rates. The low abundance virus taxa are K-selected with larger genomes, longer-lived, and potentially forming stable associations with their host.

Blooms of phytoplankton exemplify situations where there is an unbalance between growth and loss processes, either because the algal species is able to outgrow infecting (at least temporarily) or because it is (largely) resistant to infection. Blooms of *E. huxleyi* illustrate the concept of KtW well, because during the development of the bloom the percentage of infected cells is increasing to a point that causes the collapse of the bloom (Bratbak et al. 1993; Brussaard et al. 1996a; Martínez-Martínez et al. 2007). Interestingly, the cell morphotype of *E. huxleyi* seems to affect the chance of infection. A laboratory study showed that the diploid calcifying coccolith-bearing cells were virally infected but the haploid cells were not (Frada et al. 2008). This finding led to the proposition of the Cheshire Cat escape strategy, where viral infection indirectly promotes a resistant haploid phenotype and thus sexual cycling (Frada et al. 2008). However, the genetically distinct communities of the *E. huxleyi* host and the co-occurring viruses under natural conditions were stable over several years, which seems to be inconsistent with the Cheshire Cat hypothesis (sexual reproduction would actually reshuffle the genes in *E. huxleyi*; Morin 2008). Moreover, a study of a natural bloom of *E. huxleyi* showed that the haploid organic scale-bearing cells (haploid by nature, Laguna et al. 2001) were also visibly infected by viruses (using transmission electron microscopy; Brussaard et al. 1996a). It needs to be seen whether these differences between the studies (1) indicate that sex does not act as an antiviral strategy (Morin 2008), (2) are due to growth-controlling factors (i.e., the natural bloom was nitrogen depleted while the laboratory study used nutrient replete cultures), (3) different virus types (the natural study showed co-infection of two different viruses), (4) genetically distinct algal host strains (and thus potentially different sensitivity to infection), or (5) the haploid *E. huxleyi* cells in the laboratory study being different than the natural scale-bearing cells.

Another example of how blooms are formed by (largely) viral infection-resistant cell morphotypes is the algal blooms of the phytoplankton genus *Phaeocystis* (Brussaard et al. 2005). *Phaeocystis* can make colonies provided that there is enough light available to allow the excess organic carbon to be excreted in the form of carbohydrates forming a colonial matrix (Schoemann et al. 2005). The colonial cell morphotype has a much lower chance of becoming infected (or grazed) than a single cell (Jacobsen et al. 2007; Ruardij et al. 2005) and therefore a considerable increase in biomass is possible. At the moment, the cells become shed from the colonial matrix (due to low light or nutrient depletion; Brussaard et al. 2005) the liberated single cells are readily infected and eaten and the bloom crashes (Baudoux et al. 2006).

Viruses can be so specific that they only infect certain strains. This provides the potential for intraspecies succession of host and virus within a phytoplankton bloom (Baudoux et al. 2006; Tarutani et al. 2000). For example, the different sensitivities of *Heterosigma akashiwo* clones to viral infection resulted not only in changes in the total abundance of the virus infecting *H. akashiwo* but also the clonal diversity of the algal host. Another example of high differential sensitivity to viral infection within a species is the blooms of *Synechococcus* spp. that were composed of many (genetically) different populations of host and cyanophage (Mühling et al. 2005; Suttle and Chan 1994; Waterbury and Valois 1993). Differences in infectivity properties of the host strains may be directly due to changes in the composition of the attachment sites on the host's cell surface (Marston et al. 2012), or may be the indirect result of altered virus proliferation as a consequence of variation in host cell metabolic traits between the host strains (Brussaard, unpublished data). Resistance is thought to come with a cost, i.e., a resource competitive disadvantage compared to viral infection susceptible strains (Lenski 1988; Thingstad et al. 2015). Furthermore, host growth-controlling factors have been shown to affect virus latent period, viral burst size, the level of infectivity of the viral offspring, and even viral life strategy (Baudoux and Brussaard 2008; Bratbak et al. 1998; Maat et al. 2014; Wilson et al. 1996). Altogether, variations in host susceptibility to viral infection and viral production rate and yield add not only to biodiversity but also to spatial and temporal variability of microbial hosts and viruses.

Although many studies have indicated that viral infection is a significant loss factor, accurate measurements of viral lysis rates of microorganisms in the field are still few (Suttle 2005). Furthermore, rates of virally mediated mortality are assessed using different approaches (results are therefore not necessarily comparable), and most methods rely on different assumptions and conversion factors. The first approach, using visibly infected cells obtained from transmission electron microscopy analysis, is dependent on good estimation of the proportion of the lytic cycle during which virus particles are visible in the infected host cells (Brussaard et al. 1996a; Proctor and Fuhrman 1990). This factor is difficult to assess as the lytic virus growth cycle is highly variable between viruses and it can also differ within a virus depending on host metabolism. Estimating viral lysis rates using (fluorescent) virus tracers requires estimates of burst sizes (Heldal and Bratbak 1991; Noble and Fuhrman 2000; Suttle and Chen 1992). Using viral decay rates to estimate production rates also requires the assumption that there is a steady-state situation. The use of synthesis rates of viral DNA requires many conversion factors (Steward et al. 1992). Measuring viral production rates is most widely used but does need an estimate of burst size and comes with substantial sample handling (Weinbauer et al. 2010). Many of these methods, except the visibly infected host cell assay, measure viral lysis of the total community, because the different virus (and often also host) populations cannot be discriminated. Because algal viruses typically represent a few percent of the total virus abundance, the same methods are not well suited for phytoplankton viral lysis rate estimates unless the algal virus can be separately counted. For phytoplankton, viral lysis rate can be assessed using a modified dilution assay (Baudoux et al. 2006; Evans et al. 2003; Kimmance and Brussaard 2010), but here the restriction is that the algal host cells are newly infected and lysed within the time of incubation (typically 24 h because of synchronized cell division of many phytoplankton species).

The available viral lysis rate measurements in the field demonstrate that viral lysis is similar to predation rates (Baudoux et al. 2006, 2007; Mojica et al. 2015). The pathways of cellular organic matter and energy through the food web will, therefore, follow different ways (Suttle 2005). Upon predation organic carbon and nutrients are transferred to higher trophic levels, while upon viral lysis the host's cellular content is released into the surrounding waters as dissolved and detrital particulate organic matter. Viruses are as mortality agents causing accelerated transformation of organic matter from the particulate (lysed host cells) state to the dissolved phase, upon which heterotrophic microbial communities mineralize the largely (semi-) labile organic matter directly (Lønborg et al. 2013). The diversion of organic matter towards the microbial loop (microbial recycling) results in enhanced respiration (Suttle 2005). Cell lysis of microorganisms has been reported to sustain the heterotrophic bacterial carbon demand (Brussaard et al. 1995, 1996b, 2005). Globally, viral lysis seems to negatively affect the efficiency of the biological pump (i.e., the transfer of photosynthetically fixed carbon to the deep ocean as dead organisms or feces). The rapid release and remineralization of nutrients in the photic zone may increase the ratio of carbon relative to nutrients exported to the deep. Conversely, viral lysis may potentially short-circuit the biological pump by stimulating the formation of transparent exopolymer particles (TEP) and aggregates (Brussaard et al. 2008; Mari et al. 2005), as well as release elements that could as such stimulate primary production (e.g., ligand-bound iron and organic phosphorus; Gobler et al. 1997; Løvdal et al. 2008; Poorvin et al. 2004). With respect to the global anthropogenic climate change (greenhouse effect but also alterations in the physicochemical characteristics of the water masses, e.g., temperature, salinity, light, nutrients), it is urgent to improve our understanding of how viral activity affects the processes involved in the natural sequestration of carbon dioxide. A study in the northeast Atlantic Ocean showed for instance that the share of viral lysis (as compared to grazing) of phytoplankton increased under conditions of temperature-induced vertical stratification (Mojica et al. 2015).

5.6 Isolation, Culture, and Characterization of Marine Viruses

Almost from their first discovery in 1917, cultured virus–host systems (VHS) were viewed as relevant models for microbiologists and geneticists who investigated the nature of genes and heredity. This early era of virus research constituted the foundations of molecular biology. It was not until the mid-1950s that marine viruses were discovered with the isolation and the description of a phage lytic to the fish gut associated bacterium, *Photobacterium phosphoreum* (Spencer 1955). Although the potential of bacteriophages to control host populations emerged quite rapidly, virus isolation from oceanic settings was only reported sporadically and most studies focused on the biology rather than the ecological implications of these model systems (Hidaka and Fujimura 1971; Hidaka 1977; Spencer 1960). The breakthrough discovery that viruses are the most abundant biological entities in the ocean marked a turning point in the field of virus ecology and literally renewed interest in isolating, culturing, and characterizing marine viruses.

Several protocols describing procedures for the isolation of lytic viruses have been reported from various laboratories. These protocols include three main steps: (1) A natural sample is added to a selection of prospective host cultures, (2) the lysis of this host is visualized and, (3) in case of positive lysis, the lytic agent is cloned and stored appropriately until characterization. In the following section, we provide a thorough description of these steps and recommendations to optimize successful virus isolation. Although these features are applicable to most viruses, viral strains may differ in their development, and therefore isolation protocols may have to be adapted accordingly. In theory, at least one, and usually multiple, viruses, can infect any given organism. However, empirical data have shown that some strains are more permissive to viral infection than others. It is, therefore, preferable to use several clonal host lineages in order to increase the chance of successful isolation. When possible, one should isolate the host strains just prior to virus isolation from the same location in order to increase success rate in case the virus has a narrow host range. Furthermore, owing to their parasitic nature, viruses largely rely on the metabolism of their hosts for production. In most cases, well-growing hosts tend to be more susceptible to viral infection.

Prior to the isolation, it is recommended to remove particles larger than viruses by low-speed centrifugation and/or pre-filtration through low protein binding membrane filters (e.g., polycarbonate, polyethylsulfone, polyvinylidene fluoride) with a preferable pore size of 0.45–0.8 μm (filtration through 0.2 μm could exclude larger virus particles). However, this may result in loss of viruses and therefore it is recommended to also use a non-treated sample. For sediment or any other type of porous samples, viral particles need to be transferred into a buffer solution prior to centrifugation and/or pre-filtration (Danovaro and Middelboe 2010). If the targeted virus population is expected to be low in abundance, tangential flow filtration/ultrafiltration or ultra-centrifugal units with a retention cutoff of at least 100 kDa can concentrate viral particles. As an alternative to these techniques that can become time consuming and costly, the host-enrichment technique also increases the probability of successful isolation considerably. This enrichment step consists of amending the viral community to be screened with 1–5 prospective host strains and host growth medium. Upon incubation this mixture, viruses lytic to these hosts should propagate and hence this procedure facilitates isolation. After incubation, the suspension can be clarified by centrifugation or pre-filtration and, if necessary, further concentrated by ultrafiltration as described above. Once prepared, the extracted viral community can be stored at 4 °C until the isolation procedure.

5.6.1 Detection of Lytic Viral Lysis and Virus Purification

There are several approaches to visualize cell lysis of the targeted host due to lytic viral infection. One common technique is the plaque assay, which has been successfully applied to the isolation of viruses of bacteria and microalgae (Fig. 5.2). This technique relies on combining an inoculum of the viral sample and the targeted host culture in soft agar overlay on agar plates. On incubation, the host develops a homogenous lawn, except where virus lysis occurs, resulting in localized translucent plaques, referred to as a plaque-forming unit (PFU, Fig. 5.2). PFU typically originates from one infectious virus and displays a well-defined morphology. Different viruses infecting the same host can coexist in the same sample. It is therefore recommended to pick as many individual plaques as possible and to repeat the plaque assay for another two times to produce as many clonal lineages of these isolates. Alternatively, the lysis of infected host can be detected in liquid medium. In this case, the propagation of viruses is not limited by the diffusion in semi-solid agar media. Upon attachment on a susceptible host, the viral progeny is released in liquid medium and the newly produced virions can initiate another lysis cycle and propagate until complete lysis of the cultured host. Lysis is detected by complete

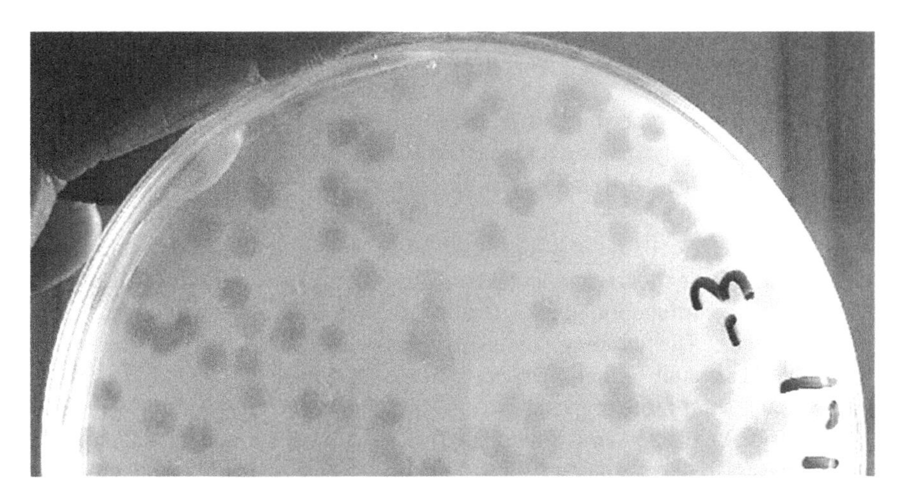

Fig. 5.2 Petri dish showing a plaque assay for viruses infecting the picoeukaryotic phytoplankton *Micromonas* sp. (*photo courtesy* Nigel Grimsley)

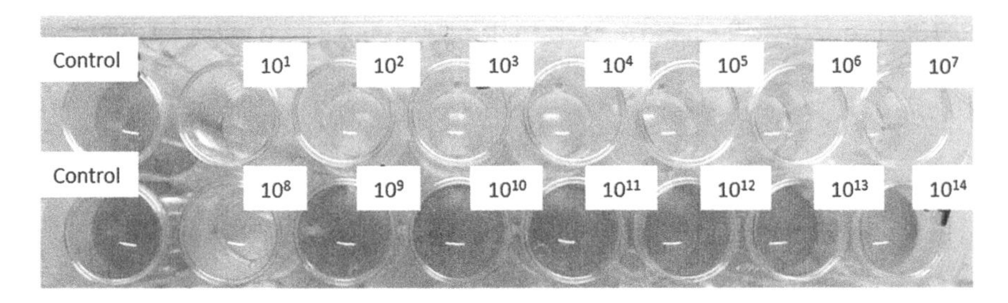

Fig. 5.3 Aquatic serial dilution set-up using microwell plates whereby a suspension of *Micromonas* virus RCC4265 was diluted from $\times 10^{1}$ down to $\times 10^{-14}$ in *Micromonas* host culture RCC829. Lysis is scored positive until dilution $\times 10^{8}$. Non-infected controls (*first column*) were taken along for reference

clearing of the host suspension as shown in Fig. 5.3. The resulting lysate can comprise a mixture of viruses that need to be separated as soon as possible to obtain clonal viral lineages. This can be achieved by extinction dilution methods in multiwell-plates (Fig. 5.3). The principle of this procedure is to serially dilute (usually using tenfold dilution increment) the lytic agents in vigorously growing host culture until complete extinction (i.e., no lysis detected anymore). The lysed culture from the most diluted dilution is selected and the complete procedure is repeated twice to ensure a clonal virus. While the plaque assay can lead to the isolation of several viral strains infecting the same host, isolation in liquid medium generally selects for the most abundant viral strain lytic for the targeted hosts. However, this latter procedure requires only non-specialized equipment, consumables and minimal sample handling.

5.6.2 Isolation of Temperate Viruses

The most common procedure to induce lysogens and isolate temperate viruses consists of inducing lysogens with the antibiotic mitomycin-C (Paul and Weinbauer 2010). The recommended mitomycin-C concentration varies depending on the lysogen strains but usually ranges between 0.1 and 2 μg mL^{-1}. This treatment usually induces the collapse of lysogenized culture and the release of temperate virus within 12 h. Cellular debris and larger particles can be clarified from the suspension by low-speed centrifugation and/or filtration through 0.45 μm low protein binding membrane filters. The filtrate containing temperate viral isolates can then be stored at 4 °C in the dark until further characterization. Alternatively, lysogens can also be induced upon short (30 s) exposure to UV-C radiation (Jiang and Paul 1998; Lohr et al. 2007; Weinbauer and Suttle 1996). These procedures demonstrated that lysogeny was frequent among cultured marine bacteria (Jiang and Paul 1998; Lossouarn et al. 2015; Stopar et al. 2004) yet it has been seldom reported in microalgae (Lohr et al. 2007) and archaea. However, one should bear in mind that lysogeny represents a complex and stable relationship between the virus genome and that of its hosts. The induction of lysogens upon chemical or physical induction does not take into account the biological principle of this symbiosis (Paul 2008). It is thus likely that some lysogens are not induced following these treatments.

5.6.3 Isolation of Chronic Viruses

Chronic viruses that are released by the host cell by continuous or intermittent budding or extrusion are less investigated than their lytic and temperate counterparts, yet they have been reported in marine eukaryotic photoautotrophs *Emiliania huxelyi* (Mackinder et al. 2009) and *Ostreococcus* sp. (Thomas et al. 2011), but also for viruses of the hyperthermophilic euryarchaeote *Pyrococcus abyssi* (Geslin et al. 2003) and *Thermococcus prieurii* (Gorlas et al. 2012). Chronic viruses are typically detected by the examination of host culture by transmission electron microscopy. The isolation of these viruses mostly relies on extracting the virus particles from the host culture. This can be done either by low-speed centrifugation and/or filtration through low protein binding membrane filters with pore size ranging from 0.2 to 0.45 μm.

5.6.4 Maintenance and Storage

The most common procedure to store virus stocks is to keep them refrigerated (4 °C) in the dark. However, viral strains differ considerably with respect to their stability in the cold. While some isolates can be kept for years, others decay when stored refrigerated. Therefore, appropriate storage conditions should be designed for each individual type of virus. For the most challenging isolates, repeated transfers into fresh host culture can be considered for maintaining viruses. Finally, cryopreservation can be an alternative for, at least, some isolates. Bacteriophages affiliated to the order *Caudovirales* are often stored at −80 °C in 10–20 % sucrose, 10-20 % dimethyl sulfoxide (DMSO). These concentrations of cryoprotectant were also appropriate to store viruses of phytoplankton; however, storage temperature has to be modified according to the isolate stability (Nagasaki and Bratbak 2010). Conversely, to viruses pathogenic to human, animals or plants, marine virus isolates are rarely deposited in culture collections. Except for the Roscoff Culture Collection (RCC), which curates and supplies viruses of marine microbes, most VHS are maintained in personal culture collections of individual laboratories.

5.7 Marine Virus-Host Model Systems in Culture

During the past three decades few hundreds of viruses have been isolated from marine systems and brought into culture. Cultured viral isolates have been reported for many species emblematic of heterotrophic and phototrophic marine bacteria such as Proteobacteria, Bacteroidetes, Cyanobacteria (e.g., Holmfeldt et al. 2013, 2014; Moebus and Nattkemper 1983; Sullivan et al. 2003, 2005, 2009) and algae (Brussaard and Martínez-Martínez 2008; Short 2012; Tomaru et al. 2015 and references therein), whereas thus far only few viruses infecting archaea (Geslin et al. 2003; Gorlas et al. 2012), or heterotrophic protozoa (Arslan et al. 2011; Fischer et al. 2010; Garza and Suttle 1995) were brought into culture. Among the latter, the virus of the protozoa *Cafeteria roenbergensis* (CroV) was even parasitized by a small virophage (Mavirus) during CroV host infection (Fischer and Suttle 2011).

The isolation and characterization of these VHS unveiled an extraordinarily high level of morphological, taxonomical, and functional novelty. The characterization of 31 cellulophages demonstrated considerable divergence with known counterparts (Holmfeldt et al. 2013). The morphological and genomic analysis of these isolates showed that they represent 12 novel genera globally distributed in the ocean and together they comprise the largest diversity of phages associated to a single marine host (Holmfeldt et al. 2013). Likewise, marine protists host a wide diversity of viral pathogens. The majority of these viruses is complex, possesses a large dsDNA genome, and belongs to the Nucleo-Cytoplasmic Large DNA Viruses (NCLDV), for which the order Megavirales has been proposed (Colson et al. 2013). NCLDV from marine environments belong to 2 main families. The Phycodnaviridae family includes the majority of viruses that infects photosynthetic protists among which the haptophytes *E. huxleyi* (Schroeder et al. 2002; Wilson et al. 2005), *P. globosa* (Baudoux and Brussaard 2005; Brussaard et al. 2004b), the prasinophytes *Micromonas*, *Bathycoccus,* and *Ostreococcus* (Baudoux et al. 2015; Derelle et al. 2008, 2015; Martínez-Martínez et al. 2015; Zingone et al. 2006), or the raphidophyte *H. akashiwo* (Nagasaki et al. 1999).

Marine NCLDV also classify within the Mimiviridae family. This clade, named after the discovery of Mimivirus (Raoult et al. 2004), originally included viruses with exceptionally large dsDNA genomes. With a sophisticated shell 0.7 μm and a huge genome of 1.2 Mb, *Megavirus chilensis,* isolated from coastal waters of Chile, represents the largest Mimiviridae and it surprisingly propagates on a freshwater acanthamoeba (Arslan et al. 2011). Viruses with smaller genome sizes but sharing several genetic features with these giant viruses were added to this virus family; Mimiviridae includes now the virus CroV that infects the marine protozoa *Cafeteria roenbergensis* (750 kb, Fischer et al. 2010), but also viruses that infect photosynthetic protists such as the haptophytes *Phaeocystis pouchetii* (485 kb, Jacobsen et al. 1996), *Phaeocystis globosa* (460 kb PgV-16T, Santini et al. 2013), *Haptolina ericina* (formely named *Chrysochromulina ericina*, 530 kb, Sandaa et al. 2001; Johannessen et al. 2015), *Prymnesium kappa* (507 kb, Johannessen et al. 2015), the prasinophyte *Pyramimonas orientalis* (560 kb, Sandaa et al. 2001) and the pelagophyte *Aureococcus anophagefferens* (371 kb, Moniruzzaman et al. 2014) are awaiting assignment to this virus family. The discovery of Mimiviridae initiated intensive debates on the concept virus. Their huge and astonishingly complex genome was found to encode trademark cellular functions, which literally blurs the boundaries between viruses and cellular organisms. The upper limits of the viral world both in terms of particle size and genome complexity have been pushed out with the isolation of Pandoraviruses from coastal marine waters and a freshwater pond (Philippe et al. 2013). These amoeba viruses exhibit dsDNA genome of 2.5 Mb lacking similarities with known NCLDV.

Not all protist viruses are of the dsDNA type. The prasinophyte *Micromonas*, which hosts many Phycodnaviridae, can also be infected by dsRNA virus (Brussaard et al. 2004b). This reovirus displays unusual morphology and genetic features compared to known Reoviridae, which led to the proposition of a new genus, *Mimoreovirus*, within the family Reoviridae (Attoui et al. 2006). Moreover, diatom viruses exhibit tiny particle diameter (< 40 nm) compared to Phycod naviridae and belong to two main groups: the Bacilladnavirus (ssDNA viruses, Kimura and Tomaru 2013, 2015; Nagasaki et al. 2005; Tomaru et al. 2008, 2011a, b, 2013b) and the Bacillarnavirus (ssRNA, Kimura and Tomaru 2015; Nagasaki et al. 2004; Shirai et al. 2008; Tomaru et al. 2009, 2012, 2013a).

Together, these novel VHS demonstrate that we have barely scratched the surface of viral diversity. Characterizing these model systems is essential to gaining knowledge about the functional role of this tremendous taxonomic diversity. Importantly, marine environments have been inadequately sampled. Most VHS cultured thus far originate from temperate coastal marine environments, while open-ocean, tropical and extreme marine systems have been largely under-explored. Likewise, VHS mostly include dsDNA virus that infect emblematic marine bacteria or protists. By comparison archaeal viruses, RNA viruses, or ssDNA are under-represented while some of these groups seem to comprise an important fraction of the viral community (Culley et al. 2006; Labonté and Suttle 2013a, b; Labonté et al. 2015). Hence, there is no question that more efforts for isolating novel viruses need to be pursued.

5.8 Marine Viruses and Biotechnological Applications

Metagenomic studies of marine viruses reveal the presence of many new genes and novel proteins and as such the vast amount of genetically diverse marine viruses is a largely untapped genetic resource for biotechnological applications. Interesting findings from studying marine viruses thus far involve the algal virus EhV infecting *E. huxleyi*, which has a unique cluster of seven genes that are involved in the sphingolipid biosynthesis, leading to ceramide formation (Wilson et al. 2005). Ceramides are the key structural lipids of skin, nails and hair as an effective epidermal barrier against water evaporation and entry of microorganisms. Use of ceramides by the cosmetic industry depends still largely on biomolecules obtained from natural sources, and thus the expression and production of viral ceramides may be a promising alternative source of natural ceramides (Wilson et al. 2005).

Other exciting recent discoveries are the viruses infecting viruses which showed that the infection with such virophage ('eaters of viruses') not only helps the survival of the cellular organism by decreasing the yield of the larger virus, but it also illustrated the potential of the virophages as source of biotechnologically interesting features (Fischer and Suttle 2011; La Scola et al. 2008). For example, the name of the virophage Mavirus is derived from Maverick virus as its genome encoding 20 predicting coding sequences of which seven have homology to the Maverick/Polinto family of DNA transposons. Furthermore, the DNA genome of the virophage Sputnik contains genes related to those in viruses infecting the three domains of life, and may as such serve the transfer of genetic material among viruses.

In the marine aquaculture industry, viruses are typically directly linked to (lethal) diseases and therefore are considered a major economic burden (Suttle 2007). However, the beneficial use of lytic viruses as mortality agents and source of lytic enzymes has been explored for multiple purposes. In aquaculture, *Chlorella* viruses were exploited to improve the extraction efficiency of algal lipids for biodiesel production (Sanmukh et al. 2014). The stringent host specificity of virus is another feature that has been exploited in aquaculture. Cultured fish, shellfish, or crustaceans, like any animal, are the target of bacterial infections that can have a dramatic economic impact. The use of chemotherapy is a rapid and efficient way to limit these infections. Yet these treatments are also lethal for beneficial bacteria and, importantly, they have led to the evolvement of multidrug-resistant bacterial strains that are a serious threat for farming industries. By comparison, lytic viruses are efficient antimicrobial agents that specifically infect and propagate on a limited number of host strains, leaving the co-occurring microbial community untouched. They have thus emerged as potential therapeutic agents to prevent and treat bacterial infections. Phage therapy has already been implemented to treat several fish pathogens and its effectiveness is currently tested for the treatment of other cultured animals in experimental conditions (Karunasagar et al. 2007; Martínez-Díaz and Hipólito-Morales 2013; Nakai and Park 2002; Oliveira et al. 2012 and references therein).

The application of viruses as therapeutic agents goes beyond aquaculture. In natural ecosystems, the combined effect of anthropogenic pressure and global change severely threatens the health of coral reefs. The increased surface temperature and pollution has been associated with the development of infectious diseases that lead to the demise of infected reefs. Phage therapy reached a new milestone with promising tests accomplished on the scleractinian corals *Favia favus* infected by *Thalassomonas loyana*, the causative agent of the white plague disease (Atad et al. 2012). Inoculation of the marine phage BA3 (Efrony et al. 2007, 2009) with diseased corals prevented the progression of the white plague disease and its transmission to healthy corals in the Gulf of Aqba, Red Sea. This promising study constitutes the first application of phage therapy to cure diseased corals in situ.

Another emerging field of interest is the detection and identification of viruses in ballast water. Ballast water discharges from ships are responsible for introducing invasive species and their viruses to other marine regions. The knowledge of marine microorganisms spread by ballast water discharge is still largely understudied but has potential of spreading disease agents. There are a few examples (already) of specific viruses that were introduced via ships' ballast water, i.e., the Infectious Salmon Anemia Virus (ISAV) in salmon farms of Scotland were linked to vessel visits, and marine cyanophages were introduced to the Great Lakes, USA (Murray et al. 2002; Wilhelm et al. 2006). Given the increasing volume of shipping and rapidly growing aquaculture industry, introduction of pathogenic viruses to near-coast ecosystems and aquaculture farms is a realistic threat. At the same time, increasingly strict regulations on ballast water discharge force the shipping industry to make use of on-board ballast water treatment systems in order to decrease the abundance of organisms and viruses. Effective methods of virus concentration and virus detection are needed to comply with the new quality standards.

The high yield genes found as AMG can be used by synthetic biology to create more efficient artificial cells in the future. It is clear that the vast genetic diversity contained in the viral biosphere compartment will eventually be used for biotech applications or nanotechnology (Fischlechner and Donath 2007).

5.9 Future Perspectives

Viruses play a major role in the natural history and population dynamics of most living organisms and this is even truer for Bacteria, Archaea, and protists. Marine viruses might contribute decisively to the marine ecosystem functioning and to the stability and performance of the microbial communities. Therefore, it is important to strive to get a meaningful representation of all the groups involved in the community behavior and that includes viruses. If we are to understand the biology of these microbes we have to advance in the knowledge of their viruses. Culture-independent approaches are allowing fast advances, but do not exclude the need for in situ viral lysis rate measurements, as well as experimental studies of viral–host interactions.

Viruses are probably the largest reservoir of novel genes in the biosphere and this novelty is important for their host biology. The role that viruses play in the evolution of microbes is considered to be fundamental. Cells can outsource their most innovative evolutionary experiments to their viruses that are not under the pressure of optimizing growth parameters to compete. The number of genes shared by (microbial) hosts and their viruses has been continuously increasing. As more viral genomes become available, the list of AMG genes increases and is now including multiples aspects of the host physiology (Breitbart 2012).

It has been proposed that viruses are a part of the bacterial and archaeal pan-genomes (Rodriguez-Valera and Ussery 2012) and that as such are impossible to extricate from the cellular lineages from an evolutionary perspective, i.e., to form a single selection unit. In the formidable task that microbiologists face when struggling to describe microbial diversity and its ecological function, viral diversity and activity can be considered the last but important frontiers that need to be overcome.

Acknowledgments The research leading to these results has received funding from the European Union Seventh Framework Programme (FP7/2007-2013) under grant agreement n° 311975. This publication reflects the views only of the author, and the European Union cannot be held responsible for any use which may be made of the information contained therein.

References

Abedon ST (2000) The murky origin of Snow White and her T-even dwarfs. Genetics 155:481–486

Angly FE, Felts B, Breitbart M, Salamon P, Edwards RA, Carlson C, Chan AM, Haynes M, Kelley S, Liu H, Mahaffy JM, Mueller JE, Nulton J, Olson R, Parsons R, Rayhawk S, Suttle CA, Rohwer F (2006) The marine viromes of four oceanic regions. PLoS Biol 4:e368

Arslan D, Legendre M, Seltzer V, Abergel C, Claverie JM (2011) Distant Mimivirus relative with a larger genome highlights the fundamental features of Megaviridae. Proc Natl Acad Sci 108:17486–17491

Atad I, Zvuloni A, Loya Y, Rosenberg E (2012) Phage therapy of the white plague-like disease of *Favia favus* in the Red Sea. Coral Reefs 31:665–670

Attoui H, Mohd Jaafar F, Belhouchet M, de Micco P, de Lamballerie X, Brussaard CPD (2006) *Micromonas pusila* reovirus: a new member of the family Reoviridae assigned to a novel proposed genus (Mimoreovirus). J Gen Virol 87:1375–1383

Avrani S, Schwartz DA, Lindell D (2012) Virus-host swinging party in the oceans: Incorporating biological complexity into paradigms of antagonistic coexistence. Mobile Genetic Elements 2:88–95

Baudoux A-C, Brussaard CPD (2005) Characterization of different viruses infecting *Phaeocystis globosa*. Virology 341:80–90

Baudoux A-C, Brussaard CPD (2008) Influence of irradiance on viral-algal host interactions. J Phycol 44:902–908

Baudoux AC, Noordeloos AAM, Veldhuis MJW, Brussaard CPD (2006) Virally induced mortality of *Phaeocystis globosa* during two spring blooms in temperate coastal waters. Aquat Microb Ecol 44:207–217

Baudoux A-C, Veldhuis MJW, Witte HJ, Brussaard CPD (2007) Viruses as mortality agents of picophytoplankton in the deep chlorophyll maximum layer during IRONAGES III. Limnol Oceanogr 52:2519–2529

Baudoux A-C, Hendrix RW, Lander GC, Bailly X, Podell S, Paillard C, Johnson JE, Potter CS, Carragher B, Azam F (2012) Genomic and functional analysis of vibrio phage SIO-2 reveals novel insights into ecology and evolution of marine siphoviruses. Environ Microbiol 14:2071–2086

Baudoux A-C, Lebredonchel H, Dehmer H, Latimier M, Edern R, Rigaut-Jalabert F, Latimier M, Moreau H, Simon N (2015) Interplay between the genetic clades of micromonas and their viruses in the Western English Channel. Environmental Microbiology Reports 7:765–773

Bratbak G, Egge JK, Heldal M (1993) Viral mortality of the marine alga *Emiliania huxleyi* (Haptophyceae) and termination of algal blooms. Mar Ecol Prog Ser 93:39–48

Bratbak G, Jacobsen A, Heldal M, Nagasaki K, Thingstad F (1998) Virus production in *Phaeocystis pouchetii* and its relation to host cell growth and nutrition. Aquat Microb Ecol 16:1–9

Breitbart M (2012) Marine viruses: truth or dare. Ann Rev Mar Sci 4:425–448

Breitbart M, Felts B, Kelley S, Mahaffy JM, Nulton J, Salamon P, Rohwer F (2004) Diversity and population structure of a near–shore marine–sediment viral community. Proc R Soc Lond B Biol Sci 271:565–574

Brum JR, Sullivan MB (2015) Rising to the challenge: accelerated pace of discovery transforms marine virology. Nat Rev Microbiol 13:147–159

Brum JR, Ignacio-Espinoza JC, Roux S, Doulcier G, Acinas SG, Alberti A, Sullivan MB (2015) Patterns and ecological drivers of ocean viral communities. Science 348:1261498

Brussaard CPD (2004a) Optimization of procedures for counting viruses by flow cytometry. Appl Environ Microbiol 70:1506–1513

Brussaard CPD, Martínez-Martínez J (2008) Algal bloom viruses. Plant Viruses 2(1):1–13

Brussaard CPD, Riegman R, Noordeloos AAM, Cadee GC, Witte H, Kop AJ, Nieuwland G, van Duyl FC, Bak RPM (1995) Effects of grazing, sedimentation and phytoplankton cell lysis on the structure of a coastal pelagic food web. Mar Ecol Prog Ser 123:259–271

Brussaard CPD, Gast GJ, Van Duyl FC, Riegman R (1996a) Impact of phytoplankton bloom magnitude on pelagic microbial food web. Mar Ecol Prog Ser 144:211–221

Brussaard CPD, Kempers RS, Kop AJ, Riegman R, Heldal M (1996b) Virus-like particles in a summer bloom of *Emiliania huxleyi* in the North Sea. Aquat Microb Ecol 10:105–113

Brussaard CPD, Marie D, Bratbak G (2000) Flow cytometric detection of viruses. J Virol Methods 85:175–182

Brussaard CPD, Noordeloos AAM, Sandaa R-A, Heldal M, Bratbak G (2004a) Discovery of a dsRNA virus infecting the marine photosynthetic protist *Micromonas pusilla*. Virology 319:280–291

Brussaard CPD, Short SM, Frederickson CM, Suttle CA (2004b) Isolation and phylogenetic analysis of novel viruses infecting the phytoplankter *Phaeocystis globosa* (Prymnesiophyceae). Appl Environ Microbiol 70:3700–3705

Brussaard CPD, Mari X, Van Bleijswijk JDL, Veldhuis MJW (2005) A mesocosm study of *Phaeocystis globosa* population dynamics. II. Significance for the microbial community. Harmful Algae 4:875–893

Brussaard CPD, Wilhelm SW, Thingstad F, Weinbauer MG, Bratbak G, Heldal M, Kimmance SA, Middelboe M, Nagasaki K, Paul JH, Schroeder DC, Suttle CA, Vaque D, Wommack KE (2008) Global-scale processes with a nanoscale drive: the role of marine viruses. The ISME Journal 2:575–578

Chen F, Suttle CA (1995) Amplification of DNA polymerase gene fragments from viruses infecting microalgae. Appl Environ Microbiol 61:1274–1278

Clokie MR, Shan J, Bailey S, Jia Y, Krisch HM, West S, Mann NH (2006) Transcription of a 'photosynthetic' T4-type phage during infection of a marine cyanobacterium. Environ Microbiol 8:827–835

Colson P, De Lamballerie X, Yutin N, Asgari S, Bigot Y, Bideshi DK, Cheng X-W, Federici BA, Van Etten JL, Koonin EV, La Scola B, Raoult D (2013) Megavirales, a proposed new order for eukaryotic nucleocytoplasmic large DNA viruses. Arch Virol 158:2517–2521

Culley AI, Lang AS, Suttle CA (2006) Metagenomic analysis of coastal RNA virus communities. Science 312:1795–1798

Danovaro R, Middelboe M (2010) Separation of free virus particles from sediments in aquatic systems. In: Wilhelm SW, Weinbauer MG, Suttle CA, Waco TX (eds) Manual of aquatic viral ecology. ASLO, pp 74–81

Danovaro R, Corinaldesi C, Filippini M, Fischer UR, Gessner MO, Jacquet S, Magagnini M, Velimirov B (2008) Viriobenthos in freshwater and marine sediments: a review. Freshw Biol 53:1186–1213

Derelle E, Ferraz C, Escande ML, Eychenié S, Cooke R, Piganeau G, Desdevises Y, Bellec L, Moreau H, Grimsley N (2008) Life-cycle and genome of OtV5, a large DNA virus of the pelagic marine unicellular green alga *Ostreococcus tauri*. PlosOne 3(5):e2250. doi:10.1371/journal.pone.0002250

Derelle E, Monier A, Cooke R, Worden AZ, Grimsley NH, Moreau H (2015) Diversity of Viruses Infecting the Green Microalga *Ostreococcus lucimarinus*. J Virol 89:5812–5821

Dutilh BE, Cassman N, McNair K, Sanchez SE, Silva GG, Boling L, Barr JJ, Speth DR, Seguritan V, Aziz RK, Felts B, Dinsdale EA, Mokili JL, Edwards RA (2014) A highly abundant bacteriophage discovered in the unknown sequences of human faecal metagenomes. Nature Communications 5:4498

Efrony R, Loya Y, Bacharach E, Rosenberg E (2007) Phage therapy of coral disease. Coral Reefs 26:7–13

Efrony R, Atad I, Rosenberg E (2009) Phage therapy of coral white plague disease: properties of phage BA3. Curr Microbiol 58:139–145

Ellis EL, Delbrück M (1939) The growth of bacteriophage. J Gen Physiol 22:365–384

Evans C, Archer SD, Jacquet S, Wilson WH (2003) Direct estimates of the contribution of viral lysis and microzooplankton grazing to the decline of a *Micromonas* spp. population. Aquat Microb Ecol 30:207–219

Fischer MG, Suttle CA (2011) A virophage at the origin of large DNA transposons. Science 332 (6026):231–234

Fischer MG, Allen MJ, Wilson WH, Suttle CA (2010) Giant virus with a remarkable complement of genes infects marine zooplankton. Proc Natl Acad Sci 107:19508–19513

Fischlechner M, Donath E (2007) Viruses as building blocks for materials and devices. Angew Chem Int Ed 46:3184–3193

Flores CO, Meyer JR, Valverde S, Farr L, Weitz JS (2011) Statistical structure of host-phage interactions. Proc Nat Acad Sci USA 108:E288–E297

Flores CO, Valverde S, Weitz JS (2013) Multi-scale structure and geographic drivers of cross-infection within marine bacteria and phages. ISME J 7:520–532

Forterre P (2010) Defining life: the virus viewpoint. Orig Life Evol Biosph 40:151–160

Frada M, Probert I, Allen MJ, Wilson WH, de Vargas C (2008) The "Cheshire Cat" escape strategy of the coccolithophore *Emiliania huxleyi* in response to viral infection. Proc Natl Acad Sci 105:15944–15949

Fraser JS, Maxwell KL, Davidson AR (2007) Immunoglobulin-like domains on bacteriophage: weapons of modest damage? Curr Opin Microbiol 10:382–387

Garcia-Doval C, van Raaij MJ (2013) Bacteriophage receptor recognition and nucleic acid transfer. Structure and physics of viruses. Springer, Berlin, pp 489–518

Garza DR, Suttle CA (1995) Large double-stranded DNA viruses which cause the lysis of a marine heterotrophic nanoflagellate (*Bodo* sp.) occur in natural marine viral communities. Aquat Microb Ecol 9:203–210

Geslin C, Le Romancer M, Erauso G, Gaillard M, Perrot G, Prieur D (2003) PAV1, the first virus-like particle isolated from a hyperthermophilic euryarchaeote,"*Pyrococcus abyssi*". J Bacteriol 185:3888–3894

Ghai R, Martin-Cuadrado AB, Molto AG, Heredia IG, Cabrera R, Martin J, Verdu M, Deschamps P, Moreira D, Lopez-Garcia P, Mira A, Rodriguez-Valera F (2010) Metagenome of the Mediterranean deep chlorophyll maximum studied by direct and fosmid library 454 pyrosequencing. ISME J 4:1154–1166

Ghai R, Mizuno CM, Picazo A, Camacho A, Rodriguez-Valera F (2013) Metagenomics uncovers a new group of low GC and ultra-small marine Actinobacteria. Sci Rep 3:2471

Glud RN, Middelboe M (2004) Virus and bacteria dynamics of a coastal sediment: implication for benthic carbon cycling. Limnol Oceanogr 49:2073–2081

Gobler CJ, Hutchins DA, Fisher NS, Cosper EM, Sanudo Wilhelmy SA (1997) Release and bioavailability of C, N, P, Se, and Fe following viral lysis of a marine chrysophyte. Limnol Oceanogr 42:1492–1504

Gorlas A, Koonin EV, Bienvenu N, Prieur D, Geslin C (2012) TPV1, the first virus isolated from the hyperthermophilic genus *Thermococcus*. Environ Microbiol 14:503–516

Heldal M, Bratbak G (1991) Production and decay of viruses in aquatic environments. Mar Ecol Prog Ser 72:205–2012

Hewson I, O'Niel JM, Fuhrman JA, Dennison WC (2001) Virus-like particle distribution and abundance in sediments and overlying waters along eutrophication gradients in two subtropical estuaries. Limnol Oceanogr 46:1734–1746

Hidaka T (1977) Detection and isolation of marine bacteriophage systems in the southwestern part of the Pacific Ocean. Mem Fac Fish Kagoshima Univ 26:55–62

Hidaka T, Fujimura T (1971) A morphological study of marine bacteriophages. Mem Fac Fish Kagoshima Univ 20:141–154

Holmfeldt K, Solonenko N, Shah M, Corrier K, Riemann L, Verberkmoes NC (2013) Twelve previously unknown phage genera are ubiquitous in global oceans. Proc Nat Acad Sci USA 110:12798–12803

Hurwitz BL, Hallam SJ, Sullivan MB (2013) Metabolic reprogramming by viruses in the sunlit and dark ocean. Genome Biol 14:R123

Hutchinson GE (1961) Theparadoxoftheplankton. Am Nat 95:137–145

Jacobsen A, Bratbak G, Heldal M (1996) Isolation and characterization of a virus infecting *Phaeocystis pouchetii* (Prymnesiophyceae). J Phycol 32:923–927

Jacobsen A, Larsen A, Martínez-Martínez J, Verity PG, Frischer ME (2007) Susceptibility of colonies and colonial cells of *Phaeocystis pouchetii* (Haptophyta) to viral infection. Aquat Microb Ecol 48:105–112

Jiang SC, Paul JH (1998) Significance of lysogeny in the marine environment: studies with isolates and a model of lysogenic phage production. Microb Ecol 35:235–243

Johannessen TV, Bratbak G, Larsen A, Ogata H, Egge ES, Edvardsen B, Eikrem W, Sandaa RA (2015) Characterisation of three novel giant viruses reveals huge diversity among viruses infecting Prymnesiales (Haptophyta). Virology 476:180–188

Kang I, Oh HM, Kang D, Cho JC (2013) Genome of a SAR116 bacteriophage shows the prevalence of this phage type in the oceans. Proc Natl Acad Sci 110:12343–12348

Karunasagar I, Shivu MM, Girisha SK, Krohne G, Karunasagar I (2007) Biocontrol of pathogens in shrimp hatcheries using bacteriophages. Aquaculture 268:288–292

Kimmance SA, Brussaard CPD (2010) Estimation of viral-induced phytoplankton mortality using the modified dilution method. Limnol Oceanogr Methods 7:65–73

Kimura K, Tomaru Y (2013) Isolation and characterization of a single-stranded DNA virus infecting the marine diatom *Chaetoceros* sp. strain SS628–11 isolated from western Japan. PLoS ONE 8:e82

Kimura K, Tomaru Y (2015) Discovery of two novel viruses expands the diversity of ssDNA and ssRNA viruses infecting a cosmopolitan marine diatom. Appl Environ Microbiol 81:1120–1131

La Scola B, Audic S, Robert C (2003) A giant virus in amoebae. Science 299:2033

La Scola B, Desnues C, Pagnier I, Robert C, Barrassi L, Fournous G, Merchat M, Suzan-Monti M, Forterre P, Koonin E, Raoult D (2008) The virophage as a unique parasite of the giant mimivirus. Nature 455:100–104

Labonté JM, Suttle CA (2013a) Metagenomic and whole-genome analysis reveals new lineages of gokushoviruses and biogeographic separation in the sea. Front Microbiol 4:404
Labonté JM, Suttle CA (2013b) Previously unknown and highly divergent ssDNA viruses populate the oceans. ISME J 7:2169–2177
Labonté JM, Swan BK, Poulos B, Luo H, Koren S, Hallam SJ, Sullivan MB, Woyke T, Wommack KE, Stepanauskas R (2015) Single-cell genomics-based analysis of virus-host interactions in marine surface bacterioplankton. ISME J 9:2386–2399
Laguna R, Romo J, Read BA, Wahlund TM (2001) Induction of phase variation events in the life cycle of the marine coccolithophorid *Emiliania huxleyi*. Appl Environ Microbiol 67:3824–3831
Leggett HC, Buckling A, Long GH, Boots M (2013) Generalism and the evolution of parasite virulence. Trends Ecol Evol 28(10):592–596
Lenski RE (1988) Dynamics of interactions between bacteria and virulent bacteriophage. In: Advances in microbial ecology. Springer, New York, pp 1–44
Lohr J, Munn CB, Wilson WH (2007) Characterization of a latent virus-like infection of symbiotic zooxanthellae. Appl Environ Microbiol 73:2976–2981
Lønborg C, Middelboe M, Brussaard CPD (2013) Viral lysis of *Micromonas pusilla*: impacts on dissolved organic matter production and composition. Biogeochemistry 116:231–240
Lossouarn J, Nesbø CL, Mercier C, Zhaxybayeva O, Johnson MS, Charchuck R, Farasin J, Bienvenu N, Baudoux A-C, Michoud G, Jebbar M, Geslin C (2015) 'Ménage à trois': a selfish genetic element uses a virus to propagate within Thermotogales. Environ Microbiol 17:3278–3288
Løvdal T, Eichner C, Grossart H-P, Carbonnel V, Chou L, Martin-Jezequel V, Thingstad TF (2008) Competition for inorganic and organic forms of nitrogen and phosphorous between phytoplankton and bacteria during an *Emiliania huxleyi* spring bloom. Biogeosciences 5:371–383
Maat DS, Crawfurd KJ, Timmermans KR, Brussaard CPD (2014) Elevated CO_2 and phosphate limitation favor *Micromonas pusilla* through stimulated growth and reduced viral impact. Appl Environ Microbiol 80:3119–3127
Mackinder LC, Worthy CA, Biggi G, Hall M, Ryan KP, Varsani A, Harper GM, Wilson WH, Brownlee C, Schroeder DC (2009) A unicellular algal virus, *Emiliania huxleyi* virus 86, exploits an animal-like infection strategy. J Gen Virol 90:2306–2316
Mann NH, Cook A, Millard A, Bailey S, Clokie M (2003) Marine ecosystems: bacterial photosynthesis genes in a virus. Nature 424:741
Mari X, Rassoulzadegan F, Brussaard CPD, Wassmann P (2005) Dynamics of transparent exopolymeric particles (TEP) production by *Phaeocystis globosa* under N- or P-limitation: a controlling factor of the retention/export balance? Harmful Algae 4:895–914
Marston MF, Pierciey FJ Jr, Shepard A, Gearin G, Qi J, Yandava C (2012) Rapid diversification of coevolving marine Synechococcus and a virus. Proc Nat Acad Sci USA 109:4544–4549
Martínez-Díaz SF, Hipólito-Morales A (2013) Efficacy of phage therapy to prevent mortality during the vibriosis of brine shrimp. Aquaculture 400–401:120–124
Martínez-Martínez J, Schroeder DC, Larsen A, Bratbak G, Wilson WH (2007) Molecular dynamics of *Emiliania huxleyi* and cooccurring viruses during two separate mesocosm studies. Appl Environ Microbiol 73:554–562
Martínez-Martínez J, Swan BK, Wilson WH (2014) Marine viruses, a genetic reservoir revealed by targeted viromics. ISME J 8:1079–1088
Martínez-Martínez J, Boere A, Gilg I, van Lent JW, Witte HJ, van Bleijswijk JDL, Brussaard CPD (2015) New lipid envelope-containing dsDNA virus isolates infecting *Micromonas pusilla* reveal a separate phylogenetic group. Aquat Microb Ecol 74:17–28
Matsuzaki S, Tanaka S, Koga T, Kawata T (1992) A broad-host-range vibriophage, KVP40, isolated from sea water. Microbiol Immunol 36:96–97
McDaniel L, Houchin LA, Williamson SJ, Paul JH (2002) Lysogeny in marine *Synechococcus*. Nature 451:496

Mizuno CM, Rodriguez-Valera F, Kimes NE, Ghai R (2013) Expanding the marine virosphere using metagenomics. PLoS Genet 9:e1003987

Moebus K, Nattkemper H (1981) Bacteriophage sensitivity patterns among bacteria isolated from marine waters. Helgoländer Meeresuntersuchungen 34:375–385

Moebus K, Nattkemper H (1983) Taxonomic investigations of bacteriophage sensitive bacteriaisolated from marine waters. Helgolhnder Meeresunters 36:357–373

Mojica KDA, Brussaard CPD (2014) Factors affecting virus dynamics and microbial host–virus interactions in marine environments. FEMS Microbiol Ecol 89:495–515

Mojica KDA, Huisman J, Wilhelm SW, Brussaard CPD (2015) Latitudinal variation in virus-induced mortality of phytoplankton across the North Atlantic Ocean. ISME J. doi:10.1038/ismej.2015.130

Moniruzzaman M, LeCleir GR, Brown CM, Gobler CJ, Bidle KD, Wilson WH, Wilhelm SW (2014) Genome of brown tide virus (AaV), the little giant of the Megaviridae, elucidates NCLDV genome expansion and host–virus coevolution. Virology 466:60–70

Morin PJ (2008) Sex as an algal antiviral strategy. Proc Natl Acad Sci 105:15639–15640

Mühling M, Fuller NJ, Millard A, Somerfield PJ, Marie D, Wilson WH, Scanlan DJ, Post AF, Joint I, Mann NH (2005) Genetic diversity of marine Synechococcus and co-occurring cyanophage communities: evidence for viral control of phytoplankton. Environ Microbiol 7:499–508

Murray AG, Smith RJ, Stagg RM (2002) Shipping and the spread of infectious Salmon Anemia in Scottish aquaculture. Emerg Infect Dis 8:1–5

Nagasaki K, Bratbak G (2010) Isolation of viruses infecting photosynthetic and nonphotosynthetic protists. In: Wilhelm SW, Weinbauer MG, Suttle CA, Waco TX (eds) Manual of aquatic viral ecology. ASLO, pp 92–101

Nagasaki K, Tarutani K, Yamaguchi M (1999) Cluster analysis on algicidal activity of HaV clones and virus sensitivity of *Heterosigma akashiwo* (Raphidophyceae). J Plankton Res 21:2219–2226

Nagasaki K, Tomaru Y, Katanozaka N, Shirai Y, Nishida K, Itakura S, Yamaguchi M (2004) Isolation and characterization of a novel single-stranded RNA virus infecting the bloom-forming diatom *Rhizosolenia setigera*. Appl Environ Microbiol 70:704–711

Nagasaki K, Tomaru Y, Takao Y, Nishida K, Shirai Y, Suzuki H, Nagumo T (2005) Previously unknown virus infects marine diatom. Appl Environ Microbiol 71:3528–3535

Nakai T, Park SC (2002) Bacteriophage therapy of infectious diseases in aquaculture. Res Microbiol 153:13–18

Noble RT, Fuhrman JA (1998) Use of SYBR Green I for rapid epifluorescence counts of marine viruses and bacteria. Aquat Microb Ecol 14:113–118

Noble RT, Fuhrman JA (2000) Rapid virus production and removal as measured with fluorescently labeled viruses as tracers. Appl Environ Microbiol 66:3790–3797

Oliveira J, Castilho F, Cunha A, Pereira MJ (2012) Bacteriophage therapy as a bacterial control strategy in aquaculture. Aquacult Int 20:879–910

Paul JH (2008) Prophages in marine bacteria: dangerous molecular time bombs or the key to survival in the seas? ISME J 2:579–589

Paul JH, Weinbauer MG (2010) Detection of lysogeny in marine environments. In: Wilhelm SW, Weinbauer MG, Suttle CA, Waco TX (eds) Manual of aquatic viral ecology. ASLO, pp 30–33

Paul JH, Cochran PK, Jiang SC (1999) Lysogeny and transduction in the marine environment. In: Bell CR, Brylinsky M, Johnson-Green P (eds) Microbial biosystems: new frontiers. Proceedings of the 8th international symposium on microbial ecology. Atlantic Canada Society for Microbial Ecology, Halifax, Canada

Philippe N, Legendre M, Doutre G, Couté Y, Poirot O, Lescot M, Arslan D, Seltzer V, Bertaux L, Bruley C, Garin J, Claverie J-M, Abergel C (2013) Pandoraviruses: amoeba viruses with genomes up to 2.5 Mb reaching that of parasitic eukaryotes. Science 341:281–286

Poorvin L, Rinta-Kanto JM, Hutchins DA, Wilhelm SW (2004) Viral release of Fe and its bioavailability to marine plankton. Limnol Oceanogr 49:1734–1741

Pope WH, Bowman CA, Russell DA, Jacobs-Sera D, Asai DJ, Cresawn SG, Jacobs WR, Hendrix RW, Lawrence JG, Hatfull GF (2015) Science education alliance phage hunters advancing, S. Evolutionary, R. Phage Hunters Integrating, Education and C. Mycobacterial Genetics. Whole genome comparison of a large collection of mycobacteriophages reveals a continuum of phage genetic diversity. Elife 4:e06416

Proctor LM, Fuhrman JA (1990) Viral mortality of marine bacteria and cyanobacteria. Nature 343:60–62

Raoult D, Audic S, Robert C, Abergel C, Renesto P, Ogata H, La Scola B, Suzan M, Claverie JM (2004) The 1.2-megabase genome sequence of Mimivirus. Science 306:1344–1350

Rodriguez-Brito B, Li L, Wegley L, Furlan M, Angly F, Breitbart M, Buchanan J, Desnues C, Dinsdale E, Edwards R, Felts B, Haynes M, Liu H, Lipson D, Mahaffy J, Martin-Cuadrado AB, Mira A, Nulton J, Pasic L, Rayhawk S, Rodriguez-Mueller J, Rodriguez-Valera F, Salamon P, Srinagesh S, Thingstad TF, Tran T, Thurber RV, Willner D, Youle M, Rohwer F (2010) Viral and microbial community dynamics in four aquatic environments. ISME J 4:739–751

Rodriguez-Valera F, Ussery DW (2012) Is the pan-genome also a pan-selectome? F1000Res 1:16

Rodriguez-Valera F, Martin-Cuadrado AB, Rodriguez-Brito B, Pasic L, Thingstad TF, Rohwer F, Mira R (2009) Explaining microbial population genomics through phage predation. Nat Rev Microbiol 7:828–836

Rodriguez-Valera F, Mizuno CM, Ghai R (2014) Tales from a thousand and one phages. Bacteriophage 4:e28265

Rohwer F, Segall A, Steward G, Seguritan V, Breitbart M, Wolven F, Farooq Azam F (2000) The complete genomic sequence of the marine phage Roseophage SIO1 shares homology with nonmarine phages. Limnol Oceanogr 45:408–418

Rosenwasser S, Mausz MA, Schatz D, Sheyn U, Malitsky S, Aharoni A, Weinstock E, Tzfadia O, Ben-Dor S, Feldmesser E, Pohnert G, Vardi A (2014) Rewiring host lipid metabolism by large viruses determines the fate of *Emiliania huxleyi*, a bloom-forming alga in the Ocean. Plant Cell 26:2689–2707

Ruardij P, Veldhuis MJW, Brussaard CPD (2005) Modelling the development and termination of a *Phaeocystis* bloom. Harmful Algae 4:941–963

Sánchez-Paz A, Muhlia-Almazan A, Saborowski R, García-Carreño F, Sablok G, Mendoza-Cano F (2014) Marine viruses: the beneficial side of a threat. Appl Biochem Biotechnol 174:2368–2379

Sandaa R-A, Larsen A (2006) Seasonal variations in virus-host populations in Norwegian coastal waters: focusing on the cyanophage community infecting marine *Synechococcus* spp. Appl Environ Microbiol 72:4610–4618

Sandaa R-A, Heldal M, Castberg T, Thyrhaug R, Bratbak G (2001) Isolation and characterization of two viruses with large genome size infecting *Chrysochromulina ericina* (Prymneriophyceae) and *Pyramimonas orientalis* (Prasinophyceae). Virology 290:272–280

Sanmukh SG, Khairnar K, Chandekar RH, Paunikar WN (2014) Increasing the extraction efficiency of algal lipid for biodiesel production: novel application of algal viruses. Afr J Biotechnol 13:1666–1670

Santini S, Jeudy S, Bartoli J, Poirot O, Lescot M, Abergel C, Berbe V, Wommack KE, Noordeloos AAM, Brussaard CPD, Claverie J-M (2013) Genome of *Phaeocystis globosa* virus PgV-16T highlights the common ancestry of the largest known DNA viruses infecting eukaryotes. Proc Natl Acad Sci 110:10800–10805

Schoemann V, Becquevort S, Stefels J, Rousseau V, Lancelot C (2005) Phaeocystis blooms in the global ocean and their controlling mechanisms: a review. J Sea Res 57:43–66

Schroeder DC, Oke J, Malin G, Wilson WH (2002) Coccolithovirus (Phycodnaviridae): characterisation of a new large dsDNA algal virus that infects *Emiliana huxleyi*. Arch Virol 147:1685–1698

Shirai Y, Tomaru Y, Takao Y, Suzuki H, Nagumo T, Nagasaki K (2008) Isolation and characterization of a singlestranded RNA virus infecting the marine planktonic diatom *Chaetoceros tenuissimus* Meunier. Appl Environ Microbiol 74:4022–4027

Short SM (2012) The ecology of viruses that infect eukaryotic algae. Environ Microbiol 14:2253–2271
Short CM, Suttle CA (2005) Nearly identical bacteriophage structural gene sequences are widely distributed in both marine and freshwater environments. Appl Environ Microbiol 71:480–486
Sintes E, Del Giorgio PA (2014) Feedbacks between protistan single-cell activity and bacterial physiological structure reinforce the predator/prey link in microbial foodwebs. Front Microbiol 5:453
Spencer R (1955) A marine bacteriophage. Nature 175:690–691
Spencer R (1960) Indigenous marine bacteriophages. J Bacteriol 79:614
Stern A, Sorek R (2011) The phage-host arms race: shaping the evolution of microbes. BioEssays 33:43–51
Steward GF, Wikner J, Cochlan WP, Smith DC, Azam F (1992) Estimation of virus production in the sea. II: field results. Mar Microb Food Webs 6:79–90
Stopar D, Černe A, Žigman M, Poljšak-Prijatelj M, Turk V (2004) Viral abundance and a high proportion of lysogens suggest that viruses are important members of the microbial community in the Gulf of Trieste. Microb Ecol 47:1–8
Sullivan MB, Waterbury JB, Chisholm SW (2003) Cyanophages infecting the oceanic cyanobacterium *Prochlorococcus*. Nature 424:1047–1051
Sullivan MB, Coleman ML, Weigele P, Rohwer F, Chisholm SW (2005) Three *Prochlorococcus* cyanophage genomes: signature features and ecological interpretations. PLoS Biol 3:e144
Sullivan MB, Krastins B, Hughes JL, Kelly L, Chase M, Sarracino D, Chisholm SW (2009) The genome and structural proteome of an ocean siphovirus: a new window into the cyanobacterial 'mobilome'. Environ Microbiol 11:2935–2951
Suttle CA (2005) Viruses in the sea. Nature 437:356–361
Suttle CA (2007) Marine viruses–major players in the global ecosystem. Nat Rev Microbiol 5:801–812
Suttle CA, Chan AM (1994) Dynamics and distribution of cyanophages and their effect on marine *Synechococcus* spp. Appl Environ Microbiol 60:3167–3174
Suttle CA, Chen F (1992) Mechanisms and rates of decay of marine viruses in seawater. Appl Environ Microbiol 58:3721–3729
Suttle CA, Chan AM, Cottrell MT (1995) Viruses infecting the marine Prymnesiophyte *Chrysochromulina* spp.: isolation, preliminary characterization and natural abundance. Mar Ecol Prog Ser 118:275–282
Taruntani K, Nagasaki K, Yamaguchi M (2000) Viral impacts on total abundance and clonal composition of the harmful bloom-forming phytoplankton *Heterosigma akashiwo*. Appl Environ Microbiol 66:4916–4920
Thingstad TF (2000) Elements of a theory for the mechanisms controlling abundance, diversity, and biogeochemical role of lytic bacterial viruses in aquatic systems. Limnol Oceanogr 45:1320–1328
Thingstad TF, Pree B, Giske J, Våge S (2015) What difference does it make if viruses are strain-, rather than species-specific? Front Microbiol 6:320
Thomas R, Grimsley N, Escande ML, Subirana L, Derelle E, Moreau H (2011) Acquisition and maintenance of resistance to viruses in eukaryotic phytoplankton populations. Environ Microbiol 13:1412–1420
Tomaru Y, Shirai Y, Suzuki H, Nagumo T, Nagasaki K (2008) Isolation and characterization of a new single-stranded DNA virus infecting the cosmopolitan marine diatom *Chaetoceros debilis*. Aquat Microb Ecol 50:103–112
Tomaru Y, Takao Y, Suzuki H, Nagumo T, Nagasaki K (2009) Isolation and characterization of a single-stranded RNA virus infecting the bloom-forming diatom *Chaetoceros socialis*. Appl Environ Microbiol 75:2375–2381
Tomaru Y, Fujii N, Oda S, Toyoda K, Nagasaki K (2011a) Dynamics of diatom viruses on the western coast of Japan. Aquat Microb Ecol 63:223–230

Tomaru Y, Shirai Y, Toyoda K, Nagasaki K (2011b) Isolation and characterisation of a single-stranded DNA Virus infecting the marine planktonic diatom *Chaetoceros tenuissimus* Meunier. Aquat Microb Ecol 64:175–184

Tomaru Y, Takao Y, Suzuki H, Nagumo T, Koike K, Nagasaki K (2011c) Isolation and characterization of a single-stranded DNA virus infecting *Chaetoceros lorenzianus* Grunow. Appl Environ Microbiol 77:5285–5293

Tomaru Y, Toyoda K, Kimura K, Hata N, Yoshida M, Nagasaki K (2012) First evidence for the existence of pennate diatom viruses. ISME J 6:1445–1448

Tomaru Y, Toyoda K, Kimura K, Takao Y, Sakurada K, Nakayama N, Nagasaki K (2013a) Isolation and characterization of a single-stranded RNA virus that infects the marine planktonic diatom *Chaetoceros* sp. (SS08-C03). Phycol Res 61:27–36

Tomaru Y, Toyoda K, Suzuki H, Nagumo T, Kimura K, Takao Y (2013b) New single-stranded DNA virus with a unique genomic structure that infects marine diatom *Chaetoceros setoensis*. Scientific Reports 3:3337

Tomaru Y, Toyoda K, Kimura K (2015) Marine diatom viruses and their hosts: Resistance mechanisms and population dynamics. Perspect Phycol 2:69–81

Våge S, Storesund JE, Thingstad TF (2013) SAR11 viruses and defensive host strains. Nature 499: E3–E4

Waterbury JB, Valois FW (1993) Resistance to co-occurring phages enables marine Synechococcus communities to coexist with cyanophages abundant in seawater. Appl Environ Microbiol 59:3393–3399

Weinbauer MG, Suttle CA (1996) Potential significance of lysogeny to bacteriophage production and bacterial mortality in coastal waters of the Gulf of Mexico. Appl Environ Microbiol 62:4374–4380

Weinbauer MG, Brettar I, Hofle MG (2003) Lysogeny and virus-induced mortality of bacterioplankton in surface, deep, and anoxic marine waters. Limnol Oceanogr 48:1457–1465

Weinbauer MG, Rowe JM, Wilhelm S (2010) Determining rates of virus production in aquatic systems by the virus reduction approach. In: Wilhelm SW, Weinbauer MG, Suttle CA, Waco TX (eds) Manual of aquatic viral ecology. ASLO, pp 1–8

Weitz JS, Poisot T, Meyer JR, Flores CO, Valverde S, Sullivan MB, Hochberg ME (2013) Phage–bacteria infection networks. Trends Microbiol 21:82–91

Wilhelm SW, Carberry MJ, Eldridge ML, Poorvin L, Saxton MA, Doblin MA (2006) Marine and freshwater cyanophages in a Laurentian Great Lake: evidence from infectivity assays and molecular analyses of g20 genes. Appl Environ Microbiol 72:4957–4963

Wilson WH (2005) The versatility of giant algal viruses: from shunting carbon to antiwrinkle cream. Ocean Challenge 14:8–9

Wilson WH, Carr NG, Mann NH (1996) The effect of phosphate status on the kinetics of cyanophage infection in the oceanic cyanobacterium *Synechococcus* sp. WH7803. J Phycol 32:506–516

Wilson WH, Schroeder DC, Allen MJ, Holden MT, Parkhill J, Barrell BG, Churcher C, Hamlin N, Mungall K, Norbertczak H, Quail MA, Price C, Rabbinowitsch E, Walker D, Craigon M, Roy D, Ghazal P (2005) Complete genome sequence and lytic phase transcription profile of a coccolithovirus. Science 309:1090–1092

Zeigler Allen L, Ishoey T, Novotny MA, McLean JS, Lasken RS, Williamson SJ (2011) Single virus genomics: a new tool for virus discovery. PLoS ONE 6:e17722

Zhao Y, Temperton B, Thrash JC, Schwalbach MS, Vergin KL, Landry ZC, Ellisman M, Deerinck T, Sullivan MB, Giovannoni SJ (2013) Abundant SAR11 viruses in the ocean. Nature 494:357–360

Zingone A, Natale F, Biffali E, Borra M, Forlani G, Sarno D (2006) Diversity in morphology, infectivity, molecular characteristics and induced host resistance between two viruses infecting *Micromonas pusilla*. Aquat Microb Ecol 45:1–14

Part II
Marine Habitats, Their Inhabitants, Ecology and Biogeochemical Cycles

Chapter 6
Biogeography of Marine Microorganisms

Viggó Þór Marteinsson, René Groben, Eyjólfur Reynisson and Pauline Vannier

Abstract Marine microbial biogeography describes the occurrence and abundance of microbial taxa and aims to understand the mechanisms by which they are dispersed and then adapt to their environment. The development of novel technologies, such as Next-Generation Sequencing (NGS) in combination with large-scale ocean sampling campaigns, generated a vast amount of taxonomic data that allowed for in-depth analyses of biogeographic patterns. Globally occurring groups of microorganisms were detected that dominate the marine environment (e.g., SAR11, SAR86, *Roseobacter,* and *Vibrio*), however, NGS data revealed the presence of distinct eco- and phylotypes inside these clades and genera that showed clear ecological niche adaptation and different biogeographic distributions. Genome analyses of these marine microorganisms helped to understand potential adaptive mechanisms that could explain why certain taxa are occurring ubiquitously and others are limited to certain regions and ecosystems. Marine microorganisms can employ a vast variety of adaptive mechanisms to deal with environmental parameters such as temperature, light or nutrient availability, for example through exploitation of specific energy sources or protective mechanisms against UV radiation or viruses. The availability or lack of physiological pathways and traits in ecotypes is then responsible for shaping the marine microbial biogeography.

V.Þ. Marteinsson (✉) · R. Groben · E. Reynisson · P. Vannier
Matís ohf./Food Safety, Environment and Genetics, Vínlandsleið 12,
113 Reykjavík, Iceland
e-mail: viggo@matis.is

V.Þ. Marteinsson
University of Iceland, Sæmundargata 2, 101 Reykjavík, Iceland

V.Þ. Marteinsson
Agricultural University of Iceland, Hvanneyri, IS-311 Borgarnes, Iceland

L.J. Stal and M.S. Cretoiu (eds.), *The Marine Microbiome*,
DOI 10.1007/978-3-319-33000-6_6

6.1 Introduction

Biogeography can be defined as the spatial and temporal distribution of taxa, or more elaborate and detailed: "Biogeography is the study of the distribution of biodiversity over space and time. It aims to reveal where organisms live, at what abundance, and why" (Martiny et al. 2006). While this topic has been studied for all kinds of larger, multicellular organisms, like higher plants and animals [review articles include, e.g. (Davies et al. 2011; Cowman 2014; Wang and Ran 2014)], investigations have more recently also included microorganisms, like bacteria, archaea and protists, e.g. (Angel et al. 2010; Foissner 2006; King et al. 2010; Nemergut et al. 2011). More than 70 % of Earth's surface is covered by the ocean with an average depth of 3800 m and a maximum depth of 10,790 m at the Marianas Trench. The microbial communities in this enormous water column and the connected sediments have a preponderant role in marine and even global biogeochemical cycles such as for carbon, nitrogen, oxygen availability, phosphorus, sulfur, and iron (Azam and Malfatti 2007; Fuhrman 1999; Huber et al. 2007). Marine microorganisms include both primary producers like photoautotrophs and chemoautotrophs and also secondary producers such as heterotrophs. The understanding of life-sustaining biochemical fluxes, such as global carbon and nitrogen cycles, is very much linked to the identification and characterization of the microorganisms present in the community and of their physiological capabilities. Therefore, marine microbial biogeography is an important aspect of understanding local and global biochemical cycles. In addition to their importance on these cycles, marine microorganisms also play an important role in marine food webs and in shaping ecological niches and ecosystems (Azam et al. 1983; Azam and Malfatti 2007), and again, knowledge about the distribution of marine microorganisms is paramount for our understanding of ecosystem functions.

Biogeography studies of marine microorganisms are complicated by a number of specific methodological problems. First, the environment of study is highly dynamic and sampling and biogeographic analyses require from scientists to deal with a three-dimensional water column that is constantly changing due to large and small scale impacts, like ocean currents, mixing of water layers and turbulences. This open and dynamic system is therefore at risk to be undersampled (Shade et al. 2009), because the physical movements of currents, eddies, fronts, up- and down-welling regions can cause different prevailing environmental conditions and give rise to different microbial communities, even at small spatial scales (Seymour et al. 2012) or at a single location during short time spans (short temporal scales) (Needham et al. 2013). On larger scales, surface currents could constrain microbial community biogeographic structure by affecting their physical limits and therefore contribute to microbial dissemination in the upper layers of the oceans (Selje et al. 2004; Wilkins et al. 2013a). Second, microorganisms are less easy to study than multicellular organisms due to their small size and often lack of easily identifiable morphologic features. These methodological challenges were the reason for the previous lack of marine microbial biogeographic research, and the surge of new publications on that topic (Brown et al. 2014) appeared when these hurdles were overcome by the availability of new technologies. In the early days of marine microbiology, cell identification was normally based on the isolation and culturing of strains and their characterization through morphological traits and metabolic abilities. Most studies of the diversity of marine microorganisms focused therefore on the culturing and identification of specific groups of organisms, those that were possible to culture and those that showed well-defined morphological or metabolic characteristics that allowed for an easy taxonomic identification. For example, among the first described marine bacteria were members of the genus *Photobacterium*, as they were easy to culture on complex media and easy to identify (Beijerinck 1889). Following this methodology, many more marine bacterial taxa were characterized and formally described in the nineteenth and twentieth century which covered a good part of the—cultivable—marine microbiome. ZoBell and Upham (1944) performed one of the first surveys of the diversity of marine microbes by describing 60 new bacterial species isolated from the marine environment. This early study generated bacterial strains that were used to describe many of the genera of marine bacterial cultivated in later studies. These studies focused on particular groups of Bacteria using numerical taxonomy to rigorously define genera and species. Later studies included work on, e.g., *Alteromonas* (Gauthier 1976), *Oceanospirillum*, *Pseudomonas*, *Deleya* (now *Halomonas*), *Alteromonas*, *Vibrio*, *Photobacterium* (Baumann et al. 1971, 1972; Baumann and Baumann 1977; Baumann et al. 1983; Reichelt and Baumann 1973), *Cytophaga*, *Microscilla*, and *Flavobacterium* (Lewin 1969; Lewin and Lounsbery 1969). During this period other groups of bacteria were isolated and classified on the basis of specific metabolic abilities or morphologic traits, like methylotrophs (Janvier et al. 1985; Sieburth et al. 1987), budding bacteria (Poindexter 1964; Staley 1973) and nitrifying bacteria (Watson 1965; Watson and Mandel 1971). Eukaryotic microorganisms, especially phytoplankton, were often easier to identify and describe as they show more diverse and distinct morphologic characteristics that can be seen by standard light microscopy, such as silicate cell walls (diatoms), calcium carbonate scales (coccolithophores) or flagella (e.g., dinoflagellates). However, subsequent investigations sometimes revealed cryptic speciation, emphasizing the limits of morphology-based identifications (Saez et al. 2003).

The dependence on culturing and simple light microscopy largely limited the possibilities of a thorough analysis of the marine microbial diversity and therefore a wide-spanning microbial biogeography. Technological and methodological advancements, like e.g. epifluorescence microscopy and direct counting methods, flow cytometry and molecular biology, opened up the field of microbiology beyond culture-based identification methods for bacteria and archaea and microscopy for microbial Eukarya. The advent of novel technologies, especially of "Next-Generation-Sequencing" (NGS) methodologies that are capable of generating millions of sequences from an environmental sample, and the corresponding bioinformatics tools that allow for their analysis, made it possible to obtain a picture of the marine microbial biogeography (Tseng and Tang 2014; Wood et al. 2013). Today, nearly all publications in the field of marine microbial biogeography are based on sequences of the small subunit (SSU) of the ribosomal RNA gene (16S rRNA gene for Bacteria and Archaea, 18S rRNA gene for Eukarya). These ribosomal genes are one of the standards for molecular taxonomic identification and phylogenetic analyses, although further improvements in NGS technologies will allow for the use of whole genomes and metagenomes for taxonomic and phylogenetic analyses.

Another way of analysis of microbial biogeography, remote sensing, is set to become a key component in biological oceanography and marine microbial ecology studies in the same way as it has long been the case in physical and chemical oceanography. It is now possible to estimate biomass, composition and even some community trait characteristics, such as diversity of cell size in the photoautotrophic community (Alvain et al. 2008). However, it is difficult to foresee how such technique could also cover heterotrophic microorganisms.

6.2 The Concept of "Microbial Biogeography"

The study of microbial biogeography is of course only logical and necessary if such a feature really exists. The often cited statement '*alles is overall*: maar *het milieu selecteert*' (Everything is everywhere: *but* the environment selects) (Baas Becking 1934) has for a long time guided microbiologists with respect to their view in regard to the distribution of microbial taxa and communities. In contrast to larger organisms, dispersal of small, unicellular microorganisms was seen as ubiquitous through, for example wind and, in the case of marine microorganisms, currents. Finlay and coworkers described globally ubiquitous distributions of certain protist morphospecies (Finlay 2002; Finlay and Fenchel 2004; Finlay et al. 2006). Based on their observations and taking into account the even larger population sizes of bacteria and archaea, Finlay and coworkers extrapolated that these globally ubiquitous distributions would also apply to these organisms, and that distribution of microorganisms in general would be driven by random dispersal and therefore microbial species would be globally ubiquitous (Finlay and Fenchel 2004). If this is true then microbial ecology would be very much different from general ecological principles derived from larger, "higher" organisms. Only part of the biogeography definition of Martiny et al. (2006) would apply (abundance, but not the "where", would be relevant to microbial organisms) and not all of the general mechanisms that have been identified to determine biogeographic distributions of organisms (see below) would apply to microorganisms. However, one should remember that we can address the concept of biogeography at different taxonomic levels, from morphospecies, as Finley and coworkers did (Finlay 2002), down to subgroups and phylotypes, as we see for example among the SAR11 clade (Brown et al. 2012). Therefore, what seems to be a globally ubiquitous distribution of a taxon can show clearly defined biogeographic patterns when more information, e.g. specific DNA sequences, like internal transcribed spacers (ITS), are taken into account that characterize organisms at lower taxonomic levels.

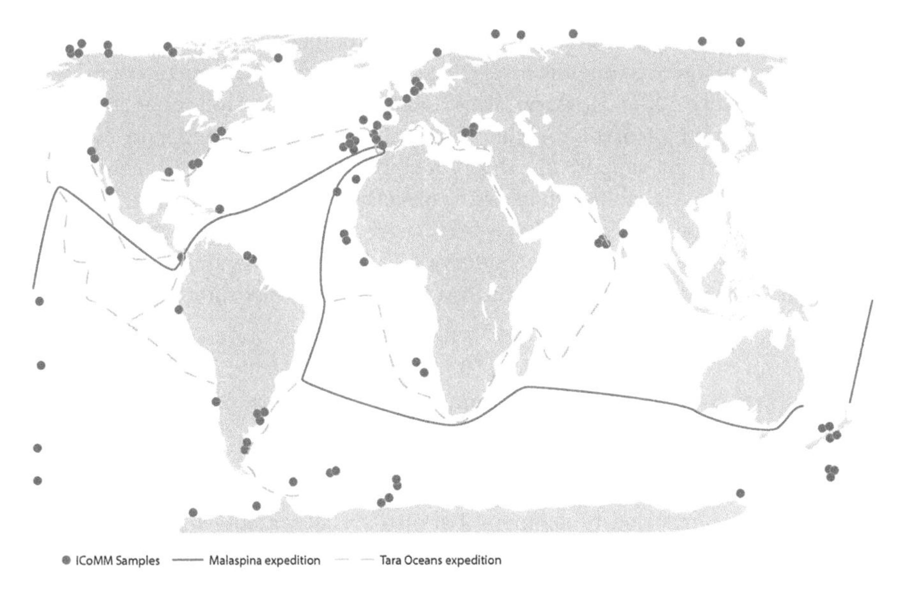

Fig. 6.1 Overview of selected large-scale marine microbial sampling and data collection initiatives showing sampling sites and routes

In an effort to give insights into the great microbial diversity in different geographic locations of the sea, various national and international sampling expeditions have been launched (Karsenti et al. 2011; Kopf et al. 2015; Rusch et al. 2007; Sunagawa et al. 2015). A large step forward in the understanding of biogeographic patters of marine microorganisms has been made due to these large-scale sampling and data collection initiatives (Fig. 6.1) such as the "The Sorcerer II Global Ocean Sampling expedition" (Nealson and Venter 2007; Rusch et al. 2007), the International Census of Marine Microbes (ICOMM; http://icomm.mbl.edu/) (Amaral-Zettler et al. 2010), Tara Oceans (http://oceans.taraexpeditions.org/) (Pesant et al. 2015), Malaspina expedition (http://scientific.expedicionmalaspina.es/), or one of the latest events, the Ocean Sampling Day (OSD) (Kopf et al. 2015). Another portal has also been established (http://www.megx.net/) for geo-referenced databases and tools for the analysis of marine bacterial, archaeal and phage genomes and metagenomes (Kottmann et al. 2010). As an example, data consisting of more than 7.7 million sequencing reads from 41 sampling sites on a 8000 km route, from the North Atlantic southwards through the Panama Canal and onwards toward the South Pacific, has revealed more new genes, proteins and biodiversity than might have been thought possible (Nealson and Venter 2007). Yet this is only a fraction of the total biological diversity that is present in the ocean at greater depths and in other geographical areas and that awaits discovery.

These initiatives and other research projects greatly benefit from the use of NGS technologies for the identification and characterization of microbial taxa. Sequencing data have shown that previously identified morphospecies of microorganisms often consist of genetically different organisms (Saez et al. 2003) and that this cryptic diversity means that the idea of cosmopolitan microorganisms can be artificial and disappears when lower taxonomic levels (e.g., ecotypes) are taken into account (Bass et al. 2007; Paasche 2001). However, while the scientific discussion about ubiquitous versus local distribution of microorganisms is still ongoing, more and more evidence has been gathered that supports the theory that microorganisms display similar biogeographic patterns as multicellular organisms (Brown et al. 2012, 2014; Fuhrman et al. 2008; Martiny et al. 2006). It has been suggested (van der Gast 2015) to use a combined model of microbial biogeography, in which some organisms show a cosmopolitan distribution, while others are restricted to certain locations.

6.3 Environmental Factors Shaping Microbial Biogeography

Biogeography is impacted by temporal but also spatial variations (Fuhrman et al. 2015). Communities can be altered by physicochemical parameters such as temperature, UV light, high hydrostatic pressures, pH, and salinity, but also by nutrient supply, interspecies competition, and viral infection. The microbial dispersion in the upper layer of the water column or the photic zone ($\sim$10–400 m), which is approximately 10 % of the ocean by volume, depends on the season and availability of sunlight. Ocean features, such as light intensity, may influence the distribution of a community, such as the vertical distribution of organisms in the water column based on their UV tolerance (DeLong et al. 2006). Below the photic zone or in the deeper waters, which make up 90 % of the ocean system, there is a physical separation of water bodies, driven by temperature and density, called thermocline. The different, separated water bodies caused by the thermocline contain each their own circulation and different microbial communities (Agogue et al. 2011).

An interesting case in regard to the dispersal of marine microorganisms and their biogeography comes from polar research and ocean currents (Fig. 6.2). Pole-to-pole studies revealed that dissemination of microbial communities by surface ocean currents is unlikely but that these microbial communities could possibly disseminate with deep water currents between the poles on a timescale of hundreds or possibly several thousands of years (Ballarotta et al. 2014; Cavicchioli 2015; Matsumoto 2007; Rahmstorf 2003; Zika et al. 2012). Although the marine environment is continuous between Antarctica and the Arctic, environmental factors exist that affect the dissemination of microbes. Surface water has not only little connectivity by major currents, it is also characterized by large changes in temperature (<0 °C at the poles but >30 °C at the equator), salinity, nutrients, and sunlight hours. Moreover, both poles are enclosed. The Antarctic Circumpolar Current isolates Antarctica and the continents Eurasia and North America enclose the Arctic Ocean (Fig. 6.2). Several factors seem to impact surface versus deep ocean communities, with surface communities being more influenced by local environmental factors and deep ocean communities more influenced by dispersal (Ghiglione et al. 2012).

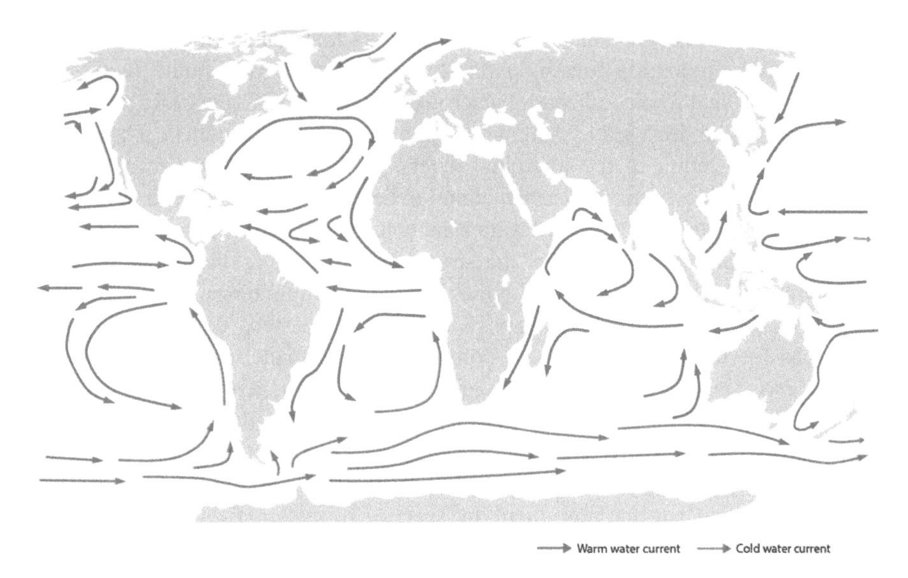

Fig. 6.2 Overview of major global ocean currents

Arctic Ocean communities differ noticeably from those in the Southern Ocean, having only ~15 % of microbial species in common which are represented by members of the typical marine taxonomic groups Alphaproteobacteria, Gammaproteobacteria, and Flavobacteria (Ghiglione et al. 2012). It seems that the limited dispersal of the marine communities at the poles exert more distinction on surface water community composition versus deep ocean communities. This indicates rather more influence of surface water dynamics on community composition than environmental selection (Sul et al. 2013b). It has also been demonstrated that physical transport with ocean currents affects microbial community composition in the Southern Ocean (Wilkins et al. 2013b). All the major Southern Ocean water masses, from the surface to ~6 km depth and latitudinal separation of up to ~3000 km, were covered in these analyses and the effect of dissemination or advection was tested by methods that included an ocean model using environmental factors and distance (Wilkins et al. 2013b). The effects of distance and environment on advection have been demonstrated in other marine studies (Fuhrman et al. 2008; Galand et al. 2009; Ghiglione et al. 2012; Giebel et al. 2009; Lauro et al. 2007; Sul et al. 2013a) but without testing the role of advection in shaping the composition of marine microbial communities. These studies concluded that dispersal in the ocean has physical limitations and ocean currents increase the opportunity for microorganisms to colonize new locations. However, it should be mentioned that single-cell genomics studies of cultured and uncultured marine bacteria tend to differ markedly and led to the conclusion that the surface ocean distribution is unlikely to limit dispersal (Swan et al. 2013).

Nutrient and energy availability are also factors that shape microbial biogeography. Primary production done by phototrophic microorganisms in the photic zone is quickly used and remineralized by heterotrophs. The resulting organic matter can be exported in the deep sea by sinking of aggregates and fecal pellets. Those steps induce a modification of the microbial diversity depending on the nutrients and oxygen availabilities. Thus, microbial communities are different close to the surface than in deeper water. The deep sea waters are considered oligotrophic environments due to the rapid degradation of the organic matter along the water column. Exceptions are for instance hydrothermal vents, upwelling zones, cold seeps, and carcasses on the seafloor (see Chaps. 8 and 9). Three steps exist under anoxic conditions for the mineralization of complex organic matter (Schink and Stams 2013). First, oligomers and monomers are formed from the hydrolysis of complex molecules and then these substrates are fermented in reduced organic compounds (short chain fatty acids and alcohols). These reduced compounds can be used by sulfate-reducing bacteria in sulfate-rich sediments, whereas in sulfate-depleted sediments strains can ferment them to acetate, formate, H_2, and CO_2. Then, methanogenic archaea can use those last substrates to produce CH_4 (Schink and Stams 2013; Stams and Plugge 2009). Consequently, microbial diversity changes along the water column from the sea surface to the deep sea waters (Brown et al. 2009; DeLong 2006; DeLong et al. 2006; Zinger et al. 2011). Depending on the study and on the site where it was conducted, some research has demonstrated less taxonomical diversity in the deep sea (Agogue et al. 2011; Brown et al. 2009), whereas other studies have shown higher diversity (Kembel et al. 2011; Pommier et al. 2010). These differences seemed to be related to the oxygen and nutrients content at the studied sites. Microbial diversity fluctuations were also detected in the water column and in sediments. Due to lack of nutrients and oxygen, it appears that sediments from continental margins and estuaries harbor a higher microbial diversity than deep sea sediments (Feng et al. 2009; Lozupone and Knight 2007).

6.4 Dominant Groups of Microorganisms in the Ocean

Originally, the dominating bacterial species of the ocean were thought to be members of *Vibrio*, *Alteromonas* and *Pseudomonas* to name a few, based on the culturing of the microbiota from seawater and marine fish. However, molecular biology, genomics, and information technology have changed our understanding and have revealed unforeseen levels of microbial diversity and metabolic potential in the seawater column and sediments (DeLong 1997; Venter et al. 2004). NGS technology has, for example, revealed unknown microbial diversity by deep sequencing the unexplored “rare biosphere” (Sogin et al. 2006). Evidence has been published that indicates a complex and diversified community composition in the ocean. The research unveils a large microbial diversity and led to spectacular discoveries of previously unknown marine microorganisms with new metabolic processes (Karsenti et al. 2011; Sogin et al. 2006; Sunagawa et al. 2015; Venter et al. 2004; Whitman et al. 1998). Although microbial diversity is huge, in most natural assemblages only a small number of phylogenetic groups dominate the community (Giovannoni and Stingl 2005). Among bacterial and archaeal taxa, Schattenhofer et al. (2009) reported the dominance of about 11 marine phyla in the upper water layers along a transect across the Atlantic Ocean from South Africa (32.9°S) to the UK (46.4°N). The alphaproteobacterial clade SAR11 is one such dominant bacterial group that may comprise between 24 and 55 % of microbial cells in the Atlantic Ocean (Malmstrom et al. 2004; Mary et al. 2006; Schattenhofer et al. 2009) and the coastal North Sea (Alonso and Pernthaler 2006). The clade was first observed and described from the Sargasso Sea (Giovannoni et al. 1990) and probably represents the most abundant heterotrophic bacterial group in pelagic marine environments (Morris et al. 2002). Another example of a globally abundant and ecologically important taxon is the cyanobacterium *Synechococcus* (see Chap. 3). The question of dominance is an important aspect of biogeography and the determination of factors that cause this dominance is of utmost importance in understanding marine microbial biogeography.

6.4.1 SAR11

The most dominant heterotrophic bacterial taxon in the ocean, SAR11, was originally discovered by sequencing 16S rRNA gene clone libraries of environmental samples without previous culturing or microscopic identification (Britschgi and Giovannoni 1991; Giovannoni et al. 1990; Mullins et al. 1995). These early molecular analyses of marine bacterial diversity found a number of novel linages that were widely distributed in ocean surface waters, of which the most abundant one was called SAR11. Phylogenetic analyses identified this new lineage as a deeply branching alphaproteobacterial clade (Britschgi and Giovannoni 1991; Field et al. 1997; Giovannoni et al. 1990; Mullins et al. 1995; Rappe et al. 2000). After its discovery by molecular methods in the 1990s, the culturing of strains from this clade was first described in 2002 (Rappe et al. 2002).

Inside the SAR11 clade, Morris et al. (2005) identified four subgroups based on distinct 16S rRNA gene sequences which were later further divided into phylotypes using internal transcribed spacer (ITS) sequences (Brown and Fuhrman 2005; Brown et al. 2012; Garcia-Martinez and Rodriguez-Valera 2000). The biogeography of SAR11, its subgroups and phylotypes shows ecological niche adaptation as well as a strong correlation with temperature and/or latitude (Brown et al. 2012). SAR11 subgroups 1a, 1b, and 2 are most abundant in ocean surface waters, whereas subgroup 3 seems to be mainly found in coastal or brackish environments, including polar and warm waters (Brown et al. 2012; Carlson et al. 2009; Morris et al. 2005). While subgroup 1b seems to be found only in warm waters (>18 °C), subgroups 1a and 2 have a cosmopolitan distribution. However, different phylotypes in those two subgroups occur in different environments according to temperature: the three phylotypes in subgroup 1a each occur either in tropical, temperate or polar waters, while subgroup 2 contains two phylotypes, each of which can dominate ocean surface waters depending on a temperature above or below 10 °C (Brown and Fuhrman 2005; Brown et al. 2012; Carlson et al. 2009; Morris et al. 2005). A third phylotype of subgroup 2 shows a more latitude-independent distribution at lower depths (Field et al. 1997; Garcia-Martinez and Rodriguez-Valera 2000). While temperature seems to be the most important factor in shaping the biogeography of SAR11 subgroups and phylotypes, other environmental parameters, such as water mixing, influence the abundance of these clades as well (Brown and Fuhrman 2005; Brown et al. 2012; Carlson et al. 2009).

While it is not clear why SAR11 is so successful and globally ubiquitous, it has been suggested, that its small size and its streamlined genome, both decreasing nutrient requirements, play a proponent role (Giovannoni et al. 2005). These and other traits may also give competitive advantages in nutrient uptake and defense against viruses (Zhao et al. 2013). Another finding was that the expression of proteorhodopsin, for which genes were found in all sequenced SAR11 genomes, is part of a mechanism that helps the cell under low carbon conditions (Sun et al. 2011). Proteorhodopsin is a light-harvesting pigment and it has been suggested that it may help covering part of the energy demand of the organism. In general, it seems that the ecological success and widespread distribution and dominance of SAR11 is linked to its ability to handle limitations in nutrient availability, something that might be found in many oceanic environments and conditions.

6.4.2 SAR86

The second most common group of marine heterotrophic bacteria according to the Global Ocean Sampling (GOS) datasets is the gammaproteobacterial SAR86 clade (Yooseph et al. 2010). SAR86 strains show similar adaptation to an oligotrophic, nutrient-limited environment as SAR11 (Lauro et al. 2009), including metabolic streamlining (Dupont et al. 2012; Swan et al. 2013) and the presence of proteorhodopsin genes (Sabehi et al. 2004). These similarities between the two globally most dominant groups of heterotrophic marine microorganisms strongly support the importance of these features for the success of the organisms. Differences between SAR86 and SAR11 include their use of various carbon sources. While SAR11 utilizes dicarboxylic acids and simple peptides, SAR86 seems to grow mainly on lipids and polysaccharides as a carbon source, and it is speculated that these different mechanisms allow for the co-occurrence of the two clades, possibly avoiding competition for resources (Dupont et al. 2012).

While sequence analyses showed some phylogenetic sub-clades within SAR86, it is not clear yet whether they represent distinct ecotypes (Treusch et al. 2009). However, some research indicates that this might be the case (Dupont et al. 2012). SAR86 sequences from sampling sites representing the open ocean, colder coastal areas, and warmer coastal areas were mainly recruited into different genomes during assembly and might represent three different ecotypes (Dupont et al. 2012). Further research is needed on this ubiquitous bacterial clade, which is hampered by the fact that there are currently no cultured representatives.

6.4.3 *Roseobacter*

Different lineages among marine bacteria use different strategies for adaptation to their environment, which can nevertheless be equally successful (Luo and Moran 2014). The ubiquitous clades SAR11/SAR86 and the also globally widespread marine alphaproteobacterial genus *Roseobacter* are examples for successful strategies of adaptation. While SAR11 and SAR86 use genomic and metabolic streamlining to decrease nutritional requirements, *Roseobacter* strains have large genomes that code for a wide variety of traits, including chemotaxis, motility, and other functions for finding and utilizing nutrients (Newton et al. 2010). These two ecological strategies are often called "free living" (SAR11 and SAR86) and "patch associated" (*Roseobacter*) (Luo and Moran 2014). The genus *Roseobacter* consists of a phylogenetically broad group of species/strains and the differences in their genomes are also reflected in their versatility with specific metabolic pathways only available to certain strains (Azam and Malfatti 2007; Newton et al. 2010). These different metabolic capacities lead then to different trophic strategies, niche adaptation and therefore ecotypes that are weakly but nevertheless significantly correlated to genome sequences and the phylogeny of the group (Newton et al. 2010). An example for the link between specific metabolic pathways and the biogeography of *Roseobacter* strains is the abundance of high affinity phosphorus uptake systems known to function at low phosphate concentrations in Atlantic strains, while uptake systems known to operate in high phosphorous conditions occur more prominently in Pacific strains, which reflects phosphorous concentrations in these oceans (Newton et al. 2010).

6.5 Alternative Vectors of Marine Microbial Dispersal—The Hitchhikers

Although the free-floating currents in the ocean are the main way of dispersion of marine microorganisms, there are also other means of dissemination. One possible vector of distribution, and a niche for various kinds of marine microbial communities, is the association with marine animals and plants, in the form of attached biofilms or as gut content of marine animals. Microbial biofilms can also be attached to and distributed by abiotic supports, like ship hulls or plastic waste. A large number of species have been isolated from various marine environments, which have been distributed among four different families, i.e., *Enterovibrionaceae*, *Photobacteriaceae*, *Salinivibrionaceae*, and *Vibrionaceae* (Thompson et al. 2004).

The genus *Vibrio*, belonging to the family *Vibrionaceae* is an example of a genus containing many marine bacteria that are dispersed both by oceanic currents and other vectors such as marine animals. The genus is ubiquitous in the aquatic environment, both planktonic and associated to multicellular hosts, and is easy to culture. *Vibrio* is, therefore, very well suited as a model of global biogeography of culturable bacteria versus bacterial clades that are only or mainly known based on NGS data. *Vibrio* is highly abundant worldwide. This genus is found in estuaries, marine coastal waters, and sediments and it has also been detected in tissues or organs of various marine algae and animals, e.g., corals, sponges, shrimp, zooplankton, abalones and other bivalves, and fish (Barbieri et al. 1999; Heidelberg et al. 2002; Ortigosa et al. 1989, 1994; Rehnstam et al. 1993; Thompson et al. 2004; Urakawa et al. 2000). One of the most studied species of *Vibrio*, is *V. cholera*, a marine bacterium that enters into a dormant "viable but non-culturable" state when conditions are unfavorable for growth. Indeed, the first identified *Vibrio* species was *V. cholera*, the causative agent of the deadly disease cholera. Subsequently, and after Robert Koch (1843–1910) obtained pure cultures on plates, numerous studies have been performed on this group of bacteria (Thompson et al. 2004). Colwell and colleagues have researched *Vibrio* species extensively and demonstrated that various vectors, including zooplankton (e.g., copepods), chironomid insects and cyanobacteria, play major roles in its dissemination (Colwell and Huq 1999; Islam et al. 2004; Nicolas et al. 1996). The association of *V. cholera* with plankton has established the link between plankton blooms and epidemic patterns of cholera in countries where it is epidemic. The sporadic and erratic nature of cholera epidemics can now be related to climate and specific climate events, such as El Nino (Colwell and Huq 1999). Thus, cholera outbreaks can be correlated to environmental factors that influence microbial dissemination, such as high tides and raising sea temperatures. In Bangladesh, data has shown the existence of two annual peaks of *Vibrio* abundance in coastal waters that are associated to the number of cholera outbreaks occurring each year in the country.

Biofilms on fish skin, in gills, and in the gut content of marine fish contain naturally high numbers of bacteria, which are dispersed into the water column. Large shoals of pelagic fish migrate seasonally between oceans and in situ studies of fish microbiota have shown evidence that the composition of skin community is specific both for the host species, e.g., for cod and killifish (Larsen et al. 2013) and for their sampling locations (Larsen et al. 2015; Wilson et al. 2008). It has been estimated that herring shoals in the North Atlantic can occupy up to 4.8 km^3 with fish densities between 0.5 and 1.0 fish m^{-3}, totaling in about three billion fish in a single shoal (Radakov 1973). Microbes are therefore likely to hitchhike in these travels of larger animals and spread to different geographical locations. It has also been proposed that large pelagic fish shoals can contribute indirectly to altered microbial processes, e.g., by decreasing oxygen availability (Gilly et al. 2013). The microbiome of lionfishes showed similar variations in diversity when compared between different geographic locations (Stevens and Olson 2015).

There is data available in the literature on microbial communities associated with fish that was caught in different geographical locations and from different species. This data might give insight into host-specific microbial ecology. However, the primary objective of these studies was to study the microbiome of fish with regard to food production, shelf life and quality of the final product, rather than to document marine microbial diversity per se. The microbes that became dominant during various storage conditions of caught fish were mostly rare species in situ when attached to living fish in the ocean. Examples of such studies are from, e.g., cod (Reynisson et al. 2009), halibut (Hovda et al. 2007), mackerel (Svanevik and Lunestad 2011) and many more.

Besides being associated with mobile animals for dispersal, bacteria are also associated with abiotic pollutants such as plastic waste, which has become an environmental threat. The term "plastisphere" has been proposed for microbial communities on marine plastic debris and it has been shown that it harbors distinct microbial communities from surrounding surface water (Zettler et al. 2013). The plastic debris may not be one of the main dispersal driving forces but certainly adds another vector for microorganisms that are attached to these surfaces as biofilm. Further anthropogenic vectors for microbial dispersal are ship hulls and the ballast waters they carry (Al-Yamani et al. 2015; Ruiz et al. 2015).

6.6 Conclusion

Marine microbial biogeography has come a long way since the first marine microorganisms were characterized in the nineteenth century and is only currently developing its full potential through the availability of novel technologies, particularly NGS that allows for extensive full genome analyses and metagenomic studies of microbial communities. While some marine microorganisms exhibit a global distribution, recent molecular analyses showed nevertheless biogeographic patterns and niche adaptations at lower taxonomic levels, such as eco- and phylotypes inside these ubiquitous groups. Genome analyses of these eco- and phylotypes groups displaying biogeographic distributions can help us to understand which genes and physiological pathways are responsible for the adaptation to a particular global region or ecological niche. Our understanding of the factors that are shaping marine microbial biogeography, both through dispersal mechanisms (for instance currents, wind, anthropogenic transport) and microbial adaptation to environmental conditions (for instance nutrients, temperature, light), are still growing at steady pace. Given the importance of marine microorganisms for global biogeochemical cycles and food webs, it is of paramount importance for us to understand the mechanisms of adaptation, dispersion, and the resulting biogeographic patterns, in order to be able to predict and possibly even mitigate negative effects of climate change and other anthropogenic influences on marine microbial communities.

Acknowledgments The research leading to these results has received funding from the European Union Seventh Framework Programme (FP7/2007-2013) under grant agreement No. 311975. This publication reflects the views only of the author, and the European Union cannot be held responsible for any use which may be made of the information contained therein.

References

Agogue H, Lamy D, Neal PR, Sogin ML, Herndl GJ (2011) Water mass-specificity of bacterial communities in the North Atlantic revealed by massively parallel sequencing. Mol Ecol 20 (2):258–274. doi:10.1111/j.1365-294X.2010.04932.x

Alonso C, Pernthaler J (2006) *Roseobacter* and SAR11 dominate microbial glucose uptake in coastal North Sea waters. Environ Microbiol 8(11):2022–2030. doi:10.1111/j.1462-2920.2006.01082.x

Alvain S, Moulin C, Dandonneau Y, Loisel H (2008) Seasonal distribution and succession of dominant phytoplankton groups in the global ocean: a satellite view. Glob Biogeochem Cycles 22(3). doi:10.1029/2007gb003154

Al-Yamani FY, Skryabin V, Durvasula SRV (2015) Suspected ballast water introductions in the Arabian Gulf. Aquat Ecosyst Health Manage 18(3):282–289. doi:10.1080/14634988.2015.1027135

Amaral-Zettler L, Artigas LF, Baross J, Bharathi PAL, Boetius A, Chandramohan D, Herndl G, Kogure K, Neal P, Pedrós-Alió C, Ramette A, Schouten S, Stal L, Thessen A, de Leeuw J, Sogin M (2010) A global census of marine microbes. In: Life in the world's oceans. Wiley-Blackwell, pp 221–245. doi:10.1002/9781444325508.ch12

Angel R, Soares MIM, Ungar ED, Gillor O (2010) Biogeography of soil archaea and bacteria along a steep precipitation gradient. ISME J 4(4):553–563. doi:10.1038/ismej.2009.136

Azam F, Malfatti F (2007) Microbial structuring of marine ecosystems. Nat Rev Microbiol 5 (10):782–791

Azam F, Fenchel T, Field JG, Gray J, Meyer-Reil L, Thingstad F (1983) The ecological role of water-column microbes in the sea. Mar Ecol Prog Ser 10:257–263

Baas Becking LGM (1934) Geobiologie of inleiding tot de milieukunde. W.P. Van Stockum & Zoon, The Hague, the Netherlands

Ballarotta M, Falahat S, Brodeau L, Döös K (2014) On the glacial and interglacial thermohaline circulation and the associated transports of heat and freshwater. Ocean Sci 10(6):907–921

Barbieri E, Falzano L, Fiorentini C, Pianetti A, Baffone W, Fabbri A, Matarrese P, Casiere A, Katouli M, Kühn I (1999) Occurrence, diversity, and pathogenicity of halophilic *Vibrio* spp. and Non-O1 *Vibrio cholerae* from estuarine waters along the Italian Adriatic coast. Appl Environ Microbiol 65(6):2748–2753

Bass D, Richards TA, Matthai L, Marsh V, Cavalier-Smith T (2007) DNA evidence for global dispersal and probable endemicity of protozoa. BMC Evol Biol 7:162. doi:10.1186/1471-2148-7-162

Baumann P, Baumann L (1977) Biology of the marine enterobacteria: genera *Beneckea* and *Photobacterium*. Annu Rev Microbiol 31(1):39–61

Baumann P, Baumann L, Mandel M (1971) Taxonomy of marine bacteria: the genus *Beneckea*. J Bacteriol 107(1):268–294

Baumann L, Baumann P, Mandel M, Allen RD (1972) Taxonomy of aerobic marine eubacteria. J Bacteriol 110(1):402–429

Baumann P, Baumann L, Woolkalis MJ, Bang SS (1983) Evolutionary relationships in *Vibrio* and *Photobacterium*: a basis for a natural classification. Annu Rev Microbiol 37(1):369–398

Beijerinck M (1889) Le *Photobacterium luminosum*, bactérie lumineuse de la Mer du Nord. Archives Néerlandaises des Sciences Exactes et Naturelles 23:401–427

Britschgi TB, Giovannoni SJ (1991) Phylogenetic analysis of a natural marine bacterioplankton population by ribosomal-RNA gene cloning and sequencing. Appl Environ Microbiol 57 (6):1707–1713

Brown MV, Fuhrman JA (2005) Marine bacterial microdiversity as revealed by internal transcribed spacer analysis. Aquat Microb Ecol 41(1):15–23. doi:10.3354/ame041015

Brown MV, Philip GK, Bunge JA, Smith MC, Bissett A, Lauro FM, Fuhrman JA, Donachie SP (2009) Microbial community structure in the North Pacific ocean. ISME J 3(12):1374–1386. doi:10.1038/ismej.2009.86

Brown MV, Lauro FM, DeMaere MZ, Muir L, Wilkins D, Thomas T, Riddle MJ, Fuhrman JA, Andrews-Pfannkoch C, Hoffman JM, McQuaid JB, Allen A, Rintoul SR, Cavicchioli R (2012) Global biogeography of SAR11 marine bacteria. Mol Syst Biol 8:595. doi:10.1038/msb.2012.28

Brown MV, Ostrowski M, Grzymski JJ, Lauro FM (2014) A trait based perspective on the biogeography of common and abundant marine bacterioplankton clades. Mar Genomics 15:17–28. doi:10.1016/j.margen.2014.03.002

Carlson CA, Morris R, Parsons R, Treusch AH, Giovannoni SJ, Vergin K (2009) Seasonal dynamics of SAR11 populations in the euphotic and mesopelagic zones of the northwestern Sargasso Sea. ISME J 3(3):283–295. doi:10.1038/ismej.2008.117

Cavicchioli R (2015) Microbial ecology of Antarctic aquatic systems. Nat Rev Microbiol 13 (11):691–706

Colwell RR, Huq A (1999) Global microbial ecology: biogeography and diversity of Vibrios as a model. J Appl Microbiol 85:134S–137S

Cowman PF (2014) Historical factors that have shaped the evolution of tropical reef fishes: a review of phylogenies, biogeography, and remaining questions. Front genet 5:394. doi:10.3389/fgene.2014.00394

Davies TJ, Buckley LB, Grenyer R, Gittleman JL (2011) The influence of past and present climate on the biogeography of modern mammal diversity. Philos Trans R Soc B Biol Sci 366 (1577):2526–2535. doi:10.1098/rstb.2011.0018

DeLong EF (1997) Marine microbial diversity: the tip of the iceberg. Trends Biotechnol 15 (6):203–207. doi:10.1016/s0167-7799(97)01044-5

DeLong EF (2006) Archaeal mysteries of the deep revealed. Proc Natl Acad Sci USA 103 (17):6417–6418. doi:10.1073/pnas.0602079103

DeLong EF, Preston CM, Mincer T, Rich V, Hallam SJ, Frigaard NU, Martinez A, Sullivan MB, Edwards R, Brito BR, Chisholm SW, Karl DM (2006) Community genomics among stratified microbial assemblages in the ocean's interior. Science 311(5760):496–503. doi:10.1126/science.1120250

Dupont CL, Rusch DB, Yooseph S, Lombardo M-J, Richter RA, Valas R, Novotny M, Yee-Greenbaum J, Selengut JD, Haft DH, Halpern AL, Lasken RS, Nealson K, Friedman R, Venter JC (2012) Genomic insights to SAR86, an abundant and uncultivated marine bacterial lineage. ISME J 6(6):1186–1199. doi:10.1038/ismej.2011.189

Feng B-W, Li X-R, Wang J-H, Hu Z-Y, Meng H, Xiang L-Y, Quan Z-X (2009) Bacterial diversity of water and sediment in the Changjiang estuary and coastal area of the East China Sea. FEMS Microbiol Ecol 70(2):236–248. doi:10.1111/j.1574-6941.2009.00772.x

Field KG, Gordon D, Wright T, Rappe M, Urbach E, Vergin K, Giovannoni SJ (1997) Diversity and depth-specific distribution of SAR11 cluster rRNA genes from marine planktonic bacteria. Appl Environ Microbiol 63(1):63–70

Finlay BJ (2002) Global dispersal of free-living microbial eukaryote species. Science 296 (5570):1061–1063. doi:10.1126/science.1070710

Finlay BJ, Fenchel T (2004) Cosmopolitan metapopulations of free-living microbial eukaryotes. Protist 155(2):237–244. doi:10.1078/143446104774199619

Finlay BJ, Esteban GF, Brown S, Fenchel T, Hoef-Emden K (2006) Multiple cosmopolitan ecotypes within a microbial eukaryote morphospecies. Protist 157(4):377–390. doi:10.1016/j.protis.2006.05.012

Foissner W (2006) Biogeography and dispersal of micro-organisms: a review emphasizing protists. Acta Protozoologica 45(2):111–136

Fuhrman JA (1999) Marine viruses and their biogeochemical and ecological effects. Nature 399 (6736):541–548

Fuhrman JA, Steele JA, Hewson I, Schwalbach MS, Brown MV, Green JL, Brown JH (2008) A latitudinal diversity gradient in planktonic marine bacteria. Proc Natl Acad Sci 105 (22):7774–7778

Fuhrman JA, Cram JA, Needham DM (2015) Marine microbial community dynamics and their ecological interpretation. Nat Rev Microbiol 13(3):133–146. doi:10.1038/nrmicro3417
Galand PE, Casamayor EO, Kirchman DL, Lovejoy C (2009) Ecology of the rare microbial biosphere of the Arctic Ocean. Proc Natl Acad Sci USA 106(52):22427–22432. doi:10.1073/pnas.0908284106
Garcia-Martinez J, Rodriguez-Valera F (2000) Microdiversity of uncultured marine prokaryotes: the SAR11 cluster and the marine Archaea of Group I. Mol Ecol 9(7):935–948. doi:10.1046/j.1365-294x.2000.00953.x
Gauthier M (1976) Morphological, physiological, and biochemical characteristics of some violet-pigmented bacteria isolated from seawater. Can J Microbiol 22(2):138–149
Ghiglione JF, Galand PE, Pommier T, Pedros-Alio C, Maas EW, Bakker K, Bertilson S, Kirchman DL, Lovejoy C, Yager PL, Murray AE (2012) Pole-to-pole biogeography of surface and deep marine bacterial communities. Proc Natl Acad Sci USA 109(43):17633–17638. doi:10.1073/pnas.1208160109
Giebel HA, Brinkhoff T, Zwisler W, Selje N, Simon M (2009) Distribution of *Roseobacter* RCA and SAR11 lineages and distinct bacterial communities from the subtropics to the Southern Ocean. Environ Microbiol 11(8):2164–2178
Gilly WF, Beman JM, Litvin SY, Robison BH (2013) Oceanographic and biological effects of shoaling of the oxygen minimum zone. In: Carlson CA, Giovannoni SJ (eds) Annual review of marine science, vol 5. Annual reviews, Palo Alto, pp 393–420. doi:10.1146/annurev-marine-120710-100849
Giovannoni SJ, Stingl U (2005) Molecular diversity and ecology of microbial plankton. Nature 437(7057):343–348. doi:10.1038/nature04158
Giovannoni SJ, Britschgi TB, Moyer CL, Field KG (1990) Genetic diversity in Sargasso Sea bacterioplankton. Nature 345(6270):60–63. doi:10.1038/345060a0
Giovannoni SJ, Tripp HJ, Givan S, Podar M, Vergin KL, Baptista D, Bibbs L, Eads J, Richardson TH, Noordewier M, Rappe MS, Short JM, Carrington JC, Mathur EJ (2005) Genome streamlining in a cosmopolitan oceanic bacterium. Science 309(5738):1242–1245. doi:10.1126/science.1114057
Heidelberg J, Heidelberg K, Colwell R (2002) Seasonality of Chesapeake Bay bacterioplankton species. Appl Environ Microbiol 68(11):5488–5497
Hovda MB, Sivertsvik M, Lunestad BT, Lorentzen G, Rosnes JT (2007) Characterisation of the dominant bacterial population in modified atmosphere packaged farmed halibut (*Hippoglossus hippoglossus*) based on 16S rDNA-DGGE. Food Microbiol 24(4):362–371. doi:10.1016/j.fm.2006.07.018
Huber JA, Welch DBM, Morrison HG, Huse SM, Neal PR, Butterfield DA, Sogin ML (2007) Microbial population structures in the deep marine biosphere. Science 318(5847):97–100
Islam M, Mahmuda S, Morshed M, Bakht HB, Khan MN, Sack R, Sack D (2004) Role of cyanobacteria in the persistence of *Vibrio cholerae* O139 in saline microcosms. Can J Microbiol 50(2):127–131
Janvier M, Frehel C, Grimont F, Gasser F (1985) *Methylophaga marina* gen. nov., sp. nov. and *Methylophaga thalassica* sp. nov., marine methylotrophs. Int J Syst Bacteriol 35(2):131–139
Karsenti E, Acinas SG, Bork P, Bowler C, De Vargas C, Raes J, Sullivan M, Arendt D, Benzoni F, Claverie J-M, Follows M, Gorsky G, Hingamp P, Iudicone D, Jaillon O, Kandels-Lewis S, Krzic U, Not F, Ogata H, Pesant S, Reynaud EG, Sardet C, Sieracki ME, Speich S, Velayoudon D, Weissenbach J, Wincker P, Tara Oceans C (2011) A holistic approach to marine eco-systems biology. PLoS Biol 9(10). doi:10.1371/journal.pbio.1001177
Kembel SW, Eisen JA, Pollard KS, Green JL (2011) The phylogenetic diversity of metagenomes. PLoS ONE 6(8):e23214–e23214
King AJ, Freeman KR, McCormick KF, Lynch RC, Lozupone C, Knight R, Schmidt SK (2010) Biogeography and habitat modelling of high-alpine bacteria. Nat Commun 1. doi:10.1038/ncomms1055
Kopf A, Bicak M, Kottmann R, Schnetzer J, Kostadinov I, Lehmann K, Fernandez-Guerra A, Jeanthon C, Rahav E, Ullrich M, Wichels A, Gerdts G, Polymenakou P, Kotoulas G, Siam R,

Abdallah RZ, Sonnenschein EC, Cariou T, O'Gara F, Jackson S, Orlic S, Steinke M, Busch J, Duarte B, Cacador I, Canning-Clode J, Bobrova O, Marteinsson V, Reynisson E, Loureiro CM, Luna GM, Quero GM, Loscher CR, Kremp A, DeLorenzo ME, Ovreas L, Tolman J, LaRoche J, Penna A, Frischer M, Davis T, Katherine B, Meyer CP, Ramos S, Magalhaes C, Jude-Lemeilleur F, Aguirre-Macedo ML, Wang S, Poulton N, Jones S, Collin R, Fuhrman JA, Conan P, Alonso C, Stambler N, Goodwin K, Yakimov MM, Baltar F, Bodrossy L, Van De Kamp J, Frampton DM, Ostrowski M, Van Ruth P, Malthouse P, Claus S, Deneudt K, Mortelmans J, Pitois S, Wallom D, Salter I, Costa R, Schroeder DC, Kandil MM, Amaral V, Biancalana F, Santana R, Pedrotti ML, Yoshida T, Ogata H, Ingleton T, Munnik K, Rodriguez-Ezpeleta N, Berteaux-Lecellier V, Wecker P, Cancio I, Vaulot D, Bienhold C, Ghazal H, Chaouni B, Essayeh S, Ettamimi S, Zaid EH, Boukhatem N, Bouali A, Chahboune R, Barrijal S, Timinouni M, El Otmani F, Bennani M, Mea M, Todorova N, Karamfilov V, Ten Hoopen P, Cochrane G, L'Haridon S, Bizsel KC, Vezzi A, Lauro FM, Martin P, Jensen RM, Hinks J, Gebbels S, Rosselli R, De Pascale F, Schiavon R, Dos Santos A, Villar E, Pesant S, Cataletto B, Malfatti F, Edirisinghe R, Silveira JAH, Barbier M, Turk V, Tinta T, Fuller WJ, Salihoglu I, Serakinci N, Ergoren MC, Bresnan E, Iriberri J, Nyhus PAF, Bente E, Karlsen HE, Golyshin PN, Gasol JM, Moncheva S, Dzhembekova N, Johnson Z, Sinigalliano CD, Gidley ML, Zingone A, Danovaro R, Tsiamis G, Clark MS, Costa AC, El Bour M, Martins AM, Collins RE, Ducluzeau A-L, Martinez J, Costello MJ, Amaral-Zettler LA, Gilbert JA, Davies N, Field D, Glockner FO (2015) The ocean sampling day consortium. GigaScience 4:27. doi:10.1186/s13742-015-0066-5

Kottmann R, Kostadinov I, Duhaime MB, Buttigieg PL, Yilmaz P, Hankeln W, Waldmann J, Gloeckner FO (2010) Megx.net: integrated database resource for marine ecological genomics. Nucleic Acids Res 38:D391–D395. doi:10.1093/nar/gkp918

Larsen A, Tao Z, Bullard SA, Arias CR (2013) Diversity of the skin microbiota of fishes: evidence for host species specificity. FEMS Microbiol Ecol 85(3):483–494. doi:10.1111/1574-6941.12136

Larsen AM, Bullard SA, Womble M, Arias CR (2015) Community structure of skin microbiome of gulf killifish, *Fundulus grandis*, is driven by seasonality and not exposure to oiled sediments in a Louisiana salt marsh. Microb Ecol 70(2):534–544. doi:10.1007/s00248-015-0578-7

Lauro FM, Chastain RA, Blankenship LE, Yayanos AA, Bartlett DH (2007) The unique 16S rRNA genes of piezophiles reflect both phylogeny and adaptation. Appl Environ Microbiol 73 (3):838–845

Lauro FM, McDougald D, Thomas T, Williams TJ, Egan S, Rice S, DeMaere MZ, Ting L, Ertan H, Johnson J, Ferriera S, Lapidus A, Anderson I, Kyrpides N, Munk AC, Detter C, Han CS, Brown MV, Robb FT, Kjelleberg S, Cavicchioli R (2009) The genomic basis of trophic strategy in marine bacteria. Proc Natl Acad Sci USA 106(37):15527–15533. doi:10.1073/pnas.0903507106

Lewin R (1969) A classification of flexibacteria. J Gen Microbiol 58(2):189–206

Lewin RA, Lounsbery DM (1969) Isolation, cultivation and characterization of flexibacteria. J Gen Microbiol 58(2):145–170

Lozupone CA, Knight R (2007) Global patterns in bacterial diversity. Proc Natl Acad Sci 104 (27):11436–11440. doi:10.1073/pnas.0611525104

Luo H, Moran MA (2014) Evolutionary ecology of the marine *Roseobacter* clade. Microbiol Mol Biol Rev 78(4):573–587. doi:10.1128/mmbr.00020-14

Malmstrom RR, Kiene RP, Cottrell MT, Kirchman DL (2004) Contribution of SAR11 bacteria to dissolved dimethylsulfoniopropionate and amino acid uptake in the North Atlantic ocean. Appl Environ Microbiol 70(7):4129–4135. doi:10.1128/aem.70.7.4129-4135.2004

Martiny JBH, Bohannan BJM, Brown JH, Colwell RK, Fuhrman JA, Green JL, Horner-Devine MC, Kane M, Krumins JA, Kuske CR, Morin PJ, Naeem S, Ovreas L, Reysenbach AL, Smith VH, Staley JT (2006) Microbial biogeography: putting microorganisms on the map. Nat Rev Microbiol 4(2):102–112. doi:10.1038/nrmicro1341

Mary I, Cummings DG, Biegala IC, Burkill PH, Archer SD, Zubkov MV (2006) Seasonal dynamics of bacterioplankton community structure at a coastal station in the western English Channel. Aquat Microb Ecol 42(2):119–126. doi:10.3354/ame042119

Matsumoto K (2007). Radiocarbon-based circulation age of the world oceans. J Geophys Res Oceans (1978–2012) 112(C9). doi:10.1029/2007JC004095

Morris RM, Rappe MS, Connon SA, Vergin KL, Siebold WA, Carlson CA, Giovannoni SJ (2002) SAR11 clade dominates ocean surface bacterioplankton communities. Nature 420(6917):806–810. doi:10.1038/nature01240

Morris RM, Vergin KL, Cho JC, Rappe MS, Carlson CA, Giovannoni SJ (2005) Temporal and spatial response of bacterioplankton lineages to annual convective overturn at the Bermuda Atlantic Time-series Study site. Limnol Oceanogr 50(5):1687–1696

Mullins TD, Britschgi TB, Krest RL, Giovannoni SJ (1995) Genetic comparisons reveal the same unknown bacterial lineages in Atlantic and Pacific bacterioplankton communities. Limnol Oceanogr 40(1):148–158

Nealson KH, Venter JC (2007) Metagenomics and the global ocean survey: what's in it for us, and why should we care? ISME J 1(3):185–187. doi:10.1038/ismej.2007.43

Needham DM, Chow C-ET, Cram JA, Sachdeva R, Parada A, Fuhrman JA (2013) Short-term observations of marine bacterial and viral communities: patterns, connections and resilience. ISME J 7(7):1274–1285. doi:10.1038/ismej.2013.19

Nemergut DR, Costello EK, Hamady M, Lozupone C, Jiang L, Schmidt SK, Fierer N, Townsend AR, Cleveland CC, Stanish L, Knight R (2011) Global patterns in the biogeography of bacterial taxa. Environ Microbiol 13(1):135–144. doi:10.1111/j.1462-2920.2010.02315.x

Newton RJ, Griffin LE, Bowles KM, Meile C, Gifford S, Givens CE, Howard EC, King E, Oakley CA, Reisch CR, Rinta-Kanto JM, Sharma S, Sun S, Varaljay V, Vila-Costa M, Westrich JR, Moran MA (2010) Genome characteristics of a generalist marine bacterial lineage. ISME J 4(6):784–798. doi:10.1038/ismej.2009.150

Nicolas J-L, Corre S, Gauthier G, Robert R, Ansquer D (1996) Bacterial problems associated with scallop *Pecten maximus* larval culture. Dis Aquat Org 27(1):67–76

Ortigosa M, Esteve C, Pujalte M-J (1989) *Vibrio* species in seawater and mussels: abundance and numerical taxonomy. Syst Appl Microbiol 12(3):316–325

Ortigosa M, Garay E, Pujalte M-J (1994) Numerical taxonomy of *Vibrionaceae* isolated from oysters and seawater along an annual cycle. Syst Appl Microbiol 17(2):216–225

Paasche E (2001) A review of the coccolithophorid *Emiliania huxleyi* (Prymnesiophyceae), with particular reference to growth, coccolith formation, and calcification-photosynthesis interactions. Phycologia 40(6):503–529. doi:10.2216/i0031-8884-40-6-503.1

Pesant S, Not F, Picheral M, Kandels-Lewis S, Le Bescot N, Gorsky G, Iudicone D, Karsenti E, Speich S, Trouble R, Dimier C, Searson S, Tara Oceans Consortium C (2015) Open science resources for the discovery and analysis of Tara Oceans data. Sci Data 2:150023. doi:10.1038/sdata.2015.23

Poindexter JS (1964) Biological properties and classification of the Caulobacter group. Bacteriol Rev 28(3):231

Pommier T, Neal PR, Gasol JM, Coll M, Acinas SG, Pedrós-Alió C (2010) Spatial patterns of bacterial richness and evenness in the NW Mediterranean Sea explored by pyrosequencing of the 16S rRNA. Aquat Microb Ecol 61(3):221–233. doi:10.3354/ame01484

Radakov DV (1973) Schooling in the ecology of fish. Wiley, Hoboken, NJ, USA

Rahmstorf S (2003) Thermohaline circulation: the current climate. Nature 421(6924):699

Rappe MS, Vergin K, Giovannoni SJ (2000) Phylogenetic comparisons of a coastal bacterioplankton community with its counterparts in open ocean and freshwater systems. FEMS Microbiol Ecol 33(3):219–232. doi:10.1016/s0168-6496(00)00064-7

Rappe MS, Connon SA, Vergin KL, Giovannoni SJ (2002) Cultivation of the ubiquitous SAR11 marine bacterioplankton clade. Nature 418(6898):630–633. doi:10.1038/nature00917

Rehnstam A-S, Bäckman S, Smith DC, Azam F, Hagström Å (1993) Blooms of sequence-specific culturable bacteria in the sea. FEMS Microbiol Lett 102(3):161–166

Reichelt JL, Baumann P (1973) Taxonomy of the marine, luminous bacteria. Arch für Mikrobiol 94(4):283–330

Reynisson E, Lauzon HL, Magnusson H, Jonsdottir R, Olafsdottir G, Marteinsson VT, Hreggvidsson GO (2009) Bacterial composition and succession during storage of

North-Atlantic cod (*Gadus morhua*) at superchilled temperatures. BMC Microbiol 9(1):250. doi:10.1186/1471-2180-9-250

Ruiz GM, Fofonoff PW, Steves BP, Carlton JT (2015) Invasion history and vector dynamics in coastal marine ecosystems: a North American perspective. Aquat Ecosyst Health Manage 18 (3):299–311. doi:10.1080/14634988.2015.1027534

Rusch DB, Halpern AL, Sutton G, Heidelberg KB, Williamson S, Yooseph S, Wu D, Eisen JA, Hoffman JM, Remington K, Beeson K, Tran B, Smith H, Baden-Tillson H, Stewart C, Thorpe J, Freeman J, Andrews-Pfannkoch C, Venter JE, Li K, Kravitz S, Heidelberg JF, Utterback T, Rogers Y-H, Falcon LI, Souza V, Bonilla-Rosso G, Eguiarte LE, Karl DM, Sathyendranath S, Platt T, Bermingham E, Gallardo V, Tamayo-Castillo G, Ferrari MR, Strausberg RL, Nealson K, Friedman R, Frazier M, Venter JC (2007) The Sorcerer II global ocean sampling expedition: northwest Atlantic through eastern tropical Pacific. PLoS Biol 5 (3):398–431. doi:10.1371/journal.pbio.0050077

Sabehi G, Beja O, Suzuki MT, Preston CM, DeLong EF (2004) Different SAR86 subgroups harbour divergent proteorhodopsins. Environ Microbiol 6(9):903–910. doi:10.1111/j.1462-2920.2004.00676.x

Saez AG, Probert I, Geisen M, Quinn P, Young JR, Medlin LK (2003) Pseudo-cryptic speciation in coccolithophores. Proc Natl Acad Sci USA 100(12):7163–7168. doi:10.1073/pnas.1132069100

Schattenhofer M, Fuchs BM, Amann R, Zubkov MV, Tarran GA, Pernthaler J (2009) Latitudinal distribution of prokaryotic picoplankton populations in the Atlantic Ocean. Environ Microbiol 11(8):2078–2093. doi:10.1111/j.1462-2920.2009.01929.x

Schink B, Stams AJ (2013) Syntrophism among prokaryotes. Springer

Selje N, Simon M, Brinkhoff T (2004) A newly discovered *Roseobacter* cluster in temperate and polar oceans. Nature 427(6973):445–448. doi:10.1038/nature02272

Seymour JR, Doblin MA, Jeffries TC, Brown MV, Newton K, Ralph PJ, Baird M, Mitchell JG (2012) Contrasting microbial assemblages in adjacent water masses associated with the East Australian Current. Environ Microbiol Rep 4(5):548–555. doi:10.1111/j.1758-2229.2012.00362.x

Shade A, Carey CC, Kara E, Bertilsson S, McMahon KD, Smith MC (2009) Can the black box be cracked? The augmentation of microbial ecology by high-resolution, automated sensing technologies. ISME J 3(8):881–888. doi:10.1038/ismej.2009.56

Sieburth JN, Johnson PW, Eberhardt MA, Sieracki ME, Lidstrom M, Laux D (1987) The first methane-oxidizing bacterium from the upper mixing layer of the deep ocean: *Methylomonas pelagica* sp. nov. Curr Microbiol 14(5):285–293

Sogin ML, Morrison HG, Huber JA, Mark Welch D, Huse SM, Neal PR, Arrieta JM, Herndl GJ (2006) Microbial diversity in the deep sea and the underexplored "rare biosphere". Proc Natl Acad Sci USA 103(32):12115–12120. doi:10.1073/pnas.0605127103

Staley JT (1973) Budding bacteria of the Pasteuria-Blastobacter group. Can J Microbiol 19 (5):609–614

Stams AJ, Plugge CM (2009) Electron transfer in syntrophic communities of anaerobic bacteria and archaea. Nat Rev Microbiol 7(8):568–577

Stevens JL, Olson JB (2015) Bacterial communities associated with lionfish in their native and invaded ranges. Mar Ecol Prog Ser 531:253–262. doi:10.3354/meps11323

Sul WJ, Oliver TA, Ducklow HW, Amaral-Zettler LA, Sogin ML (2013a) Marine bacteria exhibit a bipolar distribution. Proc Natl Acad Sci USA 110(6):2342–2347. doi:10.1073/pnas.1212424110

Sul WJ, Oliver TA, Ducklow HW, Amaral-Zettler LA, Sogin ML (2013b) Marine bacteria exhibit a bipolar distribution. Proc Natl Acad Sci 110(6):2342–2347

Sun J, Steindler L, Thrash JC, Halsey KH, Smith DP, Carter AE, Landry ZC, Giovannoni SJ (2011) One carbon metabolism in SAR11 pelagic marine bacteria. PLoS One 6(8). doi:10.1371/journal.pone.0023973

Sunagawa S, Coelho LP, Chaffron S, Kultima JR, Labadie K, Salazar G, Djahanschiri B, Zeller G, Mende DR, Alberti A, Cornejo-Castillo FM, Costea PI, Cruaud C, d'Ovidio F, Engelen S, Ferrera I, Gasol JM, Guidi L, Hildebrand F, Kokoszka F, Lepoivre C, Lima-Mendez G, Poulain J, Poulos BT, Royo-Llonch M, Sarmento H, Vieira-Silva S, Dimier C, Picheral M, Searson S, Kandels-Lewis S, Bowler C, de Vargas C, Gorsky G, Grimsley N, Hingamp P, Iudicone D, Jaillon O, Not F, Ogata H, Pesant S, Speich S, Stemmann L, Sullivan MB, Weissenbach J, Wincker P, Karsenti E, Raes J, Acinas SG, Bork P, Tara Oceans C (2015) Structure and function of the global ocean microbiome. Science 348(6237). doi:10.1126/science.1261359

Svanevik CS, Lunestad BT (2011) Characterisation of the microbiota of Atlantic mackerel (*Scomber scombrus*). Int J Food Microbiol 151(2):164–170. doi:10.1016/j.ijfoodmicro.2011.08.016

Swan BK, Tupper B, Sczyrba A, Lauro FM, Martinez-Garcia M, Gonzalez JM, Luo HW, Wright JJ, Landry ZC, Hanson NW, Thompson BP, Poulton NJ, Schwientek P, Acinas SG, Giovannoni SJ, Moran MA, Hallam SJ, Cavicchioli R, Woyke T, Stepanauskas R (2013) Prevalent genome streamlining and latitudinal divergence of planktonic bacteria in the surface ocean. Proc Natl Acad Sci USA 110(28):11463–11468. doi:10.1073/pnas.1304246110

Thompson FL, Iida T, Swings J (2004) Biodiversity of vibrios. Microbiol Mol Biol Rev 68 (3):403–431

Treusch AH, Vergin KL, Finlay LA, Donatz MG, Burton RM, Carlson CA, Giovannoni SJ (2009) Seasonality and vertical structure of microbial communities in an ocean gyre. ISME J 3 (10):1148–1163. doi:10.1038/ismej.2009.60

Tseng C-H, Tang S-L (2014) Marine microbial metagenomics: from individual to the environment. Int J Mol Sci 15(5):8878–8892. doi:10.3390/ijms15058878

Urakawa H, Yoshida T, Nishimura M, Ohwada K (2000) Characterization of depth-related population variation in microbial communities of a coastal marine sediment using 16S rDNA-based approaches and quinone profiling. Environ Microbiol 2(5):542–554

van der Gast CJ (2015) Microbial biogeography: the end of the ubiquitous dispersal hypothesis? Environ Microbiol 17(3):544–546. doi:10.1111/1462-2920.12635

Venter JC, Remington K, Heidelberg JF, Halpern AL, Rusch D, Eisen JA, Wu DY, Paulsen I, Nelson KE, Nelson W, Fouts DE, Levy S, Knap AH, Lomas MW, Nealson K, White O, Peterson J, Hoffman J, Parsons R, Baden-Tillson H, Pfannkoch C, Rogers YH, Smith HO (2004) Environmental genome shotgun sequencing of the Sargasso Sea. Science 304 (5667):66–74. doi:10.1126/science.1093857

Wang X-Q, Ran J-H (2014) Evolution and biogeography of gymnosperms. Mol Phylogenet Evol 75:24–40. doi:10.1016/j.ympev.2014.02.005

Watson SW (1965) Characteristics of a marine nitrifying bacterium, *Nitrosocystis oceanus* sp. nov. Limnol Oceanogr 10(suppl):R274–R289

Watson SW, Mandel M (1971) Comparison of the morphology and deoxyribonucleic acid composition of 27 strains of nitrifying bacteria. J Bacteriol 107(2):563–569

Whitman WB, Coleman DC, Wiebe WJ (1998) Prokaryotes: the unseen majority. Proc Natl Acad Sci USA 95(12):6578–6583. doi:10.1073/pnas.95.12.6578

Wilkins D, Lauro FM, Williams TJ, Demaere MZ, Brown MV, Hoffman JM, Andrews-Pfannkoch C, McQuaid JB, Riddle MJ, Rintoul SR, Cavicchioli R (2013a) Biogeographic partitioning of Southern Ocean microorganisms revealed by metagenomics. Environ Microbiol 15(5):1318–1333. doi:10.1111/1462-2920.12035

Wilkins D, van Sebille E, Rintoul SR, Lauro FM, Cavicchioli R (2013b) Advection shapes Southern Ocean microbial assemblages independent of distance and environment effects. Nat Commun 4. doi:10.1038/ncomms3457

Wilson B, Danilowicz BS, Meijer WG (2008) The diversity of bacterial communities associated with Atlantic cod *Gadus morhua*. Microb Ecol 55(3):425–434. doi:10.1007/s00248-007-9288-0

Wood SA, Smith KF, Banks JC, Tremblay LA, Rhodes L, Mountfort D, Cary SC, Pochon X (2013) Molecular genetic tools for environmental monitoring of New Zealand's aquatic habitats, past, present and the future. NZ J Mar Freshwat Res 47(1):90–119. doi:10.1080/00288330.2012.745885

Yooseph S, Nealson KH, Rusch DB, McCrow JP, Dupont CL, Kim M, Johnson J, Montgomery R, Ferriera S, Beeson K, Williamson SJ, Tovchigrechko A, Allen AE, Zeigler LA, Sutton G, Eisenstadt E, Rogers Y-H, Friedman R, Frazier M, Venter JC (2010) Genomic and functional adaptation in surface ocean planktonic prokaryotes. Nature 468(7320):60–66. doi:10.1038/nature09530

Zettler ER, Mincer TJ, Amaral-Zettler LA (2013) Life in the "plastisphere": microbial communities on plastic marine debris. Environ Sci Technol 47(13):7137–7146. doi:10.1021/es401288x

Zhao Y, Temperton B, Thrash JC, Schwalbach MS, Vergin KL, Landry ZC, Ellisman M, Deerinck T, Sullivan MB, Giovannoni SJ (2013) Abundant SAR11 viruses in the ocean. Nature 494(7437):357–360. doi:10.1038/nature11921

Zika JD, England MH, Sijp WP (2012) The ocean circulation in thermohaline coordinates. J Phys Oceanogr 42(5):708–724

Zinger L, Amaral-Zettler LA, Fuhrman JA, Horner-Devine MC, Huse SM, Welch DBM, Martiny JBH, Sogin M, Boetius A, Ramette A (2011) Global patterns of bacterial beta-diversity in seafloor and seawater ecosystems. PLoS ONE 6(9):e24570. doi:10.1371/journal.pone.0024570

ZoBell CE, Upham HC (1944) A list of marine bacteria including descriptions of sixty new species. University of California Press, Oakland, CA, USA

Chapter 7
The Euphotic Realm

Lucas J. Stal

Abstract The euphotic realm of the ocean is defined as the zone that receives enough light to allow photosynthesis. The bottom of the euphotic zone is often set as the depth at which 1 % of the incident sunlight is still available, which is in the open oligotrophic ocean approximately 200 m below the surface. In more turbid coastal waters, the euphotic zone ends at much shallower waters. Whether or not photosynthesis occurs at 1 % of the incident light intensity depends first on the actual value of the latter, on the sun angle, and not in the last place on the organism considered. Even among the same species low and high-light adapted ecotypes exist. The euphotic realm is important because it provides the primary production for the food web of the whole ocean. The microbiome of this realm is therefore characterized by microorganisms that use light. The composition of the microbiome and the ecology is intimately associated with the physicochemical characteristics of the euphotic realm.

7.1 Introduction

The oceanic water column ('the pelagic') is subdivided in five major realms. The epipelagic extends from the surface to a depth of 200 m and is defined by the layer that receives enough sunlight for photosynthesis. This is followed by the mesopelagic that extends to 1000 m depth, the bathypelagic that covers the depths between 1000–4000 m, and everything deeper is called the abyssal or abyssopelagic. The mesopelagic zone is also known as the twilight or disphotic zone, which

L.J. Stal (✉)
Department of Marine Microbiology and Biogeochemistry, NIOZ Royal Netherlands Institute for Sea Research and Utrecht University, PO Box 59, 1790 AB Den Burg, Texel, The Netherlands
e-mail: Lucas.stal@nioz.nl

L.J. Stal
Department of Aquatic Microbiology, University of Amsterdam, Amsterdam, The Netherlands

L.J. Stal and M.S. Cretoiu (eds.), *The Marine Microbiome*,
DOI 10.1007/978-3-319-33000-6_7

receives insufficient light for photosynthesis but plays a role for organisms that migrate in a day–night rhythm to the surface (Orcutt et al. 2011). Below the mesopelagic zone there is no sunlight at all. The bathypelagic is defined as the zone that starts at the edge of the continental shelf to the depth where the water temperature is constant 4 °C. The uniform environmental conditions below the bathypelagic define the abyssal where the temperature is 4 °C. The trenches are even deeper and are called the Hadal zone. The epipelagic is also known as the euphotic or photic zone, which means "well-lit." As a rule of thumb the euphotic depth is taken as the depth at which still 1 % of the incident sunlight reaches (Cullen 2015). The euphotic depth is also often taken as equal to the compensation depth where the oxygen produced by photosynthesis equals the oxygen consumption by heterotrophic activity. The depth of 200 m as the bottom of the euphotic zone is a bit arbitrary because it depends largely on the turbidity of the water. Turbidity is obvious much higher in coastal areas, seas, bays, and estuaries when compared to the oligotrophic open ocean. Because photosynthetic primary production takes place in the euphotic realm of the marine ecosystem and forms the basis of the food web, most of the sea life is concentrated here.

Organic matter produced in the euphotic realm is generally biologically labile, which means that it is degraded within hours to days while it is residing inside the euphotic realm (Carlson et al. 2008). Degradation of this biologically labile material leaves behind semi-labile organic matter (degraded in weeks to months) and compounds that are more recalcitrant to degradation, which degrade from decades, to thousands of years, and even to ten thousands of years. These compounds sink in the water column and with depth the composition of the organic material increases towards more recalcitrant molecules.

Mixing of the surface water column determines its productivity. When surface waters receive nutrients by an upward flux caused, e.g., by upwelling, the region may be termed eutrophic and is distinguished from oligotrophic regions that lack such nutrient input. Primary productivity and phytoplankton abundances are usually high in eutrophic regions and low in oligotrophic surface waters. However, situations have been reported in which the chlorophyll content in the euphotic layer was lower than expected from the high nutrient concentrations. These regions are known as HNLC (high nutrient low chlorophyll) regions. These phenomena have been attributed to iron limitation (Nolting et al. 1998; Wagener et al. 2008).

In this Chap. I will briefly review the characteristics and importance of the euphotic realm in the marine environment.

7.2 Light

Photosynthetic organisms use light and in many cases this represents their only source of energy. In the ocean's water column, light is absorbed by the water and by the living and non-living particles that are suspended in the water column causing light attenuation with depth in an exponential way. Therefore, light penetrates only to a certain depth below which the environment is permanently veiled in darkness. The euphotic realm is the water column in which enough light penetrates to allow photosynthesis and is as a rule of thumb set at 1 % of the incident sunlight irradiation at the surface, which marks the bottom of this water layer (Cullen 2015). However, photosynthetic organisms vary widely with respect to their photosynthetic capacities and the light levels that allow net photosynthesis differ between species and even ecotypes. Some ecotypes are low-light adapted while others are high-light adapted as for instance is the case of *Prochlorococcus*, a group of unicellular cyanobacteria (Hess et al. 2001). At night, when the euphotic realm is also veiled in darkness, phototrophic organisms switch to chemoheterotrophic metabolism. Most species utilize internally stored carbohydrate, such as starch, glycogen and chrysolaminarin, which is synthesized during the daytime, respire it aerobically, and use the generated energy to cover their maintenance requirements (Deschamps et al. 2008; Kromkamp 1987). Some organisms may even continue to grow uninterrupted and at the same rate as in the light, using the intermediates of carbohydrate degradation as building blocks for synthesis of structural cell material (van Liere et al. 1979). Obviously, the integrated day–night balance of photosynthesis and dark respiration of an obligate photoautotrophic organism needs to be positive when the organism wants to maintain itself. The daily integral of light where photosynthesis and respiration are in equilibrium is called the compensation point. The daily-integrated euphotic depth equals the depth of the compensation point. The critical depth is where the cumulative daily photosynthetic integral becomes negative (Stal and Walsby 2000). These authors noted that neither the euphotic nor the critical depths were dependent on temperature. However, it should be noted that organisms differ widely in the light intensities at which they reach their compensation point and their critical depths. These depths can also be defined based on the whole photosynthetic community activity rather than for a single population.

Sunlight intensity follows a sinus during the day, starting with sunrise in the morning and sunset in the afternoon. While this will not change the attenuation of the light in the water column, it will affect the actual available light. Hence, the 1 % rule as a measure for the euphotic realm (defined as the realm in which photosynthesis takes place) will not stand up. Moreover, the sinus course of light during the day is in many places on Earth an exception rather than a rule. Overcast and clouds cause large fluctuations of the incident light and therefore cause large changes in the availability of light in the water column. Another important variable is the day length. Except for the equator, where the day and night both last ~12 h during the whole year, at the higher latitudes the day length and thus the daily-integrated incident light changes seasonally. Also, in winter the sun angle (winter solstice) is low, which causes a large part of the sunlight reflecting at the water surface rather than penetrating the water column. In summer the sun angle (summer solstice) is high resulting in a much higher proportion of the light penetrating the water column. Finally, stratification of the water column and turbulence are important factors that determine the amount of light that photosynthetic plankton will be able to absorb. When the mixing depth exceeds the critical depth, the photosynthetic community will not be able to grow (Stal and Walsby 2000). It is likely that the daily-integrated sunlight in the euphotic realm is the critical factor for photosynthesis and primary production (Stal and Walsby 2000). This is supported by extensive measurements of phytoplankton biomass and rates of primary production in the South Atlantic Ocean that showed a bloom in the austral summer that collapsed although nutrients were not depleted (Holm-Hansen and Mitchell 1991). The phytoplankton in that study was low-light adapted and the compensation point was only 1 $\mu mol\ m^{-2}\ s^{-1}$.

Due to molecular vibrations, water molecules absorb light of specific wavelengths. Although this absorption is barely visible in the absorption spectrum of pure water, it produces large gaps in the light spectrum with greater depth. This results in a set of spectral niches underwater that correspond with the absorption spectra of photosynthetic microorganisms (Stomp et al. 2007a). These authors concluded that the absorption of light by the water molecule was a major factor for the selection of pigmentation of phototrophic microorganisms.

Water turbidity is also an important factor that selects for the pigmentation of picocyanobacteria (Stomp et al. 2007b). Waters with low concentrations of gilvin and tripton (dissolved and particulate organic matter, respectively) and therefore low turbidity such as in the oligotrophic ocean select for red (phycoerythrin) pigmented picocyanobacteria, while waters with very high content of these suspended particles such as in peat lakes select for green (phycocyanin) pigmented cyanobacteria. Waters with intermediate turbidity have coexisting populations of red and green picocyanobacteria. The fraction of each of these groups agrees with the degree of turbidity.

Although coastal waters are characterized by high concentrations of gilvin and tripton and therefore high turbidity relative to the open ocean, the seafloor of a substantial (33 %) part of the global shelf area receives enough light to support net community production by benthic primary producers (Gattuso et al. 2006). Net primary production can even happen deeper than 200 m and the minimum daily-integrated photon irradiance varies from 0.4 to 5.1 $mol\ m^{-2}$. The compensation point in these communities ranges from 0.24 to 4.4 $mol\ m^{-2}$.

7.3 Subsurface Chlorophyll Maximum Layers

In many cases the highest concentration of chlorophyll in the water column is not at the surface where the light intensity is highest but in the subsurface. The highest chlorophyll concentration in the subsurface is called the 'subsurface chlorophyll maximum layer' or SCML and also known as 'deep chlorophyll maximum' or DCM (Cullen 2015). This phenomenon is usually present in tropical waters and is more variable elsewhere, although DCMs have also been reported from the Antarctic (Holm-Hansen and Hewes 2004). The DCM is not necessary coinciding with the subsurface biomass maximum layer (SBML) because shade-adapted phytoplankton may accumulate a high content of chlorophyll. Hence, in that case a relatively low amount of biomass may contain a large amount of chlorophyll. The mechanisms that lead to an SBML include subsurface production of phytoplankton and surface nutrient depletion by sinking living or dead particles (Beckmann and Hense 2007). The depth of the SBML can be described in terms of optimum growth rate (Cullen 2015). Growth is nutrient limited above the SBML and light limited below.

The DCM or SBML may be located close to the bottom of the euphotic zone where a thermocline exists, beneath which the nutricline exists. An SBML forms only in stable, stratified water columns. In oligotrophic waters deep mixing of the water column usually prevents the formation of an SBML. Differences in light and mixing may cause very different phytoplankton bloom developments; even when the compensation point and other (geographical) factors were the same (Lacour et al. 2015). The availability of nutrients in the deep water and shade-adapted phytoplankton results in the accumulation of chlorophyll. The location of the nutricline is therefore determined by the SBML and its thickness (Beckmann and Hense 2007). The upper nutricline is the area with the highest productivity in a stratified water column. Feedback loops lead to a steady state of an SBML. A change in light intensity will affect the growth rate, which will affect the nutricline, which will affect growth rate, which will eventually stabilize the SMBL. When a DCM is found below the euphotic zone it usually is formed by sinking cells or by water transport phenomena.

Herbland and Voituriez (1979) reported strong correlations between the depth of nitracline (above which the concentrations of nitrate are very low) and the depth of the chlorophyll maximum in the tropical Atlantic Ocean. They also observed a negative correlation of the chlorophyll concentration at the chlorophyll maximum and the depth where it occurred (greater depth, less chlorophyll) and, consequently, a lower depth integrated productivity with increasing depth of the nitracline.

7.4 UV Light

Although seawater absorbs UV light, UV-B (280–315 nm) can penetrate as deep as 60 m particularly in the Southern Ocean where the atmosphere seasonally contains very low levels of ozone (Smith et al. 1992). UV-B light is deleterious for life as it causes damage to DNA especially by forming pyrimidine dimers (Buma et al. 1996, 2001). During the Archean (4.0–2.5 Ga) the atmosphere was devoid of oxygen and ozone and UV light could reach unhindered the surface of the ocean causing an estimated three orders of magnitude more DNA damage than at the present day (Cockell 2000). This might have seriously limited the viability of microorganisms and thereby diversity of life in the upper part of the sunlit ocean, although it might also have contributed to genetic variation. However, at 30 m depth the UV level was predicted to be similar as today's ocean surface, which would have allowed developing a diverse microbial community below the mixed layer and above the depth of the compensation point, i.e., the SBML. Moreover, some microorganisms, including phototrophic species, have developed efficient DNA repair mechanisms allowing them to survive and thrive at high levels of UV light. Another consequence of UV-B incidence is the reduction of ferric iron (Fe^{3+}) to ferrous iron (Fe^{2+}) (Rijkenberg et al. 2005). In the present-day oxygenated ocean, iron is mainly present in the oxidized (ferric) form, which is virtually insoluble (Liu and Millero 2002). Reduced (ferrous) iron is soluble but quickly oxidized to the insoluble ferric form. UV-B light reduces ferric iron to ferrous iron thereby overcoming iron limitation of the microorganisms (Rijkenberg et al. 2005). Particles of ferric iron may also be attached to the extracellular layers of some microorganisms (composed of polysaccharide-rich extracellular polymeric substances, EPS) and this may have a double function: it protects against UV light and it provides bio-available ferrous iron. A phototrophic organism that is dependent on light cannot always avoid exposure to UV light. Such organisms sometimes produce specific pigments that absorb UV light, which protect the organism from UV damage (Gao and Garcia-Pichel 2011). Some cyanobacteria are known to produce an extracellular pigment with a unique dimeric indole-phenolic structure named scytonemin which acts as sunscreen by absorbing the UV-A. Other cyanobacteria produce secondary metabolites such as microsporin-like amino acids (MAA) that absorb in a wide range of wavelengths. Many eukaryotes produce MAA but except in cyanobacteria they have not been found in Bacteria or Archaea (Gao and Garcia-Pichel 2011). Melanines are UV-B sunscreens found in fungi. All of these sunscreens are found in terrestrial microorganisms and not in aquatic environments. Carotenoids may also protect against UV-B and are in some aquatic organisms induced upon exposure to UV-B.

7.5 Biogeochemical Cycling

7.5.1 Carbon

The carbon cycle in the euphotic realm is dominated by the fixation of inorganic carbon (CO_2, bicarbonate) into organic carbon by photosynthetic microorganisms (cyanobacteria and microalgae). These organisms use the Calvin–Benson–Bassham pathway in which the enzyme ribulose-1,5-bisphosphate carboxylase/oxygenase (RuBisCO) is the key enzyme and responsible for the fixation of CO_2. To some extent chemosynthesis may also contribute to CO_2 fixation and the production of organic material (Orcutt et al. 2011).

The organic matter produced by the photosynthetic organisms in the form of their own biomass (primary production) serves as the basis of the (microbial) food web. There are a number of ways by which this organic matter becomes available to the community. Grazing by protists and viral lysis represent the most important processes for the recycling of organic matter (Worden et al. 2015). However, exudation of organic matter by phototrophs can also be considerable, especially under low CO_2 and high O_2, when RuBisCO's oxygenase activity dominates and glycolate is excreted (Latifi et al. 2009). Large amounts of extracellular polymeric substances are also formed under nutrient limitation when no structural cell material can be synthesized (Wotton 2004). Dissolved organic matter that is produced by phytoplankton under nutrient limited conditions may coagulate to large particles known as transparent exopolymer particles (TEP) (Mari et al. 2001). TEP may measure several millimeters in size and are found in the euphotic realm of many marine environments.

The chemotrophic community degrades the organic matter that becomes available after viral lysis of the phototrophs or by the production of fecal pellets by the grazers. Most of the (easily) degradable organic matter is decomposed within the euphotic zone. The more recalcitrant material sinks and is more slowly degraded in the deeper realms of the ocean. The growth and activity of chemotrophic microorganisms in the euphotic realm strongly depends on the availability of nutrients.

Polymeric compounds need first to be decomposed to low-molecular weight compounds that can be taken up and used as substrate for the chemoheterotrophic microorganisms. This process requires the activity of exo- or ectoenzymes that decompose the polymers (Sass et al. 2001). These enzymes are produced by some but not all chemoheterotrophs and those that do not possess them may benefit nevertheless from the activity of the others. Exoenzymes are exuded into the medium where they attack the polymers. Ectoenzymes remain associated with the exterior layers of the organism, which may be beneficial when other organisms do not benefit from their activity. Besides the enzymatic activity, UV light may also degrade polymers (Wetzel et al. 1995).

The free-living chemoheterotrophic microorganisms decompose the organic matter aerobically to CO_2 (Del Giorgio and Duarte 2002). Anaerobic decomposition of organic matter using alternative electron acceptors (nitrate/nitrite) or fermentation may be confined to particles such as marine snow and fecal pellets (Ploug et al. 1997).

Calcification is the precipitation of calcite and other related minerals when the activities of calcium and carbonate ions exceed its solubility product (Zavarzin 2002). Coccolithophoric microalgae form coccoliths that are composed of calcite (O'Brien et al. 2016). The production of these coccoliths is continuously going on. The coccoliths that are of an amazing beautiful morphology fall off and sink to the seafloor. It is thought that by producing carbonate from bicarbonate, which is the dominant form of inorganic carbon in seawater, the organism raises the concentration of CO_2, the substrate of RuBisCO, according to the following reactions (Borman et al. 1982):

$$2HCO_3^- \rightarrow CO_2 + CO_3^{2-} + H_2O \text{ and } Ca^{2+} + CO_3^{2-} \rightarrow CaCO_3$$

For example, chalk cliffs such as those that are found in Dover (UK) find their origin in these calcifying algae. Another form of calcite formation is the so-called whitings. Whitings are massive calcite formations in the euphotic zone caused by blooms of picocyanobacteria in warm shallow waters (Obst et al. 2009). The picocyanobacteria withdraw the CO_2 from the seawater, raising the pH to above 9 and then producing such high concentrations of carbonate ion that it exceeds the solubility product of calcite, resulting in a massive precipitation of small calcite needles that float in the water and rendering the surface a milky white appearance.

7.5.2 Nitrogen

Nitrogen occurs in various oxidation states from the most oxidized in nitrate to the most reduced in ammonia. Nitrogen is a quantitatively and qualitatively important element for organisms. Its share in the dry weight biomass amounts to ~10 % and it can be found in proteins, nucleic acids, pigments, and cell walls. In these cell components nitrogen is present in the reduced (amino) form. Microorganisms can take up nitrogen in different forms such as nitrate, nitrite, ammonium, urea, organic nitrogen, and dinitrogen gas (N_2), depending on the organism. Nitrate, nitrite, and N_2 need to be reduced in the cell to ammonium (assimilatory nitrogen reduction). Nitrate and nitrite require special uptake transporters/uptake mechanisms and enzymes to reduce (nitrate- and nitrite reductase) and assimilate it. N_2 is a gas that diffuses into the cell, where it is fixed by the nitrogenase enzyme complex to ammonium, which is subsequently assimilated. Reduced nitrogen sources need to be transported into cell and can be directly assimilated into cell material. Ammonia is a gas and diffuses in (and out, futile cycle, Kleiner 1985) the cell.

Cellular nitrogen becomes available to other organisms due to grazing or viral lysis or other processes that lead to the death and disintegration of the cell. This reduced organic nitrogen may be directly recycled or deaminated. The latter results in the formation of free ammonium that is directly recycled by the primary producers in the euphotic zone. Nitrogen released by organisms that sink below the euphotic zone is oxidized to nitrite and subsequently to nitrate in dissimilatory processes (nitrification). Below the euphotic zone, recycling by assimilation of ammonium or nitrate is less important because primary production by photoautotrophs does not take place. Some production, however, will take place by chemoautotrophic organisms. The oxidation of ammonium to nitrite (nitritification) is performed mainly by ammonium oxidizing Thaumarchaeota (Santoro and Casciotti 2011). The fate of the nitrite may be several. It can be further nitrified to nitrate (nitratification), it can be denitrified to N_2, or used as electron acceptor in anaerobic ammonium oxidation (anammox), also leading to the formation of N_2 (Lam and Kuypers 2011). Denitrification and anammox take place in oxygen minimum zones that can be found in various locations in the ocean, or in aggregates (marine snow) that comprise anoxic zones.

Below the euphotic zone and above the oxygen minimum zone (if present) nitrate accumulates. Upwelling or other physical processes such as vertical fluxes, local stratification and temperature differences bring this nutrient-rich water into the euphotic zone where it fuels primary production (Zehr and Kudela 2011). The nitrate that is transported from the deep waters is also known as "new nitrogen" to distinguish it from nitrogen that is recycled in the euphotic zone.

Denitrification and anammox catalyze the transformation of nitrate/nitrite to N_2 (and to some extend to N_2O), which is dissolved as gas in the water or escapes to the atmosphere. N_2 is inaccessible for all but a few specialized bacteria (and even fewer archaea). The triple bond that bind the two nitrogen atoms makes it almost inert and the biological fixation requires a large amount of energy and low-potential electrons as well as a dedicated enzyme complex (nitrogenase). These prerequisites force the bulk of oceanic N_2 fixation in the euphotic zone, where it is carried out by a number of different cyanobacteria that combine the fixation of N_2 with that of CO_2 and are really the drivers of new production (Zehr 2011). However, this cyanobacteria-mediated N_2 fixation in the ocean seems to be confined to the warmer (tropical and subtropical) ocean (Stal 2009). It is not clear how primary production in the temperate and cold ocean is supplied by nitrogen. This may be by transport from the warm ocean by surface gyres, by recycling in the euphotic zone or by upwelling (Hansell et al. 2004). The contribution of N_2 fixation by heterotrophic diazotrophs is debated. Although these organisms have been detected by their *nifH* genes (structural gene for nitrogenase), their activity is uncertain and they belong to (facultative) anaerobic organisms that only express nitrogenase activity under anoxic conditions (Riemann et al. 2010). N_2 fixation does occur in oxygen minimum zones and it is likely that it also plays a role in marine aggregates. However, these aggregates sink and hence will play a minor role in the euphotic realm.

7.5.3 Phosphorus

Phosphorus is after nitrogen the second-most important nutrient for primary production in the euphotic zone (ignoring carbon). Although phosphorus may exist in a few oxidation states, dissimilatory redox reactions are not known to support energy metabolism in microorganisms. Most of the phosphorus is present in the oxidized form: free phosphate or bound to organic matter. Since N_2 fixation in theory gives access to an infinite source of this element (most of the nitrogen on Earth is present as N_2 in the atmosphere), it has been suggested that phosphorus is the final limiting nutrient for primary producers (Tyrrell 1999). This is, however, still debated because the fixation of N_2 in the ocean is globally not equally distributed and even in the tropical and subtropical areas of the ocean where N_2 fixation takes place, this process may not be running optimally because of the sensitivity of the process for oxygen. Hence, nitrogen or iron may still be the limiting nutrient in vast areas of the ocean.

The source of phosphorus in the ocean is mostly riverine and to a lesser extent atmospheric. The cycle of phosphate is one of the uptake and assimilation in structural cell material and its release after cell death. The released phosphate is bound to organic matter but alkaline phosphatases convert them to free phosphates, which is the preferred source of phosphorus by most (phototrophic) microorganisms (Dyhrman et al. 2007). In the euphotic realm of the ocean phosphate is usually present at such low concentrations that some microorganisms also utilize other sources of phosphorus. Phosphonates are compounds in which phosphorus is chemically bound to carbon and have been proven to be a key component of organic carbon in the ocean. In structural cell material phosphonates may occur in lipids and proteins. Several organisms including picocyanobacteria possess genes that encode for enzymes that break the P–C bond and have been shown to be capable of utilizing phosphonates. The source of phosphonates is microorganisms that produce them among which some *Trichodesmium* species (Dyhrman et al. 2009) and some archaea (Metcalf et al. 2012).

7.6 The Microbial Loop

The phytoplankton thriving in the euphotic realm photosynthesizes and fixes CO_2 into organic structural cell material. These cells are grazed by zooplankton that produce fecal pellets or the phytoplankton cells are lysed by viral attack. The dissolved organic matter that is liberated in this way into the environment is decomposed by bacteria and archaea and respired back to CO_2 (Azam and Malfatti 2007). The bacteria and archaea are also exposed to viral attack and grazed by protozoa that are subsequently grazed by zooplankton. These processes mineralize the organic matter and recycle nutrients that are taken up by the phytoplankton. This short cut in the euphotic realm is also known as the microbial loop. The viral loop is a component of this microbial loop, basically after a lytic cycle directly returning the dissolved organic matter back to the bacteria/archaea. The microbial loop is recycling much faster than the classical food web where zooplankton serves as food for higher trophic levels.

7.7 The Carbon Pump

Organisms, particulate organic matter, fecal pellets, and other suspended material may aggregate to larger particles that sink rapidly below the euphotic zone and eventually part of it will reach the seafloor. During its journey to the seafloor the organic matter in the aggregates will be further degraded. Most of the organic matter is decomposed and the nutrients recycled within the euphotic zone. The material that eventually is deposited on the seafloor is extremely recalcitrant and depleted in nitrogen and phosphorus (Carlson et al. 2008). Hence, there is a net transport of fixed carbon from the euphotic zone to the deep sea where it becomes buried and subsequent diagenetic processes take place at the geological timescale. The removal of fixed CO_2 to the deep sea is known as the carbon pump and is one of the mechanisms by which the anthropogenic induced increase of atmospheric CO_2 is counteracted (Azam and Malfatti 2007).

7.8 The Microorganisms in the Euphotic Realm

The Sorcerer II Global Ocean Sampling Expedition (GOS) revealed that the number of dominant genera of free-living microorganisms in the surface waters of the ocean is only 10–20 (Rusch et al. 2007). Three of those have been cultured *Synechococcus*, *Prochlorococcus*, and *Pelagibacter ubique*. Other common groups were related to Alphaproteobacteria, Acidimicrobidae, Cytophaga, *Rhodospirillum*, *Roseobacter*, SAR86, and SAR116. However, the DNA from which these conclusions were drawn was extracted from water samples that were filtered in the size range of 0.1–0.8 μm, which would contain only the smallest bacteria (and Archaea and a few Eukarya). A common group of marine bacteria such as *Vibrio* appears in very low numbers (if at all) in seawater but reaches high abundances in aggregates and particles or as symbionts and pathogens (Nealson and Venter 2007). To make a census of microbial life one has to take into account the sampling strategy that was chosen and realize that perhaps only a fraction of marine microbial life is freely suspended in the water. Moreover, the census of marine microorganisms consists of countless populations that occur in very low abundances, which has been termed the 'rare biosphere' (Pedrós-Alío 2006, 2007). This rare biosphere may serve as a 'seed bank' presenting an almost infinite source of natural genetic innovation, insuring an optimum function of the ecosystem under fluctuating environmental conditions.

The euphotic realm is an environment with many peculiar conditions and properties and therefore it comes as no surprise that it is inhabited with a specific community of microorganisms. Since it is the realm where light is available and photosynthesis is possible, phototrophic microorganisms are a key functional group.

Cyanobacteria are oxygenic phototrophic bacteria that have a plant-like photosynthetic apparatus with two photosystems in series. They use water as electron donor and evolve oxygen as a consequence. Light is harvested by phycobiliproteins and chlorophyll *a* and the light energy is used to fix CO_2 through the Calvin–Benson–Bassham cycle. Many species are capable of fixing N_2 (diazotrophs; diazote = dinitrogen, troph = feeding on) (Stal 1995). *Trichodesmium* spp. are filamentous, non-heterocystous, diazotrophs which are widespread in the warmer tropical ocean (Bergman et al. 2013). They possess gas vesicles, which makes them buoyant. They can form surface blooms that are visible from space. *Trichodesmium* spp. occur as single trichomes or as aggregates (tufts and puffs). They are responsible for the bulk of the fixation of N_2 in the euphotic realm of the ocean. Nevertheless, the contribution of unicellular diazotrophic cyanobacteria may also be considerable. They form three phylogenetic groups indicated as UCYN-A, UCYN-B, and UCYN-C (unicellular nitrogen-fixing cyanobacteria) (Moisander et al. 2010; Mulholland et al. 2012). UCYN-A (*Candidatus Atelocyanobacterium thalassa*) is hitherto uncultured but known by its genome sequence and lives symbiotic with a picoeukaryotic unicellular Prymnesiophyte alga (Zehr 2013). This symbiosis seems to be widespread and although also confined to warmer waters, it has a wider temperature distribution than *Trichodesmium*. UCYN-A is an unusual cyanobacterium because it lacks among others photosystem II and does not fix CO_2. UCYN-B cyanobacteria are free-living (related to *Crocosphaera*) and UCYN-C is related to *Cyanothece* and are probably benthic or epiphytic rather than free-living in suspension. Heterocystous cyanobacteria occur in the open ocean only as symbionts of certain diatoms (Foster et al. 2011). In the marine environment, free-living heterocystous cyanobacteria are only found in brackish waters such as the Baltic Sea or as benthic and epiphytic organisms in microbial mats, coral reefs, seagrass meadows and rocky shores.

The most abundant cyanobacteria in the euphotic realm are the picocyanobacteria with cell sizes ranging from 0.5 to 3.0 μm and can reach cell numbers as high as 10^5–10^6 cells per milliliter of seawater (Scanlan et al. 2009). *Prochlorococcus* is the genus with the smallest cell sizes among the picocyanobacteria. They are unusual cyanobacteria because they do not use phycobiliproteins for light harvesting and contain instead divinyl derivatives of chlorophyll *a* and *b*. *Prochlorococcus* spp. come in a variety of ecotypes, roughly divided into high-light adapted ecotypes that thrive at the surface layers and low-light adapted ecotypes that thrive at the bottom of the euphotic realm (Moore and Chisholm 1999). *Prochlorococcus* spp. are all confined to warmer extremely eutrophic ocean waters. *Synechococcus* spp. are picocyanobacteria that have a wider distribution and can also be found in temperate seas, more turbid waters, and at higher nutrient concentrations. They have evolved an enormous diversity and adapted to the prevailing light conditions by changing the content and ratio of phycobiliproteins with different chromophores (Six et al. 2007). Several strains are capable of changing their pigmentation (various forms of complementary chromatic adaptation).

Although their numbers are much smaller compared to cyanobacteria, microalgae are an important component of the microbial community in the euphotic realm (Worden et al. 2015). Diatoms (Heterokontophyta) and dinoflagellates (Dinoflagellata) are characterized by a much higher turnover because they are heavily grazed by protists. This keeps their numbers low but their productivity rival that of the cyanobacteria. Also, eukaryotic phototrophs are extremely diverse and cover a wide range of life strategies (Worden et al. 2015). Diatoms are a highly diverse and important group of primary producers with approximately 200,000 species described (Armbrust 2009). It has been estimated that 20 % of the total primary production on Earth is attributed to this group of organisms. Diatoms are surrounded by two valves that are made of silicate (frustules). Diatoms are either single cells or form chains of cells. Diatoms have a complex reproductive cycle depending on their morphology (pennate or centric) and many of them alternate vegetative reproduction (mitosis) with the sexual one (meiosis). They divide mitotically but sexual reproduction occurs as well and is a necessary event in the life cycle of the diatom. Diatoms also vary in size from a few micrometers to a few millimeters. They grow rapidly and the cells sink quickly out the euphotic realm. A considerable number may reach the seafloor where it is stored over geological periods of time. Some species can control buoyancy and use this property to migrate from the bottom of the euphotic zone where nutrients concentrations are high to the surface for photosynthesis. Diatoms are distributed worldwide from the poles to the tropics, in marine as well as in freshwater settings.

Coccolithophorids are a clade within the Haptophytes. These algae form coccoliths that are made of calcite. Coccolithophorid algae occur worldwide in the euphotic zone of the ocean. In the temperate ocean, they may form massive surface blooms that due to the scattering of the calcite give a milky appearance to the water. The coccoliths sink to the seafloor and form locally large calcite deposits. The highest diversity is found in the low latitude areas of the ocean. Light and temperature are the environmental factors that best predict their growth (O'Brien et al. 2016). Due to the fact that these algae depend on calcite formation it has been suggested that this group of organisms may be strongly affected by ocean acidification (Riebesell et al. 2000).

Dinoflagellates are another group of algae that thrive in the euphotic zone of the ocean. They represent the largest group of eukaryotic phytoplankton after the diatoms and belong to the phylum Alveolata. More than 2000 species have been described (Caron et al. 2009). Some species are toxic and produce saxitoxin that is also known as paralytic shellfish poisoning (PSP) and blooms of such organisms may cause serious health and environmental problems. The behavior may be photoautotrophic, heterotrophic or mixotrophic. Dinoflagellates possess two flagella, which allow them to swim at rates up to 0.5 mm s^{-1}. They migrate in a daily rhythm and swim down at night and up to the surface during the day (Bresolin de Souza et al. 2014; Follows and Dutkiewicz 2011). This allows these organisms to take advantage of the nutrients available at the bottom of the euphotic zone and light during the day.

Picoeukaryotes are the smallest protists and are similar in size as bacteria. By definition their size class ranges from 0.2 to 2.0 μm, although sometimes investigators prefer 3.0 μm as the size limit (Worden et al. 2015). These organisms are found worldwide in the euphotic zone of the ocean and exhibit a high diversity. Just like the picocyanobacteria, picoeukaryotes do not sink. Picoeukaryotes may be photoautotrophic, heterotrophic, or mixotrophic. Phototrophic picoeukaryotes in the ocean belong to three divisions: Chlorophyta, Heterokonta, and Haptophyta. Common genera belonging to the Chlorophyta include *Bathyoccus*, *Micromonas*, and *Ostreococcus*. Some of these possess flagella and are able to swim. Representatives of the Heterokonta include *Pelagomonas*, *Pelagococcus*, *Aureococcus*, *Bolidomonas*, *Nannochloropsis*, and *Pinguiochrysis*. Haptophyta include *Imantonia* and *Phaeocystis*. A detailed review of the ecology and diversity of picoeukaryotes can be found in Worden and Not (2008).

Aerobic anoxygenic phototrophic bacteria (AAnPB) are abundantly present in the euphotic realm and may account to ~10 % of the total microbial community (Kolber et al. 2001), but in some cases 30 % has been reported (Shiba 1991). These organisms belonging to Alpha-, Beta-, and Gammaproteobacteria possess bacteriochlorophyll *a*, which is located in chromatophores in the cytoplasmic membrane and are facultative photoheterotrophs (Eiler 2006). The best-known marine strains are the Alphaproteobacteria *Erythrobacter* and *Roseobacter*. They metabolize organic substrates but can use light energy as an additional energy source. The genome sequences of some representatives suggest that they may live as autotrophs or as mixotrophs because CO_2 fixation pathways were found, but none of the isolates have been shown to grow autotrophically. RuBisCO is absent in all cultured isolates so far.

Other microorganisms that use light without being autotrophic are those that have the pigment proteorhodopsin. Proteorhodopsin has been shown in Alpha- (SAR11) and Gammaproteobacteria (SAR86 and *Vibrio* sp. AND4) and in Flavobacteria. This protein is a light-driven proton pump that is common in marine Proteobacteria, including the Alphaproteobacterium SAR11 (*Pelagibacter ubique*). Bacteria belonging to SAR11 may make up to 35 % of the bacterial community and this indicates that light harvesting by proteorhodopsin may represent an important advantage. There is a great diversity of proteorhodopsins that possess distinct light absorption spectra and there is evidence that the organisms are tuned to the quality of light that they experience in their environment (Béjà et al. 2001). A single amino acid residue change switches the absorption maximum of the pigment from green (525 nm) to blue (490 nm) wavelengths (Man et al. 2003). Genes coding for proteorhodopsin are ubiquitous in the euphotic realm, emphasizing the important fact that light may help to overcome energy limitation in otherwise heterotrophic microorganisms (Akram et al. 2013; Gómez-Consarnau et al. 2010). Nevertheless, in only few cases it has been shown that proteorhodopsin harvested light indeed accelerated growth.

Acknowledgments The research leading to these results has received funding from the European Union Seventh Framework Programme (FP7/2007–2013) under grant agreement No. 311975. This publication reflects the views only of the author, and the European Union cannot be held responsible for any use which may be made of the information contained therein.

References

Akram N, Palovaara J, Forsberg J, Lindh MV, Milton DL, Luo H, González JM, Pinhassi J (2013) Regulation of proteorhodopsin gene expression by nutrient limitation in the marine bacterium *Vibrio* sp. AND4. Environ Microbiol 15:1400–1415

Armbrust EV (2009) The life of diatoms in the world's oceans. Nature 459:185–192

Azam F, Malfatti F (2007) Microbial structuring of marine ecosystems. Nat Rev Microbiol 5: 782–791

Beckmann A, Hense I (2007) Beneath the surface: characteristics of oceanic ecosystems under weak mixing conditions—a theoretical investigation. Prog Oceanogr 75:771–796

Béjà O, Spudich EN, Spudich JL, Leclerc M, DeLong EF (2001) Proteorhodopsin phototrophy in the ocean. Nature 411:786–789

Bergman B, Sandh G, Lin S, Larsson J, Carpenter EJ (2013) *Trichodesmium*—a widespread marine cyanobacterium with unusual nitrogen fixation properties. FEMS Microbiol Rev 37:286–302

Borman AH, de Jong EW, Huizinga M, Kok DJ, Westbroek P, Bosch L (1982) The role in $CaCO_3$ crystallization of an acid Ca^{2+}-binding polysaccharide associated with coccoliths of *Emiliania huxleyi*. Eur J Biochem 129:179–183

Bresolin de Souza K, Jephson T, Berg Hasper T, Carlsson P (2014) Species-specific dinoflagellate vertical distribution in temperature-stratified waters. Mar Biol 161:1725–1734

Buma AGJ, van Hannen EJ, Veldhuis MJW Gieskes WWC (1996) UB-B induces DNA damage and DNA synthesis delay in the marine diatom *Cyclotella* sp. Sci Mar 60(suppl. 1):101–106

Buma AGJ, de Boer MK, Boelen P (2001) Depth distribution of DNA damage in Antarctic marine phyto- and bacterioplankton exposed to summertime radiation. J Phycol 37:200–208

Carlson CA, Del Giorgio PA, Herndl GJ (2008) Microbes and the dissipation of energy and respiration: from cells to ecosystems. Oceanography 20:89–100

Caron DA, Worden AZ, Countway PD, Demir E, Heidelberg KB (2009) Protists are microbes too: a perspective. ISME J 3:4–12

Cockell CS (2000) Ultraviolet radiation and the photobiology of earth's early oceans. Orig Life Evol Biosph 30:467–499

Cullen JJ (2015) Subsurface chlorophyll maximum layers: enduring enigma or mystery solved? Annu Rev Mar Sci 7:207–239

Del Giorgio PA, Duarte CM (2002) Respiration in the open ocean. Nature 420:379–384

Deschamps P, Haferkamp I, d'Hulst C, Neuhaus HE, Ball SG (2008) The relocation of starch metabolism to chloroplasts: when, why and how. Trends Plant Sci 11:574–582

Dyhrman ST, Ammerman JW, Van Mooy BAS (2007) Microbes and the marine phosphorus cycle. Ocenography 20:110–116

Dyhrman ST, Benitez-Nelson CR, Orchard ED, Haley ST, Pellechia PJ (2009) A microbial source of phosphonates in oligotrophic marine systems. Nat Geosci 2:696–699

Eiler A (2006) Evidence for the ubiquity of mixotrophic bacteria in the upper ocean: implications and consequences. Appl Environ Microbiol 72:7431–7437

Follows MJ, Dutkiewicz S (2011) Modeling diverse communities of marine microbes. Annu Rev Mar Sci 3:427–451

Foster RA, Kuypers MMM, Vagner T, Paerl RW, Musat N, Zehr JP (2011) Nitrogen fixation and transfer in open ocean diatom-cyanobacterial symbioses. ISME J 5:1484–1493

Gao Q, Garcia-Pichel F (2011) Microbial ultraviolet sunscreens. Nat Rev Microbiol 9:791–802

Gattuso JP, Gentili B, Duarte CM, Kleypas JA, Middelburg JJ, Antoine D (2006) Light availability in the coastal ocean: impact on the distribution of benthic photosynthetic organisms and contribution to primary production. Biogeosciences 3:489–513

Gómez-Consarnau L, Akram N, Lindell K, Pedersen A, Neutze R, Milton DL, González JM, Pinhassi J (2010) Proteorhodopsin phototrophy promotes survival of marine bacteria during starvation. PLoS Biol 8(4):e1000358

Hansell DA, Bates NR, Olson DB (2004) Excess nitrate and nitrogen fixation in the North Atlantic Ocean. Mar Chem 84:243–265

Herbland A, Voituriez B (1979) Hydrological structure-analysis for estimating the primary production in the tropical Atlantic Ocean. J Mar Res 37:87–101

Hess WR, Rocap G, Ting CS, Larimer F, Stilwagen S, Lamerdin J, Chisholm SW (2001) The photosynthetic apparatus of *Prochlorococcus*: insights through comparative genomics. Photosynth Res 70:53–71

Holm-Hansen O, Hewes CD (2004) Deep chlorophyll-*a* maxima (DCMs) in Antarctic waters. Polar Biol 27:699–710

Holm-Hansen O, Mitchell BG (1991) Spatial and temporal distribution of phytoplankton and primary production in the western Bransfield Strait region. Deep-Sea Res 38:961–980

Kleiner D (1985) Bacterial ammonium transport. FEMS Microbiol Rev 32:87–100

Kolber ZS, Plumley FG, Lang AS, Beatty JT, Blankenship RE, VanDover CL, Vetriani C, Koblizek M, Rathgeber C, Falkowski PG (2001) Contribution of aerobic photoheterotrophic bacteria to the carbon cycle in the ocean. Science 292:2492–2495

Kromkamp J (1987) Formation and functional significance of storage products in cyanobacteria. NZ J Mar Freshw Res 21:457–465

Lacour L, Claustre H, Prieur L, d'Ortenzio F (2015) Phytoplankton biomass cycles in the North Atlantic subpolar gyre: a similar mechanism for two different blooms in the Labrador Sea. Geophys Res Lett 42:5403–5410

Lam P, Kuypers MMM (2011) Microbial nitrogen cycling processes in oxygen minimum zones. Annu Rev Mar Sci 3:317–345
Latifi A, Ruiz M, Zhang C-C (2009) Oxidative stress in cyanobacteria. FEMS Microbiol Rev 33:258–278
Liu X, Millero FJ (2002) The solubility of iron in seawater. Mar Chem 77:43–54
Man D, Wang W, Sabehi G, Aravind L, Post AF, Massana R, Spudich EN, Spudich JL, Béjà O (2003) Diversification and spectral tuning in marine proteorhodopsins. EMBO J 22:1725–1731
Mari X, Beauvais S, Lemée R, Pedrotti L (2001) Non-redfield C: N ratio of transparent exopolymeric particles in the northwestern Mediterranean Sea. Limnol Oceanogr 46: 1831–1836
Metcalf WW, Griffin BM, Cicchillo RM, Gao J, Janga SC, Cooke HA, Circello BT, Evens BS, Martens-Habbena W, Stahl DA, van der Donk WA (2012) Synthesis of methylphosphonic acid by marine microbes: a source for methane in the aerobic ocean. Science 337:1104–1107
Moisander PH, Beinart RA, Hewson I, White AE, Johnson KS, Carlson DJ, Montoya JP, Zehr JP (2010) Unicellular cyanobacterial distributions broaden the oceanic N_2 fixation domain. Science 327:1512–1514
Moore LR, Chisholm SW (1999) Photophysiology of the marine cyanobacterium *Prochlorococcus*: ecotypic differences among cultured isolates. Limnol Oceanogr 44:628–638
Mulholland MR, Bernhardt PW, Blanco-Garcia JL, Mannino A, Hyde K, Mondragon E, Turk K, Moisander PH, Zehr JP (2012) Rates of dinitrogen fixation and the abundance of diazotrophs in North American coastal waters between Cape Hatteras and Georges Bank. Limnol Oceanogr 57:1067–1083
Nealson KH, Venter JC (2007) Metagenomics and the global ocean survey: what's in it for us, and why should we care? ISME J 1:185–187
Nolting RF, Gerringa LJA, Swagerman MJW, Timmermans KR, de Baar HJW (1998) Fe (III) speciation in the high nutrient, low chlorophyll Pacific region of the Southern Ocean. Mar Chem 62:335–352
O'Brien CJ, Vogt M, Gruber N (2016) Global coccolithophore diversity: drivers and future change. Prog Oceanogr 140:27–42
Obst M, Wehrli B, Dittrich M (2009) $CaCO_3$ nucleation by cyanobacteria: laboratory evidence for a passive, surface-induced mechanism. Geobiology 7:324–347
Orcutt BN, Sylvan JB, Knab NJ, Edwards KJ (2011) Microbial ecology of the dark ocean above, at, and below the seafloor. Microbiol Mol Biol Rev 75:361–422
Pedrós-Alió C (2006) Marine microbial diversity: can it be determined? Trends Microbiol 14:257–263
Pedrós-Alió C (2007) Dipping into the rare biosphere. Science 315:192–193
Ploug H, Kühl M, Buchholz-Cleven B, Jørgensen BB (1997) Anoxic aggregates—an ephemeral phenomenon in the pelagic environment? Aquat Microb Ecol 13:285–294
Riebesell U, Zondervan I, Rost B, Tortell PD, Zeebe RE, Morel FMM (2000) Reduced calcification of marine plankton in response to increased atmospheric CO_2. Nature 407: 364–367
Riemann L, Farnelid H, Stewards GF (2010) Nitrogenase genes in non-cyanobacterial plankton: prevalence, diversity and regulation in marine waters. Aquat Microb Ecol 61:235–247
Rijkenberg MJA, Fischer AC, Kroon JJ, Gerringa LJA, Timmermans KR, Wolterbeek HTh, de Baar HJW (2005) The influence of UV irradiation on the photoreduction of iron in the Southern Ocean. Mar Chem 93:119–129
Rusch DB, Halpern AL, Sutton G, Heidelberg KB, Williamson S, Yooseph S, Wu D, Eisen JA, Hoffman JM, Remington K, Beeson K, Tran B, Smith H, Baden-Tillson H, Stewart C, Thorpe J, Freeman J, Andrews-Pfannkoch C, Venter JE, Li K, Kravitz S, Heidelberg JF, Utterback T, Rogers Y-H, Falcón LI, Souza V, Bonilla-Rosso G, Eguiarte LE, Karl DM, Sathyendranath S, Platt T, Bermingham E, Gallardo V, Tamayo-Castillo G, Ferrari MR, Strausberg RL, Nealson K, Friedman R, Frazier M, Venter JC (2007) The *Sorcerer II* global ocean sampling expedition: Northwest Atlantic through Eastern Tropical Pacific. PLoS Biol 5(3):e77

Santoro AE, Casciotti KL (2011) Enrichment and characterization of ammonia-oxidizing archaea from the open ocean: phylogeny, physiology and stable isotope fractionation. ISME J 5: 1796–1808

Sass AM, Sass H, Coolen MJL, Cypionka H, Overmann J (2001) Microbial communities in the chemocline of a hypersaline deep-sea basin (Urania Basin, Mediterranean Sea). Appl Environ Microbiol 67:5392–5402

Scanlan DJ, Ostrowski M, Mazard S, Dufresne A, Garczarek L, Hess WR, Post AF, Hagemann M, Paulsen I, Partensky F (2009) Ecological genomics of marine picocyanobacteria. Microbiol Mol Biol Rev 73:249–299

Shiba T (1991) *Roseobacter litoralis* gen. nov., sp. nov., and *Roseobacter denitrificans* sp. nov., aerobic pink-pigmented bacteria which contain bacteriochlorophyll *a*. Syst Appl Microbiol 14:140–145

Six C, Thomas J-C, Garczarek L, Ostrowski M, Dufresne A, Blot N, Scanlan DJ, Partensky F (2007) Diversity and evolution of phycobilisomes in marine *Synechococcus* spp.—a comparative genomics study. Genome Biol 8(12):R259

Smith RC, Prezelin BB, Baker KS, Bidigare RR, Boucher NP, Coley T, Karentz D, Macintyre S, Matlick HA, Menzies D, Ondrusek M, Wan Z, Waters KJ (1992) Ozone depletion—ultraviolet radiation and phytoplankton biology in Antarctic waters. Science 255:952–959

Stal LJ (1995) Physiological ecology of cyanobacteria in microbial mats and other communities. New Phytol 131:1–32

Stal LJ (2009) Is the distribution of nitrogen-fixing cyanobacteria in the oceans related to temperature? Environ Microbiol 11:1632–1645

Stal LJ, Walsby AE (2000) Photosynthesis and nitrogen fixation in a cyanobacterial bloom in the Baltic Sea. Eur J Phycol 35:97–108

Stomp M, Huisman J, Stal LJ, Matthijs HCP (2007a) Colourful niches of phototrophic microorganisms shaped by vibrations of the water molecule. ISME J 1:271–282

Stomp M, Huisman J, Vörös L, Pick FR, Laamanen M, Haverkamp T, Stal LJ (2007b) Colourful coexistence of red and green picocyanobacteria in lakes and seas. Ecol Lett 10:290–298

Tyrrell T (1999) The relative influences of nitrogen and phosphorus on oceanic primary production. Nature 400:525–531

van Liere L, Mur LR, Gibson CE, Herdman M (1979) Growth and physiology of *Oscillatoria agardhii* Gomont cultivated in continuous culture with a light-dark cycle. Arch Microbiol 123:315–318

Wagener T, Guieu C, Losno R, Bonnet S, Mahowald N (2008) Revisiting atmopsheric dust export to the Southern Hemisphere Ocean: biogeochemical implications. Glob Biogeochem Cycles GB2006 22

Wetzel RG, Hatcher PG, Bianchi TS (1995) Natural photolysis by ultraviolet irradiance of recalcitrant dissolved organic matter to simple substrates for rapid bacterial metabolism. Limnol Oceanogr 40:1369–1380

Worden AZ, Not F (2008) Ecology and diversity of picoeukaryotes. In: Kirchman DL (ed) Microbial ecology of the oceans. Wiley, NY, pp 159–205

Worden AZ, Follows MJ, Giovannoni SJ, Wilken S, Zimmerman AE Keeling PJ (2015) Rethinking the marine carbon cycle: factoring in the multifarious lifestyles of microbes. Science 347(6223):1257594 (1-10)

Wotton RS (2004) The ubiquity and many roles of exopolymers (EPS) in aquatic systems. Sci Mar 68:13–21

Zavarzin GA (2002) Microbial geochemical calcium cycle. Microbiology 71:1–17

Zehr JP (2011) Nitrogen fixation by marine cyanobacteria. Trends Microbiol 19:162–173

Zehr JP (2013) Interactions with partners are key for oceanic nitrogen-fixing cyanobacteria. Microbe 8:117–122

Zehr JP, Kudela RM (2011) Nitrogen cycle of the open ocean: from genes to ecosystems. Ann Rev Mar Sci 3:197–225

Chapter 8
Exploring the Microbiology of the Deep Sea

Mohamed Jebbar, Pauline Vannier, Grégoire Michoud and Viggó Thór Marteinsson

Abstract In this chapter the current knowledge of the diversity of piezophiles isolated so far is reviewed. The isolated piezophiles originated from high-pressure environments such as the cold deep sea, hydrothermal vents, and crustal rocks. Several "stress" conditions can be experienced in these environments, in particular high hydrostatic pressure (HHP). Discoveries of abundant life in diverse high-pressure environments (deep biosphere) support the existence and an adaptation of life to HHP. At least 50 piezophilic and piezotolerant Bacteria and Archaea have been isolated from different deep-sea environments but these do not by far cover the large metabolic diversity of known microorganisms thriving in deep biospheres. The field of biology of piezophiles has suffered essentially from the requirements for high-pressure retaining sample containments and culturing laboratory equipment, which is technically complicated and expensive. Only a few prototypes of HHP bioreactors have been developed by a number of research groups and this could explain the limited number of piezophiles isolated up till now.

M. Jebbar (✉)
Univ Brest, CNRS, Ifremer, UMR 6197-Laboratoire de Microbiologie des Environnements Extrêmes (LM2E), Institut Universitaire Européen de La Mer (IUEM), Rue Dumont D'Urville, 29280 Plouzané, France
e-mail: mohamed.jebbar@univ-brest.fr

P. Vannier · V.T. Marteinsson
Food Safety, Environment and Genetics, Matís Ohf., Vinlandsleið 12, 113 Reykjavik, Iceland

G. Michoud
Biological and Environmental Sciences and Engineering Division (BESE), King Abdullah University of Science and Technology (KAUST), Thuwal 23955-6900, Kingdom of Saudi Arabia

V.T. Marteinsson
Agricultural University of Iceland, Hvanneyri, 311 Borgarnes, Iceland

L.J. Stal and M.S. Cretoiu (eds.), *The Marine Microbiome*,
DOI 10.1007/978-3-319-33000-6_8

8.1 Introduction

The ocean accounts for 70 % of the Earth's surface with an average depth of ~3800 m, i.e., an average hydrostatic pressure of 38 MPa (380 bars). Indeed, hydrostatic pressure increases with depth with 10 MPa per km in the ocean and 30 MPa per km in the oceanic crust. The deepest site in ocean is the Deep Challenger that is located in the Mariana trench at 10,998 m where the high hydrostatic pressure (HHP) reaches 110 MPa. However, on Earth, the ocean, and more particularly the deep ocean and the subsurface biosphere, constitutes the last microbiological frontier to be studied and it still remains one of the unexplored biospheres due to severe technological constraints (Oger and Jebbar 2010).

The study and characterization of marine biodiversity are essential to gain a better insight into the functioning of the Earth system. Bacteria constitute the largest diversity reservoir, the main primary producers, as well as the creatures that have evolved for the longest period (Whitman et al. 1998). From global estimates of volume and numbers of cells, the upper 200 m of the ocean contains a total of 3.6×10^{28} bacterial and archaeal cells of which 2.9×10^{27} cells are autotrophs. Ocean water below 200 m depth contains 6.5×10^{28} Bacteria and Archaea (Whitman et al. 1998). Although the number of bacterial and archaeal species is estimated from as low as 10^4–10^5 to as high as 10^6–10^7 (Hammond 1995) there are only 12,260 species (451 Archaea and 11,809 Bacteria) whose taxonomy has been validated and that have been described (Euzeby 2013).

Despite the remarkable work carried out by the "Census of Marine Life" over the last decade to highlight numerous prospects, the current knowledge about the biodiversity in the marine environment is still limited with respect to the one in terrestrial habitats, and this is more crucial for the deep ocean and the deep-sea biosphere (below the sea-floor). The discovery of living microorganisms at the deepest location in the ocean (Mariana Trench), in deep-sea hydrothermal vents, cold seeps, and in the deep-sea subsurface (1600–2500 mbsf) has extended the geographical limits of life on Earth, and has driven scientists to assume that microbial biomass in the deep-sea biosphere might exceed the one in the water column and Earth surface (Ciobanu et al. 2014; Inagaki et al. 2015; Roussel et al. 2008).

In the biological sciences history, the discovery of deep-sea ecosystems, more than 30 years ago, whose food web relies on microbial chemosynthesis, is of a recent date (Corliss and Ballard 1977; Jannasch and Taylor 1984). These deep-sea ecosystems constitute extreme environments with a strong endemism composed of communities frequently dominated by complex biological assemblages where primary producer microorganisms (e.g., chemolithoautotrophs) are often associated with one or several producers that have coevolved in these pressurized habitats. Being mainly situated in accretion zones (ocean ridges), passive continental margins and subduction zones, these microbial communities are associated to sources of reduced chemical compounds (hydrogen, sulfide, methane, or simple hydrocarbons). These unique spots on the ocean floor have high emission of biologically rich media that turns them into an oasis in an otherwise extremely oligotrophic desert. These isolated islands are densely populated by a variety of communities thriving in this HHP environment. These communities are dependent on a time-varying resource and unaffected by seasonal variations. This resource is composed of fluids that are expelled towards the ridges and continental margins at the limit of ocean plates. Other deep-sea ecosystems, more dispersed and short-lived, display close features and depend on unpredictable supplies of organic matter (corpses of large animals, e.g., whales, wood falls, and other organic substrates), which, while decaying, feed a food chain based on the degradation of sulfides (and/or methane) by microorganisms.

Until recently, only the deep oceans were known to harbor microorganisms that grow at these depths and love HHP. Initially, these microbes were called "barophiles" but recently renamed to piezotolerant or piezophilic ("piezo" is derived from the Greek 'piezein' which means to squeeze or to press, and "phile" means love) microorganisms that are able to grow at HHP. The first records for deep-sea bacteria were made by Certes (Certes 1884) who cultured microorganisms from water samples collected at depths down to 5000 m during the cruises of the ships «Le Travailleur» and «Talisman» in 1882–1883. Although Certes' experiments were not performed under HHP, his report unlocked the way of deep-sea microbiology. Indeed, HHP tools for collecting samples at great depth and then culture microorganisms from these samples under high hydrostatic pressure did not exist at that time.

One of the first questions addressed in the second half of the twentieth century concerned the assimilation of organic matter under conditions of low temperature and high pressure. Zobell, Morita, Jannasch (see Prieur and Marteinsson 1998 for review) and others tackled this problem and after hundreds of experiments it was concluded that heterotrophic metabolism of marine microorganisms isolated from surface waters as well as from deep waters proceeded at a considerably lower rate under low temperature and high hydrostatic pressure. These early investigators also concluded that probably no free-living bacteria adapted to deep-sea conditions exist. However, Schwarz, Colwell, Deming and Yayanos (see Prieur and Marteinsson 1998 for review) demonstrated the existence of deep-sea piezophilic psychrophiles mainly in free-living state or associated to the digestive tracts of deep-sea invertebrates. Later, many hyperthermophilic microorganisms ($T_{opt} \geq 80$ °C) were also isolated from deep-sea hydrothermal vents. Some of them appeared to be piezotolerant, piezophiles, and even obligate piezophiles, depending on the depth from which they originated.

This chapter focuses on deep-sea microbiology and in particular on the latest available data about piezophilic microorganisms that have been isolated and characterized from various deep biospheres, which are still poorly explored.

8.2 What are the Technical Limits of Isolation and Culturing Piezophiles?

The field of biology of piezophiles has suffered largely from the need of high-pressure retaining sample containments and laboratory culturing equipment. The first HHP-adapted bacteria were isolated from deep-sea sediments in 1949 by ZoBell and Johnson (1949). The first obligate piezophiles, i.e., organisms that cannot develop at atmospheric pressure and ambient temperature, were isolated in 1981 (Yayanos et al. 1981). Since the time of Zobell and other pioneering researchers in the field of piezomicrobiology researchers have been aware that piezophiles are sensitive to hydrostatic pressure shifts and that special equipment is required.

Culturing piezophiles under aerobic or anaerobic conditions at various temperatures requires specific equipment. At the laboratory scale, batch cultures are usually carried out in syringes that are pressurized in stainless steel bioreactors (Marteinsson et al. 1997). Typically, culture of hyperthermophilic Archaea or Bacteria is prepared under anaerobic conditions. Selecting medium is prepared along with the addition of a reducing agent ($Na_2S{\cdot}9H_2O$) in order to remove traces of oxygen. The culture medium is inoculated with 1–2 % exponential phase growing cells. Culturing is performed in sterile gas-tight glass syringes sealed with needles inserted in a rubber stopper. Subsequently, the syringes are transferred into a high-pressure and high-temperature incubation system. The temperature and pressure of the vessel are controlled independently. A hydraulic pump typically generates the required pressure and cold tap water serves as the hydraulic fluid. In the case of the HP/HT (high pressure/high temperature) reactor described by Marteinsson et al. (1997) the maximum working pressure is 60 MPa and the maximum temperature is up to 200–300 °C (Fig. 8.1a).

However, these devices do not allow for production of a large amount of biomass, or subsampling for time series experiments. HP/HT bioreactors can be categorized into two types: hydrostatic and hyperbaric. Hydrostatic bioreactors, in which the reactor contains only a liquid phase, are usually easy to build and operate and these devices have been used to perform many types of laboratory analyses and experiments including microbial enrichment and isolation and measurement of microbial activity under HHP (Zhang et al. 2015b). In contrast to the hydrostatic bioreactor, the hyperbaric bioreactor allows an adequate supply of substrate gases for the culture of autotrophs that need H_2, CO_2, or other gases (Miller et al. 1988; Park and Clark 2002).

Complex devices such as the "DEEPBATH" have been designed for the above purposes by the Japanese Agency for Marine-Earth Science and Technology (JAMSTEC). This DEEP-sea BAro-piezophile and THermophile isolation and culturing system consists of four separate devices: (1) a pressure-retaining sampling device, (2) dilution device under pressure conditions, (3) isolation device, and (4) culturing device (Kato 2006). The DEEPBATH system is controlled by central regulation systems and the pressure and temperature ranges from 0.1 to 65 MPa and

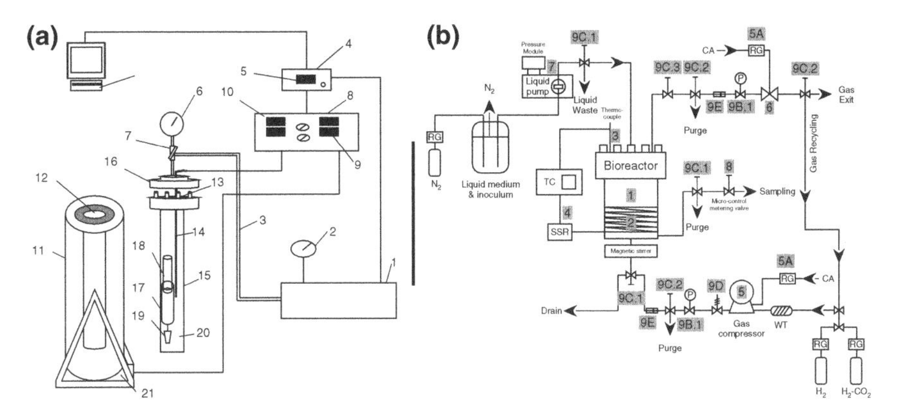

Fig. 8.1 Description of a low and a high-throughput HP/HT reactor (Marteinsson et al. 1997; Park et al. 2006). **a** Schematic representation of the high-pressure apparatus "hot bucket" for cultivating microorganisms at high temperature and pressure. Key: *1* hydraulic pressure generator; *2* Bourdon gauge (100 MPa); *3* water tube (inox); *4* pressure indicator; *5* digital pressure indicator; *6* Bourdon gauge; *7* valve; *8* computerized independent thermoregulators; *9* digital temperature indicator in oven; *10* digital temperature indicator inside the pressure vessel; *11* heating element (oven); *12* bucket for the pressure vessel; *13* O-ring; *14* thermocouple in jacket; *15* stainless steel pressure vessel; *16* the head for the pressure vessel; *17* culture syringe; *18* cut piston; *19* needle embedded in rubber stopper; *20* water; *21* thermocouple for heating unit; *22* computer (Marteinsson et al. 1997). **b** Schematic diagram the high T–P bioreactor system. Key: *1* high-pressure vessel; *2* heating belt; *3* thermocouple; *4* proportional-integral-derivative controller; *5* gas compressor; *6* back-pressure regulator; *7* liquid pump; *8* micro-control metering value; *9* high-pressure valves and rupture disks; *P* pressure gauge; *CA* compressed air; *TC* temp controller; *SSR* solid state relay; *WT* water trap

from 0 to 150 °C, respectively. The capacity of the culturing devices (2 sets) is 1.5 l each and, therefore, cultures of up to 3 l can be obtained. This sophisticated equipment may overcome the issue of repeated decompression and repressurization processes, which might affect the quality of the growth results. DEEPBATH is currently the only system available that permits sampling without a change in pressure and temperature. Using this system, growth experiments of hyperthermophiles piezotolerants and piezophiles have been performed under high pressure and temperature conditions (HP-HT). Other systems have been developed to allow culturing of microorganisms in hyperbaric conditions. Indeed, many hyperthermophiles are autotrophic microorganisms that require gaseous substrates such as H_2 and CO_2. In this regard, hydrostatic HP/HT bioreactors can be inadequate to grow autotrophic hyperthermophiles because initially there is a limited amount of gaseous substrate dissolved in the liquid medium (Park et al. 2006). The group of Douglas Clark developed a bioreactor that is equipped with a thermocouple and a heating belt for temperature control and has a total volume of 1.15 l. It can operate at up to 88 MPa and 200 °C, and pressurization from 0.78 to 50 MPa can be achieved within 10 min. The main difference between the two kinds of HP bioreactors is the utilization of a gas compressor in the hyperbaric bioreactor, which is needed to achieve the pressurization of the reactor (Park et al. 2006) (Fig. 8.1b). Also, Parkes et al. (2009) has proposed a device for recovering and processing sediments under in situ pressure conditions. This equipment allows performing experiments without depressurization under mild temperature and low hydrostatic pressures.

8.3 Piezophilic Microorganisms from the Cold Ocean

To decipher the functional diversity of deep-sea microorganisms from the ocean water column, culture-dependent approaches were combined with culture-independent methods, i.e., the use of molecular approaches including metagenomics and single cell genomics (Eloe et al. 2011a; Lasken and McLean 2014; León-Zayas et al. 2015).

Most of the deep-sea microorganisms isolated so far are psychrophiles (Jebbar et al. 2015). In fact, except for a few sites such as the deep-sea hydrothermal vents, the deep ocean has an average temperature of 2 °C. Only few microorganisms have been isolated from the deepest site on Earth: the Deep Challenger, and the first one was *Pseudomonas bathycetes* (Morita 1976). After this discovery of life existing in such an extreme environment, more studies focused on piezophilic microorganisms. *Colwellia* sp. strain MT41 was one of the first piezophile that was isolated and studied. This strain originated from an amphipod sampled at 10,476 m (Yayanos et al. 1981). Most of the psychro-piezophilic microorganisms isolated so far belong to three phyla: Actinobacteria, Firmicutes and Proteobacteria (Fig. 8.2a).

8.3.1 Bacteria

8.3.1.1 Actinobacteria

Despite the fact that numerous studies have shown the presence of Actinobacteria in deep sediments (Zhang et al. 2015b) or in the deep-water column, only one strain has been isolated so far from the deep biosphere. This isolate is *Dermacoccus abyssi* MT1.1T and was isolated from sediment sampled from the deepest site on Earth: the Deep Challenger in the Mariana Trench (Pathom-Aree et al. 2006). This strain was cultivated at 28 °C and atmospheric pressure but it still grew at 40 MPa.

8.3.1.2 Firmicutes

Firmicutes represent at least 6 % of the microbial diversity found in the sediments from the Okinawa trench (Zhang et al. 2015b). *Carnobacterium* sp. strain AT12 and AT7 were isolated from Aleutian Trench from an amphipod kept at HHP

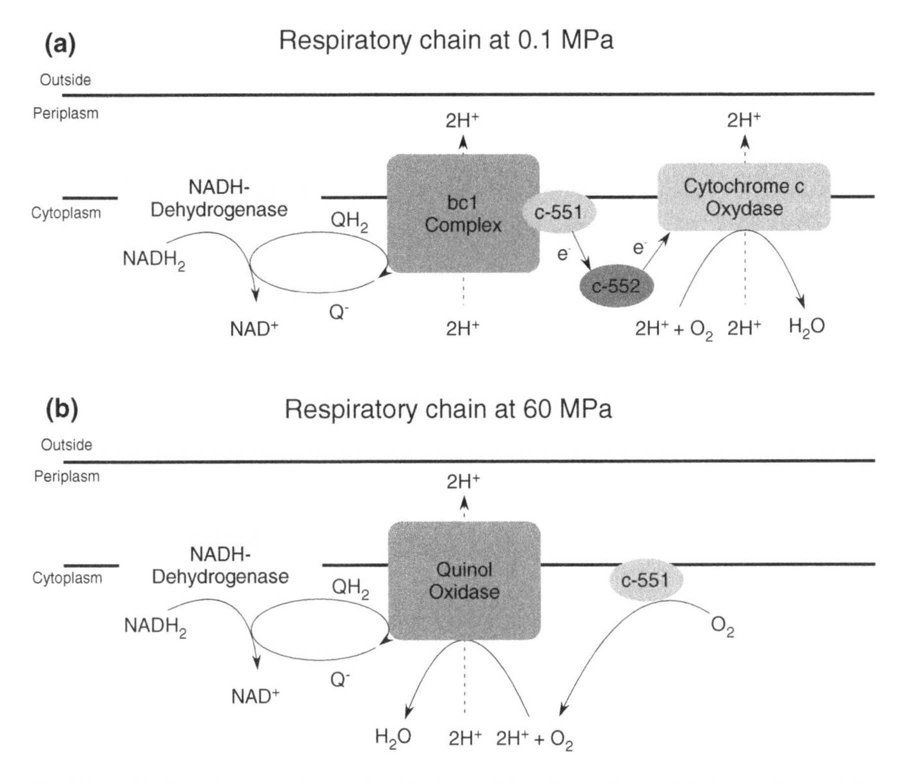

Fig. 8.2 **a** Phylogenic tree of psychropiles/mesophile piezophile and **b** hyper/thermophiles piezophiles. The trees were created using the SILVA database (Quast et al. 2013) and the Phylogeny.fr website with default parameters (Dereeper et al. 2008)

(Lauro et al. 2007). Their optimum growth temperatures are 2 and 20 °C, respectively, with a pressure range from 0.1 to 60 MPa and an optimum pressure of 15 MPa (Yayanos and DeLong 1987). These strains are chemoorganoheterotrophic and were the first Gram-positive piezophilic bacteria described.

8.3.1.3 Proteobacteria

Proteobacteria represent the largest microbial diversity (more than 35 %) discovered in deep sediments from the Okinawa Trench as was concluded from culture-independent methods (Zhang et al. 2015a). Rhodobacterales strain PRT1 was the first psychro-piezophilic organism isolated by using the dilution-extinction method using a natural seawater medium (Eloe et al. 2011b). This study showed that increasing the concentration of the carbon source might decrease the ability of a microorganism to grow. For instance, PRT1 has a higher growth rate with 0.001 % of a defined mixture of organic carbon compounds than with 0.01 or 0.1 %.

8.3.1.4 Colwellia Genus

Colwellia spp. are defined as facultative anaerobes and psychrophilic bacteria (Deming et al. 1988). So far, only five piezophilic species belonging to *Colwellia* have been isolated in pure culture and described.

Colwellia sp. strain MT41 was the first obligate piezophilic microorganism that was isolated in 1981 from the Mariana Trench from the amphipod *Hirondellea gigas* (Yayanos et al. 1981). *Colwellia piezophila* strains Y223GT and Y251ET are obligate piezophiles with a pressure range for growth from 10 to 80 MPa at 4 °C and from 40 to 80 MPa at 10 °C. The higher growth (yield and rate) was observed at 10 °C and 60 MPa. These strains are chemoorganotrophic facultative anaerobes (Nogi et al. 2004). *Colwellia hadaliensis* BNL-1T can grow from 30 to 102 MPa at 2 °C and from 50 to 102 MPa at 10 °C. This strain is an obligate piezophile and lysis happens at 2 °C and HHP <20 MPa and at 10 °C with a HHP <40 MPa (Deming et al. 1988). *Colwellia* sp. strain KT27 (Lauro et al. 2007) is closely related to strain MT41 and was isolated from the Kermadec Trench in the Southern hemisphere at 9858 m at 1.8 °C. This strain is still able to grow at the in situ pressure of 98 MPa.

8.3.1.5 Psychromonas Genus

Most of the *Psychromonas* species were isolated from polar environments and also from deep-sea environments. All members are psychrophilic Gram-negative and only five strains were described as piezophilic.

Psychromonas profunda 2825T (Xu et al. 2003a) was isolated from Atlantic sediments and grows optimal at 15–20 MPa and 6 °C and about 25 MPa at 10 °C. *Psychromonas kaikoae* strains JT7301 and JT7304T are obligate piezophiles with an optimum growth at 10 °C and 50 MPa (Nogi et al. 2002). *Psychromonas* sp. strain CNPT3 was isolated from decaying amphipods sampled at 5700 m in the Central North Pacific Ocean and grows optimal at 50 MPa at 2 and 4 °C (Yayanos et al. 1979). *Psychromonas hadalis* K41GT is an obligate piezophile isolated from 7542 m depth in the Japan Trench and grows optimal at 6 °C and 60 MPa (Nogi et al. 2007).

8.3.1.6 Moritella Genus

Moritella is a genus name gathering the most common psychrophilic microorganisms isolated from marine environments. Only five piezophilic strains belonging to *Moritella* were isolated so far. *Moritella japonica* DSK1 was isolated from the Japan Trench at 6356 m with optimal temperature and pressure conditions for growth at, respectively, 15 °C and 50 MPa (Nogi et al.1998a). This strain was the first to be described as a piezophile belonging to the genus *Moritella*. *Moritella profunda* strain 2674^T isolated from Atlantic sediment at 2815 m close to the

African coast showed an optimal growth at 10 °C and 20–24 MPa (Xu et al. 2003b). *Moritella abyssi* strain 2693^{T} was isolated from Atlantic sediment and is also a piezophile with an optimal growth at 30 MPa and 10 °C (Xu et al. 2003b). *Moritella* sp. strain PE36 is a facultative piezophile isolated from the California coast at 3584 m showing optimal growth at 10 °C (Yayanos 1986). *Moritella yayanosii* strain DB21MT-5 was isolated from sediment sampled at the deepest ocean site: Challenger Deep. Its optimal parameters for growth are 10 °C and 80 MPa (Nogi and Kato 1999).

8.3.1.7 Shewanella Genus

A large number of *Shewanella* species were isolated from the deep-sea because this genus is widespread in aquatic and marine environments (DeLong 1997). *Shewanella* is divided into two groups: Group 1 harboring psychrophilic or psychrotolerant piezophiles and producing eicosapentanoic acid (EPA) and comprises the species: *Shewanella benthica*, *Shewanella piezotolerans*, *Shewanella profunda* and *Shewanella violacea* and Group 2 comprising the mesophilic pressure sensitive species, producing no or only small amounts of EPA (Kato and Nogi 2001). *S. benthica* has been isolated from different sites from which at least eight strains have been described: KT99, F1A (Wirsen et al. 1987), DB6101, DB5501, DB6705, DB6909 (Kato et al. 1995), DB172R and DB172F (Kato et al. 1996), and DB21MT-2 (Kato et al. 1998). *S. benthica* KT99 was the first *Shewanella* species that was found in the Southern hemisphere and acquired from a depth of 9856 m with an in situ temperature of 1.8 °C in the Kermadec Trench (southwest Pacific Ocean).

S. piezotolerans WP2T and WP3T were isolated from deep-sea sediment located in the west Pacific (Xiao et al. 2007). These strains are facultative anaerobic rods, Gram-negative and motile. They are able to grow from 0.1 to 50 MPa with an optimal growth at 20 MPa at 10–15 °C for WP2 and 15–20 °C for WP3. *S. profunda* LT13a was isolated from a deep sediment sample in the Nankai Trough and has an optimal growth at 10 MPa at 25–30 °C Toffin et al. (2004). *S. violacea* DSS12 was isolated from the Ryukyu Trench at 5110 m. This strain is a Gram-negative rod using a polar flagellum to move and able to produce a violet pigment. This strain can grow from 0.1 to 70 MPa and from 4 to 15 °C with an optimal growth at 30 MPa and 8 °C (Kato et al. 1995). It has been found that the d-type cytochrome, a respiratory protein, plays a crucial role for the growth at HHP of this strain (Tamegai et al. 2011).

8.3.1.8 Photobacterium Genus

Photobacterium is a genus of marine bacteria found in all oceans (Ruby and Nealson 1978). Two strains from the species *Photobacterium profundum* are piezophilic: DSJ4 and SS9. *P. profundum* DSJ4 was also isolated from the Ryukyu trench at 5110 m. This strain grows optimally at 10 MPa and 10 °C (Nogi et al. 1998b). *P. profundum* SS9 is one of the piezo-psychrophilic bacteria most described with more than 20 publications and the first of which the genome has been sequenced (Vezzi et al. 2005). Optimum growth parameters of this strain are 25 MPa and 15 °C (Bartlett et al. 2008).

8.3.1.9 Profundimonas Genus

Profundimonas piezophila strain YC-1 is a psychro-piezophilic member of the Oceanospirillales (Gammaproteobacteria) (Cao et al. 2014). This strain changes its morphology depending on the hydrostatic pressure. *P. piezophila* strain YC-1 grows optimal at 50 MPa at 8 °C. It is unable to grow below 20 MPa or above 80 MPa. MUFAs are the predominant fatty acid of its membrane.

8.3.1.10 Piezophilic Bacteria and Archaea from Deep-Sea Hydrothermal Vents

More than one hundred species of Bacteria (17 phyla, 72 genera, and 113 species) and Archaea (3 phyla, 17 genera, and 62 species) have been isolated from deep-sea hydrothermal vents (Jebbar et al. 2015). However, in contrast to the cold ocean only a few of the microorganisms that have been isolated and characterized are in fact piezophilic (Fig. 8.2b).

8.3.1.11 Proteobacteria

Amongst the bacterial phylum of Proteobacteria, only four piezophilic species were isolated from hydrothermal vents, *Piezobacter thermophilus* 108, *Desulfovibrio hydrothermalis* AM13, *Desulfovibrio piezophilus,* and *Thioprofundum lithtotrophica* 106 (Alazard et al. 2003; Khelaifia et al. 2011; Takai et al. 2009).

P. thermophilus 108 was isolated from a black smoker chimney of the TAG field in the Mid-Atlantic Ridge (MAR) (Takai et al. 2009). This strain is a facultative anaerobic and chemolithoautotrophic, piezophilic, thermophilic, and neutrophilic marine bacterium. The genus *Piezobacter* is related to members of the family Rhodobacteraceae within the Alphaproteobacteria. This non-motile, short oval rod can grow optimally at 50 °C, 35 MPa, at pH 6.5–7.0 at an NaCl concentration of 2 %.

D. hydrothermalis AM13^{T} was isolated from a hydrothermal chimney at a depth of 2600 m at 13° N from the East Pacific Rise (Alazard et al. 2003). The organism is motile, Gram-negative, vibrio-shaped or sigmoid. It grows optimal at 35 °C, at 26 MPa and at pH 7.8.

T. lithtotrophica 106 was isolated from a black smoker chimney of the TAG field in the MAR (Takai et al. 2009). This strain is a facultative anaerobic, thermophilic, and piezophilic rod that grows optimal at 50 °C, pH 7, and 15 MPa. The genus *Thioprofundum* is related to the endosymbionts of the deep-sea animals within the Gammaproteobacteria.

8.3.1.12 Firmicutes

The only piezophilic species isolated from deep-sea hydrothermal vents that belongs to the Firmicutes order is *Anoxybacter fermentans* DY22613 (Zeng et al. 2015). The type strain, DY22613^T was isolated from a hydrothermal sulfide sample collected from an East Pacific Ocean hydrothermal field at a depth of 2891 m (Zeng et al. 2015). The organism is motile and the cells are round-ended rods with flagella. The organism grows optimal at 60 °C, 20 MPa, pH 7, and 35 g l^{-1} of sea salts. *A. fermentans* DY22613 reduces insoluble Fe(III) compounds, including amorphous Fe(III) oxyhydroxide (pH 7.0), amorphous iron (III) oxide (pH 9.0), goethite (α-FeOOH; pH 12.0), and Fe(III) citrate to Fe(II).

8.3.1.13 Thermotogales

The bacterial order of Thermotogales is broadly represented in deep biosphere ecosystems, especially in deep-sea hydrothermal vents. This order consists of mesophilic and hyperthermophilic, anaerobic, chemoorganotrophic microorganisms (Bonch-Osmolovskaya 2008; Nesbø et al. 2012). Among this order, the only three known piezophilic species are *Thermosipho japonicus* IHB1^T, *Marinitoga piezophila* KA3^T, and *Kosmotoga pacifica* SLHLJ1^T. *T. japonicus* IHB1^T is a Gram-negative rod with optimal growth at 72 °C, pH 7.0–7.5, 20 MPa, and 40 g l^{-1} of sea salts. The type species was isolated from a deep-sea hydrothermal vent chimney at the Iheya Basin, in the Okinawa area, Japan (Takai and Horikoshi 2000). *M. piezophila* KA3^T (Alain et al. 2002) (= DSM 14283T = JCM11233T) was isolated from a hydrothermal vent sample, under 30 MPa hydrostatic pressure at 65 °C, from the East Pacific Ridge, 13° N, at a depth of 2630 m. This strain is an obligate anaerobe, whose optimal growth is at 30 g l^{-1} of sea salts, pH 6.0, 40 MPa, and 65 °C. Under optimal conditions the cells appear as short rods (1–1.5 μm long, 0.5 μm wide), single, or in short chains of fewer than five cells. This organism shows cellular deformations, becoming twisted and elongated, when cultured at low hydrostatic pressure. It is the only known piezophilic microorganism available in pure culture that harbors a provirus in its genome (Lossouarn et al. 2015). High-pressure conditions do not seem to affect the virus production. A virus-mediated plasmid exchange was reported presenting the first molecular piracy system in deep-sea hydrothermal vents. *K. pacifica* SLHLJ1^T was isolated from sediments of an active hydrothermal vent on the East Pacific Rise (L'Haridon et al. 2014). Cells are Gram-negative non-motile short rods or ovoid cocci. Optimal growth occurs at 70 °C and 30 MPa (unpublished data). Growth occurs under strictly anaerobic and obligate chemoorganoheterotrophic conditions.

8.3.2 Archaea

Methanocaldococcus jannaschii JAL-1^T renamed by Stackebrandt et al. (2002), was formally described as *M. jannaschii* and isolated from sedimentary material from a "white smoker" submarine hydrothermal vent, East Pacific Rise, at a depth of 2600 m (Jones et al. 1983). This strain has been reported to grow between 50 °C (Jones et al. 1983) and 90 °C (Miller et al. 1988). In previous experiments employing hyperbaric helium, *M. jannaschii* exhibited faster growth at 86 and 90 ° C with increasing pressure up to 75 MPa (Miller et al. 1988). These cocci are osmotically fragile and easily lysed in solutions of low osmolality. Optimal growth occurs at 85 °C and pH 6.0. Being an obligate anaerobe, the organism grows autotrophically using $H_2 + CO_2$ as substrate while forming CH_4.

Methanothermococcus thermolithotrophicus is an autotrophic thermophilic motile coccoid methanogen and was isolated from geothermally heated sea sediments close to Naples, Italy. Growth occurs at the expense of H_2/CO_2 or formate and the organism grows optimally at 65 °C, 4 % NaCl (Huber et al. 1982). High hydrostatic pressure of up to 50 MPa increased the growth rate without extending the temperature range of viability (Bernhardt et al. 1988).

Methanopyrus kandleri 116 is a hyperthermophilic methanogen that is characterized by growth under high temperature and hydrostatic pressures. Elevated hydrostatic pressure extends the temperature maximum for proliferation from 116 ° C at 0.4 MPa to 122 °C at 20 MPa, providing the potential for growth even at 122 °C under an in situ high pressure. This strain was isolated from a water column sample of Aleutian Trench, at 2500 m depth (Takai et al. 2008).

Palaeococcus ferrophilus DMJ^T was isolated from a black smoker sample obtained from the Myojin Knoll in the Ogasawara-Bonin Arc, Japan, at a depth of 1,338 m (Takai et al. 2000). The irregular cocci grow optimally at 83 °C, pH 6.0, and a sea salts concentration of 43 g l^{-1} at 30 MPa.

Palaeococcus pacificus DY20341^T was isolated from a sediment sample collected from an East Pacific Ocean hydrothermal field at a depth of 2737 m (Zeng et al. 2013). Cells are irregular cocci and grow optimally at 80 °C, pH 7.0, 3.0 % NaCl, and at 30 MPa.

Pyrococcus abyssi GE5 was isolated from hot fluid from an active chimney at a newly discovered deep-sea vent with a temperature up to 296 °C at 2000 m depth in the North Fiji Basin (SW Pacific) (Marteinsson et al. 1997). Cells are Gram-negative slightly irregular cocci. Elevated hydrostatic pressure (20 MPa) increased the maximal growth temperature of strain GE5 by at least 3 °C, i.e., from 102 °C (determined under about 0.2 MPa of N_2) to 105 °C.

Pyrococcus yayanosii CH1^T was isolated from an active deep-sea hydrothermal chimney at the mid-Atlantic Ridge (Ashadze site at 4100 m depth (Birrien et al. 2011). Optimal growth occurs at 95 °C, 52 MPa, at pH 7.5–8 and 3.5 % NaCl. No growth was recorded at any temperature at hydrostatic pressures below 20 MPa meaning that CH1 is the only anaerobic hyperthermophilic archaeon isolated so far to be strictly piezophilic. The pH and NaCl ranges for growth are 6–9.5 and 2.5–5.5 %, respectively.

Pyrococcus kukulkanii NCB100^T was isolated from the flange of an active deep-sea hydrothermal chimney at the depth of 1997 m at the Rebecca's Roost site on the Guaymas Basin (Callac et al., unpublished). The organism is a motile irregular rod with an optimal growth at 105 °C, pH 7.0, and an NaCl concentration of 2.5–3 %. The strain NCB100^T is also able to grow at 20 and 30 MPa at 110 °C.

Thermococcus aggregans TY^T is a coccoid Archaea isolated from marine sediments at the Guaymas Basin hydrothermal vent site (2000 m depth) in the Gulf of California (Canganella et al. 1998). This strain is a Gram-positive, non-motile, chemoorganotrophic, and strictly anaerobic organism. Optimum growth occurs at pH 7.0, 88 °C with an NaCl concentration of 2.5 %, and 20 MPa.

Thermococcus guaymasensis TYS^T was isolated from marine sediments at the Guaymas Basin hydrothermal vent site (2000 m depth) in the Gulf of California (Canganella et al. 1998). This coccoid Archaea is Gram-positive, non-motile, chemoorganotrophic, and strictly anaerobic. Optimum growth occurs at pH 7.2, 88 °C, 3.0 % NaCl, and 20–35 MPa.

Thermococcus peptonophilus OG-1^T was isolated from a marine hydrothermal vent in the western Pacific Ocean, in the Izu-Bonin and South Mariana Trough areas (González et al. 1995). The organism is a coccus, motile, strictly anaerobic with an optimum growth at 85 °C, pH 6, and 3 % NaCl. The results of HP-HT growth studies showed that the temperature for growth of *T. peptonophilus* could be increased under higher-pressure conditions and its growth profile appeared piezophilic (optimum hydrostatic pressure 30–50 MPa).

Thermococcus eurythermalis A501^T was isolated from the chimney sample of a deep-sea hydrothermal vent site located at a depth of 2007 m in the Guaymas Basin (Zhao and Xiao 2015). Cells are irregular cocci and motile. Optimal growth occurs at pH 7.0, at 2.5 % NaCl, at 85 °C, and with a hydrostatic pressure range of 0.1–70 MPa (optimal growth at 0.1–30 MPa), which narrows down to 0.1–50 MPa (optimal growth at 10 MPa) at 95 °C. This strain is strictly anaerobic and chemoorganotrophic.

Thermococcus barophilus MP^T was isolated from a chimney wall, under 40 MPa hydrostatic pressure at 95 °C, from the Mid-Atlantic Ridge Snakepit at a depth of 3550 m (Marteinsson et al. 1999b). Cells are regular to irregular cocci that are obligate anaerobic and chemoorganotrophic. Its optimal growth was obtained at 2–3 % NaCl, pH 7.0, at 85 °C, and 40 MPa. This type strain becomes a strict piezophile between 95 and 100 °C and requires 15.0–17.5 MPa in order to be able to grow at these temperatures.

8.4 Adaptations to HHP

The effect of high hydrostatic pressure on the microbial physiology was initially largely studied in mesophilic model microorganisms (*Escherichia coli* and *Saccharomyces cerevisiae*). This showed that high pressure affects most of the cellular functions (Bartlett 2002). Later, the exploration of these mechanisms was extended to a few piezophilic and piezotolerant microorganisms. Currently, *P. profundum* SS9, a moderate piezophilic and psychrophilic bacterium, represents, so far the most well studied model for the characterization of the molecular and physiological mechanisms involved in the response to HHP (Simonato et al. 2006).

The first function modified by high hydrostatic pressure is the transport of molecules through the membrane due to a loss of membrane fluidity. Furthermore, it seems that psychrophilic piezophilic strains have a larger genome size than their surface counterparts (DeLong et al. 2006; Eloe et al. 2011a; Konstantinidis et al. 2009; Lauro and Bartlett 2008; Thrash et al. 2014). A new method to counteract the need of culturing using HHP incubators is the single cell genomic technique (Lasken 2012; Lasken and McLean 2014). This technique permitted the sequencing and analyzing of four psychrophilic strains from the hadal zone (6000 m to the bottom of the ocean) belonging to Alpha- and Gammaproteobacteria, Bacteriodetes, and Planctomycetes (León-Zayas et al. 2015). Nitrosopumilus of the group I.1a of Thaumarchaeota, SAR11, Marinosulfonomonas, and Psychromonas are also groups found using the single amplified genomes (SAG) method in Puerto Rico Trench (PRT) at 8000 m (León-Zayas et al. 2015).

The first study that analyzed the evolutionary behavior of 20 psychro-piezophiles established that both piezophiles and their closest relative non-piezophiles share high similarity of their 16S rRNA gene sequences (Lauro et al. 2007). *Colwellia*, *Psychromonas,* and *Moritella* are genera found only in cold waters (Lauro et al. 2007). Cold piezophile representatives could be the descendants of a mesophilic strain.

One of the most important effects of HHP on microorganisms is the higher rigidity of the plasma membrane, which is caused by the compaction of the constituent lipids. DeLong and Yayanos (1985) have shown that deep-sea organisms harbor an unusually high proportion of mono- and polyunsaturated fatty acids. This leads to highly disordered phospholipid bilayers that are less permeable to water molecules. Hence, this maintains the membrane in a functional liquid crystalline state despite the effect of pressure. The genes responsible for unsaturated lipid synthesis are up regulated by HHP in the moderate piezophile *P. profundum* strain SS9 and induced as part of the HHP induced stress response in yeast.

Lipid membrane composition is different among psychro-piezophilic strains. Indeed, different long-chain polyunsaturated fatty acids (PUFAs) were found in different genera. *Photobacterium*, *Shewanella*, *Colwellia,* and *Moritella* piezophilic strains have a high proportion of eicosapentaenoic acid (EPA, C20:5) and docosahexaenoic acid (C22:6) suggesting an important role of these PUFAs in the maintenance of membrane fluidity (DeLong et al. 1997). Nevertheless, PUFAs are not required for high-pressure growth in *P. profundum* SS9, even though these fatty acids are produced in larger amounts, in contrast to the monounsaturated fatty acid (MUFA) cis-vaccenic acid (C18:1) (Allen et al. 1999). However, in *S. violacea* strain DSS12, EPA is required for high-pressure growth (Usui et al. 2012).

As a confirmation that the membrane plays an important role in HHP adaption *P. profundum* SS9 possesses two distinct flagella, polar and lateral. These flagella show fine-tuning of the high-pressure adaptation. Considering that motility is one of the most pressure sensitive cellular processes in mesophilic microorganisms, it does not come as a surprise that piezophilic bacteria possess uniquely adapted motility systems to maintain movement under high pressure (Eloe et al. 2008). Indeed, the lateral flagella seem to be activated when the cell is placed under HHP conditions. One of the hypotheses is that contrary to polar flagella lateral flagella would allow the cells to migrate on a surface. In deep-sea environments where the amount of organic carbon is low microorganisms may attach to particles or animals, such as the amphipod from which *P. profundum* SS9 was isolated (Nogi et al. 1998b). Similarly to *P. profundum* SS9, the piezotolerant *S. piezotolerans* WP3 also possesses two types of flagella. However, unlike *P. profundum* SS9, the genes coding for the two flagella in *S. piezotolerans* WP3 are differentially expressed as a function of temperature and pressure. The genes encoding the lateral flagella are overexpressed at low temperature (4 vs. 20 °C) and under-expressed at high pressure, while it is the opposite for the polar flagella (Wang et al. 2008). This shows that even though both organisms do not have the same regulating mechanism, they appear to have developed mobility strategies enabling them to adapt to their environment, especially HHP. Interestingly, in *T. barophilus*, a piezophilic Archaea, which only possesses one type of flagella (polar one), the genes responsible for its formation are down regulated at atmospheric pressure, which is considered to be stressful for this organism (Vannier et al. 2015).

Another cellular process that is strongly affected by HHP in piezophilic microorganisms is respiration. Most piezophilic bacteria studied are aerobes or facultative anaerobes and possesses cytochromes that are affected by HHP. As an example, *S. benthica* possesses two respiratory systems, one that is activated at low hydrostatic pressure and the other under HHP conditions. These two respiratory systems involve three enzymatic complexes (NADH-dehydrogenase, a bc1 complex, and a cytochrome c oxidase). At low pressure the NADH-dehydrogenase oxidizes NADH to NAD resulting in two electrons that are transferred to the bc1 complex by a quinone Q. The transfer by the bc1 complex of electrons to a membrane-bound cytochrome (c-551) gives rise to the formation of a proton gradient across the membrane. The subsequent transfer of electrons to the cytochrome c oxidase will also result in a proton gradient due to the reduction of oxygen (Kato and Qureshi 1999) (Fig. 8.3). Under HHP conditions, the bc1 complex and the cytochrome c oxidase are under-expressed, whereas the quinol oxidase is overexpressed (Qureshi et al. 1998; Tamegai et al. 1998). In this case, the electrons coming from the NADH-dehydrogenase are transferred by quinone Q to quinol oxidase, which possesses the same function as the cytochrome c (Kato and Qureshi 1999) (Fig. 8.3). *S. violacea* DSS 12 seems to possess another adaptive system and

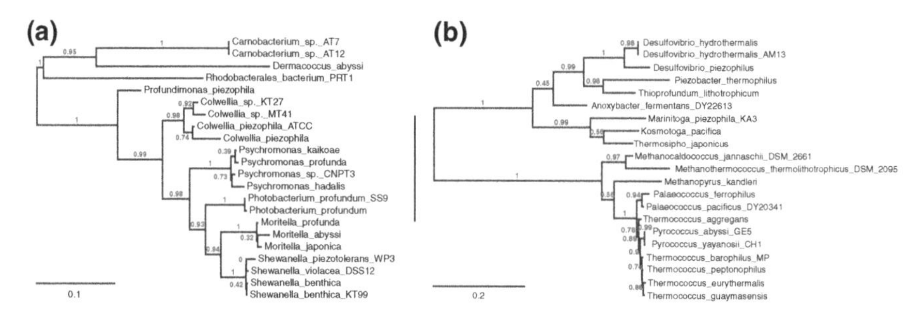

Fig. 8.3 Electron transport system in the piezophilic *Shewanella benthica* (Kato and Qureshi 1999). Proposed electron transport system in piezophilic *Shewanella benthic*a strain DB172F grown at 0.1 MPa (**a**) and 60 MPa (**b**). Abbreviations used are: *Q* quinone; *QH2* quinol; *C-551* cytochrome c-551; *C-552* cytochrome c-552

expresses a specific cytochrome protein complex at high pressure (Tamegai et al. 1998). This suggests that *S. violacea* DSS 12 adapts its respiratory mechanism during growth. Under HHP, the cytosolic cytochrome c is repressed while the membrane-bound cytochrome d is expressed. This suggests a shift in the electron transport chain and the respiration system even though there is no difference in oxygen consumption (Chikuma et al. 2007). For *P. profundum* SS9, a higher piezotolerant terminal oxidase activity was observed although, contrary to *Shewanella,* pressure did not alter the cytochrome content and expression of the genes encoding terminal oxidases, suggesting different adaptation mechanisms by these two species (Tamegai et al. 2011). piezophilic Archaea such as *T. barophilus* hydrogenases and sulfhydrogenases are the main constituents of anaerobic respiration. A transcriptomic study has shown that when *T. barophilus* is placed under stressful conditions, compared to the optimum pressure of 40 MPa, numerous genes coding for hydrogenases and sulfhydrogenases are overexpressed suggesting that the rigidity of the membrane affects the activity of theses enzymes and therefore an increase of these enzymes may be one way to counterbalance the loss of activity (Vannier et al. 2015). These authors proposed that adaptation to HHP mainly involves the fine-tuning of gene expression and protein synthesis of a common gene set, or function, (e.g., energy conservation, amino acids metabolism) in piezophiles.

Several membrane proteins whose regulation changes drastically under different hydrostatic pressure such as the proteins ToxR/S and OmpL/OmpH have been studied (Welch and Bartlett 1996, 1998). The OmpL/OmpH porines-like proteins are differentially expressed under HHP. It has been suggested that OmpH, which is expressed under HHP, could serve as a nutrient transporter in the deep sea (Oger and Jebbar 2010). In *P. profundum* SS9 transporters are mainly up regulated at 0.1 MPa (compared to the pressure optimum of 28 MPa). It was speculated that SS9 transporters might have evolved a particular protein structure to adapt to elevated pressure. The up-regulation of these transporters at 0.1 MPa could compensate for a decrease of functionality (Campanaro et al. 2005; Vezzi et al. 2005). However, because the cells possess several copies of the same transporter and since some of them are down regulated, it is difficult to generalize. Similarly, in *T. barophilus* or *D. piezophilus*, a piezophilic sulfate-reducing anaerobe, many transporter genes were up regulated under stressful pressure (Pradel et al. 2013; Vannier et al. 2015). Among those, many are involved in the transport and the metabolism of amino acids, like those transporting glutamine in *D. piezophilus* or *D. hydrothermalis* (Amrani et al. 2014). Indeed, the regulation of the transporters under HHP seems to be correlated with the capacity of the cell to produce amino acids and to take up substrate inside its cytoplasm. The first example studied was a strain of the piezotolerant eukaryote *S. cerevisiae* that lost the capacity to synthesize tryptophan. Abe and Horikoshi (2000) showed that growth of *S. cerevisiae* was greatly affected by HHP and that the addition of a plasmid containing the TAT2 gene coding for a high-affinity tryptophan permease enabled growth under HHP. HHP has also a major impact on amino acid metabolism in *T. barophilus*, which increased from three amino acids required at atmospheric pressure to 17 at 40 MPa, the optimal pressure for growth in this organism (Cario et al. 2015).

8.5 Conclusions and Trends

In order to decipher the universal HHP adaptation mechanisms from genetic, genomic, evolutionary, and physiological point of view it is mandatory to put more efforts on the isolation and characterization of new piezophilic microorganisms possessing a larger diversity of metabolic features. The phylogenetic and metabolic diversity of piezophilic microorganisms isolated so far from both cold and hot deep-sea environments are limited and do not at all reflect the large metabolic diversity that is expected in the deep biosphere. Autochthonous microorganisms of the deep biosphere are inherently "adapted" to the extreme conditions of their environment, i.e., to the high-pressure, low- or high temperature, low or high pH, anoxic or aerobic, oligotrophic conditions found throughout deep-sea ecosystems. In order to isolate new piezophiles from deep-sea ecosystems, it requires (i) development of adapted and specialized high-pressure equipment to retain and transfer samples at in situ pressure and temperature, (ii) enhance knowledge and imagination about the geochemistry of the environment in which the organisms thrive and to recreate these conditions in the laboratory, and (iii) to develop high-throughput culture equipment that works under a large range of temperature and hydrostatic pressure. All these requirements are mandatory when one wants to grow and isolate true piezophilic microorganisms. It has been shown that piezophiles do not tolerate atmospheric pressures or at best only for a short period of time. Piezophiles can be resilient when they are kept under good growth conditions i.e., temperature but not kept under their in situ hydrostatic pressure during sample collection or enrichments (Marteinsson et al. 1997, 1999a). Indeed, Yayanos (1986) speculated that microorganisms collected from deeper than 6000 m would be resilient and the true piezophiles may thrive at this and greater depth. Indeed, all piezophilic psychrophiles were isolated from depths greater than 4000 m and obligate piezophiles deeper than 6000 m (Prieur 1997). For thermophilic piezophiles, it has been established that increasing temperature permits a higher optimum HHP. Indeed. *T. barophilus*, which for long time was the only known true thermophilic piezophile, originated from 3550 m depth and is an obligate piezophile at 95 °C (Marteinsson et al. 1999b). At that time this was the greatest depth of known hydrothermal vents and the hypothesis of Yayanos (1986) was verified a few years later for thermophilic piezophiles. Indeed, the discovery of *P. yayanosi* CH1 originated from 4200 m and this organism is an obligate piezophilic hyperthermophile (Zeng et al. 2009) and this confirms that true piezophiles are found below 4000 m. Hence, it is obvious that depth of sampling, the in situ pressure of the microorganism, and containments of the sample conditions are crucial criteria for successful isolation of piezophiles.

Acknowledgments The research leading to these results has received funding from the European Union Seventh Framework Programme (FP7/2007-2013) under grant agreement n° 311975. This publication reflects the views only of the author, and the European Union cannot be held responsible for any use which may be made of the information contained therein.

References

Abe F, Horikoshi K (2000) Tryptophan permease gene TAT2 confers high-pressure growth in *Saccharomyces cerevisiae*. Mol Cell Biol 20:8093–8102

Alain K, Marteinsson VT, Miroshnichenko ML et al (2002) *Marinitoga piezophila* sp nov., a rod-shaped, thermo-piezophilic bacterium isolated under high hydrostatic pressure from a deep-sea hydrothermal vent. Int J Syst Evol Microbiol 52:1331–1339

Alazard D, Dukan S, Urios A et al (2003) *Desulfovibrio hydrothermalis* sp. nov., a novel sulfate-reducing bacterium isolated from hydrothermal vents. Int J Syst Evol Microbiol 53:173–178

Allen EE, Facciotti D, Bartlett DH (1999) Monounsaturated but not polyunsaturated fatty acids are required for growth of the deep-sea bacterium *Photobacterium profundum* SS9 at high pressure and low temperature. Appl Environ Microbiol 65:1710–1720

Amrani A, Bergon A, Holota H et al (2014) Transcriptomics reveal several gene expression patterns in the Piezophile *Desulfovibrio hydrothermalis* in response to hydrostatic pressure. PLoS ONE 9(9):e106831

Bartlett DH (2002) Pressure effects on in vivo microbial processes. Biochim et Biophys Acta-Protein Struct Mol Enzymol 1595:367–381

Bartlett DH, Ferguson G, Valle G (2008) Adaptations of the psychrotolerant piezophile *Photobacterium profundum* strain SS9. High-Pressure Microbiol 319–337

Bernhardt G, Jaenicke R, Lüdemann HD et al (1988) High pressure enhances the growth rate of the thermophilic archaebacterium *Methanococcus thermolithotrophicus* without extending its temperature range. Appl Env Microbiol 54:1258–1261

Birrien JL, Zeng X, Jebbar M et al (2011) *Pyrococcus yayanosii* sp nov., an obligate piezophilic hyperthermophilic archaeon isolated from a deep-sea hydrothermal vent. Int J Syst Evol Microbiol 61:2827–2831

Bonch-Osmolovskaya EA (2008) Thermotogales. In Encyclopedia of Life Sciences (ELS). Wiley, Chichester

Campanaro S, Vezzi A, Vitulo N et al (2005) Laterally transferred elements and high pressure adaptation in *Photobacterium profundum* strains. BMC Genom 6:122

Canganella F, Jones WJ, Gambacorta A, Antranikian G (1998) *Thermococcus guaymasensis* sp. nov. and *Thermococcus aggregans* sp. nov., two novel thermophilic archaea isolated from the Guaymas Basin hydrothermal vent site. Int J Syst Evol Bacteriol 48(4):1181–1185

Cao Y, Chastain RA, Eloe EA et al (2014) Novel psychropiezophilic *Oceanospirillales* species *Profundimonas piezophila* gen. nov., sp. nov., isolated from the deep-sea environment of the Puerto Rico trench. Appl Environ Microbiol 80:54–60

Cario A, Lormières F, Xiang X, Oger P (2015) High hydrostatic pressure increases amino acid requirements in the piezo-hyperthermophilic archaeon *Thermococcus barophilus*. Res Microbiol 166:710–716

Certes A (1884) Sur la culture, a l'abri des germes atmospheriques, des eaux et des sediments rapportes par les expeditions du Travailleur et du Talisman; 1882–1883. Compt Rend Acad Sci 98:690–693

Chikuma S, Kasahara R, Kato C, Tamegai H (2007) Bacterial adaptation to high pressure: a respiratory system in the deep-sea bacterium *Shewanella violacea* DSS12. FEMS Microbiol Lett 267:108–112

Ciobanu M-C, Burgaud G, Dufresne A et al (2014) Microorganisms persist at record depths in the subseafloor of the Canterbury Basin. ISME J 8:1370–1380

Corliss JB, Ballard RD (1977) Oases of life in cold abyss. Nat Geogr 152:441–453

DeLong EF (1997) Marine microbial diversity: the tip of the iceberg. Trends Biotechnol 15:203–207

DeLong EF, Yayanos AA (1985) Adaptation of the membrane-lipids of a deep-sea bacterium to changes in hydrostatic-pressure. Science 228:1101–1102

DeLong EF, Franks DG, Yayanos AA (1997) Evolutionary relationships of cultivated psychrophilic and barophilic deep-sea bacteria. Appl Environ Microbiol 63:2105–2108

DeLong EF, Preston CM, Mincer T et al (2006) Community genomics among stratified microbial assemblages in the ocean's interior. Science 311:496–503

Deming JW, Somers LK, Straube WL et al (1988) Isolation of an obligately barophilic bacterium and description of a new genus, *Colwellia* gen-nov. Syst Appl Microbiol 10:152–160

Dereeper A, Guignon V, Blanc G, Audic S, Buffet S, Chevenet F, Dufayard JF, Guindon S, Lefort V, Lescot M, Claverie JM, Gascuel O (2008) Phylogeny.fr: robust phylogenetic analysis for the non-specialist. Nucleic Acids Res 36(Web Server issue):W465–W469

Eloe EA, Lauro FM, Vogel RF, Bartlett DH (2008) The deep-sea bacterium *Photobacterium profundum* SS9 utilizes separate flagellar systems for swimming and swarming under high-pressure conditions. Appl Environ Microbiol 74:6298–6305

Eloe EA, Fadrosh DW, Novotny M et al (2011a) Going deeper: metagenome of a hadopelagic microbial community. PLoS ONE 6:e20388

Eloe EA, Malfatti F, Gutierrez J et al (2011b) Isolation and characterization of a psychropiezophilic alphaproteobacterium. Appl Environ Microbiol 77:8145–8153

Euzeby J (2013) List of prokaryotic names with standing in nomenclature-Genus. *Staphylococcus*. http://www.bacterio.cict.fr. Accessed April 2010

González JM, Kato C, Horikoshi K (1995) *Thermococcus peptonophilus* sp. nov., a fast-growing, extremely thermophilic archaebacterium isolated from deep-sea hydrothermal vents. Arch Microbiol 164:159–164

Hammond P (1995) Described and estimated species numbers: an objective assessment of current knowledge. Microb Divers Ecosyst Funct 29–71

Huber H, Thomm M, Knig H et al (1982) *Methanococcus thermolithotrophicus*, a novel thermophilic lithotrophic methanogen. Arch Microbiol 132:47–50

Inagaki F, Hinrichs K-U, Kubo Y et al (2015) Exploring deep microbial life in coal-bearing sediment down to ~2.5 km below the ocean floor. Science 349:420–424

Jannasch HW, Taylor CD (1984) Deep-sea microbiology. Ann Rev Microbiol 38:487–514

Jebbar M, Franzetti B, Girard E, Oger P (2015) Microbial diversity and adaptation to high hydrostatic pressure in deep-sea hydrothermal vents prokaryotes. Extremophiles 1–20

Jones WJ, Leigh JA, Mayer F et al (1983) *Methanococcus jannaschii* sp. nov., an extremely thermophilic methanogen from a submarine hydrothermal vent. Arch Microbiol 136:254–261
Kato C (2006) Handling of piezophilic microorganisms. Methods Microbiol 35:733–741
Kato C, Nogi Y (2001) Correlation between phylogenetic structure and function: examples from deep-sea *Shewanella*. FEMS Microbiol Ecol 35:223–230
Kato C, Qureshi MH (1999) Pressure response in deep-sea piezophilic bacteria. J Mol Microbiol Biotechnol 1:87–92
Kato C, Sato T, Horikoshi K (1995) Isolation and properties of barophilic and barotolerant bacteria from deep-sea mud samples. Biodivers Conserv 4:1–9
Kato C, Inoue A, Horikoshi K (1996) Isolating and characterizing deep-sea marine microorganisms. Trends Biotechnol 14:6–12
Kato C, Li L, Nogi Y et al (1998) Extremely barophilic bacteria isolated from the Mariana Trench, Challenger Deep, at a depth of 11,000 meters. Appl Environ Microbiol 64:1510–1513
Khelaifia S, Fardeau ML, Pradel N et al (2011) *Desulfovibrio piezophilus* sp nov., a piezophilic, sulfate-reducing bacterium isolated from wood falls in the Mediterranean Sea. Int J Syst Evol Microbiol 61:2706–2711
Konstantinidis KT, Braff J, Karl DM, DeLong EF (2009) Comparative metagenomic analysis of a microbial community residing at a depth of 4,000 meters at station ALOHA in the North Pacific subtropical gyre. Appl Environ Microbiol 75:5345–5355. doi:10.1128/AEM.00473-09
L'Haridon S, Jiang L, Alain K et al (2014) *Kosmotoga pacifica* sp. nov., a thermophilic chemoorganoheterotrophic bacterium isolated from an East Pacific hydrothermal sediment. Extremophiles 18:81–88
Lasken RS (2012) Genomic sequencing of uncultured microorganisms from single cells. Nat Rev Microbiol 10:631–640
Lasken RS, McLean JS (2014) Recent advances in genomic DNA sequencing of microbial species from single cells. Nat Rev Genet 15:577–584
Lauro FM, Bartlett DH (2008) Prokaryotic lifestyles in deep sea habitats. Extremophiles 12:15–25
Lauro FM, Chastain RA, Blankenship LE et al (2007) The unique 16S rRNA genes of piezophiles reflect both phylogeny and adaptation. Appl Environ Microbiol 73:838–845
León-Zayas R, Novotny M, Podell S et al (2015) Single cells within the Puerto Rico trench suggest hadal adaptation of microbial lineages. Appl Environ Microbiol 81:8265–8276
Lossouarn J, Nesbø CL, Mercier C et al (2015) "Ménage à trois": a selfish genetic element uses a virus to propagate within *Thermotogales*. Environ Microbiol 17:3278–3288
Marteinsson VT, Moulin P, Birrien J et al (1997) Physiological responses to stress conditions and barophilic behavior of the hyperthermophilic vent archaeon *Pyrococcus abyssi*. Appl Environ Microbiol 63:1230–1236
Marteinsson VT, Reysenbach AL, Birrien JL, Prieur D (1999a) A stress protein is induced in the deep-sea barophilic hyperthermophile *Thermococcus barophilus* when grown under atmospheric pressure. Extremophiles 3:277–282
Marteinsson VT, Birrien JL, Reysenbach AL, Vernet M, Marie D, Gambacorta A, Messner P, Sleytr UB, Prieur D (1999b) *Thermococcus barophilus* sp. nov., a new barophilic and hyperthermophilic archaeon isolated under high hydrostatic pressure from a deep-sea hydrothermal vent. Int J Syst Evol Bacteriol 49:351–359
Miller JF, Shah NN, Nelson CM et al (1988) Pressure and temperature effects on growth and methane production of the extreme thermophile *Methanococcus jannaschii*. Appl Environ Microbiol 54:3039–3042
Morita R (1976) Survival of bacteria in cold and moderate hydrostatic pressure environments with special reference to psychrophilic and barophilic bacteria. Soc Gen Microbiol Symp Ser 17:279–298
Nesbø CL, Bradnan DM, Adebusuyi A et al (2012) *Mesotoga prima* gen. nov., sp. nov., the first described mesophilic species of the *Thermotogales*. Extremophiles 16:387–393
Nogi Y, Kato C (1999) Taxonomic studies of extremely barophilic bacteria isolated from the Mariana Trench and description of *Moritella yayanosii* sp. nov., a new barophilic bacterial isolate. Extremophiles 3:71–77

Nogi Y, Kato C, Horikoshi K (1998a) *Moritella japonica* sp. nov., a novel barophilic bacterium isolated from a Japan Trench sediment. J Gen Appl Microbiol 44:289–295
Nogi Y, Masui N, Kato C (1998b) *Photobacterium profundum* sp. nov., a new, moderately barophilic bacterial species isolated from a deep-sea sediment. Extremophiles 2:1–7
Nogi Y, Kato C, Horikoshi K (2002) *Psychromonas kaikoae* sp nov., a novel piezophilic bacterium from the deepest cold-seep sediments in the Japan Trench. Int J Syst Evol Microbiol 52:1527–1532
Nogi Y, Hosoya S, Kato C, Horikoshi K (2004) *Colwellia piezophila* sp nov., a novel piezophilic species from deep-sea sediments of the Japan Trench. Int J Syst Evol Microbiol 54:1627–1631
Nogi Y, Hosoya S, Kato C, Horikoshi K (2007) *Psychromonas hadalis* sp nov., a novel plezophilic bacterium isolated from the bottom of the Japan Trench. Int J Syst Evol Microbiol 57:1360–1364
Oger PM, Jebbar M (2010) The many ways of coping with pressure. Res Microbiol 161:799–809
Park CB, Clark DS (2002) Rupture of the cell envelope by decompression of the deep-sea methanogen *Methanococcus jannaschii*. Appl Environ Microbiol 68:1458–1463
Park CB, Boonyaratanakornkit BB, Clark DS (2006) Toward the large scale cultivation of hyperthermophiles at high-temperature and high-pressure. Methods Microbiol 35:109–126
Parkes RJ, Sellek G, Webster G et al (2009) Culturable prokaryotic diversity of deep, gas hydrate sediments: first use of a continuous high-pressure, anaerobic, enrichment and isolation system for subseafloor sediments (DeepIsoBUG). Environ Microbiol 11:3140–3153
Pathom-Aree W, Stach JE, Ward AC et al (2006) Diversity of actinomycetes isolated from Challenger Deep sediment (10,898 m) from the Mariana Trench. Extremophiles 10:181–189
Pradel N, Ji B, Gimenez G et al (2013) The first genomic and proteomic characterization of a deep-sea sulfate reducer: insights into the piezophilic lifestyle of *Desulfovibrio piezophilus*. PLoS ONE 8:e55130
Prieur D (1997) Microbiology of deep-sea hydrothermal vents. Trends Biotechnol 15:242–244. doi:10.1016/S0167-7799(97)01052-4
Prieur D, Marteinsson VT (1998) Prokaryotes living under elevated hydrostatic pressure. In Biotechnology of extremophiles. Springer, Heidelberg, pp 23–35
Quast C, Pruesse E, Yilmaz P, Gerken J, Schweer T, Yarza P, Peplies J, Glöckner FO (2013) The SILVA ribosomal RNA gene database project: improved data processing and web-based tools. Nucleic Acids Res 41:590–596
Qureshi MH, Kato C, Horikoshi K (1998) Purification of two pressure-regulated c-type cytochromes from a deep-sea barophilic bacterium, *Shewanella* sp. strain DB-172F. FEMS Microbiol Lett 161:301–309
Roussel EG, Bonavita M-AC, Querellou J et al (2008) Extending the sub-sea-floor biosphere. Science 320:1046
Ruby E, Nealson K (1978) Seasonal changes in the species composition of luminous bacteria in nearshore seawater. Limnol Oceanogr 23:530–533
Simonato F, Campanaro S, Lauro FM et al (2006) Piezophilic adaptation: a genomic point of view. J Biotechnol 126:11–25
Stackebrandt E, Frederiksen W, Garrity GM et al (2002) Report of the ad hoc committee for the re-evaluation of the species definition in bacteriology. Int J Syst Evol Microbiol 52:1043–1047
Takai K, Horikoshi K (2000) *Thermosipho japonicus* sp. nov., an extremely thermophilic bacterium isolated from a deep-sea hydrothermal vent in Japan. Extremophiles 4(1):9–17. PMID:10741832
Takai K, Sugai A, Itoh T, Horikoshi K (2000) *Palaeococcus ferrophilus* gen. nov., sp. nov., a barophilic, hyperthermophilic archaeon from a deep-sea hydrothermal vent chimney. Int J Syst Evol Microbiol 50:489–500
Takai K, Nakamura K, Toki T et al (2008) Cell proliferation at 122 °C and isotopically heavy CH_4 production by a hyperthermophilic methanogen under high-pressure cultivation. Proc Natl Acad Sci USA 31:10949–10954

Takai K, Miyazaki M, Hirayama H et al (2009) Isolation and physiological characterization of two novel, piezophilic, thermophilic chemolithoautotrophs from a deep-sea hydrothermal vent chimney. Environ Microbiol 11:1983–1997

Tamegai H, Kato C, Horikoshi K (1998) Pressure-regulated respiratory system in barotolerant bacterium, *Shewanella* sp. strain DSS 12. J Biochem Mol Biol Biophys 1:213–220

Tamegai H, Ota Y, Haga M et al (2011) Piezotolerance of the respiratory terminal oxidase activity of the piezophilic *Shewanella violacea* DSS12 as compared with non-piezophilic *Shewanella* species. Biosci Biotechnol Biochem 75:919–924

Thrash JC, Temperton B, Swan BK et al (2014) Single-cell enabled comparative genomics of a deep ocean SAR11 bathytype. ISME J 8:1440–1451

Toffin L, Bidault A, Pignet P et al (2004) *Shewanella profunda* sp. nov., isolated from deep marine sediment of the Nankai Trough. Int J Syst Evol Microbiol 54:1943–1949

Usui K, Hiraki T, Kawamoto J et al (2012) Eicosapentaenoic acid plays a role in stabilizing dynamic membrane structure in the deep-sea piezophile *Shewanella violacea*: a study employing high-pressure time-resolved fluorescence anisotropy measurement. Biochim Biophys Acta-Biomembr 1818:574–583

Vannier P, Michoud G, Oger P et al (2015) Genome expression of *Thermococcus barophilus* and *Thermococcus kodakarensis* in response to different hydrostatic pressure conditions. Res Microbiol 166:717–725

Vezzi A, Campanaro S, D'Angelo M et al (2005) Life at depth: *Photobacterium profundum* genome sequence and expression analysis. Science 307:1459–1461

Wang F, Wang J, Jian H et al. (2008) Environmental adaptation: genomic analysis of the piezotolerant and psychrotolerant deep-sea iron reducing bacterium *Shewanella piezotolerans* WP3. PLoS ONE 3(4):e1937

Welch TJ, Bartlett DH (1996) Isolation and characterization of the structural gene for OmpL, a pressure-regulated porin-like protein from the deep-sea bacterium *Photobacterium* species strain SS9. J Bacteriol 178:5027–5031

Welch TJ, Bartlett DH (1998) Identification of a regulatory protein required for pressure-responsive gene expression in the deep-sea bacterium *Photobacterium* species strain SS9. Mol Microbiol 27:977–985

Whitman WB, Coleman DC, Wiebe WJ (1998) Prokaryotes: the unseen majority. Proc Natl Acad Sci USA 95:6578–6583

Wirsen CO, Jannasch HW, Wakeham SG, Canuel EA (1987) Membranes lipids of a psychrophilic and barophilic deep-sea bacterium. Curr Microbiol 14:319–322

Xiao X, Wang P, Zeng X et al (2007) *Shewanella psychrophila* sp. nov. and Shewanella piezotolerans sp. nov., isolated from west Pacific deep-sea sediment. Int J Syst Evol Microbiol 57:60–65

Xu Y, Nogi Y, Kato C et al (2003a) *Psychromonas profunda* sp nov., a psychropiezophilic bacterium from deep Atlantic sediments. Int J Syst Evol Microbiol 53:527–532

Xu Y, Nogi Y, Kato C et al (2003b) *Moritella profunda* sp nov and *Moritella abyssi* sp nov., two psychropiezophilic organisms isolated from deep Atlantic sediments. Int J Syst Evol Microbiol 53:533–538

Yayanos AA (1986) Evolutional and ecological implications of the properties of deep-sea barophilic bacteria. Proc Natl Acad Sci USA 83:9542–9546

Yayanos A, DeLong EF (1987) Deep-sea bacterial fitness to environmental temperatures and pressure. In: Jannasch HW, Marquis RE, Zimmerman AM (eds) Current perspectives in high pressure biology. Academic Press, Toronto, pp 17–32

Yayanos AA, Dietz AS, Vanboxtel R (1979) Isolation of a deep sea barophilic bacterium and some of its growth-characteristics. Science 205:808–810

Yayanos AA, Dietz AS, Vanboxtel R (1981) Obligately barophilic bacterium from the Mariana Trench. Proc Natl Acad Sci USA 78:5212–5215

Zeng X, Birrien JL, Fouquet Y, Cherkashov G, Jebbar M, Querellou J, Oger P, Cambon-Bonavita MA, Xiao X, Prieur D (2009) *Pyrococcus* CH1, an obligate piezophilic hyperthermophile: extending the upper pressure-temperature limits for life. ISME J 3:873–876

Zeng X, Zhang X, Jiang L et al (2013) *Palaeococcus pacificus* sp. nov., a novel archaeon from a deep-sea hydrothermal sediment. Int J Syst Evol Microbiol 63:2155–2159

Zeng X, Zhang Z, Li X et al (2015) *Anoxybacter fermentans* gen. nov., sp. nov., a piezophilic, thermophilic, anaerobic, fermentative bacterium isolated from a deep-sea hydrothermal vent. Int J Syst Evol Microbiol 65:710–715

Zhang J, Sun Q, Zeng Z et al (2015a) Microbial diversity in the deep-sea sediments of Iheya North and Iheya Ridge, Okinawa Trough. Microbiol Res 177:43–52

Zhang Y, Li X, Bartlett DH, Xiao X (2015b) Current developments in marine microbiology: high-pressure biotechnology and the genetic engineering of piezophiles. Curr Opin Biotechnol 33:157–164

Zhao W, Xiao X (2015) Complete genome sequence of *Thermococcus eurythermalis* A501, a conditional piezophilic hyperthermophilic archaeon with a wide temperature range, isolated from an oil-immersed deep-sea hydrothermal chimney on Guaymas Basin. J Biotechnol 193:14–15

ZoBell CE, Johnson FH (1949) The influence of hydrostatic pressure on the growth and viability of terrestrial and marine bacteria. J Bacteriol 57:179–189

Chapter 9
Extreme Marine Environments (Brines, Seeps, and Smokers)

Francesca Mapelli, Elena Crotti, Francesco Molinari, Daniele Daffonchio and Sara Borin

Abstract Several different extreme environments—characterized by geochemical and physical extremes—are found in the ocean and in seas and many of them appeared to be hot spots for microbial abundance and diversity, thanks to the overwhelming presence of substrates and energy sources that support microbial metabolism. The most studied oceanic extreme environments are the vent ecosystems, such as the hot deep-sea hydrothermal vents (DSHVs) or cold seeps and mud volcanoes, and the hypersaline ecosystems such as the deep anoxic hypersaline lakes, brine lakes on mud volcanoes, and brines contained within sea ice. However, new fascinating extreme habitats for microbial life in the ocean are being discovered continuously such as water droplets entrapped in oil deposits. These environments comprise a large variety of extreme physicochemical conditions, which contribute importantly to the composition and shaping of the residing microbial communities and select for extremophile populations of microorganisms. These extremophiles are the key players of the element cycles in these environments, often responsible for primary productivity, and endemic. Many of the extremophiles have

F. Mapelli · E. Crotti · F. Molinari · D. Daffonchio · S. Borin (✉)
Department of Food, Environmental and Nutritional Sciences (DeFENS), University of Milan, 20133 Milan, Italy
e-mail: sara.borin@unimi.it

F. Mapelli
e-mail: francesca.mapelli@unimi.it

E. Crotti
e-mail: elena.crotti@unimi.it

F. Molinari
e-mail: francesco.molinari@unimi.it

D. Daffonchio
e-mail: daniele.daffonchio@unimi.it

D. Daffonchio
Biological and Environmental Sciences and Engineering Division, King Abdullah University of Science and Technology, Thuwal 23955-6900, Kingdom of Saudi Arabia

L.J. Stal and M.S. Cretoiu (eds.), *The Marine Microbiome*,
DOI 10.1007/978-3-319-33000-6_9

not yet been obtained in pure culture. The study of the microbiota associated with extreme marine environments confirmed that they constitute an important source of Bacteria and Archaea with biotechnological potential, producing enzymes and metabolites—"extremozymes" and "extremolytes"—that might find industrial application.

9.1 Extreme Marine Environments: Hot Spots of Microbial Diversity Possessing Biotechnological Potential

The ocean and adjacent seas host habitats characterized by geochemical and physical extremes at the edge of compatibility with life, which have been assigned as "extreme environments." The assignment "extreme" is anthropocentric since microbiologists have discovered that these environments host communities of specialized microorganisms, the so-called "extremophiles" (Macelroy 1974). Extremophiles are physiologically adapted to grow under what would be for humans and for many other organisms harsh physicochemical conditions. Many extreme marine environments are the result of a particularly tight geosphere–biosphere interaction. For instance, vents are environments in which gases and reduced chemicals from the deep subsurface geoactive layers reach the seafloor, and deep hypersaline anoxic lakes (DHALs) came into being when deeply stratified evaporitic rocks were exposed to seawater. Moreover, DHALs are environments separated from the overlying seawater. This separation is due to the difference in density of the brine and the overlying seawater that prevents their mixing. Some extreme environments are considered to be analogues for the conditions on early Earth and it has been hypothesized that life arose in environments such as hydrothermal vents (Cockell 2006; Shock 1992). The specific features of extreme environments contribute to the selection and enrichment of peculiar bacterial and archaeal communities that are highly adapted to the prevailing physicochemical conditions. Life is divided into three domains: the Bacteria, the Archaea, and the Eukarya. The former two have a simpler cell structure compared to the latter, but are metabolically highly diverse and able to use almost any form of chemical energy-rich compounds. Autotrophic microorganisms are pioneers that colonize the barren environments, because they utilize atmospheric gases as C and N sources, oxidize inorganic reduced chemicals as energy source through chemosynthesis (the so-called "black energy", which is crucial in dark ecosystems), and use alternative electron acceptors when oxygen is not available. All environments, even the most extreme in terms of physicochemical conditions, which provide sufficient amounts of reduced substrate as energy source, appear to be hot spots of microbial diversity and give rise to microbial communities with highly abundant populations of microorganisms. Therefore, the study of these extremophile microbial communities is of special interest in order to (i) understand the limits of life on our planet, (ii) speculate about the origin of life on Earth and elsewhere in the universe (astrobiology), and (iii) access to new biotechnological applications.

Despite the recent and remarkable technological and molecular advances in the description of microbial communities associated to extreme marine environments, the relevance of culture-dependent methods should not be underestimated (Joint et al. 2010; Prakash et al. 2013). Although culture-dependent methods have an important drawback when used to describe the biodiversity associated with a complex ecosystem, these methods are necessary to assess the microbial metabolic potential, the ecology of the microorganisms and their function in the geochemical element cycles of that particular habitat. Isolation and culturing of the microorganisms of a given ecosystem is also indispensable when the exploitation of microbes or the products they generate is the aim. Marine extremophile microorganisms have a great potential in industrial biotechnology because their metabolic and physiological adaptation to habitats characterized by extreme and harsh conditions hold promise for enzymes and metabolites that are active when those from other organisms fail. Extremophiles indeed withstand extreme physicochemical conditions such as low or high temperatures, low or high pH, high ionic strength, and high hydrostatic pressure. These adaptations are the result of a natural directed molecular evolution aiming at structural and chemical adaptations to the specific extreme environment. Hence, marine extremophile microorganisms may show improved efficiency for producing valuable products (mostly enzymes and metabolites) of practical interest for industrial processes. For this reason, the extensive study of the microbiology of extreme marine environments offers to biotechnologists new resources of inestimable potential.

9.2 Environmental Challenges and Microbial Diversity in Extreme Marine Habitats

9.2.1 Vent Environments

9.2.1.1 Hot Environments: Deep-Sea Hydrothermal Vents

DSHVs represent one of the most fascinating extreme environments on Earth. Since their discovery in late 70s, microbiologists have been attracted to this particular habitat aiming at investigating the microbial diversity associated to these highly productive ecosystems and characterizing the functioning of the entire system. In these oases of life on a deserted ocean floor, the primary role is played by microorganisms, which provide the basis for a rich ecosystem that supports also macroorganisms (Sievert and Vetriani 2012).

Hydrothermal vents originate from volcanic activity below the ocean floor. Water seeping through fissures of the Earth's crust becomes enriched with metals and minerals and, heated to high temperature, rises back through the ocean floor.

The contact of this fluid with cold and oxidized seawater results in the formation of hydrothermal plumes and in the precipitation of the dissolved minerals and metals, creating chimneys around the vents (Dick et al. 2013). DSHVs are globally distributed in many submarine habitats, such as mid-ocean ridges, back-arc basins, hot spot volcanoes, seamounts, and off-axis locations. Up to now, more than 500 hydrothermal vents have been discovered. The East Pacific Rise and the Mid-Atlantic Ridge are the preferential targets of scientific cruises (Godet et al. 2011). Unfortunately, a large part of the mid-ocean ridges, mainly localized in the Artic and Southern ocean, remains still unexplored (German et al. 2010).

Deep-sea vents influence the chemistry of global oceans considerably (Elderfield and Schultz 1996). They comprise a wide variety of niches and energy sources for microorganisms that through chemolithoautotrophy (chemosynthesis) and heterotrophy sustain benthic communities. Various endemic organisms such as giant tubeworms, shrimps, crabs, and fishes thrive in these environments, thanks to the existence of microbes and their metabolic activities (Sievert and Vetriani 2012). In these deep-sea vents, chemolithoautotrophic microorganisms produce organic matter from inorganic carbon sources (chemosynthetic CO_2 fixation), while obtaining energy from the oxidation of reduced inorganic compounds (Sievert and Vetriani 2012). The different reduced compounds in the hydrothermal fluid, such as H_2, H_2S, Fe^{2+}, CH_4, and NH_4^+, determine the microbial diversity (Takai and Nakamura 2011). Microbial life encompasses free-living organisms within the hydrothermal plume, biofilm communities attached to the sediment, and obligate symbionts within invertebrate hosts such as tubeworms (Takai et al. 2006). Steep physicochemical gradients of temperature, salinity, hydrostatic pressure, and availability of oxygen are the main drivers of the DSHV microbiota. Samples collected from DSHVs allowed the isolation of extremophile bacteria and archaea, including hyper/thermophiles, psychrophiles, acidophiles, piezophiles, and moderate halophiles (Jebbar et al. 2015). As an example, hyperthermophilic and barophilic microorganisms isolated from these environments recorded the upper limit for life ever detected so far in terms of high temperature (122 °C) and pressure (120 MPa), respectively (Takai et al. 2008; Zeng et al. 2009).

Although the accessibility of the sampling sites is extremely difficult, archaeal and bacterial communities associated to DSHVs have been thoroughly investigated (Huber et al. 2007; Sievert and Vetriani 2012). Epsilonproteobacteria and Aquificales are the predominant bacterial taxa in DSHVs and are abundantly present. These taxa are known for their versatile metabolism and therefore they are ubiquitously found in deep-sea vents worldwide. Epsilonproteobacteria oxidize hydrogen and sulfur compounds, reduce oxygen, nitrate, and sulfur compounds and use the reverse TCA (tricarboxylic acid) cycle for carbon fixation. They inhabit structures build of sulfide minerals and hydrothermal sediments and fluids. Moreover, Epsilonproteobacteria colonize the surface of deep-sea vent metazoans in episymbiotic associations. Endosymbiotic associations have been also described: an example is the association between Epsilonproteobacteria and a gastropod mollusk, *Alviniconcha hessleri* (Sievert and Vetriani 2012; Suzuki et al. 2006). Bacteria of the order Aquificales utilize electron donors and electron acceptors similar to those used by Epsilonproteobacteria. Aquificales are globally widespread in deep-sea vents, but differ in their distribution from Epsilonproteobacteria, which flourish at temperatures between 20 and 60 °C, while the former prefer higher temperatures, i.e, 60–80 °C. Other bacteria found in deep-sea vent environments include members of the Firmicutes, the Verrucomicrobia, the Thermales, and the Cytophaga–Flexibacter–Bacteroides (CFB) group (Takai et al. 2006).

Archaea in DSHVs comprise lineages such as the Thermococcales, Archaeoglobales, Desulfurococcales, Ignicoccales, Methanococcales, Methanopyrales, Halobacteriales, Thaumarchaeota, DHVE2 (Deep-sea Hydrothermal Vent Euryarchaeota group 2), and the MCG (Miscellaneous Crenarchaeota Group), which constitute the major part of the microbial communities in DSHVs (Jebbar et al. 2015; Takai et al. 2006; Takai and Nakamura 2011). DSHVs, together with freshwater habitats, have been pointed out as the main reservoirs of archaeal diversity worldwide (Auguet et al. 2010). In deep-sea vent ecosystems, Archaea colonize specific portions of the chimney structures, such as the inner parts of sulfide chimneys, as well as the fluids and sediments (Takai and Nakamura 2011).

Steep physicochemical gradients create in DSHVs a number of niches for microorganisms that possess peculiar characteristics that ultimately influence the richness and distribution of microbial communities. Anderson et al. (2015), using metagenomics, investigated the distribution of abundant and rare microorganisms associated to several hydrothermal vent systems. The vast majority of archaeal lineages were rare and, together with the bacterial ones, were geographically restricted. Abundant archaeal operational taxonomic units (OTUs) were widely distributed and many of them belonged to Marine Groups I and II. These microorganisms are typically found in the deep sea and are pointed out as organisms that are able to exploit deep-sea currents and colonize new DSHVs systems (Anderson et al. 2015).

The ability of microorganisms to form biofilm in deep-sea vent ecosystems has attracted much attention (Sievert and Vetriani 2012). At present, little is known about the ability to form biofilms in deep-sea vents, nor about the mechanisms that underpin this process. Biofilms have been observed in many niches in deep-sea vents, e.g., the white microbial mats on basalt close to deep-sea vents or the microbial matrices covering invertebrates. Attention has been also given to microbial genes expressed when growing as a biofilm, with particular reference to genes related to quorum sensing and to those involved in the production of polysaccharides (Campbell et al. 2009; Giovannelli et al. 2011; Nakagawa et al. 2007). However, whether or not a microbial life in a biofilm is advantageous compared to a planktonic lifestyle in these extreme environments is not known and has received little attention.

A biofilm growing on the surface of a black smoker chimney in the Loki's Castle Vent Field has been studied in order to unravel the functional interactions at the base of the colonizing microbial consortium (Stokke et al. 2015). Taking advantage of a multidisciplinary approach (based on microscopy and -omics techniques, including metagenomics, metatranscriptomics, and metaproteomics), Stokke et al. (2015) investigated the structure and the metabolic potential associated to the biofilm. In Loki's Castle Vent Field, lithoautotrophic Epsilonproteobacteria of the genus *Sulfurovum* produce long and sheathed threads (>100 μm) made up of stable sugar polymers, on which organotrophic rod-shaped Bacteroidetes grow as episymbionts. The filamentous nature of these organisms facilitates their attachment to the chimney wall, thereby contributing to the biofilm formation. This biofilm consortium of microorganisms represents a link between the primary producers (*Sulfurovum*) and consumers (Bacteroidetes). Moreover, the detection of an acetyl-CoA synthetase expressed by *Sulfurovum* suggests the ability of this organism to grow also heterotrophically, recycling organic matter through acetate assimilation. The metabolism of *Sulfurovum* and Bacteroidetes is also linked in regard to nitrogen cycling. *Sulfurovum* assimilates nitrogen mainly as ammonia, but both bacterial taxa cooperate in the denitrification process (Stokke et al. 2015).

In the marine environment the microbial contribution to the nitrogen cycle is highly relevant. In DSHVs, nitrate is present in the deep-seawater, while the hydrothermal fluids are deprived of it, resulting in a nitrate concentration gradient at the interface between these two habitats. Nitrate reductase is responsible for nitrate reduction. Besides the cytoplasmic nitrate reductase, which is not linked to respiratory electron transfer, this enzyme can be divided into two types, according to its affinity to nitrate and to its cellular localization: (i) the membrane-bound nitrate reductase (Nar), which is preferentially expressed in nitrate-rich environments, because it possesses a low affinity for nitrate and (ii) the nitrate reductase complex (Nap), located in the periplasmic space, which has a higher affinity for nitrate. DSHVs microorganisms benefit from a shorter doubling time and gain energy through the coupling of nitrate reduction to the oxidation of hydrogen, rather than using sulfur (Pérez-Rodríguez et al. 2013; Vetriani et al. 2004). Pérez-Rodríguez et al. (2013) found that in deep-sea vents the *nar* gene is likely expressed in temperate and less reducing environments, enriched with Alpha- and Gammaproteobacteria and with a nitrate concentration similar to that in deep-seawater. In vent fluids where Epsilonproteobacteria are abundant, the *nap* gene is expressed to overcome the low nitrate availability. Specifically, vent Epsilonproteobacteria have been shown to encode a conserved and widespread *nap* gene cluster (Vetriani et al. 2014). Furthermore, *nap* diversity in bacterial communities associated with deep-sea vents with different temperature and redox status has been investigated (Vetriani et al. 2014). The distribution of the *nap* gene showed that Epsilonproteobacteria were different in vents with moderate temperature when compared to those associated to black smokers, which are characterized by high temperatures, or associated to substrate-attached communities at low temperatures (Vetriani et al. 2014).

Hydrothermal plumes of deep-sea vents represent an environment that is special but yet little explored. Hydrothermal plumes have indeed received less attention in regard to the characterization of the microbial diversity and, consequently, to the ecological role of the microorganisms in this habitat (Dick et al. 2013). Plumes are formed by the injection of hydrothermal fluids into the cold seawater and can be transported even kilometers away from the vent (Dick et al. 2013). The study of the plume microbial diversity was aiming at the elucidation of the origin of its microbiome by evaluating as potential bank the plume itself, the surrounding water, and the seafloor and subsurface communities (Dick et al. 2013). The plume microbiome showed a dynamic behavior in terms of time and space, differing from those of the seafloor or subsurface, but also maintained some features in common with them (Dick et al. 2013; Dick and Tebo 2010; Lesniewski et al. 2012). For example, the microbial communities of Guaymas Basin hydrothermal plumes in the Gulf of California has been investigated in detail and compared to those of the background waters above the plume and in the nearby Carmen Basin (a basin located 100 km away from Guaymas Basin and without vent activity) (Lesniewski et al. 2012). The active microbiomes of plumes and background waters were similar, with the presence of the same groups of methanotrophs and chemolithoautotrophs as abundant microorganisms. Microorganisms typical of seafloor environments were not found. This led to the conclusion that microbial groups typically inhabiting the water column rather than the seafloor are active in these hydrothermal plumes (Lesniewski et al. 2012). Furthermore, plumes have been pointed out as vectors for dispersal of microbial communities between seafloor vents (Dick et al. 2013).

Several research groups spent much effort in the isolation of microorganisms from DSHVs, and as a result many archaeal and bacterial extremophiles have been obtained in pure culture (Takai and Nakamura 2011). The DSHV bacterial community is dominated by Epsilonproteobacteria and Aquificales and soon after their detection by molecular tools efforts to isolate members of these two groups were successful (reviewed by Sievert and Vetriani 2012). One of the first enrichments of Epsilonproteobacteria from DSHV samples was performed in 2001 (Campbell et al. 2001) and since then many different new Epsilonproteobacteria such as *Sulfurovum lithotrophicum*, *Thioreductor micantisoli*, *Nautilia lithotrophica*, and *Caminibacter profundus* have been isolated. From the Aquificales *Persephonella marina*, *Desulfurobacterium thermolithotrophum* and *Thermovibrio ammonificans* were obtained in culture (Sievert and Vetriani 2012), just to mention a few examples. Besides Epsilonproteobacteria and Aquificales several other bacteria have been isolated. The new anaerobic, piezophilic, and thermophilic bacterial strain *Anoxybacter fermentans* within the class Clostridia was isolated from a DSHV (Zeng et al. 2015). Among the Archaea, several extremophiles were isolated from DSHVs that have very high temperature optima (90–105 °C), i.e., *Ignicoccus pacificus*, *Methanopyrus kandleri*, and *Pyrolobus fumarii* (Sievert and Vetriani 2012).

9.2.1.2 Cold Environments: Seeps and Mud Volcanoes

On the seafloor a peculiar extreme environment is represented by cold seeps (CSs). CSs are characterized by the seepage into the ocean of dissolved and gaseous phase compounds (methane, petroleum, other hydrocarbon gases and gas hydrates) at the same temperature of the nearby seawater (Suess 2014). By pressure, these fluids are forced to pass through the seafloor sediments and their release results in a wide array of geological and sedimentary structures such as pockmarks, mounds, brine pools, and mud volcanoes. The simplest manifestation of a cold seep is bubbles escaping from the seafloor (Levin 2005). CSs are typically found at geologically active and passive continental margins with fluids released more slowly than in hydrothermal vents (Suess 2014). The deepest cold seep habitat ever found is located in the Sea of Japan at the depth of 7300 m. Since its first description 30 years ago CSs have attracted the attention of scientists and studies aimed at unraveling their microbial diversity have been performed. Similar as in hydrothermal vents, CSs comprise a rich benthic community that is sustained by chemolithoautotrophic processes fueled by reduced chemical species like sulfides, methane, and hydrogen (Levin 2005; Niemann et al. 2013).

Methane seeps are found worldwide, patchy distributed, and are characterized by highly reduced sediments in which oxygen is limited to the first few millimeters or centimeters of the sediment layer. In this oxygenated layer aerobic oxidation of methane (AeOM) takes place. Below this layer, microbial consortia of anaerobic methane-oxidizing archaea (ANME) and sulfate-reducing bacteria (SRB) oxidize the methane anaerobically (AOM) coupled to sulfate reduction (Suess 2014). Several studies aimed at elucidating the carbon and sulfur cycling pathways in CSs. For instance, Ruff et al. (2015) analyzed the microbiomes associated with 23 methane seeps distributed worldwide and compared these with the bacterial and archaeal communities associated with 54 other habitats on the seafloor. Microbial communities inhabiting methane seeps were less diverse and characterized by a high level of endemism when compared to other seafloor habitats. The authors showed that methane seeps are inhabited by few OTUs that resulted more abundantly and widely distributed, while the majority of OTUs was rare and locally restricted. Moreover, it was suggested that the abundant microorganisms play an important role in the ecosystem, mediating methane oxidation and impacting methane emission in the ocean (Ruff et al. 2015).

Metagenomics and metaproteomics have also been used to explore the microbial community associated to the Nyegga cold seep field, localized on the mid-Norwegian margin, and to investigate its microbial physiology (Stokke et al. 2012). Using 454-deep sequencing a high abundance of Euryarchaeota was discovered in the CS microbial community. More than half of the sequences were related to ANME-1 (a lineage included into ANME, distantly affiliated to the Methanosarcinales and Methanomicrobiales). Using a metaproteomic approach the interactions of microorganisms and their adaptation to the environment was evaluated. It revealed that the vast majority of the metabolic pathways is involved in sulfate-dependent AOM. Proteins were related to cold adaptation and to production of gas vesicles, underling the ability of ANME-1 to adapt to cold habitats and to position themselves at specific depths of sediment (Stokke et al. 2012).

Besides carbon and sulfur cycling, microbes in CSs participate also in the cycling of nitrogen. Anaerobic ammonium oxidizing (anammox) bacteria were found to be abundant and diverse in CSs, whereas ammonia-oxidizing archaea (AOA) were present but at low diversity (Cao et al. 2015; Shao et al. 2014). Cao et al. (2015) used the ammonia monooxygenase subunit A (*amoA*) as the genetic marker to investigate AOA from samples collected at the Thuwal seepage site in the

Red Sea. These authors included into the analysis samples from the seepage point as well as from the adjacent area. Hence, the analyses included marine sediments of the brine pool, marine sediments and water outside the brine pool, and benthic microbial mats. The aim was to elucidate the changes in the microbial community along this gradient from seepage sites to its margins. Deep sequencing of the 16S rRNA gene has also been performed resulting in the detection of a large number of reads that were unclassified at genus level. This suggests the presence of potentially novel microbial groups in these CSs (Cao et al. 2015). *AmoA* analysis showed that AOA from seeps and marine waters were different from those associated with microbial mats and marine sediments. It was therefore conceived that variations in ammonia-oxidizing archaeal community might depend on modification of environmental factors along the environmental gradient such as salinity and oxygen concentration (Cao et al. 2015).

Seepage manifestations at the seafloor include mud volcanoes (MVs) (Boetius and Wenzhöfer 2013). These marine structures, with variable diameters of 1–10 km, are typically elevated on the seafloor and originate from eruptions of muds and gases from the Earth's subsurface. They can be pointed, flat or show a crater-like structure on the top (Boetius and Wenzhöfer 2013). Fluids released by MVs comprise gases among which methane is predominant, and material components, known as "mud breccia." Besides methane, other compounds commonly found in the eruptions are wet gases, hydrogen sulfide, carbon dioxide, and petroleum products. MVs are considered dynamic features, subjected to changes in space and time, in consequence to modifications of fluids' fluxes and eruptions. This could thus influence the structure and function of the associated microbial community.

Microbial investigations on sediments of MVs showed high bacterial and low archaeal diversity (Pachiadaki et al. 2011). For instance, an active site of the Amsterdam mud volcano, located in the East Mediterranean Sea, has been characterized through the establishment of 16S rRNA gene clone libraries (Pachiadaki et al. 2011). These authors studied the stratification of the bacterial and archaeal communities along a spatial gradient of depth and pointed out the presence of low number of archaeal phylotypes and a high number of bacterial ones, at all depths. Especially, anaerobic methanotrophs of ANME-1, -2, and -3 groups have been retrieved supporting the importance of anaerobic oxidation of methane in this environment, which is particularly enriched in methane. 16S rRNA gene clone libraries revealed a more complex bacterial community than that of the Archaea. Indeed, in contrast to the Archaea, the bacterial diversity was not fully recovered and this suggested that other metabolisms may occur in this peculiar habitat, such as those related to nitrogen and sulfur conversions. The main bacterial phylotypes were recovered and revealed a community dominated by Proteobacteria, in particular Gammaproteobacteria (Pachiadaki et al. 2011). Comparison of the microbial community of the Amsterdam MV with data obtained from the Kazan MV of the same field showed that a high number of archaeal phylotypes and a low number of bacterial phylotypes were shared between the two sites, suggesting that some phylotypes are endemic to MVs (Pachiadaki et al. 2011).

Methanogens represent another important microbial group in the MV environment. In the center of the Amsterdam MV, methanogenesis and the archaeal community were investigated along a vertical transect using measurements with ^{14}C-labeled substrates, archaeal 16S rRNA gene PCR-Denaturing Gradient Gel Electrophoresis (DGGE) and *mcrA* gene clone libraries (*mcrA* encodes for the alpha subunit of methyl coenzyme M reductase, which catalyzes the final reaction of methane formation) (Lasar et al. 2012). The main component of the MV fluid is methane, which originates partially from the activity of methanogenic archaea. It was demonstrated that this methanogenesis is relevant for MVs. However, contrary to what has been reported for other CSs, in the Amsterdam MV sediments acetate methanogenesis was more important than hydrogenotrophic methanogenesis. This is probably due to the availability of acetate that is produced abiotically in the deep part of the mud volcano (Lasar et al. 2012).

Besides the application of molecular methods to describe the microbial assemblages colonizing CSs and MVs, also cultivation-dependent methods have been applied on samples collected from these environments. The moderate halophiles, *Halobacillus profundi* and *Halobacillus kuroshimensis*, have been cultivated from a carbonate rock at a methane cold seep in Japan (Hua et al. 2007) and *Altererythrobacter epoxidivorans*, a bacterial strain showing epoxide hydrolase activity (with potential application in enantioselective biocatalysis) has been isolated from the sediment of another Japanese cold seep (Kwon et al. 2007). *Marinobacter alkaliphilus* was isolated from samples collected from the Mariana forearc serpentinite mud volcanoes (Takai et al. 2005) and piezophilic strains belonging to the species *Psychromonas kaikoae* were obtained from the deepest cold seep sediments in the Japan Trench, at a depth of 7434 m (Nogi et al. 2002).

9.2.2 Hypersaline Environments

9.2.2.1 Deep Anoxic Hypersaline Lakes

Brine lakes located in the deep sea are among the most challenging environments described on Earth, distinguished by multiple physical and chemical parameters, which values were once considered so harsh that they would hamper their habitability. DHALs have been discovered few decades ago on the seafloor in the Gulf of Mexico (Joye et al. 2005; LaRock et al. 1979), in the Red Sea (Hartmann et al. 1998), and the Eastern Mediterranean Sea (Camerlenghi 1990; Jongsma et al. 1983). It is intriguing that the discovery of new DHALs in the Mediterranean Sea is still ongoing as for instance the recent exploration of the Thetis and Kryos basins (La Cono et al. 2011; Yakimov et al. 2015). The formation of DHALs is driven by tectonic events that cause the dissolution of evaporitic rocks trapped in the subsurface for millions of years. Since the 1970s, DHALs are known to host active microbes (LaRock et al. 1979). Despite their location in different geographical areas, DHALs share some common features such as the presence of brines with salinity 7–10 times higher than the overlying seawater (La Cono et al. 2011). The brine and the seawater column above do not mix due to their different densities and are separated by a well-defined interface. The brine–seawater interface hosts stable gradients of salinity, oxygen, and temperature (e.g., Daffonchio et al. 2006; Eder et al. 2001; LaRock et al. 1979) and are normally 1–3 m deep (Borin et al. 2009; Daffonchio et al. 2006; Eder et al. 2001; Yakimov et al. 2015). Besides containing sharp oxy- and haloclines, within the thick interface it is possible to measure gradients of nutrients concentrations, indicating their production/consumption within this ecological niche (Borin et al. 2009; Bougouffa et al. 2013; Daffonchio et al. 2006; Eder et al. 2001). A high precision sampling was designed allowing separating the interface of DHALs in fractions of increasing salinity and subsequently the geochemistry and microbiology were studied (Daffonchio et al. 2006). The results showed that the interfaces are hot spots of microbiological abundance and activity. The specific conditions in these interfaces selected for unusual assemblages of Bacteria and Archaea (Borin et al. 2009; Bougouffa et al. 2013; Daffonchio et al. 2006; La Cono et al. 2011; Yakimov et al. 2013) as exemplified by the molecular fingerprints shown in Fig. 9.1. The bacterial and archaeal communities thriving in the brine of DHALs are different from those of the layers above. van der Wielen et al. (2005) focused on Eastern Mediterranean Sea DHALs investigating four basins (i.e., Bannock, Discovery, L'Atalante, and Urania), which are characterized by diverse geochemical settings. They unraveled that specific and different microbial communities colonized each basin. The four brines were indeed colonized by different microbial assemblages and the cluster analysis that was performed on the microbial dataset showed that the distribution of microorganisms perfectly fitted the geochemical data (van der Wielen et al. 2005). Based on the 16S rRNA signatures the authors also demonstrated the possibility of life in the Discovery brine, which according to the high concentration of $MgCl_2$ (5 M) is considered as one of the most extreme environments on Earth (Wallmann et al. 2002). Because of its chaotropicity such high concentration of manganese in the absence of other compensating ions has been postulated to be unsuitable for life (Hallsworth et al. 2007). Another study displayed nearly saturated $MgCl_2$ brine in the newly discovered Lake Kryos, which made it the biggest athalassohaline formation on our planet (Yakimov et al. 2015). Differently from Discovery, Kryos also contains sodium and sulfate ions. The first evidence has been reported that microbial life exists and is active in lake Kryos (Yakimov et al. 2015).

Similarly to what is occurring in many extreme ecosystems, most of the studies performed on DHALs during the last two decades were based on cultivation-independent approaches and aimed at unveiling their hidden ecological processes. The main research focus was on the seawater–brine interface that acts as a trap for nutrients, minerals, and microbial cells, and creates environmental gradients of paramount interest for clarifying the limits of life. These studies tested whether specific metabolic pathways occur at the conditions typical of these oxic/anoxic boundaries (e.g., high pressure and high salinity). In this regard, DHAL interfaces are precious natural laboratories that are being colonized by microbial communities that occur sharply stratified and are actively involved in

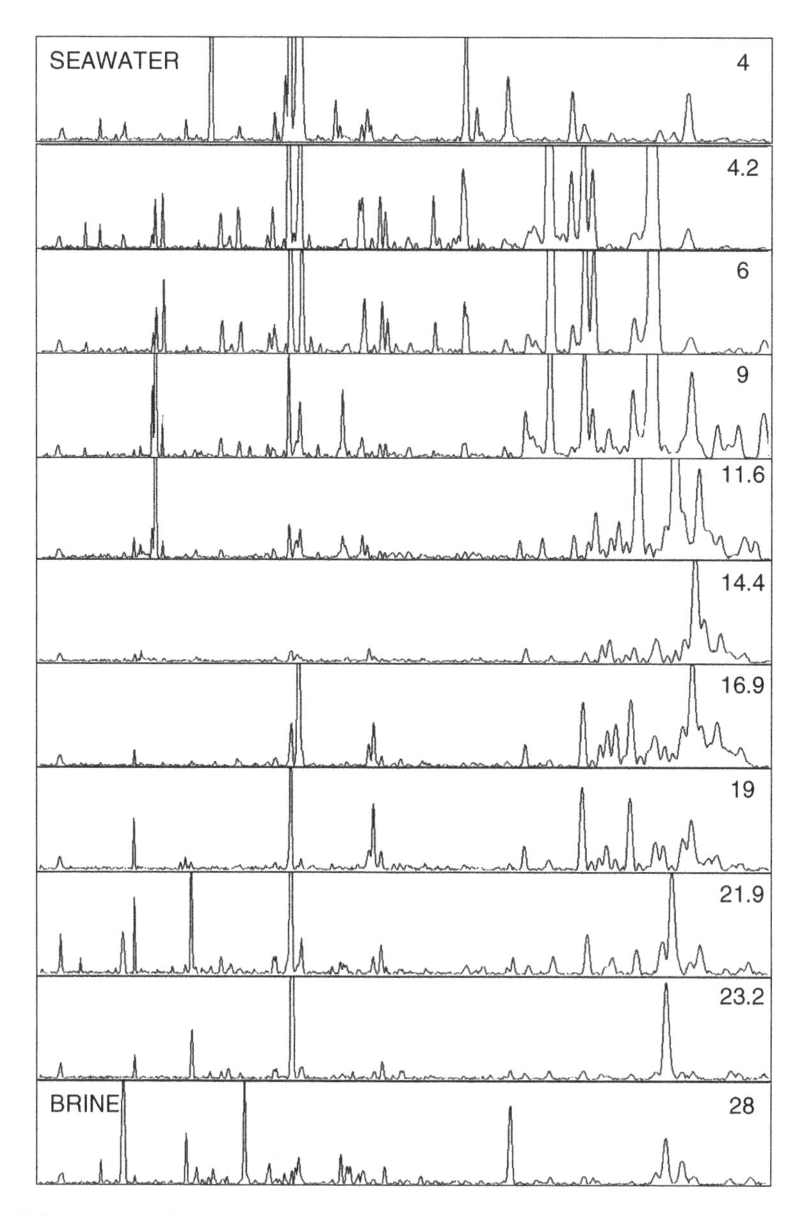

Fig. 9.1 Automated ribosomal intergenic spacer analysis (ARISA) fingerprinting of the 2-m thick seawater–brine interface in the Bannock DHAB. The interface was fractionated immediately after recovery in layers exhibiting increasing salinity values, reported as %salinity (as estimated by hand refractometer) on the right side of each panel. The fingerprints show that the bacterial communities are stratified along the interface, with specific populations confined at defined salinity intervals (Daffonchio et al. 2006)

biogeochemical processes (Daffonchio et al. 2006). Sulfur and carbon cycles have been widely recognized as crucial metabolisms in DHAL interfaces and brines, and this was confirmed by activity measurements and the detection of phylogenetic groups involved in such processes (Borin et al. 2009; Daffonchio et al. 2006; Joye et al. 2009; La Cono et al. 2011; Yakimov et al. 2007a, b). Novel phylogenetic groups were detected in DHALs located both in the Mediterranean and the Red Sea. The novel bacterial phylogenetic lineage KB1 from the Kebrit lake in the Red Sea was described by Eder et al. (1999), and it was later also detected in the Shaban deep (Eder et al. 2002) and in the Mediterranean DHALs (van der Wielen et al. 2005; Yakimov et al. 2015). New archaeal and bacterial lineages, named MSBL-1 and MSBL-2 (Mediterranean Sea Brine Lakes), respectively, were detected at Bannock, Urania, L'Atalante, and Discovery (van der Wielen et al. 2005). The bacterial groups MSBL 3–6 were subsequently identified at Bannock interface (Daffonchio et al. 2006). MSBL-2 is phylogenetically related to the division SB1 (Daffonchio et al. 2006) and detected in the interface of the Shaban lake (Eder et al. 2002). Besides these DHAL-specific lineages, Proteobacteria of the epsilon, gamma, and delta subclasses are widely spread in the DHAL interfaces and brines, where they occur stratified according to the gradients of different sulfur compounds. The Proteobacteria are involved in sulfate reduction (Deltaproteobacteria) and in sulfur oxidation (Epsilon- and Gammaproteobacteria) (Borin et al. 2009; Daffonchio et al. 2006; Ferrer et al. 2012; Yakimov et al. 2007a, b). The archaea MSBL-1 was hypothesized to possess a methanogenic metabolism (van der Wielen et al. 2005) and was found to be highly abundant in different brine lakes such as Urania and Medee (Borin et al. 2009; Yakimov et al. 2013). Additional prevailing archaeal populations in DHALs belong to the Crenarchaeota and are involved in sulfur metabolism, and ANME lineages, which are responsible for anaerobic methane oxidation (Borin et al. 2009; Daffonchio et al. 2006; Yakimov et al. 2013).

Besides overall community studies, bacterial and archaeal populations were characterized using specific gene markers such as *dsrA* for sulfate-reducing bacteria (Joye et al. 2009; van der Wielen and Heijs 2007), *aprA* for both sulfate-reducing bacteria and sulfide-oxidizing bacteria (La Cono et al. 2011) and *mcrA* for methanogenic and methane-oxidizing archaea (Joye et al. 2009; La Cono et al. 2011; Yakimov et al. 2013). The expression of the *cbbL* gene, encoding for the large subunit of ribulose-1,5-biphosphate carboxylase/oxygenase (RuBisCo), was measured in L'Atalante Lake and detected across the halocline, providing evidence for CO_2 fixation (Yakimov et al. 2007b). The *cbbL* sequences were highly similar to those detected in the body of the brine of other lakes and of various other hydrothermal vents and, combined with 16S rRNA data, pointed to the possible role of Gammaproteobacteria in CO_2 fixation at L'Atalante (Yakimov et al. 2007b). The nitrogen cycle in DHALs was also investigated to some extent (Borin et al. 2013; Ngugi et al. 2015; Yakimov et al. 2007b). The distribution of AOA and bacteria was analyzed at L'Atalante Lake by PCR-amplifying the *amo*A gene, which encodes the ammonia monoxygenase alpha subunit. This showed the prevalence of sequences belonging to Crenarchaea (Yakimov et al. 2007b). The distribution of *amoA* sequences along water column above the L'Atalante interface was related to depth although the abundance of the *amoA* gene in the interface was low (Yakimov et al. 2007b). In the Bannock and L'Atalante lakes the gradients of ammonium concentration plotted versus salinity were concave, differently from what observed for chemical species with a conservative behavior, indicating its active consumption by microorganisms (Daffonchio et al. 2006; Yakimov et al. 2007b). Anaerobic ammonium oxidation (anammox) is a process that takes place at oxic–anoxic boundaries and the bacteria of the order Planctomycetales responsible for this process were identified for the first time less than 20 years ago (Strous et al. 1999). The occurrence of anammox bacteria was investigated in DHALs located at the Eastern Mediterranean Sea (Borin et al. 2013). Anammox had never been reported to occur at salinity values higher than seawater. The presence of anammox bacteria was assessed at the Bannock and L'Atalante haloclines from clone libraries and Fluorescent In Situ Hybridization (FISH) using specific primers and probes, and by detecting anammox activity. Analysis of the 16S rRNA dataset demonstrated that the anammox community in the two lakes was diverse and stratified according to the increase of salinity (Borin et al. 2013). The majority of the anammox sequences retrieved in the Bannock Lake was affiliated to *Scalindua brodae*, a known anammox bacterium. The remaining sequences in Bannock and all sequences detected at L'Atalante belonged to a cluster that does not match with any known anammox bacteria but were previously reported in samples from deep hydrothermal vents where anammox occurred (Borin et al. 2013). Anammox activity was detected in the halocline of L'Atalante with salinity of 6–9.2 %, suggesting that the anammox-related phylotypes retrieved here may be new branches of anammox bacteria, specifically adapted to high salinity (Borin et al. 2013). Previously, the presence of anammox was also suggested in the Urania interface at salinity values ranging from 5.1 to 13.9 % (Borin et al. 2009). Moreover, in Bannock and L'Atalante, the functional gene *hszA*, encoding for the hydrazine synthase involved in anammox was detected (Borin et al. 2013). The 16S rRNA signatures of anammox bacteria were also detected by an extensive pyrosequencing study in the Red Sea, at the Atlantis II and Discovery DHALs (Bougouffa et al. 2013) showing that anammox is possible at salinities well above that of seawater.

Although molecular surveys demonstrated that DHALs (i) contain habitats (i.e, interfaces) with high abundance and diversity of Bacteria and Archaea, (ii) show evidence of complete biogeochemical cycles, and (iii) host novel phylogenetic lineages that are often highly DHAL-specific, few strains belonging to those novel microbial taxa have been brought into culture. Eder et al. (2001) were the first to report the isolation of microorganisms from DHALs, which presence was until then inferred exclusively through biochemical data (Dickins and Van Vleet 1992; LaRock et al. 1979). The halophilic bacterium *Halanaerobium* sp. strain KT-8-13 was isolated from the interface and brine of the Kebrit lake in the Red Sea (Eder et al. 2001). A few months later in 2001, the bacteria *Alteromonas macleodii* and *Halomonas aquamarina* were isolated also from the interface of Urania Lake in the Mediterranean Sea, where they were also detected by molecular methods (Sass et al. 2001). The new species *Haloplasma contractile* was isolated from the Shaban lake located in Northern Red Sea (Antunes et al. 2008). Culturing approaches were applied during the last years leading to the selection of different bacteria and archaeal species (Borin et al. 2009; Daffonchio et al. 2006; Sorokin et al. 2015). Genomic studies are now available for some of the bacteria and archaea isolated from DHALs, such as the gammaproteobacterium *Salinisphaera shabanensis* (Antunes et al. 2011) and the euryarchaeon *Halorhabdus tiamatea* (Werber et al. 2014), shedding a new light on traits potentially involved in habitat-specific adaptation and opening the way toward their biotechnological exploitation.

9.2.2.2 Mud Volcanoes Brine Lakes

During the last decades geologists described a number of mud volcanoes (MVs) filled with muddy brine in the Eastern Mediterranean Sea, within the Mediterranean Ridge (Dupré et al. 2007, 2014; Zitter et al. 2005). The muddy brine of MVs represents a distinctive ecosystem colonized by active extremophiles. Microbial investigations of MV-associated brines have been performed on the mud fluids collected below the Urania DHAL (Yakimov et al. 2007a), at the Napoli MV (Lazar et al. 2011), and at the Menes Caldera located in the Nile deep-sea fan (Omoregie et al. 2008).

The Urania DHAL has a horseshoe shape and hosts on one of its edges a hypersaline anoxic brine-filled lake called “mud pit” which has been indicated as one of the most sulfidic water bodies on Earth (Yakimov et al. 2007a). The presence of mud underneath the lake is caused by the activity of a mud volcano, which releases fluids from the sedimentary reservoir. The Urania mud is rich of methane and shows a higher temperature and lower salinity than the overlying brine from which it is separated due to the clay content and the differences in density (Yakimov et al. 2007a). Molecular analysis based on reverse transcription of 16S rRNA genes from the Urania mud revealed active bacterial and archaeal communities. Using clone libraries, three taxonomic groups belonging to euryarchaeota and ten classes/subclasses within the Bacteria were retrieved (Yakimov et al. 2007a). The archaeal sequences were all affiliated to uncultured organisms and included representatives of the Anoxic Methane-oxidizing Euryarchaeota group 1 (ANME-1), suggesting the occurrence of anaerobic methane oxidation (Yakimov et al. 2007a). Moreover, the archaeal MSBL-1 group, typically found in DHABs (Borin et al. 2009; Daffonchio et al. 2006; van der Wielen et al. 2005), was also detected by Yakimov et al. (2007a) in the muddy fluid below the Urania Lake. The bacterial communities of the warm and oxygenated mud of the Urania Lake were highly diverse and differed from those occurring in the above hypersaline anoxic brine (Yakimov et al. 2007a). The Proteobacteria, particularly the epsilon and delta subclasses, were the prevalent divisions in the clone libraries, but Alpha- and Betaproteobacteria, Actinobacteria, Bacteroides, Deinococcus-Thermus, and KB1 and OP-11 candidate divisions were also detected (Yakimov et al. 2007a).

Archaeal sequences related to uncultured lineages were also detected in the RNA-based analyses of the mud samples collected at the Napoli volcano, a MV located on saline deposit from which the brine moves up to the overlying seawater

(Lazar et al. 2011). These authors investigated the microbiology of orange-pigmented mats of the Napoli MV, which ejects a mixture of brine and mud enriched with biogenic methane, focusing particularly on the microorganisms involved in methane cycling. Clone libraries of the *mcrA* gene, encoding for the methyl coenzyme M reductase, a key enzyme in methanogenesis and in anaerobic oxidation of methane (AOM), suggested the occurrence of AOM in the sediments of Napoli MV (Lazar et al. 2011). These results, together with the enrichment of coccus-shaped methanogens from the sediments, led to infer the presence of an active microbial assemblage involved in methane production and consumption under hypersaline conditions in the mats of Napoli's brine.

Besides methane, sulfide is another key substrate for bacterial and archaeal growth in the MV ecosystem. Sulfide is the result of sulfate reduction coupled to AOM or anaerobic oxidation of higher hydrocarbons. Mat-forming gammaproteobacteria of the genera *Beggiatoa*, *Thiomargarita,* and *Arcobacter* oxidize these hydrocarbons in the sediment layers where sulfate and hydrocarbons meet (Grünke et al. 2011; Omoregie et al. 2008). The brine pool at the Chefren active mud volcano contains a hypersaline fluid rich of methane and sulfide (Dupré et al. 2014) and white and orange mat patches were identified at a brine emission point located along a steep slope of the Chefren MV, about 3000 m below sea level (Omoregie et al. 2008). The biogeochemical cycles driven by the microorganisms in the muddy saline fluid fuel a complex ecosystem that flourishes as pigmented microbial mats, which are populated by higher organisms such as tubeworms, crabs, and polychaetes (Omoregie et al. 2008). The bacteria detected in these mats by culture-independent approaches were mostly represented by the Deltaproteobacteria involved in sulfate reduction, and by the Gammaproteobacteria able to oxidize iron, sulfide, and methane (Omoregie et al. 2008). The archaeal community was mainly constituted of anaerobic methane oxidizers (ANME) of the clusters ANME-2 and ANME-3, which accounted for 55 and 74 % of the total detected sequences in the white and orange mats, respectively (Omoregie et al. 2008). The microbiological investigations performed on these warm muddy brine lakes on the seafloor of the Eastern Mediterranean Sea provided consistent results, indicating the presence of microbial communities involved in the metabolic pathways that, according to the lake geochemistry, are of primary importance in this ecosystem. However, the isolation of the key microbial players from MV brine lakes is still lacking.

9.2.2.3 Sea Ice Brines

At first sight, sea ice appears as a homogeneous solid phase. However, in fact sea ice includes a wide network of pores and channels containing fluid brine. The brine is the result of the freezing process during sea ice growth, through which saline water undergoes cryoconcentration, hence resulting into salinities of up to three times those typical of seawater (Thomas and Dieckmann 2002). The brine-filled channels represent a habitat where well-adapted extremophile organisms flourish. These extremophiles may include all three domains of life and are resistant to multiple stressors such as high salinity, subzero temperature, and dark and anoxic conditions. Among the Eukarya, diatoms are abundant components of this peculiar ecosystem, together with metazoans and other protists. Their numbers exceed those of seawater (Thomas and Dieckmann 2002). Besides microalgae, metabolically active bacterial communities have been studied in Arctic and Antarctic sea ice brines, where they play key roles in the biogeochemical cycles (Díez et al. 2012; Mikucki and Priscu 2007; Mikucki et al. 2009; Møller et al. 2010; Murray et al. 2012). Various bacterial genera have been isolated from sea ice brines (Junge et al. 1998; Mikucki and Priscu 2007). This resulted in the identification of a new species within the *Planococcus* genus (Junge et al. 1998) and even to the characterization of a mercury-resistant bacterial strain (Møller et al. 2010). Other studies that focused on the bacterial diversity of sea ice brines are based on culture-independent approaches. Díez et al. (2012) reported for the first time the occurrence of diazotrophic bacteria in the brine of the Greenland Sea and Fram Strait, where the number of viable cells was 0.6×10^9 cells/L. The investigation of the *nifH* gene, coding for dinitrogenase reductase, one of the components of the enzyme complex nitrogenase and responsible for nitrogen fixation, allowed the detection of several sequences affiliated to marine bacteria, including cyanobacteria (Díez et al. 2012). About 52 % of the clones obtained from the arctic brine were related to Alpha- and Gammaproteobacteria, while the rest of the sequences belonged to cyanobacteria. Although the authors did not assess the metabolic activity of the detected diazotrophs, they reported the possibility of N_2 fixation in Arctic seawater and ice brine.

The most fascinating story about sea ice brine takes place at the McMurdo Dry Valleys (Antarctica) and tells about the episodically outflow events at the Taylor Glacier, on which surface the so-called Blood Falls appear. The name of this unique phenomenon is due to the drainage, from below the Taylor Glacier, of subglacial brine rich of salt and the iron mineral goethite, which swiftly precipitates after meeting the atmosphere (Mikucki et al. 2009). It is estimated that the brine emerging at the Blood Falls remained isolated for 1.5 million years before its release. However, the hydrology of the system is poorly understood (Mikucki et al. 2009). Using an electromagnetic sensor system, designed for hydrogeophysical research, it was demonstrated that a highly saline liquid is present within the permafrost below glaciers and lakes of the McMurdo Dry Valleys, implying the presence of a widespread habitat suitable for life of extremophiles (Mikucki et al. 2015). The presence of extremophiles was also shown by the microbiological studies performed on the brine emerging at the Blood Falls. Heterotrophic activity was demonstrated through the measurement of ^{3}H-thymidine incorporation (Mikucki et al. 2009), and the detection of about 6×10^4 microbial cells per milliliter (Mikucki et al. 2015). A culturing approach allowed the isolation of strains belonging to phyla belonging to the Proteobacteria and Bacteroidetes from the discharged brine (Mikucki and Priscu 2007). Molecular studies indicated that bacteria inhabiting the Blood Falls brine are related to phylotypes of marine origin. These studies supported the presence of microorganisms able to grow chemoautotrophically and chemoorganotrophically, and revealed the high abundance of the genus *Thiomicrospira*, which is also an important player in hydrothermal vent ecosystems (Mikucki and Priscu 2007). Further analyses confirmed the occurrence of metabolically active bacterial assemblages involved in the sulfur cycle, which is crucial in order to allow the oxidation of organic matter using ferric iron as the electron terminal acceptor (Mikucki et al. 2009).

It is widely recognized that global warming is severely hampering the cryosphere, eventually influencing the time and space available for life of sea ice organisms thriving in the brine (Vincent 2010). The role of compounds of microbial origin affecting the physical structure of the sea ice is still poorly known. Exopolymers, mainly constituted of polysaccharides, and proteins are of paramount interest from the biotechnological perspective, as well as for the ecological implications of their production and accumulation in the brine. An experimental study performed on the sea ice of the Arctic and Antarctic showed the high level of expression of eukaryotic ice-binding proteins (Uhlig et al. 2015). These proteins, which are encoded by genes occurring in distantly related microbes, enhance the retention of liquid brine inside the ice. Exopolymers alter the sea ice microstructure and they have been found in brine inclusions where they likely operate as cryoprotectants (Krembs et al. 2002). The impact of exopolymers produced by the diatom *Melosira arctica* var. *krembsii* on the ice permeability was demonstrated in laboratory experiments (Krembs et al. 2011). The exopolymers probably acted by clogging of pores inside the ice and increase brine viscosity.

9.2.3 Low Water Extreme Marine Environments: Life in Oil

Various marine extreme ecosystems are rich in hydrocarbons (HCs) and due to plate tectonics sometimes represent a source of HCs in the seabed. Oil reservoirs can be abundant in deepwater areas worldwide and represent an extreme habitat with an unknown microbiology. Innovative research on deep subsurface oil basins has been performed in the framework of microbiological oil degradation in the hope to discover and design new strategies for oil recovery (Head et al. 2003). Microorganisms have been isolated from oil well producing waters, but there were doubts as to the indigenous origin of these microbes. About a decade ago the common opinion was that oil degradation takes place only at the oil–water interface (Head et al. 2003). However, a groundbreaking study analyzed samples collected at the largest asphalt lake of natural origin on Earth, the Pitch lake in Trinidad and Tobago, and showed life in water droplets dispersed in oil (Meckenstock et al. 2014). Salinity and isotopic data ascertained the origin of the water droplets and indicated their provenience from ancient seawater and brine. Motile cells were microscopically visible and the resident microbial community was studied in single droplets showing active microorganisms thriving in a HCs-saturated fluid at seawater salinity (Meckenstock et al. 2014). 16S rRNA pyrosequencing analysis detected Bacteria and Archaea, mainly halophiles and phylotypes typical of oil environments. Moreover, the characterization of HCs in water droplets demonstrated the occurrence of microbial populations transforming oil components, which implied that this newly detected space for life is of paramount importance for oil degradation in subsurface reservoirs (Meckenstock et al. 2014). This study discovered a new habitat for microorganisms, and its exploration promises exciting discoveries about the limits of life under saline and hydrocarbon rich conditions. The extremophiles retrieved in this habitat may hold promise for their biotechnological exploitation.

9.3 Biotechnological Exploitation of Marine Extremophile Microorganisms: Products and Perspectives

Marine extremophile microorganisms produce a wide range of enzymes and metabolites with attractive biological and chemical diversity; this structural diversity has proved to be potentially useful in various fields of application (Litchfield 2011). Enzymes derived from marine extremophile microorganisms are often effective biocatalysts with higher activity, greater stability, and higher salt and solvent tolerance. Different strategies of adaptation to the salinity of marine environments have evolved. By accumulating inorganic solutes (K^+ and Na^+) in the cytoplasm osmotic stress is relieved and osmotic equilibrium is achieved ("salt-in" strategy); this strategy is common among many groups of marine microorganisms, such as many *Halobacteriaceae*. Alternatively, halophiles accumulate low molecular weight organic molecules (osmolytes) in the cytoplasm in order to compensate for high ionic strength; these molecules can be charged (anionic or zwitterionic), or even uncharged ("salt-out" strategy). Based on these general arrangements, different bioproducts can be expected: metabolites produced for balancing the osmotic equilibrium will be the primary products of strains adopting the "salt-out" strategy, whereas molecules adapted to high ionic stress (i.e., halophilic proteins) will be mostly found within microorganisms utilizing the "salt-in" strategy.

An interesting perspective concerns the use of marine extremophile microorganisms as cell factories for improving the performance of bioproduct preparation by metabolic engineering; the ongoing availability of new genomes will provide new tools for specific genetic manipulation and heterologous expression and this will open the possibility of improving the performances of marine extremophile microorganisms for industrial biotechnology. Hence, marine extremophile microorganisms can be industrially attractive for:

- discovery of new bioactive agents (mostly secondary metabolites)
- discovery of new enzymes
- development of strains and/or enzymes with improved tolerance toward harsh conditions, similar to industrial environments

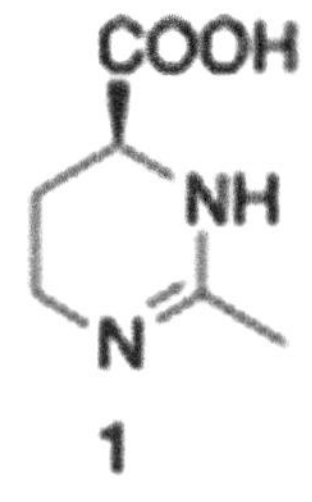

Fig. 9.2 Chemical structure of ectoine (*1*) and 5-(S)-hydroxyectoine (*2*)

9.3.1 Extremolytes

Secondary metabolites from marine extremophiles are often known as "extremolytes" and, in the case of marine extremophile microorganisms, salinity may be crucial for the biosynthesis of these molecules. We refer to Pettit (2011) for a review on the state of the art of the discovery of new bioactive secondary metabolites from cultured extreme-tolerant and deep-sea microbes. Secondary metabolites have been isolated from marine extremophile microorganisms, showing a wide array of chemical structures having often interesting properties as chemotherapeutic agents.

Osmolytes are metabolites synthesized and intracellularly accumulated to avoid shrinking of the cell due to external osmotic pressure. Intracellular osmolytes can reach concentrations up to 2 M. Ectoines (ectoine and 5-(S)-hydroxyectoine, Fig. 9.2) are intracellular solutes produced and accumulated by halophilic and halotolerant bacteria in response to osmotic and heat stress. Ectoines have been applied as an ingredient of cosmetics. These extremolytes are accumulated to a high concentration and may make up 25 % of cell dry mass (Lentzen and Schwarz 2006). This allows for large-scale production and industrial production of ectoines has been accomplished using *Halomonas elongata* ATCC 33173 under high salt conditions (15–20 % w/v NaCl) (Sauer and Galinski 1998). Ectoines were easily recovered by osmotic shock, which causes the rupture of the inner membrane of the bacterium and subsequent release of the product into the medium. Alternatively, ectoines were produced at a large scale under anaerobic conditions using *Pyrococcus furiosus* at 98 °C (Lentzen and Schwarz 2006).

Ectoine production from lignocellulosic biomass-derived sugars by engineered *Halomonas elongata* has also been developed (Tanimura et al. 2013). Optimization of the preparation of ectoine and hydroxyectoine was achieved using the halophile bacterium *Chromohalobacter salexigens*. A recombinant strain of *C. salexigen* was constructed in which the three genes involved in ectoine biosynthesis were overexpressed and fermentation was carried out at 4.35 % NaCl (Rodríguez-Moya et al. 2013). Two continuous bioreactors were employed in order to produce ectoines. In the first bioreactor high cell density was achieved at 37 °C in the presence of 108 g/L of NaCl. Ectoines (2.1 g/L) were recovered after osmotic downshock using distilled water in the second reactor (Van-Thuoc et al. 2010).

Haloferax mediterranei produces polyhydroxyalkanoate (PHA) a polymer that can be used for the production of biodegradable plastics (Quillaguaman et al. 2010). Distinctive advantages linked to the use of the archaeal halophilic *H. mediterranei* include the possibility of transforming agricultural wastes and lignocellulose residues in general (i.e., rice bran, vinasse and stillage) with high yields into PHA. The halophilic characteristics of *H. mediterranei* makes this strain suitable for performing fermentations with poorly diluted biomass hydrolysates and high salt concentration, which minimizes medium contamination. An excellent example of biorefinery is the waste stillage of rice-based ethanol industry for the production of poly-3-hydroxybutyrate-hydroxyvalerate (PHBV) by *H. mediterranei* (Bhattacharyya et al. 2014). This process was carried out in a plug flow reactor of the activated sludge process, giving a final yield of 63 % PHA of dry cell weight, corresponding to 13.12 PHA/L. An interesting feature of this bioprocess is the cost of sludge desalinization: PHBV production was estimated as US$2.05 /kg using the current methods of desalination, whereas the cost of PHBV production could be lowered to US$1.30 /kg, using innovative and cheaper desalination technologies (Bhattacharyya et al. 2015). Moreover, cell lysis of *H. mediterranei* can be obtained by simple treatment with water (like many haloarchaea), making PHA recovery easy and cheap, and potentially decreasing the recovery costs up to 50 %, compared to other extractions in conventional fermentations (bacteria, yeasts).

Besides for the production of PHA, *H. mediterranei* has been also exploited for other processes such as the preparation of anionic polysaccharides (applied as gelling agents and emulsifiers, or for enhanced oil recovery) (Anton et al. 1988). Marine extremophile microorganisms are particularly attractive for oil recovery by microbial remediation (Banat et al. 2000). Petroleum reservoirs are featured by high ionic strength conditions and halophilic bacteria able to secrete high concentrations of surfactants help lowering the tension at the interface between oil and solid materials (Martinez-Checa et al. 2007).

Metatranscriptomic analyses have revealed a substantial presence of fungi in the deep marine subsurface (Orsi et al. 2013). In particular, marine fungal secondary metabolites have been investigated (Rateb and Ebel 2011). Screening for genes involved in secondary metabolite synthesis among marine fungi from deep subseafloor sediments revealed their potential for the production of bioactive metabolites (Rédou et al. 2015). Interestingly, marine yeasts proved to be effective in the production of astaxanthin (Fig. 9.3), a carotene derivative, which is approved by the United States Food and Drug Administration (USFDA) as food colorant in animal and fish feed and by the European Commission as a food dye. A marine strain of *Rhodotorula glutinis* was isolated and was capable of producing 2.7 mg/L of astaxanthin under optimized conditions. This concentration is similar to what is reported from terrestrial strains (Ambati et al. 2014).

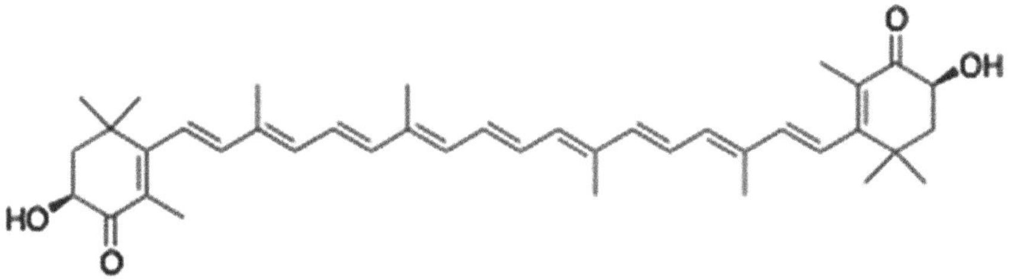

Fig. 9.3 Chemical structure of astaxanthin

9.3.2 *Extremozymes*

Because of their ability of catalyzing chemical reactions under nonstandard conditions proteins produced by marine extremophile microorganisms (extremozymes) are of great interest for industrial application, where aggregation, precipitation, and denaturation decrease the activity of most nonextremophilic enzymes (Trincone 2011). Haloarchaea, which adopt a "salt-in" strategy, have evolved intracellular halophilic enzymes able to tolerate salt concentrations of up to 5 M. Variation of protein charges and increased hydrophobicity are the two most common features of halophilic proteins compared to regular proteins. The proteomic analysis of *Halobacterium* sp. NRC-1 showed an overall preponderance of proteins with negative charges, which is the result of the predominance of acidic over basic amino acids (Kennedy et al. 2001). The analysis of a number of genomes of halophilic bacteria isolated from marine hypersaline environments confirmed that these organisms produce proteins with an overall negative charge (Capes et al. 2012). Another general property of many halophilic proteins is the occurrence of hydrophobic residues (phenylalanine, isoleucine, and leucine) on their surfaces. The extent of negative charge of the proteome is somehow associated with the degree of halophilicity of the microorganisms; moderate halophilic microorganisms generally contain proteomes, which are less acidic than more extreme halophiles (Oren 2013). In conditions of high salinity/low water activity, interactions between water and negatively charged residues become crucial for establishing functional hydration of the protein. Extremophilic enzymes often remain active in solution even when liquid water is limiting by maintaining a tight hydration shell, even in the presence of high ionic concentrations and/or with water close to the freezing point. Hence, halophilic proteins are often characterized by slow refolding, resistance to organic solvents, and high aqueous solubility. These properties may be advantageous for industrial applications.

Seawater has been suggested as an alternative medium for sustainable bioprocesses and halophilic enzymes seem to be suited for these applications (Domínguez de María 2013; Zambelli et al. 2015). In particular, hydrolases have been thoroughly studied (e.g., glycosidases, carboxylester hydrolase, amidases, and proteases), since they catalyze valuable reactions, while working without organic cofactor under mild conditions of pH, temperature and pressure. These hydrolases are also often chemo-, regio-, and stereoselective. Marine-derived hydrolases are particularly interesting for their robustness under harsh conditions, such as high organic solvent content. Dalmaso et al. (2015) have reviewed the use of marine extremophiles as a source of hydrolases for biotechnological applications.

Organic-solvent-tolerant bacteria are an interesting group of marine extremophile microorganisms of potential interest for the biotechnology industry. Organic-solvent-tolerant bacteria are over 100 times more abundant in deep-sea samples than in terrestrial soils. The β-galactosidase from the cold-adapted halophilic *Halorubrum lacusprofundi*, which is active over a large temperature range of 5–70 °C and at a salt concentration of 5 M NaCl, is also active in concentrated methanolic and ethanolic solutions (Karan et al. 2013). Typical industrial applications of halophilic enzymes include the use of cellulase for hydrolysing lignocellulosic biomass (Ogan et al. 2012) or chitin. Chitin is the second most abundant natural polymer after cellulose and is the major waste from the seafood industry. *H. mediterranei* is able to use colloidal and powdered chitin for the production of PHBV, having powerful and highly stable chitinases (Hou et al. 2014).

Deep-sea hypersaline anoxic basins (DHALs) extremophile microbiome showed to be an innovative source of marine-derived enzymes with interesting application in stereoselective biocatalysis. A screening among bacteria isolated from DHABs water–brine interface of the Eastern Mediterranean allowed the selection of various strains able to enantioselectively hydrolyse racemic esters of anti-2-oxotricyclo [2.2.1.0]heptan-7-carboxylic acid, a key intermediate for the synthesis of prostaglandins. An esterase from *Halomonas aquamarina* 9B catalyzed the kinetic resolution of the racemic substrate, while the strain *Bacillus hornekiae* 15A gave highly stereoselective reduction of the same substrate. The two approaches furnished enantiomerically pure unreacted (R)-substrate, which could be easily recovered and purified (De Vitis et al. 2015).

9.3.3 Extremophiles Potential in Bioenergy Processes

Microbial fuel cells (MFC) have been applied using seawater microorganisms found in sediments for bioelectricity production. MFCs have been reported to operate in closed and in open marine systems (Logan 2005), showing higher performances when operated in high salinity environments. Hence, halophilic archaea have been exploited as biocatalysts for establishing MFC under conditions of critical high ionic strength for having high efficient electrofermentations (Abrevaya et al. 2011). Mathis et al. (2008) showed that marine sediments inhabit thermophilic electricity-generating bacteria and that a thermophilic microbial fuel cell was able to generate electricity using cellulose. The Gram-positive thermophile, *Thermincola carboxydophila* was isolated from biofilms formed on the graphite anodes of fuel cells, constructed with marine sediment. This microorganism proved to be efficient as electrode-reducing bacteria. Moreover, the use of this marine thermophile bacterium under thermophilic conditions provided higher electrical currents than what is normally obtained using mesophilic microorganisms (Mathis et al. 2008).

Marine yeasts have been studied as alternative cell factories for the production of bioethanol (Kathiresan et al. 2011). These authors were attracted by the use of marine biomass hydrolysates (e.g., seaweed and sea lettuce) as substrate for bioethanol fermentation. These hydrolysates contain high concentrations of salts and thus marine yeasts that are able to perform the conversion of sugars into ethanol under high salinity are the best suited for these bioprocesses, since no pretreatment involving desalinization is necessary, which saves energy. Zaky et al. (2014) summarize the best results obtained with a marine strain of conventional (*Saccharomyces cerevisiae*) and nonconventional yeasts (*Candida, Debaryomyces, Rhodotorula*). These yeasts were able to convert different hydrolysates into ethanol with satisfactory yields, albeit lower than the best producer strains described in literature. It should be pointed out that also nonmarine hydrolysates of lignocellulosic residues often contain high amounts of salts causing severe osmotic stress. This is because it contains inorganic ions from the lignocellulose hydrolysates derived from the original biomass, chemicals added during pretreatment, hydrolysis, and conditioning. A good bioethanol producer strain of *S. cerevisiae* grows in synthetic media up to 1.5 M NaCl, but an even more crucial parameter is the ratio between intracellular $[Na^+]/[K^+]$, which should be as low as possible.

9.4 Conclusions

Many studies performed during the last decades on extreme environments are discussed in this chapter. These studies represent exciting proofs that the seafloor is still having surprises for oceanographers and marine microbiologists. The discovery of new extreme habitats hosting rich microbial communities and complex ecological networks calls for more research because of the potential impact on the functioning of global biogeochemical cycles and the possibility of finding new compounds or processes that can be applied in biotechnology.

Despite the great efforts to elucidate the microbial diversity, ecology, and functioning of the various extreme marine ecosystems presented in this chapter, many questions remain open. Particularly, questions await answers referring to the main microbial players of key reactions, the metabolic pathways, the reaction rates, and the overall microbial productivity and role in these ecosystems. Many powerful techniques are now available and will contribute to answer to the aforementioned questions, including "omic" methodologies, single-cell tools, thermodynamic modeling, and the possibility to use in situ instrumentation to measure reaction rates and nutrient concentrations. Moreover, metagenomic techniques are providing an incredible amount of data and the same can be said for metatranscriptomics and metaproteomics.

More efforts are still required to bring into culture the majority of the key extremophile microorganisms and explore their biotechnological potential. Many of these microorganisms have been identified by molecular surveys. Now new methods for the isolation, culturing, and genetic manipulation of marine extremophile microorganisms have been developed. This will allow expanding the use of these organisms and the quest for more robust and sustainable bioprocesses. The combined use of novel marine extremophile microorganisms and evolved molecular techniques will furnish tailored microorganisms for bioprocesses to be performed under harsh conditions.

Acknowledgments The research leading to these results has received funding from the European Union Seventh Framework Programme (FP7/2007–2013) under grant agreement n° 311975. This publication reflects the views only of the author, and the European Union cannot be held responsible for any use which may be made of the information contained therein.

References

Abrevaya XC, Sacco N, Mauas PJD, Corton E (2011) Archaea-based microbial fuel cell operating at high ionic strength conditions. Extremophiles 15:633–642

Ambati RR, Phang SM, Ravi S, Aswathanarayana RG (2014) Astaxanthin: sources, extraction, stability, biological activities and its commercial applications. Mar Drugs 12:128–152

Anderson RE, Sogin ML, Baross JA (2015) Biogeography and ecology of the rare and abundant microbial lineages in deep-sea hydrothermal vents. FEMS Microbiol Ecol 91:1–11

Anton J, Meseguer I, Rodriguez-Valera F (1988) Production of an extracellular polysaccharide by *Haloferax mediterranei*. Appl Environ Microbiol 54:2381–2386

Antunes A, Rainey FA, Wanner G, Taborda M, Patzold J, Nobre MF, da Costa MS, Huber R (2008) A new lineare of halophilic, wall-less, contractile bacteria from a brine-filled deep of the Red Sea. J Bacteriol 190:3580–3587

Antunes A, Alam I, Bajic VB, Stingl U (2011) Genome sequence of *Salinisphaera shabanensis*, a gammaproteobacterium from the harsh, variable environment of the brine-seawater interface of the Shaban deep in the Red Sea. J Bacteriol 193:4555–4556

Auguet J-C, Barberan A, Casamayor EO (2010) Global ecological patterns in uncultured Archaea. ISME J 4:182–190

Banat IM, Makkar RS, Cameotra SS (2000) Potential commercial applications of microbial surfactants. Appl Microbiol Biotechnol 53:459–508

Bhattacharyya A, Saha J, Haldar S, Bhowmic A, Mukhopadhyay UK, Mukherjee J (2014) Production of poly-3-(hydroxybutyrate-co-hydroxyvalerate) by Haloferax mediterranei using rice-based ethanol stillage with simultaneous recovery and re-use of medium salts. Extremophiles 18:463–470

Bhattacharyya A, Jana K, Haldar S, Bhowmic A, Mukhopadhyay UK, Mukherjee J (2015) Integration of poly-3-(hydroxybutyrate-co-hydroxyvalerate) production by Haloferax mediterranei through utilization of stillage from rice-based ethanol manufacture in India and its techno-economic analysis. World J Microbiol Biotechnol 31:717–727

Boetius A, Wenzhöfer F (2013) Seafloor oxygen consumption fuelled by methane from cold seeps. Nat Geosci 6:725–734

Borin S, Brusetti L, Mapelli F, D'Auria G, Brusa T, Marzorati M, Rizzi A, Yakimov M, Marty D, De Lange GJ, Van der Wielen PWJJ, Bolhuis H, McGenity TJ, Polymenakou PN, Malinverno E, Giuliano L, Corselli C, Daffonchio D (2009) Sulfur cycling and methanogenesis primarily drive microbial colonization of the highly sulfidic Urania deep hypersaline basin. Proc Natl Acad Sci USA 106:9151–9156

Borin S, Mapelli F, Rolli E, Song B, Tobias C, Schmid MC, De Lange GJ, Reichart GJ, Schouten S, Jetten M, Daffonchio D (2013) Anammox bacterial populations in deep marine hypersaline gradient systems. Extremophiles. doi:10.1007/s00792-013-0516-x

Bougouffa S, Yang JK, Lee OO, Wang Y, Batang Z, Al-Suwailem A et al (2013) Distinctive microbial community structure in highly stratified deep-sea brine water columns. Appl Environ Microbiol 79:3425–3437

Camerlenghi A (1990) Anoxic basins of the eastern Mediterranean: geological framework. Mar Chem 31:1–19

Campbell BJ, Jeanthon C, Kostka JE, Luther GW 3rd, Cary SC (2001) Growth and phylogenetic properties of novel bacteria belonging to the epsilon subdivision of the Proteobacteria enriched from *Alvinella pompejana* and deep-sea hydrothermal vents. Appl Environ Microbiol 67:4566–4572

Campbell BJ, Smith JL, Hanson TE, Klotz MG, Stein LY, Lee CK, Wu D, Robinson JM, Khouri HM, Eisen JA, Cary SC (2009) Adaptations to submarine hydrothermal environments exemplified by the genome of *Nautilia profundicola*. PLoS Genet 5:e1000362

Cao H, Zhang W, Wang Y, Qiari P-Y (2015) Microbial community changed along the active seepage site of one cold seep in the Red Sea. Front Microbiol 6. doi:10.3389/fmicb.2015.00739

Capes MD, DasSarma P, DasSarma S (2012) The core and unique proteins of haloarchaea. BMC Genom 13:39–45

Cockell CS (2006) The origin and emergence of life under impact bombardment. Philos Trans R Soc B 361:1845–1856

Daffonchio D, Borin S, Brusa T, Brusetti L, van der Wielen PWJJ, Bolhuis H, Yakimov M, D'Auria G, Giuliano L, Marty D, Tamburini C, McGenity T, Hallsworth J, Sass A, Timmis KN, Tselepides A, de Lange G, Hübner H, Thomson J, Varnavas S, Gasparoni F, Gerber H, Malinverno E, Corselli C, Party The Biodeep Scientific (2006) Stratified prokaryote network in the oxic-anoxic transition of a deep sea halocline. Nature 440:203–207

Dalmaso GZL, Ferreira D, Vermelho AB (2015) Marine extremophiles: a source of hydrolases for biotechnological applications. Mar Drugs 13:1925–1965

De Vitis V, Guidi B, Contente ML, Granato T, Conti P, Molinari F, Crotti E, Mapelli F, Borin S, Daffonchio D, Romano D (2015) Marine microorganisms as source of stereoselective esterases and ketoreductases: kinetic resolution of a prostaglandin intermediate. Mar Biotechnol 17:144–152

Dick GJ, Tebo BM (2010) Microbial diversity and biogeochemistry of the Guaymas Basin deep-sea hydrothermal plume. Environ Microbiol 12:1334–1347

Dick GJ, Anantharaman K, Baker BJ, Li M, Reed DC, Sheik CS (2013) The linkages to seafloor and water microbiology of deep-sea hydrothermal vent plumes: ecological and biogeographic column habitats. Front Microbiol 4:124

Dickins HD, Van Vleet ES (1992) Archaebacterial activity in the Orca basin determined by the isolation of characteristic isopranyl ether-linked lipids. Deep-Sea Res 39:521–536

Díez B, Bergman B, Pedrós-Alió C et al (2012) High cyanobacterial nifH gene diversity in Arctic seawater and sea ice brine. Environ Microbiol Rep 4(3):360–366

Domínguez de María P (2013) On the use of seawater as reaction media for large-scale applications in biorefineries. ChemCatChem 5:1643–1648

Dupré S, Woodside J, Foucher J-P, de Lange G, Mascle J, Boetius A, Mastalerz V, Stadnitskaia A, Ondréas H, Huguen C, Harmegnies F, Gontharet S, Loncke L, Deville E, Niemann H, Omoregie E, Olu-Le Roy K, Fiala-Medioni A, Dählmann A, Caprais J-C, Prinzhofer A, SibuetM, Pierre C, SinningheDamsté J & NAUTINIL scientific party (2007) Seafloor geological studies above active gas chimneys off Egypt (Central Nile Deep Sea Fan). Deep Sea Res Part I Oceanogr Res Pap 54(7):1146–1172

Dupré S, Mascle M, Foucher JP, Harmegnies F, Woodside J, Pierre C (2014) Warm brine lakes in craters of active mud volcanoes, Menes caldera off NW Egypt: evidence for deep-rooted thermogenic processes. Geo-Mar Lett 34:153–168

Eder W, Ludwig W, Huber R (1999) Novel 16S rRNA gene sequences retrieved from highly saline brine sediments of Kebrit Deep, Red Sea. Arch Microbiol 172:213–218

Eder W, Jahnke LL, Schmidt M, Huber R (2001) Microbial diversity of the brine-seawater interface of the Kebrit Deep, Red Sea, studied via 16S rRNA gene sequences and cultivation methods. Appl Environ Microbiol 67:3077–3085

Eder W, Schmidt M, Koch M, Garbe-Shonberg D, Huber R (2002) Prokaryotic phylogenetic diversity and corresponding geochemical data of the brine-seawater interface of the Shaban Deep, Red Sea. Environ Microbiol 4:758–763

Elderfield H, Schultz A (1996) Mid-ocean ridge hydrothermal fluxes and the chemical composition of the ocean. Annu Rev Earth Planet Sci 24:191–224

Ferrer M, Werner J, Chernikova TN, Bargiela R, Fernández L, La Cono V et al (2012) Unveiling microbial life in the new deep-sea hypersaline Lake Thetis. Part II: a metagenomic study. Environ Microbiol 14:268–281

German CR, Bowen A, Coleman ML, Honig DL, Huber JA, Jakuba MV, Kinsey JC, Kurz MD, Leroy S, McDermott JM, de Lepinay BM, Nakamura K, Seewald JS, Smith JL, Sylva SP, van Dover CL, Whitcomb LL, Yoerger DR (2010) Diverse styles of submarine venting on the ultra-slow spreading Mid-Cayman rise. Proc Natl Acad Sci USA 107:14020–14025

Giovannelli D, Ferriera S, Johnson J, Kravitz S, Pérez-Rodriguez I, Ricci J, O'Brien C, Voordeckers JW, Bini E, Vetriani C (2011) Draft genome sequence of *Caminibacter mediatlanticus* strain TB-2T, an epsiloproteobacterium isolated from a deep-sea hydrothermal vent. Stan Genomic Sci 5:135–143

Godet L, Zelnio KA, Van Dover CL (2011) Scientists as stakeholders in conservation of hydrothermal vents. Conserv Biol 25:214–222

Grünke S, Felden J, Lichtschlag A, Girnth A, De Beer D, Wenzhöfer F, Boetius A (2011) Niche differentiation among mat-forming, sulfide-oxidizing bacteria at cold seeps of the Nile Deep Sea Fan (Eastern Mediterranean Sea). Geobiology 9:330–348

Hallsworth JE, Yakimov MM, Golyshin PN, Gillion JL, Auria D, de Lima G et al (2007) Limits of life in MgCl2-containing environments: chaotropicity defines the window. Environ Microbiol 9:801–813

Hartmann M, Scholten JC, Stoffers P, Wehner F (1998) Hydrographic structure of brine-filled deeps in the Red Sea—new results from the Shaban Kebrit, Atlantis II, and Discovery Deep. Mar Geol 144:311–330

Head IM, Jones DM, Larter SR (2003) Biological activity in the deep subsurface and the origin of heavy oil. Nature 426:344–352

Hou J, Han J, Cai L, Zhou J, Lü Y, Jin C, Liu J, Hua Xiang H (2014) Characterization of genes for chitin catabolism in *Haloferax mediterranei*. Appl Microbiol Biotechnol 98:1185–1194

Hua NP, Kanekiyo A, Fujikura K, Yasuda H, Naganuma T (2007) *Halobacillus profundi* sp. nov. and *Halobacillus kuroshimensis* sp. nov., moderately halophilic bacteria isolated from a deep-sea methane cold seep. Int J Syst Evol Microbiol 57:1243–1249

Huber JA, Welch DBM, Morrison HG, Huse SM, Neal PR, Butterfield DA, Sogin ML (2007) Microbial population structures in the deep marine biosphere. Science 318:97–100

Jebbar M, Franzetti B, Girard E, Oger P (2015) Microbial diversity and adaptation to high hydrostatic pressure in deep-sea hydrothermal vents prokaryotes. Extremophiles 19:721–740

Joint I, Mühling M, Querellou J (2010) Culturing marine bacteria—an essential prerequisite for biodiscovery. Microb Biotechnol 3(5):564–575

Jongsma D, Fortuin AR, Huson W, Troelstra SR, Klaver GT, Peters JM, van Harten D, de Lange GJ, ten Haven L (1983) Discovery of an anoxic basin within the Strabo trench, eastern Mediterranean. Nature 305:795–797

Joye SB, MacDonald IR, Montoya JP, Piccini M (2005) Geophysical and geochemical signatures of Gulf of Mexico seafloor brines. Biogeosciences 2:295–309

Joye SB, Samarkin VA, Orcutt BN, MacDonald IR, Hinrichs KU, Elvert M, Teske AP, Lloyd KG, Lever MA, Montoya JP, Meile CD (2009) Metabolic variability in seafloor brines revealed by carbon and sulphur dynamics. Nat Geosci 2:349–354

Junge K, Gosink JJ, Hoppe HG et al (1998) *Arthrobacter*, *Brachybacterium* and *Planococcus* isolates identified from Antarctic Sea ice brine. Description of *Planococcus mcmeekinii*, sp. nov. Syst Appl Microbiol 21:306–314

Karan R, Capes MD, DasSarma P, DasSarma S (2013) Cloning, overexpression, purification, and characterization of a polyextremophilic β-galactosidase from the Antarctic haloarchaeon *Halorubrum lacusprofundi*. BMC Biotechnol 13:3–9
Kathiresan K, Saravanakumar K, Senthilraja P (2011) Bio-ethanol production by marine yeasts isolated from coastal mangrove sediment. Int Multi Res J 1:19–24
Kennedy SP, Ng WV, Salzberg SL, Hood L, DasSarma S (2001) Understanding the adaptation of Halobacterium species NRC-1 to its extreme environment through computational analysis of its genome sequence. Genome Res 11:1641–1650
Krembs C, Eicken H, Junge K, Deming JW (2002) High concentrations of exopolymeric substances in Arctic winter sea ice: implications for the polar ocean carbon cycle and cryoprotection of diatoms. Deep Sea Res I 49:2163–2181
Krembs C, Eicken H, Deming JW (2011) Exopolymer alteration of physical properties of sea ice and implications for ice habitability and biogeochemistry in a warmer Arctic. Proc Natl Acad Sci USA 108(9):3653–3658
Kwon KK, Woo JH, Yang SH, Kang JH, Kang SG, Kim SJ, Sato T, Kato C (2007) *Altererythrobacter epoxidivorans* gen. nov., sp. nov., an epoxide hydrolase-active, mesophilic marine bacterium isolated from cold-seep sediment, and reclassification of *Erythrobacter luteolus* Yoon et al. 2005 as *Altererythrobacter luteolus* comb. nov. Int J Syst Evol Microbiol 57:2207–2211
La Cono V, Smedile F, Bortoluzzi G, Arcadi E, Maimone G, Messina E et al (2011) Unveiling microbial life in new deep-sea hypersaline Lake Thetis. Part I: prokaryotes and environmental settings. Environ Microbiol 13:2250–2268
LaRock PA, Lauer RD, Schwarz JR, Watanabe KK, Wiesenburg DA (1979) Microbial biomass and activity distribution in an anoxic hypersaline basin. Appl Environ Microbiol 37:466–470
Lazar CS, L'Haridon S, Pignet P, Toffin L (2011) Archaeal populations in hypersaline sediments underlying orange microbial mats in the Napoli mud volcano. Appl Environ Microbiol 77 (9):3120–3131
Lazar CS, Parkes JR, Cragg BA, L'Haridon S, Toffin L (2012) Methanogenic activity and diversity in the centre of the Amsterdam mud volcano, Eastern Mediterranean Sea. FEMS Microbiol Ecol 81:243–254
Lentzen G, Schwarz T (2006) Extremolytes: natural compounds from extremophiles for versatile applications. Appl Microbiol Biotechnol 72:623–634
Lesniewski RA, Jain S, Anantharaman K, Schloss PD, Dick GJ (2012) The metatranscriptome of a deep-sea hydrothermal plume is dominated by water column methanotrophs and lithotrophs. ISME J 6:2257–2268
Levin LA (2005) Ecology of cold seep sediments: interactions of fauna with flow, chemistry and microbes. Oceanogr Mar Biol Annu Rev 43:1–46
Litchfield CD (2011) Potential for industrial products from the halophilic Archaea. J Ind Microbiol Biotechnol 38:1635–1647
Logan BE (2005) Simultaneous wastewater treatment and biological electricity generation. Water Sci Technol 52:31–37
Macelroy RD (1974) Some comments on the evolution of the extremophiles. Biosystems 6:74–75
Martinez-Checa F, Toledo FL, El Mabrouki K, Quesada E, Calvo C (2007) Characteristics of bioemulsiWer V2-7 synthesized in culture media added of hydrocarbons: chemical composition, emulsifying activity and rheological properties. Bioresour Technol 98:3130–3135
Mathis BJ, Marshall CW, Milliken CE, Makkar RS, Creager SE, May HD (2008) Electricity generation by thermophilic microorganisms from marine sediment. Appl Microbiol Biotechnol 78:147–155
Meckenstock RU, von Netzer F, Stumpp C, Lueders T, Himmelberg AM, Hertkorn N, Schmitt-Kopplin P, Harir M, Hosein R, Haque S, Schulze-Makuch D (2014) Water droplets in oil are microhabitats for microbial life. Science 345:673–676
Mikucki JA, Priscu JC (2007) Bacterial diversity associated with Blood Falls, a subglacial outflow from the Taylor Glacier, Antarctica. Appl Environ Microbiol 73:4029–4039

Mikucki JA, Pearson A, Johnston DT et al (2009) A contemporary microbially maintained subglacial ferrous "ocean". Science 324:397–400
Mikucki JA, Auken E, Tulaczyk S et al (2015) Deep groundwater and potential subsurface habitats beneath an Antarctic dry valley. Nat Commun 6:6831. doi:10.1038/ncomms7831
Møller AK, Barkay T, Al-Soud WA et al (2010) Diversity and characterization of mercury-resistant bacteria in snow, freshwater and sea-ice brine from the high Arctic. FEMS Microbiol Ecol 75(2011):390–401
Murray AE, Kenig F, Fritsen CH et al (2012) Microbial life at −13 & #xB0;C in the brine of an ice-sealed Antarctic lake. Proc Natl Acad Sci USA 109(50):20626–20631
Nakagawa S, Takaki Y, Shimamura S, Reysenbach AL, Takai K, Horikoshi K (2007) Deep-sea vent ε-proteobacterial genomes provide insights into emergence of pathogens. Proc Natl Acad Sci USA 104:12146–12150
Ngugi DK, Blom J, Alam I, Rashid M, Ba-Alawi W, Zhang G, Hikmawan T, Guan Y, Antunes A, Siam R, El Dorry H, Bajic V, Stingl U (2015) Comparative genomics reveals adaptations of a halotolerant thaumarchaeon in the interfaces of brine pools in the Red Sea. ISME J 9:396–411
Niemann H, Linke P, Knittel K, MacPherson E, Boetius A, Brückmann W, Larvik G, Wallmann K, Schacht U, Omoregie E, Hilton D, Brown K, Rehder G (2013) Methane-carbon flow into the benthic food web at cold seeps—a case study from the Costa Rica subduction zone. PLoS ONE 8:e74894
Nogi Y, Kato C, Horikoshi K (2002) Psychromonas kaikoae sp nov., a novel piezophilic bacterium from the deepest cold-seep sediments in the Japan Trench. Int J Syst Evol Microbiol 52:1527–1532
Ogan A, Danis O, Gozuacik A, Cakmar E, Birbir M (2012) Production of cellulase by immobilized whole cells of Haloarcula. Appl Biochem Microbiol 48:440–453
Omoregie EO, Mastalerz V, de Lange G, Straub KL, Kappler A, Røy H, Stadnitskaia A, Foucher J-P, Boetius A (2008) Biogeochemistry and community composition of iron- and sulfur-precipitating microbial mats at the Chefren mud volcano (Nile Deep Sea Fan, Eastern Mediterranean). Appl Environ Microbiol 74(10):3198–3215
Oren A (2013) Life at high salt concentrations, intracellular KCl concentrations, and acidic proteomes. Front Microbiol 4:315–321
Orsi WD, Edgcomb VP, Christman GD, Biddle JF (2013) Gene expression in the deep biosphere. Nature 499:205–208
Pachiadaki MG, Kallionaki A, Dählmann A, De Lange GJ, Kormas KA (2011) Diversity and spatial distribution of prokaryotic communities along a sediment vertical profile of a deep-sea mud volcano. Microb Ecol 62:655–668
Pérez-Rodríguez I, Bohnert KA, Cuebas M, Keddis R, Vetriani C (2013) Detection and phylogenetic analysis of the membrane-bound nitrate reductase (Nar) in pure cultures and microbial communities from deep-sea hydrothermal vents. FEMS Microbiol Ecol 86:256–267
Pettit RK (2011) Culturability and secondary metabolite diversity of extreme microbes: expanding contribution of deep sea and deep-sea vent microbes to natural product discovery. Mar Biotechnol 13:1–11
Prakash O, Shouche Y, Jangid K, Kostka JE (2013) Microbial cultivation and the role of microbial resource centers in the omics era. Appl Microbiol Biotechnol 97:51–62
Quillaguaman J, Guzman H, Van-Thuoc D, Hatti-Kaul R (2010) Synthesis and production of polyhydroxyalkanoates by halophiles: current potential and future prospects. Appl Microbiol Biotechnol 85:1687–1696
Rateb ME, Ebel R (2011) Secondary metabolites of fungi from marine habitats. Nat Prod Rep 28:290–344
Rédou V, Navarri M, Meslet-Cladière L, Barbier G, Burgaud G (2015) Species richness and adaptation of marine fungi from deep-subseafloor sediments. Appl Environ Microbiol 81:3571–3583

Rodríguez-Moya J, Argandoña M, Iglesias-Guerra F, Nieto JJ, Vargas C (2013) Temperature- and salinity-decoupled overproduction of hydroxyectoine by Chromohalobacter salexigens. Appl Environ Microbiol 79:1018–1023

Ruff SE, Biddle JF, Teske AP, Knittel K, Boetius A, Ramette A (2015) Global dispersion and local diversification of the methane seep microbiome. Proc Natl Acad Sci USA 112:4015–4020

Sass AM, Sass H, Coolen MJL, Cypionka H, Overmann J (2001) Microbial communities in the chemocline of a hypersaline deep-sea basin (Urania Basin, Mediterranean Sea). Appl Environ Microbiol 67:5392–5402

Sauer T, Galinski EA (1998) Bacterial milking: a novel bioprocess for production of compatible solutes. Biotechnol Bioeng 57:306–313

Shao S, Luan X, Dang H, Zhou H, Zhao Y, Liu H, Zhang Y, Dai L, Ye Y, Klotz MG (2014) Deep-sea methane seep sediments in the Okhotsk Sea sustain diverse and abundant anammox bacteria. FEMS Microbiol Ecol 87:503–516

Shock EL (1992) Chemical environments of submarine hydrothermal systems. Orig Life Evol Biosph 22:67–107

Sievert SM, Vetriani C (2012) Chemoautotrophy at deep-sea vents: past, present, and future. Oceanography 25:218–233

Sorokin DY, Kublanov IV, Gavrilov SN, Rojo D, Roman P, Golyshin PN, Slepak VZ, Smedile F, Ferrer M, Messina E, La Cono V, Yakimov MM (2015) Elemental sulfur and acetate can support life of a novel strictly anaerobic haloarchaeon. ISME J. doi:10.1038/ismej.2015.79

Stokke R, Roalkvam I, Lanzen A, Haflidason H, Steen IH (2012) Integrated metagenomics and metaproteomic analyses of an ANME-1-dominated community in marine cold seep sediments. Environ Microbiol 14:1333–1346

Stokke R, Dahle H, Roalkvam I, Wissuwa J, Daae FL, Tooming-Klunderud A, Thorseth IH, Pedersen RB, Steen IH (2015) Functional interactions among filamentous Epsilonproteobacteria and Bacteroidetes in a deep-sea hydrothermal vent biofilm. Environ Microbiol. doi:10.1111/1462-2920.12970

Strous M, Fuerst JA, Kramer EHM, Logemann S, Muyzer G, van de Pas-Schoonen KT, Webb R, Kuenen JG, Jetten MSM (1999) Missing lithotroph identifed as new planctomycete. Nature 400:446–449

Suess E (2014) Marine cold seeps and their manifestations: geological control, biogeochemical criteria and environmental conditions. Int J Earth Sci (Geol Rundsch) 103:1889–1916

Suzuki Y, Kojima S, Sasaki T, Suzuki M, Utsumi T, Watanabe H, Urakawa H, Tsuchida S, Nunoura T, Hirayama H, Takai K, Nealson KH, Horikoshi K (2006) Host-symbiont relationships in hydrothermal vent gastropods of the genus *Alviniconcha* from the Southwest Pacific. Appl Environ Microbiol 72:1388–1393

Takai K, Nakamura K (2011) Archaeal diversity and community development in deep-sea hydrothermal vents. Curr Opin Microbiol 14:282–291

Takai K, Moyer CL, Miyazaki M, Nogi Y, Hirayama H, Nealson KH, Horikoshi K (2005) *Marinobacter alkaliphilus* sp. nov., a novel alkaliphilic bacterium isolated from subseafloor alkaline serpentine mud from Ocean drilling program site 1200 at South Chamorro seamount, Mariana Forearc. Extremophiles 9:17–27

Takai K, Nakagawa S, Reysenbach A-L, Hoek J (2006) Microbial ecology of mid-ocean ridges and back-arc basins. In: Christie DM, Fisher CR, Lee S-M, Givens S (eds) Back-arc spreading systems: geological, biological, chemical, and physical interactions. American Geophysical Union, Washington, DC. doi:10.1002/9781118666180.ch9

Takai K, Nakamura K, Toki T, Tsunogai U, Miyazaki M, Miyazaki J, Hirayama H, Nakagawa S, Nunoura T, Horikoshi K (2008) Cell proliferation at 122 °C and isotopically heavy CH_4 production by a hyperthermophilic methanogen under high-pressure cultivation. Proc Natl Acad Sci USA 105:10949–10954

Tanimura K, Nakayama H, Tanaka T, Kondo A (2013) Ectoine production from lignocellulosic biomass-derived sugars by engineered *Halomonas elongata*. Bioresour Technol 142:523–529

Thomas DN, Dieckmann GS (2002) Antarctic sea ice—a habitat for extremophiles. Science 295:641–644

Trincone A (2011) Marine biocatalysts: enzymatic features and applications. Mar Drugs 9:478–499

Uhlig C, Kilpert F, Frickenhaus S et al (2015) In situ expression of eukaryotic ice-binding proteins in microbial communities of Arctic and Antarctic sea ice. ISME J. doi:10.1038/ismej.2015.43

van der Wielen PWJJ, Bolhuis H, Borin S, Daffonchio D, Corselli C, Giuliano L, de Lange GJ, Varnavas SP, Thompson J, Tamburini C, Marty D, McGenity TJ, Timmis KN, Party The BioDeep Scientific (2005) The enigma of prokaryotic life in deep hypersaline anoxic basins. Science 307:121–123

van der Wielen PW, Heijs SK (2007) Sulfate-reducing prokaryotic communities in two deep hypersaline anoxic basins in the Eastern Mediterranean deep sea. Environ Microbiol 9:1335–1340

Van-Thuoc D, Guzman H, Quillaguaman J, Hatti-Kaul R (2010) High productivity of ectoines by *Halomonas boliviensis* using a combined two-step fed-batch culture and milking process. J Biotechnol 147:46–51

Vetriani C, Speck MD, Ellor SV, Lutz RA, Starovoytov V (2004) *Thermovibrio ammonificans* sp. nov., a thermophilic, chemolithotrophic, nitrate ammonifying bacterium from deep-sea hydrothermal vents. Int J Syst Evol Microbiol 54:175–181

Vetriani C, Voordeckers JW, Crespo-Medina M, O'Brien CE, Giovannelli D, Lutz RA (2014) Deep-sea hydrothermal vent Epsilonproteobacteria encode a conserved and widespread nitrate reduction pathway (Nap). ISME J 8:1510–1521

Vincent WF (2010) Microbial ecosystem responses to rapid climate change in the Arctic. ISME J 4:1089–1091

Wallmann K, Aghib FS, Castradori D, Cita MB, Suess E, Greinert J, Rickert D (2002) Sedimentation and formation of secondary minerals in the hypersaline Discovery basin, eastern Mediterranean. Mar Geol 186:9–28

Werber J, Ferrer M, Michel G, Mann AJ, Huang S, Juarez S, Ciordia S, Albar JP, Alcaide M, La Cono V, Yakimov, MM, Antunes A, Taborda M, da Costa MS, Hai T, Glöckner FO, Golyshina OV, Golyshin PN, Teeling H, The MAMBA Consortium (2014) *Halorhabdus tiamatea*: proteogenomics and glycosidase activity measurements identify the first cultivated euryarchaeon from a deep-sea anoxic brine lake as potential polysaccharide degrader. Environ Microbiol 16(8):2525–2537

Yakimov MM, Giuliano L, Cappello S, Denaro R, Golyshin PN (2007a) Microbial community of a hydrothermal mud vent underneath the deep-sea anoxic brine lake Urania (Eastern Mediterranean). Orig Life Evol Biosp 37:177–188

Yakimov MM, La Cono V, Denaro R, D'Auria G, Decembrini F, Timmis KN, Golyshin PN, Giuliano L (2007b) Primary producing prokaryotic communities of brine, interface and seawater above the halocline of deep anoxic lake L'Atalante, Eastern Mediterranean Sea. ISME J 0:1–13

Yakimov MM, La Cono V, Slepak VZ, La Spada G, Arcadi E, Messina E et al (2013) Microbial life in the Lake Medee, the largest deep-sea salt-saturated formation. Sci Rep 3:3554

Yakimov MM, La Cono V, La Spada G, Bortoluzzi G, Messina E, Smedile F, Arcadi E, Borghini M, Ferrer M, Schmitt-Kopplin P, Hertkorn N, Cray JA, Hallsworth JE, Golyshin PN, Giuliano L (2015) Microbial community of the deep-sea brine Lake Kryos seawater–brine interface is active below the chaotropicity limit of life as revealed by recovery of mRNA. Environ Microbiol 17(2):364–382

Zaky AS, Tucker GA, Daw ZY, Du C (2014) Marine yeast isolation and industrial application. FEMS Yeast Res 1–13

Zambelli P, Serra I, Fernandez-Arrojo L, Plou FJ, Tamborini L, Conti P, Contente ML, Molinari F, Romano D (2015) Sweet-and-salty biocatalysis: fructooligosaccharides production using *Cladosporium cladosporioides* in seawater. Process Biochem 50:1086–1090

Zeng X, Birrien J-L, Fouquet Y, Cherkashow G, Jebbar M, Querellou J, Oger P, Cambon-Bonavita M-A, Xiao X, Prieur D (2009) *Pyrococcus* CH1, an obligate piezophilic hyperthermophile: extending the upper pressure-temperature limits for life. ISME J 3:873–876

Zeng X, Zhang Z, Li X, Zhang X, Cao J, Jebbar M, Alain K, Shao Z (2015) Anoxybacter fermentans gen. nov., sp. nov., a piezophilic, thermophilic, anaerobic, fermentative bacterium isolated from a deep-sea hydrothermal vent. Int J Syst Evol Microbiol 65:710–715

Zitter TAC, Huguen C, Woodside JM (2005) Geology of mud volcanoes in the Eastern Mediterranean from combined sidescan sonar and submersible surveys. Deep-Sea Res I 52:457–475

Chapter 10
Coastal Sediments: Transition from Land to Sea

Lucas J. Stal

Abstract Coastal habitats differ greatly from the open sea and the ocean. The coast is a transition zone from the land to the sea and is therefore influenced by both. Coastal habitats are also very productive environments and therefore vulnerable to disturbances. The coast protects the land behind from flooding and it often keeps pace with sea-level changes. The coasts of most seas and the ocean are exposed to the tides and waves, and locally to ice cover and drifting ice. The tidal influence in some enclosed seas is less or even negligible. On the other hand, tidal influence may reach more than 100 km upstream of estuaries and rivers. The ecosystems of coastal habitats are complex and as everywhere microorganisms play a crucial role in maintaining ecosystem functions and keep the biogeochemical cycles going. There is a great variety of different coastal habitats. They may be intertidal or sublittoral. It is impossible to review all those different coastal habitats and therefore this chapter treats only three different intertidal coastal microbial ecosystems as representative cases for the microbiology of coastal sediments. Intertidal mudflats are dominated by diatoms, which are important primary producers and exude extracellular polymeric substances (EPS). These exopolymers may increase the erosion threshold of the mud and also forms the basis of the microbial food web. The second case is microbial mats that are formed by benthic cyanobacteria on sandy beaches. They are thought to be the modern representatives of the world's earliest ecosystem, the Precambrian stromatolites. The third case is mangrove forests that are important ecosystems in tropical and subtropical regions. The mangrove trees and shrubs form the basis of productive ecosystems that provide food for a plethora of local fauna, and protect the land behind. Mangrove

L.J. Stal (✉)
Department of Marine Microbiology and Biogeochemistry,
NIOZ Royal Netherlands Institute for Sea Research and Utrecht University,
PO Box 59, 1790 AB Utrecht, Texel, The Netherlands
e-mail: Lucas.stal@nioz.nl

L.J. Stal
Department of Aquatic Microbiology, University of Amsterdam,
Amsterdam, The Netherlands

L.J. Stal and M.S. Cretoiu (eds.), *The Marine Microbiome*,
DOI 10.1007/978-3-319-33000-6_10

ecosystems export a considerable amount of carbon to the ocean. The activity of diverse microbial communities is responsible for maintaining the major ecosystem functions and the cycling of the major element.

10.1 Introduction

Coastal seas and coastal environments are in general distinguished from open ocean by higher nutrient levels and consequently by higher primary productivity (Heip et al. 1995). These environments receive the nutrients mainly from run-off of the land and from river discharge and to a lesser extend from wet and dry deposition. The high primary productivity results in a high input of organic matter in coastal zones. Because of the shallowness of most coastal waters, much of the organic matter deposits on the seafloor, where microorganisms decompose it, while microbial decomposition in the shallow water column is less important.

Most coastal areas are exposed to strong tidal movements and currents, although there are also examples of closed seas where the tide is negligible such as the Mediterranean-, Black-, and Baltic Seas. Tidal movements cause salinity gradients in river estuaries and deltas. At high tide, the saline seawater enters the river upstream and at low tide the fresh river water reaches the mouth of the river and, depending on the volume of the discharge, dilutes the seawater well beyond the river mouth. Tidal movement may reach the river much farther upstream than the seawater, leading to purely freshwater tidal areas, such as is the case in the Scheldt river (Belgium, The Netherlands), where the tides are present 180 km upstream (up to the city of Ghent, Belgium). Other tidal estuaries in Europe include the Elbe and Weser (Germany), and the Humber (UK). The more dense seawater also causes a salinity wedge in which the seawater dives underneath the lighter freshwater, causing a more or less strongly stratified vertical salinity gradient. Tides also cause tidal influenced areas along the coast and along the borders of estuaries and river deltas (de Boer et al. 1988).

Life may have started in the ocean, possibly near a deep-sea hydrothermal vent (Cockell 2006). DNA sequence analysis points to a hyperthermophilic aquatic microorganism as the common ancestor of life. The reduced chemicals that may have provided the energy for the earliest forms of life and the steep temperature gradients from the vent to the cold seawater may have provided excellent conditions for life to develop. However, the earliest fossil evidence of life is found in stromatolites that developed in shallow coastal areas about 3.5 billion years before present (Schopf 2006). Stromatolites are laminated rocks that were formed by photosynthetic microbial mats that fixed CO_2 and precipitated silicate (and carbonate). The lamination originates from different layers of microbial mats of which growth and activity may have varied seasonally. The growth of some stromatolites followed sea-level rises. Modern stromatolites are formed in extreme environments, such as alkaline environments where they are protected from grazing. Today, stromatolites in shallow coastal environments can be found in the Exuma Cays

(Bahamas) and Shark Bay (Western Australia) (Reid et al. 2000). Marine microalgae (Eukarya) evolved in the ocean and have given rise to the evolution of land plants and perhaps most terrestrial animals may have originated from the sea. Hence, coastal areas have always functioned as the area of transition of life from the sea to the land and it should therefore not surprise that these areas still harbor life that is versatile and adapted to extreme variations in environmental conditions.

The littoral zone is usually defined as the intertidal zone, covering the part of the coast that is rarely inundated at spring tide to rarely exposed at neap tide. This so-called eulittoral (meaning 'real littoral') zone can be divided in an upper-, mid-, and lower littoral. Above the littoral is the area called supralittoral, which is influenced by the sea and by seawater through splash, or during storms or exceptional high tide. Likewise, the sublittoral starts where the eulittoral stops at the low shore and is never exposed, apart from exceptional cases (storms, exceptional low tide, tsunamis), but still under influence of the tide (waves and currents). In many coastal areas the sublittoral inhabit seagrass meadows. Seagrasses such as *Posidonia*, *Thalassia*, and *Zostera*, are marine flowering plants that returned from the land to the sea (Klap et al. 2000). Some species such as *Zostera noltii* may thrive in intertidal sediments. Seagrasses are associated with abundant microbial life occurring as epiphytes on the leaves and in the rhizosphere (Isaksen and Finster 1996; Welsh et al. 1996). This chapter is limited to a few examples of microbial ecosystems in the eulittoral with tidal sediment flats. Tidal flats can be situated on the continental shelf, in delta- or estuarine areas or on exposed coast. In this chapter, I present three different tidal sediment ecosystems, the intertidal mudflats, microbial mats, and mangrove forests.

10.2 Intertidal Mudflats

Intertidal sediments are coastal areas that reach from the low tide mark to the high tide mark and the sea therefore irregularly inundates parts of these areas. Irregularly, because inundation depends on several factors such as the tidal range, which is not constant and varies from neap tide and spring tide in a month's cycle, and also varies in a year cycle. Another important factor is the wind direction and force, which may push the water far landward or the opposite. When inundated, the seawater brings nutrients, sand, silt and clays, and marine life to the sediment. When exposed, the sediment may experience desiccation, with the consequence of increase in salinity of the pore water, large temperature changes, and nutrient depletion. Rain (wet deposition) may cause considerable salinity changes but also add nutrients (especially nitrogen). Therefore, intertidal areas are dynamic systems where sediment may be deposited from the seawater or eroded by currents or wind. Intertidal areas may be composed of mud and silt (mudflats) or of fine sand or any mixture of these. Beaches are often composed of fine and coarse sand. The sediment properties of intertidal areas depend on the physical forces that exert on them.

In high-energy areas silt and clays will remain in suspension while sand is deposited, while in low energy areas the smaller particles will deposit.

Mudflats are composed of a mixture of mud and sand. Mud is a term that collectively addresses silt and clays and other fine mineral compounds. These particles are charged and tend to stick together and therefore these sediments are said to be cohesive. Sand is usually composed of silicate (except for carbonate sand that is found on beaches of coral reefs) and is composed of uncharged grains that are noncohesive. Mudflats contain enough mud to be cohesive, but the amount of sand may vary between but also within mudflats. The upper mudflats usually contain the highest proportion of very fine clays, the middle mudflats are composed of silt and the lower mudflats contain the highest amount of fine sand (Shi and Chen 1996). Fine or coarse sandy sediments as found on beaches are noncohesive.

Intertidal mudflats are found globally and are geographically subdivided in those found at low-latitude (arid and tropical areas), mid-latitude (temperate areas), and high-latitude (influenced by ice cover) (Dionne 1988). The most important discriminator of a tidal flat is tidal range (the difference between neap and spring tide or highest tide and lowest tide), with subdivisions of exposure to waves and the mudflat slope, the latter is low when <0.04, steep when >0.04 and very steep when ~ 0.16 (Dyer et al. 2000). A tidal range of 2–6 m is characteristic of meso- and macrotidal mudflats that can be further subdivided according to their sediment density. Low-density mudflats contain <600 kg m^{-3}, medium density 600–1000 kg m^{-3} and high-density >1000 kg m^{-3}.

Sediment type and grain size are the best descriptors for the biology of the upper mudflats while no descriptors can be given for the biology of the middle and lower mudflats (Dyer et al. 2000). Microbial communities on mudflats favor fine-grained sediments, which in turn stabilize the sediment surface and increase porosity and water content. Hence, organic content and nutrients are inversely correlated with sediment grain size, i.e., increase with decreasing grain size.

Intertidal mudflats in estuaries and deltas or exposed coasts are usually characterized by the occurrence of biofilms of benthic (bottom dwelling) diatoms. Diatoms are microalgae that possess a silica frustule as cell envelope. Diatoms are photosynthetic and use light as the source of energy with which they fix inorganic carbon (CO_2) to synthesize organic matter for growth. Benthic diatoms are well adapted to the dynamic regime of mudflats. They are motile and this allows the organism to migrate fast. When the mudflat is exposed, diatoms migrate to the surface, while during inundation they migrate down into the sediment. This migration into the sediment serves probably two goals. By doing so, the diatoms escape grazing by marine life that comes with the water. But also, nutrient concentrations are higher deeper into the sediment. At low tide, the diatoms move to the sediment surface allowing them to photosynthesize, at least when low tide is at daytime. Hence, benthic diatoms will have only a short window of time during each 24-h day when low tide and daylight co-occur. This can be as short as a few hours during which they will have to photosynthesize and store carbohydrate to survive the rest of the day in darkness. The dense diatom biofilm will deplete rapidly the available nutrients present in the very limited volume of pore water in the biofilm, making the balanced synthesis of cell material quickly impossible. This will result in so-called unbalanced growth during which the photosynthetic products of the diatom will be diverted to carbohydrate. The intracellular pool of the storage compound of diatoms, chrysolaminarin, a polymer of glucose is filled up. Excess carbohydrate is exuded as EPS. This EPS can be taken up again by the diatom and used as carbon- and energy source in the dark (De Brouwer et al. 2002; Staats et al. 2000).

There are usually two operational fractions of EPS distinguished in diatom biofilms. These fractions depend very much on the protocol used to extract them and are therefore called 'operational'. Warm (30 °C) water extraction of the biofilm, or intertidal sediment, yields 'water-soluble' EPS, also termed mucilage (less often 'slime'), or colloidal EPS, and is usually loosely associated with the diatoms and is the matrix in which the diatoms are embedded. This EPS is glucose rich. The other operational fraction is usually extracted from the sediment using EDTA, and is also known as EDTA-extractable EPS. This EPS is apparently tightly bound to the sediment. It is low in glucose and rich in acid residues such as uronic acids, sulfated sugars, pyruvate, and other acid organic groups. These charged groups are interacting with each other and with the charged silt and mud particles. This EPS therefore renders stability to the sediment. It has been argued that the EDTA-extractable EPS is in fact derived from the colloidal fraction after it has been depleted of glucose (Stal 2010). The diatoms themselves are metabolizing their own EPS in the dark (De Brouwer et al. 2002). It is unclear whether the intracellular storage compound, chrysolaminarin, is metabolized in the dark in addition to the EPS since this intracellular reserve compound did not show such dynamics in an axenic culture (Staats et al. 2000).

But this colloidal EPS should also be available as carbon- and energy source to other microorganisms in the mudflat sediment. Also these microorganisms may utilize preferentially the glucose component of EPS that is hydrolyzed from the polymer by exo- or ectoenzymes, leaving a complex, highly branched acidic polymer behind, which might be recalcitrant to further degradation, or at least is degraded at much lower rate. Degradation of this recalcitrant material is more efficient at higher temperature (Middelburg et al. 1996) and therefore more important in summer and in (sub)tropical mudflats. However, it cannot be excluded that EDTA-extractable EPS is directly synthesized as such by the diatoms (as well as perhaps by other microorganisms in the mudflat) rather than being the product of degradation of colloidal EPS as was speculated by Stal (2010). It can also not be excluded that diatoms produce other types of EPS. There may be a polysaccharide sheath tightly associated with the cell frustule. Moreover, benthic diatoms are motile and migration goes together with the exudation of EPS.

The operational fractions of EPS from diatom cultures are different from those from sediments. When the diatom culture is centrifuged, the acidic fraction is found in the supernatant. This highly charged EPS is water-soluble in the absence of charged sediment particles. In the muddy sediment, this fraction seems to be strongly bound to the silt and clay particles and can only be removed using EDTA. The warm (30 °C) water extracts from axenic diatom cultures and from natural diatom biofilms seem to yield the same operational fraction of glucose-rich EPS.

Hence, in the intertidal mudflat, EPS fractions serve various functions. They form the matrix in which the biofilm organisms are embedded and render stability to this biofilm. There are several other important aspects connected to this matrix. The EPS may bind and provide the microorganisms with rare but essential trace elements or scavenge toxic compounds (heavy metals, xenobiotics, antibiotics) (Gutierrez et al. 2012; Tourney and Ngwenya 2014). The matrix may protect the biofilm from grazing and from desiccation, when exposed at low tide. EPS may also serve as substrate for other microorganisms, although its food quality is low, since it does not contain nitrogen or phosphorus. Hence, only when nutrients are available, bacteria may utilize EPS as substrate for growth.

Diatom biofilms may cover large areas on intertidal mudflats but have a patchy distribution. This patchiness is obvious at different scales, from the micro- to the macroscale (Benyoucef et al. 2014). The highest standing stock density of microphytobenthos is usually seen in the upper- and mid-littoral and the seasonality seems to be controlled by temperature.

The bacterial component of intertidal mudflats has not received much attention. The mostly unknown bacteria in mudflat sediments may possess unexpected properties with respect to the production of novel antimicrobial compounds and new bioactive non-ribosomal peptides (Tambadou et al. 2014). It has also been conceived that the bacterial component and the interactions of diatoms with bacteria are fundamental to the formation of benthic diatom biofilms on intertidal mudflats (Bruckner et al. 2011).

The copious amounts of EPS produced by the benthic diatoms on intertidal mudflats form an excellent substrate as carbon- and energy source for other microorganisms. This EPS is also transported to deeper sediment layers due to the vertical migration of diatoms in the tidal- and day–night rhythm. Taylor et al. (2013) estimated that a turnover of 52–369 % of EPS was required in order to explain its content in the mudflats taking into account the microphytobenthic biomass, its photosynthetic activity, and the production rate of EPS. Degradation is rapid and experiments with ^{13}C-labeled EPS showed that it ended up in phospholipid fatty acids of Gram-positive and Gram-negative bacteria but also in diatoms (Taylor et al. 2013). The latter could be taken as support for the idea that diatoms take up (their own) EPS as was demonstrated in axenic cultures, although the label could also have been taken up from degradation products of EPS. It is anyway likely that these polymers are partly degraded extracellular by exo- or ectoenzymes. Alpha- and Gammaproteobacteria seemed to be specialists for growth at the expense of EPS as was concluded from their high labeling of rRNA (Taylor et al. 2013). The addition of colloidal EPS to mudflat samples resulted in an increase of beta-glucosidase activity and in a shift of the bacterial community towards Gammaproteobacteria, notably *Acinetobacter* (Haynes et al. 2007), but it seems likely that a complex community of heterotrophic mudflat bacteria are required for the hydrolysis of the polymers and the subsequent degradation of EPS (Hofmann et al. 2009). Also, anaerobic conditions seem to be required for the degradation of high-molecular EPS. *Clostridium* and *Bacteroidetes* were held responsible for the hydrolysis of the polymers and a joint venture of sulfate-reducing bacteria and methanogens further decomposed the hydrolyzed and fermented EPS (McKew et al. 2013). Under oxygenated conditions degradation did not occur.

The organic matter content depends on the sediment composition and porosity and this in turn determines the sulfate reduction rate, which also strongly depends on temperature. Highest activities are thus observed in muddy sediments with the highest content of organic matter and in spring, summer, and autumn (Al-Raei et al. 2009). The abundance of functional genes for sulfate reduction (*dsrAB*) follows exactly this pattern (Leloup et al. 2005).

Using Mag-SIP (stable isotope probing of 16S rRNA) Miyatake et al. (2013) also showed that cyanobacteria and diatoms took up glucose under dark anoxic conditions and assumed that these organisms survived by anaerobic fermentation. In the top 2 cm layer of the sediment, diatoms and cyanobacteria took up all offered substrates (acetate, propionate, glucose, and amino acids), while deeper in the sediment it was mainly glucose and propionate that were used by these organisms. In the deeper (2–5 cm) sediment layer, it was mainly sulfate-reducing bacteria that were active. In the top layer, all major groups of bacteria were using the offered substrates (Miyatake et al. 2013). The sulfate-reducing bacteria did not take up glucose. Gammaproteobacteria fermented glucose and subsequently the sulfate-reducing bacteria end-oxidized the fermentation products.

Different groups of microorganisms in the mudflat are resuspended into the water at different shear forces (friction velocities) and the threshold of resuspension was a function of the organism's association with or binding to the sediment. Hence, free-living viruses, bacteria, and some nanoflagellates are resuspended at low shear force, while those attached to the sediment come into suspension at higher forces, together with the muddy sediment bed and the diatoms smaller than 60 μm. Larger diatoms need higher velocities of 5.5–6.5 cm s^{-1} (Dupuy et al. 2014; Montanié et al. 2014a, b). The resuspension of the sediment microphytobenthos has been considered an important supply to the food web of the (estuarine) water column. However, Guizien et al. (2014) demonstrated that grazing predominated over resuspension and that even import of picocyanobacteria from offshore waters was grazed before erosion transfer could happen.

Diatoms may accumulate dimethylsulfoniopropionate (DMSP). It is still debated whether this compound serves as osmoprotectant to compensate the osmotic pressure of the cytoplasmic to the prevailing salinity, cryoprotectant, or as storage for excess reducing equivalents and energy in a situation of unbalanced growth (Stefels 2000). Upon a salinity down-shock (e.g., when a rain shower occurs while the sediment is exposed at low tide), the diatoms must quickly exude the superfluous DSMP. The first step in the fate of this DMSP in the sediment is its conversion to acrylate and the gaseous dimethyl sulfate (DMS) (which renders the sea its typical 'smell'), which is emitted into the atmosphere where it has been suggested to play a role in climate change feedbacks (Gypens et al. 2014). There is a variety of bacteria that possess DMSP lyase and degrade this rather unstable compound. Some planktonic algae such as *Phaeocystis* also possess DMSP lyase and may thereby regulate intracellular DMSP levels. It is not known whether benthic diatoms contain DMSP lyase. Acrylate is also known to serve as an antioxidant for the microalgae, deterrent of grazers, inhibit viral infection. When DMSP lyase is activated as a response on grazing, the production of acrylate would protect the benthic diatoms. Acrylate is further metabolized by a variety of heterotrophic bacteria.

The nitrogen (N) cycle is prominently present in any microbial ecosystem. It consists of oxidation and reduction reactions, transforming nitrogen from its most oxidized state (nitrate, NO_3^-) to its most reduced state (ammonium, NH_4^+). Denitrification and anaerobic ammonium oxidation lead to the formation of dinitrogen (N_2) and nitrogen fixation transforms it back to ammonium. Until recently, it was thought that Bacteria and Archaea are predominantly responsible for running the nitrogen cycle but it has now become clear that Eukarya play a far more important role. Many marine microbial Eukarya, among which benthic diatoms, are capable of storing nitrate intracellular, and are able to use it as an electron donor under anaerobic or oxygen-limiting conditions in the dark (Kamp et al. 2015). In intertidal sediments, diatoms are the major component of the microphytobenthos and also the organisms that are predominantly storing nitrate in these environments. Other Eukarya such as foraminifers and gromiids or large sulfur bacteria such as *Beggiatoa* or *Thioploca*, all known to store nitrate, were not important (Stief et al. 2014). Diatoms store up to 274 mM nitrate in their cells and respire this to ammonium, probably using the dissimilatory nitrate reduction to ammonium (DNRA) pathway. Benthic diatoms on intertidal sediments are forced to live chemotrophically for a considerable part of the day. When migrating into the sediment they will experience anoxic conditions and nitrate respiration would increase their metabolic efficiency considerably and be close to aerobic respiration. However, it is not clear whether this nitrate respiration is really part of the dark metabolism in these diatoms or only serves short-term survival. The genes coding for the dissimilatory nitrate reduction have not yet been identified in these diatoms. Nitrate uptake, however, occurs only under aerobic conditions and is temperature-dependent in the way that the rate decreases with increasing temperature. The nitrate is stored in vacuoles.

10.3 Microbial Mats and Stromatolites

Fine sandy intertidal sediments, beaches, and coastal wetlands are often characterized by colonization of cyanobacteria (Bolhuis et al. 2014). Cyanobacteria are oxygenic phototrophic bacteria that exhibit a plant-like photosynthesis. They use water as the electron donor, splitting it and evolve O_2. The sun light energy and the electrons from water are used to fix CO_2 into organic matter that is used for growth. Many species are also capable of fixing atmospheric nitrogen gas (N_2), making these organisms both carbon and nitrogen autotrophs (Severin and Stal 2008). Most cyanobacteria are low-light-adapted organisms and have modest nutrient requirements. Apart from being able to use the ubiquitous dinitrogen from the air, they are capable of luxury uptake of phosphate storing it intracellular as polyphosphate (Gomez-Garcia et al. 2013), and possess carbon concentration mechanism (CCM) that accumulates inorganic carbon in the carboxysomes, allowing them to fix CO_2 when this is in low supply (Kaplan et al. 1991). The sediment precipitates and accumulates iron and the seawater is rich in sulfate, which further satisfies the demand of cyanobacteria. Many cyanobacteria resist extreme environmental fluctuations such as temperature, salinity, alkalinity, and desiccation (Stal 2007). Cyanobacteria are metabolically versatile, flexible, and reactive. Besides being oxygenic photoautotrophic they may also live as anoxygenic photoautotroph, and in the dark perform aerobic respiration or fermentation under anaerobic conditions and switch rapidly between these different modes. These properties make cyanobacteria preeminently the perfect organism to colonize low-nutrient coastal fine sandy sediments (63–200 μm grain size) and cope with the physicochemical gradients that are characteristic for microbial mats.

Cyanobacteria are usually not found on muddy sediments probably because of the strong attenuation on light and because they will lose the competition with diatoms who will take better advantage of the higher nutrient concentrations and there efficient chemotrophic metabolism that is important in the highly dynamic mudflats (Watermann et al. 1999). Cyanobacteria are also not found on coarse sand because the large silicate sand grains may damage the cells. Another factor is temperature. Benthic diatoms grow better at lower temperature and cyanobacteria thrive best at higher temperature. In annual microbial mats normally diatoms appear early in the year, subsequently being taken over by cyanobacteria when the temperature increases (Watermann et al. 1999).

The cyanobacteria enrich the sediment with organic matter that forms the substrate for a diverse microbial community. The organic matter becomes available through a variety of mechanisms (Stal 2001). The cyanobacteria may excrete glycolate as a product of photorespiration and exude EPS. The latter forms the matrix of the mat in which the microorganisms are embedded, gives stability to the sediment, protect the organisms from grazing and desiccation, and immobilize toxins or accumulate nutrients (Stal 2012). In addition, it can serve as organic substrate for a variety of microorganisms. In the dark under anoxic conditions, cyanobacteria may excrete low-molecular fermentation products such as acetate, lactate, and ethanol, which serve as substrate for sulfate-reducing bacteria. Last but not least, organisms may lyse, being grazed or subjected to viral attack, all of which result in a liberation of a plethora of organic compounds and nutrients.

The organic matter is degraded by a large variety of microorganisms. While during the day, photosynthesis supersaturates the top few millimeters of the mat with oxygen, this disappears within the mere time of minutes, rendering anoxic sediment up to the surface. Any oxygen diffusing into the sediment from the air or overlying water is immediately consumed and aerobic metabolism is diffusion limited and anaerobic metabolism is dominant. Fermentative processes provide low-molecular organic acids and alcohols that are end-oxidized by sulfate-reducing bacteria that produce sulfide. In the light, the sulfide is oxidized back to sulfate by anoxygenic phototrophic bacteria. Purple sulfur bacteria often form a distinct purple layer beneath the layer of the cyanobacteria, although sometimes also green sulfur bacteria are found. These bacteria use light from the far infrared spectrum that is not absorbed by the cyanobacteria and least attenuated by the sediment. Colorless sulfur bacteria may oxidize sulfide independent on light, either using oxygen or nitrate. Moreover, also chemical oxidation of sulfide is possible. The sulfur cycle in the mat is complex. A mini sulfur cycle of reduction of elemental sulfur (or polysulfide) to sulfide and oxidation back to elemental sulfur is probably cycling much faster. There are many redox stages between the most oxidized state of sulfur, sulfate (SO_4^{2-}) and the most reduced form sulfide (S^{2-}) and they all play a role in the sulfur cycle in the microbial mat. Several sulfur compounds with intermediate redox states can be disproportionated, i.e., one part is oxidized while the other is reduced. Disproportionation is a form of fermentation and yields energy, but since sulfur disproportionation is a process involving an inorganic substrate, the disproportionating organisms are usually autotrophs and use CO_2 as carbon source. Chemotrophic CO_2 fixation in mats can be considerable since a variety of chemoautotrophic microorganisms are active in these microbial mats. Sulfide can also chemically react with iron and form iron sulfide (FeS), which renders the anoxic sediment the typical black color. When sulfide reacts with oxidized (ferric, Fe^{3+}) iron, iron will be reduced to ferrous (Fe^{2+}) iron and sulfide oxidized to elemental sulfur, which may further react to form the stabile mineral pyrite FeS_2. Hence, the iron cycle is intimately associated with the sulfur cycle and cycling between the ferrous and ferric form, both chemically as well as biologically. Ferrous iron may also serve as electron donor in anoxygenic photosynthesis by some specialized purple bacteria.

Bolhuis et al. (2013) investigated the diversity of a coastal mat developing on the North Sea beach of the Dutch barrier island Schiermonnikoog. They sampled with high spatial resolution a transect perpendicular to the coast and representing a natural salinity gradient. The highest salinity and least developed mats were found near the low water mark, while mature mats were found towards the dunes. Using denaturing gradient gel electrophoresis (DGGE) the 16S rRNA diversity was determined. Bolhuis et al. (2013) found three structurally different types of microbial mats along this transect. Stal et al. (1985) already described the structure, physiology, and cyanobacterial composition of these three types of mats from a German barrier island. The high spatial resolution along this tidal transect and the high resolution of the DGGE fingerprints confirmed the presence of three different mat types for the domains Bacteria and Archaea. The cyanobacteria as key organisms for these microbial mats and belonging to the Bacteria followed the same clustering. The Archaea split into two subgroups of mats close to the low water mark. The Eukarya clustered into four groups but in large also followed the pattern of the other two domains. The diversity of all groups (Shannon diversity index) increased with distance from the low water mark and was usually highest in the mature microbial mats found in the intermediate or upper tidal ranges. The photosynthetic pigments followed the same pattern with a dominance of diatoms at the low water mark, cyanobacteria in the intermediate, and green algae in the upper littoral. Bolhuis et al. (2013) found strong evidence that the clustering was salinity driven.

Bolhuis and Stal (2011) revealed the diversity of Bacteria and Archaea in these three mats during three different seasons using 454-pyrosequencing of 16S rRNA gene tag sequencing. They concluded that these coastal mats were among the most diverse marine ecosystems studied so far. Coastal microbial mats were also strikingly more diverse than their counterparts from more extreme environments such as hypersaline and hot spring microbial mats (Bolhuis et al. 2014). Obviously, diversity decreases with increasing environmental extremes. The differences between the three structurally different mats were again confirmed, but there were no major differences between the seasons. This means that the community composition at one site does not change during the year, although this does not exclude that different groups of microorganisms are active at different times of the year. Since this investigation did not look at ribosomal RNA and proteins, this remains to be seen. Another remarkable result was that these microbial mats are predominantly bacterial systems and that Archaea play a minor role. Average Chao richness was 2853 for Bacteria and 182 for Archaea. Proteobacteria were the most dominant group. Alphaproteobacteria dominated in the upper littoral while in the lower littoral Delta- and Gammaproteobacteria were more important. In the mid-littoral zone, these distributions were also intermediate. Betaproteobacteria were virtually absent in the lower littoral. Other important groups included Bacteroidetes, Actinobacteria, Planctomycetes, and Cyanobacteria. The latter were especially important in the upper and intermediate mats. Cyanobacteria catch the eye when observing these mats in the field and under the microscope and it is striking why these organisms are not more represented in this sequence database. This has also been observed in other microbial mats systems such as the hypersaline microbial mats of Guerrero Negro, Mexico (Ley et al. 2006). Microbial mat cyanobacteria are much larger than most other bacteria (or archaea) and are more conspicuous when doing microscopy, however, they do not possess more copies of 16S rRNA gene compared with the smaller microorganisms. Also, mat-forming cyanobacteria are notorious difficult to extract DNA from and this may also give a bias against cyanobacteria. Fact is that, these microbial mats are built and maintained by cyanobacteria but it is striking that so many other microorganisms contribute to the mat system's function.

Although microbial mats are mainly bacterial systems, Archaea are certainly present as well and predominantly represented by Euryarchaeota. The mid-littoral mat was dominated by Halobacteria, which can be explained by high salinity of the pore water after desiccation. Another important group is the methanogenic bacteria, which are especially dominant in the upper littoral. Sulfate-reducing bacteria are responsible for most of the end oxidation of the organic matter as a consequence of the seawater sulfate. However, the upper littoral is more influenced by freshwater and the availability of sulfate may be less. Moreover, there are also non-competitive substrates in the marine environment that are used by methane bacteria and cannot be degraded by sulfate-reducing bacteria such as methylamines.

When analyzing the functional groups of bacteria in the microbial mats it strikes that chemoorganotrophic bacteria represent half of all functional groups. Photoautotrophs and photoheterotrophs are other large functional groups in addition to sulfate-reducing bacteria and sulfur-oxidizing bacteria. It should be noted that a considerable number of sequences belong to hitherto unknown Bacteria and Archaea, and, consequently, also their functions are unknown.

All microbial mats that have been investigated for it are capable of fixing atmospheric nitrogen (dinitrogen, N_2) (Severin and Stal 2010b). Nitrogen is an important element for organisms as it represents $\sim$10 % of the dry weight of living biomass. Most organisms obtain this essential food supply as bound nitrogen. Seawater is usually low in supply of bound nitrogen (i.e., nitrate/nitrite, ammonium, organic nitrogen). The bulk of nitrogen on Earth is present as N_2 in the atmosphere, but this molecule is inaccessible for most organisms, except some specialized Bacteria and even fewer Archaea. No representative of the Eukarya is known that is capable of using N_2, except in symbiosis with bacteria. In order to form a mat with a dense biomass, it should not surprise that the supply of bound nitrogen from the seawater is limiting and that N_2 fixation is an important process for the formation of a microbial mat. Cyanobacteria are well known as a group with many diazotrophic (N_2-fixing) species, but many other bacteria possess this capacity as well, including anoxygenic phototrophic bacteria and sulfate-reducing bacteria (Zehr et al. 2003). Since nitrogenase is sensitive to oxygen and cyanobacteria are oxygenic bacteria, it has been a debate for some time which organisms were actually responsible for the observed N_2 fixation. Measurements that followed the day–night variations of nitrogenase activity in microbial mats gave many different patterns (Severin and Stal 2008; Villbrandt et al. 1990). Sometimes, nitrogenase activity was confined to the dark (to avoid oxygenic photosynthesis and oxygen super saturation), sometimes activity peaks were observed at sunrise and sunset that was explained as benefiting from the presence of light (as energy source) while oxygen saturation as low. But also, nitrogenase activity during daytime has been observed, especially when heterocystous cyanobacteria were involved. Severin and Stal (2008) showed a clear light-dependency of nitrogenase activity in the Dutch barrier islands coastal mats, even when most of the activity was measured during the night. It was also concluded that the mats maximized nitrogenase activity and that the total daily-integrated nitrogenase activity was independent on the daily-integrated photon flux density.

Severin et al. (2010) showed that cyanobacterial *nifH* genes contributed for 50 % of the total pool of this gene. Gamma- and Deltaproteobacteria contributed for 25 % and the rest was unidentified. Cyanobacterial *nifH* transcripts dominated during most of the day–night cycle (60–80 %), while Gammaproteobacteria were the second-most important organisms. Especially during the early morning, Gammaproteobacteria seemed to be more important than the cyanobacteria. It is likely that the diazotrophic Gammaproteobacteria were anoxygenic phototrophic bacteria that equally as the cyanobacteria use light as the source of energy for the fixation of N_2. The contribution of diazotrophic Deltaproteobacteria may be by sulfate-reducing bacteria, some of which are known to be diazotrophs. However, caution must be taken since the discovery of a deltaproteobacterial *nif* cluster in the genome of the cosmopolitan mat-forming cyanobacterium *Microcoleus* (*Coleofasciculus*) *chthonoplastes* (Bolhuis et al. 2010). Studies of the expression of the *nifH* gene in these mats confirmed the important role of cyanobacteria for N_2 fixation. Salinity is an important factor. Increased salinity worked against N_2 by cyanobacteria mostly through a lower proportion of *nif* gene expression (Severin et al. 2012). A complicating factor was, however, the fact that day–night gene expression patterns did not follow actual nitrogenase activity (Severin and Stal 2010a). Such a relationship is perhaps also not logic or necessary since gene expression is not necessarily one-to-one related to the presence of enzyme activity. Even when an active enzyme is present, environmental and physiological factors may prevent it from exerting activity.

The fate of this fixed N in microbial mats has not been extensively investigated as N_2 fixation. Denitrification is certainly an important loss factor, although the rates seem to be considerably lower (order of magnitude) than N_2 fixation (Fan et al. 2015a). Denitrification occurs more or less at the same time as N_2 fixation with highest rates during summer, although seasonality of denitrification is much less pronounced than is the case with N_2 fixation. Anaerobic ammonium oxidation (anammox) does not seem to play a role. The denitrifier community was diverse with more or less equal contributions of *nirS* and *nirK* genotypes and an average Chao richness of $\sim$75 (Fan et al. 2015a). The diversity of the denitrifying community in the three mat types along the tidal gradient followed the same pattern as the whole microbial community (Bolhuis and Stal 2011). This was also the case with the nitrifying (ammonium-oxidizing) community. Although the functional gene *amoA*, the alpha subunit of ammonia monooxygenase, of both ammonium-oxidizing bacteria (AOB) and ammonium-oxidizing archaea (AOA) were detected, it was concluded that the latter did not play a major role in nitrification (Fan et al. 2015b). Contrary to denitrification, ammonium oxidation was highest in autumn and winter and therefore temporally separated from N_2 fixation and denitrification. It is likely that nitrifiers are competing for ammonia when the primary producers of the mat are most active.

Nitrogen and the nitrogen cycle are key factors for the development microbial mats. The fact that all microbial mats fix dinitrogen emphasizes that these systems are basically limited in this essential nutrient. Moreover, the mostly non-heterocystous cyanobacteria, and the oxygen super saturation of the mat during the day and anoxic conditions during most of the night are seriously limiting the possibilities of efficient nitrogen fixation. Stal (2003) calculated that these coastal mats fix 80 % of carbon in excess to what can be accommodated for balanced growth, i.e., synthesis of structural cell material. Most of this excess carbon is exuded as EPS, which forms the matrix in which the mat microorganisms are embedded. This EPS serves many important purposes. It protects against grazing and desiccation, while EPS may immobilize toxic elements and scavenge nutrients. EPS also renders stability to the sediment and is therefore important for the establishment of the microbial mat. It has also been conceived that EPS binds calcium (Ca^{2+}) ions to the charged moieties of the polymer. This would lead to a considerable lower calcium concentration and prevent the precipitation of calcium carbonate (Westbroek et al. 1994). This might be the reason that modern coastal microbial mats do not form carbonate rock and do not form stromatolites, their presumed fossil counterparts. Kremer et al. (2008) discovered that the coastal mats of the Dutch barrier island Schiermonnikoog showed strong calcifying activity. This calcification was associated with degrading EPS. Apparently, the massive amount of calcium carbonate that precipitated in these mats dissolved subsequently due to microbial activity and therefore these annual mats did not form carbonate rock. Modern coastal stromatolites that are found in the Bahama's Exuma Cays are formed following the same principles. Calcium ion is initially bound to EPS. This EPS is degraded by microorganisms and calcium ion is massively liberated causing calcium carbonate precipitation. Dissolution and re-precipitation of the calcium carbonate forms micrite that cement the calcium sand grains, forming a solid calcium carbonate layer. The process repeats resulting in a layered lithified structure, the stromatolite (Reid et al. 2000). It is questionable whether these modern coastal stromatolites are true analogs of those of the Archaean fossil record (Kazmierczak and Kempe 2004).

10.4 Mangroves

Mangrove forests or swamps are found in intertidal areas in the tropics and subtropics. They are important coastal ecosystems that can cope with sea-level rise and coastal protection. Most of the mangroves are found between the 5°N and 5°S latitudes in Asia, Africa, America, and Oceania and comprise about 60 species of halotolerant species of trees (Holguin et al. 2001). Mangroves are also found in estuaries and arid environments with little supply of freshwater. Plants of the family among others, Rhizophoraceae, comprising shrubs and short trees, form mangrove forests along the coast or the borders of estuaries where they are protected from wave action. The environment is therefore depositional and the sediment contains high amounts of organic matter. These sediments are usually anoxic except for the top few millimeters (Kristensen et al. 2008). The plant roots are adapted to high saline and anoxic conditions. Mangroves are highly productive systems that provide food for a variety of invertebrates such as crabs and other animals such as fish and fish larvae (Bouillon et al. 2002; Hsieh et al. 2002; Kieckbusch et al. 2004). The removal of litter by these animals is important for clearing the surface and allowing benthic photosynthesis. Kristensen et al. (2008) estimated a total net primary production of mangroves of 149 mol C m^{-2} $year^{-1}$, including the leaf litter, wood, and root production. Leaf litter probably represents only 25 % of the total net primary production. Gross production is obviously much higher and losses through root exudation may be considerable. Microphytobenthic production ranges between 7 and 73 mol C m^{-2} $year^{-1}$ and of macroalgae 110–118 mol C m^{-2} $year^{-1}$ (see Kristensen et al. 2008). The contribution of phytoplankton to primary production in mangrove ecosystems varies considerably depending on the type and environment of the mangrove ecosystem, but may be substantial. From data of $\delta^{13}C$ of the organic matter, Kristensen et al. (2008) concluded that mangrove litter is an important source of organic matter but that there may also be a considerable import from external sources such as from phytoplankton or seagrass or from the dense microbial mats of microphytobenthos that are inherent of many mangrove ecosystems (Wooller et al. 2003).

Although mangrove ecosystems occupy only 0.5 % of the global coastal area, they store 10–15 % of the coastal carbon (24 Tg C $year^{-1}$) and the tides cause a draining of the mangrove area resulting in an export of 10–11 % of the particulate terrestrial derived carbon to the ocean (Alongi 2014). The most important microbial processes involved in the decomposition of the organic material are aerobic respiration and sulfate reduction each of which makes up approximately 50 % (Alongi et al. 1998).

Invertebrates and infauna are important components driving the microbial remineralization processes (Dye and Lasiak 1986; Schrijvers et al. 1998). Burrowing of these organisms introduce oxygen into the otherwise anoxic sediments. This allows the re-oxidation of reduced compounds such as sulfide and iron. Oxygen availability also helps to degrade recalcitrant polymers such as cellulose and lignin (Marchand et al. 2005).

Mangroves are hot spots of microbial diversity and among the most-productive ecosystems in the world (Andreote et al. 2012; Thatoi et al. 2013). Mangroves inhabit a variety of different functional groups of bacteria such as phosphate solubilizing organisms (*Bacillus*, *Paenibacillus*, *Xanthobacter*, *Vibrio proteolyticus*, *Enterobacter*, *Kluyvera*, *Chryseomonas*, and *Pseudomonas* sp.) and anoxygenic photosynthetic bacteria (*Chloronema*, *Chromatium*, *Beggiatoa*, *Thiopedia*, *Leucothiobacteria* sp.). Microorganisms are found as biofilms on the exposed roots and other plant surfaces of the mangrove trees such as the leaves (phyllosphere), inside and on the soft soil as microbial mats, and in the water, where they are involved in the biogeochemical cycling of elements, notably, carbon, nitrogen, sulfur, phosphorus, and iron. Nevertheless, few studies reported on the microbial diversity of mangrove ecosystems and actually investigated their roles in biochemical cycling and ecosystem function (Holguin et al. 2001).

Marine algae belonging to groups such as Chlorophyta, Chrysophyta, Phaeophyta, Rhodophyta, and Cyanobacteria are common in mangrove ecosystems. However, surprisingly little information is available on cyanobacteria from mangrove ecosystems although they are known to be abundantly present, e.g., on the roots of mangrove trees but also as mats on the soft surface in the upper intertidal, depending on the shadowing of the plants. Cyanobacteria are also reported from the mangrove phyllosphere (Rigonato et al. 2012). Cyanobacteria are photoautotrophic bacteria that contribute to carbon and nitrogen fixation. These organisms also may store large amounts of phosphate as polyphosphate granules and as such represent a storage and retention for this important nutrient. The composition of the cyanobacterial community in mangrove ecosystems depends on environmental parameters (Rigonato et al. 2013). Rigonato et al. (2013) reported picocyanobacteria (*Synechococcus* and *Prochlorococcus*) inhabiting near-shore sites but it is more likely that these are not native residents, as these cyanobacteria are known from the oceanic phytoplankton. Rigonato et al. (2013) also reported slightly acidic conditions in the mangroves they studied, which is usually not promoting cyanobacterial growth. Alvarenga et al. (2015) gave an extensive list of cyanobacteria reported from mangrove ecosystems, which included all five sections of this group, including many known diazotrophic species. Many species seem to be related to hitherto unknown (i.e., not described by their 16S rRNA gene sequence) cyanobacteria.

In general, only low rates of N_2 fixation have been reported from mangrove forests. This is remarkable, since in many environments diazotrophic cyanobacteria provide fixed nitrogen to the microbial community and even to the plants. Many chemotrophic diazotrophic bacteria have been reported from the soils (*Azospirillum*, *Azotobacter*, *Rhizobium*, *Clostridium*, *Klebsiella*). They may fix N_2 using the abundantly present organic matter in the soil as energy substrate but their actual contribution to the nitrogen budget varies among different mangrove ecosystems and among various sub-habitats within the ecosystem (Holguin et al. 2001). However, the ecosystem may also receive sufficient nitrogen from terrestrial run-off.

Fungi are another important component of the microbiome of mangrove ecosystems that are considered 'hot-spots' for this group of organisms (Shearer et al. 2007). These organisms produce extracellular enzymes that help in the degradation of plant polymers such as lignin and cellulose. A variety of fungal enzymes have been found such as those with pectinolytic, proteolytic, and amylolytic activities (Holguin et al. 2001; Thatoi et al. 2013).

The reconstruction of metabolic pathways for the carbon-, nitrogen- and sulfur cycles from the metagenomics information of mangrove sediments show that methanogenesis is important and that the methanogenic organisms use formaldehyde and CO_2 as substrate. For the nitrogen cycle, the dissimilatory reduction of nitrate, denitrification, and ammonia oxidation (ammonia-oxidizing bacteria) are predominantly present. For the sulfur cycle, sulfate reduction and oxidation are important. The organisms involved in these metabolic processes include Burkholderiaceae, Planctomycetaceae, Rhodobacteraceae, and Desulfobacteraceae. Microbial activity in the anoxic soil cause loss of nitrogen, iron, and phosphate by denitrification, iron reduction, and solubilizing of phosphate, respectively, which tends to make the soil nutrient poor. Sulfate reduction and methanogenesis decrease the organic content of the soil.

Proteobacteria make up approximately 50 % of the bacterial community with a dominance of Delta- and Gammaproteobacteria. But also Alpha- Beta and Epsilonproteobacteria are quite common. Firmicutes, Actinobacteria, Bacteroides, and Chloroflexi each may account for 5–10 %. Minor groups are represented by Planctomycetes, Cyanobacteria, Acidobacteria. Archaea play a minor role. As in any marine frequently anoxic sedimentary environment, the biogeochemical sulfur cycle is prominently present in the mangrove microbiome. This explains the prominence of the Deltaproteobacteria that comprise the sulfate-reducing bacteria (e.g., *Desulfovibrio*, *Desulfotomaculum*, *Desulfosarcina*, and *Desulfococcus*). Sulfate-reducing bacteria are capable of oxidizing more complex and high-molecular substrates. Mangrove sediments contain high amounts of pyrite and elemental sulfur but low amounts of iron sulfide (Holmer et al. 1994).

Iron respiration may be more important for anaerobic decomposition of organic carbon than sulfate reduction, especially when infauna burrows and mangrove roots oxidize iron, and as such contribute to high amounts of this potent electron acceptor (Nielsen et al. 2003). Iron respiration may account for as much as 80 % of the anaerobic decomposition of organic carbon in mangrove ecosystems as in other marine sediments (Jensen et al. 2003).

The oxygen concentration in the top layer of the sediments may be highly variable and support nitrification by mainly ammonium-oxidizing bacteria. These may provide the nitrite that serves as the substrate for anaerobic ammonium oxidation.

Our knowledge of the diversity and ecology of Archaea in mangroves is limited (Bhattacharyya et al. 2015). These authors reported Thaumarchaeota (Marine Group I) and Euryarchaeota (Marine Group II). The former group is known for aerobic chemolithotrophic ammonia oxidation (nitrification) in the marine environment. Dias et al. (2011) found ammonia-oxidizing archaea but whether these aerobic organisms are important for ammonium oxidation remains to be seen. Luo et al. (2014) showed that dissolved oxygen was a main factor controlling ammonium oxidation in an estuarine wetland. Ammonium-oxidizing bacteria were more affected by dissolved oxygen than ammonium-oxidizing archaea. Luo et al. (2014) found higher nitrification rates in the mangrove sediments compared to bare mud and this correlated with a higher abundance and diversity of ammonium-oxidizing archaea. From the group of Euryarchaeota, representatives of the Thermoplasmatales and Halobacteria are common as well as Methanomicrobia and Methanobacteria. The latter groups are methanogens but may also be involved in sulfate reduction. The presence of the methanogenic archaea *Methanococcoides methylutens* has been reported from mangrove forest sediments. The importance of methanogenesis in mangrove sediments is uncertain but it is most likely low. Often this process cannot be detected at all (Alongi et al. 2005), although this does not necessarily mean that it is absent since methane may be removed by anaerobic methane oxidation (Canfield et al. 2005). Most of the organic matter is probably end-oxidized by sulfate-reducing bacteria, but the presence of non-competitive substrates such as methanol, mono-, di-, and trimethylamine may allow some methanogenesis. Also, in case sulfate is depleted, methane may be produced from acetate or CO_2 and H_2.

Considering the high microbial diversity and the small portion of this diversity being known and described, mangrove ecosystems may be hot spots of novel bioactive compounds, enzymes, new drugs and antimicrobial substances, and source of food and feed. Particularly, actinomycetes and fungi are among the most promising organisms for biotechnological applications (Hong et al. 2009; Polizeli et al. 2005), but also algae are producers of polymers such as agar, carrageenan, and alginate (Shanmugam and Mody 2000).

Acknowledgments The research leading to these results has received funding from the European Union Seventh Framework Programme (FP7/2007–2013) under grant agreement no 311975. This publication reflects the views only of the author, and the European Union cannot be held responsible for any use which may be made of the information contained therein.

References

Alongi DM (2014) Carbon cycling and storage in mangrove forests. Ann Rev Mar Sci 6:195–219

Alongi DM, Sasekumar A, Tirendi F, Dixon P (1998) The influence of stand age on benthic decomposition and recycling of organic matter in managed mangrove forests of Malaysia. J Exp Mar Biol Ecol 225:197–218

Alongi DM, Pfitzner J, Trott LA, Tirendi F, Dixon P, Klumpp DW (2005) Rapid sediment accumulation and microbial mineralization in forests of the mangrove *Kandelia candel* in the Jiulongjiang estuary, China. Estuar Coast Shelf Sci 63:605–618

Al-Raei AM, Bosselmann K, Böttcher ME, Hespenheide B, Tauber F (2009) Seasonal dynamics of microbial sulfate reduction in temperate intertidal surface sediments: controls by temperature and organic matter. Ocean Dyn 59:351–370

Alvarenga DO, Rigonato J, Branco LHZ, Fiore MF (2015) Cyanobacteria in mangrove ecosystems. Biodivers Conserv 24:799–817

Andreote FD, Jiménez DJ, Chaves D, Dias ACF, Luvizotto DM, Dini-Andreote F, Fasanella CC, Lopez MV, Baena S, Taketani RG, de Melo IS (2012) The microbiome of Brazilian mangrove sediments as revealed by metagenomics. PLoS ONE 7(6):e38600

Benyoucef I, Blandin E, Lerouxel A, Jesus B, Rosa P, Méléder V, Launeau P, Barillé L (2014) Microphytobenthos interannual variations in a north-European estuary (Loire estuary, France) detected by visible-infrared multispectral remote sensing. Estuar Coast Shelf Sci 136:43–52

Bhattacharyya A, Majumder NS, Basak P, Mukherji S, Roy D, Nag S, Haldar A, Chattopadhyay D, Mitra S, Bhattacharyya M, Ghosh A (2015) Diversity and distribution of Archaea in the mangrove sediment of Sundarbans. Archaea 2015(968582):1–14

Bolhuis H, Stal LJ (2011) Analysis of bacterial and archaeal diversity in coastal microbial mats using massive parallel 16S rRNA gene tag sequencing. ISME J 5:1701–1712

Bolhuis H, Severin I, Confurius-Guns V, Wollenzien UIA, Stal LJ (2010) Horizontal transfer of the nitrogen fixation gene cluster in the cyanobacterium *Microcoleus chthonoplastes*. ISME J 4:121–130

Bolhuis H, Fillinger L, Stal LJ (2013) Coastal microbial mat diversity along a natural salinity gradient. PLoS ONE 8(5):e63166

Bolhuis H, Cretoiu MS, Stal LJ (2014) Molecular ecology of microbial mats. FEMS Microbiol Ecol 90:335–350

Bouillon S, Raman AV, Dauby P, Dehairs F (2002) Carbon and nitrogen stable isotope ratios of subtidal benthic invertebrates in an estuarine mangrove ecosystem (Andhra Pradesh, India). Estuar Coast Shelf Sci 54:901–913

Bruckner CG, Rehm C, Grossart H-P, Kroth PG (2011) Growth and release of extracellular organic compounds by benthic diatoms depend on interactions with bacteria. Environ Microbiol 13:1052–1063

Canfield DE, Kristensen E, Thamdrup B (2005) The methane cycle. In: Aquatic geomicrobiology. Advances in marine biology, vol 48, pp 383–418

Cockell CS (2006) The origin and emergence of life under impact bombardment. Philos Trans R Soc Lond B 361:1845–1856

de Boer PL, van Gelder A, Nio SD (1988) Tide-influenced sedimentary environments and facies. D. Reidel Publishing Company, Dordrecht, p 529

De Brouwer JFC, Wolfstein K, Stal LJ (2002) Physical characterization and diel dynamics of different fractions of extracellular polysaccharides in an axenic culture of a benthic diatom. Eur J Phycol 37:37–44

Dias ACF, Dini-Andreote F, Taketani RG, Tsai SM, Azevedo JL, de Melo IS, Andreote FD (2011) Archaeal communities in the sediments of three contrasting mangroves. J Soils Sedim 11:1466–1476

Dionne JC (1988) Characteristic features of modern tidal flats in cold regions. In: de Boer PL, van Gelder A, Nio SD (eds) Tide-influenced sedimentary environments and facies. D. Reidel Publishing Company, Dordrecht, pp 301–332

Dupuy C, Mallet C, Guizien K, Montanié H, Bréret M, Mornet F, Fontaine C, Nérot C, Orvain F (2014) Sequential resuspension of biofilm components (viruses, prokaryotes and protists) as measured by erodimetry experiments in the Brouage mudflat (French Atlantic coast). J Sea Res 92:56–65

Dye AH, Lasiak TA (1986) Microbenthos, meiobenthos and fiddler crabs: trophic interactions in a tropical mangrove sediment. Mar Ecol Prog Ser 32:259–264

Dyer KR, Christie MC, Wright EW, Shi Z (2000) The classification of intertidal mudflats. Cont Shelf Res 20:1039–1060

Fan H, Bolhuis H, Stal LJ (2015a) Denitrification and the denitrifier community in coastal microbial mats. FEMS Microbiol Ecol 91:fiu033

Fan H, Bolhuis H, Stal LJ (2015b) Nitrification and nitrifying bacteria in a coastal microbial mat. Front Microbiol 6:1367

Gomez-Garcia MR, Fazeli F, Grote A, Grossman AR, Bhaya D (2013) Role of polyphosphate in thermophilic *Synechococcus* sp. from microbial mats. J Bacteriol 195:3309–3319

Guizien K, Dupuy C, Ory P, Montanié H, Hartmann H, Chatelain M, Karpytchev M (2014) Microorganism dynamics during a rising tide: disentangling effects of resuspension and mixing with offshore waters above an intertidal mudflat. J Mar Syst 129:178–188

Gutierrez T, Biller DV, Shimmield T, Green DH (2012) Metal binding properties of the EPS produced by *Halomonas* sp. TG39 and its potential in enhancing trace element bioavailability to eukaryotic phytoplankton. Biometals 25:1185–1194

Gypens N, Borges AV, Speeckaert G, Lancelot C (2014) The dimethylsulfide cycle in the eutrophied southern North Sea: a model study integrating phytoplankton and bacterial processes. PLoS ONE 9(1):e85862

Haynes K, Hofmann TA, Smith CJ, Ball AS, Underwood GJC, Osborn AM (2007) Diatom-derived carbohydrates as factors affecting bacterial community composition in estuarine sediments. Appl Environ Microbiol 73:6112–6124

Heip CHR, Goosen NK, Herman PMJ, Kromkamp J, Middelburg JJ, Soetaert K (1995) Production and consumption of biological particles in temperate tidal estuaries. Oceanogr Mar Biol Annu Rev 33:1–149

Hofmann T, Hanlon ARM, Taylor JD, Ball AS, Osborn AM, Underwood GJC (2009) Dynamics and compositional changes in extracellular carbohydrates in estuarine sediments during degradation. Mar Ecol Prog Ser 379:45–58

Holguin G, Vazquez P, Bashan Y (2001) The role of sediment microorganisms in the productivity, conservation, and rehabilitation of mangrove ecosystems: an overview. Biol Fertil Soils 33:265–278

Holmer M, Kristensen E, Banta G, Hansen K, Jensen MH, Bussawarit N (1994) Biogeochemical cycling of sulfur and iron in sediments of a south-east Asian mangrove, Phuket Island, Thailand. Biogeochemistry 26:145–161

Hong K, Gao A-H, Xie Q-Y, Gao H, Zhuang L, Lin H-P, Yu H-P, Li J, Yao X-S, Goodfellow M, Ruan J-S (2009) Actinomycetes for marine drug discovery isolated from mangrove soils and plants in China. Mar Drugs 7:24–44

Hsieh H-L, Chen C-P, Chen Y-G, Yang H-H (2002) Diversity of benthic organic matter flows through polychaetes and crabs in a mangrove estuary: $\delta^{13}C$ and $\delta^{34}S$ signals. Mar Ecol Prog Ser 227:145–155

Isaksen MF, Finster K (1996) Sulphate reduction in the root zone of the seagrass *Zostera noltii* on the intertidal flats of a coastal lagoon (Arcachon, France). Mar Ecol Prog Ser 137:187–194

Jensen MM, Thamdrup B, Rysgaard S, Holmer M, Fossing H (2003) Rates and regulation of microbial iron reduction in sediments of the Baltic-North Sea transition. Biogeochemistry 65:295–317

Kamp A, Høgslund S, Risgaard-Petersen N, Stief P (2015) Nitrate storage and dissimilatory nitrate reduction by eukaryotic microbes. Front Microbiol 6:1492

Kaplan A, Schwarz R, Liemanhurwitz J, Reinhold L (1991) Physiological and molecular aspects of the inorganic carbon-concentrating mechanism in cyanobacteria. Plant Physiol 97:851–855

Kazmierczak J, Kempe S (2004) Microbialite formation in seawater of increased alkalinity, Satonda Crater Lake, Indonesia—discussion. J Sedim Res 74:314–317
Kieckbusch DK, Koch MS, Serafy JE, Anderson WT (2004) Trophic linkages among primary producers and consumers in fringing mangroves of subtropical lagoons. Bull Mar Sci 74:271–285
Klap VA, Hemminga MA, Boon JJ (2000) Retention of lignin in seagrasses: angiosperms that returned to the sea. Mar Ecol Prog Ser 194:1–11
Kremer B, Kazmierczak J, Stal LJ (2008) Calcium carbonate precipitation in cyanobacterial mats from sandy tidal flats of the North Sea. Geobiology 6:46–56
Kristensen E, Bouillon S, Dittmar T, Marchand C (2008) Organic carbon dynamics in mangrove ecosystems: a review. Aquat Bot 89:201–219
Leloup J, Petit F, Boust D, Deloffre J, Bally G, Clarisse O, Quillet L (2005) Dynamics of sulfate-reducing microorganisms (*dsrAB* genes) in two contrasting mudflats of the Seine estuary (France). Microb Ecol 50:307–314
Ley RE, Harris JK, Wilcox J, Spear JR, Miller SR, Bebout BM, Maresca JA, Bryant DA, Sogin ML, Pace NR (2006) Unexpected diversity and complexity of the Guerrero Negro hypersaline microbial mat. Appl Environ Microbiol 72:3685–3695
Luo Z, Qiu Z, Wei Q, Laing GD, Zhao Y, Yan C (2014) Dynamics of ammonia-oxidizing archaea and bacteria in relation to nitrification along simulated dissolved oxygen gradient in sediment–water interface of the Jiulong river estuarine wetland, China. Environ Earth Sci 72:2225–2237
Marchand C, Disnar JR, Lallier-Vergès E, Lottier N (2005) Early diagenesis of carbohydrates and lignin in mangrove sediments subject to variable redox conditions (French Guiana). Geochim Cosmochim Acta 69:131–142
McKew BA, Dumbrell AJ, Taylor JD, McGenity TJ, Underwood GJC (2013) Differences between aerobic and anaerobic degradation of microphytobenthic biofilm-derived organic matter within intertidal sediments. FEMS Microbiol Ecol 84:495–509
Middelburg JJ, Klaver G, Nieuwenhuize J, Wielemaker A, de Haas W, Vlug T, van der Nat FJWA (1996) Organic matter mineralization in intertidal sediments along an estuarine gradient. Mar Ecol Prog Ser 132:157–168
Miyatake T, MacGregor BJ, Boschker HTS (2013) Depth-related differences in organic substrate utilization by major microbial groups in intertidal marine sediment. Appl Environ Microbiol 79:389–392
Montanié H, Ory P, Orvain F, Delmas D, Dupuy C, Hartmann HJ (2014) Microbial interactions in marine water amended by eroded benthic biofilm: a case study from an intertidal mudflat. J Sea Res 92:74–85
Nielsen OI, Kristensen E, Holmer M (2003) Impact of *Arenicola marina* (Polychaeta) on sediment sulfur dynamics. Aquat Microb Ecol 33:95–105
Polizeli MLTM, Rizzatti ACS, Monti R, Terenzi HF, Jorge JA, Amorim DS (2005) Xylanases from fungi: properties and industrial applications. Appl Microbiol Biotechnol 67:577–591
Reid RP, Visscher PT, Decho AW, Stolz JF, Bebout BM, Dupraz C, Macintyre IG, Paerl HW, Pinckney JL, Prufert-Bebout L, Steppe TF, DesMarais DJ (2000) The role of microbes in accretion, lamination and early lithification of modern marine stromatolites. Nature 406: 989–992
Rigonato J, Alvarenga DO, Andreote FD, Dias ACF, Melo IS, Kent A, Fiore MF (2012) Cyanobacterial diversity in the phyllosphere of a mangrove forest. FEMS Microbiol Ecol 80:312–322
Rigonato J, Kent AD, Alvarenga DO, Andreote FD, Beirigo RM, Vidal-Torrado P, Fiore MF (2013) Drivers of cyanobacterial diversity and community composition in mangrove soils in south-east Brazil. Environ Microbiol 15:1103–1114
Schopf JW (2006) Fossil evidence of Archaean life. Philos Trans R Soc Lond B 361:869–885
Schrijvers J, Camargo MG, Pratiwi R, Vincx M (1998) The infaunal macrobenthos under East African *Ceriops tagal* mangroves impacted by epibenthos. J Exp Mar Biol Ecol 222:175–193
Severin I, Stal LJ (2008) Light dependency of nitrogen fixation in a coastal cyanobacterial mat. ISME J 2:1077–1088

Severin I, Stal LJ (2010a) Temporal and spatial variability of *nifH* expression in three filamentous *Cyanobacteria* in coastal microbial mats. Aquat Microb Ecol 60:59–70
Severin I, Stal LJ (2010b) Diazotrophic microbial mats. In: Seckbach J, Oren A (eds) Microbial mats. Modern and ancient microorganisms in stratified systems, vol 14. Springer Sciences, Heidelberg, pp 321–339
Severin I, Acinas SG, Stal LJ (2010) Diversity of nitrogen-fixing bacteria in cyanobacterial mats. FEMS Microbiol Ecol 73:514–525
Severin I, Confurius-Guns V, Stal LJ (2012) Effect of salinity on nitrogenase activity and composition of the active diazotrophic community in intertidal microbial mats. Arch Microbiol 194:483–491
Shanmugam M, Mody KH (2000) Heparinoid-active sulphated polysaccharides from marine algae as potential blood anticoagulant. Curr Sci 79:1672–1683
Shearer CA, Descals E, Kohlmeyer B, Kohlmeyer J, Marnanová L, Padgett D, Porter D, Raja HA, Schmit JP, Thorton HA, Voglymayr H (2007) Fungal biodiversity in aquatic habitats. Biodivers Conserv 16:49–67
Shi Z, Chen JY (1996) Morphodynamics and sediment dynamics on intertidal mudflats in China (1961–1994). Cont Shelf Res 16:1909–1926
Staats N, Stal LJ, de Winder B, Mur LR (2000) Oxygenic photosynthesis as driving process in exopolysaccharide production of benthic diatoms. Mar Ecol Prog Ser 193:261–269
Stal LJ (2001) Coastal microbial mats: the physiology of a small-scale ecosystem. S Afr J Bot 67:399–410
Stal LJ (2003) Nitrogen cycling in marine cyanobacterial mats. In: Krumbein WE, Paterson DM, Zavarzin GA (eds) Fossil and recent biofilms, a natural history of life on earth. Kluwer Academic Publishers, Dordrecht, pp 119–140
Stal LJ (2007) Cyanobacteria: diversity and versatility, clues to life in extreme environments. In: Seckbach J (ed) Algae and cyanobacteria in extreme environments. Springer, Dordrecht, pp 659–680
Stal LJ (2010) Microphytobenthos as a biogeomorphological force in intertidal sediment stabilization. Ecol Eng 36:236–245
Stal LJ (2012) Microbial mats and stromatolites. In: Whitton BA (ed) The ecology of cyanobacteria II. Springer, Dordrecht, pp 65–125
Stal LJ, van Gemerden H, Krumbein WE (1985) Structure and development of a benthic marine microbial mat. FEMS Microbiol Ecol 31:111–125
Stefels J (2000) Physiological aspects of the production and conversion of DMSP in marine algae and higher plants. J Sea Res 43:183–197
Stief P, Kamp A, de Beer D (2014) Role of diatoms in the spatial-temporal distribution of intracellular nitrate in intertidal sediment. PLoS ONE 8(9):e73257
Tambadou F, Lanneluc I, Sablé S, Klein GL, Doghri I, Sopéna V, Didelot S, Barthélémy C, Thiéry V, Chevrot R (2014) Novel *nonribosomal peptide synthetase* (*NRPS*) genes sequenced from intertidal mudflat bacteria. FEMS Microbiol Lett 357:123–130
Taylor JD, McKew BA, Kuhl A, McGenity TJ, Underwood GJC (2013) Microphytobenthic extracellular polymeric substances (EPS) in intertidal sediments fuel both generalist and specialist EPS-degrading bacteria. Limniol Oceanogr 58:1463–1480
Thatoi H, Behera BC, Mishra RR, Dutta SK (2013) Biodiversity and biotechnological potential of microorganisms from mangrove ecosystems: a review. Ann Microbiol 63:1–19
Tourney J, Ngwenya BT (2014) The role of bacterial extracellular polymeric substances in geomicrobiology. Chem Geol 386:115–132
Villbrandt M, Stal LJ, Krumbein WE (1990) Interactions between nitrogen fixation and oxygenic photosynthesis in a marine cyanobacterial mat. FEMS Microbiol Ecol 74:59–72
Watermann F, Hillebrand H, Gerdes G, Krumbein WE, Sommer U (1999) Competition between benthic cyanobacteria and diatoms as influenced by different grain sizes and temperatures. Mar Ecol Prog Ser 187:77–87

Welsh DT, Bourguès S, de Wit R, Herbert RA (1996) Sulphate reduction in the root zone of the seagrass *Zostera noltii* on the intertidal flats of a coastal lagoon (Arcachon, France). Mar Biol 125:619–628

Westbroek P, Buddemeier B, Coleman M, Kok DJ, Fautin D, Stal LJ (1994) Strategies for the study of climate forcing by calcification. In: Doumenge F (ed) Past and present biomineralization processes. Musee Oceanographique, Monaco, pp 37–60

Wooller M, Smallwood B, Jacobson M, Fogel M (2003) Carbon and nitrogen stable isotopic variation in *Laguncularia racemosa* (L.) (white mangrove) from Florida and Belize: implications for trophic level studies. Hydrobiologia 499:13–23

Zehr JP, Jenkins BD, Short SM, Steward GF (2003) Nitrogenase gene diversity and microbial community structure: a cross-system comparison. Environ Microbiol 5:539–554

Chapter 11
Photosymbiosis in Marine Pelagic Environments

Fabrice Not, Ian Probert, Catherine Gerikas Ribeiro, Klervi Crenn, Laure Guillou, Christian Jeanthon and Daniel Vaulot

Abstract Photosymbiosis is a symbiotic relationship between two or more organisms, one of which is capable of photosynthesis. Like other forms of symbiosis, photosymbioses can involve the full spectrum of trophic interactions from mutualism through commensalism to parasitism. As in marine benthic environments (e.g., coral reef ecosystems), photosymbiotic associations are frequently encountered in marine pelagic environments and can involve various combinations of microalgae with bacteria, protists, or metazoans. Here, we aim to provide a brief overview of current knowledge on the diversity of the organisms involved in pelagic photosymbioses, their ecological role, and their relevance for the ecosystem. This chapter focuses on mutualistic interactions occurring between photosynthetic protists and bacteria, between two protists and between microalgae and metazoans, as well as on photosymbiotic interactions involving parasitic protists. A section reviewing the most common and recent approaches used to study pelagic photosymbioses and presenting general perspectives in the field concludes the chapter.

11.1 Introduction

While studying the formation of lichens in the 19th century, H.A. de Bary first coined the term "symbiosis" as "the living together of unlike organisms" (de Bary 1879). This definition is broad and technically includes any distinct taxa, from any kingdom of life that are physically in contact and that have an enduring relationship

F. Not (✉) · I. Probert · K. Crenn · L. Guillou · C. Jeanthon · D. Vaulot
UPMC Univ Paris 06, CNRS, UMR 7144, Station Biologique, Sorbonne Universités, Place Georges Teissier, 29680 Roscoff, France
e-mail: not@sb-roscoff.fr

C. Gerikas Ribeiro
Departamento de Oceanografia Biológica, Instituto Oceanográfico, Universidade de São Paulo, São Paulo, Brazil

L.J. Stal and M.S. Cretoiu (eds.), *The Marine Microbiome*,
DOI 10.1007/978-3-319-33000-6_11

over multiple generations. Symbiosis therefore includes the full spectrum of trophic interactions, from mutualism through commensalism to parasitism.

Photosymbiosis is a symbiotic relationship between two (or more) organisms, one of which is capable of photosynthesis. Photosynthesis originated in cyanobacteria and has since spread across the eukaryotic tree of life by multiple serial endosymbiotic events, leading to the evolution of multiple lineages of algae, one of which (the Chlorobionta) was at the origin of the 'higher' terrestrial plants. Photosymbiosis has thus been, and still is, a highly relevant evolutionary process, but is also a key ecological interaction for ecosystem functioning both on land and in the ocean (Thompson 1999). Terrestrial plants are involved in many well-known photosymbiotic relationships, both mutualistic (e.g., with nitrogen-fixing bacteria in root nodules) and parasitic (e.g., with the oomycete *Phytophthora infestans* causing the disease known as potato blight). Unicellular algae are also involved in some prominent symbiotic relationships in terrestrial environments, notably in partnership with filamentous fungi in lichens.

The best-known photosymbiotic relationship in the marine environment is the association of cnidarian corals with unicellular algae from the dinoflagellate genus *Symbiodinium*. This photosymbiotic relationship structures and sustains benthic reef ecosystems and has been extensively studied, notably in relation to the negative impact of stresses linked to environmental change ('coral bleaching', e.g., Sampayo et al. 2008). Unicellular algae are involved in mutualistic symbiotic relationships with a number of other benthic hosts in the marine environment, including other cnidarians such as sea anemones, molluscs such as the giant clam *Tridacna*, and acoel flatworms (Bailly et al. 2014). Benthic seaweeds are known to have a number of bacterial, unicellular eukaryotic (='protistan') and macroalgal parasites, and some unicellular algae have been reported to parasitize benthic invertebrates (Trench 1993).

Photosymbiotic associations are also frequently encountered in the marine pelagic environment and can involve various combinations of microalgae with bacteria, protists, or metazoans (Anderson 2012; Decelle et al. 2015; Jephcott et al. 2015; Nowack and Melkonian 2010; Stoecker et al. 2009; Taylor 1982). Despite, the independently recognized key roles of oceanic plankton on the one hand and symbiosis on the other hand, the nature, diversity, and importance of pelagic photosymbioses are still poorly understood. In this chapter, we aim to provide a brief overview of the current knowledge of the diversity of the organisms involved in pelagic photosymbioses and their ecological role and importance in the ecosystem. The review will focus on mutualistic interactions occurring between photosynthetic protists and bacteria, between two protists and between microalgae and metazoans, as well as on photosymbiotic interactions involving parasitic protists. An overview of the most common approaches used to study pelagic photosymbioses and general perspectives in the field will conclude this chapter.

11.2 Symbioses Between Phytoplankton and Cyanobacteria

11.2.1 Symbiotic Nitrogen Fixation

Nitrogen is a major limiting factor in oceanic ecosystems (Moore et al. 2013). Eukaryotes can only obtain nitrogen through the uptake of dissolved forms (mainly nitrates and ammonia), whereas some bacteria and a few archaea have the ability to fix dinitrogen (N_2) and convert it into particulate organic nitrogen. Land plants have developed symbioses with N_2-fixing bacteria such as *Rhizobium* (Franche et al. 2009; Santi et al. 2013) and similar symbioses exist in eukaryotic phytoplankton. The earliest reports were from diatom-diazotroph associations (DDAs), with the cyanobacterial symbionts ('cyanobionts') *Richelia* (Ostenfeld and Schmidt 1902) and *Calothrix* (Lemmerman 1905). More recently, the unicellular N_2-fixing cyanobacteria UCYN-A has been shown to form an unusual symbiosis with a unicellular haptophyte alga (Thompson et al. 2012). Diazotrophic cyanobacteria have also been documented to form symbiotic partnerships with a wide variety of eukaryotic marine organisms, like sponges, ascidians (although N_2 fixation in ascidians can be linked to Rhizobiales, see Erwin et al. 2014), flagellated protists, dinoflagellates, radiolarians, macroalgae, and tintinnids (Carpenter 2002; Foster et al. 2006, and references therein).

11.2.2 Symbioses Between Cyanobacteria and Diatoms

DDAs involve either filamentous heterocystous (e.g., *Calothrix rhizosolenia* and *Richelia intracellularis*) or unicellular (e.g. *Cyanothece* sp.) nitrogen-fixing cyanobacteria (Rai et al. 2002). DDAs are non-obligate endosymbioses between diatoms from several different genera (notably including *Hemiaulus*, *Rhizosolenia*, and *Chaetoceros*) and diazotrophic cyanobacteria. The diatom hosts and the cyanobacterial symbionts can be found free living in the ocean, and horizontal transfer between cells and vertical transmission from host to daughter cell are both common. In diatom–*Richelia* associations, cyanobiont *hetR* sequences from the same host species vary by less than 1 % which suggests a high degree of specificity, probably linked to vertical transmission of the cyanobiont during the host division process (Janson et al. 1999). When in association, the diazotrophs appear to be localized in different regions of the diatom depending on the host species (Foster and O'Mullan 2008). After a long period in isolation, *Calothrix* trichomes start to change their morphological features, indicating host control of cyanobiont characteristics (Foster and O'Mullan 2008).

The metabolic influence of DDA symbioses on the cyanobiont has been observed in recent studies. Foster et al. (2011) estimated that symbiotic *Richelia* fixes up to 651 % more N_2 than required for its own growth. Symbiont genome reduction can be an evolutionary consequence of long-term nutrient exchanges pointing to an increasing dependency between symbiont and host. Genome streamlining of cyanobionts has been reported for *Richelia* in intracellular association with *Hemiaulus* (Hilton et al. 2013), with genome reduction mainly affecting genes related to nitrogen metabolism: symbionts have a decreased capability to assimilate urea or nitrate (lack of ammonium transporters, nitrate and nitrite reductases and glutamine: 2-oxoglutarate aminotransferase), thus favoring N_2 fixation (Hilton et al. 2013). Diazotrophic cyanobacteria have evolved several mechanisms (both spatial and temporal) to overcome the deleterious effect of oxygen, a photosynthetic by-product, for the nitrogenase enzyme (Berman-Frank et al. 2001; Fay 1992; Thompson and Zehr 2013). In *Richelia intracellularis* in symbiosis with *Rhizosolenia clevei*, nitrogenase is protected by spatial separation, being confined to the heterocysts, the thick-walled, specialized N_2-fixation cells (Janson et al. 1995). In addition to spatial separation, a pronounced day-night periodicity of N_2 fixation was observed for *Richelia-Rhizosolenia* associations at the ALOHA station (Church et al. 2005, Foster and Zehr 2006). Unicellular *Cyanothece* sp. separate temporally the processes of carbon and nitrogen fixation (Reddy et al. 1993), and were found in association with the diatom *Climacodium frauenfeldianum* (Carpenter and Janson 2000).

11.2.3 Symbioses Between Cyanobacteria and Haptophytes

Using HISH-SIMS (halogenated in situ hybridization nanometer-scale secondary ion mass spectrometry) imaging, Thompson et al. (2012) observed a loose cell-surface association between the diazotrophic cyanobacterium UCYN-A and an apparently non-calcifying microalgal host. The host partial 18S rRNA gene sequences were >99 % identical to sequences obtained from sorted picoeukaryotic cells from South Pacific Ocean samples (BIOSOPE T60.34) (Shi et al. 2009) related to sequences of *Braarudosphaera bigelowii* (an atypical coccolithophore that produces pentalith-shaped coccoliths) and the non-calcifying haptophyte *Chrysochromulina parkeae* (Thompson et al. 2012). Using transmission electron microscopy Hagino et al. (2013) observed spheroidal bodies within *B. bigelowii* which were determined to be intracellular cyanobacterial symbionts belonging to the UCYN-A clade. Hagino et al. (2013) suggested that *C. parkeae* might be an alternate life-cycle stage of *B. bigelowii*, the former being an elongate, motile, unicellular organism with non-calcified organic scales (Green and Leadbeater 1972). *B. bigelowii* seems to comprise a set of pseudo-cryptic species, consisting of at least five 18S rRNA genotypes that correspond to morphotypes that differ slightly in size (Hagino et al. 2009). As *B. bigelowii* has a coastal distribution and the haptophyte related to BIOSOPE T60.34 was recovered from an open ocean site (Shi et al. 2009; Thompson et al. 2012), it has been hypothesized that the intracellular UCYN-A symbiosis in *B. bigelowii* was acquired after separation of those coastal/open ocean haptophyte ancestors (Hagino et al. 2013). Adding further complexity, three clades of UCYN-A, with distinct but overlapping distributions, can be distinguished based on *nifH* sequences (Thompson et al. 2014), forming a monophyletic group with the marine cyanobacteria *Crocosphaera* sp. and *Cyanothece* sp. (Bombar et al. 2014). UCYN-A1 is mostly found in the open ocean (Thompson et al. 2012) and its host is smaller than that of UCYN-A2, which has coastal distribution and whose host is *B. bigelowii* (Hagino et al. 2013). Little is known about the host and spatial distribution of UCYN-A3. A global study by Cabello et al. (2016) has provided evidence that these cyanobacterium-haptophyte symbioses are mandatory for the hosts. UCYN-A cells were reported to transfer up to 95 % of newly fixed nitrogen to their hosts (Thompson et al. 2012). UCYN-A has a reduced genome that lacks the genes involved in carbon fixation, such as those for RuBisCO (ribulose-1,5-bisphosphate carboxylase-oxygenase) (Zehr et al. 2008) and the tricarboxylic acid (TCA) cycle responsible for the biosynthesis of amino acids (Bombar et al. 2014; Tripp et al. 2010). Such modifications in the genome of symbionts are analogous to the situation for cellular organelles with specific metabolic functions such as the chloroplast or the mitochondrion, although there are still no reports on the existence of a "diazoplast" (Thompson and Zehr 2013). Tripp et al. (2010) observed that the reduced genome of UCYN-A (1.44 Mb) structurally resembles those found in most chloroplasts (as well as in some bacteria), which may indicate a similar evolutionary path. In addition, the lack of the oxygen-evolving pathway (Tripp et al. 2010; Zehr et al. 2008) de facto prevents nitrogenase damage.

11.2.4 Symbioses Between Cyanobacteria and Dinoflagellates

Little is known about symbioses between cyanobacteria and dinoflagellates despite the fact that they were first observed more than 100 years ago (Schütt 1895). In most known cases, such as for the dinoflagellates *Ornithocercus* and *Histoneis* (Farnelid et al. 2010), the cyanobacteria are ectosymbionts (i.e., associated to the cell surface) located in the cingulum of the dinoflagellate cell. These cyanobacteria appear to be nitrogen-fixers (Foster et al. 2006), but more than one type can occur in association with a single dinoflagellate cell (Farnelid et al. 2010; Foster et al. 2006). Surprisingly, sequences recovered from dinoflagellate symbionts corresponded to cyanobacteria that are not known to fix nitrogen such as *Prochlorococcus* or to other types of bacteria, suggesting the complexity of the associations between dinoflagellates and bacteria.

11.2.5 Ecological Relevance of Symbioses Involving Diazotrophs

New production in oligotrophic areas is largely dependent on N_2 fixation, since upward nutrient fluxes are limited in these regions. Several studies have highlighted the importance of symbiosis between diazotrophic bacteria and photosynthetic eukaryotes in the marine environment, both in terms of the abundance of the organisms involved and of the impact on overall N_2 fixation (Foster et al. 2009; Goebel et al. 2010; Montoya et al. 2004; Turk et al. 2011). Goebel et al. (2010) and Foster et al. (2007) found high abundances of *Richelia-Hemiaulus* symbiosis in the western equatorial Atlantic under the influence of the Amazon River plume, while during the circumnavigating Malaspina expedition, *Richelia*-diatom associations were mostly found in the South Atlantic Gyre and Indian South Subtropical Gyre (Fernández-Castro et al. 2015). The nitrogen fixed by DDAs may be an important source of nutrients to other, non-diazotrophic planktonic groups. Villareal (1990) reported evidence of release of newly fixed N to the environment in *Rhizosolenia-Richelia* symbiosis under culture conditions. Due to their size and aggregation capability, diatoms sink rapidly. Therefore, DDAs might account for an important part of the downward flux of carbon linked to new production (Scharek et al. 1999), representing an important link between nitrogen and carbon cycles in the ocean (Foster and O'Mullan 2008). Goebel et al. (2010) observed that UCYN-A was the second most abundant diazotrophic organism in tropical Atlantic waters. UCYN-A N_2 fixation was the highest among diazotrophic groups in both coastal and oligotrophic waters of the eastern North Atlantic (Turk et al. 2011). The widespread distribution of UCYN-A cells throughout the tropical and subtropical ocean observed by Cabello et al. (2016) indicates that the symbioses involving these unicellular cyanobacteria may have an important, and thus, far underestimated impact on the nitrogen cycle in these environments. This unicellular cyanobacteria-Prymnesiophyceae association may also be responsible for important contributions to vertical carbon fluxes.

11.3 Symbioses Between Phytoplankton and Heterotrophic Bacteria

11.3.1 Diversity and Dynamics of Microalgal-Bacterial Interactions

Interactions between phytoplankton and heterotrophic bacteria in marine environments are numerous, varied and often complex (Amin et al. 2012; Bell and Mitchell 1972; Ramanan et al. 2015). Some bacteria are loosely associated with algae, while others are associated more closely and colonize algal surfaces (Kaczmarska et al. 2005). Interactions range from obligate to facultative, as well as from mutualistic to parasitic, and can be mediated by cell-to-cell attachment or through the release of allelopathic compounds (Doucette 1995; Geng and Belas 2010; Seyedsayamdost et al. 2011).

The development of molecular biology tools has facilitated the study of links between phytoplankton and bacteria in natural communities (Grossart et al. 2005; Rooney-Varga et al. 2005) and from culture collections (Abby et al. 2014; Green et al. 2004; Jasti et al. 2005; Sapp et al. 2007). A molecular survey of bacterial diversity from cultures of six diatom genera (*Ditylum*, *Thalassiosira*, *Asterionella*, *Chaetoceros*, *Leptocylindrus*, and *Coscinodiscus*) revealed distinct bacterial phylotypes associated with each genus. Alphaproteobacteria related to the genera *Sulfitobacter*, *Roseobacter*, *Ruegeria*, and *Erythrobacter*, members of the Bacteroidetes and to a lesser extent Betaproteobacteria were the most prominent bacteria in the diatom cultures examined (Schäfer et al. 2002). Of these, members of the *Roseobacter* clade are commonly found in natural assemblages with marine algae, and have been shown to increase in abundance during phytoplankton blooms (Allgaier et al. 2003; Buchan et al. 2014; Mayali et al. 2008). Several molecular microbial surveys using the 16S rRNA gene marker have shown that key bacterial phylogenetic groups such as Bacteroidetes and Alpha- and Gammaproteobacteria actively respond to the decay of algal blooms (Pinhassi and Hagstrom 2000; Pinhassi et al. 2004; Riemann et al. 2000). Succession of bacterial taxa was observed during a bloom of centric diatoms in the North Sea and their occurrence patterns were linked to their capacity to degrade algal-derived organic matter (Teeling et al. 2012). The final phase of the bloom favored the dominance of Bacteroidetes with *Ulvibacter* and *Formosa* during early and mid-stages of the decline, and *Polaribacter* in the final stages. The latter metagenomic analysis demonstrated that the bacterial response to coastal phytoplankton blooms was more dynamic than previously thought and consisted of a succession of different bacterial populations with distinct functional and transporter profiles.

11.3.2 Parasitic Interactions

Bacteria can control microalgal populations by inhibiting growth or by active lysis of algal cells. Reports of algicidal bacteria have mainly focused on bacteria acting against bloom forming algae known to produce toxins that can affect human health (Mayali and Azam 2004; Paul and Pohnert 2011). The most common algicidal bacteria belong to the Gammaproteobacteria (mainly the genera *Alteromonas* and *Pseudoalteromonas*) and the Bacteroidetes (mainly the genera *Cytophaga* and *Saprospira*) (Mayali and Azam 2004). The algicidal activity can be caused either by the release of dissolved algicidal compounds or by the lysis of microalgal cells after attachment. Only few compounds or enzymes responsible for the algicidal effect have been identified. Different levels of specificity have been reported from algicidal bacteria. Selective activity against one algal species and universal activity against all tested species in a given taxon have been reported as well as all intermediate forms of specificity (Mayali and Azam 2004). Several studies indicate that some algicidal bacteria can kill their algal prey by releasing proteases (Lee et al. 2000; Paul and Pohnert 2011). Other algicidal bacteria directly attach to the microalgal cells in order to lyse them (Furusawa et al. 2003).

11.3.3 Mutualistic Interactions

Mutualistic partnerships between bacteria and marine microalgae based on the exchange of metabolites and nutrients are common (see Cooper and Smith 2015 for a recent review). Identifying chemical compounds involved in these trophic interactions between bacteria and phytoplankton is essential for our understanding of marine elemental cycles. Amin et al. (2009) found that several clades of the gammaproteobacterial genus *Marinobacter* provide an enhanced supply of Fe(III) to the dinoflagellate *Scripsiella trochoidea*, and, in return, the bacterium depends on organic matter produced by the alga. Durham et al. (2015) established a model microbial system in which the marine alphaproteobacterium *Ruegeria pomeroyi* had an obligate trophic dependency on the diatom *Thalassiosira pseudonana* for carbon while the diatom obtained vitamin B_{12} from the bacterium. A transcriptional analysis of cocultures of *T. pseudonana* and *R. pomeroyi* using RNA-seq revealed that many transcripts up-regulated in *R. pomeroyi* were involved in the transport and metabolism of 2,3-dihydroxypropane-1-sulfonate (DHPS), a sulfur compound produced by the diatom with no currently recognized role in marine microbial food webs, but which, like dimethylsulfoniopropionate (DMSP), is produced in large amounts by many marine algae. Amin et al. (2015) combined transcriptomic analysis with microbiological and biochemical experiments to study the mutualistic interactions between the coastal diatom *Pseudonitzschia multiseries* and its associated bacteria. Among 49 bacterial strains isolated from *P. multiseries* cultures, members of the genus *Sulfitobacter* (Rhodobacterales) had the largest positive effect on the growth of the alga. A *Sulfitobacter* species promoted diatom cell division via secretion of the auxin indole-3-acetic acid (IAA), while this bacterium used both diatom-secreted and endogenous tryptophan. This study also detected levels of IAA in five coastal North Pacific sites equivalent to that found in laboratory cocultures and presented transcriptomic evidence from natural samples for multiple IAA biosynthesis pathways. Amin et al. (2015) proposed that tryptophan and IAA are signaling molecules to recognize and sustain beneficial partners. Another study of *Phaeobacter inhibens* BS107, a member of the *Roseobacter* clade, and *Emiliania huxleyi*, a dominant marine phytoplankton found in large algal blooms, revealed that interaction between algae and *Roseobacter* could be mutualistic, antagonistic, or shift between both (Seyedsayamdost et al. 2011). The bacterium initially provided a growth enhancing effect by producing an auxin and an antibiotic that protected the alga from other bacteria. This mutualistic relationship shifted to a pathogenic relationship when the algal senescence signal p-coumaric acid released by aging *E. huxleyi* cells elicited the production by the bacterium of algicidal compounds termed roseobacticides that increase cell death of *E. huxleyi*. A similar effect was also observed in co-cultures of the dinoflagellate *Prorocentrum minimum* and *Dinoroseobacter shibae*, suggesting that a shift from mutualism to parasitism is a common feature in *Rhodobacterales*-based symbiosis (Wang et al. 2014).

11.4 Mutualistic Photosymbioses Between Eukaryotes

In pelagic environments, photosymbiotic interactions between eukaryotes include relationships that involve microalgae with other protists or with metazoans. Often assumed to be mutually beneficial or commensal because of the presumed trophic exchanges and recycling of nutrients between the host and symbionts, the exact nature of the partnership is often difficult to formally demonstrate. Eukaryotic epibionts (i.e., cells living on the surface of other organisms) are common in benthic environments and are also encountered in pelagic ecosystems, such as the association between the centric diatom *Thalassiosira* sp. and the coccolithophore *Reticulofenestra sessilis* (Decelle et al. 2015; Taylor 1982). However, planktonic photosymbioses between eukaryotes most often involve a photosynthetic symbiont that lives intracellularly within a heterotrophic host (Anderson 2012; Decelle et al. 2015). The most common host taxa in marine plankton are Radiolaria, Foraminifera, ciliates and dinoflagellates (Stoecker et al. 2009). Microalgal symbionts, often collectively referred to as "zooxanthellae," have long been thought to all be rather similar, but recent studies have revealed more diversity in this group.

11.4.1 Radiolarian Hosts

Based on current knowledge, Radiolaria is the most diverse group of planktonic hosts harboring eukaryotic microalgal symbionts. All main radiolarian lineages (Spumellaria, Collodaria, Nassellaria, Acantharia) include numerous species harboring obligate eukaryotic microalgal symbionts (Suzuki and Not 2015). It is assumed that these symbiotic species have to specifically acquire their symbionts from the environment at each host generation (i.e., horizontal transmission). In the Spumellaria, Collodaria, and Nassellaria, the most commonly occurring symbiont appears to be the dinoflagellate *Brandtodinium nutricula* that was first described (as *Zooxanthella nutricula*) over a century ago (Brandt 1881), but which was only recently cultured and morphologically characterized, leading to placement in the new genus *Brandtodinium* (Probert et al. 2014). The exact identity of the microalgal symbionts of the main monophyletic clade of symbiotic Acantharia was recently revealed to be members of the well-known haptophyte genus *Phaeocystis* (Decelle et al. 2012). In apparent contrast to the symbionts of other radiolarians, based on phylogenies performed on the 18S rRNA and D1–D2 region of the 28S rRNA gene sequences, acantharian symbionts have the exact same genetic identity as species that are abundant in the plankton in their free-living stage, and display a lack of species-level host specificity (e.g., symbiont geography rather than host taxonomy is the main determinant of the association). *Acanthochiasma*, an early branching clade of Acantharia, has been found to simultaneously harbor multiple symbiotic microalgae, including distantly related dinoflagellates (*Heterocapsa* sp., *Pelagodinium* sp., *Azadinium* sp., and *Scrippsiella* sp.) as well as a haptophyte (*Chrysochromulina* sp.) (Decelle et al. 2013).

Acantharia is widely distributed throughout the world's ocean and typically outnumber planktonic Foraminifera and other Radiolaria in oligotrophic open ocean waters. Environmental molecular diversity surveys of protistan communities in pelagic ecosystems have demonstrated the ubiquitous occurrence of radiolarian sequences and notably those of Collodaria (de Vargas et al. 2015; Not et al. 2009). The Collodaria are large, fragile, colony-forming Radiolaria that have been estimated, using in situ imaging tools, to contribute significantly to total oceanic carbon standing stock in the upper 200 m of the water column (Biard et al. 2016). Along with other heterotrophic protists harboring microalgal endosymbionts, their predominance in surface waters of the intertropical ocean is likely linked to their photosymbiotic character, illustrating the significance of acquired phototrophy for global marine ecology (Stoecker et al. 2009).

11.4.2 Foraminiferal Hosts

Only 5 of the nearly 50 species of planktonic Foraminifera described to date harbor microalgal symbionts, yet these five species correspond to 50–90 % of Foraminifera individuals found in surface waters of the tropical and subtropical ocean (Caron et al. 1995; Stoecker et al. 1996). Each host cell can contain up to 20,000 symbionts. These five species, belonging to the genera *Globigerinoides*, *Globigerinella*, and *Orbulina*, form a monophyletic clade within the Foraminifera based on 18S rRNA gene phylogenies and they all possess spines along which symbionts are positioned during the day (Spero 1987). In contrast to benthic Foraminifera that have a wide diversity of microalgal symbionts, all planktonic symbiotic species form associations with the recently described dinoflagellate genus *Pelagodinium* (Siano et al. 2010), which is related to *Symbiodinium* in the order Suessiales. Other microalgal symbionts belonging to the haptophyte genus *Chrysochromulina* have been reported (Gast et al. 2000), but this relationship is less well characterized.

11.4.3 Ciliate Hosts

Symbiotic associations between ciliates and eukaryotic microalgae (e.g., *Paramecium bursaria* and *Chlorella* sp.) are well known and abundant in freshwater ecosystems (Kodama et al. 2014). In marine environments ciliates preferentially associate with cyanobionts (e.g., *Codonella* sp.) or perform klepto-plastidy (retention of plastids only rather than the whole cell) such as the well-known *Mesodinium rubrum* which sequesters plastids from a cryptophyte algae and can form massive blooms (Johnson and Stoecker 2005), or Oligotrichida ciliates which harbor klepto-chloroplasts from green algae in estuarine environments (Stoecker et al. 1989a). An original pelagic photosymbiosis between a calcifying ciliate host and the dinoflagellate *Symbiodinium* was recently described from surface ocean waters (Mordret et al. 2015). The host is a new ciliate species closely related to *Tiarina fusus* (Colepidae) and phylogenetic analysis of the symbionts revealed that they are novel genotypes of *Symbiodinium*, closely related to clade A, that do not seem to associate with any benthic host. Based on molecular diversity surveys, this symbiotic partnership occurs globally, in particular in nutrient-poor surface waters.

11.4.4 Dinoflagellate Hosts

Photosynthetic dinoflagellates can be symbionts of other large protists (e.g., Foraminifera or Radiolaria), but heterotrophic dinoflagellates can also harbor photosynthetic symbionts. These symbionts are mainly cyanobionts (see above), but in some cases can be eukaryotic microalgae. For instance, the genus *Amphisolenia* has been described to simultaneously harbor both cyanobionts and pelagophyte microalgae (Daugbjerg et al. 2013). The bioluminescent dinoflagellate species *Noctiluca scintillans* lives in symbiosis with a green prasinophyte alga, described from its morphology as *Pedinomonas noctilucae* (Sweeney 1976), and can harbor up to 10,000 symbionts that swim freely within large vacuoles in the host cell. The *Noctiluca–Pedinomonas* association is common in tropical and subtropical areas of southeast Asia, in the Indian Ocean, the Pacific Ocean, and the Red Sea where it regularly forms extensive blooms (called “green tides”) reaching densities of up to 5×10^6 cells L^{-1} (Harrison et al. 2011). Diatoms (“dinotoms”) and other symbionts of uncertain affiliation can be found in symbiosis with dinoflagellate hosts, but these are less well described (Imanian et al. 2010).

11.4.5 Metazoan Hosts

Endosymbiotic microalgae are also found in association with large multicellular metazoan plankton, such as jellyfish and acoel flatworms. Among the most studied jellyfish, the scyphozoan *Cassiopea* has been described in symbiosis with the dinoflagellate *Symbiodinium microadriaticum*, but the specificity of the relationships between host and symbiont are currently unclear as morphological, biochemical, and physiological differences between strains cultured from different hosts have been observed (Arai 1997). Other dinoflagellates, namely *Gymnodinium linuchae* and *Scrippsiella vellelae*, have been isolated and described from the scyphozoan *Linuche unguiculata* and the hydrozoan *Vellela vellela*, respectively (Trench and Thinh 2007). *S. vellelae* was redefined as *B. nutricula* and is the same symbiont found in the majority of radiolarians (Probert et al. 2014). All species of pelagic acoel flatworms collected over a 13 year sampling effort in surface waters of the open oceans harbor microalgal endosymbionts (Stoecker et al. 1989b). From this latter study, three types of oceanic flatworms were discriminated: a "bright green" and a "dark brown" acoel presumably belonging to the host genus *Convoluta* and both harboring a green prasinophyte-like symbiont identified based on ultrastructure. The "dark brown" flatworm was mostly observed on the surface of colonial radiolarians and other gelatinous plankton. The third type of acoel was referred to as "golden" and harbors a dinoflagellate symbiont of uncertain taxonomic affiliation. Dinoflagellate endosymbionts of the pelagic acoel *Amphiscolops* sp. were identified as belonging to the genus *Amphidinium* (Lopes and Silveira 1994). Acoel flatworms with algal endosymbionts depend on both autotrophic and heterotrophic nutrition and are a widespread, though sporadic component of the plankton in the upper water column in warm, oceanic waters. Jellyfish harboring photosymbionts are frequently observed in the environment and besides a trophic relationship that is presumably advantageous in oligotrophic environments (i.e., recycling of nutrients between the host and symbionts), the exact role of the photosymbiosis is not well understood. It has been suggested that symbiosis enhances the rate of strobilation, being potentially involved in the host cell cycle (Arai 1997).

11.5 Parasitic Photosymbioses Between Eukaryotes

Parasitism is a non-mutual symbiotic relationship that can be neutral to lethal (i.e., never beneficial) for the host. The parasite has an obligate physical association with its host, at least during a part of its life cycle. Parasitic photosymbioses include heterotrophic protists infecting microalgae and microalgae infecting larger animals. All marine protists involved in parasitic photosymbioses have a similar life cycle that typically includes three stages which allow the parasite to fulfill three essential functions: infection of the host via an actively swimming zoospore, acquisition of energy via a feeding stage (the trophont), and sporulation by a sporocyst which produce zoospores used for propagation (i.e., they are all zoosporic parasites). These parasites are thus horizontally transmitted, meaning that the host is newly infected from the surrounding environment at each generation.

All of these parasites can be classified based on their impact on their host, their localization on their host and their mode of transmission (Lafferty and Kuris 2002, Poulin 2011). Parasites infecting microalgae generally kill their host and these highly virulent parasites are called parasitoids. The impact of microalgal parasites infecting larger animals is generally lower. Beside their negative impact on host populations, zoospores of parasites are actively grazed by predators and should also be considered as an important trophic resource in marine pelagic systems.

11.5.1 Heterotrophic Parasites Infecting Microalgae

All known protistan parasites of marine microalgae infect either diatoms or dinoflagellates. Protistan parasites infecting other important marine microalgal lineages exist in freshwater, e.g., the perkinsoan *Rastrimonas subtilis* infecting the cryptophyte *Chilomonas paramecium* (Brugerolle 2002), but have never been reported in marine habitats. Parasites of marine diatoms include chytrids, aphelids, stramenopiles (including the genus *Pirsonia*, oomycetes, labyrinthuloids, and hyphochytrids), dinoflagellates, cercozoans, and phytomyxids (for a review see Scholz et al. 2016). Parasites of dinoflagellates include chytrids (different from those infecting diatoms), Syndiniales (Amoebophryidae) and Perkinsozoa (for a review see Jephcott et al. 2015). The trophont of these parasites develops either outside (ectoparasites) or inside (endoparasites) their host. Ectoparasites may partially penetrate inside the host (part of the cytoplasm, and even the mitochondrion, Lepelletier et al. 2014a), but the nucleus always remains outside the host. Ectoparasites of microalgae use different strategies to penetrate their host. Fungi produce a germ tube that penetrates into the dinoflagellate host through gaps between thecal plates (Lepelletier et al. 2014a, b). Most ectoparasites of microalgae, however, are active phagotrophs and feed either by endocytosis, pinocytosis or phagocytosis. The heterokont *Pirsonia* spp. and the cercozoans *Pseudopirsonia mucosa* and *Cryothecomonas longipes* infect diatoms using a pseudopodium-like cytoplasmic strand that either pierces the diatom frustule, generally in the girdle region, or passes through natural orifices of centric diatoms (Schnepf and Kuhn 2000; Schweikert and Schnepf 1997). *Paulsenella* is a dinoflagellate ectoparasite of diatoms that pierces the host plasmalemma by a feeding tube (called the pedoncule) and gradually sucked out the host cytoplasm, resembling drinking through a straw. The prey's cytoplasm is deposited in a food vacuole where it is digested. This mode of endocytosis (called myzocytosis, Schnepf and Deichgräber 1984) is a feeding mode exclusively observed in alveolate parasitoids.

When the nucleus of the parasite enters inside the host, the parasite is considered as an endoparasite. There are several advantages to being an endoparasite. First, the endoparasite remains protected by the external envelope of its host during its whole maturation. Second, an endoparasite more efficiently exploits its host than an ectoparasite. While in the case of ectoparasites, the host nucleus and enough of the cytoplasm can be left to allow the host to survive the infection (Kühn et al. 1996; Schnepf and Melkonian 1990; Schweikert and Schnepf 1997), endoparasites always kill their host and can digest them entirely, including the nucleus. However, endoparasites need to overcome two major difficulties: bypassing the natural defenses of the host to enter the cell and finding a way to leave the cell after maturation. Different strategies are used by endoparasites to enter and develop safely inside their host. Like ectoparasites, the aphelid *Pseudaphelidium drebessi* (Karpov et al. 2014) and the cercozoan *Cryothecomonas aestivalis* (Drebes et al. 1996) produce a pseudopodium-like structure and squeeze into the interior of the diatom frustule. Inside the diatom frustule, these parasites are in intimate contact with the host plasma membrane, but never pierce it (Schweikert and Schnepf 1997). Apicomplexans and relatives (Syndiniales and Perkinsozoa) use an apical complex derived structure to penetrate their host (Leander and Keeling 2003). The host of *Parvilucifera* spp. is rapidly immobilized at the penetration of the parasite that then feeds by osmotrophy by producing external enzymes that totally digest its host. In contrast, *Amoebophrya ceratii* is an endoparasitic phagotroph that preserves its host alive (swimming in the water column) until the very end of the maturation process.

The intracellular trophont often distorts the host cell. Dinoflagellates infected by *A. ceratii* become much larger than healthy cells (Hanic et al. 2009; Kim et al. 2004). For sporulation, *C. aestivalis* forms slightly amoeboid flagellate spores that are discharged by slipping with their posterior pole foremost through the diatom frustule (Drebes et al. 1996). Oomycetes and chytrids infecting the diatom *Pseudo-nitzschia pungens* (Hanic et al. 2009) produce similarly shaped discharge tubes through the host cell wall. In *A. ceratii* the sporocyst makes a complex evagination to leave its host (Cachon 1964) and once outside the cell becomes an elongated multicellular flagellated structure (the vermiform stage). Within hours, each cell forming this vermiform is released from this multicellular structure and is available to infect a novel host.

Several parasites are known to produce resting stages, e.g., *Pirsonia* spp. (Drebes and Schnepf 1988; Kühn et al. 1996), aphelids (Karpov et al. 2014) or *C. aestivalis* (Drebes et al. 1996). Whether these resting stages are the result of a sexual reproduction is unknown. Mature zoospores will be released via several opercules that will be opened possibly after activation by water-borne signals (Garcés et al. 2013). *A. ceratii* may enter into dormancy in its host resting cyst and new infections are initiated after germination of the host cyst (Chambouvet et al. 2011a). This strategy allows a perfect physical coupling in time and space of the parasite with its host.

11.5.2 Microalgal Parasites Infecting Larger Organisms

Microalgae may be obligatory and/or facultative parasites of marine metazoans in benthic and pelagic ecosystems (Rodriguez et al. 2008). Members of the dinoflagellate genus *Blastodinium* are endocommensal of copepods (Skovgaard et al. 2012). They are directly ingested by their hosts as food particles and once inside the copepod gut, they develop and produce one to several large (several hundreds of microns) trophont or sporocyst individuals surrounded by the same outer mother membrane. Infected copepods are generally smaller, less fit, and less fecund than healthy ones as a consequence of a physical blocking of the alimentary tract in the gut of the animal and competition for food uptake.

11.5.3 Detection of Parasites in Environmental Genetic Surveys

In molecular surveys of plankton diversity, Amoebophryidae, also known as Marine ALVeolate group II or MALV II (López-García et al. 2001), always represent a large fraction of the sequences retrieved from marine habitats, from surface waters to deep hydrothermal vents, but have never been found in freshwater. This family includes a single genus, *Amoebophrya*, with seven species, all described by Cachon (1964). Environmental surveys based on sequencing the SSU rRNA gene have revealed an impressive diversity within this lineage, including more than 40 genetic clusters identified, most composed of several distinct subclusters that potentially correspond to separate species (Guillou et al. 2008). All *Amoebophrya* are parasites, infecting either radiolarians (*A. acanthometrae* and *A. sticholonchae*), ciliates (*A. tintinni*), dinoflagellates (*A. ceratii* and *A. leptodisci*) or other parasites (i.e., the hyperparasites *A. grassei* and *A. rosei*). *A. ceratii* infects most, if not all, phototrophic dinoflagellates (Cachon 1964; Siano et al. 2011). This wide host range may explain the ecological success of this parasitic group. MALV II are predominantly detected in the smallest picoplanktonic size fractions by metabarcoding (Massana et al. 2015) and from environmental DNA rather than RNA (Massana et al. 2015; Not et al. 2009). Environmental genetic surveys are likely to detect preferentially zoospores, which are actively swimming propagules that do not reproduce mitotically. Their viability in marine water is at most a few days, even in culture (Cachon 1964; Coats and Park 2002). One infection potentially releases hundreds of zoospores that, like spermatozoids, have a high nucleus/cytoplasm ratio. The genetic trace of these zoospores is likely detectable long after they die as a part of free environmental genetic material. Not all marine parasites are preferentially detected in the smallest size fractions. *Blastodinium* spp. environmental sequences are more prevalent within the mesoplanktonic fraction (180–2000 μm; de Vargas et al. 2015), leading to the conclusion that these parasites are mainly detected within their hosts. Less destructive methods, such as Fluorescent In Situ Hybridization (FISH) techniques, have been used to confirm that infections of dinoflagellates by *A. ceratii* and of copepods by *Blastodinium* spp. occur from coastal environments to the most oligotrophic areas of the planet (Alves-de-Souza et al. 2011; Siano et al. 2011).

11.6 Methods for Studying Pelagic Symbioses

11.6.1 Microscopy and Related Approaches

Marine pelagic photosymbiotic associations were first discovered soon after light microscopes started to become widely available in the latter half of the 19th century. Only two years after de Bary (1879) coined the term symbiosis, K. Brandt used light microscopy to recognize that the "yellow cells" inside radiolarians, actinian corals, and hydrozoans were in fact symbiotic microalgae (Brandt 1881). Symbiotic associations between diatoms and cyanobacteria were first described from light microscope observations by Karsten (1907). In the second half of the 20th century, increasingly sophisticated microscopy-related methods were employed to discover and describe pelagic symbiotic relationships. The development of transmission electron microscopy in the 1960s allowed more precise localization and sometimes taxonomic identification of symbionts inside host cells (e.g., Hollande and Carré 1974; Taylor 1971). Characterization by electron microscopy of the morphology and/or ultrastructure of symbionts either within the host cell '*in hospite*' (Hagino et al. 2013; Miller et al. 2012; Yuasa et al. 2012) or outside, '*ex hospite*' or 'free living' (Probert et al. 2014), remains central to the study of these associations. When coupled with immuno-labeling, electron microscopy allows precise intracellular localization within host and/or symbiont cells of specific proteins such as nitrogenase (Foster et al. 2006). From the 1980s onwards, epifluorescence microscopy has been widely used to assess the type of pigment present in the symbionts (and thus general taxonomic assignation). For example, fluorescence microscopy permits easy distinction of chlorophyll-containing eukaryotes from phycoerythrin-containing cyanobacteria (Stoecker et al. 1987) or Syndiniales parasites within dinoflagellates based on their specific green fluorescence (Chambouvet et al. 2011a). The subsequent development of molecular probes coupled with fluorescent labels (FISH) using amplification approaches, such as Tyramide Signal Amplification (TSA), necessary in many cases because of the low ribosomal signal of the symbionts or parasites compared to their hosts, allows determination of the taxonomical affiliation of hosts and/or symbionts (e.g., Biegala et al. 2002; Cabello et al. 2016; Chambouvet et al. 2008). Nanoscale secondary ion mass spectrometry (Nano-SIMS; Musat et al. 2012) is a powerful emerging technique which allows assessment of the cellular localization and metabolic fluxes of compounds such as nitrogen or carbon (Foster et al. 2011; Thompson et al. 2012). Flow cytometry allows characterization and physical separation of cells based on their size and fluorescence and has been used, for example, to sort small eukaryotes associated to nitrogen-fixing cyanobacteria in order to determine their taxonomic affiliation (Thompson et al. 2012) or to study pico- and nano-phytoplankton associations with fungi (Lepère et al. 2015).

11.6.2 Ex Situ Laboratory Culture

The successful maintenance of planktonic organisms in ex situ laboratory culture greatly facilitates in depth morphological, genetic and physiological studies. Culturing of organisms that are capable of living in isolation from other species is a challenge in itself (Stewart 2012). The co-culture of organisms involved in symbiotic associations tends to be even more complex. For pelagic photosymbioses, most success to date has come from separating the partners and culturing one (or more rarely both) as a free-living organism. For mutualistic symbioses involving unicellular photosynthetic organisms as symbionts, separation of the partners by manual micropipetting has been increasingly successful (e.g. Decelle et al. 2012; Probert et al. 2014; Siano et al. 2010). This method involves disintegration (crushing) of the host cell with a micropipette under an inverted microscope to release the symbionts, which are subsequently individually isolated into an appropriate culture medium in which they develop as a culture of the free-living form. Disintegration methods are often used to release microalgal symbionts from corals and anemones, but induction of the release of symbionts by physical (e.g. heat, light, or salinity shock) and/or chemical (e.g. 3-(3,4-dichlorophenyl)-1,1-dimethylurea (DCMU) or menthol) treatment has also been reported for these benthic metazoan hosts (Wang et al. 2012). The viability of the symbionts released using these physical or chemical treatments may vary for a given method between host species (Wang et al. 2012). The fact that microalgal symbionts are typically maintained within host cells in a simplified ('protoplasmic') state with considerably altered phenotype (for example, lack of flagella and/or theca, abnormally large cell size) facilitates single-cell isolation once the cells are released, but can mean that cells are more prone to deleterious physicochemical shock when released from the host. There is undoubtedly considerable scope for better mimicking the physicochemical environment within the host in order to increase the success of isolation of symbionts, particularly for those with specific physiological capacities such as N_2-fixation. It has been possible to maintain the N_2-fixing cyanobacterial symbiont *Calothrix* in culture outside its host diatom *Chaetoceros* (Foster et al. 2010), but *Richelia* can only be maintained within its host. Neither the host diatom *Rhizosolenia* (Villareal 1990) or the haptophyte hosts of UCYN-A N_2-fixing cyanobacterial symbiont have been cultivated to date (Fig. 11.1).

Co-culturing of host and symbiont (the 'holobiont') is required to conduct experiments to develop a mechanistic understanding of the functioning of symbiotic relationships. Many symbiotic coral species can be maintained in culture for long periods of time with colonies growing by asexual reproduction. However, mastering the sexual reproduction of corals in ex situ culture has proven much more difficult, particularly for species that have separate mating types and that expel (rather than brood) their gametes. Some pelagic organisms harboring microalgal symbionts can be maintained for short periods (up to a few weeks) in laboratory conditions, particularly if kept in circulating seawater aquarium systems that maintain them in suspension. However, the almost total lack of knowledge about asexual and sexual reproduction processes of host organisms from the pelagic environment has prevented successful long-term culture of these organisms. Most heterotrophic hosts are known to undergo sexual reproduction and release aposymbiotic (i.e., without symbiont) gametes at some point in their life cycle, but nothing is known about the processes of gamete recognition, fusion, and subsequent establishment of a daughter generation that would need to reacquire symbionts from the environment (horizontal transmission). The culture of pelagic photosymbiotic holobionts will require considerable advances not only in the technology of culture

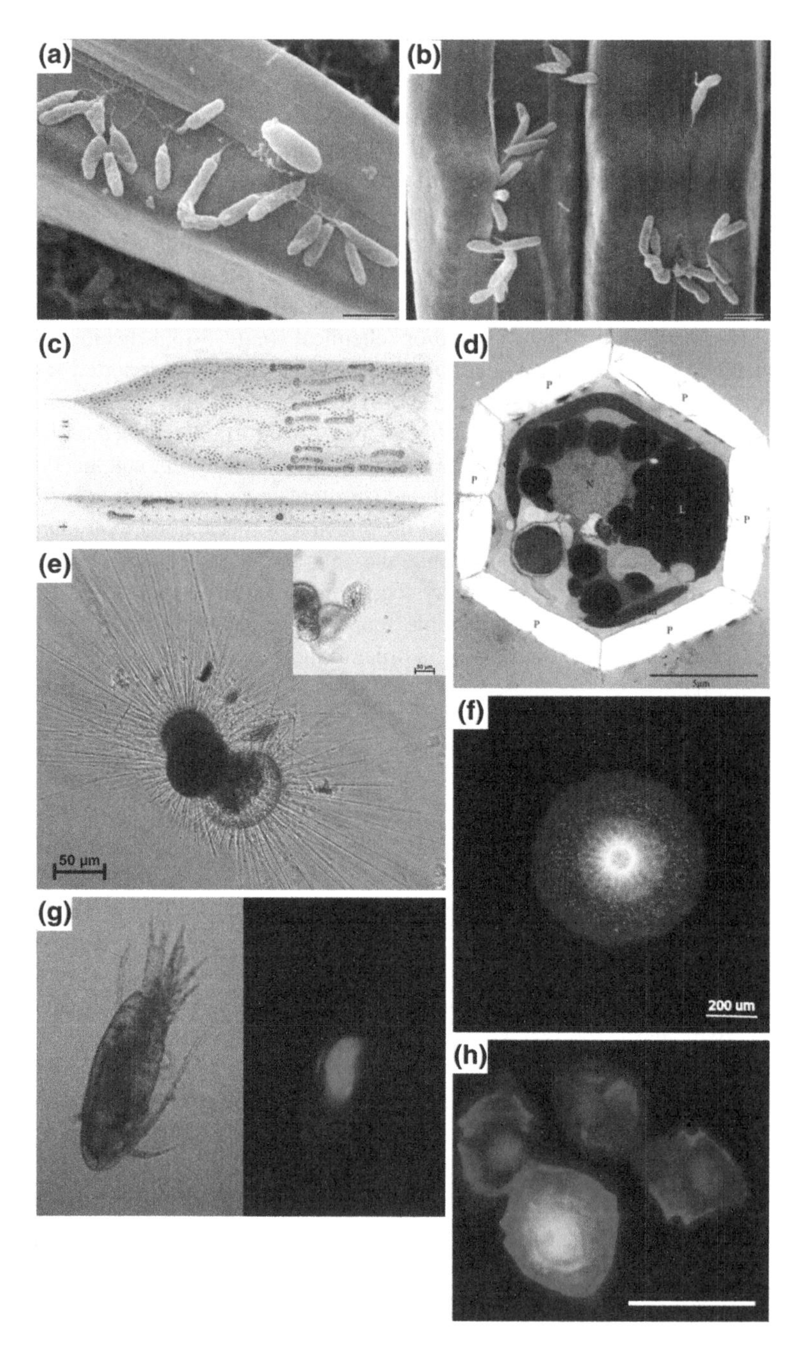
(a)
(b)
(c)
(d)
P
N
L
(e)
50 µm
(f)
200 um
(g)
(h)

◀ **Fig. 11.1** Illustrations of pelagic photosymbioses. **a** and **b** Scanning electron microscopy images of heterotrophic bacteria associated to the diatom *Pseudonitzschia multiseries* (from Kaczmarska et al. 2005, scale bars = 1 μm). **c** Symbiosis between the cyanobacterium *Richelia* and the diatom *Rhizosolenia* drawn from microscopic observations (Karsten 1907). **d** TEM images of the prymnesiophyte algae *Braarudosphaera bigelowii* showing nucleus (N), chloroplasts (Chl), lipid globules (L), pentaliths (P), mitochondria (mt) and cyanobacterial symbiont (S) (from Hagino et al. 2013, under CC BY license). **e** Planktonic Foraminifera in association with its dinoflagellate symbiont *Pelagodinium beii*, insert shows the Foraminifera test broken and symbiotic algae released (small golden dots). **f** One large Radiolaria cell (Collodaria), displaying its dinoflagellate symbionts (*Brandtodinium nutricula*) on the outer part (numerous small golden dots). **g** Left, optical microscopy image of a copepode (*Clausocalanus* type) infected by the microalgae *Blastodinium contortum*, and right, same specimen observed under epifluorescence showing the chlorophyll autofluorescence (*red*) of its algal parasitic endosymbiont. **h** The dinoflagellate species *Heterocapsa triquetra* infected and noninfected cells from a natural sample collected in the Penzé estuary, France. The parasites is detected by a FISH using the ALV01 probe (*green*), the host parasite is stained in red by propidium iodine and the host theca stained in blue by calcofluor (photo credit. C. Alves-de-souza, scale bar: 20 μm)

systems, but also in knowledge of the undoubtedly complex life cycles of these organisms.

The majority of parasites cannot be maintained in culture without their host. Generalist parasites (infecting a large range of hosts) are typically much easier to isolate than specialist parasites (having a narrow host range). For specialists, it is recommended to first establish the host strain in culture from the locality where the parasite will be isolated. The main bottleneck for their cultivation remains the labor intensiveness of their maintenance, as parasites of phytoplankton typically have rapid life cycles and have to be regularly transferred into a fresh host culture (as frequently as twice per week). Some parasites of microalgae, such as *Parvilucifera* spp., can be stored for longer periods at 4 °C and/or cryopreserved (Lepelletier et al. 2014a, b).

11.6.3 Molecular Approaches

The introduction of molecular techniques into plankton research has allowed much better characterization of the nature and diversity of hosts as well as symbionts using marker 'barcode' genes such as 18S or 16S rRNA (e.g., Chambouvet et al. 2011b; Decelle et al. 2012; Thompson et al. 2014) or functional genes linked to the key role of the symbiont such as *nifH* or *hetR* involved in N_2-fixation (Foster and Zehr 2006). In light of the difficulty of culturing pelagic photosymbiotic associations, one big advantage of molecular techniques is that they can usually be employed in culture-independent studies. In recent years, new "omics" approaches (genomics, transcriptomics, and their meta- declinations when dealing with uncultured organisms) have increasingly been employed to study the nature of symbiotic relationships. For example, determination from flow cytometry sorted cells of the genome sequence of the symbiotic cyanobacterium UCYN-A highlighted the absence of photosystem II in this organism and therefore its inability to fix carbon (Zehr et al. 2008) for which it has to rely on its host (Thompson et al. 2012). Genome sequencing also revealed that *Richelia*, a cyanobacterial symbiont of diatoms, lacks key N metabolism genes (Hilton et al. 2013). Transcriptomic approaches are currently more accessible than full genome sequencing for eukaryotes and these have been used, for example, to identify genes potentially involved in symbiosis or parasitic attack such as those coding for lectins (Balzano et al. 2015; Lu et al. 2014).

Interactions between marine protists and bacteria have been demonstrated using single-cell sorting by flow cytometry and further sequencing SSU rRNA genes of the individual protist and the bacteria physically associated with it (Martinez-Garcia et al. 2012). In particular, the latter pilot study suggested the discovery of novel symbionts, distantly related to Rickettsiales and the candidate divisions ZB3 and TG2, associated with cercozoan, and chrysophyte hosts. Although further studies are required to unequivocally determine whether these newly discovered associations represent parasitic or mutualistic relationships, single-cell sequencing is a promising approach for the analysis of ecological interactions between uncultured protists and bacteria.

11.7 Concluding Remarks

Only in recent years scientists have started to realize the full extent of the critical roles and services provided by symbioses across ecosystems and scales, from molecular to ecological (McFall-Ngai 2008). It has long been recognized that symbiotic interactions exist in the marine pelagic environment, but the pace of discovery has increased in recent years through the application of both classical techniques and novel methodologies such as high-throughput sequencing associated to bioinformatic analysis of interaction networks (Guidi et al. 2016; Lima-Mendez et al. 2015; Thompson et al. 2012; Worden et al. 2015). Several new, ecologically important pelagic photosymbiotic associations have been discovered and at least partially characterized and it would not be surprising to see this trend continue and even intensify in the near future. It is clear that future studies aiming to model nutrient and energy budgets in the ocean must take into account the importance of pelagic symbiotic associations for the input of new nitrogen, as well as for the downward flux of carbon in the water column.

In order to progress toward a holistic understanding of the marine microbiome (Dubilier et al. 2015), it is important to further complement descriptive studies of the nature of photosymbiotic interactions with understanding of the physiological and molecular mechanisms involved. Despite promising developments in culture-independent methods (e.g., single-cell approaches), ex situ culturing, and experimentation remains a critical step to comprehensively understand any biological system. The establishment of new, ecologically relevant, culturable biological model systems to study pelagic photosymbioses is one of the main challenges facing researchers in this field in coming years.

Acknowledgments The research leading to these results has received funding from the European Union Seventh Framework Programme (FP7/2007–2013) under grant agreement n° 311975. This publication reflects the views only of the author, and the European Union cannot be held responsible for any use which may be made of the information contained therein. The authors also acknowledge the ANR HAPAR (ANR-14-CE02-0007), the ANR IMPEKAB (ANR-15-CE02-0011), the CNRS GDRI "Diversity, Evolution and Biotechnology of Marine Algae", the CAPES-COFECUB "PicoBras" project (Te 871-15), and the MICROMAR project funded by a CNRS grant (INSU-EC2CO program). Klervi Crenn received a doctoral grant half funded by Région Bretagne.

References

Abby SS, Touchon M, De Jode A, Grimsley N, Piganeau G (2014) Bacteria in *Ostreococcus tauri* cultures—friends, foes or hitchhikers? Front Microbiol 5:505

Allgaier M, Felske A, Wagner-do I (2003) Aerobic anoxygenic photosynthesis in *Roseobacter* clade bacteria from diverse marine habitats. Appl Environ Microbiol 69:5051–5059

Alves-de-Souza C, Cornet C, Nowaczyk A, Gasparini S, Skovgaard A, Guillou L (2011) *Blastodinium* spp. infect copepods in the ultra-oligotrophic marine waters of the Mediterranean Sea. Biogeosciences 8:2125–2136

Amin SA, Green DH, Hart MC, Küpper FC, Sunda WG, Carrano CJ (2009) Photolysis of iron–siderophore chelates promotes bacterial–algal mutualism. Proc Natl Acad Sci USA 106:17071

Amin SA, Parker MS, Armbrust EV (2012) Interactions between diatoms and bacteria. Microbiol Mol Biol Rev 76:667–684

Amin SA, Hmelo LR, van Tol HM, Durham BP, Carlson LT, Heal KR, Morales RL et al (2015) Interaction and signalling between a cosmopolitan phytoplankton and associated bacteria. Nature 522:98–101

Anderson OR (2012) Living together in the plankton: a survey of marine protist symbioses. Acta Protozool 52:1–10

Arai MN (1997) A functional biology of scyphozoa, vol XVI. Springer, Netherlands, p 316

Bailly X, Laguerre L, Correc G, Dupont S, Kurth T, Pfannkuchen A, Entzeroth R et al (2014) The chimerical and multifaceted marine acoel *Symsagittifera roscoffensis*: from photosymbiosis to brain regeneration. Front Microbiol 5:498

Balzano S, Corre E, Decelle J, Sierra R, Wincker P, da Silva C, Poulain J et al (2015) Transcriptome analyses to investigate symbiotic relationships between marine protists. Front Microbiol 6:98

Bell W, Mitchell R (1972) Chemotactic and growth responses of marine bacteria to algal extracellular products. Biol Bull 143:265–277

Berman-Frank I, Lundgren P, Chen YB, Küpper H, Kolber Z, Bergman B, Falkowski P (2001) Segregation of nitrogen fixation and oxygenic photosynthesis in the marine cyanobacterium *Trichodesmium*. Science 294:1534–1537

Biard T, Stemmann L, Picheral M, Mayot N, Vandromme P, Hauss H, Gorsky G, Guidi L, Kiko R, Not F (2016) In situ imaging reveals the biomass of giant protists in the global ocean. Nature 532:504–507

Biegala IC, Kennaway G, Alverca E, Lennon JF, Vaulot D, Simon N (2002) Identification of bacteria associated with dinoflagellates (Dinophyceae) *Alexandrium* spp. using tyramide signal amplification-fluorescent in situ hybridization and confocal microscopy. J Phycol 38:404–411

Bombar D, Heller P, Sanchez-Baracaldo P, Carter BJ, Zehr JP (2014) Comparative genomics reveals surprising divergence of two closely related strains of uncultivated UCYN-A cyanobacteria. ISME J 8:2530–2542

Brandt K (1881) Uber das Zusammenleben von Thieren und Algen. Verh Physiol Ges 524–527
Brugerolle G (2002) *Colpodella vorax*: ultrastructure, predation, life-cycle mitosis, and phylogenetic relationships. Eur J Protistol 38:113–125
Buchan A, Lecleir GR, Gulvik CA, González JM (2014) Master recyclers: features and functions of bacteria associated with phytoplankton blooms. Nat Publ Gr 12:686–698
Cabello AM, Cornejo-Castillo FM, Raho N, Blasco D, Vidal M, Audic S, de Vargas C et al (2016) Global distribution and vertical patterns of a prymnesiophyte-cyanobacteria obligate symbiosis. ISME J 10:693–706
Cachon J (1964) Contribution à l'étude des péridiniens parasites. Cytologie, cycles évolutifs. Ann des Sci Nat Zool Paris VI:1–158
Caron DA, Michaels AF, Swanberg NR, Howse FA (1995) Primary productivity by symbiont-bearing planktonic sarcodines (Acantharia, Radiolaria, Foraminifera) in surface waters near Bermuda. J Plankton Res 17:103–129
Carpenter EJ (2002) Marine cyanobacterial symbioses. Biol Environ 102:15–18
Carpenter EJ, Janson S (2000) Intracellular cyanobacterial symbionts in the marine diatom *Climacodium frauenfeldianum* (Bacillariophyceae). J Phycol 36:540–544
Chambouvet A, Morin P, Marie D, Guillou L (2008) Control of toxic marine dinoflagellate blooms by serial parasitic killers. Science 322:1254–1257
Chambouvet A, Alves-de-Souza C, Cueff V, Marie D, Karpov S, Guillou L (2011a) Interplay between the parasite *Amoebophrya* sp. (Alveolata) and the cyst formation of the red tide dinoflagellate *Scrippsiella trochoidea*. Protist 162:637–649
Chambouvet A, Laabir M, Sengco M, Vaquer A, Guillou L (2011b) Genetic diversity of Amoebophryidae (Syndiniales) during the *Alexandrium catenella/tamarense* (Dinophyceae) blooms in Thau lagoon (Mediterranean Sea, France). Res Microbiol 162:959–968
Church MJ, Short CM, Jenkins BD, Karl DM, Zehr JP (2005) Temporal patterns of nitrogenase gene (*nifH*) expression in the oligotrophic North Pacific Ocean. Appl Environ Microbiol 71:5362–5370
Coats DW, Park MG (2002) Parasitism of photosynthetic dinoflagellates by three strains of *Amoebophrya* (Dinophyta): parasite survival, infectivity, generation time, and host specificity. J Phycol 38:520–528
Cooper MB, Smith AG (2015) Exploring mutualistic interactions between microalgae and bacteria in the omics age. Curr Opin Plant Biol 26:147–153
Daugbjerg N, Jensen MH, Hansen PJ (2013) Using nuclear-encoded LSU and SSU rDNA sequences to identify the eukaryotic endosymbiont in *Amphisolenia bidentata* (Dinophyceae). Protist 164:411–422
de Bary A (1879) Die erscheinungder symbiose. Verlag von Karl J. Trubner, Strassburg
de Vargas C, Audic S, Henry N, Decelle J, Mahé F, Logares R, Lara E et al (2015) Eukaryotic plankton diversity in the sunlit ocean. Science 348:1261605
Decelle J, Probert I, Bittner L, Desdevises Y, Colin S, de Vargas C, Gali M et al (2012) An original mode of symbiosis in open ocean plankton. Proc Natl Acad Sci USA 109:18000–18005
Decelle J, Martin P, Paborstava K, Pond DW, Tarling G, Mahé F, de Vargas C et al (2013) Diversity, ecology and biogeochemistry of byst-forming Acantharia (Radiolaria) in the oceans. PLoS ONE 8:e53598
Decelle J, Colin S, Foster RA (2015) Photosymbiosis in marine planktonic protists. In: Ohtsuka S, Suzaki T, Horiguchi T, Suzuki N, Not F (eds) Marine protists: diversity and dynamics. Springer, Japan, pp 465–500
Doucette GJ (1995) Interactions between bacteria and harmful algae: a review. Nat Toxins 3:65–74
Drebes G, Schnepf E (1988) *Paulsenella* Chatton (Dinophyta), ectoparasites of marine diatoms: development and taxonomy. Helgoländer Meeresunters 42:563–581
Drebes G, Kühn SF, Gmelch A, Schnepf E (1996) *Cryothecomonas aestivalis* sp. nov., a colourless nanoflagellate feeding on the marine centric diatom *Guinardia delicatula* (Cleve) Hasle. Helgolander Meeresunters 50:497–515

Dubilier N, McFall-Ngai M, Zhao L (2015) Create a global microbiome effort. Nature 526:631–634
Durham BP, Sharma S, Luo H, Smith CB, Amin SA, Bender SJ, Dearth SP et al (2015) Cryptic carbon and sulfur cycling between surface ocean plankton. Proc Natl Acad Sci USA 112:453–457
Erwin PM, Pineda MC, Webster N, Turon X, López-Legentil S (2014) Down under the tunic: bacterial biodiversity hotspots and widespread ammonia-oxidizing archaea in coral reef ascidians. ISME J 8:575–588
Farnelid H, Tarangkoon W, Hansen G, Hansen PJ, Riemann L (2010) Putative N2-fixing heterotrophic bacteria associated with dinoflagellate-cyanobacteria consortia in the low-nitrogen Indian Ocean. Aquat Microb Ecol 61:105–117
Fay P (1992) Oxygen relations of nitrogen fixation in cyanobacteria. Microbiol Rev 56:340–373
Fernández-Castro B, Mouriño-Carballido B, Marañón E, Chouciño P, Gago J, Ramírez T, Vidal M, et al. (2015). Importance of salt fingering for new nitrogen supply in the oligotrophic ocean. Nat Commun 6:8002
Foster RA, O'Mullan GD (2008). Nitrogen-fixing and nitrifying symbioses in the marine environment. In: Capone DG, Bronk DA, Mulholland MR, Carpenter EJ (eds) Nitrogen in the marine environment. Academic Press, pp 1197–1218
Foster RA, Zehr JP (2006) Characterization of diatom-cyanobacteria symbioses on the basis of nifH, hetR and 16S rRNA sequences. Environ Microbiol 8:1913–1925
Foster RA, Carpenter EJ, Bergman B (2006) Unicellular cyanobionts in open ocean dinoflagellates, radiolarians, and tintinnids: ultrastructural characterization and immuno-localization of phycoerythrin and nitrogenase. J Phycol 42:453–463
Foster RA, Subramaniam A, Mahaffey C, Carpenter EJ, Capone DG, Zehr JP (2007) Influence of the Amazon River plume on distributions of free-living and symbiotic cyanobacteria in the western tropical north Atlantic Ocean. Limnol Oceanogr 52:517–532
Foster RA, Subramaniam A, Zehr JP (2009) Distribution and activity of diazotrophs in the eastern equatorial Atlantic. Environ Microbiol 11:741–750
Foster RA, Goebel NL, Zehr JP (2010) Isolation of *Calothrix rhizosoleniae* (Cyanobacteria) strain SC01 from *Chaetoceros* (Bacillariophyta) spp. diatoms of the subtropical North Pacific Ocean. J Phycol 46:1028–1037
Foster RA, Kuypers MMM, Vagner T, Paerl RW, Musat N, Zehr JP (2011) Nitrogen fixation and transfer in open ocean diatom-cyanobacterial symbioses. ISME J 5:1484–1493
Franche C, Lindström K, Elmerich C (2009) Nitrogen-fixing bacteria associated with leguminous and non-leguminous plants. Plant Soil 321:35–59
Furusawa G, Yoshikawa T, Yasuda A, Sakata T (2003) Algicidal activity and gliding motility of *Saprospira* sp. SS98-5. Can J Microbiol 49:92–100
Garcés E, Alacid E, Reñé A, Petrou K, Simó R (2013) Host-released dimethylsulphide activates the dinoflagellate parasitoid *Parvilucifera sinerae*. ISME J 7:1065–1068
Gast RJ, McDonnell TA, Caron DA (2000) srDna-based taxonomic affinities of algal symbionts from a planktonic foraminifer and a solitary radiolarian. J Phycol 36:172–177
Geng H, Belas R (2010) Molecular mechanisms underlying *Roseobacter*–phytoplankton symbioses. Curr Opin Biotechnol 21:332–338
Goebel NL, Turk KA, Achilles KM, Paerl R, Hewson I, Morrison AE, Montoya JP et al (2010) Abundance and distribution of major groups of diazotrophic cyanobacteria and their potential contribution to N2 fixation in the tropical Atlantic Ocean. Environ Microbiol 12:3272–3289
Green JC, Leadbeater BSC (1972) *Chrysochromulina parkeae* sp. nov. (Haptophyceae) a new species recorded from S. W. England and Norway. J Mar Biol Assoc UK 52:469–474
Green DH, Llewellyn LE, Negri AP, Blackburn SI, Bolch CJS (2004) Phylogenetic and functional diversity of the cultivable bacterial community associated with the paralytic shellfish poisoning dinoflagellate *Gymnodinium catenatum*. FEMS Microb Ecol 47:345–357
Grossart HP, Levold F, Allgaier M, Simon M, Brinkhoff T (2005) Marine diatom species harbour distinct bacterial communities. Environ Microbiol 7:860–873
Guidi L, Chaffron S, Bittner L, Eveillard D (2016). Plankton networks driving carbon export in the oligotrophic ocean. Nature 532:465–470

Guillou L, Viprey M, Chambouvet A, Welsh RM, Kirkham AR, Massana R, Scanlan DJ et al (2008) Widespread occurrence and genetic diversity of marine parasitoids belonging to Syndiniales (Alveolata). Environ Microbiol 10:3349–3365

Hagino K, Takano Y, Horiguchi T (2009) Pseudo-cryptic speciation in *Braarudosphaera bigelowii* (Gran and Braarud) Deflandre. Mar Micropaleontol 72:210–221

Hagino K, Onuma R, Kawachi M, Horiguchi T (2013) Discovery of an endosymbiotic nitrogen-fixing cyanobacterium UCYN-A in *Braarudosphaera bigelowii* (Prymnesiophyceae). PLoS ONE 8:e81749

Hanic LA, Sekimoto S, Bates SS (2009) Oomycete and chytrid infections of the marine diatom *Pseudo-nitzschia pungens* (Bacillariophyceae) from Prince Edward Island Canada. Botany 87:1096–1105

Harrison PJ, Furuya K, Glibert PM, Xu J, Liu HB, Yin K, Lee JHW et al (2011) Geographical distribution of red and green *Noctiluca scintillans*. Chin J Ocean Limnol 29:807–831

Hilton JA, Foster RA, Tripp HJ, Carter BJ, Zehr JP, Villareal TA (2013) Genomic deletions disrupt nitrogen metabolism pathways of a cyanobacterial diatom symbiont. Nat Commun 4:1767

Hollande A, Carré D (1974) Les xanthelles des radiolaires sphaerocollides, des acanthaires et de *Velella velella*: Infrastructure-cytochimie-taxonomie. Protistolog 10:573–601

Imanian B, Pombert JF, Keeling PJ (2010) The complete plastid genomes of the two "Dinotoms" *Durinskia baltica* and *Kryptoperidinium foliaceum*. PLoS ONE 5:e10711

Janson S, Rai AN, Bergman B (1995) Intracellular cyanobiont *Richelia intracellularis*: ultrastructure and immuno-localisation of phycoerythrin, nitrogenase, Rubisco and glutamine synthetase. Mar Biol 124:1–8

Janson S, Wouters J, Bergman B, Carpenter EJ (1999) Host specificity in the *Richelia*—diatom symbiosis revealed by *hetR* gene sequence analysis. Environ Microbiol 1:431–438

Jasti S, Sieracki ME, Poulton NJ, Giewat MW, Rooney-Varga JN (2005) Phylogenetic Diversity and specificity of bacteria closely associated with *Alexandrium* spp. and other phytoplankton. Appl Environ Microbiol 71:3483

Jephcott TG, Alves-de-Souza C, Gleason FH, van Ogtrop F, Sime-Ngando T, Karpov S, Guillou L (2015) Ecological impacts of parasitic chytrids, Syndiniales and perkinsids on populations of marine photosynthetic dinoflagellates. Fungal Ecol 19:47–58

Johnson MD, Stoecker DK (2005) Role of feeding in growth and photophysiology of *Myrionecta rubra*. Aquat Microb Ecol 39:303–312

Kaczmarska I, Ehrman JM, Bates SS, Green DH, Léger C, Harris J (2005) Diversity and distribution of epibiotic bacteria on *Pseudo-nitzschia multiseries* (Bacillariophyceae) in culture, and comparison with those on diatoms in native seawater. Harmful Algae 4:725–741

Karpov SA, Mamkaeva MA, Benzerara K, Moreira D, López-García P (2014) Molecular phylogeny and ultrastructure of *Aphelidium* aff. *melosirae* (Aphelida, Opisthosporidia). Protist 165:512

Karsten G (1907) Das indische phytoplankton. G. Fischer, Jena

Kim S, Park MG, Yih W, Coats DW (2004) Infection of the bloom-forming thecate dinoflagellates *Alexandrium affine* and *Gonyaulax spinifera* by two strains of *Amoebophrya* (Dinophyta). J Phycol 40:815–822

Kodama Y, Suzuki H, Dohra H, Sugii M, Kitazume T, Yamaguchi K, Shigenobu S et al (2014) Comparison of gene expression of *Paramecium bursaria* with and without *Chlorella variabilis* symbionts. BMC Genom 15:1–8

Kühn S, Drebes G, Schnepf E (1996) Five new species of the nanoflagellate *Pirsonia* in the German Bight, North Sea, feeding on planktic diatoms. Helgol Mar Res 50:205–222

Lafferty KD, Kuris AM (2002) Trophic strategies, animal diversity and body size. Trends Ecol Evol 17:507–513

Leander BS, Keeling PJ (2003) Morphostasis in alveolate evolution. Trends Ecol Evol 18:395–402

Lee S, Kato J, Takiguchi N, Kuroda A, Ikeda T (2000) Involvement of an extracellular protease in algicidal activity of the marine bacterium involvement of an extracellular protease in algicidal activity of the marine bacterium *Pseudoalteromonas* sp. strain A28. Appl Environ Microbiol 66:4334–4339

Lemmerman E (1905) Sandwich-islen. Ergebnisse einer reise nach dem Pacific. H. Schauinsland 1896/97. Bot Jahrb Syst Pflanzengesch Planzengeogr 34:607–663

Lepelletier F, Karpov SA, Alacid E, Le Panse S, Bigeard E, Garcés E, Jeanthon C et al (2014a) *Dinomyces arenysensis* gen. et sp. nov. (Rhizophydiales, Dinomycetaceae fam. nov.), a chytrid infecting marine dinoflagellates. Protist 165:230–244

Lepelletier F, Karpov SA, Le Panse S, Bigeard E, Skovgaard A, Jeanthon C, Guillou L (2014b) *Parvilucifera rostrata* sp. nov., a novel parasite in the phylum Perkinsozoa that infects the toxic dinoflagellate *Alexandrium minutum* (Dinophyceae). Protist 165:31–49

Lepère C, Ostrowski M, Hartmann M, Zubkov MV, Scanlan DJ (2015) In situ associations between marine photosynthetic picoeukaryotes and potential parasites—a role for fungi? Environ Microbiol Rep doi:10.1111/1758-2229.12339 (in press)

Lima-Mendez G, Faust K, Henry N, Decelle J, Colin S, Carcillo F, Chaffron S et al (2015) Determinants of community structure in the global plankton interactome. Science 348:1262073-1–1262073-9

Lopes RM, Silveira M (1994) Symbiosis between a pelagic flatworm and a dinoflagellate from a tropical area—structural observations. Hydrobiologia 287:277–284

López-García P, Rodriguez-Valera F, Pedrós-Alió C, Moreira D (2001) Unexpected diversity of small eukaryotes in deep-sea Antarctic plankton. Nature 409:603–607

Lu Y, Wohlrab S, Glockner G, Guillou L, John U (2014) Genomic insights into processes driving the infection of *Alexandrium tamarense* by the parasitoid *Amoebophrya* sp. Eukaryot Cell 13:1439–1449

Martinez-Garcia M, Brazel D, Poulton NJ, Swan BK, Gomez ML, Masland D, Sieracki ME et al (2012) Unveiling in situ interactions between marine protists and bacteria through single cell sequencing. ISME J 6:703

Massana R, Gobet A, Audic S, Bass D, Bittner L, Boutte C, Chambouvet A et al (2015) Marine protist diversity in European coastal waters and sediments as revealed by high-throughput sequencing. Environ Microbiol 17:4035–4049

Mayali X, Azam F (2004) Algicidal bacteria in the sea and their impact on algal blooms. J Eukaryot Microbiol 51:139–144

Mayali X, Franks PJS, Azam F (2008) Cultivation and ecosystem role of a marine Roseobacter clade-affiliated cluster bacterium. Appl Environ Microbiol 74:2595–2603

McFall-Ngai M (2008) Are biologists in 'future shock'? Symbiosis integrates biology across domains. Nat Rev Microbiol 6:789–792

Miller JJ, Delwiche CF, Coats DW (2012) Ultrastructure of *Amoebophrya* sp. and its changes during the course of infection. Protist 163:720–745

Montoya JP, Holl CM, Zehr JP, Hansen A, Villareal TA, Capone DG (2004) High rates of N-2 fixation by unicellular diazotrophs in the oligotrophic Pacific Ocean. Nature 430:1027–1031

Moore CM, Mills MM, Arrigo KR, Berman-Frank I, Bopp L, Boyd PW, Galbraith ED et al (2013) Processes and patterns of oceanic nutrient limitation. Nat Geosci 6:701–710

Mordret S, Romac S, Henry N, Colin S, Carmichael M, Berney C, Audic S et al (2015) The symbiotic life of Symbiodinium in the open ocean within a new species of calcifying ciliate (*Tiarina* sp.). ISME J doi:10.1038/ismej.2015.211 (in press)

Musat N, Foster R, Vagner T, Adam B, Kuypers MMM (2012) Detecting metabolic activities in single cells, with emphasis on nanoSIMS. FEMS Microbiol Rev 36:486–511

Not F, del Campo J, Balagué V, de Vargas C, Massana R (2009) New insights into the diversity of marine picoeukaryotes. PLoS ONE 4:e7143

Nowack ECM, Melkonian M (2010) Endosymbiotic associations within protists. Philos Trans R Soc B Biol Sci 365:699–712

Ostenfeld CH, Schmidt J (1902) Plankton fra det Røde Hav of Adenbugten (Plankton from the Red Sea and the Gulf of Eden.). Videnskabelige Meddelelser fra Dansk Naturhistorisk Forening 1901:141–182

Paul C, Pohnert G (2011) Interactions of the algicidal bacterium *Kordia algicida* with diatoms: regulated protease excretion for specific algal lysis. PLoS ONE 6:e21032

Pinhassi J, Hagstrom A (2000) Seasonal succession in marine bacterioplankton. Aquat Microb Ecol 21:245–256
Pinhassi J, Sala MM, Havskum H, Peters F, Guadayol Ò, Malits A, Marrasé C (2004) Changes in bacterioplankton composition under different phytoplankton regimens. Appl Environ Microbiol 70:6753–6766
Poulin R (2011). Evolutionary ecology of parasites. Princeton University press. 360 pp
Probert I, Siano R, Poirier C, Decelle J, Biard T, Tuji A, Suzuki N et al (2014) *Brandtodinium* gen. nov. and *B. nutricula* comb. nov. (Dinophyceae), a dinoflagellate commonly found in symbiosis with polycystine radiolarians. J Phycol 50:388–399
Rai AN, Bergman B, Rasmussen U (2002) Cyanobacteria in symbiosis. Kluwer Academic Publishers. 319 pp
Ramanan R, Kim BH, Cho DH, Oh HM, Kim HS (2015) Algae–bacteria interactions: evolution, ecology and emerging applications. Biotechnol Adv 34:14–29
Reddy KJ, Haskell JB, Sherman DM, Sherman LA (1993) Unicellular, aerobic nitrogen-fixing cyanobacteria of the genus Cyanothece. J Bacteriol 175:1284–1292
Riemann L, Steward GF, Azam F (2000) Dynamics of bacterial community composition and activity during a mesocosm diatom bloom. Appl Environ Microbiol 66:578–587
Rodriguez F, Feist SW, Guillou L, Harkestad LS, Bateman K, Renault T, Mortensen S (2008) Phylogenetic and morphological characterization of the green algae infesting blue mussel *Mytilus edulis* in the North and South Atlantic. Dis Aquat Organ 81:231–240
Rooney-Varga JN, Giewat MW, Savin MC, Sood S, Legresley M, Martin JL (2005) Links between phytoplankton and bacterial community dynamics in a coastal marine environment. Microb Ecol 49:163–175
Sampayo EM, Ridgway T, Bongaerts P, Hoegh-Guldberg O (2008) Bleaching susceptibility and mortality of corals are determined by fine-scale differences in symbiont type. Proc Natl Acad Sci USA 105:10444–10449
Santi C, Bogusz D, Franche C (2013) Biological nitrogen fixation in non-legume plants. Ann Bot 111:743–767
Sapp M, Wichels A, Gerdts G (2007) Impacts of cultivation of marine diatoms on the associated bacterial community. Appl Environ Microbiol 73:3117–31120
Schäfer H, Abbas B, Witte H, Muyzer G (2002) Genetic diversity of "satellite" bacteria present in cultures of marine diatoms. FEMS Microbiol Ecol 42:25–35
Scharek R, Tupas LM, Karl DM (1999) Diatom fluxes to the deep sea in the oligotrophic North Pacific gyre at station ALOHA. Mar Ecol Prog Ser 182:55–67
Schnepf E, Deichgräber G (1984) "Myzocytosis", a kind of endocytosis with implications to compartmentation in endosymbiosis. Naturwissenschaften 71:218–219
Schnepf E, Kuhn SF (2000) Food uptake and fine structure of *Cryothecomonas longipes* sp nov., a marine nanoflagellate incertae sedis feeding phagotrophically on large diatoms. Helgol Mar Res 54:18–32
Schnepf E, Melkonian M (1990) Bacteriophage-like particles in endocytic bacteria of *Cryptomonas* (Cryptophyceae). Phycologia 29:338–343
Scholz B, Guillou L, Marano AV, Neuhauser S, Sullivan BK, Karsten U, Küpper FC et al (2016) Zoosporic parasites infecting marine diatoms—a black box that needs to be opened. Fungal Ecol 19:59–76
Schütt F (1895) Die peridineen der plankton-expedition. Lipsius & Tischer, Kiel
Schweikert M, Schnepf E (1997) Electron microscopical observations on *Pseudaphelidium drebesii* Schweikert and Schnepf, a parasite of the centric diatom *Thalassiosira punctigera*. Protoplasma 199:113–123
Seyedsayamdost MR, Case RJ, Kolter R, Clardy J (2011) The Jekyll-and-Hyde chemistry of *Phaeobacter gallaeciensis*. Nat Chem 3:331–335
Shi XL, Marie D, Jardillier L, Scanlan DJ, Vaulot D (2009) Groups without cultured representatives dominate eukaryotic picophytoplankton in the oligotrophic South East Pacific Ocean. PLoS ONE 4:e7657

Siano R, Montresor M, Probert I, Not F, de Vargas C (2010) *Pelagodinium* gen. nov. and *P. beii* comb. nov., a dinoflagellate symbiont of planktonic foraminifera. Protist 161:385–399
Siano R, Alves-de-Souza C, Foulon E, Bendif EM, Simon N, Guillou L, Not F (2011) Distribution and host diversity of Amoebophryidae parasites across oligotrophic waters of the Mediterranean Sea. Biogeoscience 8:267–278
Skovgaard A, Karpov SA, Guillou L (2012) The parasitic dinoflagellates *Blastodinium* spp. inhabiting the gut of marine, planktonic copepods: morphology, ecology, and unrecognized species diversity. Front Microbiol 3:305
Spero HJ (1987) Symbiosis in the planktonic foraminifer, *Orbulina universa*, and the isolation of its symbiotic dinoflagellate Gymnodinium beii sp. nov. J Phycol 23:307–317
Stewart EJ (2012) Growing unculturable bacteria. J Bacteriol 194:4151–4160
Stoecker DK, Michaels AE, Davis LH (1987) Large proportion of marine planktonic ciliates found to contain functional chloroplasts. Nature 326:790–792
Stoecker DK, Sliver MW, Michaels AE, Davis LH (1989a) Enslavement of algal chloroplasts by four *Strombidium* spp. (Ciliophora, Oligotrichida). Mar Microb Food Webs 3:79–100
Stoecker DK, Swanberg N, Tyler S (1989b) Oceanic mixotrophic flatworms. Mar Ecol Ser 58:41–51
Stoecker DK, Gustafson DE, Verity PG (1996) Micro- and mesoprotozooplankton at 140°W in the equatorial Pacific: heterotrophs and mixotrophs. Aquat Microb Ecol 10:273–282
Stoecker DK, Johnson MD, de Vargas C, Not F (2009) Acquired phototrophy in aquatic protists. Aquat Microb Ecol 57:279–310
Suzuki N, Not F (2015) Biology and ecology of radiolaria. In: Ohtsuka S, Suzaki T, Horiguchi T, Suzuki N, Not F (eds) Marine Protists. Spinger, Japan, pp 179–222
Sweeney BM (1976) *Pedinomonas noctilucae* (Prasinophyceae), the flagellate symbiotic in *Noctiluca* (Dinophyceae) in Southeast Asia. J Phycol 12:460–464
Taylor DL (1971) Ultrastructure of the 'Zooxanthella' *Endodinium chattonii* in situ. J Mar Biol Assoc U.K. 51:227–234
Taylor FJR (1982) Symbioses in marine microplankton. Ann L Inst Oceanogr 58:61–90
Teeling H, Fuchs BM, Becher D, Klockow C, Gardebrecht A, Bennke CM, Kassabgy M et al (2012) Substrate-controlled succession of marine bacterioplankton populations induced by a phytoplankton bloom. Science 336:608–611
Thompson JN (1999) The evolution of species interactions. Science 284:2116–2118
Thompson AW, Zehr JP (2013) Cellular interactions: lessons from the nitrogen-fixing cyanobacteria. J Phycol 49:1024–1035
Thompson AW, Foster RA, Krupke A, Carter BJ, Musat N, Vaulot D, Kuypers MMM et al (2012) Unicellular cyanobacterium symbiotic with a single-celled eukaryotic alga. Science 337:1546–1550
Thompson AW, Carter BJ, Turk-Kubo K, Malfatti F, Azam F, Zehr JP (2014) Genetic diversity of the unicellular nitrogen-fixing cyanobacteria UCYN-A and its prymnesiophyte host. Environ Microbiol 16:3238–3249
Trench RK (1993) Microalgal-invertebrate symbioses—a review. Endocytobiosis Cell Res 9:135–175
Trench RK, Thinh L (2007) *Gymnodinium linucheae* sp. nov.: the dinoflagellate symbiont of the jellyfish *Linuche unguiculata*. Eur J Phycol 30:149–154
Tripp HJ, Bench SR, Turk KA, Foster RA, Desany BA, Niazi F, Affourtit JP et al (2010) Metabolic streamlining in an open-ocean nitrogen-fixing cyanobacterium. Nature 464:90–94
Turk KA, Rees AP, Zehr JP, Pereira N, Swift P, Shelley R, Lohan M et al (2011) Nitrogen fixation and nitrogenase (nifH) expression in tropical waters of the eastern North Atlantic. ISME J 5:1201–1212
Villareal TA (1990) Laboratory culture and preliminary characterization of the nitrogen-fixing *Rhizosolenia-Richelia* symbiosis. Mar Ecol 11:117–132
Wang JT, Chen YY, Tew KS, Meng PJ, Chen CA (2012) Physiological and biochemical performances of menthol-induced aposymbiotic corals. PLoS ONE 7:e46406
Wang H, Tomasch J, Jarek M, Wagner-Döbler I (2014) A dual-species co-cultivation system to study the interactions between *Roseobacters* and dinoflagellates. Front Microbiol 5:1–11

Worden AZ, Follows MJ, Giovannoni SJ, Wilken S, Zimmerman AE, Keeling PJ (2015) Rethinking the marine carbon cycle: factoring in the multifarious lifestyles of microbes. Science 347:1257594

Yuasa T, Horiguchi T, Mayama S, Matsuoka A, Takahashi O (2012) Ultrastructural and molecular characterization of cyanobacterial symbionts in *Dictyocoryne profunda* (Polycystine Radiolaria). Symbiosis 57:51–55

Zehr JP, Bench SR, Carter BJ, Hewson I, Niazi F, Shi T, Tripp HJ et al (2008) Globally distributed uncultivated oceanic N2-fixing cyanobacteria lack oxygenic photosystem II. Science 322:1110–1112

Part III
Marine Resources—The Hidden Treasure

Chapter 12
Marine Microbial Systems Ecology: Microbial Networks in the Sea

Gerard Muyzer

Abstract Next-generation sequencing of DNA has revolutionized microbial ecology. Using this technology, it became for the first time possible to analyze hundreds of samples simultaneously and in great detail. 16S rRNA amplicon sequencing, metagenomics and metatranscriptomics became available to determine the diversity and activity of microbial communities. Moreover, the huge amount of data that is obtained made it possible to build statistically significant networks from which ecological (or metabolic) interactions amongst microbes and between microbes and their environment could be inferred. Here I give an overview of the use of next-generation sequencing and network analysis in marine microbial ecology.

12.1 Introduction

Bacteria are everywhere and they are present in large numbers. For instance, there are more bacteria in the human gut (10^{13}; Sears 2005), than stars in the sky (4×10^{11}; Cain 2013). It has been estimated that the total number of bacterial cells on Earth is between 4×10^{30} and 6×10^{30} (Whitman et al. 1998). To study these bacteria, microbiologists have tried to isolate them in pure culture. But unfortunately only less than 1 % of the bacteria that could be seen and counted by microscopy were growing as colonies on solid media, a phenomenon which is known as the "Great Plate Count Anomaly" (Staley and Konopka 1985). To answer important questions on the diversity and activity of microorganisms, molecular techniques, such as cloning and community fingerprinting of PCR-amplified 16S rRNA gene fragments (e.g., SSCP, Lee et al. 1996; T-RFLP, Marsh 1999; DGGE, Muyzer 1999) were used. The results of investigations using these molecular techniques showed that microbial diversity was enormous, and that most of the

G. Muyzer (✉)
Microbial Systems Ecology, Department of Aquatic Microbiology,
Institute for Biodiversity and Ecosystem Dynamics, University of Amsterdam,
Science Park 904, 1098 XH Amsterdam, The Netherlands
e-mail: g.muijzer@uva.nl

L.J. Stal and M.S. Cretoiu (eds.), *The Marine Microbiome*,
DOI 10.1007/978-3-319-33000-6_12

bacteria that could be detected in nature had not been isolated in pure culture and so were not described. However, although successful, these techniques were mainly detecting predominant community constituents leaving rare members, the "rare biosphere" (Sogin et al. 2006), undetected (Pedrós-Alió 2006). In addition, these methods were limited in the number of samples that could be studied simultaneously, which made it difficult to perform statistical analysis. The application of next-generation sequencing technologies (Maclean et al. 2009) revolutionized the field of microbial ecology; now it became, for the first time, possible to study hundreds of samples of microbial communities at the same time (Hamady et al. 2008) and in great detail. This made it possible to perform co-occurrence analysis resulting in microbial association networks consisting of nodes and edges from which ecological relationships between different microorganisms, and between microorganisms and their environment could be inferred (Fig. 12.1; Faust and Raes 2012; Faust et al. 2015; Fuhrman et al. 2015). Positive relations between different microorganisms could be interpreted as mutualism, while negative relations could point to competition. Nodes with many links are assumed to be "keystone" species. Co-occurrence analysis has been used to study the ecological interactions of microbes in lakes (Eiler et al. 2012; Peura et al. 2015), soils (Barberán et al. 2011), streams (Widder et al. 2014), the human microbiome (Faust et al. 2012), and the marine environment (see below).

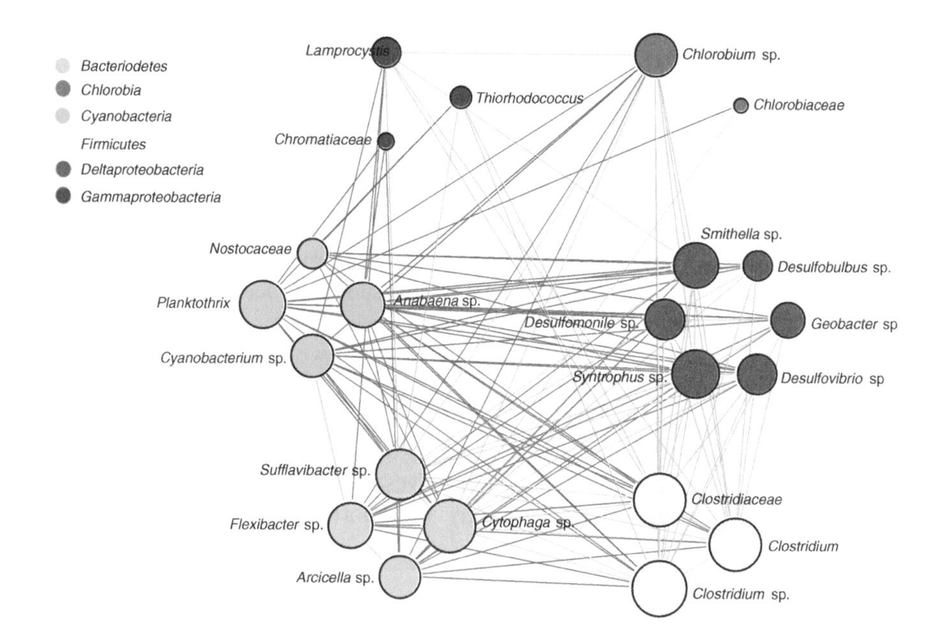

Fig. 12.1 Co-occurrence network showing positive (*green*) and negative (*red*) relations between different bacteria. The size of the nodes is an indication of the number of links. The network has been calculated from 16S rRNA amplicon data using the software program CoNET (Faust and Raes 2012) and drawn in Cytoscape (Shannon et al. 2003)

Here I will give an overview on the use of next-generation sequencing and will argue for the use of a systems biology approach to study marine microbial ecosystems. I will limit this review to microorganisms (Bacteria, Archaea, protists, and viruses) in the water column that play an essential role in the cycling of carbon and nutrients in marine ecosystems (Falkowski et al. 2008). These microorganisms are part of the "microbial loop," whereby dissolved organic carbon (DOC) is processed and consumed and returned into the marine food web, where it becomes available to higher trophic levels (Azam et al. 1983). Although much research has been done on the microbial loop, and key species have been identified, little is known about the interactions between the different community members, and the environmental factors that might influence the diversity and activity of these microorganisms. However, this information is essential for a comprehensive understanding of the consequences of short- and long-term perturbations such as eutrophication and global warming on the microbial loop and the marine ecosystem as a whole. Only by using an integrated or holistic approach with knowledge and technologies of different disciplines this information can be obtained (Karsenti et al. 2011). Such a systems biology approach allows the possibility to model these ecosystems and finally to predict the effects of perturbations (Raes and Bork 2008).

12.2 16S rRNA Amplicon Sequencing

Fuhrman and colleagues (Steele et al. 2011) used network analysis to reveal ecological relationships between Bacteria, Archaea, and protists and the environmental parameters that might influence these relationships. They took monthly samples over a period of three years from the deep chlorophyll maximum layer at the San Pedro Ocean Time Series (SPOT) site off the coast of California. Different molecular techniques (i.e., ARISA and T-RFLP combined with cloning and qPCR) were used to identify the community members and to monitor their dynamics over time. In addition, several environmental parameters, such as temperature, salinity, dissolved oxygen, nutrients, pigments, and total abundance of bacteria and viruses were measured. Steele et al. (2011) found different networks in which particular OTUs, such as for instance SAR11, played a central role. These highly connected OTUs can be regarded as "keystone" species. Positive correlations between microorganisms and water density indicated a preference of these organisms for cold water. Positive correlations without time delay showed that organisms were growing under similar conditions, while positive correlations with a time delay indicated a succession. Negative correlations between protists and bacteria pointed to grazing, and a negative correlation between the cyanobacterium *Synechococcus* and the green alga *Ostreococcus* indicated competition. In a subsequent study, Chow et al. (2014) used network analysis to determine the effect of viral lysis and protistan grazing ('top-down" controls) on the bacterial community structure. These authors found more associations between viruses and bacteria than between protists and bacteria, indicating host specificity and nonselective interactions, respectively.

In another publication, they described the differences in microbial associations within and between different water depths (Cram et al. 2015).

Gilbert et al. (2012) studied the seasonal dynamics of microbial communities over a period of 6 years. They collected water samples between January 2003 and December 2008 from a long-time series station in Western English Channel. Different environmental parameters (e.g., temperature, salinity, length of the day, nutrient concentrations, abundance of phyto- and zooplankton) were measured, and 16S rRNA gene tag pyrosequencing was used to identify the community members and to monitor their succession over time. Members of the *Rickettsiales*, such as SAR11 (currently placed in the order *Pelagibacterales* (Ferla et al. 2013)), and the *Rhodobacterales*, both within the *Alphaproteobacteria*, were the most dominant community constituents. These populations showed strong seasonal dynamics; SAR11 was most abundant in winter, when primary productivity was low, while members of the *Roseobacter* clade were abundantly present in spring and autumn, when primary productivity was high. Gilbert et al. (2012) found, to their surprise, that the most explanatory parameter for bacterial richness was day length. Co-occurrence network analysis showed that interactions within domains (e.g., *Bacteria*) were stronger than between domains (i.e., *Bacteria* and *Eukarya*), or between bacterial taxa and environmental factors.

12.3 Metagenomics

Apart from using one gene as a molecular marker (e.g., 16S rRNA gene), scientists also used the complete collection of genes, the "metagenome," to study the structure and function of marine microbial communities. Within the *Tara Oceans* project (Karsenti et al. 2011) a consortium of scientists from different disciplines studied the bacterial and archaeal (Sunagawa et al. 2015), picoeukaryal (Lima-Mendez et al. 2015) and viral (Brum et al. 2015) communities using a holistic or systems biology approach. About 35,000 water samples from different depths and across the globe were collected and used for morphological, genetic, and environmental analysis. Sunagawa et al. (2015) used metagenomics to study the microbial communities in 243 water samples from 68 locations and three different depths. From this information they generated an Ocean Microbial Reference Gene Catalog consisting of more than 40 million mostly novel sequences. Moreover, they showed in 139 bacteria-enriched samples that (i) the most dominant community members were affiliated to the *Alphaproteobacteria* (SAR11) and the *Gammaproteobacteria* (SAR86), (ii) that richness increased with sample depth, and (iii) that temperature was the main driver of community composition in the epipelagic layer (up to 200 m depth). The ocean core metagenome, i.e., present in all 139 samples, consisted of 5755 clusters of orthologous groups (OGs). Comparative analysis of the ocean core OGs with the human gut core OGs showed a large overlap of functions (i.e., 73 % for the ocean core OGs and 63 % for the gut core OGs). Nevertheless, differences in the abundance of different functional groups were present, such as OGs involved in defense mechanisms and carbohydrate metabolism, which were more abundant in the gut core, while those for transport mechanism and energy production were more abundant in the ocean core. Although the taxonomic composition between different ocean regions was variable, the functional variability was more or less stable, indicating functional redundancy. Furthermore, nearly all of the OGs present in the epipelagic layer were also present in the mesopelagic layer (between 200 and 1000 m depth, although some functions were more enriched in the mesopelagic layer, such as those involved in aerobic respiration for remineralization of organic matter produced in the epipelagic layer, and flagella assembly and chemotaxis to colonize sinking particles and marine snow.

In the same *Tara Oceans* project Lima-Mendez et al. (2015) studied the influence of biotic and abiotic factors on the interaction network ("interactome") of oceanic plankton. They collected 313 plankton samples (from viruses to small metazoans) from two water depths at 68 stations across eight oceanic provinces around the globe. They used different network inference methods and machine-learning techniques to construct interactions networks from metagenomic ($_{mi}$tags, Logares et al. 2014) and 18S rRNA amplicon sequences, and environmental variables. The results showed that environmental variables only had a limited effect on community structure. The integrated network contained 81,590 biotic interactions, for which 72 % were positive and 23 % were negative interactions. Most interactions resulted from dinoflagelates (i.e., *Syndiniales* and *Dinophyceae*) and from arthropods. Examples were the positive interaction of *Syndiniales* with tintinnid ciliates, and the association of diatoms and *Flavobacteria*. The authors emphasized the role of top-down biotic interactions, such as parasitic or phage-microbe interactions, in the epipelagic zone. They concluded that the presented interactome could be used to predict the structure and dynamics of ocean ecosystems.

Apart from *Bacteria*, *Archaea*, and micro-eukaryotes the *Tara Oceans* project also studied viruses. Brum et al. (2015) used metagenomics of double-stranded DNA (dsDNA) viruses ('viromes') and virus morphology to determine the patterns and ecological drivers of ocean viral communities. They established a dataset of 43 viromes from the epipelagic layer of seven oceans and seas (*Tara Oceans* Viromes (TOV) dataset). Analysis of sequence similarity-based protein clusters (PC) showed a pan virome of a million PCs, which only slightly increased after the analysis of 20 viromes. Brum et al. (2015) found that oceanic currents transported viral communities and that local environmental factors stimulating microbial hosts indirectly also structured these viral communities, thereby confirming the "seed-bank" hypothesis posed by Breitbart and Rohwer (2005).

The results obtained by Brum et al. (2015) nicely complemented the results found by Hurwitz and Sullivan (2013) who studied, over different seasons, the viromes of different stations, and various depths in the Pacific Ocean. These authors obtained a dataset of 456,000 PCs (Pacific Ocean Virome (POV) dataset) and found that functional diversity decreased from deep to surface water, from winter to summer, and with distance from the shore. In a subsequent study Hurwitz et al. (2014) used the POV dataset for comparative metagenomics and network analysis to determine the ecological drivers of viral community structure. The results showed that the 32 viromes clustered in eight groups, three in the photic and five in the aphotic zone, with geographic location, depth, and proximity to shore as the main predictors of the network. However, analysis of sub-networks of viromes, such as those obtained from samples along the LineP transect, showed that depth, season, and oxygen were significant drivers of the viral community structure.

12.4 Metatranscriptomics

So far, most research has been focused on DNA, i.e., on the presence of microorganisms using 16S rRNA amplicon sequencing or metagenomics. However, for a comprehensive understanding of the complete marine microbial ecosystem the expression of genes using metatranscriptomics and metaproteomics needs to be studied as well. DeLong and coworkers were pioneers in gene expression analysis of marine microbial communities. In their first study, they amplified small amounts of RNA to produce enough cDNA for pyrosequencing. Subsequently, the reads were compared to the metagenome of the same sample (Frias-Lopez et al. 2008). The results showed expression of genes involved in photosynthesis, carbon fixation and nitrogen acquisition, genes encoding hypothetical proteins, as well as many genes (ca. 50 %) that had never been detected before. In subsequent studies the authors designed a strategy to collect and preserve samples automatically by a robotic system, and investigated gene expression of bacterioplankton over time (Ottesen et al. 2011, 2013). Using transcriptomic network analysis, Aylward et al. (2015) demonstrated that gene expression of microbial plankton communities was orchestrated by the 24-h metabolic cycle of the photoautotrophs *Ostreococcus* and *Prochlorococcus*.

Baker et al. (2013) used a slightly different approach in studying gene expression of microbial populations from a hydrothermal plume in the Guaymas Basin, Gulf of California. They performed de novo assembly of the reads into "mRNA" contigs that were subsequently annotated; this approach has the advantage that it might detect gene expression in less abundant, but highly active populations. Most transcripts belonged to genes involved in the oxidation of sulfur, ammonia, and methane, as well as genes encoding ribosomal proteins. In addition, they found novel genes involved in nitrite oxidation (*nxr*) in novel *Nitrospira*-like organisms.

12.5 Metaproteomics

Metaproteomics or community proteomics (Verberkmoes et al. 2009) is another powerful tool to determine the activity of microbial community members. In metaproteomics proteins are extracted from microbial communities, digested with trypsin, separated by liquid chromatography, and analyzed by tandem mass spectroscopy (LC-MS/MS). Subsequently, the peptides are identified after comparison with genes in a database. Although this approach is still in its infancy several studies have used this approach to study marine microbial communities (see Wang et al. 2014 and Williams and Cavicchioli 2014). Teeling et al. (2012) studied the response of bacterioplankton to a diatom bloom in the North Sea using a suite of molecular tools. CARD-FISH and 16S rRNA amplicon sequencing showed the dominance of different members of the *Flavobacteria*, *Alphaproteobacteria*, and *Gammaproteobacteria*, and their succession during the different stadia of the diatom bloom. Furthermore, metaproteogenomics (i.e., metaproteomics combined with metagenomics) showed the expression of different enzymes for degrading algal polysaccharides, different transporters for taking up the produced oligo- and monosaccharides, and diverse phosphate acquisition strategies. Metatranscriptomic analysis by RNA sequencing confirmed these results and showed that *Flavobacteria* were involved in degrading algal polymers during the diatom bloom (Klindworth et al. 2014) while they were not active before the bloom started (Knopf et al. 2015). These results nicely showed the niche differentiation of these bacterial populations in a homologous habitat as the marine water column.

Metaproteomics was also used to study the survival of bacteria, in particular SAR11, in the surface layer of the Sargasso Sea during summer and autumn, when nutrient (nitrogen and phosphorus) concentrations were extremely low (Sowell et al. 2009). The results showed an increase of periplasmic proteins for transport of amino acids, phosphate, phosphonate, sugars, and spermidine, as well as proteins involved in protein refolding and protection to oxidative stress. In a subsequent study, Sowell et al. (2011) used the same approach to study gene expression of bacterioplankton under nutrient-rich conditions. Water samples were collected from surface water off the coast of Oregon during summer when nutrient concentrations were high. Genes involved in transportation of amino acids, glycine-betaine, polyamines, and taurine were highly expressed, but genes involved in phosphate acquisition were not expressed, which is in contrast with the results found for the Sargasso Sea. Apart from bacterial proteins also viral proteins were detected, indicating the importance of viral infection of bacterioplankton members in this coastal region. Morris et al. (2010) performed comparative metaproteomics of membrane proteins from microbial communities of 10 different places along a natural gradient of nutrient concentrations, i.e., from the low-nutrient South Atlantic gyre to the high-nutrient Benguela upwelling region. Dominant proteins were transporters, such as TonB-dependent transporters (TBDT), involved in transporting nutrients across the outer membrane, and microbial rhodopsins (light-driven proteins that pump protons and other ions (Na^+, Cl^-) across the membrane). Detection of TBDT proteins and rhodopsins in the same taxonomic lineage might suggest the presence of light-driven transport activities, which might be an excellent adaptation to nutrient-limited conditions. TBDT proteins from low-nutrient sites were related to *Alpha-* and *Deltaproteobacteria*, and from high-nutrient sites to *Gammaproteobacteria* and *Bacteroidetes*. Apart from these proteins, viral proteins were detected in all samples, and archaeal amoA was detected in nutrient-rich samples.

12.6 Concluding Remarks

We have seen in the preceding sections that next-generation sequencing of DNA/RNA and high-resolution mass spectroscopy of peptides can be used to determine the diversity and activity of microbiology communities, and to infer ecological interactions between community members and their environment. However, for obtaining a comprehensive understanding of the structure and function of microbial communities and their role in nature we need to use a systems biology approach, in which experiments are combined with mathematical modeling (Hanemaaijer et al. 2015; Raes and Bork 2008; Thiele et al. 2013; Zengler and Palsson 2012). The resulting models will generate novel questions and hypotheses that will be answered and tested in new experiments. To reach this goal integration of datasets obtained from different "omics" approaches is essential (Franzosa et al. 2015; Muller et al. 2013). Although this is a big challenge it will not take long before it is possible to produce the first detailed models that can predict the response of microbial communities to different environmental conditions and to get insight into the complex interactions between community constituents.

Acknowledgments The research leading to these results has received funding from the European Union Seventh Framework Programme (FP7/2007–2013) under grant agreement no 311975. This publication reflects the views only of the author, and the European Union cannot be held responsible for any use which may be made of the information contained therein. Gerard Muyzer is financially supported by the Research Priority Area *Systems Biology* of the University of Amsterdam. Muhe Diao is acknowledged for calculating the co-occurrence network, and Tim Bush for correcting the English.

References

Aylward FO, Eppley JM, Smith JM, Chavez FP, Scholin CA, DeLong EF (2015) Microbial community transcriptional networks are conserved in three domains at ocean basin scales. PNAS 112:5443–5448

Azam A, Fenchel T, Field JG, Gray JS, Meyer-Reil LA, Thingstad F (1983) The ecological role of water-column microbes in the sea. Mar Ecol 10:257–263

Baker BJ, Sheik CS, Taylor CA et al (2013) Community transcriptomic assembly reveals microbes that contribute to deep-sea carbon and nitrogen cycling. ISME J 7:1962–1973

Barberán A, Bates ST, Casamayor EO, Fierer N (2011) Using network analysis to explore co-occurrence patterns in soil microbial communities. ISME J 6:343–351

Breitbart M, Rohwer F (2005) Here a virus, there a virus, everywhere the same virus? Trends Microbiol 13:278–284

Brum JR, Ignacio-Espinoza JC, Roux S et al (2015) Patterns and ecological drivers of ocean viral communities. Science 348:1261498-1–1261498-10

Cain F (2013) How many stars in the universe? Universe Today. http://www.universetoday.com/102630/how-many-stars-are-there-in-the-universe/

Chow C-ET, Kim DY, Sachdeva R et al (2014) Top-down controls on bacterial community structure: microbial network analysis of bacteria, T4-like viruses and protists. ISME J 8:816–829

Cram JA, Xia LC, Needham DM, Sachdeva R, Sun F, Fuhrman JA (2015) Cross-depth analysis of marine bacterial networks suggests downward propagation of temporal changes. ISME J 9:2573–2586

Eiler A, Heinrich F, Bertilsson S (2012) Coherent dynamics and association networks among lake bacterioplankton taxa. ISME J 6:330–342

Falkowski PG, Fenchel T, DeLong EF (2008) The microbial engines that drive the Earth's biogeochemical cycles. Science 320:1034–1039

Faust K, Raes J (2012) Microbial interactions: from networks to models. Nat Rev Microbiol 10:538–550

Faust K, Sathirapongsasuti JF, Izard J, Segata N, Gevers D, Raes J et al (2012) Microbial co-occurrence relationships the human microbiome. PLoS Comput Biol 8:e1002606

Faust K, Lima-Mendez G, Lerat J-S, Sathirapongsasuti JF et al (2015) Cross-biome comparison of microbial association networks. Front Microbiol 6:1200

Ferla MP, Thrash J, Giovannoni SJ, Patrick WM (2013) New rRNA gene-based phylogenies of the *Alphaproteobacteria* provide perspective on major groups, mitochondrial ancestry and phylogenetic instability. PLoS ONE 8:e83383

Franzosa EA, Hsu T, Sirota-Madi A et al (2015) Sequencing and beyond: integrating molecular 'omics' for microbial community profiling. Nat Rev Microbiol 13:360–372

Frias-Lopez J, Shi Y, Tyson GW, Coleman ML et al (2008) Microbial community gene expression in ocean surface waters. Proc Natl Acad Sci 105:3805–3810

Fuhrman JA, Cram JA, Needham DM (2015) Marine microbial community dynamics and their ecological interpretation. Nat Rev Microbiol 13:133–146

Gilbert JA, Steele JA, Caporaso JG et al (2012) Defining seasonal marine microbial community dynamics. ISME J 6:298–308

Hamady M, Walker JJ, Harris JK et al (2008) Error-correcting barcoded primers for pyrosequencing hundreds of samples in multiplex. Nat Methods 5:235–237

Hanemaaijer M, Röling WFM, Olivier BG et al (2015) Systems modeling approaches for microbial community studies: from metagenomics to inference of the community structure. Front Microbiol 6:213

Hurwitz BL, Sullivan MB (2013) The Pacific Ocean Virome (POV): a marine viral metagenomic dataset and associated protein clusters for quantitative viral ecology. PLoS ONE 8:e57355

Hurwitz BL, Westveld AH, Brum JR, Sullivan MB (2014) Modeling ecological drivers in marine viral communities using comparative metagenomics and network analyses. Proc Natl Acad Sci 111:10714–10719

Karsenti E, Acinas SG, Bork P et al (2011) A holistic approach to marine eco-system biology. PLoS Biol 9:e1001177

Klindworth A, Mann AJ, Huang S, Wichels A et al (2014) Diversity and activity of marine bacterioplankton during a diatom bloom in the North Sea assessed by total RNA and pyrotag sequencing. Mar Genomics 18:185–192

Knopf A, Kostadinov I, Wichels A, Quast C, Glöckner FO (2015) Metatranscriptomics of marine bacterioplankton during winter time in the North Sea by total RNA sequencing. Mar Genomics 19:45–46

Lee D-H, Zo Y-G, Kim S-J (1996) Nonradioactive method to study genetic profiles of natural bacterial communities by PCR-single-strand-conformation-polymorphism. Appl Environ Microbiol 62:3112–3120

Lima-Mendez G, Faust K, Henry N et al (2015) Determinants of community structure in the global plankton interactome. Science 348:1262073-1–1262073-9

Logares R, Sunagawa S, Salazar G et al (2014) Metagenomic 16S rDNA Illumina tags are a powerful alternative to amplicon sequencing to explore diversity and structure of microbial communities. Environ Microbiol 16:2659–2671

Maclean D, Jones JDG, Studholme DJ (2009) Application of 'next-generation' sequencing technologies to microbial genetics. Nat Rev Microbiol 7:287–296

Marsh TL (1999) Terminal restriction fragment length polymorphism (T-RFLP): an emerging method for characterizing diversity among homologous populations of amplified products. Curr Opin Microbiol 2:323–327

Morris RM, Nunn BL, Frazar C, Goodlett DR, Ting YS, Rocap G (2010) Comparative metaproteomics reveals ocean-scale shifts in microbial nutrient utilization and energy transduction. ISME J 4:673–685

Muller EEL, Glaab E, May P, Vlassis N, Wilmes P (2013) Condensing the omics fog of microbial communities. Trends Microbiol 21:325–333

Muyzer G (1999) DGGE/TGGE a method for identifying genes from natural ecosystems. Curr Opin Microbiol 2:317–322

Ottesen EA, Marin R III, Preston CM, Young CR et al (2011) Metatranscriptomic analysis of autonomously collected and preserved marine bacterioplankton. ISME J 5:1881–1895

Ottesen EA, Young CR, Eppley JM, Ryan JP et al (2013) Pattern and synchrony of gene expression among sympartic marine microbial populations. PNAS 110:E488–E497

Pedrós-Alió C (2006) Marine microbial diversity: can it be determined? Trends Microbiol 14: 257–263

Peura S, Bertilsson S, Jones RI, Eiler A (2015) Resistant microbial co-occurrence patterns inferred by network topology. Appl Environ Microbiol 81:2090–2097

Raes J, Bork P (2008) Molecular eco-systems biology: towards an understanding of community function. Nat Rev Microbiol 6:693–699

Sears CL (2005) A dynamic partnership: celebrating our gut flora. Anaerobe 11:247–251

Shannon P, Markiel A, Ozier O et al (2003) Cytoscape: a software environment for integrated models of biomolecular interaction networks. Genome Res 13:2498–2504

Sogin ML, Morrison HG, Huber JA, Welch DM et al (2006) Microbial diversity in the deep sea and the underexplored "rare biosphere". Proc Natl Acad Sci 103:12115–12120

Sowell SM, Wilhelm LJ, Norbeck AD, Lipton MS et al (2009) Transport functions dominate the SAR11 metaproteome at low-nutrient extremes in the Sargasso Sea. ISME J 3:93–105

Sowell SM, Abraham PE, Shah M, Verberkmoes NC et al (2011) Environmental proteomics of microbial plankton in a highly productive coastal upwelling system. ISME J 5:856–865

Staley JT, Konopka A (1985) Measurements of *in situ* activities of nonphotosynthetic microorganisms in aquatic and terrestrial habitats. Ann Rev Microbiol 39:321–346

Steele JA, Countway PD, Xia L et al (2011) Marine bacterial, archaeal and protistan association networks reveal ecological linkages. ISME J 5:1414–1425

Sunagawa S, Coelho LP, Chaffron S et al (2015) Structure and function of the global ocean microbiome. Science 348:1261359-1–1261359-9

Teeling H, Fuchs BM, Becher D, Klockow C et al (2012) Substrate-controlled succession of marine bacterioplankton populations induced by a phytoplankton bloom. Science 336:608–611

Thiele I, Heinken A, Fleming RMT (2013) A systems biology approach to studying the role of microbes in human health. Curr Opin Biotechnol 24:4–12

Verberkmoes NC, Denef VJ, Hettich RL, Banfield JF (2009) Functional analysis of natural microbial consortia using community proteomics. Nat Rev Microbiol 7:96–205

Wang D-Z, Xie Z-X, Zhang S-F (2014) Marine metaproteomics; current status and future directions. J Proteomics 97:27–35

Whitman WB, Coleman DC, Wiebe WJ (1998) Prokaryotes: the unseen majority. Proc Natl Acad Sci 95:6578–6583

Widder S, Besemer K, Singer GA, Ceola S et al (2014) Fluvial network organization imprints on microbial co-occurrence networks. PNAS 111:12799–12804

Williams TJ, Cavicchioli R (2014) Marine metaproteomics: deciphering the microbial metabolic food web. Trends Microbiol 22:248–260

Zengler K, Palsson BO (2012) A road map for the development of community systems (CoSy) biology. Nat Rev Microbiol 10:366–372

Chapter 13
Screening Microorganisms for Bioactive Compounds

Sonia Giubergia, Carmen Schleissner, Fernando de la Calle, Alexander Pretsch, Dagmar Pretsch, Lone Gram and Mariane Schmidt Thøgersen

Abstract Novel bioactive compounds are in high demand due to the development of microbial antibiotic resistance, increase in age-related diseases, and requirements for optimized manufacturing processes. The bioactive compounds can act as specific antitumor, anti-inflammatory, and antifungal compounds as well as pharmaceuticals against metabolic diseases. Bioprospecting from marine microorganisms has a tremendous potential for the discovery of novel bioactive compounds as enzymes and complex secondary metabolites for industrial as well as for biotechnological and therapeutic applications. A bioprospecting process usually begins with the isolation of a native microorganism, extraction of compounds from a culture sample, or with the isolation of environmental DNA for either heterologous expression or sequencing. Extracts from live microorganisms can be screened for bioactivity in function-based screening assays, while DNA from either the isolated

S. Giubergia
Novo Nordisk Foundation Center for Biosustainability, Technical University of Denmark, Kogle Allè 6, 2970 Hørsholm, Denmark
e-mail: sogiu@biosustain.dtu.dk

C. Schleissner · F. de la Calle
PharmaMar, Avenida de los Reyes 1, 28770 Colmenar Viejo, Madrid, Spain
e-mail: cschleissner@pharmamar.com

F. de la Calle
e-mail: fdelacalle@pharmamar.com

A. Pretsch · D. Pretsch
SeaLife Pharma GmbH, Technopark 1, 3430 Tulln, Austria
e-mail: pretsch@sealifepharma.com

D. Pretsch
e-mail: dagmar.pretsch@sealifepharma.com

S. Giubergia · L. Gram (✉) · M.S. Thøgersen
Department of Systems Biology, Technical University of Denmark, Matematiktorvet 301, 2800 Kgs. Lyngby, Denmark
e-mail: gram@bio.dtu.dk

M.S. Thøgersen
e-mail: marsm@bio.dtu.dk

L.J. Stal and M.S. Cretoiu (eds.), *The Marine Microbiome*,
DOI 10.1007/978-3-319-33000-6_13

microorganisms or from the environment can be screened with sequence-based screening methods or genome mining. Once a desired activity has been detected, the bioactive compound should be purified, the structure should be elucidated, and preferably its mechanism of action should be described. In this chapter, we give an overview of the bioprospecting process with special focus on compounds with therapeutic properties from marine microorganisms, and we evaluate some of the most commonly used strategies that have been used at different steps in the bioprospecting processes when searching for novel bioactive compounds.

13.1 The Need for Novel Bioactive Compounds

During the last century, microorganisms have emerged as a valuable source of bioactive compounds with a wide range of biotechnological applications. Two major groups of such compounds are enzymes and compounds with bioactivity on bacterial, archaeal, or eukaryotic cells. Enzymes of microbial origin are used as biocatalyzers in many different industries such as feed, detergent, textile, chemical, and biofuel industries. Polymer-degrading enzymes such as cellulases and xylanases can be used for bioremediation and food processing purposes (Adrio and Demain 2014; Schmid et al. 2001). Compounds exhibiting bioactivity on bacterial, archaeal, and eukaryotic cells (from now on referred to as bioactive compounds) include microbial secondary metabolites with therapeutic properties such as antibiotic, antitumor, or anti-inflammation activity. Secondary metabolites belong to several different chemical classes of compounds, which among others include polyketides, non-ribosomal peptides, terpenes, and terpenoids. Bioactive compounds of microbial origin also include bacteriocins and bacterial polysaccharides. Bacteriocins are ribosomal peptides produced by bacteria to kill or inhibit the growth of closely related bacteria and are used in for example dairy and food industries to prevent food spoilage (Cleveland et al. 2001), however bacteriocins are also increasingly being recognized as potential alternatives to classical antibiotics (reviewed in Reen et al. 2015). Bacterial polysaccharides are used in food, cosmetic, and pharmaceutical industries as well as in bioremediation and for the development of biomaterials and bioplastics (Nicolaus et al. 2010; Sutherland 2001).

The United Nations' Convention on Biological Diversity (CBD) defines biological prospecting, or just bioprospecting, as "the exploration of biodiversity for commercially valuable genetic and biochemical resources" (https://www.cbd.int). In other words, bioprospecting is the search for any product from nature, such as genes, enzymes, chemical compounds, or organisms for commercial purposes. In this chapter, we will focus on bioprospecting for small molecules of microbial origin with therapeutic properties.

The need for novel bioactive compounds is best exemplified by the rapid development and spread of antibiotic resistance in bacteria caused by the widespread use of antibiotics in both the clinic and the food production sector. Infections caused by antibiotic resistant bacteria are increasing (World Health Organization

2014) and this is of great concern and a threat to human and animal health, since known antibiotics are losing their effectiveness. Novel antibiotics, preferably with new mechanisms of action, are therefore urgently needed. Many compounds with antibacterial activity are of microbial origin, and some classes of such compounds may have the desired effect without being hampered by already known resistance mechanisms (reviewed in Clardy et al. 2006). As an initiative to discover and develop novel therapeutics, the Infectious Disease Society of America (IDSA) has launched the "10 × '20 Initiative" with the aim of bringing at least 10 new antibiotics to the marked before 2020 (http://www.idsociety.org/10x20/).

With the increase in average life expectancy of the human population, age-related diseases such as Alzheimer's disease, Parkinson's disease, osteoporosis, and several types of cancer are also requiring new and improved drugs to optimize treatment. Especially for Alzheimer's and Parkinson's diseases, the drugs used today only provide symptomatic relief but no cure.

13.1.1 Bioactive Compounds from the Marine Environment

Groups of soil microorganisms such as *Streptomyces* and *Micromonospora* as well as several species of filamentous fungi have for years been the most commonly known microbial reservoirs of bioactive compounds (Bérdy 2005; Hoffmeister and Keller 2007; Keller et al. 2005; Watve et al. 2001). However, recent advances in novel bioprospecting technology are providing the opportunity to explore new habitats. These include the ocean, the deep sea, the cryosphere, and the deserts (Gerwick and Moore 2012; Molinski et al. 2009).

The ocean covers more than 70 % of the Earth's surface and represents more than 95 % of the total biosphere. Thirty-three of the thirty-five known animal phyla of life are present in the oceans and thirteen are exclusively marine phyla (Hiep and McDonough 2012), making it an invaluable source of biodiversity potentially useful for bioprospecting purposes. The environmental conditions of the oceans require selection and maintenance of unique biosynthetic pathways as a result of adaptation to pressure, salinity, temperature, oligotrophic conditions, and unique chemical compounds (Fenical and Jensen 2006; Lozupone and Knight 2007). Several marine-derived natural products therefore have new chemical features as compared to molecules isolated from terrestrial environments, exemplified by the incorporation of the halogen atoms bromide or chloride (Gerwick and Moore 2012; van Pée 1996).

Early studies of the marine environment focused on natural products from invertebrates such as sponges, bryozoans, algae, and corals, and led to the isolation of several classes of bioactive natural products including polyketides, non-ribosomal peptides, terpenes, and indoles (DeGroot et al. 2015; Proksch et al. 2002; Woodhouse et al. 2013). However, evidence is currently emerging, based on metagenomics of marine sponges and tunicates and their associate microbiota, that many of the compounds originally isolated from eukaryotic organisms are in fact of bacterial origin, as will be outlined below.

Many marine organisms have not developed an adaptive immune response and instead rely mostly on an immediate, innate immune system as the defense against pathogenic microorganisms. Also, many marine microorganisms have developed specialized strategies against predators. It is therefore expected that marine bacteria are capable of producing an array of anti-inflammatory compounds to better be able to evade the innate immune response of the host organism. For instance, several cyanobacteria produce such anti-inflammatory compounds, potentially targeting a broad range of molecular targets (Stevenson et al. 2002; Villa et al. 2010).

There are several cases of antitumor compounds originally isolated from marine invertebrates and tunicates and later identified in the genomes of bacteria (Lane and Moore 2012; Wilson and Piel 2013; Wilson et al. 2014). The similar chemical structure between safracin B and ecteinascidins (trabectedin, Yondelis®, the first European marine-derived compound approved for cancer treatments and developed by PharmaMar) has been corroborated to a genetic level by the identification of a NRPS gene cluster for the putative biosynthesis of ecteinascidins into a bacterial symbiont living into the ascidian cells (Rath et al. 2011). Another example is the anticancer and anti-Alzheimer's compound bryostatin, which is currently undergoing clinical trials. Bryostatin was first identified in extracts from the bryozoan *Bugula neritina*, but the polyketide synthase (PKS) cluster putatively responsible for its biosynthesis has been identified in the bacterial symbiont *Candidatus* Endobugula sertula (Davidson et al. 2001). Similarly, the antitumor depsipeptide dolastatin 10 was first isolated from the sea hare *Dollabella auricularia* (Pettit et al. 1993) and later from the marine cyanobacterium *Symploca* sp. VP642 (Luesch et al. 2001). Also, the Gram-positive *Bacillus pumilus*, isolated from the marine sponge *Acanthella acuta*, produces diglucosyl-glycerolipids with antitumor properties (Ramm et al. 2004).

The Gram-positive actinobacteria, which are already known as an excellent reservoir of natural products in the terrestrial environment (Bérdy 2005), have also emerged as producers of novel bioactive compounds in the marine environment. Although several compounds have been isolated from marine species of *Streptomyces* (Khan et al. 2011; Manivasagan et al. 2014; Schleissner et al. 2011), probably the best-known example is the novel actinomycete genus *Salinispora*. This genus includes only three species, but has already provided a number of bioactive compounds, including the anticancer compound salinosporamide A and derivatives of the antibiotic rifamycin (Fenical and Jensen 2006). Other interesting actinomycete genera for drug discovery are *Saccharopolyspora* (Pérez et al. 2009), *Micromonospora* (Romero et al. 1997) or *Nocardiopsis* (PharmaMar, personal communication), which contain several putative biosynthetic gene clusters.

Marine Gram-negative bacteria are currently emerging as potential producers of such compounds and novel natural products have been discovered in marine cyanobacteria (Burja et al. 2001; Calteau et al. 2014; Tan 2007) and proteobacteria. For instance, myxobacteria belonging to the class of Deltaproteobacteria as well as the gammaproteobacterial genera *Pseudomonas*, *Pseudoalteromonas*, and *Vibrio* have provided several compounds with bioactivities including antibacterial and anti-virulence activity (Bowman 2007; Chellaram et al. 2012; Gram et al. 2010;

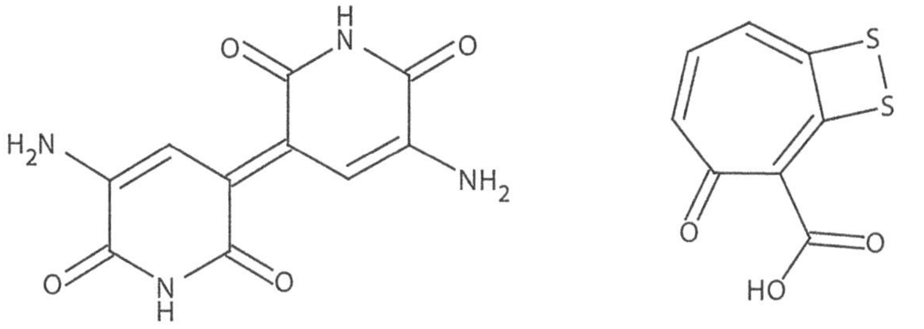

Indigoidine

Tropodithietic acid (TDA)

Fig. 13.1 Structures of the antimicrobial compounds indigoidine and tropodithietic acid (TDA) derived from members of the marine *Roseobacter* clade

Mansson et al. 2011; Nielsen et al. 2012, 2014; Proksch et al. 2002; Schäberle et al. 2010; Vynne et al. 2011; Whalen et al. 2015; Wietz et al. 2010). Furthermore, members of the alphaproteobacterial *Roseobacter* clade produce bioactive compounds, including the antimicrobial compounds indigoidine (Cude et al. 2012) and tropodithietic acid (TDA) (Brinkhoff et al. 2004; Bruhn et al. 2007) (Fig. 13.1).

13.2 Bioprospecting for Bioactive Compounds

Bioprospecting of microbial natural products can include the following steps:

1. Isolation and molecular characterization of microorganisms
2. Culture and extract preparation
3. Construction of metagenomic libraries
4. Screening for bioactivities
5. Chemical dereplication, compound purification, and structure elucidation.

A bioprospecting process is not linear, and many different approaches can be used. Sometimes the starting material is a live microorganism, wild type or recombinant, in other situations it is a crude extract from cultures of microorganisms, and in some cases it is a library of pure compounds. Here we will describe a classical methodology following the steps described above, well knowing that many researchers choose alternative strategies, and that one often has to revisit a screen or an assay several times before a novel, pure bioactive compound can be presented.

It is important to note that the strategy chosen for bioprospecting is independent of the origin of the samples; hence, here we will provide examples that have been used in the discovery of natural products from microorganisms of both terrestrial and marine origin with a focus on the latter.

13.3 Isolation of Microorganisms

The first step in bioprospecting for microbial natural products is to isolate microorganisms or to purify microbial DNA from environmental samples. Any biological sample collection should be carried out in accordance with the Convention on Biological Diversity, the Cartagena Protocol and the Nagoya Protocol (https://www.cbd.int). These documents have been created to ensure the conservation of biological resources, their sustainable use, and the equitable and fair sharing of benefits deriving from the use of biological and genetic resources. Despite the fact that several countries in the world have signed the convention and the protocols, they have not been ratified by all of them. But nontheless there should always be an agreement between those collecting the biological material and the country where the sampling takes place.

The culturability of microorganisms is an important consideration in bioprospecting. With the current standard laboratory culture conditions, only approximately 1 % of all bacteria have been brought into culture. For marine bacteria this number is thought to be even lower; between 0.01 and 0.1 % (Gram et al. 2010; Kogure et al. 1979). The bioprospecting potential of microorganisms is therefore potentially overlooked since they simply have never been cultured in the laboratory. However, efforts are now being made to develop devices and techniques in order to increase microbial culturability or to use microbial DNA as the starting material.

In traditional isolation techniques, microorganisms from for example soil, water, or invertebrate samples are cultured on artificial substrates and strains able to grow into colonies are selected for further analyses. Selective media are often used to enrich for a desired group of bacteria. Thaker et al. (2013) succeeded in the isolation of scaffold-specific natural product producers; the assumption was that producers of antibiotic compounds must have a mechanism of self-resistance to the produced compounds to avoid suicide. Therefore, the addition of a specific antibiotic to the medium will select for the resistant microorganisms, which would also be likely to produce other compounds with the same scaffold. Indeed, the addition of the glycopeptide antibiotic (GPA) vancomycin to the isolation medium led to the isolation of the novel GPA pekiskomycin.

Several strategies are currently being exploited to increase microbial culturability. When in their natural habitat, microorganisms are exposed to a multitude of factors that most often are not present in traditional laboratory culturing, including the presence of other microorganisms, specific substrates, and growth factors (Marmann et al. 2014; Stewart 2012). Connon and Giovannoni (2002) succeeded in the high-throughput isolation of previously uncultured marine bacteria using the "dilution-to-extinction" approach based on the work of Button et al. (1993). The dilution of seawater samples to a concentration of one to five bacterial cells per ml led to an increase in the number of isolated strains of up to three orders of magnitude compared to direct inoculum of water samples onto solid medium. In contrast, after plating serial dilutions of a microbial suspension obtained from marine environmental samples, D'Onofrio et al. (2010) observed that there were significantly more colonies on densely inoculated plates as compared to diluted ones. They hypothesized and proved that the growth of some of the isolates from the densely inoculated plates depended on neighboring colonies providing growth factors, in this case siderophores.

An alternative strategy to increase bacterial culturability is to introduce an inoculum of environmental cells into culture chambers spatially delimited by semipermeable membranes, whose pore size prevents cells to pass through but enables free exchange of molecules with the external environment. The chamber is then incubated either at the original sampling location or in an artificial environment simulating it. Several variants of such chambers are available, such as the isolation Chip (ichip) (Nichols et al. 2010) which is miniaturized and optimized for high-throughput isolation of hundreds of strains in parallel, and the MicroDish Culture Chip, which consists of up to 180,000 culture areas 20 μm across on top of a porous base that allows for passage of nutrients from below when the chip is placed on for example an agar substrate (http://www.microdish.nl). The use of culture chambers has recently resulted in the isolation of the novel, previously uncultured species *Eleftheria terrae* (provisional name), which produces teixobactin, an antibiotic with a novel mode of action active against both *Staphylococcus aureus* and *Mycobacterium tuberculosis* (Ling et al. 2015).

In order to construct a collection of unique culturable strains, biotechnological methods for molecular dereplication, such as DNA fingerprinting, and identification based on sequencing of the 16S rRNA gene must be used. In the future, genome sequence-based identification and dereplication is likely to be the method of choice.

13.4 Culture and Extract Preparation

When native organisms are successfully brought into culture, the next challenge is to unlock their bioactive potential. The artificial in vitro culture conditions are most often not identical to the original environment of the microorganisms, and chemical substances, signaling molecules, and other compounds important for induction of gene expression might be missing.

In bioprospecting, screening for bioactivity is carried out using either live organisms or extracts from whole cultures or sub-fractions hereof (i.e., separation of biomass from the supernatant). In preparation of an extract, the choice of the solvent depends on the availability of information about the nature of the compound to be extracted. For nonpolar compounds, organic solvents such as hexane and chloroform can be used, while for polar compounds water, ethanol, or methanol are suitable. When there is no information about the nature of the compounds, extraction will have to be carried out by a trial and error approach: often medium polarity extractions with ethyl acetate or dichloromethane can be applied, or alternatively a mixture of solvents. In the case of large liquid cultures where liquid–liquid extraction would be challenging due to the big volumes to be handled, it is possible to add an adsorptive resin (e.g., DIAION HP 20®) to the culture: after 24–48 h of incubation, the resin is separated from the cultures and the secreted metabolites can be extracted from its surface.

Microbial crude extracts are usually complex mixtures of compounds including cellular and media components. The most commonly used strategy to narrow down the complexity of the crude extract is bioassay-guided isolation, where the complexity is reduced by fractionation, and bioactive fractions are identified by means of a bioassay (Gerwick et al. 1994; Wietz et al. 2010). An alternative to working with crude extracts is to build libraries containing pure compounds, which are then screened for bioactivity. This approach, however, requires more resources as compared to the bioassay-guided isolation (Wagenaar 2008).

When it is not possible to bring the native microorganisms into culture, an alternative strategy is to extract DNA directly from the environment for metagenomic analyses. The isolated DNA can either be used to construct clone libraries to be tested in bioactivity screening assays, or alternatively, the DNA sequences can be mined using software developed to search for biosynthetic gene clusters or other structures indicating potential production of bioactive compounds.

13.5 Construction of Metagenomic Libraries

The metagenomic approaches bypasses the microbial isolation and culturing steps, however, isolation of DNA directly from an environmental sample is a major challenge, especially when dealing with samples from extreme environments. Extremophilic microorganisms are often reluctant to standard lysis protocols developed for mesophilic microbes and, when lysed, they often release stable nucleases that will degrade the purified DNA (Simon and Daniel 2011).

When constructing a metagenomic library, DNA is purified directly from an environmental sample, fragmented, and cloned into a host organism using a selected vector. This vector can be either an expression plasmid for direct gene expression from small DNA fragments (Lynch and Gill 2006; Schmitz et al. 2008) or a larger plasmid such as BAC (bacterial artificial chromosome) plasmids. Larger plasmids are commonly used to insert and sequence large fragments of environmental DNA (eDNA) to search *in silico* for known or hypothetical genes or gene clusters (O'Connor et al. 1989; Shizuya and Kouros-Mehr 2001). However, it is important to keep in mind that many gene clusters encoding e.g., polyketide synthase/non-ribosomal peptide synthase (PKS/NRPS) hybrids are too large for classical heterologous expression and that novel expression systems should be developed for expression of large gene clusters (>50 kb).

In the preparation of expression libraries for smaller genes or gene clusters, both the vector and the host for direct gene expression have to be chosen with great care to ensure maximal chance of successful expression of the desired gene(s). In case the host has an expression system or a secretion system too different from that of the native donor strain (e.g., the simple issue of Gram type), no active compounds will

Table 13.1 Natural products discover through metagenomic approaches

Compound	Source	Type of screening
Onnamide A	*Theonella swinhoei*, bacterial symbiont	Sequence-based screening
Bryostatin	*Bugula neritina*, bacterial symbiont	Sequence-based screening
Minimide	*Didemnum molle*, microbiome	Sequence-based screening
Apratoxin A	*Lyngbya bouillonii*	Sequence-based screening
Patellamides	*Lissoclinum patella*	Function-based screening
Zn-coproporphyrin III	*Discodermia calyx*	Function-based screening

Modified from Barone et al. (2014)

be produced even though the genetic potential was actually there. Several commercial host organisms as well as broad-host-range vectors are available, but there is no universal expression system available that covers all bacterial species yet (Aakvik et al. 2009; Lale et al. 2011).

Promoter trap or gene trap libraries have now become more commonly used since they rely on only small fragments of eDNA and can be easily screened for the presence of a desired gene or promoter by using a standardized reporter molecule like green fluorescent protein (GFP) or luciferase (Izallalen et al. 2002; Kondo et al. 1993; Rediers et al. 2005).

A variety of bioactive compounds including primarily enzymes but also therapeutics from the marine environment have in the last decade been discovered via metagenomic approaches (reviewed in Barone et al. 2014). In a study by Piel et al. (2004), DNA was extracted from the marine sponge *Theonella swinhoei*, cloned into cosmids, and screened using primers specific for PKS and NRPS clusters. This resulted in the identification of two gene clusters originating from a bacterial endosymbiont encoding onnamides and theopederins, a class of polyketides with antitumor activity (Piel et al. 2004). Other examples of marine compounds found through metagenomic approaches can be seen in Table 13.1.

13.6 Screening for Bioactivity

When screening for bioactivity, two overall approaches are possible: sequence-based screening and function-based screening. In sequence-based screening, sequence analysis is performed to identify genes or gene clusters potentially encoding molecules or biosynthetic pathways of interest. In function-based screening, bioassays are performed to identify the desired bioactivity directly in the live microorganisms or the extracts. The two strategies should not be considered as independent and incompatible, instead they are often combined to accelerate and optimize the bioprospecting process (Mansson et al. submitted).

13.6.1 Sequence-Based Screening

In sequence-based screening, DNA sequences are analyzed to identify conserved regions deriving from known gene or protein families. This can be achieved in vitro by using for example DNA probes or degenerated primers (Ayuso-Sacido and Genilloud 2005; Ehrenreich et al. 2005) or phage display (Yin et al. 2007). However, with recent advances in DNA sequencing techniques, such analyses are most often done by in silico homology search based on whole genome sequencing of an isolated strain or sequencing of metagenomic libraries.

13.6.1.1 Genome Mining

In silico sequence-based screening, often referred to as genome mining, is an attractive approach in the natural product discovery pipeline. Several tools have been developed for the identification of conserved domains that are likely to be part of biosynthetic gene clusters. Some of them are publicly available and include NaPDoS for the identification of keto-synthase and condensation domains, which are core enzymes in PKS and NRPS, respectively (Ziemert et al. 2012), BAGEL for the identification of bacteriocins (de Jong et al. 2006), and AntiSMASH for the identification of a wide range of biosynthetic clusters, ranging from PKS and NRPS to siderophores (Medema et al. 2011). For a complete review of the available in silico tools for the genome mining of natural products, see Weber (2014).

These sequence analysis tools are based on homology searches, meaning that only previously characterized families of genes and gene clusters are detected, and that the discovery of novel classes of bioactive compounds is almost impossible. However, efforts are being made to overcome this problem, as demonstrated by the recently developed algorithm ClusterFinder (Cimermancic et al. 2014). This tool enables the identification of both known and unknown biosynthetic gene cluster by converting nucleotide sequences into protein family (Pfam) domains and calculating for each domain, the probability of it being part of a gene cluster. A comprehensive analysis of 1154 genome sequences using ClusterFinder has revealed the presence of large families of biosynthetic gene clusters of unknown function, indicating that the microbial biosynthetic potential is far from exhausted (Cimermancic et al. 2014).

Information obtained with genome mining can accelerate the natural product bioprospecting process by identifying isolates or clones that produce already known compounds, enabling the prediction of expected classes of compounds as well as structural predictions (Jensen et al. 2014). For example, mining of the genome of the marine bacterium *Salinispora pacifica* strain CNT-133 enabled the identification of a truncated gene cluster related to the salinosporamide A biosynthetic gene cluster in *Salinispora tropica* strain CNB-440 (Eustáquio et al. 2011). The anticancer compound salinosporamide A has a chloroethyl group on C-2 (Fig. 13.2),

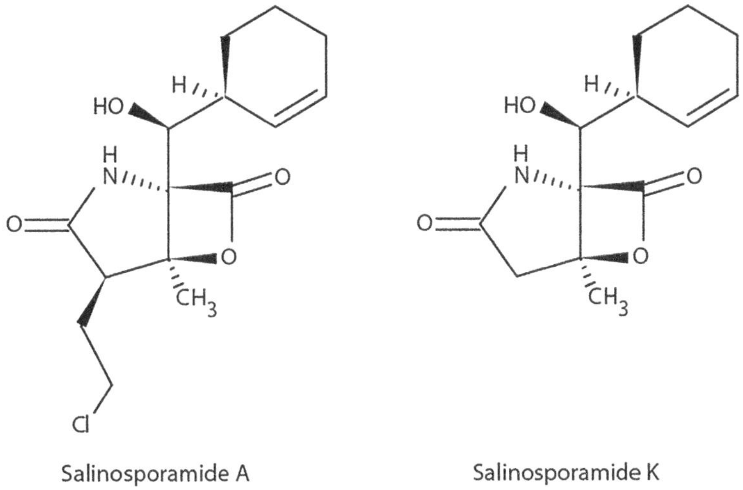

Salinosporamide A

Salinosporamide K

Fig. 13.2 Structures of the antitumor compounds salinosporamide A and salinosporamide K derived from the marine bacteria *Salinispora tropica* and *Salinispora pacifica*, respectively

and the comparison of the two gene clusters led to the hypothesis that, if a salinosporamide were to be produced by *S. pacifica*, it would not be halogenated. Indeed, the compound that later was isolated and structurally characterized, salinosporamide K, lacks the C-2 chloroethyl group (Fig. 13.2).

Silent biosynthetic gene clusters

Genome mining has revealed that microorganisms often have the potential to produce other natural products than identified using extractions and bioassay-guided fractionation (Gram 2015; Helfrich et al. 2014; Jensen et al. 2014; Machado et al. 2015). This has opened up a new branch in natural products research, where molecular biology, microbial ecology, and physiology are merged to elucidate how the silent or cryptic biosynthetic gene clusters can be elicited. The "One Strain-MAny Compounds (OSMAC)" approach (Schiewe and Zeeck 1999) was successfully used in several studies, where the variation of culture parameters (e.g., media composition, culture vessel, aeration, addition of enzyme inhibitors, or rare earth elements, or coculturing with other microorganisms) elicited previously silent biosynthetic gene clusters (Bode et al. 2002). Culture parameters can be modified to simulate the environmental niches of microorganisms. For example, antibacterial activity was observed in the supernatant of cultures of two marine *Bacillus* strains, only when they were grown into an agar-coated roller bottle mimicking the intertidal environment of isolation and not when they were cultured in standard shaking flasks (Yan et al. 2002). Similar results have been obtained with "air-membrane surface" (Yan et al. 2003) and "rotating disc" bioreactors (Sarkar et al. 2008). The exposure to low concentrations of antibiotics and other small molecules produced by microorganisms elicited two cryptic biosynthetic gene clusters in the soil bacterium *Burkholderia thailandensis* (Seyedsayamdost 2014), whilst coculturing setups caused variations in the secondary metabolism profile of *Streptomyces coelicolor*, as captured by imaging mass spectrometry (Traxler et al. 2013). Access to a carbon source typical of the niche of isolation can influence the biosynthesis of a given compound. For instance, it has been observed that the addition of chitin to the growth medium can elicit a twofold increase in the biosynthesis of the antibiotic compound andrimid in the marine bacterium *Vibrio coralliilyticus* S2052 as compared to glucose (Wietz et al. 2011).

An alternative way to access silent biosynthetic gene clusters relies on homologous and heterologous gene expression. In homologous gene expression, transcriptional, translational, or metabolic elements are manipulated to elicit the expression of the targeted biosynthetic gene cluster. However, this is possible only when the host is easily culturable and not recalcitrant to genetic manipulation. The introduction of drug resistance mutations in the ribosome and in RNA polymerase, so-called ribosomal engineering, greatly influences secondary metabolism in actinomycetes (Hosaka et al. 2009). Ribosomal engineering has also led to the isolation of the antitumor molecule fredericamycin A from the deep-sea-derived *Streptomyces somaliensis* SCSIO ZH66 (Zhang et al. 2015) and to a 180-fold higher production of the antibiotic actinorhodin in *Streptomyces coelicor* A3(2) (Wang et al. 2008). Moreover, homologous overexpression of a regulatory gene controlling the biosynthesis of the precursor of the C-2 chloroethyl group in salinosporamide A selectively doubled the yield of the compound in the natural producer *Salinispora tropica* (Lechner et al. 2011).

In heterologous expression, single genes or gene clusters are expressed in a heterologous host. A prerequisite is the availability of well-developed genetic toolboxes that enable the introduction of gene clusters up to 100 kb in size. The native promoters of biosynthetic pathways are often not strong enough to trigger the expression in heterologous hosts, and it is necessary to place the biosynthetic cluster under control of strong, inducible promoters. For instance, it has been possible to express the gene cluster encoding biosynthesis of the antibiotic polyketide oxytetracycline from *Streptomyces rimosus* in *Escherichia coli* only when one of the heterologous host's native sigma factor was overexpressed (Stevens et al. 2013). An example of heterologous expression of a large NRPS gene cluster from a marine *Micromonospora* in an industrial *Streptomyces* is described for the antitumoral peptide thiocoraline (Lombó et al. 2006). In general, several molecular tools have been used for heterologous expression of silent biosynthetic gene clusters in bacteria and fungi. Some examples to be mentioned here are the transformation associated recombination (TAR) strategy (Ross et al. 2015; Yamanaka et al. 2014), the TRansfer and EXpression of biosynthetic pathways (TREX) system (Loeschcke et al. 2013), the red/ET (Wenzel et al. 2005) and the RecE/RecT recombination systems (Fu et al. 2012). For further examples and strategies, the review from Ongley et al. (2013) exhaustively covers advances on the topic until 2013.

13.6.2 Function-Based Screening

In function-based screening, bioassays are performed to detect a desired bioactivity in collections of isolated microorganisms, on sequence-based sub-selections of such, on recombinant expression hosts, or on culture extracts. Strategies that are most commonly used in function-based screening campaigns can be subdivided into phenotypic screens and target-based screens. By definition, phenotypic screens measure the effect, or phenotype, that the tested compounds induce in target cells or organisms, whereas target-based screens investigate the ability of a compound to bind or inhibit purified targets in vitro (Kotz 2012). Phenotypic screens can be cell based (in vitro screens on single cells or tissues), or involve model organisms (in vivo screenings) such as *Saccharomyces cerevisiae* (yeast), *Drosophila melanogaster* (fruit fly), or *Caenorhabditis elegans* (nematode). In target-based screens, the target of interest is purified and in vitro biochemical assays are established to investigate the effects of a range of compounds on the target.

Traditionally, phenotypic screens were favored by the pharmaceutical industry. However, difficulties encountered in target identification as well as advancement in multiple disciplines such as molecular biology, flow cytometry, chemical proteomics, and imaging techniques led to the establishment of target-based screens. The outcome of target-based screening campaigns was not as high as expected, though, and today there is a renewed interest in phenotypic screens both in academia and in the pharmaceutical industry in companies such as Novartis AG and GlaxoSmithKline (Eggert 2013; Kell 2013; Kotz 2012; Swinney and Anthony 2011; Zheng et al. 2013). The two approaches are often combined: phenotypic screens are not completely target-agnostic and targets for target-based screens can be chosen based on results from phenotypic assays (Kell 2013; Moffat et al. 2014; Sams-Dodd 2005). Examples on how the two screening strategies can be used and combined can be seen in Sects. 6.2.1–6.2.4 on strategies for the discovery of antibacterial, antiviral, antitumor, anti-Alzheimer's, and anti-Parkinson's compounds.

13.6.2.1 Screens for antibacterial activity

Due to the increased antimicrobial drug resistance in many pathogenic microorganisms and the lack of therapeutic alternatives to the classical antibiotics, prospecting for novel antibacterial compounds has become of high priority. Classical function-based screening of live bacteria or extracts is often based on agar plate or broth-based assays where for instance growth inhibition by an antagonistic compound or a colorimetric reaction indicative of a desired activity or molecule can be observed.

Live targets

A commonly used approach to detect antibacterial compounds is the overlay method developed by Waksman (Schatz et al. 1944). In brief, a potential producer strain is grown on a solid medium, which is then overlaid with soft agar seeded with a target bacterium. After a period of incubation, the presence of antibacterial compounds, to which the target is susceptible, is indicated by a clear halo in the top agar due to the lack of growth of the target strain. Alternatively, the potential producer strain can be on top of the seeded medium. The latter option can be used directly during the isolation procedure to select for strains displaying antibacterial activity or by replica plating a master plate on the seeded medium (Gram et al. 2010). The production of antagonizing compounds will be seen as a clear zone in the agar or by the lack of growth of the target strain, respectively (Fig. 13.3).

BioMAP (antiBIOtic Mode of Action Profile) is an assay that can be used for the growth-independent function-based screening of crude extracts. The BioMAP assay consists of a panel of 15 clinically relevant bacterial pathogens and provides a function-based high-throughput platform for screening natural products in order to identify new lead compounds with unique biological profiles. The activity profile of a given extract can then be compared to profiles of known antibiotics and used to determine the structural class of the antibacterial compound in the extract (Higginbotham et al. 2014; Wong et al. 2012).

Biosensors

Another strategy to screen for novel antibiotic compounds is to apply the whole-cell antibiotic biosensors strategy described by Urban et al. (2007). Promoter regions selectively and strongly induced by bactericidal antibiotics were identified in *Bacillus subtilis* and used to construct five biosensors consisting of the *B. subtilis*

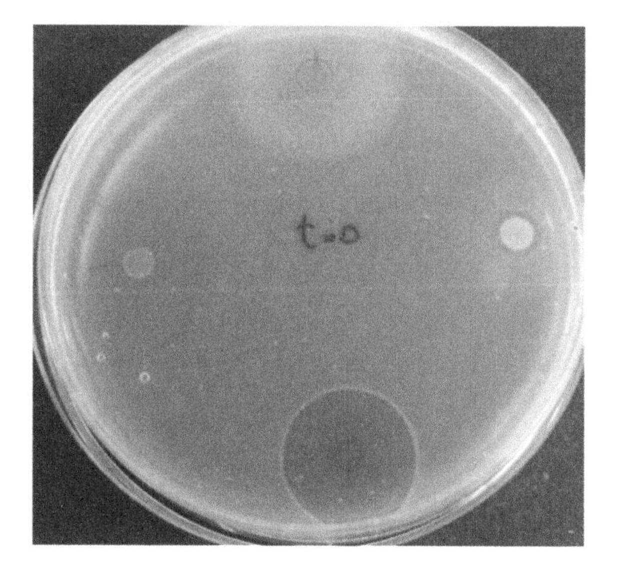

Fig. 13.3 Agar plate assay for screening for antimicrobial compounds containing a solid medium seeded with a target strain (here, the fish pathogen *Vibrio anguillarum*) on which four potential producer strains have been spotted. The clear halo surrounding the red colony in the bottom part of the picture indicates production of an antibacterial compound

promoters fused to the firefly luciferase reporter gene *lucFF*. This allowed for high-throughput detection of compounds interfering with DNA synthesis (*yorB* promoter), RNA synthesis (*yvgS* promoter), protein synthesis (*yheI* promoter), cell wall synthesis (*ypuA* promoter), and fatty acid synthesis (*fabHB* promoter). The biomarker-carrying strains of *B. subtilis* were then subjected to approximately 14,000 pure natural compounds, screened in a 384-well microtiter plate format using a luminescence detector, and the study led to the discovery of novel antibiotics in the form of DNA synthesis and translation inhibitors.

Since several bacterial virulence factors are under quorum sensing (QS) regulation, there is a great interest in compounds that can specifically block QS as potential novel classes of antibiotics (Rasmussen and Givskov 2006). Biosensors have been designed allowing for identification of such compounds. One example is the detection of the bacterial QS molecules *N*-acyl-homoserine lactones (AHLs). In *Agrobacterium tumefaciens lacZ::traG*, the *lacZ* reporter gene encoding a β-galactosidase is fused to the promoter of the QS-regulated *tra* operon. The AHL-induced expression of the operon and reporter gene is then visible as a blue precipitate due to β-galactosidase-mediated hydrolysis of X-gal present in the medium (Cha et al. 1998). High-throughput functional screening of single cells for AHL production is also possible as described by Williamson et al. (2005), who prepared a soil metagenomic library using *E. coli* cells containing a green fluorescence protein (GFP)-based AHL biosensor. Clones producing QS inducers could then be detected by fluorescence microscopy due to GFP expression, while clones producing QS inhibitors were identified using fluorescence-activated cell sorting (FACS) on cells in which GFP expression was induced by an exogenous AHL.

Molecular targets

Novel antimicrobial compounds are also identified by screens that are aimed directly at the targets of the antimicrobial compounds such as components of cellular pathways. Typically, assays are performed to screen for inhibitors of macromolecular synthesis using radioactively labeled precursors of protein, RNA, DNA, lipid, or peptidoglycan synthesis (Cotsonas King and Wu 2009; Nonejuie et al. 2013). However, these assays do not allow for determination of the mechanism of action (MOA), and often they do not distinguish between inhibitors that affect the same pathway.

Transcriptional profiling measures total gene expression to give an overall picture of the cellular functions. Even though transcriptional profiling sometimes can be used to identify the molecular targets of a bioactive compound, the method is slow and often fails to identify the target. Instead, Nonejuie et al. (2013) developed a bacterial cytological profiling (BCP) assay, which can distinguish between antibacterial compounds with different MOAs and also predict the MOA of novel compounds. In practice, the bacterial cells were treated with the potential antibacterial compounds and after incubation stained with FM4-64 which stains the membrane (Pogliano et al. 1999), with DAPI which stains DNA (Kapuscinski 1979), and with SYTOX Green Nucleic Acid Stain, which functions as a live/dead stain, since it is only able to penetrate membranes that have been made permeable

(Molecular Probes, Thermo Fisher Scientific Inc.). The cells were then subjected to fluorescence microscopy and the intensity of the stains was determined and used as a measure of the cell's cytological profile.

Several assays are available to screen for membrane-damaging compounds. However, such compounds have to be specific with respect to their target, since compounds targeting the cytoplasmic membrane would potentially not limit their activity to bacteria and ultimately cause toxicity in the mammalian host. Therefore, a range of assays to detect membrane-damaging compounds specific against e.g., *Staphylococcus aureus* have been developed. O'Neill et al. (2004) applied a β-galactosidase (BG) assay, where the *lacZ* gene from *E. coli* has been put under the control of a strong staphylococcal promoter, *cap1A*. Leakage of BG from the strain subjected to potential membrane-damaging compounds could then be detected using a fluorescence assay. Other assays to detect membrane damage in bacteria include ATP release (Johnston et al. 2003), leakage of nucleic acids (material absorbing at OD_{260}) (Carson et al. 2002), and the commercially available LIVE/DEAD® *Bac*Light™ Bacterial Viability Kit from Molecular Probes (Thermo Fisher Scientific Inc.).

Other molecular targets used in screening for novel antibacterial compounds are components of the major signal transduction pathways such as histidine kinases (Bem et al. 2015) or cell wall components like lipopolysaccharides and transporter complexes (Nayar et al. 2015). Nayar et al. (2015) applied a high-throughput phenotypic screening assay based on a *Citrobacter freundii* AmpC β-lactamase reporter system to discover novel compounds inhibiting cell wall synthesis. Inhibitors of cell wall synthesis also induce expression of AmpC β-lactamase (Sun et al. 2002). Hence, when β-lactamase was expressed, degradation of nitrocefin (a chromogenic cephalosporin substrate added to the cultures) could be detected in a plate reader, allowing for a 384-well microtiter plate format. Using this strategy, Nayar et al. (2015) successfully discovered two novel antibacterial compounds, sulfonyl piperazine and pyrazole, which are not cross-reactive with any known antibiotic and with completely novel MOAs.

13.6.2.2 Antiviral Drug Screening with CPE Inhibition Test

From the screening perspective, most antiviral test systems are based on infection of a host cell line or an animal. Many animal virus models have been established but for high-throughput screening of natural products cell lines are the better choice also because of animal ethical reasons. In most cases, a virus is highly specific, shows organ tropism, and infects defined parts of the body. The susceptibility for infection of the used viral type therefore regulates which type of host or cell line should be chosen for screening purposes.

Respiratory viruses like influenza, respiratory syncytial virus, parainfluenza, and especially human rhino virus, infect the upper respiratory tract including nose and throat and are some of the most common viruses in humans (Denny 1995). Being not highly virulent with low morbidity and weak symptoms, these types of viruses are in many cases precursors for more severe bacterial and fungal infections and can causes enormous financial damage (Gonzalez et al. 1987; Gonzales et al. 1997). With more than 100 different serotypes (Greve et al. 1989; Hofer et al. 1994), it is important to screen against as many viruses as is possible to check activity against the whole group, otherwise some serotypes will be preferred and might quickly become dominant (Andries et al. 1992).

The cytopathic effect (CPE) refers to the structural changes in a host caused by a viral infection. As screening method, the CPE inhibition assay is a fast and effective tool to verify if a natural product has antiviral potential (Schmidtke et al. 2001). Readout from the assay is the inhibition of the CPE i.e., the survival of the targeted cells. In infected and nontreated cell cultures, an increasing cytopathic effect with characteristic cell granulation, shrinking cells, plaque production, and spherical cell shapes can be observed. This cell morphology starts to change as the result of the reprogramming of the cell for viral production. At the end, most viruses replicate by destroying the host cell via lysis. In combination with plaque assay, the CPE inhibition assay could furthermore be used for definition of the detailed viral titer (Bachrach et al. 1957; Cooper 1961).

In general, there are two different types of screening in the CPE inhibition assay. In the prophylactic assay, the ability of a compound to affect the virus directly is tested. In this approach, the virus is mixed with the compound and incubated. This time allows the compound to target straight at the virus and block it or destroy it before it gets in touch with host cell line. Many antiviral blocking agents and biocides work with this mode of actions (Gonzalez et al. 1987). In contrast, in the therapeutic assay the compound attacks the virus after cell infection, and, hence, mainly stays inside the host cell. Here, the bioactive candidate is added to the cell culture media and the impact is on important switches in viral replication and packaging. Due to the fact that these approached require incubation time during the entire experiment, toxicity, solubility, and stability are critical factors to be considered. Therefore, it is recommended to collect as much as possible pre-information about the compound to be tested.

13.6.2.3 Screening for Antitumor Activity

The fundamental goal in cancer drug discovery is to kill or reprogram malignant cells while minimizing adverse effects on non-tumoral cells. Cancer encompasses a large number of molecularly and phenotypically distinct diseases, and hence it demands a much larger repertoire of drugs with distinct MOAs than most other diseases. For the past 50 years, phenotypic screenings in cancer drug discovery has been based on cytotoxicity assays using cancer cell lines that exhibit the phenotype of unrestrained fast growth. Such antiproliferative assays in cancer drug discovery have resulted in the development of a repertoire of chemotherapeutic agents (DeVita and Chu 2008) and are currently used in many oncology drug discovery programs. The National Cancer Institute 60 (NCI60) platform introduced the concept of high-throughput cell-based profiling using a panel of 60 different tumor cell lines.

The most frequently used cytotoxicity assay in a high-throughput screening platform is usually a colorimetric method, for example sulforhodamine B (SRB), for quantitative measurement of cell growth and viability (Vichai and Kirtikara 2006). Cultured cell lines derived from many different types of human cancer are used in such assays. Cytotoxicity is typically estimated using the National Cancer Institute (NCI) algorithm (Boyd and Paull 1995) which gives three end points that can be used to determine compound activity. These are GI_{50} (concentration required to inhibit 50 % of cells), TGI (concentration required for total growth inhibition), and LC_{50} (concentration required to kill 50 % of the cells) (Holbeck 2004). This screening technology requires the use of microplates, automated liquid handling and involves a high volume of data analysis.

Recent technologies that facilitate the parallel analysis of large panels of cell lines, together with genomic technologies that define their genetic constitution, have revitalized efforts to use cancer cell lines to assess the clinical utility of new investigational cancer drugs and to discover predictive biomarkers (Moffat et al. 2014). A human tumor cell line platform has been established to provide as broad a representation as possible of different cancers and includes 1200 cell lines. This panel is referred to as the Centre for Molecular Therapeutics 1000 (CMT1000) and is being used to probe the genetic basis for sensitivity to approved and investigational anticancer agents (Sharma et al. 2010).

Several bioactive anticancer compounds have been isolated from marine invertebrates, but have later been proven to be of microbial origin. The structural similarity between safracin B and ecteinascidins (trabectedin, otherwise known as Yondelis®, which was the first European marine-derived compound approved for some cancer treatments and developed by PharmaMar) has been resolved to the genetic level by the identification of a NRPS gene cluster for the putative biosynthesis of ecteinascidins in a bacterial symbiont living in the ascidian cells (Rath et al. 2011). The biosynthetic gene cluster is closely related to the NRPS genes of safracin B produced by *Pseudomonas fluorescens* A2-2 (Velasco et al. 2005) which is the starting material for the current semisynthetic manufacture of Yondelis® (Cuevas et al. 2000). The common building blocks are two units of the unusual amino acid 3-hydroxy-5-methyl-*O*-methyltyrosine (Fu et al. 2009). Another interesting example of a widespread NRPS gene cluster is the case of the antitumor compounds didemnins, initially isolated from the marine Caribbean tunicate *Trididemnum solidum* and recently described by several authors as a bacterial compound produced by free living Alphaproteobacteria (Tsukimoto et al. 2011; Xu et al. 2012). This includes the isolation of several strains belonging to the *Tistrella* genus during PharmaMar's program for isolation of marine bacteria (PharmaMar, personal communication).

Technical and biological advances, especially with the sequencing of cancer genomes and analysis of tumor cell transcriptions, have provided new insights into the molecular basis and classification of tumor phenotypes. The knowledge emerged from systematic cancer genome characterization during the last decades not only allows the discovery of new targets for target-based drug discovery

Table 13.2 Origin of new small-molecule cancer drugs approved by the FDA between 1999 and 2013

Lead discovery			
Inhibition or modulation of target		Mechanism-informed phenotypic screen	*De novo* phenotypic screen
Abiraterone*	**Afatinib**	Epirubicin	Carfilzomib
Bendamustine	**Axitinib**	Ixabepilone	Everolimus
Bexarotene*	**Bosutinib**	Nelarabine	Temsirolimus
Bortezomib*	**Cabozantinib**	Vismodegib*	Eribulin
Clofarabine	**Crizotinib***	Cabazitaxel	Omacetaxine*
Decitabine	**Dabrafenib**	Pemetrexed	Lenalidomide
Exemestane	**Dasatinib**	Azacitidine*	Pomalidomide
Temozolomide	**Erlotinib**		Romidepsin
Enzalutamide	**Ibrutinib***		Vorinostat*
Fulvestrant	**Imatinib***		**Trametinib***
Lapatinib	**Ruxolitinib***		
Nilotinib	**Sorafenib***		
Pazopanib	**Sunitinib***		
Ponatinib	**Vandetanib**		
Regorafenib	**Vemurafenib***		

Kinase inhibitors are highlighted in bold. Information on the drugs to be analyzed was obtained from the (FDA) website

*First-in-class therapeutics. Modified from Moffat et al. (2014)

programs, but also enables the definition of relevant and predictive phenotypic end points and cellular models for phenotypic screens (Moffat et al. 2014).

Some types of screening are in an intermediate category termed 'mechanism-informed phenotypic drug discovery' (MIPDD) where the mechanism of action can be detected by a specific phenotype. Moffat et al. (2014) have reviewed the screening origins of new small-molecule cancer drugs approved by the FDA between 1999 and 2013, as shown in Table 13.2.

For the 47 oncology NMEs (New Molecular Entities) approved by FDA between 1999 and 2013, the majority (30 compounds) originated from target-based drug discovery, seven originated from MIPDD and 10 originated from phenotypic screens. If the group of kinase inhibitors (21 compounds, highlighted in bold) is excluded, a higher number of drugs were discovered by mixed or phenotypic screening approaches.

Phenotypic assays have the advantage of identifying drug leads and clinical candidates that are more likely to possess therapeutically relevant molecular mechanisms of action (MMOAs) and clinical efficacy (Moffat et al. 2014; Swinney 2013). However, target-based approaches have also been prominent in the past two decades, particularly those directed against oncogenic kinases (Hoelder et al. 2012; Zhang et al. 2009) and have resulted in a new generation of anticancer agents with fewer side effects and impressive results in clinical trials.

13.6.2.4 Screens for Novel Drug Candidates for Alzheimer's and Parkinson's Disease

Neurodegenerative disorders such as Alzheimer's disease (AD) and Parkinson's disease (PD) cause serious public health problems due to the exponential increase in incidence of the diseases with age. The currently approved drugs for treatment only provide symptomatic relief to mild AD patients and do not stop progression of the disease. Therefore novel drug candidates are of great interest, and besides the chemical synthesis of compounds, natural compounds are again becoming of interest, including those of marine origin.

The difficulty in this area of drug research is the lack of validated therapeutic targets. In the absence of a target or in complex mechanisms like the human brain, a drug screening method is almost impossible to set up. An alternative may come from phenotypic chemical screens using a whole organism. The mouse *Mus musculus* is the most commonly used animal model but the costs limit its use in large-scale therapeutic screening. Instead, small invertebrate models like *Caenorhabditis elegans* or *Drosophila melanogaster* are suitable models combining genetic amenability, low cost, and culture conditions used in large-scale screenings (Giacomotto and Ségalat 2010). The invertebrate models bridge the gap between traditional high-throughput screenings and the validation in mammalian models. They allow for identification of new active compounds, targets, or molecular mechanisms, which can be further used in traditional screening assays.

Using live animals it is possible to screen bioactive compounds that are able to induce a certain phenotype (e.g., paralysis or fluorescence) or reverse an abnormal phenotype (e.g., ß-amyloid deposits, also known as plaques) to the wild-type phenotype (Arya et al. 2010). ß-amyloid and α-synuclein are aggregation-prone proteins typically associated with AD and PD, where the misfolding and accumulation of these proteins lead to neuronal cell death (Marsh and Blurton-Jones 2012). In order to discover novel drugs against AD and PD, marine-derived extracts and compounds are screened for their effect against ß-amyloid and α-synuclein toxicity, with the proteins being transgenetically expressed in different strains of *C. elegans* in a high-throughput system (Sealife Pharma, Austria, personal communication). The transgenic strains display different phenotypes allowing for direct detection of effect of the tested compound. For example, toxicity of soluble oligomers can be measured by a phenotypic readout because upshift of temperature leads to expression of ß-amyloid in muscle cells of the worms, which in turn get paralyzed. Alternatively, expression of toxic ß-amyloid or α-synuclein can be measured using a plate reader, because the proteins are coupled to fluorescent proteins like GFP or YFP. The ability of unknown compounds to prevent plaque building can be assessed because plaques can be stained with the fluorescent dye Thioflavin T and visualized using a fluorescence microscope. Fluorescence can also be used to detect expression of the reporter protein GFP in dopaminergic neurons. By using the neurotoxic molecule MPP + (1-methyl-4-phenylpyridinium), which leads to neuronal cell death as it occurs in PD it is therefore possible to detect compounds, which are able to reverse neuronal cell death. Positive controls, which are compounds protective against ß-amyloid toxicity, include coffee extract (Dostal et al. 2010), thioflavin T (Alavez et al. 2011), and reserpine (Arya et al. 2009), which are all known to protect *C. elegans* from ß-amyloid peptide toxicity.

13.7 Chemical Dereplication, Compound Purification, and Structure Elucidation

When interesting lead activities have been detected in the bioassays described above, a key challenge is to isolate and identify the compound(s) responsible for the activity. As for the screening process, several strategies are possible, depending on the starting material and to which level the compound needs to be identified.

An example of a strategy to identify bioactive fractions followed by identification of the bioactive compound is the Explorative Solid-Phase (E-SPE) strategy, where microbial extracts are loaded into columns whose stationary phases display different functionalities; the eluted fractions are tested in a bioassay and results are organized in a bioactivity matrix. The pattern of the matrix gives indications about size and functional groups of the bioactive compounds, accelerating both the dereplication and isolation process, as demonstrated for the extracts from *Pseudoalteromonas luteoviolacea* and *Penicillium roqueforti* (Mansson et al. 2010).

A key step to avoid rediscovery of already known compounds is dereplication. The identification of already known compounds occurs frequently, even in microorganisms not belonging to the same species (Egan et al. 2001; Ginolhac et al. 2005; Jin et al. 2006; Ziemert et al. 2014). Dereplication relies on analytical methods like LC-UV, LC-MS, LC-MS/MS, and LC-NMR, compound databases and, in recent years, metabolomics, molecular networking, and genome mining (El-Elimat et al. 2013; Helfrich et al. 2014; Tawfike et al. 2013; Vynne et al. 2012; Yang et al. 2013). Experimental data like UV profiles or fragmentation patterns are searched against databases such as PubChem, ChemSpider, AntiBase, or the Dictionary of Natural Products (170,000 entries as of July 2015). With increased focus on the marine environment as a reservoir of natural compounds, a number of databases containing only compounds of marine origin have become available, e.g., the Dictionary of Marine Natural Products (http://dmnp.chemnetbase.com) and MarinLit (http://pubs.rsc.org/marinlit/). However, research groups often developed their own dereplication strategies, databases, and tools, see for example Klitgaard et al. (2014), Macintyre et al. (2014), and Kildgaard et al. (2014).

When a desired bioactivity is observed in an extract or a fraction thereof, the active compound should be purified in order to proceed to structure elucidation. As in the case of extraction of crude extracts, it is important to develop an isolation protocol that considers the physical and chemical nature of the compound, particularly its lipophilic and hydrophilic characters (Ebada et al. 2008). Qualitative tests like thin-layer chromatography (TLC) can be performed to gather information about its polarity, charge, size, solubility, and acid–base properties, and such tests are able to indicate the most suitable chromatographic technique.

For low to medium polarity compounds, column chromatography (CC, normal or reverse phase) or high performance liquid chromatography (HPLC) is often preferred to produce a pure compound, whereas for high polarity compounds, reverse phase CC with elution in H_2O/MeOH followed by another round of CC with a hydrophobic matrix, size-exclusion chromatography or HPLC is preferred (Ebada et al. 2008).

Once a pure compound has been produced, the compound is most often subjected to another round of bioactivity assaying to confirm its bioactivity, before continuing to structure elucidation. Structure elucidation can be carried out by either stereochemistry methods or by spectroscopic methods (Ebada et al. 2008). The latter includes mass spectrometry (MS) and nuclear magnetic resonance spectroscopy (NMR), which is one of the most commonly used and versatile techniques for elucidation of the structure of organic compounds (Fuloria and Fuloria 2013; Kwan and Huang 2008). One of the possible approaches in natural products discovery from culturing of the microbial producer to structure elucidation by NMR of the bioactive compound is exemplified in Fig. 13.4.

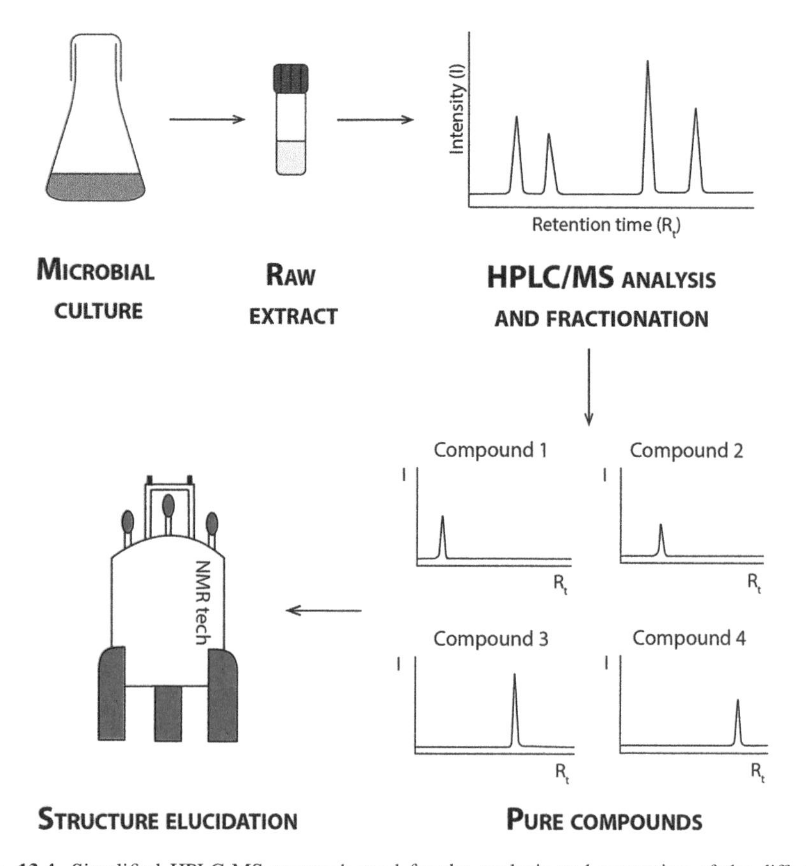

Fig. 13.4 Simplified HPLC-MS approach used for the analysis and separation of the different compounds present in the crude extract of a bacterial culture. The pure compounds can be used in bioassays and undergo several analyses for structure elucidation

13.8 Concluding Remarks

Bioprospecting for natural products is in rapid development, especially within marine microbiology. The discovery of novel bioactive compounds continues to pose innovative and (bio)technological challenges. Recent advances in microbial cultivation techniques, genomics, molecular biology, and tools for chemical analyses and dereplication, mean that an ever expanding and diverse toolbox is becoming available for bioprospecting. These new tools, in combination with biological assays and the genetic analysis of organisms, means that marine bioprospecting is entering a new era. A key question in bioprospecting is "where to search?" for novel bioactive molecules and the marine environment is, due to its chemical and physical uniqueness, a promising source of novel chemistry. Whilst marine eukaryotic macro-organisms were the first to be analyzed and provided us with several compounds with therapeutic properties, we are now realizing that many of these metabolites are actually of bacterial origin and, hence, the marine microbial world is reemerging as a promising source for bioprospecting.

Acknowledgments The research leading to these results has received funding from the European Union Seventh Framework Programme (FP7/2007–2013) under grant agreement no. 311975. This publication reflects the views only of the author, and the European Union cannot be held responsible for any use which may be made of the information contained therein (CS, FC, and MST). SG was funded by an Early Stage Researchers Grant, "BacTory", from the People Programme (Marie Curie Actions) No. 317058 of the European Union's Seventh Framework Programme FP7-People-2012-ITN.

References

Aakvik T, Degnes KF, Dahlsrud R et al (2009) A plasmid RK2-based broad-host-range cloning vector useful for transfer of metagenomic libraries to a variety of bacterial species. FEMS Microbiol Lett 296:149–158. doi:10.1111/j.1574-6968.2009.01639.x

Adrio J, Demain A (2014) Microbial enzymes: tools for biotechnological processes. Biomolecules 4:117–139. doi:10.3390/biom4010117

Alavez S, Vantipalli MC, Zucker DJS et al (2011) Amyloid-binding compounds maintain protein homeostasis during ageing and extend lifespan. Nature 472:226–229. doi:10.1038/nature09873

Andries K, Dewindt B, Snoeks J et al (1992) In vitro activity of pirodavir (R 77975), a substituted phenoxy- pyridazinamine with broad-spectrum antipicornaviral activity. Antimicrob Agents Chemother 36:100–107. doi:10.1128/AAC.36.1.100

Arya U, Dwivedi H, Subramaniam JR (2009) Reserpine ameliorates Ab toxicity in the Alzheimer's disease model in *Caenorhabditis elegans*. Exp Gerontol 44:462–466. doi:10.1016/j.exger.2009.02.010

Arya U, Das CK, Subramaniam JR (2010) *Caenorhabditis elegans* for preclinical drug discovery. Curr Sci 99:1669–1680

Ayuso-Sacido A, Genilloud O (2005) New PCR primers for the screening of NRPS and PKS-I systems in actinomycetes: detection and distribution of these biosynthetic gene sequences in major taxonomic groups. Microb Ecol 49:10–24. doi:10.1007/s00248-004-0249-6

Bachrach HL, Callis JJ, Hess WR, Patty RE (1957) A plaque assay for foot-and-mouth disease virus and kinetics of virus reproduction. Virology 4:224–236. doi:10.1016/0042-6822(57)90060-0

Barone R, De Santi C, Palma Esposito F et al (2014) Marine metagenomics, a valuable tool for enzymes and bioactive compounds discovery. Front Mar Sci 1:1–6. doi:10.3389/fmars.2014.00038

Bem AE, Velikova N, Pellicer MT et al (2015) Bacterial histidine kinases as novel antibacterial drug targets. ACS Chem Biol 10:213–224. doi:10.1021/cb5007135

Bérdy J (2005) Bioactive microbial metabolites. J Antibiot (Tokyo) 58:1–26. doi:10.1038/ja.2005.1

Bode HB, Bethe B, Hoefs R, Zeeck A (2002) Big effects from small changes: possible ways to explore nature's chemical diversity. ChemBioChem 3:619–627

Bowman JP (2007) Bioactive compound synthetic capacity and ecological significance of marine bacterial genus *Pseudoalteromonas*. Mar Drugs 5:220–241

Boyd MR, Paull KD (1995) Some practical consideration and applications of the National Cancer Institute in vitro anticancer drug discovery screen. Drug Dev Res 34:91–109

Brinkhoff T, Bach G, Heidorn T et al (2004) Antibiotic production by a *Roseobacter* clade-affiliated species from the German Wadden Sea and its antagonistic effects on indigenous isolates. Appl Environ Microbiol 70:2560–2565. doi:10.1128/AEM.70.4.2560-2565.2003

Bruhn JB, Gram L, Belas R (2007) Production of antibacterial compounds and biofilm formation by *Roseobacter* species are influenced by culture conditions. Appl Environ Microbiol 73:442–450. doi:10.1128/AEM.02238-06

Burja AM, Banaigs B, Abou-Mansour E et al (2001) Marine cyanobacteria—a prolific source of natural products. Tetrahedron 57:9347–9377. doi:10.1016/S0040-4020(01)00931-0

Button DK, Schut F, Quang P et al (1993) Viability and isolation of marine bacteria by dilution culture: theory, procedures, and initial results. Appl Environ Microbiol 59:881–891

Calteau A, Fewer DP, Latifi A et al (2014) Phylum-wide comparative genomics unravel the diversity of secondary metabolism in Cyanobacteria. BMC Genom 15:977. doi:10.1186/1471-2164-15-977

Carson CF, Mee BJ, Riley TV (2002) Mechanism of action of *Melaleuca alternifolia* (tea tree) oil on *Staphylococcus aureus* determined by time-kill, lysis, leakage, and salt tolerance assays and electron microscopy. Antimicrob Agents Chemother 46:1914–1920. doi:10.1128/AAC.46.6.1914

Cha C, Gao P, Chen Y et al (1998) Production of acyl-homoserine lactone quorum-sensing signals by Gram-negative plant-associated bacteria. Mol Plant Microbe Interact 11:1119–1129

Chellaram C, Anand TP, Shanthini CF et al (2012) Bioactive peptides from epibiotic *Pseudoalteromonas* strain P1. APCBEE Procedia 2:37–42. doi:10.1016/j.apcbee.2012.06.008

Cimermancic P, Medema MH, Claesen J et al (2014) Insights into secondary metabolism from a global analysis of prokaryotic biosynthetic gene clusters. Cell 158:412–421. doi:10.1016/j.cell.2014.06.034

Clardy J, Fischbach MA, Walsh CT (2006) New antibiotics from bacterial natural products. Nat Biotechnol 24:1541–1550. doi:10.1038/nbt1266

Cleveland J, Montville TJ, Nes IF, Chikindas ML (2001) Bacteriocins: safe, natural antimicrobials for food preservation. Int J Food Microbiol 71:1–20. doi:10.1016/S0168-1605(01)00560-8

Connon SA, Giovannoni SJ (2002) High-throughput methods for culturing microorganisms in very-low-nutrient media yield diverse new marine isolates. Appl Environ Microbiol 68:3878–3885. doi:10.1128/AEM.68.8.3878

Cooper PD (1961) The plaque assay of animal viruses. Adv Virus Res 8:319–378

Cotsonas King A, Wu L (2009) Macromolecular synthesis and membrane pertubation assay for mechanisms of action studies of antimicrobial agents. Curr Protoc Pharmacol 47:13A.7.1–13A.7.23

Cude WN, Mooney J, Tavanaei AA et al (2012) Production of the antimicrobial secondary metabolite indigoidine contributes to competitive surface colonization by the marine roseobacter *Phaeobacter* sp. strain Y4I. Appl Environ Microbiol 78:4771–4780. doi:10.1128/AEM.00297-12

Cuevas C, Pérez M, Martín MJ et al (2000) Synthesis of ecteinascidin ET-743 and phthalascidin Pt-650 from cyanosafracin B. Org Lett 2:2545–2548. doi:10.1021/ol0062502

D'Onofrio A, Crawford JM, Stewart EJ et al (2010) Siderophores from neighboring organisms promote the growth of uncultured bacteria. Chem Biol 17:254–264. doi:10.1016/j.chembiol.2010.02.010

Davidson SK, Allen SW, Lim GE et al (2001) Evidence for the biosynthesis of bryostatins by the bacterial symbiont *Candidatus* Endobugula sertula of the bryozoan *Bugula neritina*. Appl Environ Microbiol 67:4531–4537. doi:10.1128/AEM.67.10.4531-4537.2001

de Jong A, van Hijum SAFT, Bijlsma JJE et al (2006) BAGEL: a web-based bacteriocin genome mining tool. Nucleic Acids Res 34:W273–W279. doi:10.1093/nar/gkl237

DeGroot DE, Franks DG, Higa T et al (2015) Naturally occurring marine brominated indoles are aryl hydrocarbon receptor ligands/agonists. Chem Res Toxicol 28:1176–1185. doi:10.1021/acs.chemrestox.5b00003

Denny FWJ (1995) The clinical impact of human respiratory virus infections. Am J Respir Crit Care Med 152:4–12

DeVita VT, Chu E (2008) A history of cancer chemotherapy. Cancer Res 68:8643–8653. doi:10.1158/0008-5472.CAN-07-6611

Dostal V, Roberts CM, Link CD (2010) Genetic mechanisms of coffee extract protection in a *Caenorhabditis elegans* model of b-amyloid peptide toxicity. Genetics 186:857–866. doi:10.1534/genetics.110.120436

Ebada SS, Edrada RA, Lin W, Proksch P (2008) Methods for isolation, purification and structural elucidation of bioactive secondary metabolites from marine invertebrates. Nat Protoc 3:1820–1831. doi:10.1038/nprot.2008.182

Egan S, Wiener P, Kallifidas D, Wellington EMH (2001) Phylogeny of *Streptomyces* species and evidence for horizontal transfer of entire and partial antibiotic gene clusters. Antonie Van Leeuwenhoek 79:127–133. doi:10.1023/A:1010296220929

Eggert US (2013) The why and how of phenotypic small-molecule screens. Nat Chem Biol 9:206–209. doi:10.1038/nchembio.1206

Ehrenreich IM, Waterbury JB, Webb EA (2005) Distribution and diversity of natural product genes in marine and freshwater cyanobacterial cultures and genomes. Appl Environ Microbiol 71:7401–7413. doi:10.1128/AEM.71.11.7401

El-Elimat T, Figueroa M, Ehrmann BM et al (2013) High-resolution MS, MS/MS, and UV database of fungal secondary metabolites as a dereplication protocol for bioactive natural products. J Nat Prod 76:1709–1716. doi:10.1021/np4004307

Eustáquio AS, Nam SJ, Penn K et al (2011) The discovery of salinosporamide K from the marine bacterium *Salinispora pacifica* by genome mining gives insight into pathway evolution. ChemBioChem 12:61–64. doi:10.1002/cbic.201000564

Fenical W, Jensen PR (2006) Developing a new resource for drug discovery: marine actinomycete bacteria. Nat Chem Biol 2:666–673. doi:10.1038/nchembio841

Fu CY, Tang MC, Peng C et al (2009) Biosynthesis of 3-hydroxy-5-methyl-O-methyltyrosine in the saframycin/safracin biosynthetic pathway. J Microbiol Biotechnol 19:439–446. doi:10.4014/jmb.0808.484

Fu J, Bian X, Hu S et al (2012) Full-length RecE enhances linear-linear homologous recombination and facilitates direct cloning for bioprospecting. Nat Biotechnol 30:440–446. doi:10.1038/nbt.2183

Fuloria NK, Fuloria S (2013) Analytical & bioanalytical structural elucidation of small organic molecules by 1D, 2D and multi dimensional-solution NMR spectroscopy. J Anal Bioanal Tech S11:1–8. doi:10.4172/2155-9872.S1

Gerwick WH, Moore BS (2012) Lessons from the past and charting the future of marine natural products drug discovery and chemical biology. Chem Biol 19:85–98. doi:10.1016/j.chembiol.2011.12.014

Gerwick WH, Proteau PJ, Nagle DG et al (1994) Structure of curacin A, a novel antimitotic, antiproliferative and brine shrimp toxic natural product from the marine cyanobacterium *Lyngbya majuscula*. J Org Chem 59:1243–1245. doi:10.1021/jo00085a006

Giacomotto J, Ségalat L (2010) High-throughput screening and small animal models, where are we? Br J Pharmacol 160:204–216. doi:10.1111/j.1476-5381.2010.00725.x
Ginolhac A, Jarrin C, Robe P et al (2005) Type I polyketide synthases may have evolved through horizontal gene transfer. J Mol Evol 60:716–725. doi:10.1007/s00239-004-0161-1
Gonzales R, Steiner JF, Sande MA (1997) Antibiotic prescribing for adults with cold, upper respiratory tract infections, and ronchitis by ambulatory care physicians. J Am Med Assoc 278:901–904
Gonzalez ME, Alarcon B, Carrasco L (1987) Polysaccharides as antiviral agents: antiviral activity of carrageenan. Antimicrob Agents Chemother 31:1388–1393. doi:10.1128/AAC.31.9.1388
Gram L (2015) Silent clusters—speak up! Microb Biotechnol 8:13–14. doi:10.1111/1751-7915. 12181
Gram L, Melchiorsen J, Bruhn JB (2010) Antibacterial activity of marine culturable bacteria collected from a global sampling of ocean surface waters and surface swabs of marine organisms. Mar Biotechnol (NY) 12:439–451. doi:10.1007/s10126-009-9233-y
Greve JM, Davis G, Meyer AM et al (1989) The major human rhinovirus receptor is ICAM-1. Cell 56:839–847. doi:10.1016/0092-8674(89)90688-0
Helfrich EJ, Reiter S, Piel J (2014) Recent advances in genome-based polyketide discovery. Curr Opin Biotechnol 29:107–115. doi:10.1016/j.copbio.2014.03.004
Hiep C, McDonough (2012) Marine biodiversity: a science roadmap for Europe. Marine Board Future Science Brief 1, European Marine Board, Ostend, Belgium. ISBN 978-2-918428-75-6
Higginbotham S, Wong WR, Linington RG et al (2014) Sloth hair as a novel source of fungi with potent anti-parasitic, anti-cancer and anti-bacterial bioactivity. PLoS ONE. doi:10.1371/ journal.pone.0084549
Hoelder S, Clarke PA, Workman P (2012) Discovery of small molecule cancer drugs: successes, challenges and opportunities. Mol Oncol 6:155–176. doi:10.1016/j.molonc.2012.02.004
Hofer F, Gruenberger M, Kowalski H et al (1994) Members of the low density lipoprotein receptor family mediate cell entry of a minor-group common cold virus. Proc Nat Acad Sci USA 91:1839–1842. doi:10.1073/pnas.91.5.1839
Hoffmeister D, Keller NP (2007) Natural products of filamentous fungi: enzymes, genes, and their regulation. Nat Prod Rep 24:393–416. doi:10.1039/B603084J
Holbeck SL (2004) Update on NCI in vitro drug screen utilities. Eur J Cancer 40:785–793. doi:10. 1016/j.ejca.2003.11.022
Hosaka T, Ohnishi-Kameyama M, Muramatsu H et al (2009) Antibacterial discovery in actinomycetes strains with mutations in RNA polymerase or ribosomal protein S12. Nat Biotechnol 27:462–464. doi:10.1038/nbt.1538
Izallalen M, Lavesque RC, Perret X et al (2002) Broad-host-range mobilizable suicide vectors for promoter trapping in gram-negative bacteria. Biotechniques 33:1038–1043. doi:10.2144/ 000113156
Jensen PR, Chavarria KL, Fenical W et al (2014) Challenges and triumphs to genomics-based natural product discovery. J Ind Microbiol Biotechnol 41:203–209. doi:10.1007/s10295-013-1353-8
Jin M, Fischbach MA, Clardy J (2006) A biosynthetic gene cluster for the acetyl-CoA carboxylase inhibitor andrimid. J Am Chem Soc 128:10660–10661. doi:10.1021/ja063194c
Johnston MD, Hanlon GW, Denyer SP, Lambert RJW (2003) Membrane damage to bacteria caused by single and combined biocides. J Appl Microbiol 94:1015–1023. doi:10.1046/j.1365-2672.2003.01923.x
Kapuscinski J (1979) DAPI: a DNA-specific fluorescent probe. Biotech Histochem 70:220–233. doi:10.3109/10520299509108199
Kell DB (2013) Finding novel pharmaceuticals in the systems biology era using multiple effective drug targets, phenotypic screening and knowledge of transporters: where drug discovery went wrong and how to fix it. FEBS J 280:5957–5980. doi:10.1111/febs.12268
Keller NP, Turner G, Bennett JW (2005) Fungal secondary metabolism—from biochemistry to genomics. Nat Rev Microbiol 3:937–947. doi:10.1038/nrmicro1286

Khan ST, Komaki H, Motohashi K et al (2011) *Streptomyces* associated with a marine sponge *Haliclona* sp.; biosynthetic genes for secondary metabolites and products. Environ Microbiol 13:391–403. doi:10.1111/j.1462-2920.2010.02337.x

Kildgaard S, Mansson M, Dosen I et al (2014) Accurate dereplication of bioactive secondary metabolites from marine-derived fungi by UHPLC-DAD-QTOFMS and a MS/HRMS library. Mar Drugs 12:3681–3705. doi:10.3390/md12063681

Klitgaard A, Iversen A, Andersen MR et al (2014) Aggressive dereplication using UHPLC–DAD–QTOF: screening extracts for up to 3000 fungal secondary metabolites. Anal Bioanal Chem 406:1933–1943. doi:10.1007/s00216-013-7582-x

Kogure K, Simidu U, Taga N (1979) A tentative direct microscopic method for counting living marine bacteria. Can J Microbiol 25:415–420. doi:10.1139/m79-063

Kondo T, Strayer CA, Kulkarni RD et al (1993) Circadian rhythms in prokaryotes: luciferase as a reporter of circadian gene expression in cyanobacteria. Proc Nat Acad Sci USA 90:5672–5676. doi:10.1073/pnas.90.12.5672

Kotz J (2012) Phenotypic screening, take two. Sci Exch 5:1–3. doi:10.1038/scibx.2012.380

Kwan EE, Huang SG (2008). Structural elucidation with NMR spectroscopy: practical strategies for organic chemists. Eur J Org Chem 2671–2688. doi:10.1002/ejoc.200700966

Lale R, Brautaset T, Valla S (2011) Broad-host-range plasmid vectors for gene expression in bacteria. In: Williams JA (ed) Strain engineering:methods and protocols. Humana Press, Totowa, pp 327–343

Lane AL, Moore BS (2012) A sea of biosynthesis: marine natural products meet the molecular age. Nat Prod Rep 28:411–428. doi:10.1039/c0np90032j.A

Lechner A, Eustáquio AS, Gulder TAM et al (2011) Selective overproduction of the proteasome inhibitor salinosporamide A via precursor pathway regulation. Chem Biol 18:1527–1536. doi:10.1016/j.chembiol.2011.10.014

Ling LL, Schneider T, Peoples AJ et al (2015) A new antibiotic kills pathogens without detectable resistance. Nature 517:455–459. doi:10.1038/nature14098

Loeschcke A, Markert A, Wilhelm S et al (2013) TREX: a universal tool for the transfer and expression of biosynthetic pathways in bacteria. ACS Synth Biol 2:22–33. doi:10.1021/sb3000657

Lombó F, Velasco A, Castro A et al (2006) Deciphering the biosynthesis pathway of the antitumor thiocoraline from a marine actinomycete and its expression in two *Streptomyces* species. ChemBioChem 7:366–376. doi:10.1002/cbic.200500325

Lozupone CA, Knight R (2007) Global patterns in bacterial diversity. Proc Nat Acad Sci 104:11436–11440. doi:10.1073/pnas.0611525104

Luesch H, Moore RE, Paul VJ et al (2001) Isolation of dolastatin 10 from the marine cyanobacterium *Symploca* species VP642 and total stereochemistry and biological evaluation of its analogue symplostatin 1. J Nat Prod 64:907–910. doi:10.1021/np010049y

Lynch MD, Gill RT (2006) Broad host range vectors for stable genomic library construction. Biotechnol Bioeng 94:151–158. doi:10.1002/bit

Machado H, Sonnenschein EC, Melchiorsen J, Gram L (2015) Genome mining reveals unlocked bioactive potential of marine gram-negative bacteria. BMC Genom 16:158. doi:10.1186/s12864-015-1365-z

Macintyre L, Zhang T, Viegelmann C et al (2014) Metabolomic tools for secondary metabolite discovery from marine microbial symbionts. Mar Drugs 12:3416–3448. doi:10.3390/md12063416

Manivasagan P, Venkatesan J, Sivakumar K, Kim SK (2014) Pharmaceutically active secondary metabolites of marine actinobacteria. Microbiol Res 169:262–278. doi:10.1016/j.micres.2013.07.014

Mansson M, Vynne NG, Klitgaard A et al (in prep) Integrated metabolomic and genomic mining of the biosynthetic potential of bacteria

Mansson M, Phipps RK, Gram L et al (2010) Explorative solid-phase extraction (E-SPE) for accelerated microbial natural product discovery, dereplication, and purification. J Nat Prod 73:1126–1132. doi:10.1021/np100151y

Mansson M, Gram L, Larsen TO (2011) Production of bioactive secondary metabolites by marine Vibrionaceae. Mar Drugs 9:1440–1468. doi:10.3390/md9091440
Marmann A, Aly AH, Lin W et al (2014) Co-cultivation - A powerful emerging tool for enhancing the chemical diversity of microorganisms. Mar Drugs 12:1043–1065. doi:10.3390/md12021043
Marsh SE, Blurton-Jones M (2012) Examining the mechanisms that link β-amyloid and α-synuclein pathologies. Alzheimers Res Ther. doi:10.1186/alzrt109
Medema MH, Blin K, Cimermancic P et al (2011) antiSMASH: rapid identification, annotation and analysis of secondary metabolite biosynthesis gene clusters in bacterial and fungal genome sequences. Nucleic Acids Res 39:W339–W346. doi:10.1093/nar/gkr466
Moffat JG, Rudolph J, Bailey D (2014) Phenotypic screening in cancer drug discovery—past, present and future. Nat Rev Drug Discov 13:588–602. doi:10.1038/nrd4366
Molinski TF, Dalisay DS, Lievens SL, Saludes JP (2009) Drug development from marine natural products. Nat Rev Drug Discov 8:69–85. doi:10.1038/nrd2487
Nayar AS, Dougherty TJ, Ferguson KE et al (2015) Novel antibacterial targets and compounds revealed by a high throughput cell wall reporter assay. J Bacteriol 197:1726–1734. doi:10.1128/JB.02552-14
Nichols D, Cahoon N, Trakhtenberg EM et al (2010) Use of Ichip for high-throughput in situ cultivation of "uncultivable" microbial species. Appl Environ Microbiol 76:2445–2450. doi:10.1128/AEM.01754-09
Nicolaus B, Kambourova M, Oner ET (2010) Exopolysaccharides from extremophiles: from fundamentals to biotechnology. Environ Technol 31:1145–1158. doi:10.1080/09593330903552094
Nielsen A, Mansson M, Wietz M et al (2012) Nigribactin, a novel siderophore from *Vibrio nigripulchritudo*, modulates *Staphylococcus aureus* virulence gene expression. Mar Drugs 10:2584–2595. doi:10.3390/md10112584
Nielsen A, Mansson M, Bojer MS et al (2014) Solonamide B inhibits quorum sensing and reduces *Staphylococcus aureus* mediated killing of human neutrophils. PLoS ONE 9:e84992. doi:10.1371/journal.pone.0084992
Nonejuie P, Burkart M, Pogliano K, Pogliano J (2013) Bacterial cytological profiling rapidly identifies the cellular pathways targeted by antibacterial molecules. Proc Nat Acad Sci 110:16169–16174. doi:10.1073/pnas.1311066110
O'Connor M, Peifer M, Bender W (1989) Construction of large DNA segments in *Escherichia coli*. Science 244:1307–1312. doi:10.1111/j.1464-410X.1987.tb09132.x
O'Neill AJ, Miller K, Oliva B, Chopra I (2004) Comparison of assays for detection of agents causing membrane damage in *Staphylococcus aureus*. J Antimicrob Chemother 54:1127–1129. doi:10.1093/jac/dkh476
Ongley SE, Bian X, Neilan BA, Müller R (2013) Recent advances in the heterologous expression of microbial natural product biosynthetic pathways. Nat Prod Rep 30:1121. doi:10.1039/c3np70034h
Pérez M, Schleissner C, Rodríguez P et al (2009) PM070747, a new cytotoxic angucyclinone from the marine-derived *Saccharopolyspora taberi* PEM-06-F23-019B. J Antibiot (Tokyo) 62:167–169. doi:10.1038/ja.2008.27
Pettit GR, Kamano Y, Herald CL et al (1993) Isolation of dolastatins 10–15 from the marine mollusc *Dolabella auricularia*. Tetrahedron 49:9151–9170. doi:10.1016/0040-4020(93)80003-C
Piel J, Hui D, Wen G et al (2004) Antitumor polyketide biosynthesis by an uncultivated bacterial symbiont of the marine sponge *Theonella swinhoei*. Proc Nat Acad Sci 101:16222–16227. doi:10.1073/pnas.0405976101
Pogliano J, Osborne N, Sharp MD et al (1999) A vital stain for studying membrane dynamics in bacteria: a novel mechanism controlling septation during *Bacillus subtilis* sporulation. Mol Microbiol 31:1149–1159. doi:10.1046/j.1365-2958.1999.01255.x
Proksch P, Edrada RA, Ebel R (2002) Drugs from the seas—current status and microbiological implications. Appl Microbiol Biotechnol 59:125–134. doi:10.1007/s00253-002-1006-8

Ramm W, Schatton W, Wagner-Döbler I et al (2004) Diglucosyl-glycerolipids from the marine sponge-associated *Bacillus pumilus* strain AAS3: their production, enzymatic modification and properties. Appl Microbiol Biotechnol 64:497–504. doi:10.1007/s00253-003-1471-8
Rasmussen TB, Givskov M (2006) Quorum sensing inhibitors: a bargain of effects. Microbiology 152:895–904. doi:10.1099/mic.0.28601-0
Rath C, Janto B, Earl J et al (2011) Meta-omic characterization of the marine invertebrate microbial consortium that produces the chemotherapeutic natural product ET-743. ACS Chem Biol 6:1244–1256
Rediers H, Rainey PB, Vanderleyden J, De Mot R (2005) Unraveling the secret lives of bacteria: use of *in vivo* expression technology and differential fluorescence induction promoter traps as tools for exploring niche-specific gene expression. Microbiol Mol Biol Rev 69:217–261. doi:10.1128/MMBR.69.2.217
Reen F, Gutiérrez-Barranquero J, Dobson A et al (2015) Emerging concepts promising new horizons for marine biodiscovery and synthetic biology. Mar Drugs 13:2924–2954
Romero F, Espliego F, Pérez Baz J et al (1997) Thiocoraline, a new depsipeptide with antitumor activity produced by a marine *Micromonospora*. I. Taxonomy, fermentation, isolation, and biological activities. J Antibiot (Tokyo) 50:734–737
Ross AC, Gulland LES, Dorrestein PC, Moore BS (2015) Targeted capture and heterologous expression of the *Pseudoalteromonas* alterochromide gene cluster in *Escherichia coli* represents a promising natural product exploratory platform. ACS Synth Biol 4:414–420. doi:10.1021/sb500280q
Sams-Dodd F (2005) Target-based drug discovery: is something wrong? Drug Discov Today 10:139–147. doi:10.1016/S1359-6446(04)03316-1
Sarkar S, Saha M, Roy D et al (2008) Enhanced production of antimicrobial compounds by three salt-tolerant actinobacterial strains isolated from the Sundarbans in a niche-mimic bioreactor. Mar Biotechnol 10:518–526. doi:10.1007/s10126-008-9090-0
Schäberle TF, Goralski E, Neu E et al (2010) Marine myxobacteria as a source of antibiotics—comparison of physiology, polyketide-type genes and antibiotic production of three new isolates of *Enhygromyxa salina*. Mar Drugs 8:2466–2479. doi:10.3390/md8092466
Schatz A, Bugie E, Waksman SA (1944) Streptomycin, a substance ehibiting antibiotic activity against gram-negative bacteria. Exp Biol Med 55:66–69
Schiewe HJ, Zeeck A (1999) Cineromycins, gamma-butyrolactones and ansamycins by analysis of the secondary metabolite pattern created by a single strain of *Streptomyces*. J Antibiot (Tokyo) 52:635–642
Schleissner C, Pérez M, Losada A et al (2011) Antitumor actinopyranones produced by *Streptomyces albus* POR-04-15-053 isolated from a marine sediment. J Nat Prod 74:1590–1596. doi:10.1021/np200196j
Schmid A, Dordick JS, Hauer B et al (2001) Industrial biocatalysis today and tomorrow. Nature 409:258–268. doi:10.1038/35051736
Schmidtke M, Schnittler U, Jahn B et al (2001) A rapid assay for evaluation of antiviral activity against coxsackie virus B3, influenza virus A, and herpes simplex virus type 1. J Virol Methods 95:133–143
Schmitz JE, Daniel A, Collin M et al (2008) Rapid DNA library construction for functional genomic and metagenomic screening. Appl Environ Microbiol 74:1649–1652. doi:10.1128/AEM.01864-07
Seyedsayamdost MR (2014) High-throughput platform for the discovery of elicitors of silent bacterial gene clusters. Proc Nat Acad Sci USA 111:7266–7271. doi:10.1073/pnas.1400019111
Sharma SV, Haber DA, Settleman J (2010) Cell line-based platforms to evaluate the therapeutic efficacy of candidate anticancer agents. Nat Rev Cancer 10:241–253. doi:10.1038/nrc2820
Shizuya H, Kouros-Mehr H (2001) The development and applications of the bacterial artificial chromosome cloning system. Keio J Med 50:26–30. doi:10.2302/kjm.50.26
Simon C, Daniel R (2011) Metagenomic analyses: past and future trends. Appl Environ Microbiol 77:1153–1161. doi:10.1128/AEM.02345-10

Stevens DC, Conway KR, Pearce N et al (2013) Alternative sigma factor over-expression enables heterologous expression of a type II polyketide biosynthetic pathway in *Escherichia coli*. PLoS ONE 8:e64858. doi:10.1371/journal.pone.0064858

Stevenson CS, Capper EA, Roshak AK et al (2002) Scytonemin—a marine natural product inhibitor of kinases key in hyperproliferative inflammatory diseases. Inflamm Res 51:112–114. doi:10.1007/BF02684014

Stewart EJ (2012) Growing unculturable bacteria. J Bacteriol 194:4151–4160. doi:10.1128/JB.00345-12

Sun D, Cohen S, Mani N et al (2002) A pathway-specific cell based screening system to detect bacterial cell wall inhibitors. J Antibiot (Tokyo) 55:279–287

Sutherland IW (2001) Microbial polysaccharides from Gram-negative bacteria. Int Dairy J 11:663–674. doi:10.1016/S0958-6946(01)00112-1

Swinney DC (2013) Phenotypic vs. target-based drug discovery for first-in-class medicines. Clin Pharmacol Ther 93:299–301. doi:10.1038/clpt.2012.236

Swinney DC, Anthony J (2011) How were new medicines discovered? Nat Rev Drug Discov 10:507–519. doi:10.1038/nrd3480

Tan LT (2007) Bioactive natural products from marine cyanobacteria for drug discovery. Phytochemistry 68:954–979. doi:10.1016/j.phytochem.2007.01.012

Tawfike AF, Viegelmann C, Edrada-Ebel R (2013) Metabolomics and dereplication strategies in natural products. In: Roessner U, Dias DA (eds) Metabolomics tools for natural product discovery. Humana Press, Totowa, pp 227–244

Thaker MN, Wang W, Spanogiannopoulos P et al (2013) Identifying producers of antibacterial compounds by screening for antibiotic resistance. Nat Biotechnol 31:922–927. doi:10.1038/nbt.2685

Traxler MF, Watrous JD, Alexandrov T et al (2013) Interspecies interactions stimulate diversification of the *Streptomyces coelicolor* secreted metabolome. MBio 4:e00459–13–e00459–13. doi:10.1128/mBio.00459-13

Tsukimoto M, Nagaoka M, Shishido Y et al (2011) Bacterial production of the tunicate-derived antitumor cyclic depsipeptide didemnin B. J Nat Prod 74:2329–2331. doi:10.1021/np200543z

Urban A, Eckermann S, Fast B et al (2007) Novel whole-cell antibiotic biosensors for compound discovery. Appl Environ Microbiol 73:6436–6443. doi:10.1128/AEM.00586-07

van Pée KH (1996) Biosynthesis of halogenated methanes. Ann Rev Microbiol 375–399

Velasco A, Acebo P, Gomez A et al (2005) Molecular characterization of the safracin biosynthetic pathway from *Pseudomonas fluorescens* A2-2: designing new cytotoxic compounds. Mol Microbiol 56:144–154. doi:10.1111/j.1365-2958.2004.04433.x

Vichai V, Kirtikara K (2006) Sulforhodamine B colorimetric assay for cytotoxicity screening. Nat Protoc 1:1112–1116. doi:10.1038/nprot.2006.179

Villa FA, Lieske K, Gerwick L (2010) Selective MyD88-dependent pathway inhibition by the cyanobacterial natural product malyngamide F acetate. Eur J Pharmacol 629:140–146. doi:10.1021/ac901991x

Vynne NG, Mansson M, Nielsen KF, Gram L (2011) Bioactivity, chemical profiling, and 16S rRNA-based phylogeny of *Pseudoalteromonas* strains collected on a global research cruise. Mar Biotechnol 13:1062–1073. doi:10.1007/s10126-011-9369-4

Vynne NG, Mansson M, Gram L (2012) Gene sequence based clustering assists in dereplication of *Pseudoalteromonas luteoviolacea* strains with identical inhibitory activity and antibiotic production. Mar Drugs 10:1729–1740. doi:10.3390/md10081729

Wagenaar MM (2008) Pre-fractionated microbial samples—the second generation natural products library at Wyeth. Molecules 13:1406–1426. doi:10.3390/molecules13061406

Wang G, Hosaka T, Ochi K (2008) Dramatic activation of antibiotic production in *Streptomyces coelicolor* by cumulative drug resistance mutations. Appl Environ Microbiol 74:2834–2840. doi:10.1128/AEM.02800-07

Watve M, Tickoo R, Jog M, Bhole B (2001) How many antibiotics are produced by the genus *Streptomyces*? Arch Microbiol 176:386–390. doi:10.1007/s002030100345

Weber T (2014) *In silico* tools for the analysis of antibiotic biosynthetic pathways. Int J Med Microbiol 304:230–235. doi:10.1016/j.ijmm.2014.02.001

Wenzel SC, Gross F, Zhang Y et al (2005) Heterologous expression of a myxobacterial natural products assembly line in pseudomonads via Red/ET recombineering. Chem Biol 12:349–356. doi:10.1016/j.chembiol.2004.12.012

Whalen KE, Poulson-Ellestad KL, Deering RW et al (2015) Enhancement of antibiotic activity against multidrug-resistant bacteria by the efflux pump inhibitor 3,4-dibromopyrrole-2,5-dione isolated from a *Pseudoalteromonas* sp. J Nat Prod 78:402–412. doi:10.1021/np500775e

Wietz M, Mansson M, Gotfredsen CH et al (2010) Antibacterial compounds from marine *Vibrionaceae* isolated on a global expedition. Mar Drugs 8:2946–2960. doi:10.3390/md8122946

Wietz M, Mansson M, Gram L (2011) Chitin stimulates production of the antibiotic andrimid in a *Vibrio coralliilyticus* strain. Environ Microbiol Rep 3:559–564. doi:10.1111/j.1758-2229.2011.00259.x

Williamson LL, Borlee BR, Schloss PD et al (2005) Intracellular screen to identify metagenomic clones that induce or inhibit a quorum-sensing biosensor. Appl Environ Microbiol 71:6335–6344. doi:10.1128/AEM.71.10.6335-6344.2005

Wilson MC, Piel J (2013) Metagenomic approaches for exploiting uncultivated bacteria as a resource for novel biosynthetic enzymology. Chem Biol 20:636–647. doi:10.1016/j.chembiol.2013.04.011

Wilson MC, Mori T, Rückert C et al (2014) An environmental bacterial taxon with a large and distinct metabolic repertoire. Nature 506:58–62. doi:10.1038/nature12959

Wong WR, Oliver AG, Linington RG (2012) Development of antibiotic activity profile screening for the classification and discovery of natural product antibiotics. Chem Biol 19:1483–1495. doi:10.1016/j.chembiol.2012.09.014

Woodhouse JN, Fan L, Brown MV et al (2013) Deep sequencing of non-ribosomal peptide synthetases and polyketide synthases from the microbiomes of Australian marine sponges. ISME J 7:1842–1851. doi:10.1038/ismej.2013.65

World Health Organization (2014) Antimicrobial resistance: global report on surveillance

Xu Y, Kersten RD, Nam SJ et al (2012) Bacterial biosynthesis and maturation of the didemnin anti-cancer agents. J Am Chem Soc 134:8625–8632. doi:10.1021/ja301735a

Yamanaka K, Reynolds KA, Kersten RD et al (2014) Direct cloning and refactoring of a silent lipopeptide biosynthetic gene cluster yields the antibiotic taromycin A. Proc Nat Acad Sci 111:1957–1962. doi:10.1073/pnas.1319584111

Yan L, Boyd KG, Grant Burgess J (2002) Surface attachment induced production of antimicrobial compounds by marine epiphytic bacteria using modified roller bottle cultivation. Mar Biotechnol 4:356–366. doi:10.1007/s10126-002-0041-x

Yan L, Boyd KG, Adams DR, Burgess JG (2003) Biofilm-specific cross-species induction of antimicrobial compounds in bacilli. Appl Environ Microbiol 69:3719–3727. doi:10.1128/AEM.69.7.3719

Yang JY, Sanchez LM, Rath CM et al (2013) Molecular networking as a dereplication strategy. J Nat Prod 76:1686–1699. doi:10.1021/np400413s

Yin J, Straight PD, Hrvatin S et al (2007) Genome-wide high-throughput mining of natural-product biosynthetic gene clusters by phage display. Chem Biol 14:303–312. doi:10.1016/j.chembiol.2007.01.006

Zhang J, Yang PL, Gray NS (2009) Targeting cancer with small molecule kinase inhibitors. Nat Rev Cancer 9:28–39. doi:10.1038/nrc2559

Zhang Y, Huang H, Xu S et al (2015) Activation and enhancement of Fredericamycin A production in deepsea-derived *Streptomyces somaliensis* SCSIO ZH66 by using ribosome engineering and response surface methodology. Microb Cell Fact 14:64. doi:10.1186/s12934-015-0244-2

Zheng W, Thorne N, McKew JC (2013) Phenotypic screens as a renewed approach for drug discovery. Drug Discov Today 18:1067–1073. doi:10.1016/j.drudis.2013.07.001

Ziemert N, Podell S, Penn K et al (2012) The natural product domain seeker NaPDoS: a phylogeny based bioinformatic tool to classify secondary metabolite gene diversity. PLoS ONE 7:e34064. doi:10.1371/journal.pone.0034064

Ziemert N, Lechner A, Wietz M et al (2014) Diversity and evolution of secondary metabolism in the marine actinomycete genus *Salinispora*. Proc Nat Acad Sci 111:E1130–E1139. doi:10.1073/pnas.1324161111

Chapter 14
Metagenomics as a Tool for Biodiscovery and Enhanced Production of Marine Bioactives

F. Jerry Reen, Alan D.W. Dobson and Fergal O'Gara

Abstract The application of metagenomics technologies to the area of marine bioprospecting and biodiscovery has seen a major advance in our capacity to harness the bioactive potential of the ocean, not least when we consider the limitations surrounding culturability of microorganisms from this and other ecosystems. Combining genomics, bioinformatics, and systems biology, metagenomics has provided new levels of access to the rich tapestry of novel bioactivities from the marine microbiome. Notwithstanding this early promise, considerable limitations to the technology exist that currently prevent us from harnessing the full potential of marine microbial natural products. The continued growth in the number and diversity of metagenomic studies, allied with the advances in next generation sequencing platforms, has brought with it a global appreciation of the challenges that need to be addressed to ensure future developments in this applied research area. In this chapter we present the application of metagenomics for biodiscovery, discussing the potential value of this technology, and the current limitations preventing its full realization. Already, advances in bioinformatics, robotics, molecular cloning and expression, DNA sequencing and isolation, as well as the continued development of improved chemical profiling systems, have led to the discovery of new natural products and bioactivities. Successful implementation of further improvements that circumvent current bottlenecks will open new horizons for medical and industrial developments.

F. Jerry Reen · F. O'Gara (✉)
BIOMERIT Research Centre, School of Microbiology, University College Cork, National University of Ireland, Cork, Ireland
e-mail: f.ogara@ucc.ie

A.D.W. Dobson
School of Microbiology, University College Cork, National University of Ireland, Cork, Ireland

F. O'Gara
School of Biomedical Sciences, CHIRI, Curtin University, Perth, WA 6102, Australia

L.J. Stal and M.S. Cretoiu (eds.), *The Marine Microbiome*,
DOI 10.1007/978-3-319-33000-6_14

14.1 Introduction

The ocean has proven a vast reservoir of resources for human consumption and utilization for millennia. A natural and diverse cross-kingdom ecosystem, the marine environment is host to producers of a rich tapestry of compounds and molecules with huge therapeutic potential. The exponential advances in new technologies and engineering capacity have opened up the marine ecosystem to scientific exploration. As a result, new sources of the next generation of therapeutics continue to emerge. Improved discovery and mining of marine bioactivities has fed into the development of new and innovative solutions across a broad spectrum of areas including anti-inflammatories, antibiotics, anticoagulants, and anti-infectives (Reen et al. 2015a, b). The importance of continued advances in the isolation and characterization of these activities is highlighted by the emergence of resistance mechanisms to most conventional therapies.

Successful mining of the marine has already resulted in a range of natural products being discovered, some of which have made it to market. The source of bioactive material in the oceans has been diverse reflecting the natural biodiversity of what is essentially an unexplored niche. Algae and other higher order marine life have received considerable attention, with associated microbial communities gaining prominence. In fact, a significant number of natural products that have been isolated and developed from the marine, originally attributed to sponge or other eukaryotic origin, have since been shown to be produced by bacteria. Bioactive compounds from these bacterial sources have ranged from anticoagulants to anti-cancer and more recently to the next generation of antimicrobial compounds. Marine biodiscovery programs are also delivering new industrial enzymes with improved bioactivity, enantio-selectivity, and substrate specificity, leading the movement towards green chemistry.

Therefore, having established the potential of the marine ecosystem to deliver societal advances and solutions, the challenge remains to optimize our capacity to identify and mine them. Notwithstanding current successes, a growing concern is the degree to which the repeated discovery of known compounds has hampered attempts to reach the true diversity of the marine, and indeed other ecosystems. Despite advances in de-replication technologies, many screening platforms have simply rediscovered known compounds, to the extent that the number of new natural products from biological sources has diminished rapidly over the last decade. For this reason, the capacity to mine new and novel activities from the marine has been the focus of a concerted research and technological effort.

One approach has been to improve the cultivability and culturability of marine organisms (Vartoukian et al. 2010). These new initiatives are designed to enhance our capacity to produce the novel compounds that are urgently needed. Since the earliest days of marine biodiscovery, there has been an awareness of the limitations surrounding isolation of bacterial and fungal organisms from the ocean. This has mirrored a similar rate of discovery from soil and other ecosystems where transferring microbes from their natural environment onto artificial and synthetic media has not proven fruitful. Whether pelagic or benthic in origin, successful culturing of these organisms has met with limited success, and advances in this sphere have not been forthcoming. Despite efforts to supplement media with nutrients and other components, culturability of as little as 1 % has been reported (Epstein 2013). In many cases, irrespective of the marine source, similar species tend to dominate marine culture collections, with the result that the natural biodiversity is not being reached. New insights into isolation, maintenance, cell-cell communication, storage, data management, and robotics have led to the implementation of new protocols for the isolation of marine organisms. The next stage of this development is the harvesting of bioactive compounds from these new ‘growers’.

An alternative approach has been to adopt culture independent approaches, translating the genetic blueprint into active compounds. This area of applied research, most commonly referred to as metagenomics, has seen an exponential increase in activity since its broad inception, and has delivered new insights into the microbial and viral populations that inhabit the marine ecosystem (Fernandez et al. 2013; Kennedy et al. 2008, 2014; Mizuno et al. 2013; Mueller et al. 2015; Woyke et al. 2009). Derived from *meta*, the Greek word for ‘transcendent’ and genomic, the complete genetic code of an organism, metagenomics circumvents the need to culture organisms, thus circumventing a primary roadblock in harvesting their bioactive potential. Already delivering some notable successes, the ability to access the rich diversity of the marine ecosystem, independent of culture bias, has opened up exciting new potential for biodiscovery (Barone et al. 2014; Kennedy et al. 2010; Reen et al. 2015a; Rocha-Martin et al. 2014). However, important challenges remain before we can fully exploit the power of this emerging discipline of metagenomics. Issues surrounding access to DNA, its manipulation and expression (especially in the context of silent or cryptic gene clusters) remain to be resolved, while coupling with the undoubted power of informatics analysis needs continuous fine-tuning. Indeed, perhaps one of the greatest challenges will be the integration of molecular, chemical, and informatics technologies to maximize our efficiency in this field. Advances in the individual sciences have provided considerable improvements in the technologies available to identify and isolate new compounds and bioactivities from a broad spectrum of environmental ecosystems. Together, however, these technologies can be far more powerful, and there has been considerable interest in the integration of chemical and molecular technologies with the added power and discrimination offered by informatics. This systems based approach has the greatest potential for unlocking the new classes of bioactives, providing novel frameworks for the drug and enzyme design of the future.

14.2 Mining the not yet Cultured Microorganisms: Metagenomics

Combining genomics, bioinformatics and systems biology, metagenomics has already provided considerable advances in our understanding of microbial biodiversity as well as providing access to a rich tapestry of novel bioactivities from bacteria which cannot be cultured using traditional methods. Although as diverse as the ecosystems they seek to investigate, all metagenomics programs follow a classical path of DNA isolation, cloning into stable vectors and subsequent screening either by sequence or function (Fig. 14.1). Metagenomics can be broadly assigned into categories on the basis of the screening strategies employed (a) sequence-based (mass genome or next generation), (b) activity-driven screens (c) sequence-driven studies that link genome information with phylogenetic or functional marker genes of interest. Metagenomics technologies have advanced several research disciplines, including ecology, medicine, and environmental sciences, and the isolation of natural products from previously unattainable sources has proven a

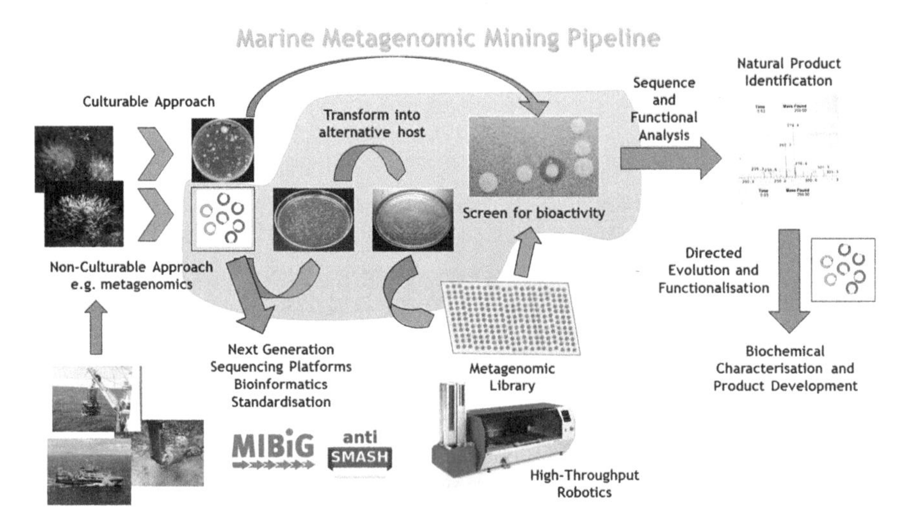

Fig. 14.1 The Marine Metagenomics Mining Pipeline. Described as the systems biology of the biosphere, metagenomics harnesses the genomic information from an environmental sample in a context that allows its expression and the capture of its bioactive potential. Isolation of metaDNA from the marine sample is followed by restriction or shearing to facilitate large insert cloning into broad-host range plasmids. Subsequent transformation is usually directed towards *Escherichia coli*, although recent advances in vector design have seen the emergence of new heterologous expression systems such as *Streptomyces* and *Pseudomonas*. High throughput libraries are generated using robotics, providing large ordered clone collections for further analysis. Finally, the detection of 'hits' can be achieved either through functional screening or sequencing and bioinformatics screening of the libraries. This results in the isolation of clones carrying the genetic information encoding bioactive potential, which can then be further characterized by natural product chemistry and functionalized by directed evolution

major step forward in harnessing the natural potential of the global microbiome. Notwithstanding this promise, limitations to the technology exist that need to be addressed in the short term. The continued growth in the number and diversity of metagenomics studies, allied with the advances in sequencing technologies, have brought with them a global appreciation of the challenges that now bottleneck future developments in this research area.

The explosion in the availability of genomic information has brought with it new insights into the diversity of bioactivities that were previously hidden or out of reach. Armed with the genetic blueprint for multiple locations within the marine and indeed other ecosystems, we can begin to infer biological function linked to changes in signature sequences, domains or motifs. Combining genomics, bioinformatics, and systems biology, the discipline of metagenomics has already provided considerable advances in our understanding of microbial biodiversity as well as providing access to a rich tapestry of novel bioactivities from bacteria which cannot be cultured using traditional methods. Metagenomics technologies have advanced several research disciplines, including ecology, medicine, and environmental sciences, and the isolation of natural products from previously unattainable sources has proven a major step forward in harnessing the natural potential of the global microbiome.

From a functional perspective, metagenomic libraries allied with innovative chemistry and state of the art robotics, provide an efficient screening mechanism to identify and isolate rich bioactives from the marine and other environments. Access to the culturable and non-culturable resources has enhanced our capacity to mine the ocean for novel and effective antimicrobials, anticancer compounds, enzymes and biocatalysts. In addition, the study of metagenomes has greatly advanced our understanding of marine ecology, and particularly the dynamics that underpin community structure in this rich aquatic environment. Beginning with the discovery of the Sargasso Sea's unprecedented diversity of Bacteria and functional genes, the abundance of metagenomes currently available has since led to the identification of habitat specific fingerprints. Comparative analysis of these metagenome datasets has uncovered correlations between functional metagenome diversity and specialized conditions of environmental niches, a feature that is beginning to inform the functional screens that seek to isolate bioactivities from the marine and other ecosystems.

Metagenomics has provided a platform for detailed biologically driven analysis of niche-specific signatures. This becomes particularly relevant where conservation of peptide-motifs underpins biological activity and biotechnological potential. Recent evidence of niche specialization and peptide motif conservation among Type Six lipase effectors in metagenome sequences underpins the need for future structural and bioactivity studies that focus on peptide domains (Egan et al. 2014). It is feasible to suggest that this motif specialization will extend to include the biosynthetic gene clusters that encode the vast majority of bioactive compounds isolated to date. Motif specialization would be expected to manifest in the functionality of these bioactives and uncovering the extent of such specialization will inform future marine mining programs.

However, the continued growth in the number and diversity of metagenomic studies, allied with the advances in sequencing technologies have brought with them a global appreciation of the challenges that now bottleneck future developments in this research area.

14.3 Bottlenecks and Solutions in Marine Metagenomics

Metagenomics is essentially a DNA-based technology that requires access to the genetic material of living organisms, its subsequent molecular manipulation, the heterologous expression of bioactive target genes, and the production of bioactive target compounds in an active and measurable form. There are several stages to the process of metagenomics, each of which presents its own challenges (summarized in Table 14.1).

Table 14.1 Metagenomic bottlenecks and challenges

Current limitations	Bottlenecks	Technical advances
1. Challenges in sampling—capturing the natural diversity and not just the domineer	– DNA size and yield	– Standardize DNA isolation, storage, sampling, and pre-enrichment
2. Limitations of metagenomics technologies—short reads and dominant species	– Insufficient clones	– Remove redundant DNA. – Pre-enrichment/selective capture of bioactive producer. – New expression systems and cloning technologies to ensure sufficient clone numbers in metagenomics libraries – Avoid loss of rare unique hits and misinterpretation of community data
3. Expression of eDNA in heterologous hosts, although possible, is fraught with limitations	– Inefficient transformation	– Heterologous expression systems: minimal interference with host biological context. – Develop new techniques to isolate, clone and stably express eDNA
4. Few standardized systems/technologies, with accurate taxonomy hampered by limitations in systematics	– Homology clutter	– Novel automated genome mining tools and pattern recognition based algorithms: large datasets – Innovative mining approaches to replace classical labor intensive screens, which deliver poor coverage with many false positives – Standardized systems and technologies to harmonize global analyses

(continued)

Table 14.1 (continued)

Current limitations	Bottlenecks	Technical advances
5. Classical screening has not delivered the full potential of the natural environment, while many gene clusters are 'silent' or regulated by insulated transcriptional regulators	– Time consuming rare clones	– New technologies for the isolation and identification of BGCs from rare species of sponge associated bacteria – Tapping into the silent potential of microbial genomes – Novel screening approaches: e.g. search for new 'antimicrobial resistance blockers'

The first stage of the metagenomics process is in many ways the most crucial. Optimization and developments downstream of this step become redundant where primary sampling is ineffective. Furthermore, bias introduced at this stage will be difficult to filter out, particularly where the genetic information from low abundant, yet lucrative, organisms is lost.

14.3.1 Deciding When and Where to Sample

Where isolation of novel bioactive natural products is the ultimate goal, the source of the metagenomic DNA is a central consideration. In order to improve the abundance of novel bioactive natural products, a pre-enrichment step might be considered.

Post-sampling, optimization of DNA extraction technologies to encompass different classes of organisms is a key consideration. The size of many biosynthetic gene clusters means that low molecular weight DNA has severely limited usefulness, particularly where bioprospecting for natural products is the ultimate goal.

14.3.2 The Ability to Isolate High Quality, High Molecular Weight DNA from Samples

The extraction of high quality and high molecular weight DNA has proven a major bottleneck to complete coverage of library construction and associated functional screens. The diversity of organisms present, the extent to which they will yield their DNA using conventional or adapted isolation protocols, the differences in abundance between the dominant potentially uninteresting species, and the rare potentially lucrative organisms, all present major headaches that need to be overcome.

Having isolated and prepared the genetic information from the environmental sample, the next step is to express that code in a meaningful way. Early studies made use of the laboratory workhorse *E. coli*, modified with plasmids to facilitate expression of foreign DNA where rare codons are frequently encountered.

14.3.3 Heterologous Expression of Environmental DNA (eDNA) in Diverse Hosts

Although considerable advances have been made in the construction and manipulation of broad-host range expression plasmids (Craig et al. 2010; Kakirde et al. 2011), marine and extreme environmental niches present a unique challenge *vis-à-vis* the novelty of the coding systems employed in their genetic blueprints. Even if this is overcome, expression in heterologous hosts can be problematic, with, e.g., the inability to fold or process the natural product. Homologous expression in the natural host, while favorable, is generally not possible, with even the culturable marine organisms proving difficult to transform. Some advances have been made with marine *Streptomyces* and *Pseudomonas*, with other organisms currently under development as metagenomic hosts. Unresolved difficulties in expressing large fragments of DNA must also be considered, bearing in mind the size of biosynthetic gene clusters such as those encoding polyketide synthases (PKS) and nonribosomal peptide synthetases (NRPSs). The gene clusters encoding the corresponding pharmaceutically relevant natural products generally exceed the insert size limitations of fosmid or even BAC vectors, making it often necessary to isolate multiple clones or perform genome walking in order to obtain the complete gene cluster. As a result, heterologous expression of modular PKS or nonribosomal peptide synthetase (NRPS) pathways can usually only be attempted after initial identification using homology-based screens.

The increased application of metagenomics and other biodiscovery technologies to the search for novel bioactives has brought with it an exponential increase in the discovery and reporting of new natural products. This has necessitated a focus on standardization, of methodologies, but more importantly also of data reporting and storage. In tandem with this, sophisticated software programs and virtual workbenches have been created to decipher the trenches of genomic data that currently exist in the public domain.

14.3.4 The Size and Complexity of Current Metagenomic Datasets

This presents an additional challenge to researchers with computational advances now urgently required to meet the explosion in available data. Automated genome mining tools and eventually pattern recognition-based algorithms are required to deal with the large datasets emerging from these studies. This is crucial in overcoming the oversampling of abundant organisms with loss of information from the lower abundant species.

Even where successful expression of rare and potentially bioactive genetic information is achieved, limitations in classical screening approaches hamper the translation into detectable natural products.

14.3.5 Limitations to Classical Screening Approaches

Many screens are based on narrow spectrum bioactivity, utilizing a single heterologous host, under defined experimental conditions that may not be favorable to the isolated activity. This often leads to rediscovery of known molecules, or at the very least, the isolation of compounds for which the framework molecule already exists. In other cases, the activities are missed using classical screening approaches. For example, mining enzymes from deep ocean organisms, while screening using temperate atmospheric conditions, may not be conducive to detecting the full bioactive repertoire from these organisms. These are some of the limitations. So, where will the solutions come from?

14.3.6 Advances in Next Generation Sequencing and Software Platforms

There is an urgent need for sequencing platforms that provide greater depth, with parallel developments in technologies that remove redundant DNA. The advent of next generation sequencing platforms such as MiSeq, HiSeq, and Genome Analyzer II platforms (Illumina), SOLiD system (Life Technologies/Applied Biosystems), Ion Torrent and Ion Proton (Life Technologies), and the PacBio RS II (Pacific Biosciences), has driven the explosion in sequencing. With shorter reads (excepting the PacBio RS II) and cheaper costs, these high throughput platforms have replaced the longer high fidelity reads of the Sanger sequencing method that were traditionally applied. In spite of the advances arising from next generation sequencing platforms, the short nature of the reads presents its own difficulties, particularly when we consider the large size of many biosynthetic gene clusters (BGCs). These clusters encode many important activities that are clinically effective such as antimicrobial, anti-inflammatory, and antitumor drugs. Typically arranged as modules, these BGCs can be quite large in size (possibly as large as 100 kb), a feature that mitigates against their expression and co-localization in functional metagenomic analyses. Here, high molecular weight DNA isolation techniques become important to cope with present and future long read sequencing approaches in order to directly obtain gene clusters from metagenomes without any need for further assemblies.

Of course the availability of large sequence datasets presents its own opportunities for data mining. The available predictive platforms (e.g., AntiSMASH, BAGEL3, and SMURF) have equipped the research community with the tools for identifying known classes of BGCs (Khaldi et al. 2010; van Heel et al. 2013; Weber et al. 2015), while further advances will be needed to expand the search to novel and as yet uncharacterized compounds (Wooley and Ye 2009). Furthermore, standardization of bioactive gene cluster isolation methods and classification are underway, with community driven platforms such as MIxS and MIBiG providing excellent frameworks (Medema et al. 2012, 2015). This will enhance the harmonization of metagenomics exploration, providing structure and improved annotation to existing and future genomic, metagenomic, and bioactive gene cluster information.

Perhaps an obvious limitation arising from the heterogeneous nature of microbial communities and the fragment sizes that classically populate metagenomic libraries is the inability to taxonomically link bioactivities to producing organisms. With rare exceptions, biosynthetic genes are not found with taxonomic markers on the same clone, hampering any accurate assignment of origin. Efforts are ongoing to overcome this limitation using what are termed classification and binning approaches. The advent of single cell genomics may also offer a solution to this problem (Macaulay and Voet 2014). It is hoped that uncovering the genome of producing organisms may provide a blueprint on which to base culturability studies. However, it is likely that many slow growing symbiotic organisms will remain elusive in this regard.

14.3.7 Strategies for Improvements in Heterologous Expression Systems

Even if bottlenecks in sequencing and bioinformatics are overcome, the expression of eDNA in heterologous hosts, although possible, is fraught with limitations, including codon usage, rare tRNAs, promoter recognition, toxicity, yield and stability. Host selection and engineering based on phylogenetic and comparative genomic approaches may assist in selecting the most appropriate universal host for a particular environmental sample. A (meta) genome-first approach may also provide a better understanding of compatible gene regulation and resistance in order to identify and engineer universal hosts that can express genes from a wide range of sources, as well as be used directly for library construction. The extent to which heterologous expression will be achieved is predicted to vary between taxonomic groups, with an average of 40 % of enzymatic activities readily recovered by random cloning in *E. coli* (Gabor et al. 2004). Analogous to the use of different environmental conditions to elicit production of new compounds, the "different host—different hit" hypothesis has been proposed (Gabor et al. 2007). Courtois and colleagues showed how expression screening of environmental DNA cloned in an *E. coli—S. lividans* shuttle cosmid led to the identification of several clones that were active in *E. coli* but silent in *S. lividans* (Courtois et al. 2003). Another study, a parallel screening of broad-host-range cosmid environmental DNA libraries maintained in six different proteobacterial hosts (*E. coli, Pseudomonas putida, Burkholderia graminis, Caulobacter vibrioides, Ralstonia metallidurans,* and *Agrobacterium tumefaciens*), described little overlap between active clones, further demonstrating the potential advantage to using multiple diverse host species (Craig et al. 2010).

Improved vectors that can carry and express entire gene clusters stably and reliably will also be required. Some success has already been achieved in this regard (Aakvik et al. 2009; Craig et al. 2010; Kakirde et al. 2011; Terron-Gonzalez et al. 2013; Troeschel et al. 2012). The plasmid RK2-based broad-host-range cloning vector developed by Aakvik and coworkers for metagenomic library transfer between species (Aakvik et al. 2009). With utility for fosmid and bacterial artificial chromosome (BAC) cloning, pRS44 can be efficiently transferred to numerous hosts by conjugation, replicating either through the PK2 or F origins. Novel pEBP shuttle vectors have been used to express heterologous DNA in *E. coli, Bacillus subtilis* and *P. putida* (Troeschel et al. 2012), while a "Gram-negative shuttle BAC" vector (pGNS-BAC) has also been described (Kakirde et al. 2011). In another approach, viral components have been used to improve vector and host strain performance, resulting in a six-fold increase in the frequency of carbenicillin resistant clones from their metagenomic library (Terron-Gonzalez et al. 2013). These systems take advantage of the phage T7 RNA-polymerase to drive metagenomic gene expression, while using the lambda phage transcription anti-termination protein N to limit transcription termination. Together with advances in host strain improvements, it is hoped these broad-host systems will open new possibilities for marine bioactive isolation.

14.3.8 Innovations in Screening Methodologies and De-replication

The successful implementation of technological developments described above are hoped to maximize the expression and production of rare and suitably bioactive natural products. The challenge then is to ensure that these important compounds are identified and detected, a challenge that should not be understated. What use in achieving production of these compounds if we do not select them from our screens?

Metagenomic libraries have generally been screened using either function or sequence homology as the primary search parameter. While homology-based screens have benefited from the development of a suite of search algorithms and software analysis tools, they remain limited by the redundancy of the genetic code, particularly where low abundant previously uncharacterized organisms are the source of the sought after activity. Functional screens using direct substrate–product conversions as readout have been widely applied in biodiscovery programs, particularly where biocatalysts have been targeted. Ideally based on growth or no growth, alternative screens without selection have relied on visible changes in colony color, morphology or clearing (halo) using agar plating methods.

More recently, the development of more targeted and sensitive assays has greatly expanded the capacity of the research community to mine metagenomic samples for bioactivity. These have included molecular based assays, also known as gene traps, chromogenic assays, and complementation assays, all of which have led to the discovery of novel activities from a range of ecosystems. For example, screening for quorum sensing inhibitors has made use of a complementation strategy whereby the presence of a signal from the test organism or clone restores pigmentation to a biosensor strain (Chu et al. 2011). Gene-trap or promoter-trap assays have been applied to the search for enzymes such as amidases, in what have been termed as PIGEX and SIGEX approaches (Uchiyama and Miyazaki 2010). This technology is particularly suited to high throughput analysis, with chromogenic or fluorogenic markers easily integrated. Indicators such as chrome azurol S (CAS), which changes color in the presence of iron, have been used to isolate clones encoding siderophores from marine metagenomic libraries (Fujita et al. 2012). Integration with fluorescence-activated cell sorting (FACS) and microfluidic systems will further enhance the high throughput capacity of these approaches. However, further developments in this area are required,, however, to fully exploit the rich diversity of the marine and other ecosystems.

Parallel developments in de-replication technologies are important where the rare bioactive is likely to be present in low abundance among a mixture of characterized natural products. The capacity to collapse the redundancy in the screening outputs, and also within the metagenomic library itself will streamline the screening process and increase the likelihood of extracting the beneficial activity. Recognizing the importance of de-replication, a significant toolkit exists to facilitate this process in natural product research (recently reviewed by Gaudencio and Pereira 2015). Integrating advances in high throughput sequencing, chromatography, bioinformatics, structural characterization, and database management (outlined in Table 14.2), the rigorous application of this process to metagenomic screening is crucial if we are to reach the true diversity of the oceans bioactivity (Eugster et al. 2014; Gaudencio and Pereira 2015; Inokuma et al. 2013; Lang et al. 2008; Ng et al. 2009; Nielsen and Larsen 2015; Wong et al. 2012; Yang et al. 2013).

Table 14.2 De-replication approaches and associated technologies

Approach	Technology	Application
High throughput sequencing	– Robotics for strain selection – HTS Platforms	Continued advances in this area have led to increased capacity for screening of e.g. NRPSs
Analytics	Separation—e.g. TLC, GC, CE, SPE, CC, FCC, HPLC, uHPLC Detection—e.g. MS, NIMS, SIMS, DESI, DART, LAESI, NMR	Based primarily on traditional chemical and chromatographic technologies, advances e.g. nanotechnology, have enhanced the utility of these approaches for de-replication of NPs
Combinatorial analytics	GC-MS, LC-PDA, LC-MS(-MS), LC-FTIR, LC-NMR(-MS), CE-MS, LC-UV-DAD, LC-CD-NMR, FIE-MS, npMALDI-I, MALDI-TOF-MS, ICM, MSn, and ESI-TOF-MS	An ever expanding repertoire of technologies that integrate components of the discovery pipeline. A balancing act between increased ease of use, high throughput capacity, and sensitivity
Bioactivity fingerprints	Cytological profiling, BioMAP	Offering high throughput screening for, e.g., novel antibiotic activity
Computational MS	Ligand guided genome-guided	Large libraries of small molecules, natural compounds, and genomic data are already available to search using computational methods. However, limitations remain in the development of suitable search algorithms, particularly for MS fragmentation data
X-ray crystallography	Classical SCD	Previous limitations with the need for single crystals have been overcome
NMR	2D-NMR, qHNMR, DOSY, ASAP-HMQC	The latter techniques offer markedly enhanced speed of analysis compared to classical NMR
CASE	SESAMI, COCOA, INFER2D, HOUDINI, COCON, SENECA, *StrucEluc*	Reliant on the availability of structural databases, this can be a complex process. Nevertheless, still a key aspect of the de-replication toolkit
NP databases	ZINC, GNPS, PubChem, ChemSpider, NAPROC, CSLS	Becoming more accessible with the move towards open source databases
'Omics'	Genomics—PyMS, RiboPrinter, FT-IR metabolomics—LC-HR-MS, LC-MS-PCA, LC-HRFTMS, proteomics—2DE, MS	Already integrated with high throughput systems, and readily aligned with downstream bioinformatics analysis, these technologies are widely applied in NP research
In silico	QSAR, INVDOCK, MZmine, SIEVE, molecular networking	Chemi-informatics is gaining traction, with structure-activity data providing key insights

14.4 Expanding the Metagenomic Horizon

Although the metagenomics workflow described above has underpinned the discovery of many valuable chemicals, nowadays it leads too often to rediscovery of known metabolites, generating a stasis in the number of new molecules described. Therefore, in order to maximize the potential from developments in metagenomic technologies, other aspects of the discovery process require parallel developments and advances.

14.4.1 Activating the Silent Gene Clusters

The parallel advent of high throughput sequencing technologies has allowed us to unlock the physiological features of a wide array of microorganisms. These data revealed that their biosynthetic abilities have been greatly unexplored. In fact, the number of biosynthetic genes in many bacteria and fungi greatly outnumbers the known metabolites described for these organisms. In many cases, the biosynthetic genes are simply not expressed, because their activation relies on environmental cues missing in laboratory conditions. These silent genes are known as "cryptic" or "silent genes". The identification of these silent gene clusters has expanded our consideration of the biosynthetic ability of microorganisms, leading to the development of multiple approaches to stimulate the production of secondary metabolites other than the known ones.

Broadly speaking, strategies to awaken silent gene clusters can be categorized as using either (i) environmental cues and co-culturing or (ii) semi-synthetic or molecular approaches (recently reviewed by Reen et al. 2015b). The former approach, which can be described as a manipulation of culture conditions, has been used for centuries as a mechanism of improving outputs from living organisms. Encompassing the well characterized "*one strain many compounds*" (OSMAC) approach first described by Bode et al. (2002), new strategies involving the use of environmental cues or signals have received attention in recent years. The advent of molecular technologies has seen the use of genetic strategies such as ribosome engineering, transcriptional factor manipulation (activating the activators and suppressing the suppressors), the use of artificial promoters, as well as epigenetic mining (Fig. 14.2).

Metagenomics has the potential to partially address the issue of silent BGCs, in so far as placing the cluster under the control of a constitutively driven promoter, thus eliciting expression. However, as described earlier, limitations surrounding the size of the DNA fragment, and the role of internal transcriptional regulators, must also be considered.

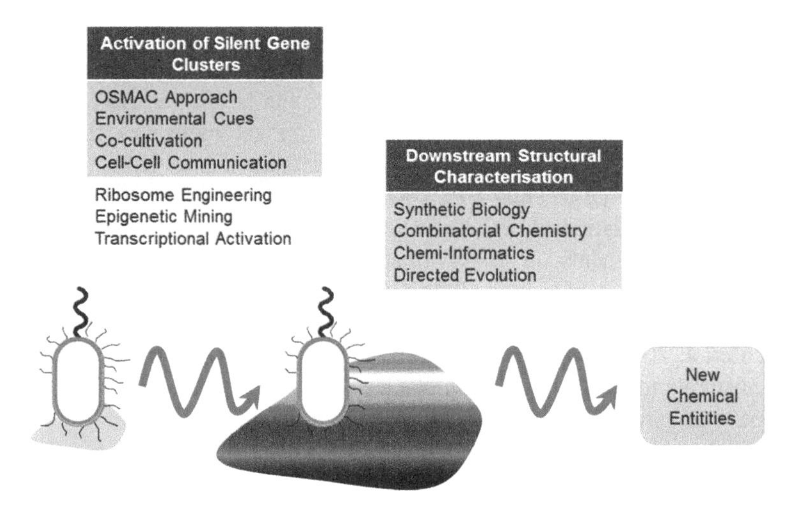

Fig. 14.2 Awakening the silent biosynthetic gene clusters. Despite the explosion on genome sequencing and the parallel developments in predictive programs, there remains a wealth of genetic information that has yet to be visualized and perhaps new paradigms for sequence analysis that have yet to be encountered. Predictive systems by their nature are predicated on prior knowledge, with platforms based on current and existing paradigms for natural product discovery. We must consider the potential for BGCs encoding previously unforeseen structures, encoded in new cluster arrangements, with enough anomalies that current predictive software cannot recognize. These unknown unknowns may be the next step in natural product discovery

14.4.2 Tandem Chemical Profiling and Structural Elucidation

Marine biodiscovery faces several challenges where the characterization of novel compound structures using chemical approaches is concerned. The chemical nature of bioactive production in the marine ecosystem can be extremely complex (de Carvalho and Fernandes 2010), making it difficult to follow standardized extraction protocols to obtain the biologically active pure compounds. The number of compounds produced by a single organism adds to the difficulty in resolving these complex samples. Broad separation of marine extracts can be resolved by successive extractions using different organic solvents (e.g., ethanol, methanol, hexane). Each fraction can then tested using appropriate biological or chemical assays linked to the bioactivity. Even when the fraction with the desired biological activity has been found, there may still be multiple compounds present in the fraction requiring additional chemical separation techniques to be applied. In this sense, chromatographic techniques play an essential role. Of these, thin-layer chromatography (TLC) and high-performance liquid chromatography (HPLC) are the most widely used techniques.

Once, the original complex sample has been fractionated, cleaned and resolved, more powerful analytical chemistry techniques are employed to elucidate the chemical structure of the bioactive compound. In this sense, the mass spectrometry (MS) and nuclear magnetic resonance (NMR) techniques offer a wide range of possibilities. Combining technologies such as MS and NMR, we can predict to an important degree the structural composition of novel molecules. In addition, combining these two techniques with the HLPC in tandem or in triplet (HPLC-MS, HPLC-MS/MS or more recently HPLC-MS-NMR), has demonstrated utility in the structural characterization of new bioactive compounds from a complex sample.

Parallel developments in bioinformatics have led to the development of MS-guided genome-mining methodologies that iteratively match de novo tandem MS (MS(n)) structures to genomic-based structures following biosynthetic logic (Kersten et al. 2011). This has enabled researchers to connect the chemotypes of peptide natural products to their biosynthetic gene clusters. In a further development, Medema et al. have introduced a new software package based on Bayesian probabilistic matching to overcome difficulties in establishing the association between tandem MS data and biosynthetic gene clusters (Medema et al. 2014). Along with other developments in the field, Pep2path is designed to optimize the pathway towards high throughput discovery of novel peptide-based natural products.

14.4.3 Bioinformatics and the Informatics Age

The development of enabling technologies, data integration and genomic standards in particular has received much attention in recent years, with bioinformatics solutions urgently required to keep pace with parallel developments in sequencing power. Recent initiatives include, the ‘Minimum Information about a Biosynthetic Gene Cluster’ (MIBiG) genomic standard, tools for the biosynthetic gene cluster (BGC) profiling in metagenomes, expanding the capabilities of the antiSMASH pipeline from genomes to metagenomes. The ‘Minimum Information about a Biosynthetic Gene Cluster’ community initiative was started in order to standardize data storage on the experimental characterization of BGCs (Medema et al. 2015). In line with the recent improvements in sequencing technologies and the development of powerful metagenomics assembly algorithms, there is an urgent unmet need to extend this work to the automated identification and analysis of BGCs in metagenomics datasets. The development of an ultra-fast BGC metagenomics profiler to screen metagenomes for BGC signatures has facilitated prioritization of samples for further sequencing and/or cloning. Furthermore, combining the profiler with a targeted assembly approach has delivered contigs and scaffolds long enough to be processed by antiSMASH (Weber et al. 2015). The antiSMASH pipeline has rapidly become the standard tool for natural product genome mining in both academia and industry, with >90,000 web server jobs processed and >75,000 software downloads. The development of bioinformatics pipelines for co-assembly and annotation of different next generation sequencing data including long reads produced by single molecule real time technology has been accompanied by parallel mobilization of metadata related to existing bacterial species via the GBIF Data

Portal (www.gbif.org) and the meta-database BacDive (bacdive.dsmz.de). The BacDive portal integrates bacterial metadata with over 400 data fields into one central database, offers easy-to-use, yet powerful search functionalities and allows users to link the annotation process to the existing information on metabolic and biosynthetic properties of related bacterial isolates. The taxonomic assignment of metagenomics fragments encoding BGCs is another important parameter for prioritizing the identified clusters/pathways related to their potential novelty. However, currently, the accurate taxonomic assignment is hampered by a number of shortcomings of microbial systematics such as the absence of a single authoritative taxonomy, inconsistencies among the different taxonomies and within them, plus incompletely annotated or incorrectly classified reference data.

14.4.4 Synthetic Biology and Metagenomics

Synthetic biology has opened new horizons for the development of improved peptide-based molecules. Comprising both enabling and core technologies, the area of synthetic biology will continue to grow and lead the genomics-based optimization of bioactive compounds for the foreseeable future (Kelley et al. 2014). The development of synthetic molecules that mimic marine products, engineering of hosts for natural product expression, and modified antimicrobial peptides are some of the pioneering advances in this exciting new era of biodiscovery (Khalil and Collins 2010; Neumann and Neumann-Staubitz 2010). Artemisinin, a botanical antimalarial compound produced by the wormwood *Artemesia annua*, and the production of taxol, were both achieved by the reconstitution of biosynthetic pathways in engineered hosts (Neumann and Neumann-Staubitz 2010; Paddon and Keasling 2014). Enhanced access to the novel sequences suggested by marine biodiversity studies has the potential to provide new starting points for synthetic development. In many cases, new natural products are effectively lead structures, which subsequently undergo synthetic development to improve activity, pharmacodynamics and pharmacokinetics, turning natural products into clinically effective drugs (Harvey et al. 2015). In addition to the modification of existing natural products and the engineering of heterologous expression hosts, the advent of 'synthetic metagenomics' has received considerable interest (Iqbal et al. 2012). This involves codon optimization of genes of interest followed by chemical synthesis, cloning, and expression in an optimized heterologous host. An important addition to the biodiscovery toolkit, the global market for synthetic biology is predicted to reach ~$11 Bn by 2016 (Kelley et al. 2014), with many challenges and ethical issues still to be resolved (Bloch and Tardieu-Guigues 2014). Foremost among those challenges will be the design and implementation of compatible broad-host vector systems and chassis organisms to expand the translation of marine genetic biodiversity into functionally novel bioactivities.

14.5 Future Trends in the Application of Metagenomics

The need for advanced solutions to societal problems is a real and current issue facing the research community. The threat of a pending 'post antibiotic era', the urgent need for greener chemistries, are all tangible problems which metagenomics can begin to address. Maximizing the potential of metagenomics to deliver on these challenges will require the integration of new technologies and complementary approaches that can underpin and enhance the quality of information mined from what are increasingly complex datasets. The emerging metaproteomics and meta-transcriptomics technologies present valuable insights into the functional and metabolic capacity of mixed microbial communities, being particularly useful in assessing the biocatalytic potential of these consortia. As the platforms supporting meta-'omics' continue to develop, new questions may be asked and previously unforeseen applications can be considered.

14.5.1 Molecular Therapeutics

The success of metagenomics approaches has coincided with the call for more targeted and personalized therapeutics, as we begin to understand at the molecular level, the pathophysiology of disease. Metagenomics has the potential to unlock a new repertoire of innovative therapeutics to challenge the continued emergence of antibiotic resistant organisms, while also delivering on the isolation and development of classical sterilizing antibiotics.

14.5.2 Cell-Cell Communication

Quorum sensing (QS) is the term used to describe coordinated behavior of bacterial cells through signal production and perception, tightly regulated at the level of a threshold or quorum (Miller and Bassler 2001). Previously thought to be uniquely controlled at the level of cell-density, these systems are now known to be more complex and exquisitely fine-tuned. A diverse range of signaling systems have been described in Gram-negative bacteria, with the broad spectrum acyl homoserine lactones (AHLs) often complemented with species specific AI-2 based systems. QS in Gram-positive organisms is controlled through the production and perception of auto-inducing peptides (AIPs). While QS molecules have been detected in Gram-negative bacteria of marine origin (Gram et al. 2002; Mohamed et al. 2008), QS peptides from Gram-positive organisms of marine origin remain to be discovered. Metagenomics has the potential to unlock the diversity of QS producing organisms in the marine, providing a new understanding of population interactions and dynamics.

14.5.3 Next Generation Antimicrobials

The rapid spread of antimicrobial resistance across all classes of antibiotics, rendering many conventional antibiotics ineffective, has focused attention on the need for alternative strategies to infection control (Cooper and Shlaes 2011). During pathogenesis, Bacteria employ a vast array of virulence factors to overcome the host defense and establish infection. The process is highly adapted and multifactorial, often requiring the temporal and coordinated expression of genes, either in tandem or in sequence, to successfully colonize the host environment. To add to this complexity, infections are rarely unique, and occur as polymicrobial communities, with interspecies and interkingdom interactions superimposed on the bi-directional signaling between microbe and host. Together, these multiple interactomes constitute what is effectively a chorus of communication, exquisitely coordinated and highly evolved. Understanding how pathogenic bacteria use virulence factors to interact with their hosts and originate the disease is a prerequisite to define new targets for vaccines and drug development.

A key approach that is developing considerable potential is the molecular disruption of virulence behavior, and particularly biofilm formation, through the use of QS signal mimics. Biofilm formation is a multicellular behavior that is almost universal among microbes, and the distinct stages of biofilm formation are coordinated through signal dependent regulatory systems. This provides researchers with the opportunity to develop smart drugs that intercept the biofilm signal, thus locking the cells into a planktonic, antibiotic sensitive state. Marine biodiscovery is providing small molecules with the potential to deliver on this next generation approach to the clinical management of infections. Advanced signal dependent screening platforms, linked to molecular biosensors, e.g., gene traps, are underpinning the discovery of new classes of antimicrobial compound (Hirakawa and Tomita 2013). This approach has already led to the identification of a suite of signal disruptive compounds with biofilm blocking potential and chemical tractability (Gutierrez-Barranquero et al. 2015; Kjaerulff et al. 2013; Nielsen et al. 2014; Teasdale et al. 2011).

14.5.4 Biocatalysts and Green Chemistry

The identification of biocatalysts from marine metagenomics libraries reinforces the largely untapped biocatalytic potential these resources present. Enzymes such as cellulases, xylanases, and esterases have broad spectrum application in the processing of paper and plant-based products (Martin et al. 2014), while there is a need for novel biodegradative enzymes for use in bioremediation (Kennedy et al. 2011).

Synergies between the chemical and biological sciences have seen a new focus on the search for enzymes to replace toxic and expensive catalysts in the drug design process. Whether lysate or whole cell bioreactors, microbe-derived enzymes have shown considerable potential for use in drug development processes, with transaminase, nitrilases, and lipases being particularly relevant. Transaminases, can be used for the production of chiral amines and non-natural amino acids for pharmaceuticals or chiral building blocks, while nitrile compounds play an important role as intermediates for the synthesis of fine chemicals and pharmaceuticals, as with the regio-selective hydrolysis of poly- and di-nitriles (Brandao and Bull 2003; Kaul and Asano 2012; Trincone 2011). However, despite considerable advances, biocatalyst still only account for a small fraction of the chemical reactions that underpin the pharmaceutical industry. The enhanced discovery of novel and effective microbial enzymes from the marine ecosystem through metagenomics has the potential to offer real alternatives to the toxic, expensive, non-selective, multi-step catalytic reactions currently in use.

14.6 Conclusion

In a world that had apparently considered infection as a solved issue, we are fast approaching a new reality, one where drugs, which had served us well in times past, are rendered ineffective against re-emerging pathogens. The battle has now shifted in favor of the invisible menace that threatens to undermine centuries of human endeavor. If the pendulum is to swing again in favor of pathogen control, new approaches and new fields of discovery are urgently needed. Perhaps, the convergence of metagenomics and marine biodiscovery will provide that momentum, hopefully underlying the next generation of bioactive molecules for societal gain.

Acknowledgments The research leading to these results has received funding from the European Union Seventh Framework Programme (FP7/2007-2013) under grant agreement no. 311975. This publication reflects the views only of the authors, and the European Union cannot be held responsible for any use which may be made of the information contained therein. This research was further supported in part by grants awarded by the European Commission (FP7-PEOPLE-2013-ITN, 607786; FP7-KBBE-2012-6, CP-TP-312184; OCEAN 2011-2, 287589; Marie Curie 256596; EU-634486), Science Foundation Ireland (SSPC-2, 12/RC/2275; 13/TIDA/B2625; 12/TIDA/B2411; 12/TIDA/B2405; 14/TIDA/2438), the Department of Agriculture and Food (FIRM/RSF/CoFoRD; FIRM 08/RDC/629; FIRM 1/F009/MabS; FIRM 13/F/516), the Irish Research Council for Science, Engineering and Technology (PD/2011/2414; GOIPG/2014/647), the Health Research Board/Irish Thoracic Society (MRCG-2014-6), the Marine Institute (Beaufort award C2CRA 2007/082) and Teagasc (Walsh Fellowship 2013).

References

Aakvik T, Degnes KF, Dahlsrud R, Schmidt F, Dam R, Yu L, Volker U, Ellingsen TE, Valla S (2009) A plasmid RK2-based broad-host-range cloning vector useful for transfer of metagenomic libraries to a variety of bacterial species. FEMS Microbiol Lett 296(2):149–158

Barone R, De Santi C, Palma Esposito F, Tedesco P, Galati F, Visone M, DiScala A, De Pascale D (2014) Marine metagenomics, a valuable tool for enzymes and bioactive compounds discovery. Front Mar Sci 1:1–6

Bloch JF, Tardieu-Guigues E (2014) Marine biotechnologies and synthetic biology, new issues for a fair and equitable profit-sharing commercial use. Mar Genomics 17:79–83
Bode HB, Bethe B, Hofs R, Zeeck A (2002) Big effects from small changes: possible ways to explore nature's chemical diversity. Chembiochem 3 (7):619–627
Brandao PF, Bull AT (2003) Nitrile hydrolysing activities of deep-sea and terrestrial mycolate actinomycetes. Antonie Van Leeuwenhoek 84(2):89–98
Chu WH, Vattem DA, Maitin V, Barnes MB, McLean RJC (2011) Bioassays of quorum sensing compounds using *Agrobacterium tumefaciens* and *Chromobacterium violaceum*. Quorum Sens Methods Protoc 692:3–19
Cooper MA, Shlaes D (2011) Fix the antibiotics pipeline. Nature 472(7341):32
Courtois S, Cappellano CM, Ball M, Francou FX, Normand P, Helynck G, Martinez A, Kolvek SJ, Hopke J, Osburne MS, August PR, Nalin R, Guerineau M, Jeannin P, Simonet P, Pernodet JL (2003) Recombinant environmental libraries provide access to microbial diversity for drug discovery from natural products. Appl Environ Microbiol 69(1):49–55
Craig JW, Chang FY, Kim JH, Obiajulu SC, Brady SF (2010) Expanding small-molecule functional metagenomics through parallel screening of broad-host-range cosmid environmental DNA libraries in diverse proteobacteria. Appl Environ Microbiol 76(5):1633–1641
de Carvalho CC, Fernandes P (2010) Production of metabolites as bacterial responses to the marine environment. Mar Drugs 8(3):705–727
Egan F, Reen FJ, O'Gara F (2014) Tle distribution and diversity in metagenomic datasets reveal niche specialization. Environ Microbiol Rep 7(2):194–203
Epstein SS (2013) The phenomenon of microbial uncultivability. Curr Opin Microbiol 16(5): 636–642
Eugster PJ, Boccard J, Debrus B, Breant L, Wolfender JL, Martel S, Carrupt PA (2014) Retention time prediction for dereplication of natural products (CxHyOz) in LC-MS metabolite profiling. Phytochemistry 108:196–207
Fernandez AB, Ghai R, Martin-Cuadrado AB, Sanchez-Porro C, Rodriguez-Valera F, Ventosa A (2013) Metagenome sequencing of prokaryotic microbiota from two hypersaline ponds of a marine saltern in Santa Pola, Spain. Genome Announcement 1(6):e00933–13
Fujita MJ, Kimura N, Yokose H, Otsuka M (2012) Heterologous production of bisucaberin using a biosynthetic gene cluster cloned from a deep sea metagenome. Mol BioSyst 8(2):482–485
Gabor EM, Alkema WB, Janssen DB (2004) Quantifying the accessibility of the metagenome by random expression cloning techniques. Environ Microbiol 6(9):879–886
Gabor E, Liebeton K, Niehaus F, Eck J, Lorenz P (2007) Updating the metagenomics toolbox. Biotechnol J 2(2):201–206
Gaudencio SP, Pereira F (2015) Dereplication: racing to speed up the natural products discovery process. Nat Prod Rep 32(6):779–810
Gram L, Grossart HP, Schlingloff A, Kiorboe T (2002) Possible quorum sensing in marine snow bacteria: production of acylated homoserine lactones by *Roseobacter* strains isolated from marine snow. Appl Environ Microbiol 68(8):4111–4116
Gutierrez-Barranquero JA, Reen FJ, McCarthy RR, O'Gara F (2015) Deciphering the role of coumarin as a novel quorum sensing inhibitor suppressing virulence phenotypes in bacterial pathogens. Appl Microbiol Biotechnol 99(7):3303–3316
Harvey AL, Edrada-Ebel R, Quinn RJ (2015) The re-emergence of natural products for drug discovery in the genomics era. Nat Rev Drug Discov 14(2):111–129
Hirakawa H, Tomita H (2013) Interference of bacterial cell-to-cell communication: a new concept of antimicrobial chemotherapy breaks antibiotic resistance. Front Microbiol 4:114
Inokuma Y, Yoshioka S, Ariyoshi J, Arai T, Hitora Y, Takada K, Matsunaga S, Rissanen K, Fujita M (2013) X-ray analysis on the nanogram to microgram scale using porous complexes. Nature 495(7442):461–466
Iqbal HA, Feng Z, Brady SF (2012) Biocatalysts and small molecule products from metagenomic studies. Curr Opin Chem Biol 16(1–2):109–116

Kakirde KS, Wild J, Godiska R, Mead DA, Wiggins AG, Goodman RM, Szybalski W, Liles MR (2011) Gram negative shuttle BAC vector for heterologous expression of metagenomic libraries. Gene 475(2):57–62

Kaul P, Asano Y (2012) Strategies for discovery and improvement of enzyme function: state of the art and opportunities. Microb Biotechnol 5(1):18–33

Kelley NJ, Whelan DJ, Kerr E, Apel A, Beliveau R, Scanlon R (2014) Engineering biology to address global problems: synthetic biology markets, needs and applications. Ind Biotechnol 10 (3):140–148

Kennedy J, Codling CE, Jones BV, Dobson AD, Marchesi JR (2008) Diversity of microbes associated with the marine sponge, *Haliclona simulans*, isolated from Irish waters and identification of polyketide synthase genes from the sponge metagenome. Environ Microbiol 10(7):1888–1902

Kennedy J, Flemer B, Jackson SA, Lejon DP, Morrissey JP, O'Gara F, Dobson AD (2010) Marine metagenomics: new tools for the study and exploitation of marine microbial metabolism. Mar Drugs 8(3):608–628

Kennedy J, O'Leary ND, Kiran GS, Morrissey JP, O'Gara F, Selvin J, Dobson AD (2011) Functional metagenomic strategies for the discovery of novel enzymes and biosurfactants with biotechnological applications from marine ecosystems. J Appl Microbiol 111(4):787–799

Kennedy J, Flemer B, Jackson SA, Morrissey JP, O'Gara F, Dobson AD (2014) Evidence of a putative deep sea specific microbiome in marine sponges. PLoS ONE 9(3):e91092

Kersten RD, Yang YL, Xu Y, Cimermancic P, Nam SJ, Fenical W, Fischbach MA, Moore BS, Dorrestein PC (2011) A mass spectrometry-guided genome mining approach for natural product peptidogenomics. Nat Chem Biol 7(11):794–802

Khaldi N, Seifuddin FT, Turner G, Haft D, Nierman WC, Wolfe KH, Fedorova ND (2010) SMURF: Genomic mapping of fungal secondary metabolite clusters. Fungal Genet Biol 47(9):736–741

Khalil AS, Collins JJ (2010) Synthetic biology: applications come of age. Nat Rev Genet 11(5):367–379

Kjaerulff L, Nielsen A, Mansson M, Gram L, Larsen TO, Ingmer H, Gotfredsen CH (2013) Identification of four new *agr* quorum sensing-interfering cyclodepsipeptides from a marine *Photobacterium*. Mar Drugs 11(12):5051–5062

Lang G, Mayhudin NA, Mitova MI, Sun L, van der Sar S, Blunt JW, Cole ALJ, Ellis G, Laatsch H, Munro MHG (2008) Evolving trends in the dereplication of natural product extracts: new methodology for rapid, small-scale investigation of natural product extracts. J Nat Prod 71(9):1595–1599

Macaulay IC, Voet T (2014) Single cell genomics: advances and future perspectives. PLoS Genet 10(1):e1004126

Martin M, Biver S, Steels S, Barbeyron T, Jam M, Portetelle D, Michel G, Vandenbol M (2014) Identification and characterization of a halotolerant, cold-active marine endo-beta-1,4-glucanase by using functional metagenomics of seaweed-associated microbiota. Appl Environ Microbiol 80(16):4958–4967

Medema MH, van Raaphorst R, Takano E, Breitling R (2012) Computational tools for the synthetic design of biochemical pathways. Nat Rev Microbiol 10(3):191–202

Medema MH, Paalvast Y, Nguyen DD, Melnik A, Dorrestein PC, Takano E, Breitling R (2014) Pep2Path: automated mass spectrometry-guided genome mining of peptidic natural products. PLoS Comput Biol 10(9):e1003822

Medema MH, Kottmann R, Yilmaz P, Cummings M, Biggins JB, Blin K, de Bruijn I, Chooi YH, Claesen J, Coates RC, Cruz-Morales P, Duddela S, Dusterhus S, Edwards DJ, Fewer DP, Garg N, Geiger C, Gomez-Escribano JP, Greule A, Hadjithomas M, Haines AS, Helfrich EJ, Hillwig ML, Ishida K, Jones AC, Jones CS, Jungmann K, Kegler C, Kim HU, Kotter P, Krug D, Masschelein J, Melnik AV, Mantovani SM, Monroe EA, Moore M, Moss N, Nutzmann HW, Pan G, Pati A, Petras D, Reen FJ, Rosconi F, Rui Z, Tian Z, Tobias NJ, Tsunematsu Y, Wiemann P, Wyckoff E, Yan X, Yim G, Yu F, Xie Y, Aigle B, Apel AK, Balibar CJ, Balskus EP, Barona-Gomez F, Bechthold A, Bode HB, Borriss R, Brady SF,

Brakhage AA, Caffrey P, Cheng YQ, Clardy J, Cox RJ, De Mot R, Donadio S, Donia MS, van der Donk WA, Dorrestein PC, Doyle S, Driessen AJ, Ehling-Schulz M, Entian KD, Fischbach MA, Gerwick L, Gerwick WH, Gross H, Gust B, Hertweck C, Hofte M, Jensen SE, Ju J, Katz L, Kaysser L, Klassen JL, Keller NP, Kormanec J, Kuipers OP, Kuzuyama T, Kyrpides NC, Kwon HJ, Lautru S, Lavigne R, Lee CY, Linquan B, Liu X, Liu W, Luzhetskyy A, Mahmud T, Mast Y, Mendez C, Metsa-Ketela M, Micklefield J, Mitchell DA, Moore BS, Moreira LM, Muller R, Neilan BA, Nett M, Nielsen J, O'Gara F, Oikawa H, Osbourn A, Osburne MS, Ostash B, Payne SM, Pernodet JL, Petricek M, Piel J, Ploux O, Raaijmakers JM, Salas JA, Schmitt EK, Scott B, Seipke RF, Shen B, Sherman DH, Sivonen K, Smanski MJ, Sosio M, Stegmann E, Sussmuth RD, Tahlan K, Thomas CM, Tang Y, Truman AW, Viaud M, Walton JD, Walsh CT, Weber T, van Wezel GP, Wilkinson B, Willey JM, Wohlleben W, Wright GD, Ziemert N, Zhang C, Zotchev SB, Breitling R, Takano E, Glockner FO (2015) Minimum information about a biosynthetic gene cluster. Nat Chem Biol 11(9):625–631

Miller MB, Bassler BL (2001) Quorum sensing in bacteria. Annu Rev Microbiol 55:165–199

Mizuno CM, Rodriguez-Valera F, Kimes NE, Ghai R (2013) Expanding the marine virosphere using metagenomics. PLoS Genet 9(12):e1003987

Mohamed NM, Cicirelli EM, Kan J, Chen F, Fuqua C, Hill RT (2008) Diversity and quorum-sensing signal production of *Proteobacteria* associated with marine sponges. Environ Microbiol 10(1):75–86

Mueller RS, Bryson S, Kieft B, Li Z, Pett-Ridge J, Chavez F, Hettich RL, Pan C, Mayali X (2015) Metagenome sequencing of a coastal marine microbial community from monterey bay, california. Genome Announcements 3(2):e00341-15

Neumann H, Neumann-Staubitz P (2010) Synthetic biology approaches in drug discovery and pharmaceutical biotechnology. Appl Microbiol Biotechnol 87(1):75–86

Ng J, Bandeira N, Liu WT, Ghassemian M, Simmons TL, Gerwick WH, Linington R, Dorrestein PC, Pevzner PA (2009) Dereplication and de novo sequencing of nonribosomal peptides. Nat Methods 6(8):596–U565

Nielsen KF, Larsen TO (2015) The importance of mass spectrometric dereplication in fungal secondary metabolite analysis. Front Microbiol 6:71

Nielsen A, Mansson M, Bojer MS, Gram L, Larsen TO, Novick RP, Frees D, Frokiaer H, Ingmer H (2014) Solonamide B inhibits quorum sensing and reduces *Staphylococcus aureus* mediated killing of human neutrophils. PLoS ONE 9(1):e84992

Paddon CJ, Keasling JD (2014) Semi-synthetic artemisinin: a model for the use of synthetic biology in pharmaceutical development. Nat Rev Microbiol 12(5):355–367

Reen F, Romano S, Dobson A, Gara F (2015a) The sound of silence: activating silent biosynthetic gene clusters in marine microorganisms. Mar Drugs 13(8):4754–4783

Reen FJ, Gutierrez-Barranquero JA, Dobson AD, Adams C, O'Gara F (2015b) Emerging concepts promising new horizons for marine biodiscovery and synthetic biology. Mar Drugs 13(5):2924–2954

Rocha-Martin J, Harrington C, Dobson AD, O'Gara F (2014) Emerging strategies and integrated systems microbiology technologies for biodiscovery of marine bioactive compounds. Mar Drugs 12(6):3516–3559

Teasdale ME, Donovan KA, Forschner-Dancause SR, Rowley DC (2011) Gram-positive marine bacteria as a potential resource for the discovery of quorum sensing inhibitors. Mar Biotechnol 13(4):722–732

Terron-Gonzalez L, Medina C, Limon-Mortes MC, Santero E (2013) Heterologous viral expression systems in fosmid vectors increase the functional analysis potential of metagenomic libraries. Sci Rep 3:1107

Trincone A (2011) Marine biocatalysts: enzymatic features and applications. Mar Drugs 9(4): 478–499

Troeschel SC, Thies S, Link O, Real CI, Knops K, Wilhelm S, Rosenau F, Jaeger KE (2012) Novel broad host range shuttle vectors for expression in *Escherichia coli*, *Bacillus subtilis* and *Pseudomonas putida*. J Biotechnol 161(2):71–79

Uchiyama T, Miyazaki K (2010) Product-induced gene expression, a product-responsive reporter assay used to screen metagenomic libraries for enzyme-encoding genes. Appl Environ Microbiol 76(21):7029–7035

van Heel AJ, de Jong A, Montalban-Lopez M, Kok J, Kuipers OP (2013) BAGEL3: Automated identification of genes encoding bacteriocins and (non-) bactericidal posttranslationally modified peptides. Nucleic Acids Res 41:W448–W453

Vartoukian SR, Palmer RM, Wade WG (2010) Strategies for culture of 'unculturable' bacteria. FEMS Microbiol Lett 309(1):1–7

Weber T, Blin K, Duddela S, Krug D, Kim HU, Bruccoleri R, Lee SY, Fischbach MA, Muller R, Wohlleben W, Breitling R, Takano E, Medema MH (2015) antiSMASH 3.0-a comprehensive resource for the genome mining of biosynthetic gene clusters. Nucleic Acids Res 43(W1): W237–W243

Wong WR, Oliver AG, Linington RG (2012) Development of antibiotic activity profile screening for the classification and discovery of natural product antibiotics. Chem Biol 19(11): 1483–1495

Wooley JC, Ye Y (2009) Metagenomics: facts and artifacts, and computational challenges. J Comput Sci Technol 25(1):71–81

Woyke T, Xie G, Copeland A, Gonzalez JM, Han C, Kiss H, Saw JH, Senin P, Yang C, Chatterji S, Cheng JF, Eisen JA, Sieracki ME, Stepanauskas R (2009) Assembling the marine metagenome, one cell at a time. PLoS ONE 4(4):e5299

Yang JY, Sanchez LM, Rath CM, Liu XT, Boudreau PD, Bruns N, Glukhov E, Wodtke A, de Felicio R, Fenner A, Wong WR, Linington RG, Zhang LX, Debonsi HM, Gerwick WH, Dorrestein PC (2013) Molecular networking as a dereplication strategy. J Nat Prod 76(9):1686–1699

Chapter 15
New Approaches for Bringing the Uncultured into Culture

Stéphane L'Haridon, Gerard H. Markx, Colin J. Ingham, Lynn Paterson, Frederique Duthoit and Gwenaelle Le Blay

Abstract It took more than 23 years to propose a defined medium to culture "*Pelagibacter ubique*" HTCC1062, one of the most dominant clades in the ocean. Although it was first identified in the 1990s by culture-independent approaches based on rRNA gene cloning and sequencing, an artificial seawater enrichment medium has only recently been proposed for this isolate. This success story is a result of the improvement of culture methods, better sensitivity of growth detection, and knowledge of metabolic activities predicted from genome sequences. The new approaches now offer a fraction of 14–40 % that can be cultured. From an optimistic point of view, all uncultured marine microorganisms could now simply be regarded as "not yet cultured". Culturing is no longer an "old fashioned" technique but an innovative and fast-moving area of research. Technological developments include micro-engineering of ichips, manipulation of single cells, community culture, high-throughput culturing (HTC) processes, and new methods for low biomass

S. L'Haridon (✉) · F. Duthoit · G. Le Blay
Laboratoire de Microbiologie des Environnements Extrêmes (LMEE),
Place Nicolas Copernic, UEB, Institut Universitaire Européen de la Mer (IUEM)—UMR
UBO, CNRS, IFREMER 6197, Université de Brest, Brest, Plouzané, France
e-mail: stephane.lharidon@univ-brest.fr

F. Duthoit
e-mail: frederique.duthoit@univ-brest.fr

G. Le Blay
e-mail: gwenaelle.leblay@univ-brest.fr

G.H. Markx · L. Paterson
School of Engineering and Physical Sciences, Institute of Biological Chemistry,
Biophysics and Bioengineering, Heriot-Watt University, Riccarton,
Edinburgh, Scotland, UK
e-mail: G.H.Markx@hw.ac.uk

L. Paterson
e-mail: L.Paterson@hw.ac.uk

C.J. Ingham
MicroDish BV, Utrecht, Netherlands
e-mail: c.ingham@microdish.nl

L.J. Stal and M.S. Cretoiu (eds.), *The Marine Microbiome*,
DOI 10.1007/978-3-319-33000-6_15

detection or targeting specific microorganisms. Culture remains a prerequisite for microbiological studies, as we need to grow microorganisms in the laboratory in order to identify their functions and validate hypotheses deduced from their genomes. The development, improvement, and combination of innovative culture techniques based on information deduced from omics will undoubtedly lead to the isolation and study of presently uncultured marine microorganisms.

15.1 Introduction

Microorganisms populate all marine habitats with an estimated abundance of 10^4 to 10^6 cells/ml of seawater. Marine microorganisms play key roles in biogeochemical cycling processes including carbon and nutrient cycling. During the past three decades, the revolution in molecular biology, combined with technological advances and a decrease in sequencing costs, has allowed genomic studies to be performed on ocean microbial communities at a worldwide scale (Kopf et al. 2015; Rusch et al. 2007; Venter et al. 2004). The 16S rRNA gene sequence-based approaches and environmental genomics have revealed the vast diversity of the microbial world. Most of the organisms detected by these approaches had never been described before, even after 120 years of culturing microorganisms. These organisms are known as "not culturable" and represent the dark matter of the microbial domain. More than 99 % of microbial diversity is not yet culturable. The establishment of pure cultures of representatives of all bacterial and archaeal divisions is still a major challenge for microbiologists today. Pure cultures continue to be essential for a true understanding of the physiology of these Bacteria and Archaea and their roles in the environment, and to enable the discovery of new products with biotechnological potentials such as new antibiotics.

During the last two decades, considerable advances have been made in culture approaches making it possible to grow some important "key-player" microorganisms in the laboratory. These advances were not brought about by a revolution or overnight breakthrough in culture methods, but by a gradual improvement in existing methods, and a renewed understanding that the "sea" is an oligotrophic medium with a low abundance of microbial cells, living together.

The aim of this chapter is to provide an overview of the new strategies and technologies used by microbiologists to enhance the isolation and culture of marine microorganisms. Some important aspects, such as the preparation of the growth medium, incubation time, and sensitivity of detection methods for cell growth are first introduced. This is followed by a description of some of the most important technological developments, i.e., high-throughput and automated dilution-to-extinction methods, microencapsulation and immobilization of microorganisms in community cultures, the development of membrane-based cultivation chips for cell enrichment and cultivation, and single-cell techniques for cell isolation.

15.2 Medium

In 1941, Claude Zobell, a pioneer in the study of marine microorganisms, introduced the marine agar 2216 medium (known as Marine Broth), a nutrient-rich medium, which made possible the isolation of many heterotrophic bacteria, mainly belonging to a small number of genera, including *Vibrio*, *Pseudomonas*, *Oceanospirillum*, *Aeromonas*, *Deleya*, *Flavobacterium*, *Alteromonas*, and *Marinomonas*. Using nutrient-rich liquid or solid medium at the enrichment step favors faster-growing bacteria, referred to as "r"-strategists, at the expense of slow-growing, "K"-strategist, bacteria (Watve et al. 2000). K-strategists represent the dominant microorganisms in the pelagic environment. Natural seawater is a complex and living environment and it is difficult to prepare an artificial sea medium that suitably mimics it, providing all the nutrients necessary to sustain microbial growth. With the exception of the precise salt composition, which can be easily achieved, the concentration and sources of organic matter, nitrogen, phosphate, sulfur, trace mineral elements, cofactors, and vitamins and the reconstruction of the carbonate buffer are important factors that often determine the success or failure of enrichment and isolation attempts. Marine Broth 2216 medium has 170 times more dissolved organic carbon than natural seawater, which inhibits the growth of the true oligotrophs that represent the majority of microorganisms present in the ocean. It was only in 1993 when D.K. Button and colleagues started using natural seawater with small amounts of inoculum for isolating marine bacteria. This work resulted in the description of two new oligotrophs, *Sphingomonas alaskensis* and *Cycloclasticus oligotrophus* (Button et al. 1998; Schut et al. 1993; Vancanneyt et al. 2001). The concept of using natural seawater as a growth medium and low cell numbers as inoculum was improved upon two decades later in Giovannoni's laboratory, allowing the isolation of members of the SAR11 clade, one of the most dominant clades of Alphaproteobacteria in the ocean (Cho and Giovannoni 2004; Connon and Giovannoni 2002; Rappe et al. 2002). "*Candidatus* Pelagibacter ubique" was cultured in an artificial seawater medium with specific nutrients (Carini et al. 2013). The medium was defined based on the metabolic reconstruction of the Ca. *P. ubique* genome (Giovannoni et al. 2005), which revealed that the genes necessary for assimilatory sulfate reduction were absent (Tripp et al. 2008). The common genes for serine and glycine biosynthesis were also absent and the bacterium was later found to be auxotrophic for the thiamin precursor HMP (Carini et al. 2014). Employing the environment itself as aiding in growing microorganisms has enabled the culture of numerous previously uncultured isolates from different clades.

Marine Crenarchaeota are now recognized as a dominant fraction of the plankton in deep ocean waters. Konneke et al. (2005) used natural seawater from an aquarium as the first step in an attempt to enrich the fraction of Marine Group I Crenarchaeota. After several transfers and the addition of bacterial antibiotics (streptomycin) the fraction of Crenarchaeota increased up to 90 % of the total community. Subsequently, "*Nitrosopumilus maritimus*" strain SCM1 was isolated in a synthetic medium.

Only a small proportion of viable cells present in microbial communities can be grown using growth media based on agar (or other gelling agents), an enigma known as the "great plate count anomaly" (Staley and Konopka 1985). The diversity of marine microorganisms obtained after streaking on solid medium may depend on the gelling agent used for their isolation (Joint et al. 2010). A hidden pitfall was revealed in the common preparation of solid medium. The simultaneous addition of phosphate and agar prior to autoclaving causes the formation of hydrogen peroxide in the medium, which inhibits the growth of microorganisms. The separation of phosphate and agar before the autoclave step allows the development of numerous colonies, among which over 30 % previously uncultured organisms (Tanaka et al. 2014).

Finally, it appears that medium based on natural seawater taken on the sampling site, in the first step of enrichment/isolation improves the growth of presently untamed microorganisms. The number of "uncultured" species domesticated in artificial media is also increased by several prior-growing periods in natural seawater (Nichols et al. 2010).

15.3 Incubation Time

An important aspect of the culturing of abundant and dominant marine microorganisms is incubation time. Long incubation times are needed to allow the growth of key players such as members of the SAR11 clade. Song et al. (2009) demonstrated that after 20–24 weeks of incubation, 64–82 % of the total number of isolates belongs to SAR11. Members of the Roseobacter clade are abundant in seawater and have been detected in many marine habitats. A new genus, *Planktomarina* gen. nov., of the Roseobacter clade was described, after it was isolated after an incubation time of 7 weeks (Giebel et al. 2013). Hahnke et al. (2015) incubated samples for three months after an algal bloom and isolated novel abundant bacterioplankton species. The autotrophic ammonia-oxidizing archaeon strain SCM1 required an enrichment period of 6 months before being isolated. Under optimal conditions, strain SCM1 reaches the stationary phase after 20 days of incubation, which indicates that long incubation times are necessary to culture members of this clade.

A long period of incubation is an important factor in the success of isolating environmentally relevant microorganisms, meaning that microbiologists have to be patient, although such time requirements are at odds with academic programs and increasing pressure to publish rapidly.

15.4 Measurement of Microbial Growth and Detection Sensitivity

Methods to measure cell density vary in terms of sensitivity and time requirements. For a long time, microbiologists assessed microbial growth based on the turbidity they observed in the inoculated liquid medium. Measurement of turbidity is fast and easy to perform. It is therefore still frequently used, but the disadvantage of this technique is the rather low sensitivity. Many cultures have been discarded because they apparently displayed no visible growth, due to poor cell density. Microscopic measurements are more sensitive but, particularly in the case of epifluorescence microscopy, they are time-consuming. Flow cytometry has a high sensitivity, the counting procedure is fast and simple, but requires strong expertise and a careful setting up of the equipment. During the last few decades, cheaper and easy-to-use flow cytometers have been introduced and are now common equipment in research laboratories.

New protocols have been developed to access high-throughput isolation and culture microorganisms on a microplate format. In Giovannoni's laboratory, two approaches have been used: first, a 48-microarray filter device was developed in order to decrease the growth detection limit and to permit cell numeration with densities as low as 10^3 cells/ml using 200-μl aliquots of culture (Connon and Giovannoni 2002). Hahnke et al. (2015) then improved this method using a 96-well blotting manifold (Bio-Dot, Bio-Rad, Munich, Germany) and a vacuum pump (Millipore, Billerica, MA, USA) under low, non-disruptive pressure (<5 mm Hg). Formaldehyde-fixed samples were filtered directly onto 4-mm polycarbonate filters with a pore size of 0.2 μm (GTTP, Millipore, Billerica, MA, USA), and stained with either 1 × SYBR Green or 1 μg/ml DAPI and mounted on glass slides with Citifluor and VectaShield (4:1).

In 2007, Stingl used a flow cytometer to enumerate cell densities after transferring samples of 200 μl into a 96-well plate and staining with SYBR Green1, the detection limit was of the same order (Stingl et al. 2007). This method is fast, as it takes around 10 min to analyze a 96-well microplate. On the Cocagne platform (described below), the SYBR Green labeling method developed in 2006 by Martens-Habbena is used to quantify cell numbers. An aliquot of 100 μl from a 96-well microplate is transferred to a black microplate containing 50 μl of a SYBR Green solution (Martens-Habbena and Sass 2006). After 3 h of incubation in the dark, a microplate reader can detect the relative fluorescence units (RFU) at 485 nm in each well of the microplate. RFU is correlated with cell density. This method is also fast that it takes less than one minute to read a microplate with a detection limit at 2×10^5 cells/ml in seawater medium.

15.5 Targeting Microorganisms of Interest

High-throughput culturing (HTC) using dilution-to-extinction, culture chip, and single-cell isolation allows large number of isolates to be recovered. In order to screen the numerous cultures, different methods can be used to dereplicate the isolates or to target specific groups of microorganisms. Dereplication using restriction fragment length polymorphism (RFLP) after amplification of the 16S rRNA gene is convenient for clustering the isolates and identifying mixed cultures (Cho and Giovannoni 2004; Connon and Giovannoni 2002). The 48-array developed in Giovannoni's laboratory allowed Rappe et al. (2002) to use labeled oligonucleotide probes targeting the SAR11 cells (FISH) directly on the array filter. Hahnke et al. (2015) developed CARD-FISH on a 96-polycarbonate membrane to quantify and to identify the presence of different taxa with specific labeled probes.

15.6 Dilution-to-Extinction Method and High-Throughput Isolation

The dilution-to-extinction method introduced by Button et al. in 1993 was improved a decade later in Giovannoni's laboratory (Connon and Giovannoni 2002). The aim was to drastically decrease the number of cells inoculated per well to around 1–5 cells in a small volume of medium, with the idea that only one cell would grow in a well of a microplate and a pure culture would thus be obtained. The theoretical number of pure cultures was estimated by the equation $\mu = -n(1 - p)\ln(1 - p)$ and the estimation of culturability was given by $V = -\ln(1 - p)/X$, where μ is an estimation of the expected number of pure cultures, n is the number of inoculated wells, V is the estimated culturability, p is the proportion of wells positive for growth (wells positive for growth/total inoculated wells), and X is the initial inoculum of cells added per well. Utilization of microplates for culturing has introduced high throughput into the culture process. In a single round, Connon and Giovannoni (2002) incubated approximately 2500 extinction cultures. A culturability of up to 14.3 % was observed after three weeks of incubation, which is 120 times higher than obtained with classical methods. Cell densities were in the range of 1.3×10^3 to 1.6×10^6 cells/ml; densities that would be impossible to detect by measurement of turbidity. Many of the microorganisms that were cultured using this approach were unique cell lineages and included cultures related to the clones SAR11 (Alphaproteobacteria), OM43 (Betaproteobacteria, related to the methylotroph *Methylophilus methylotrophus*), SAR92 (Gammaproteobacteria), and OM60/OM241 (Gammaproteobacteria).

The Laboratory of Microbiology of Extreme Environments (UMR 6197 LM2E, Plouzané, France) has automated the high-throughput dilution-to-extinction method (Cocagne platform). A pipetting robot (Starlet, Hamilton) dispenses the medium (500 µl/well) and the cells (1–3 cells/well) into the wells of 20 microplates loaded onto the deck of the pipetting robot. Filter-sterilized natural seawater is used as growth medium, amended with micromole traces of nitrogen, phosphate, and a carbon source. A column with deep wells filled with sterile medium serves as control. The microplates are sealed with silicone caps and incubated at in situ temperature for at least two months. After incubation, the detection of growth in deep-well plates is done by labeling growing cells with SYBR Green followed by the detection of relative fluorescence compared to blanks using a microplate reader (Infinite® 200 PRO, Tecan). For this purpose, deep-well plates and dark microplates (Corning) are loaded on the robot's deck and the detection method programmed in the Hamilton software is started. All steps are automatically performed according to the program instructions. The pipetting robot first distributes 50 µl/well of the SYBR Green solution into the dark microplates (Corning), and then it transfers an aliquot (100 µl/well) of cell suspension from the deep-well plates to the dark microplates. After 3 h of incubation in the dark, the dark microplates are read on the microplate reader at 485 nm. Wells are considered to be positive when their relative fluorescence is higher than the average of the blanks (sterile medium) plus three times the value of the standard deviation of the blanks. The level of the relative fluorescence correlates with cell concentration. The positive wells of the twenty inoculated microplates, i.e., the wells with microbial growth, are then redistributed into new microplates with fresh medium. These newly inoculated deep-well plates are incubated under the same conditions as described above. A DNA extraction protocol, using the NucleoSpin 96 Tissue Kit (Macherey-Nalgel) is programmed in the robot's software. The 16S rRNA gene is then amplified and sequenced. The Cocagne platform has been used on samples of coastal seawater of Brittany. Twenty microplates, in other words 1760 extinction cultures, were incubated in natural local seawater at 15 °C for three months. Two cells were inoculated per well and the culturability obtained was 14 %. DNA was extracted and partial sequences of the 16S rRNA gene from one hundred and fifty isolates were analyzed. The results indicate the presence of numerous novel genera and species affiliated to Alpha- and Gammaproteobacteria (L'Haridon, unpublished).

15.7 Cell Immobilization for Isolation and Cultivation of Marine Bacteria and Archaea

Microbial cell immobilization is a method that aims to fix microorganisms on the surface of or within specific carriers. This technique is commonly used in many fields including food, pharmaceutical, agricultural, therapeutic, environmental, and research applications (Cassidy et al. 1996). Many different strategies are employed. These include immobilization by binding (physical adsorption, ionic binding, or covalent binding) on an inert carrier, self-aggregation (flocculation or chemical cross-linking), encapsulation or entrapment (membrane entrapment or entrapment within the network of a polymer matrix). Here we present applications of whole cell immobilization by entrapment in different polymer matrices for the isolation and culture of marine heterotrophic Bacteria and Archaea.

Cell entrapment in natural polymer matrices is the most frequently used technique because it is one of the gentlest protocols, inducing the highest cell viability (Kanasawud et al. 1989). The polymer matrix allows the diffusion of small molecules that sustain the viability, activity, and growth of the entrapped cells. In addition, cells are protected against attacks by bacteriophages, and from shear forces, abiotic stresses, and potential inhibitors that may be present in the culture medium. Moreover, entrapped cells are easily recovered from their growth environment (D'Souza 2002; Nussinovitch 2010). Polymer gels can be shaped in the form of beads, spherical geometry of which increases mass transfer compared to biofilms. Various methods, such as extrusion and emulsion, can be used to produce gel beads. In the extrusion method, an aqueous mixture containing polymers and microorganisms is extruded through a syringe to form beads that fall into a hardening solution. Bead size depends on the syringe orifice diameter, the concentration, temperature, viscosity and flow rate of the polymer solution, and the distance between the orifice and the hardening solution. The emulsion technique is based on a dispersion process in a two-phase system in which an aqueous solution containing polymers and microorganisms is dispersed in an oil/organic phase under agitation to form a water-in-oil emulsion. Temperature decrease induces the polymerization of beads that are then placed in a hardening solution (Kourkoutas et al. 2004; Rathore et al. 2013). Depending on the polymer, the hardening solution contains different types of cross-linking agents. The emulsion technique allows the production of gel beads of different diameters (5–5000 μm) depending on the emulsion conditions and agitation speed. Beads can be roughly separated into microbeads and macrobeads according to their size; macrobeads being greater than 100 μm in diameter (John et al. 2011). Because of the cell immobilization, free and immobilized populations do not have the same environmental conditions. As a result, immobilized cells have a different physiology and growth capacity compared to free-living cells (Doleyres et al. 2004; Rathore et al. 2013).

Many natural polymers (e.g., κ-carrageenan, gellan gum, agarose, agar, gelatin, alginate, chitosan) can be used for cell entrapment (Cachon et al. 1997). These polymers are usually cheap and can be used alone or in combination. However, they vary in toxicity, gelling properties, rheological behavior, and mechanical stability according to the conditions during the bead formation and incubation. The selection of the right polymer and encapsulation method is therefore critical depending on the type of application and microorganisms to be immobilized, in order to ensure cell viability while preserving the beads' mechanical stability, depending on culture conditions (especially salt concentration and temperature).

15.7.1 Immobilization for the Isolation of Marine Microorganisms

An original application of whole cell immobilization is the microencapsulation method developed by Zengler et al. (2002) for high-throughput culture and isolation of marine aerobic mesophilic microorganisms. The principle behind this method is the physical separation and massive parallel culture of cells that are individually entrapped in gel microdroplets (GMDs) of agarose (60 μm diameter) and incubated as a continuous community culture (flow rate 13 ml/h) in an HPLC column under low nutrient flux conditions for long incubation periods. This GMD community culture allows the exchange of metabolites and/or signaling molecules between the micro-colonies that grow within the GMDs, which are physically separated from each other. At the end of the continuous culture, GMDs containing micro-colonies are sorted by flow cytometry and further incubated in deep-well microplates (one GMD per well) for clonal enrichment in a rich organic medium. This technique was applied to different habitats and has provided more than 10,000 bacterial and fungal isolates per natural sample, including novel marine bacteria from previously uncultured groups (Zengler et al. 2002, 2005). It is a promising technique, although the critical step is the positive GMD sorting by flow cytometry, which may decrease cell viability. The use of a suitable cell sorter that can directly disperse the positive GMDs into microplate-wells is therefore important at this stage. In Zengler's publications (Zengler et al. 2002, 2005) GMDs were directly sorted and distributed into microplates by a FACSAria® flow cytometer.

The Laboratory of Microbiology of Extreme Environments (UMR 6197 LM2E, Plouzané, France) has automated the Zengler method using a Hamilton pipetting robot (Cocagne platform) for GMD distribution into deep-well microplates and growth detection by labeling of growing cells with SYBR Green, followed by the detection of the RFU compared with blanks using a microplate reader. Moreover, a protocol for the microencapsulation (60 μm beads) and culture of thermophilic marine cells in heat-stable microbeads made of gellan and agarose was also developed (Fig. 15.1).

The same microencapsulation method was also used for the isolation of slow-growing marine bacteria from mixed samples. Akselband et al. (2006) developed a new strategy based on the growth detection of individual cells encapsulated in agarose GMDs (30–50 μm in diameter). After 6 h growth in mixed culture, the slow-growing micro-colonies were sorted from fast growing ones by measuring their size within the GMD with a flow cytometer (EPICS Elite™, Coulter) after LIVE/DEAD staining (*Bac* Light Bacterial Viability Kit), and transferred to fresh medium. Akselband et al. (2006) suggest the use of the same strategy for targeting specific activities using fluorogenic substrates for GMD sorting. In their study, they also showed that 75 % of the marine isolates grew faster in GMDs than in liquid medium.

Ben-Dov and collaborators developed another strategy for the isolation of marine aerobic mesophilic microorganisms using a macro-encapsulation technique

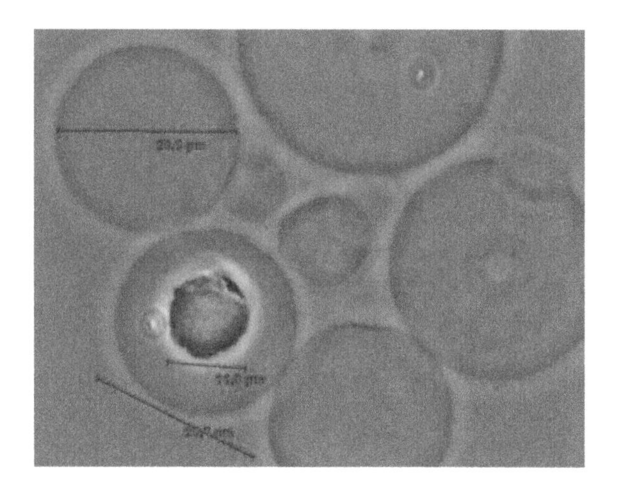

Fig. 15.1 Development of a bacterial colony inside a GMD at 65 °C

(Ben-Dov et al. 2009). After dilution of environmental samples, microbial cells were individually entrapped in agar beads (1–3 mm), which were then coated with a porous polysulfonic polymeric membrane. This type of membrane prevents agar dehydration but enables the exchange of molecules (nutrients and waste). After sterilization of the surfaces with alcohol, the beads were placed in a suitable simulated or natural environment for long periods of incubation (at least 3 weeks). Then, the beads were recovered and sterilized by flaming, and the embedded agar spheres were removed, flattened, and observed under the microscope for the presence of microbial colonies. In order to enrich and isolate microorganisms, the agar beads were diluted and remixed with warm agar to form new beads. The coated beads were then repeatedly incubated in the appropriate environment. Following every two to three transfers, colonies were streaked onto appropriate agar plates. This labor-intensive technique was successful in growing several novel microorganisms from different environments, including the mucus layer of the Red Sea coral *Fungia granulosa*.

15.7.2 Immobilization for Increasing Cell Growth Efficiency and Stability

Entrapment of mesophilic aerobic marine microorganisms within polymer matrices (usually alginate) in pure culture or as consortia is used in environmental applications (such as biosensor and in bioremediation) (Cassidy et al. 1996; Futra et al. 2014) and in biotechnological applications in bioreactors (biomass and metabolite production and waste water treatment) (Roy 2015). Cell entrapment has been used for the culture of thermophilic strains (Kanasawud et al. 1989; Klingeberg et al. 1990; Norton and Lacroix 2000) but never for marine (hyper)thermophilic anaerobic microorganisms because of the difficulties of gelling properties in a saline environment. However, cell entrapment offers numerous advantages over free-cell cultures in bioreactors (Bustard et al. 2000). Cell entrapment induces a much more stable and robust continuous culture system because it prevents cell washout, improves genetic stability, protects organisms against shear forces and possible presence of toxic compounds in the culture medium whilst increasing cell numbers and product yields (Rathore et al. 2013). Moreover, immobilization facilitates separation between biomass and products and makes possible the reusability of the beads with immobilized cells.

The Laboratory of Microbiology of the Extreme Environments (UMR 6197 LM2E, Plouzané, France) has developed a cell entrapment protocol for (hyper) thermophilic marine microorganisms in order to improve the cultivation of deep-sea hydrothermal microbial communities. An experimental design, set to evaluate different incubation conditions (pH, temperature, salt, and sulfur), showed that beads (1–2 mm diameter) made of a mixture of heat-stable polymers (gellan and xanthan gums) were resistant to most incubation conditions. After 5 weeks of incubation, beads showed good resistance to all tested conditions (Fig. 15.2), except those simultaneously including high temperature (100 °C), low NaCl concentration (<30 g/l), and extreme pH (≤ 4 or ≥ 8). Batch experiments under nitrogen gas using Ravot medium (Gorlas et al. 2013) showed that thermophilic (*Thermosipho* sp. DSM 101094) and hyperthermophilic (*Thermococcus* sp. KOD1) microorganisms were effectively immobilized in beads and grew to high cell numbers. Cell counting by regular methods is impossible in polymer beads. Therefore, a protocol based on cellular ATP measurement was developed and applied to beads and culture medium. The correlation between cell counting using a Thoma cell counting chamber and ATP values was determined for each tested sample in order to determine the number of cells/ml for each strain. The ATP content of bacterial suspensions in the liquid culture media was determined using a Kikkoman Lumitester C-110 (Isogen Life Science) in combination with the

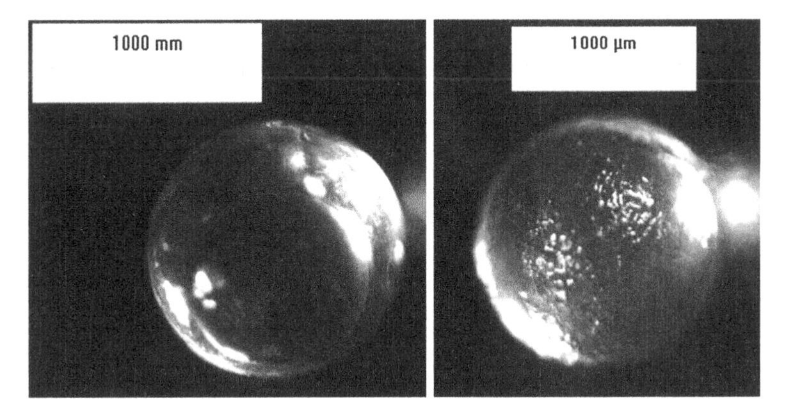

Fig. 15.2 Beads observation using a binocular magnifying glass (magnification × 2.4) after immobilization 5-week incubation

BacTiter-Glo Microbial Cell Viability assay (Promega) according to the manufacturer's instructions. In the case of beads, *ca* 100 mg of beads was placed in a pre-weighted sterile hemolysis tube (Gosselin). Subsequently, the beads were washed thrice with 100 μl of sterile degassed saline solution. For ATP measurement, 100 μl of sterile distilled water was added to the beads, which were vortexed for 10 s before adding 100 μl of BacTiter-Glo buffer. As for liquid medium, internal calibration was performed with 10 μl of a 100 nm ATP solution and maximal fluorescence emission values were considered. The number of active cells was 2.4×10^5 cells/g of beads for *Thermosipho* sp. (DSM 101094) and 2.3×10^6 cells/g of beads for *T. kodakarensis* (KOD1) just after immobilization, which corresponds to 2.7 and 54 % of cells surviving the immobilization step, respectively. After a few hours, cells grew within the beads as well as in the liquid medium. The cultures reached cell densities of 6.1×10^8 cells/ml in liquid medium and 2.9×10^7 cells/g in beads after 24 h incubation at 65 °C for *Thermosipho* sp. (DSM 101094), compared to 2.5×10^8 cells/ml for free-cell culture under the same conditions. In the case of *T. kodakarensis* (KOD1) the highest cell densities were obtained at 70 °C with 3.3×10^8 cells/ml and 4.8×10^7 cells/g in liquid medium and in beads, respectively, after 24 h incubation compared to 1.4×10^8 cells/ml in free-cell suspension medium. The high percentage of survival together with the high cell densities of both strains in beads and liquid medium showed that entrapment and culture of immobilized anaerobic thermophilic and hyperthermophilic marine strains is possible at high temperature (Landreau et al. submitted).

Using the same entrapment protocol, the proof of concept was established in a continuous culture that was performed for 42 days in a gas-lift bioreactor flushed with nitrogen and containing 40 % (v/v) of freshly inoculated beads with different thermophilic and hyperthermophilic deep-sea microorganisms. Gel beads proved to be highly resistant to mechanical and chemical degradation during the 42 days of continuous culture. Moreover, cell quantification, organic acid concentrations, and ATP monitoring showed that polymer beads and effluents were highly colonized and that the microorganisms were reactive to culture conditions.

15.8 Enrichment Chambers and Culture Chips

The environment is often an excellent growth medium (Ferrari et al. 2008; Kaeberlein et al. 2002) and this forms the basis of many laboratory cultivation techniques. We can build better and smaller culturing devices. These devices are laboratory-on-a-chip (LOC) or micro-engineering (MEMS) approaches for which the technologies are in part derived from the electronics industry that is entering life sciences (Ingham and van Hylckama 2008; Weibel et al. 2007). Therefore, microbiologists have access to improved manufacturing and design capabilities leading to the creation of sampling and culture devices for the laboratory or for implantation in the natural environment. This is both intuitively obvious and experimentally supported. Logically, we want to either move the laboratory into the environment or the environment into the laboratory in order to recreate the missing factors that act as roadblocks to cultivation. Here we deal with culture chambers (laboratory into environment) and culture chips (environment into laboratory) as examples of technological advances that use this idea.

15.8.1 Porous In Situ Cultivation Chambers

Porous cages and chambers allow the culture system to be placed in the appropriate environment. Whilst even this measure is not a perfect recreation, such systems have shown great value in culturing otherwise refractory microorganisms. One simple but effective version is a diffusion chamber bounded by two membranes (Fig. 15.3a). This chamber is filled with agar lacking nutrients and inoculated with an environmental sample and then sealed (Kaeberlein et al. 2002). The diffusion chamber can be placed in the sea or in soil, or a close simulation of the environment (e.g., sediment in an aquarium), and the external conditions rapidly equilibrate with the small volume of the interior. Micro-colonies trapped within the agar can often then be cultured conventionally, whilst closely positioned micro-colonies can be investigated for the possibility that culturing required interactions between different species. This technique has resulted in the enrichment of previously uncultured organisms, up to 30 % success in culturing microorganisms from marine and littoral habitats. A logical extension of this idea is the iChip, a highly multiplexed version of the diffusion chamber (Fig. 15.3b). The iChip allows multiple polyoxymethylene (a hydrophobic plastic) chambers to be enclosed with two polycarbonate membranes (Nichols et al. 2010). The chambers are loaded with agarose before the second membrane is put in place, dipped in the appropriate environmental sample to inoculate, sealed and implanted into soil.

A notable success of the iChip was the use of these devices in a screening of up to 10,000 micro-colonies in order to isolate an example of a new class of antibiotics, teixobactin, from *Eleftheria terrae*, a group of Bacteria not previously known to be a good source of antimicrobials (Ling et al. 2015). The iChip uses an additional strategy when looking for antimicrobials. The first line of activity screening is performed by recovering the device, then spread plating the upper membrane with indicator bacteria. In the first instance, the microorganisms grow inside the micro-wells, sandwiched between two membranes allowing entry of small molecular weight compounds such as nutrients and quorum sensing molecules that support growth. In the second stage, antimicrobials diffuse from the trapped micro-colonies to an indicator strain on the outer surface to be assayed. The high frequency of "hits" from a highly mined environment is promising and there is every reason for thinking this approach will adapt well to the less well-explored marine environment. So far, the approach is not selective—what grows and subsequently assayed. Enrichment culture, for example with poorly soluble polymers, would be one way to bias the method towards isolating microorganisms with a desired phenotype, such as polymer degradation. A variant of this diffusion chamber system that possesses

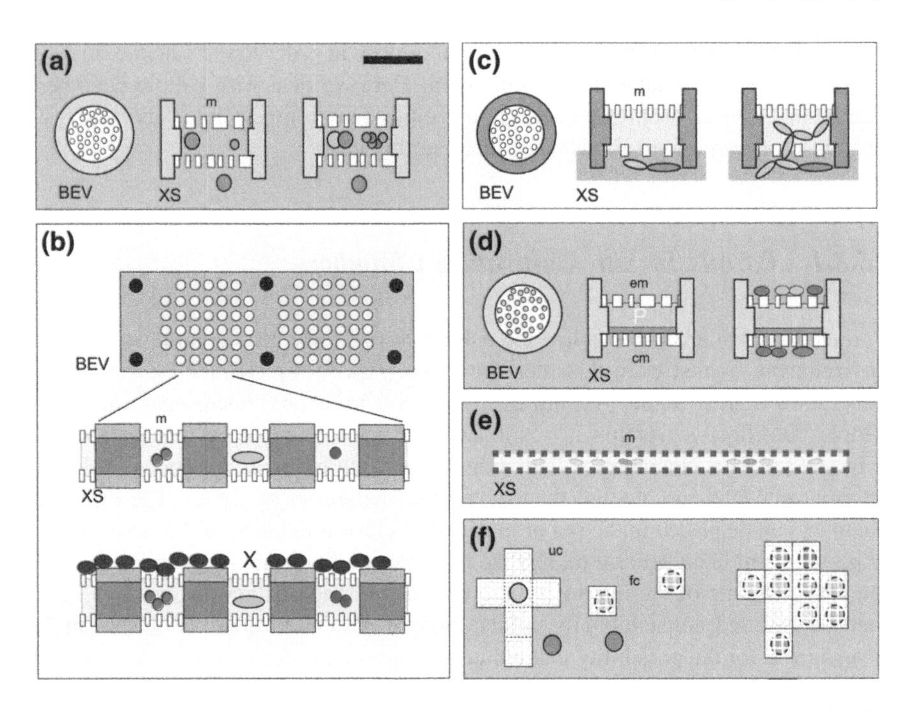

Fig. 15.3 Cultivation chambers for enrichment and near natural culture conditions. **a** In situ cultivation chamber filled with agar, then placed back in natural or simulated environment for cultivation of microorganisms in the interior. Scale bar (*top panel* ***a***) indicates 1 cm for all panels. **b** iChip, an array of cultivation chambers similar to those in panel. *Top* view of whole chip. *Middle* cross section showing microbial growth in compartments. *Bottom* Plating an indicator strain on the upper surface reveals antimicrobial activity from the bacteria contained within compartment X. **c** Selective, asymmetric capture chamber for mycelial organisms. BEV, Bird's Eye View. XS Cross section. The upper membrane (m) is a barrier for all microorganisms, the larger pore membrane can be penetrated by actinomycetes (represented by chains of green ovoids, growing from sediment) better than other bacteria, so tends to trap and enrich selectively. **d** Variant culture chamber with slowly degrading material (P, polymer) or microbial population in central chamber. Whilst bounded by two membranes, only the upper membrane (em, experimental membrane) communicates with inner chamber, the lower (cm, control membrane) does not. Therefore, the two microbial populations growing on the outer surfaces of em and cm can be expected to differ. **e** Porous tube containing bacteria (*light blue*) for exceptionally high surface area to volume ratio, allows recirculation of contents and extremely rapid exchange with the environment (*darker blue*). **f** Closing cage system of entrapment. *Left* part of this panel shows unfolded cage (uc) with a mesh hundreds of micrometers. *Center* cage now folded into a cube (fc) trapping a multicellular microbe or marine invertebrate. *Right* Assembly of multiple cages to create synthetic tissues or communities. The height in the XS of panels A, B and E is exaggerated circa two-fold to show detail. A *blue* background indicates marine use, a beige background that the main validation to date has been in the soil, but that the technology is applicable to the marine environment

selectivity uses asymmetric membranes (Fig. 15.3c). The upper membrane cannot be penetrated by microorganisms and the lower membrane has a pore size that enriches a particular group by selectively permitting growth into the chamber (Gavrish et al. 2008). When the pore size of the lower chamber is set to 0.45 μm, filamentous bacteria (particularly actinomycetes) showed an advantage in penetrating the chamber and could therefore be enriched and isolated. Precise modulation of pore size usefully excluded fungi and other microorganisms, which frequently overgrow actinomycetes in more conventional environmental screenings. Given that actinomycetes are major antibiotic producers but the number of new useful antibiotics is falling, selective entrapment of filamentous bacteria is a promising technique. If membranes can be fabricated that are selectively porous based on other properties (e.g., surface charge or hydrophobicity or surface molecules) this selective entrapment technique could become even more powerful.

Another variant of the culture chamber places the enrichment material in the central chamber and allows microorganisms to grow on the exterior of the membrane (Fig. 15.3d; MicroDish BV, unpublished). The advantage of this strategy is allowing a controlled experiment to take place. Most culturing chambers just trap what grows; they are a form of sampling rather than experimentation. In this scheme, microorganisms grow on the outer surface of a membrane exposed to the marine environment as well as the contents of an inner chamber (separated by a porous membrane, the inner chamber can contain nutrients or other microorganisms)—this acts as the experiment. A second membrane within the same chamber has equal exposure to the environment but no communication with the central chamber; this is the control. Therefore, a metagenomics experiment or targeted culturing approach should allow analysis of the differences in microorganism abundance (or nucleic acid sequence) between the experimental and control conditions. More complex geometries in culture chambers are possible, permitting multivariable experimentation in situ. An alternative format for porous enclosure is a hollow fiber (Fig. 15.3e), a tubular membrane which has an exceptionally high area in contact with the environment (Aoi et al. 2009). A porous polyvinylidene fluoride (PVDF) membrane (0.1 μm pores) contains the microorganisms of interest. This system shows a significant enhancement of culturability over agar-based methods, tested with three microbial populations derived from marine and industrial environments. Additionally, the total volume of the system can be increased (simply by extending the tubing), allowing rather larger populations of microorganisms to be enriched than is the case with micro-colony-based methods.

The diffusion chambers described above are sealed by hand under aseptic conditions and then placed in the environment. A particularly interesting development for the marine environment is the self-closing cultivation chamber (Fig. 15.3f), particularly if these could be closed in hard to reach environments, such as the deep-sea. One such approach is the creation of flat boxes that self-assemble (Leong et al. 2008). The sides to the box are all porous; when folded they form a cage. The assembled boxes can be sorted and are amenable to manipulation by magnetic fields. These properties offer the potential to stack boxes and therefore the creation of artificially structured communities of cells, again moving from sampling to experiments. Later developments have created self-folding structures, created purely from polymers that are optically transparent and capable of trapping microorganisms including eukaryotic cells (Azam et al. 2011).

15.8.2 Cultivation Chips

Starting with the Petri dish, there are a number of possible improvements that might be envisaged that would aid marine culturing. These include miniaturization, a greater potential for automation, the absence of gel polymers as a matrix (such as agar that may contain inhibitory compounds), and suitability for imaging or other detection methods. An early attempt at "the better Petri dish" is the hydrophobic grid membrane (HGM), a porous filter subdivided into hundreds of growth areas by wax barriers. HGM offers flexibility (can be placed on agar or non-agar surfaces and can be moved). The HGM has a greater dynamic range for counting colony-forming units compared to an equivalent area of agar (Sharpe and Michaud 1974). This is explained by a better segregation of colonies and effective statistics derived from the distribution of colonies segregated between a large number of compartments.

Microfabrication allows further subdivision of growth areas to the point where custom-built disposables with thousands of compartments are available (Incom, USA). Micro-engineering allows a series of configurations of what is effectively highly multiplexed multiwall plates. However, miniaturizing multiwall plates beyond a certain threshold creates problems as well as advantages. There may be issues of aeration (liquid trapped in capillaries cannot be shaken to introduce oxygen), imaging (hard to focus on cells in capillaries), assay sensitivity in low volumes and dehydration (nanoliter wells dry out unless the humidity is carefully controlled), cross contamination, and dependence on robotics that microbiology laboratories often lack. There are solutions and work-around possibilities for many of these issues. Simply optimizing geometry, shaking speed and angle, and media components within conventional multiwall plates can considerably help oxygenation (Duetz et al. 2000). Oxygen can be delivered to 96-well plates through an oxygen permeable membrane, leading to 96-well microtiter plates with oxygen transfer rates comparable to Erlenmeyer flasks (Microflask System, Applikon Biotechnology, NL). Additionally, the capillarity of micro- or nanoliter wells has been exploited to capture droplets in stackable microcapillary arrays.

Porous ceramics make a good basis for culture chips (Fig. 15.4), with the advantages of flatness, porosity, low autofluorescence, biocompatibility, and despite inertness, good ability to conjugate biomolecules (Ingham et al. 2007; Microdish BV, NL). The downside is brittleness of the ceramic, requiring reinforcement for good handling. Micro-engineering techniques can create micro-wells (7–300 μm across, 10–40 μm deep, fixed or variable geometries). Similar to culturing chambers, these can be used, in combination with sediments or other samples, to culture microorganisms that were refractory to culturing with conventional techniques. Additionally, the small volume of culture medium allows the use of high cost reagents or additives. Control of compartment size, combined with spacing allows fine-tuning of application. Furthermore, a common task in microbiology is the replication of microorganisms using a velvet pad (Lederberg and Lederberg 1952) or a simple printing device, such as a 96-pin array with a pin

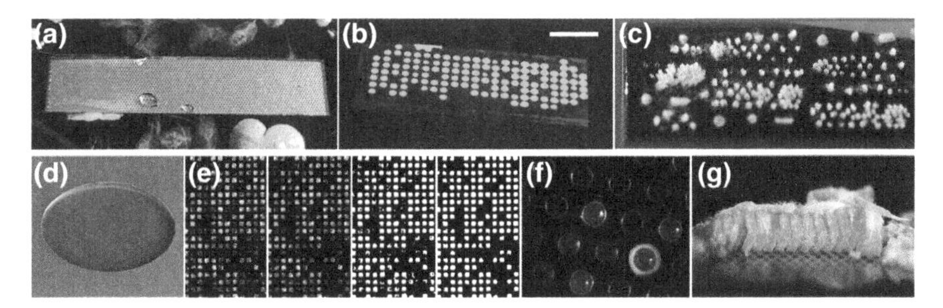

Fig. 15.4 MDCC. **a** Cultivation chip (MDCC180.10, 8 × 36 mm), floating on water, with PAO base and more than 4000 180-μm-diameter compartments in a hexagonal array coated with platinum. **b** Image of an array of 200-μm-diameter micro-colonies, visualized by green fluorescent protein and captured by a hand held digital camera. **c** Chip with variably spaced wells (MDCC180.10VAR) inoculated with fungi to look at the effect of colony density and size on growth. **d** Single 180-μm-diameter, 10-μm deep well. **e** Section of an MDCC20.10 chip with 20-μm^2 wells in an orthogonal array (125,000 wells per chip) used to culture bacteria, subsequently stained with a fluorescent dye and imaged by fluorescence microscopy. The four sections of this panel show how image processing moves the raw image from a gray scale photograph to a binary image—the final image used to score whether a compartment supports growth and therefore is scored as a CFU. **f** Image of a single colony from a screening of culture from arctic sediment. **g** Image of fungal mycelia growing out of 180-μm-diameter compartments as mycelial bundles. White scale bar (panel **b**) indicates 6 mm when applied to panel **a**, 1 mm for **b**, 2.5 mm for **c**, 40 μm for **d**, 150 μm for **e**, 420 μm for **f** and 800 μm when applied to G

spacing of ~0.5 mm. It is possible to deploy bacteria using non-contact printing devices; even ones modified from conventional inkjet printers (Flickinger et al. 2007). In terms of contact printing, miniaturized arrays of pins fabricated from elastopolymers allow a similar process but on a finer scale. A pin spacing of 80 μm is usable to replicate between micro-wells of a culture chip (Ingham et al. 2010).

Suspension cultivation is also possible in a highly multiplexed and miniaturized device. For example, arrays of microfluidic capillaries can sustain microbial growth using peristaltic pumps to move fluid and introduce oxygen (Gan et al. 2011). Moreover, micro-fabricated chips can be used to address questions such as to the organization of microbial communities (Keymer et al. 2006) or to approximate to the function of chemostats (Balagadde et al. 2005). We can expect further inventiveness in terms of microbial cultivation chips in the future (Lok 2015), and also hope that low cost manufacturing will make these devices affordable for the marine microbiologists. Therefore, these micro-fabricated chips will play a significant part in increasing the culturing members of the marine microbiome.

15.9 Single-Cell Techniques

In order to establish a pure culture, microbiologists have to isolate a viable cell and maintain its physical isolation whilst the cell divides to form a colony. Often, a pure culture is achieved by statistical means: either a sample is diluted until on average there is only a single viable cell left in the cultivation chamber, or cells are spread over a surface or in small volumes until on average there is a single isolated cell in a given location. Either way, when employing these methods there is little or no control over which individual cell in the original sample goes where. A fundamentally different approach involves the direct selective isolation of single cells from a sample, combined with a method for directing the cell to a known location, for example on an agar plate, a well in a multi-well plate, a micro-well, or micro-chamber in a microchip. Having isolated the cell and defined its position we can then alter its microenvironment to improve its culturability.

The isolation, handling, and analysis of single cells is nowadays a topic of growing interest because of the potential to target rare cells and obtain information on the heterogeneity of cultures (including gene expression) (Blainey 2013; Ishii et al. 2010; Stepanauskas 2012; Yun et al. 2013). A variety of techniques is either already available or under development to achieve this which will be discussed here. Many of the techniques were originally developed for use with mammalian cells, but could be or in some cases already have been adapted for use with marine microorganisms.

15.9.1 Cell Sorting

Sorting techniques can either be active or passive. Active systems generally use external fields (e.g. mechanical, acoustic, electric, magnetic, optical, hydrodynamic) to impose forces to displace cells, whereas passive systems use inertial forces, filters, and adhesion mechanisms to purify cell populations (Wyatt Shields IV et al. 2015). For the selective isolation of cells active sorting is generally preferred.

During the sorting process the cells can be dispersed in a static or moving fluid, or as single cells in micro-droplets or gel microbeads. To obtain a high throughput, the cells need to be suspended in a flow, sufficiently separated from each other to allow easy separation. The cells are then analyzed one at a time as they flow past a sensing system and then actively knocked out of the flow when the cell has the desirable properties. Throughput in static systems is generally lower, and the techniques used often put high demands on the dexterity of the operator. Automation is sometimes possible, for example, through the combined use of advanced control systems and robotics (Lu et al. 2010; Zhang et al. 2012), and can speed up the isolation process. However, flow-through systems can suffer from blockages.

15.9.2 Facs

The most well known technique used for flow-based cell sorting is the Fluorescence Activated Cell Sorter or FACS. Although traditionally mainly regarded as a tool for the analysis and sorting of large mammalian cells, the technique is increasingly used for microorganisms (Czechowska et al. 2008; Davey and Kell 1996; Mazard et al. 2014; Morono et al. 2013; Winson and Davey 2000). A typical FACS is shown in Fig. 15.5. It includes a stream containing the cells to be sorted, a device for hydrodynamic focusing this stream into a narrow laminar flow for cell analysis, a laser system that enables one to measure the fluorescent and light scattering properties of each single cell, a droplet generator system which produces droplets containing single cells, a method to place an electric charge on a droplet according to whether the cell is to be collected or not, and a method for selectively diverting droplets in an electric field. The droplets are sorted into tubes or wells of microtiter plates, although sorting microorganisms directly onto agar plates is also possible (Fig. 15.6). A standard FACS can achieve high cell throughputs (>10^4 cells/s), but large samples are needed, not all cells are fluorescent, it is difficult to isolate all individual cells in a sample, and it is often not possible to guarantee that the droplets contain single cells. A number of alternative cell sorters, many of them based upon miniaturization and making use of Microelectromechanical systems (MEMS) and Laboratory-on-a-Chip technologies, has been or is being developed in order to overcome these problems. Often their throughput of these miniaturized cell sorters is lower than FACS, enabling slower and more detailed cell analysis techniques such as image analysis or Raman spectroscopy to be used rather than fluorescence alone.

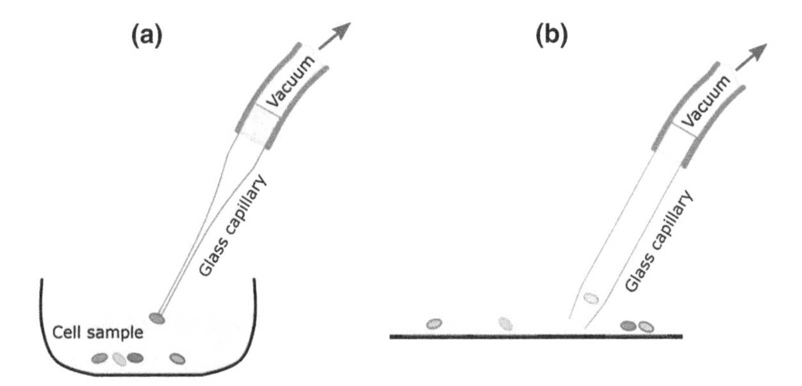

Fig. 15.5 Sketches of methods of isolation of cells with micropipettes. Cells can be isolated either by aspiration onto the tip of a narrow capillary (**a**), or into the body of the capillary itself (**b**). Aspiration onto the tip is more suitable for larger cells and filamentous organisms, but the risk of mechanical damage is greater

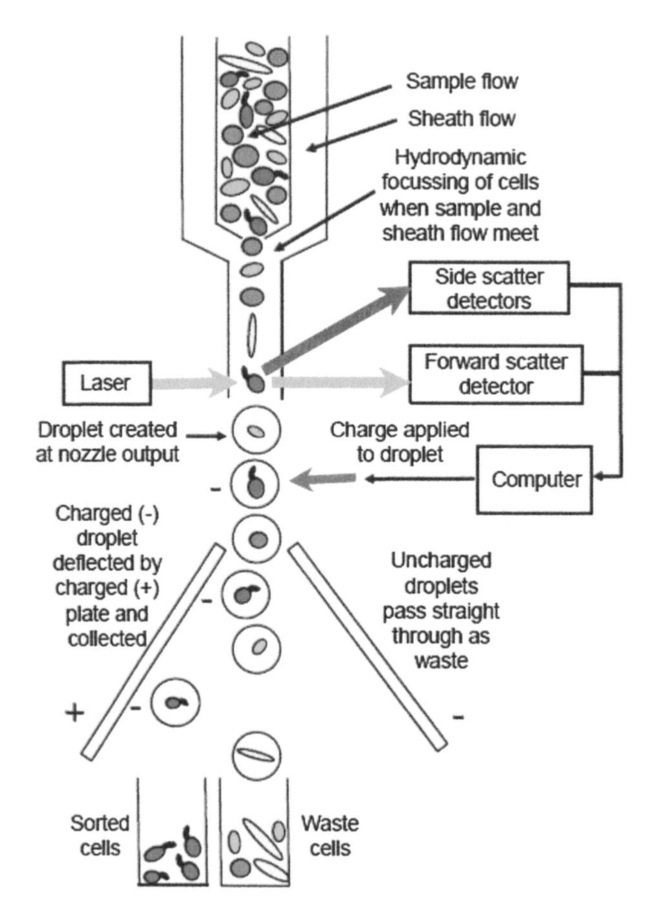

Fig. 15.6 Schematic of a fluorescence activated cell sorter (FACS). A FACS is made up of three main systems: fluidics, optics, and electronics. Fluidics: A sample of cells is hydrodynamically focused using a sheath flow. As cells flow along the stream of liquid, a laser scans them. Optics: Laser light scattered by the cell is collected by a detector such as a photo multiplier tube and is used to count cells and to measure the size and granularity of the cell. Laser light is also used to excite fluorescence from specifically labeled sub-populations of cells. Electronics: The light signals are converted into electronic signals and a computer processes the information to determine which cells are to be sorted. The electronics system controls the charging and deflection of particles. Droplets, each containing a single cell, are created at the output nozzle by vibrating the nozzle at an optimal frequency and an electrical charge (−) is applied to the droplets to be collected as they exit the nozzle. The charged droplet is then deflected left towards the positive electrode (+) into a collection tube. Droplets that contain no cells or cells that are not to be collected pass straight through into the waste tube. The collected population is a pure population for the criteria determined when setting up the experiment (for example cells with above a certain threshold of fluorescence emission)

These cell sorters include inkjet printer-based systems with integrated image analysis (Stumpf et al. 2015; Yusof et al. 2011), dielectrophoretic cell sorters (Hu et al. 2005; Zhang et al. 2015), optical force-based cell sorters (Enger et al. 2004; Keloth et al. 2015; Landenberger et al. 2012), and various valve-based microfluidic cell sorters fabricated with soft lithography (Fu et al. 2002; Robert et al. 2011).

15.9.3 Mechanical Micromanipulation Techniques

The use of mechanical micromanipulation techniques for the isolation of microbial cells has a long history, and reviews on early techniques have been given by Johnstone (1969, 1973), and more recently by Fröhlich and König (2000). Early techniques used micro-needles or microcapillaries for the isolation of microorganisms, but success was limited by the technology then available. The development of improved micromanipulators with accurate pneumatic or hydraulic pressure control systems and improved microscopy techniques have made the isolation of cells with these techniques much more straightforward, and many examples of the use of mechanical micromanipulators for the isolation of microbial cells can now be found in the literature (Fröhlich and König 1999; Ishøy et al. 2006). Either the entire cell is drawn into a micropipette that is much larger than the cell, or held at a pipette tip that has an opening that is smaller than the cell diameter (Lu et al. 2010) (Fig. 15.7). Microfabrication nowadays makes it possible to manipulate single cells with pressure force in a massively parallel way (Nagai et al. 2015), although pressure forces on mechanically isolated cells can be large, and can lead to damage to shear-sensitive cells.

An alternative approach to mechanical cell isolation using needles or pipettes involves the use of an externally applied physical field force to move the cells. Forces used for cell manipulation have been reviewed by Yun et al. (2013) and include optical, electric, magnetic and acoustic forces. To date acoustic forces have mainly lacked the resolution needed for single-cell manipulation. Most biological materials, including most cells, are diamagnetic and show little to no response to externally applied magnetic fields unless modified by the attachment of (super)-paramagnetic particles or suspended in a paramagnetic suspending fluid (Safarik and Safarikova 1999). The response of cells to electrical or optical forces is strong, however, and can be achieved without modification of the cells.

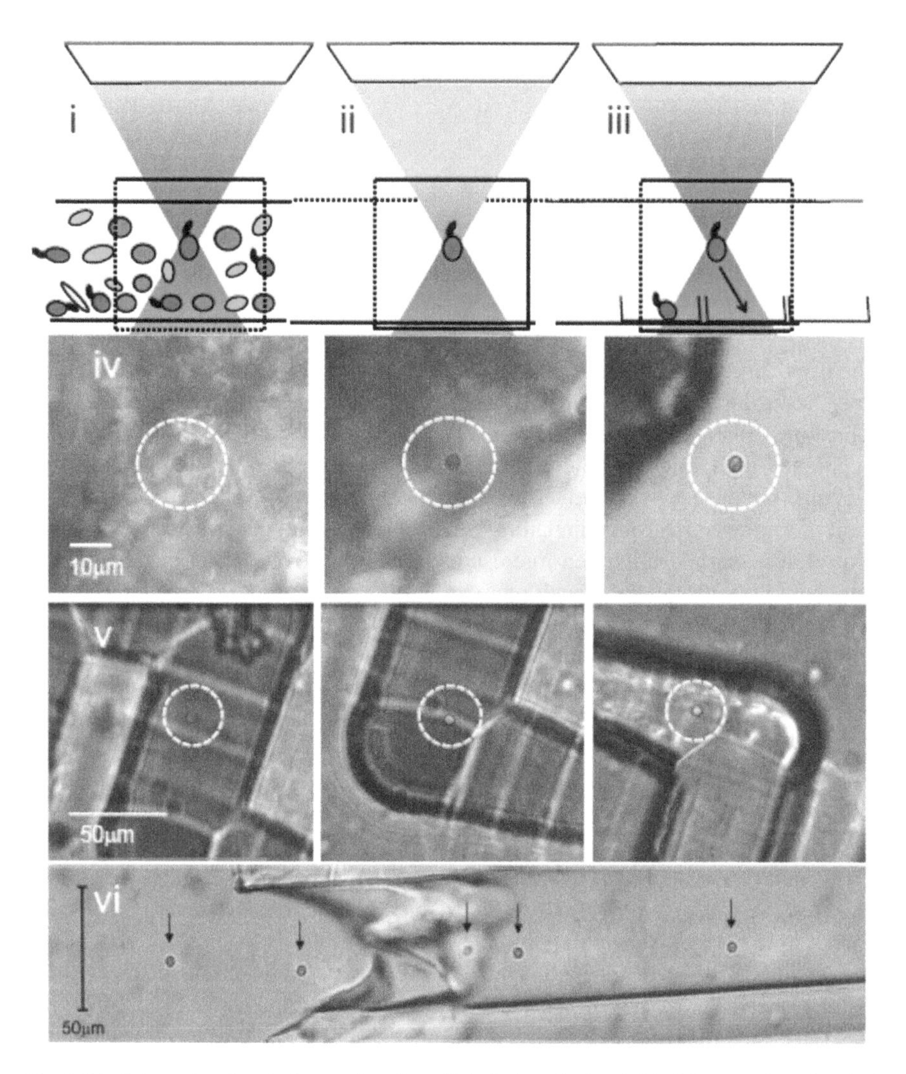

Fig. 15.7 Schematic of optical tweezing for cell isolation (*i*, *ii* and *iii*). Micrographs of optical tweezing (*iv* and *v*) and isolation (*vi*) of single cells. A tightly focused near infrared laser beam is used to trap and move a single cell in three dimensions away from surrounding cells, either in an environmental sample (schematic shown in *i*) or from a laboratory culture of cyanobacteria (*iv*). The selected cell is physically removed by optically tweezing the single cell away from the other cells via a narrow meandering channel (*ii* and *v*) to a region where there are no other contaminating cells, such as a culture chamber in a customized chip or a microcapillary (*iii* and *vi*). Cell selection and isolation can now be automated to decrease the burden on the operator. The isolated cell can be used to start a pure culture or can be joined in the chamber with a second, or any number of selected cells for co-culture

15.9.4 Cell Manipulation with Electric Fields

Cells can be moved by direct (DC) and alternating currents (AC); DC fields move cells through electrophoresis because cells have a net (negative) charge, and AC fields induce movement of the cells by the interaction between the induced dipole moment and electric field gradient (dielectrophoresis). Cell manipulation with high frequency (>10 kHz) AC fields is generally preferred over cell manipulation with DC fields as it suffers less from interference by local fluid streaming induced by the strong electric fields near the electrodes. Dielectrophoresis has been extensively used to create aggregates of microorganisms for the study of microbial interactions in biofilms, including metabolic interactions, quorum sensing, and resuscitation of dormant cells (Andrews et al. 2006; Mason et al. 2005; Zhu et al. 2010). However, to achieve the resolution needed for single-cell manipulation, structures are required of a size similar to that of a single cell. As a result, the manipulation of single cells with electric fields puts high demands on microfabrication skills. The electric field rapidly declines away from the electrode structures, and the electrode structures therefore have to be close to the cell to be isolated. Electric forces are strongly dependent on the composition of the medium, in particular its conductivity. The use of low salt media is often essential, although this can be to some extent alleviated using high frequency electric fields (Schnelle et al. 1999). Despite all this, various examples of single-cell manipulation and isolation with electric fields can be found in the literature (Graham et al. 2012; Hsiao et al. 2010; Schnelle et al. 1999; Yang et al. 2010; Zhang et al. 2015). On the whole, however, single-cell manipulation is simpler with optical techniques than with electrical techniques, and therefore preferable.

15.9.5 Optical Manipulation

Optical trapping of dielectric particles from tens of nanometers in diameter to tens of micrometers by a single-beam gradient force trap (also known as optical tweezers) (Ashkin et al. 1986; Ashkin and Dziedzic 1987) uses the phenomenon of optical force, or radiation pressure. A laser beam focused through a high numerical aperture objective lens is tightly focused and results in a three-dimensional gradient of laser intensity due to the Gaussian intensity profile of the laser in the transverse direction and the tight focusing in the longitudinal direction. A cell or any other microscopic, dielectric material with a refractive index greater than that of the surrounding medium will experience an optical pressure from transverse and longitudinal gradient forces that draws the cell towards the region of highest intensity—the laser beam focus. The scattering force also acts on the particle and the net effect can be stable trapping of the cell near the focal point. Optical tweezers are straightforward to implement, compatible with other microscopy techniques, and have been used to trap, position, and manipulate cells and molecules in a variety of experiments (Fig. 15.8). The wavelength of laser light can be selected to minimize photothermal

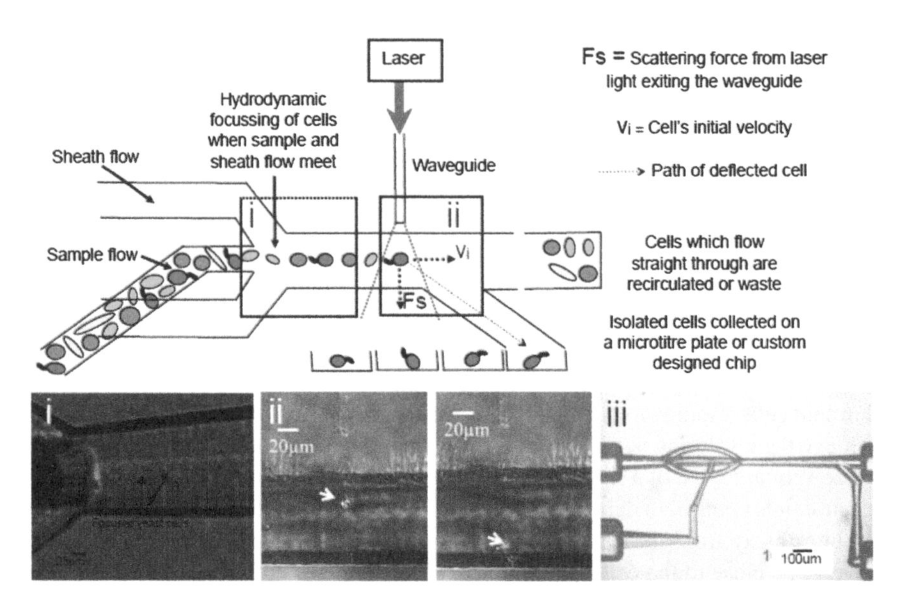

Fig. 15.8 Microfluidics with integrated optics for single-cell isolation. *Top* schematic of a working device created using ultrafast laser inscription and selective chemical etching and micrographs of the device. The cell sample is hydrodynamically focused using a sheath flow (*i*). The optical force from the laser beam emanating from the integrated waveguide deflects a single, selected cell (*ii*), which is collected from a side channel. The single isolated cell can be used to create a pure culture or other cells can be selected and deflected to create a co-culture. A micrograph of the device is shown in panel *iii*

and photochemical damage to cells (Haro-Gonzalez et al. 2013; Liang et al. 1996; Neuman et al. 1999). Cells are trapped at the focal point of a microscope objective lens, therefore working at a distance is possible, and as such experiments can be performed in enclosed, sterile sample chambers. There is no mechanical contact that can introduce risk of contamination, and subcellular organelles or endophytes can be manipulated within the cell by focusing the laser beam through the membrane (Sacconi et al. 2005). In addition, optical tweezers provide the ability to accurately measure small forces in biology, down to the level of piconewtons (Block et al. 1989).

Optical tweezers have been used for cell sorting (Ericsson et al. 2000) to select single cells and position them in patterns on a microscope slide, immobilize them, and study their viability. There have been several studies reported in the literature on the development of automated optical tweezers for manipulating and precise positioning of microspheres (Banerjee et al. 2010; Ozcan et al. 2006), diatoms (Tanaka et al. 2008), and eukaryotic cells (Grover et al. 2001; Hu and Sun 2011; Wang et al. 2013). In addition to using a tightly focused beam, laser light can be loosely focused, collimated or diverging and can exert a guiding force on a cell to push it in the direction of beam propagation (Arlt et al. 2001; Ashkin et al. 1970; Imasaka et al. 1995).

A novel, microfluidic device with integrated channels and waveguides fabricated using ultrafast laser inscription combined with selective chemical etching (Choudhury et al. 2014) is being developed in order to enable sorting and isolation of biological cells using the optical force of laser light (Fig. 15.8). The complex three-dimensional microfluidic structures within the device allow the injected cell population to focus in a hydrodynamic flow (Keloth et al. 2015; Paie et al. 2014). Continuous wave laser in the near infrared light is coupled into the integrated waveguide in the device. The laser light emerges from the waveguide into the microfluidic channel and is used to exert radiation pressure on the selected cells (Keloth et al. 2015) as these cells in the focused stream flow past the waveguide. The optical scattering force then pushes the cell from the focused stream into the sheath fluid. Thus, individual cells can be controllably deflected from the focused flow to the side channel for downstream analysis or culture.

15.9.6 Single-Cell Culture and Modification of Its Microenvironment

A large variety of methods has been or is currently under development for culture of single cells. In many cases, the work is done not with cells from the marine environment but with model organisms such as *Escherichia coli* that are easy to culture and readily modified. However, the work described can often translate well to working with (marine) uncultured microorganisms. Working with natural or artificial seawater is in itself not usually an issue when working in miniaturization; the most common materials used such as the photoresist SU8 and the polymer PDMS are highly biocompatible and resistant to seawater. The long incubation times can be an issue as dehydration is a recurrent danger when working with small volumes but can be controlled by ensuring samples remain well isolated and humidity levels are controlled. Also, PDMS is a material that is highly permeable to small molecules.

Prominent amongst single-cell culture methods are those based on micro-droplet formation in microfluidic devices (Eun et al. 2011; Joensson and Andersson Svahn 2012; Liu et al. 2009; Pan et al. 2011). Typically a stream of suspended cells is introduced into a stream on a non-miscible biocompatible fluid such as a mineral or vegetable oil or a fluorocarbon carrier fluid. Droplet formation is controlled by the interaction between hydrodynamic and interfacial forces, although electric fields may also be used to control the timing of droplet formation and droplet size (Link et al. 2006). An alternative is the formation of arrays of individual droplets by interfacial effects, for example at micro-wells in an SU8 surface (Boedicker et al. 2009). Although in each case, whether you have a single cell in a droplet or not is mainly determined by chance. The technique has great merit because it facilitates high-throughput experimentation and also automation is straightforward (Khorshidi et al., 2014). Of particular interest for the study of uncultured microorganisms is also the chemistrode, which allows sampling and formation of nanoliter volume droplets directly from the environment (Liu et al. 2009). Droplet microfluidics has allowed the investigation of the effect of cell number (Boedicker et al. 2009) on cell growth and quorum sensing responses. Confining a cell to a(n) (insulated) micro-droplet is essentially equivalent to increasing the cell density more than a thousand fold compared to having a cell in a Petri dish (Boedicker et al. 2009; Vincent et al. 2010). As a result, when confining cells to micro-droplets, Boedicker et al. (2009) found that the single cells confined in micro-droplets were able to induce a quorum sensing response. This raises the question whether quorum sensing actually is a community response. Such confinement of microorganisms to micro-droplets or micro-chambers, which have no mechanism of exchange of signaling molecules, could help inducing growth of uncultured microorganisms (Ma et al. 2014; Vincent et al. 2010). At the same time, if there is exchange, then it can be fast due the high surface to volume ratio of devices at the microscale (Boitard et al. 2015). This could be used to advantage, for example, to more effectively remove accumulated toxins.

Miniaturized methods that use compartmentalization can eliminate competition among species (Ma et al. 2014). However, as interactions between microorganisms are thought to be an important factor for cell culturability, the ability to control interactions between cells is highly attractive. A micro-droplet-based example is that developed by Park et al. (2011) for parallel co-culture of cells. Kim et al. (2008) developed a system of interconnected micro-chambers in which three of them are located close to each other but communication could only occur by diffusion. When Kim et al. (2008) constructed a community of three different species of wild-type soil bacteria with syntrophic interactions using this device they found that the spatial separation of the different species was essential for the community to survive: if the cells mixed then competition between the different species would cause the community to collapse.

The micromanipulation techniques described in the previous part of this chapter allowed one to select cells from a mixture rather than leaving the choice of cells to be investigated by chance. Yasuda et al. (Umehara et al. 2003; Wakamoto et al. 2003; Yasuda et al. 2013) have taken this approach further and used laser tweezers to move single cells of *E. coli* into individual chambers in a micro-chamber array (Wakamoto et al. 2003). After a cell had divided, the authors moved one of the two daughter cells to a vacant chamber, allowing differences between generations to be studied. In later sets of experiments, the same group also developed methods for observing the adaptation of single cells, tweezed into individual micro-chambers, to changes in nutrient concentrations (Umehara et al. 2003). The ability to change the response of cells to changes in nutrient concentration is also important for studies of culturability. Also of interest is therefore the work by Eriksson et al. (2007) who used optical tweezers to move single (yeast) cells in a gradient created within a microfluidic device, thus exposing cells to different environments and allowing the detection and analysis of rapid changes.

15.10 Conclusion

For a long time, microbiologists have judged the presence of microbial growth only by observing changes in the turbidity with the naked eye or measuring it with a spectrophotometer. Low biomass detection methods (microscopy, flow cytometry, ATPmetry, and microplate reader) now allow the observation of microbial growth with a low threshold of 10^3 cells/ml. Therefore, microbiologists have to get accustomed to working with microbial cultures with no visible turbidity. The high sensitivity of molecular methods permits the extraction of DNA from low biomass and allows one to obtain taxonomic affiliations from 16S rRNA gene amplification and sequencing. Moreover, single-cell genomic DNA can be amplified and sequenced in order to identify the metabolic functions which could help to design artificial media best suited for an optimal growth.

The dilution-to-extinction method combined with high-throughput cultivation is expected to become a widely used method in the different microbiology laboratories in order to isolate relevant microorganisms. This approach, combined with low biomass detection methods and longtime incubations, has proved to be efficient for the isolation of key microbial players. Even if a majority of the isolates that is brought into culture with this approach is later on found to be difficult to grow to high cell numbers in artificial seawater media or even under the same growth conditions, there generally should be enough cells to obtain the genomic information that may later on be useful for taming these isolates.

Culture chips and immobilization culture approaches (beads, chambers) are also promising, because they allow the maintenance of cell communication, and thus the study of syntrophic relations and interactions between cells. Moreover, many of such culture chip methods use the natural environment as the basis for the "growth medium" with advantages in culturability. It is desirable both to decrease the costs of miniaturized culture devices and automate functions such as targeting desirable micro-colonies, recovery, creating replicates, and integrating culture methods with molecular techniques.

Single-cell approaches require specific pieces of equipment (such as FACS and optical tweezers), which can be expensive and need to be handled by experts. Optical tweezers, and optical forces in general, allow one to target a specific cell in a mixture, to isolate it and study it after having isolated it in a micro-chamber.

The marine environment, the largest continuous habitat on Earth, harbors the greatest biodiversity on earth, with an untapped resource of bioactive compounds. The recent progress in the culture approaches described above could help to target marine microorganisms with high biotechnological potential, such as actinobacteria, that have been found to produce many bioactive compounds.

Acknowledgments The research leading to these results has received funding from the European Union Seventh Framework Programme (FP7/2007–2013) under grant agreement n° 311975. This publication reflects the views only of the author, and the European Union cannot be held responsible for any use which may be made of the information contained therein. We wish to thank Jochen Schuster and Anusha Keloth for useful discussions and their help with some of the illustrations. Research was supported by UBO, CNRS, and IFREMER.

References

Akselband Y, Cabral C, Castor TP, Chikarmane HM, McGrath P (2006) Enrichment of slow-growing marine microorganisms from mixed cultures using gel microdrop (GMD) growth assay and fluorescence-activated cell sorting. J Exp Mar Biol Ecol 329(2):196–205

Andrews JS, Mason VP, Thompson IP, Stephens GM, Markx GH (2006) Construction of artificially structured microbial consortia (ASMC) using dielectrophoresis: Examining bacterial interactions via metabolic intermediates within environmental biofilms. J Microbiol Methods 64(1):96–106

Aoi Y, Kinoshita T, Hata T, Ohta H, Obokata H, Tsuneda S (2009) Hollow-fiber membrane chamber as a device for in situ environmental cultivation. Appl Environ Microbiol 75(11):3826–3833

Arlt J, Garces-Chavez V, Sibbett W, Dholakia K (2001) Optical micromanipulation using a Bessel light beam. Optics Commun 197:239–245

Ashkin A (1970) Acceleration and Trapping of Particles by Radiation Pressure. Phys Rev Lett 24:156–159

Ashkin A, Dziedzic JM (1987) Optical trapping and manipulation of viruses and bacteria. Science 235(4795):1517–1520

Ashkin A, Dziedzic JM, Bjorkholm JE, Chu S (1986) Observation of a single-beam gradient force optical trap for dielectric particles. Opt Lett 11(5):288–290

Azam A, Laflin K, Jamal M, Fernandes R, Gracias D (2011) Self-folding micropatterned polymeric containers. Biomed Microdevices 13(1):51–58

Balagadde FK, You LC, Hansen CL, Arnold FH, Quake SR (2005) Long-term monitoring of bacteria undergoing programmed population control in a microchemostat. Science 309(5731):137–140

Banerjee AG, Pomerance A, Losert W, Gupta SK (2010) Developing a stochastic dynamic programming framework for optical tweezer-based automated particle transport operations. IEEE Trans Autom Sci Eng 7(2):218–227

Ben-Dov E, Kramarsky-Winter E, Kushmaro A (2009) An in situ method for cultivating microorganisms using a double encapsulation technique. FEMS Microbiol Ecol 68(3):363–371

Blainey PC (2013) The future is now: single-cell genomics of bacteria and archaea. FEMS Microbiol Rev 37(3):407–427

Block SM, Blair DF, Berg HC (1989) Compliance of bacterial flagella measured with optical tweezers. Nature 338(6215):514–518

Boedicker JQ, Vincent ME, Ismagilov RF (2009) Microfluidic confinement of single cells of bacteria in small volumes initiates high-density behavior of quorum sensing and growth and reveals its variability. Angewandte Chemie-Int Edn 48(32):5908–5911

Boitard L, Cottinet D, Bremond N, Baudry J, Bibette J (2015) Growing microbes in millifluidic droplets. Eng Life Sci 15(3):318–326

Bustard MT, Burgess JG, Meeyoo V, Wright PC (2000) Novel opportunities for marine hyperthermophiles in emerging biotechnology and engineering industries. J ChemTechnol Biotechnol 75:1095–1109

Button DK, Schut F, Quang P, Martin R, Robertson BR (1993) Viability and isolation of marine-bacteria by dilution culture theory, procedures, and initial results. Appl Environ Microbiol 59(3):881–891

Button DK, Robertson BR, Lepp PW, Schmidt TM (1998) A small, dilute-cytoplasm, high-affinity, novel bacterium isolated by extinction culture and having kinetic constants compatible with growth at ambient concentrations of dissolved nutrients in seawater. Appl Environ Microbiol 64(11):4467–4476
Cachon R, Lacroix C, Divies C (1997) Mass transfer analysis for immobilized cells of Lactococcus lactis sp. using both simulations and in-situ pH measurements. Biotechnol Tech 11(4):251–255
Carini P, Steindler L, Beszteri S, Giovannoni SJ (2013) Nutrient requirements for growth of the extreme oligotroph 'Candidatus Pelagibacter ubique' HTCC1062 on a defined medium. ISME J 7(3):592–602
Carini P, Campbell EO, Morre J, Sanudo-Wilhelmy SA, Thrash JC, Bennett SE, Temperton B, Begley T, Giovannoni SJ (2014) Discovery of a SAR11 growth requirement for thiamin's pyrimidine precursor and its distribution in the Sargasso Sea. ISME J 8(8):1727–1738
Cassidy MB, Lee H, Trevors JT (1996) Environmental applications of immobilized microbial cells: a review. J Ind Microbiol 16:79–101
Cho JC, Giovannoni SJ (2004) Cultivation and growth characteristics of a diverse group of oligotrophic marine Gammaproteobacteria. Appl Environ Microbiol 70(1):432–440
Choudhury D, Macdonald JR, Kar AK (2014) Ultrafast laser inscription: perspectives on future integrated applications. Laser Photonics Rev 8(6):827–846
Connon SA, Giovannoni SJ (2002) High-throughput methods for culturing microorganisms in very-low-nutrient media yield diverse new marine isolates. Appl Environ Microbiol 68 (8):3878–3885
Czechowska K, Johnson DR, van der Meer JR (2008) Use of flow cytometric methods for single-cell analysis in environmental microbiology. Curr Opin Microbiol 11(3):205–212
Davey HM, Kell DB (1996) Flow cytometry and cell sorting of heterogeneous microbial populations: the importance of single-cell analyses. Microbiol Rev 60(4):641
Doleyres Y, Fliss I, Lacroix C (2004) Increased stress tolerance of Bifidobacterium longum and Lactococcus lactis produced during continuous mixed-strain immobilized-cell fermentation. J Appl Microbiol 97(3):527–539
D'Souza SF (2002) Trends in immobilized enzyme and cell technology. Ind J Biotechnol 1:321–338
Duetz WA, Ruedi L, Hermann R, O'Connor K, Buchs J, Witholt B (2000) Methods for intense aeration, growth, storage, and replication of bacterial strains in microtiter plates. Appl Environ Microbiol 66(6):2641–2646
Enger J, Goksor M, Ramser K, Hagberg P, Hanstorp D (2004) Optical tweezers applied to a microfluidic system. Lab Chip 4(3):196–200
Ericsson M, Hanstorp D, Hagberg P, Enger J, Nystrom T (2000) Sorting out bacterial viability with optical tweezers. J Bacteriol 182(19):5551–5555
Eriksson E, Enger J, Nordlander B, Erjavec N, Ramser K, Goksor M, Hohmann S, Nystrom T, Hanstorp D (2007) A microfluidic system in combination with optical tweezers for analyzing rapid and reversible cytological alterations in single cells upon environmental changes. Lab Chip 7(1):71–76
Eun Y-J, Utada AS, Copeland MF, Takeuchi S, Weibel DB (2011) Encapsulating bacteria in agarose microparticles using microfluidics for high-throughput cell analysis and isolation. ACS Chem Biol 6(3):260–266
Ferrari BC, Winsley T, Gillings M, Binnerup S (2008) Cultivating previously uncultured soil bacteria using a soil substrate membrane system. Nat Protoc 3(8):1261–1269
Flickinger MC, Schottel JL, Bond DR, Aksan A, Scriven LE (2007) Painting and printing living bacteria: engineering nanoporous biocatalytic coatings to preserve microbial viability and intensify reactivity. Biotechnol Prog 23(1):2–17
Fröhlich J, König H (1999) Rapid Isolation of single microbial cells from mixed natural and laboratory populations with the aid of a micromanipulator. Syst Appl Microbiol 22(2):249–257
Fröhlich J, König H (2000) New techniques for isolation of single prokaryotic cells1. FEMS Microbiol Rev 24(5):567–572

Fu AY, Chou HP, Spence C, Arnold FH, Quake SR (2002) An integrated microfabricated cell sorter. Anal Chem 74(11):2451–2457

Futra D, Heng LY, Surif S, Ahmad A, Ling TL (2014) Microencapsulated *Aliivibrio fischeri* in alginate microspheres for monitoring heavy metal toxicity in environmental waters. Sensors (Basel) 14(12):23248–23268

Gan M, Su J, Wang J, Wu H, Chen L (2011) A scalable microfluidic chip for bacterial suspension culture. Lab Chip 11(23):4087–4092. doi:10.1039/c1lc20670b

Gavrish E, Bollmann A, Epstein S, Lewis K (2008) A trap for in situ cultivation of filamentous actinobacteria. J Microbiol Methods 72(3):257–262

Giebel H-A, Kalhoefer D, Gahl-Janssen R, Choo Y-J, Lee K, Cho J-C, Tindall BJ, Rhiel E, Beardsley C, Aydogmus O, Voget S, Daniel R, Simon M, Brinkhoff T (2013) Planktomarina temperata gen. nov., sp. nov., belonging to the globally distributed RCA cluster of the marine Roseobacter clade, isolated from the German Wadden Sea. Int J Syst Evolut Microbiol 63(Pt 11):4207–4217

Giovannoni SJ, Tripp HJ, Givan S, Podar M, Vergin KL, Baptista D, Bibbs L, Eads J, Richardson TH, Noordewier M, Rappé MS, Short JM, Carrington JC, Mathur EJ (2005) Genome streamlining in a cosmopolitan oceanic bacterium. Science 309(5738):1242–1245

Gorlas A, Alain K, Bienvenu N, Geslin C (2013) Thermococcus prieurii sp nov., a hyperthermophilic archaeon isolated from a deep-sea hydrothermal vent. Int J Syst Evol Microbiol 63:2920–2926

Graham DM, Messerli MA, Pethig R (2012) Spatial manipulation of cells and organelles using single electrode dielectrophoresis. Biotechniques 52(1):39–43

Grover SC, Skirtach AG, Gauthier RC, Grover CP (2001) Automated single-cell sorting system based on optical trapping. J Biomed Opt 6(1):14–22

Hahnke RL, Bennke CM, Fuchs BM, Mann AJ, Rhiel E, Teeling H, Amann R, Harder J (2015) Dilution cultivation of marine heterotrophic bacteria abundant after a spring phytoplankton bloom in the North Sea. Environ Microbiol 17:3515–3526

Haro-Gonzalez P, Ramsay WT, Martinez Maestro L, del Rosal B, Santacruz-Gomez K, del Carmen Iglesias-de la Cruz M, Sanz-Rodriguez F, Chooi JY, Rodriguez Sevilla P, Bettinelli M, Choudhury D, Kar AK, Garcia Sole J, Jaque D, Paterson L (2013) Quantum dot-based thermal spectroscopy and imaging of optically trapped microspheres and single cells. Small 9(12):2162–2170

Hsiao AP, Barbee KD, Huang X (2010) Microfluidic device for capture and isolation of single cells. In: Biosensing Iii 7759

Hu S, Sun D (2011) Automatic transportation of biological cells with a robot-tweezer manipulation system. Int J Robot Res 30(14):1681–1694

Hu XY, Bessette PH, Qian JR, Meinhart CD, Daugherty PS, Soh HT (2005) Marker-specific sorting of rare cells using dielectrophoresis. Proc Natl Acad Sci USA 102(44):15757–15761

Imasaka T, Kawabata Y, Kaneta T, Ishidzu Y (1995) OPTICAL CHROMATOGRAPHY. Anal Chem 67:1763–1765

Ingham CJ, van Hylckama JET (2008) MEMS and the microbe. Lab Chip 8(10):1604–1616

Ingham CJ, Sprenkels A, Bomer J, Molenaar D, van den Berg A, van Hylckama Vlieg JET, de Vos WM (2007) The micro-Petri dish, a million-well growth chip for the culture and high-throughput screening of microorganisms. Proc Natl Acad Sci 104(46):18217–18222

Ingham C, Bomer J, Sprenkels A, van den Berg A, de Vos W, van Hylckama Vlieg J (2010) High-resolution microcontact printing and transfer of massive arrays of microorganisms on planar and compartmentalized nanoporous aluminium oxide. Lab Chip 10(11):1410–1416

Ishii S, Tago K, Senoo K (2010) Single-cell analysis and isolation for microbiology and biotechnology: methods and applications. Appl Microbiol Biotechnol 86(5):1281–1292

Ishøy T, Kvist T, Westermann P, Ahring B (2006) An improved method for single cell isolation of prokaryotes from meso-, thermo- and hyperthermophilic environments using micromanipulation. Appl Microbiol Biotechnol 69(5):510–514

Joensson HN, Andersson Svahn H (2012) Droplet microfluidics—a tool for single-cell analysis. Angew Chem Int Ed 51(49):12176–12192

John RP, Tyagi RD, Brar SK, Surampalli RY, Prevost D (2011) Bio-encapsulation of microbial cells for targeted agricultural delivery. Crit Rev Biotechnol 31(3):211–226
Johnstone KI (1969) The isolation and cultivation of single organisms. In: Norris JR, Ribbons DW (eds) Methods in microbiology. Academic Press, New York, vol 1, pp 455–471
Johnstone KI (1973) Micromanipulation of bacteria. The cultivation of bacteria and their spores by the agar gel dissection technique. Churchill Livingstone, Edinburgh
Joint I, Muehling M, Querellou J (2010) Culturing marine bacteria - an essential prerequisite for biodiscovery. Microb Biotechnol 3(5):564–575
Kaeberlein T, Lewis K, Epstein SS (2002) Isolating "uncultivable" microorganisms in pure culture in a simulated natural environment. Science 296(5570):1127–1129
Kanasawud P, Hjiirleifsdottir S, Hoist O, Mattiasson B (1989) Studies on immobilization of the thermophilic bacterium Thermus aquaticus YT-1 by entrapment in various matrices. Appl Microbiol Biotechnol 31:228–233
Keloth A, Paterson L, Markx GH, Kar AK (2015) Three-dimensional optofluidic device for isolating microbes. In: Microfluidics, biomems, and medical microsystems, vol Xiii, p 9320
Keymer JE, Galajda P, Muldoon C, Park S, Austin RH (2006) Bacterial metapopulations in nanofabricated landscapes. Proc Natl Acad Sci USA 103(46):17290–17295
Khorshidi MA, Rajeswari PKP, Wahlby C, Joensson HN, Andersson Svahn H (2014) Automated analysis of dynamic behavior of single cells in picoliter droplets. Lab Chip 14(5):931–937
Kim HJ, Boedicker JQ, Choi JW, Ismagilov RF (2008) Defined spatial structure stabilizes a synthetic multispecies bacterial community. Proc Natl Acad Sci USA 105(47):18188–18193
Klingeberg M, Vorlop KD, Antranikian G (1990) Immobilization of anaerobic thermophilic bacteria for the production of cell-free thermostable alpha-amylases and pullulanases. Appl Microbiol Biotechnol 33(5):494–500
Konneke M, Bernhard A, de la Torre J, Walker C, Waterbury J, Stahl D (2005) Isolation of an autotrophic ammonia-oxidizing marine archaeon. Nature 437:543–546
Kopf A, Bicak M, Kottmann R, Schnetzer J, Kostadinov I, Lehmann K, Fernandez-Guerra A, Jeanthon C, Rahav E, Ullrich M, Wichels A, Gerdts G, Polymenakou P, Kotoulas G, Siam R, Abdallah RZ, Sonnenschein EC, Cariou T, O'Gara F, Jackson S, Orlic S, Steinke M, Busch J, Duarte B, Cacador I, Canning-Clode J, Bobrova O, Marteinsson V, Reynisson E, Loureiro CM, Luna GM, Quero GM, Loscher CR, Kremp A, DeLorenzo ME, Ovreas L, Tolman J, LaRoche J, Penna A, Frischer M, Davis T, Katherine B, Meyer CP, Ramos S, Magalhaes C, Jude-Lemeilleur F, Aguirre-Macedo ML, Wang S, Poulton N, Jones S, Collin R, Fuhrman JA, Conan P, Alonso C, Stambler N, Goodwin K, Yakimov MM, Baltar F, Bodrossy L, Van De Kamp J, Frampton DM, Ostrowski M, Van Ruth P, Malthouse P, Claus S, Deneudt K, Mortelmans J, Pitois S, Wallom D, Salter I, Costa R, Schroeder DC, Kandil MM, Amaral V, Biancalana F, Santana R, Pedrotti ML, Yoshida T, Ogata H, Ingleton T, Munnik K, Rodriguez-Ezpeleta N, Berteaux-Lecellier V, Wecker P, Cancio I, Vaulot D, Bienhold C, Ghazal H, Chaouni B, Essayeh S, Ettamimi S, Zaid EH, Boukhatem N, Bouali A, Chahboune R, Barrijal S, Timinouni M, El Otmani F, Bennani M, Mea M, Todorova N, Karamfilov V, Ten Hoopen P, Cochrane G, L'Haridon S, Bizsel KC, Vezzi A, Lauro FM, Martin P, Jensen RM, Hinks J, Gebbels S, Rosselli R, De Pascale F, Schiavon R, Dos Santos A, Villar E, Pesant S, Cataletto B, Malfatti F, Edirisinghe R, Silveira JAH, Barbier M, Turk V, Tinta T, Fuller WJ, Salihoglu I, Serakinci N, Ergoren MC, Bresnan E, Iriberri J, Nyhus PAF, Bente E, Karlsen HE, Golyshin PN, Gasol JM, Moncheva S, Dzhembekova N, Johnson Z, Sinigalliano CD, Gidley ML, Zingone A, Danovaro R, Tsiamis G, Clark MS, Costa AC, El Bour M, Martins AM, Collins RE, Ducluzeau A-L, Martinez J, Costello MJ, Amaral-Zettler LA, Gilbert JA, Davies N, Field D, Glockner FO (2015) The ocean sampling day consortium. GigaScience 4:27
Kourkoutas Y, Bekatorou A, Banat IM, Marchant R, Koutinas AA (2004) Immobilization technologies and support materials suitable in alcohol beverages production: a review. Food Microbiol 21(4):377–397
Landenberger B, Hoefemann H, Wadle S, Rohrbach A (2012) Microfluidic sorting of arbitrary cells with dynamic optical tweezers. Lab Chip 12(17):3177–3183

Lederberg J, Lederberg EM (1952) Replica plating and indirect selection of bacterial mutants. J Bacteriol 63(3):399–406
Leong TG, Randall CL, Benson BR, Zarafshar AM, Gracias DH (2008) Self-loading lithographically structured microcontainers: 3D patterned, mobile microwells. Lab Chip 8(10):1621–1624
Liang H, Vu KT, Krishnan P, Trang TC, Shin D, Kimel S, Berns MW (1996) Wavelength dependence of cell cloning efficiency after optical trapping. Biophys J 70(3):1529–1533
Ling LL, Schneider T, Peoples AJ, Spoering AL, Engels I, Conlon BP, Mueller A, Schaberle TF, Hughes DE, Epstein S, Jones M, Lazarides L, Steadman VA, Cohen DR, Felix CR, Fetterman KA, Millett WP, Nitti AG, Zullo AM, Chen C, Lewis K (2015) A new antibiotic kills pathogens without detectable resistance. Nature 517(7535):455–459
Link DR, Grasland-Mongrain E, Duri A, Sarrazin F, Cheng ZD, Cristobal G, Marquez M, Weitz DA (2006) Electric control of droplets in microfluidic devices. Angewandte Chemie-Int Edn 45(16):2556–2560
Liu W, Kim HJ, Lucchetta EM, Du W, Ismagilov RF (2009) Isolation, incubation, and parallel functional testing and identification by FISH of rare microbial single-copy cells from multi-species mixtures using the combination of chemistrode and stochastic confinement. Lab Chip 9(15):2153–2162
Lok C (2015) Mining the microbial dark matter. Nature 522(7556):270–273
Lu Z, Moraes C, Ye G, Simmons CA, Sun Y (2010) Single cell deposition and patterning with a robotic system. PLoS ONE 5(10):e13542
Ma L, Datta SS, Karymov MA, Pan Q, Begolo S, Ismagilov RF (2014) Individually addressable arrays of replica microbial cultures enabled by splitting SlipChips. Integr Biol 6(8):796–805
Martens-Habbena W, Sass H (2006) Sensitive determination of microbial growth by nucleic acid staining in aqueous suspension. Appl Environ Microbiol 72(1):87–95. doi:10.1128/aem.72.1.87-95.2006
Mason VP, Markx GH, Thompson IP, Andrews JS, Manefield M (2005) Colonial architecture in mixed species assemblages affects AHL mediated gene expression. FEMS Microbiol Lett 244(1):121–127
Mazard S, Ostrowski M, Holland R, Zubkov MV, Scanlan DJ (2014) Targeted genomics of flow cytometrically sorted cultured and uncultured microbial groups. Methods Mol Biol (Clifton, NJ) 1096:203–212
Morono Y, Terada T, Kallmeyer J, Inagaki F (2013) An improved cell separation technique for marine subsurface sediments: applications for high-throughput analysis using flow cytometry and cell sorting. Environ Microbiol 15(10):2841–2849
Nagai M, Oohara K, Kato K, Kawashima T, Shibata T (2015) Development and characterization of hollow microprobe array as a potential tool for versatile and massively parallel manipulation of single cells. Biomed Microdevices 17(2). doi:10.1007/s10544-015-9943-z
Neuman KC, Chadd EH, Liou GF, Bergman K, Block SM (1999) Characterization of photodamage to *Escherichia coli* in optical traps. Biophys J 77(5):2856–2863
Nichols D, Cahoon N, Trakhtenberg EM, Pham L, Mehta A, Belanger A, Kanigan T, Lewis K, Epstein SS (2010) Use of ichip for high-throughput in situ cultivation of "uncultivable" microbial species. Appl Environ Microbiol 76(8):2445–2450
Norton S, Lacroix C (2000) Gellan gum gel as entrapment matrix for high temperature fermentation processes: a rheological study. Biotechnol Tech 4(5):351–356
Nussinovitch A (2010) Bead formation, strengthening, and modification. Polymer macro- and micro-gel beads: fundamentals and applications. Springer, New York, pp 27–52
Ozcan M, Onal C, Akatay A (2006) A compact, automated and long working distance optical tweezer system. J Mod Opt 53(3):357–364
Paie P, Bragheri F, Vazquez RM, Osellame R (2014) Straightforward 3D hydrodynamic focusing in femtosecond laser fabricated microfluidic channels. Lab Chip 14(11):1826–1833
Pan J, Stephenson AL, Kazamia E, Huck WTS, Dennis JS, Smith AG, Abell C (2011) Quantitative tracking of the growth of individual algal cells in microdroplet compartments. Integrative Biology 3(10):1043–1051

Park J, Kerner A, Burns MA, Lin XN (2011) Microdroplet-enabled highly parallel co-cultivation of microbial communities. PLoS ONE 6(2):e17019

Rappe MS, Connon SA, Vergin KL, Giovannoni SJ (2002) Cultivation of the ubiquitous SAR11 marine bacterioplankton clade. Nature 418(6898):630–633

Rathore S, Desai PM, Liew CV, Chan LW, Lieng PWS (2013) Microencapsulation of microbial cells. J Food Eng 116(2):369–381

Robert D, Pamme N, Conjeaud H, Gazeau F, Iles A, Wilhelm C (2011) Cell sorting by endocytotic capacity in a microfluidic magnetophoresis device. Lab Chip 11(11):1902–1910

Roy D (2015) Novel bioreactors for culturing marine organisms. In: Kim S-K (ed) Springer handbook of marine biotechnology. Springer, Berlin, pp 353–358

Rusch DB, Halpern AL, Sutton G, Heidelberg KB, Williamson S, Yooseph S, Wu D, Eisen JA, Hoffman JM, Remington K, Beeson K, Tran B, Smith H, Baden-Tillson H, Stewart C, Thorpe J, Freeman J, Andrews-Pfannkoch C, Venter JE, Li K, Kravitz S, Heidelberg JF, Utterback T, Rogers Y-H, Falcon LI, Souza V, Bonilla-Rosso G, Eguiarte LE, Karl DM, Sathyendranath S, Platt T, Bermingham E, Gallardo V, Tamayo-Castillo G, Ferrari MR, Strausberg RL, Nealson K, Friedman R, Frazier M, Venter JC (2007) The sorcerer II global ocean sampling expedition: northwest atlantic through eastern tropical pacific. PLoS Biol 5(3):398–431

Sacconi L, Tolic-Norrelykke IM, Stringari C, Antolini R, Pavone FS (2005) Optical micromanipulations inside yeast cells. Appl Opt 44(11):2001–2007

Safarik I, Safarikova M (1999) Use of magnetic techniques for the isolation of cells. J Chromatogr B 722(1–2):33–53

Schnelle T, Muller T, Hagedorn R, Voigt A, Fuhr G (1999) Single micro electrode dielectrophoretic tweezers for manipulation of suspended cells and particles. Biochimica Et Biophysica Acta-Gen Subj 1428(1):99–105

Schut F, Devries EJ, Gottschal JC, Robertson BR, Harder W, Prins RA, Button DK (1993) Isolation of typical marine-bacteria by dilution culture growth-growth, maintenance and characteristics of isolates under laboratory conditions. Appl Environ Microbiol 59(7):2150–2160

Sharpe AN, Michaud GL (1974) Hydrophobic grid-membrane filters: new approach to microbiological enumeration. Appl Microbiol 28(2):223–225

Song J, Oh H-M, Cho J-C (2009) Improved culturability of SAR11 strains in dilution-to-extinction culturing from the East Sea. West Pac Ocean. FEMS Microbiol Lett 295(2):141–147

Staley JT, Konopka A (1985) Measurement of *in situ* activities of nonphotosynthetic microorganisms in aquatic and terrestrial habitats. Annu Rev Microbiol 39:321–346

Stepanauskas R (2012) Single cell genomics: an individual look at microbes. Curr Opin Microbiol 15(5):613–620

Stingl U, Tripp HJ, Giovannoni SJ (2007) Improvements of high-throughput culturing yielded novel SAR11 strains and other abundant marine bacteria from the Oregon coast and the Bermuda Atlantic Time Series study site. ISME J 1(4):361–371

Stumpf F, Schoendube J, Gross A, Rath C, Niekrawietz S, Koltay R, Roth G (2015) Single-cell PCR of genomic DNA enabled by automated single-cell printing for cell isolation. Biosens Bioelectron 69:301–306

Tanaka Y, Kawada H, Hirano K, Ishikawa M, Kitajima H (2008) Automated manipulation of non-spherical micro-objects using optical tweezers combined with image processing techniques. Opt Express 16(19):15115–15122

Tanaka T, Kawasaki K, Daimon S, Kitagawa W, Yamamoto K, Tamaki H, Tanaka M, Nakatsu CH, Kamagata Y (2014) A hidden pitfall in the preparation of agar media undermines microorganism cultivability. Appl Environ Microbiol 80(24):7659–7666

Tripp HJ, Kitner JB, Schwalbach MS, Dacey JWH, Wilhelm LJ, Giovannoni SJ (2008) SAR11 marine bacteria require exogenous reduced sulphur for growth. Nature 452(7188):741–744

Umehara S, Wakamoto Y, Inoue I, Yasuda K (2003) On-chip single-cell microcultivation assay for monitoring environmental effects on isolated cells. Biochem Biophys Res Commun 305(3):534–540

Vancanneyt M, Schut F, Snauwaert C, Goris J, Swings J, Gottschal JC (2001) Sphingomonas alaskensis sp nov., a dominant bacterium from a marine oligotrophic environment. Int J Syst Evol Microbiol 51:73–80
Venter J, Remington K, Heidelberg J, Halpern A, Rusch D, Eisen J, Wu D, Paulsen I, Nelson K, Nelson W (2004) Environmental genome shotgun sequencing of the Sargasso Sea. Science 304:66–74
Vincent ME, Liu W, Haney EB, Ismagilov RF (2010) Microfluidic stochastic confinement enhances analysis of rare cells by isolating cells and creating high density environments for control of diffusible signals. Chem Soc Rev 39(3):974–984
Wakamoto Y, Umehara S, Matsumura K, Inoue I, Yasuda K (2003) Development of non-destructive, non-contact single-cell based differential cell assay using on-chip microcultivation and optical tweezers. Sens Actuators B-Chem 96(3):693–700
Wang X, Gou X, Chen S, Yan X, Sun D (2013) Cell manipulation tool with combined microwell array and optical tweezers for cell isolation and deposition. J Micromechan Microeng 23(7):075006
Watve M, Shejval V, Sonawane C, Rahalkar M, Matapurkar A, Shouche Y, Patole M, Phadnis N, Champhenkar A, Damle K, Karandikar S, Kshirsagar V, Jog M (2000) The 'K' selected oligophilic bacteria: A key to uncultured diversity? Current Science 78:1535–1542
Weibel DB, DiLuzio WR, Whitesides GM (2007) Microfabrication meets microbiology. Nat Rev Microbiol 5(3):209–218
Winson MK, Davey HM (2000) Flow cytometric analysis of microorganisms. Method Companion Method Enzymol 21(3):231–240
Wyatt Shields Iv C, Reyes CD, Lopez GP (2015) Microfluidic cell sorting: a review of the advances in the separation of cells from debulking to rare cell isolation. Lab Chip 15(5):1230–1249
Yang S-M, Yu T-M, Huang H-P, Ku M-Y, Hsu L, Liu C-H (2010) Dynamic manipulation and patterning of microparticles and cells by using TiOPc-based optoelectronic dielectrophoresis. Opt Lett 35(12):1959–1961
Yasuda K, Hattori A, Kim H, Terazono H, Hayashi M, Takei H, Kaneko T, Nomura F (2013) Non-destructive on-chip imaging flow cell-sorting system for on-chip cellomics. Microfluid Nanofluid 14(6):907–931
Yun H, Kim K, Lee WG (2013) Cell manipulation in microfluidics. Biofabrication 5(2):022001
Yusof A, Keegan H, Spillane CD, Sheils OM, Martin CM, O'Leary JJ, Zengerle R, Koltay P (2011) Inkjet-like printing of single-cells. Lab Chip 11(14):2447–2454
Zengler K, Toledo G, Rappe M, Elkins J, Mathur EJ, Short JM, Keller M (2002) Cultivating the uncultured. Proc Natl Acad Sci U S A 99(24):15681–15686
Zengler K, Walcher M, Clark G, Haller I, Toledo G, Holland T, Mathur EJ, Woodnutt G, Short JM, Keller M (2005) High-throughput cultivation of microorganisms using microcapsules. Methods Enzymol 397:124–130
Zhang X, Leung C, Lu Z, Esfandiari N, Casper RF, Sun Y (2012) Controlled aspiration and positioning of biological cells in a micropipette. IEEE Trans Biomed Eng 59(4):1032–1040
Zhang P, Ren L, Zhang X, Shan Y, Wang Y, Ji Y, Yin H, Huang WE, Xu J, Ma B (2015) Raman-activated cell sorting based on dielectrophoretic single-cell trap and release. Anal Chem 87(4):2282–2289
Zhu K, Kaprelyants AS, Salina EG, Schuler M, Markx GH (2010) Construction by dielectrophoresis of microbial aggregates for the study of bacterial cell dormancy. Biomicrofluidics 4(2):022810

Chapter 16
Bringing New Products from Marine Microorganisms to the Market

Hywel Griffiths

Abstract The seas provide huge untapped reserves of microbial diversity and potentially useful bioactive molecules, yet their exploitation comes with many challenges as a product is brought to market. The requirements for bringing a product to market depend very much on the type of product, type of market and to some extent, geographical location of that market. Each product will face unique problems, but several themes are recurrent. These include: the need to develop a commercially viable means of production, the need to scale production to meet the demands of the market either through subcontracting or construction of production facilities, verification of the performance and safety of the product to the standards of the relevant regulatory bodies, ensuring customer acceptance of the product through market research and marketing, and above all, finding the finance to support all of these. Clearly, with such a wide variety of potential products to consider, from raw biomass, secreted products, extracts and concentrates all the way up to purified molecules, and potential markets as different as green chemistry and pharmaceuticals, it is impossible to cover all permutations but many of the challenges, while differing in the details, are surprisingly common.

16.1 Means of Production

In attempting to commercialize a product from marine microorganisms it is of course first necessary to develop methods for producing biomass in quantities sufficient to address the needs of the market, and to do so in an economically viable manner. There is significant skill involved in the transformation of a laboratory process to one which operates at large scale, is sufficiently reproducible to meet product specifications with each batch, and in which downstream processing of material to product is cost-effective and suffers minimal losses. In some cases the organism that natively produces the molecule or molecules of interest may simply

---

H. Griffiths (✉)
Fermentalg, 4 Rue Rivière, 33500 Libourne, France
e-mail: hgriffiths@fermentalg.com

L.J. Stal and M.S. Cretoiu (eds.), *The Marine Microbiome*,
DOI 10.1007/978-3-319-33000-6_16

be unsuitable for industrial exploitation and transfer of genetic material to another species or a chemical means of synthesizing the active molecule may be the only option to bring the product to market (Molinski et al. 2009).

16.1.1 Cultivation of Biomass

Marine microorganisms fall into two broad groups when considering means of production: those capable of autotrophic growth, and those that require some form of organic carbon substrate. The former group consists almost exclusively of photosynthetic eukaryotic microalgae and cyanobacteria (also known as blue-green algae), while the latter includes heterotrophic bacteria, fungi, heterotrophic protists, and some forms of heterotrophic microalgae. Many species that are autotrophic are also capable of using organic carbon, either in the absence of light or under mixotrophic conditions (where both organic carbon and light are present and used) (Kaplan et al. 1986) and this provides further options for industrial exploitation.

16.1.1.1 Cultivation in Autotrophic Conditions

Humans have been collecting photosynthetic organisms from natural lagoons for millennia. Cyanobacteria have been used in Asia, South America, and Africa as food sources (Jenson et al. 2001), in some cases as a routine foodstuff, and in other situations, a source of nutrition in times of hardship. More recently, man-made lagoons and raceways have been used to expand production area and to increase productivity (Spolaore et al. 2006). These forms of cultivation have the advantage of being easy to construct and of using a free energy-source—sunlight, and thus autotrophy is a seemingly attractive method of biomass production.

Biomass productivity in autotrophy is low, however, being limited both by the amount of solar energy that penetrates into the culture, and the availability of dissolved CO_2 for fixation into organic matter. For this reason, many modern production facilities are situated in areas of high sunlight and next to industrial sources of CO_2; Australia, Israel, and the southern states of the US are areas where the necessary level of industrialization and sunlight come together and many large-scale sites are to be found in these regions. Biomass density in these types of culture is generally low, and harvest and concentration of the biomass represents a significant part of the cost of production.

Another issue faced by production facilities using ponds and raceways is that they are open to the environment and thus susceptible to contamination by other microorganisms and predators. This can lead to so-called culture crashes, in which whole crops are lost. While some types of microorganisms can be successfully cultivated in this manner, they tend to be those that can survive in extreme environments, where competition and predation by other organisms are less of an issue. The alga *Dunaliella salina* is cultivated for the production of the pigment beta-carotene, and can be grown in a hypersaline environment, while the cyanobacterium *Spirulina* can be grown in a high-pH environment. Even then, there is no guarantee that other organisms with similar tolerances will not coexist in the culture. For example, analysis of commercially available cyanobacterial products in a number of countries has found them to be contaminated with microcystins—toxins that will have been produced by other cyanobacteria in the original culture (Jiang et al. 2008; Heussner et al. 2012).

More recently, biofuels research has allowed some companies to propose other species for use in raceways and ponds, such as the microalgae *Nannochloropsis oculata* or *Scenedesmus dimorphus*. In such cases predation and contamination have to be closely controlled by introduction of compounds such as ammonia, ozone, pesticides, or fungicides, which have a greater negative effect on predators or contaminants than on the strain of interest (McBride et al. 2014). Even with these issues managed, there are still issues of seasonal and yearly variation caused by differences in climatic conditions, and these will affect product consistency.

In response to many of these problems, closed photobioreactor systems have been developed. In these systems the issues of light penetration and contamination are less problematic, and there is a greater level of control of the culture through pH and temperature regulation. This allows higher cell densities to be reached, decreasing the harvesting costs, and also a more consistent product to be produced. These type of systems also open up the opportunity to exploit other species; for example the green alga *Chlorella* is produced for human nutrition by Roquette in Germany (Zitelli et al. 2013), the microalga *Tetraselmis* is produced for human nutrition by Fitoplancton Marino in Spain (EU 2015), and another microalga *Haematococcus* is produced by a number of companies, such as Alga Technologies in Israel or AstaReal in Sweden, again largely for the human nutrition market. As with production in ponds and raceways, climatic variation can have an effect on production and so, for extremely high-value products, completely controlled conditions with artificial light can be considered. For most products, however, the capital cost of the photobioreactors is prohibitive.

16.1.1.2 Cultivation in Heterotrophic Conditions

Most marine bacteria, all fungi and many protists are not capable of autotrophic growth and thus, to culture these organisms, some form of organic energy needs to be supplied. This is termed heterotrophic growth. While at a laboratory scale this may be performed in flasks, for efficient microbial biomass production it is preferable to use a stirred tank or airlift fermenter. Extremely large fermentation vessels with volumes of up to a million liters can be found at industrial scale. In the fermentation vessel the cells are kept well-mixed and supplied with the nutrients required to support growth while physical parameters such as dissolved oxygen levels, pH and temperature are regulated. This can allow maximal growth rates to be sustained over long periods, up to cell densities that would not naturally occur, thereby increasing productivity. In many cases a secondary set of conditions can then be imposed to induce production of the molecules of interest. For some types of fungi, solid-phase fermentation may be more appropriate, especially for production of some types of enzyme, which can be secreted directly into the substrate, avoiding the need to harvest the biomass (Bhargav et al. 2008).

Culturing under heterotrophic conditions has a number of advantages. When compared to autotrophic culture, the greatest benefits are the markedly higher productivities and higher dry weights achievable. These result from the removal of limits imposed by the energy source, such as irradiance levels, photoinhibition or mutual shading seen in autotrophic systems. These high dry weights also have a positive effect on downstream processing of the material produced. Heterotrophic culture also requires a high degree of control over culture conditions resulting in a product that is highly reproducible. Cultures are also generally carried out axenically, so the risk of product contamination is much diminished. While the capital costs of fermentation equipment are high, and there is a need to pay for the carbon substrate, the advantages of heterotrophic fermentation are such that this is usually the chosen means for production at scale. Even some types of microalgae capable of photosynthetic growth are now commercially produced heterotrophically due to the advantages in productivity and control (Barclay et al. 2013).

16.1.1.3 Cultivation Under Mixotrophic Conditions

Mixotrophy is a growth mode in which both light and organic carbon are provided. For some microalgae this provides a simple growth boost over autotrophy, but light is still a necessary ingredient for growth. For commercial exploitation the advantage comes in increased productivity (Abreu et al. 2012). For many types of marine microorganisms, however, and not just those that are capable of photosynthetic growth, light can have an effect on the behavior of cells in culture; for example metabolic pathways for the production of particular types of molecules can be induced, substrate utilization can be altered and growth rates can be changed. These effects are often wavelength-specific and may require surprisingly low levels of light for induction (Calleja 2012). The French company Fermentalg is currently using these effects in the industrial exploitation of several species of microalgae, and is convinced that it provides an option for the exploitation of many algal species that were previously limited to autotrophic production.

If the product under development is intrinsically linked with one species, then the means used for biomass production may be dictated by the biological characteristics of the organism. If circumstances allow, however, it may be wise to at least examine other similar species, or other types of organisms that produce the molecule of interest, simply to see if other types of production process are available, since each has distinct advantages and disadvantages that may be pertinent to the market one wishes to address.

16.1.2 Development of the Production Process

In the industrialization of a process to support commercial activity, one of the key objectives is to produce the biomass or extracted molecule at a price that the market will support. This generally means improving productivity of biomass and increasing the content of the molecule of interest within the biomass in a manner that is consistent with production in large volumes. For autotrophic organisms, choosing the site of the facility can have a large impact on the economics and productivity of the process as a whole, with sunlight, climate, availability of carbon dioxide, sources of brine or seawater, and materials required for downstream processing all playing an important part (Borowitzka 1990). Effects of process parameters will also need to be optimized, not only to maximize biomass production but to ensure all steps of the process through to product can be carried out effectively.

For processes carried out under more controlled conditions, process and media design are hugely important. It is worth considering appropriate scale production from the very outset of development; processes that work well at lab scale but which are dependent on expensive ingredients such as peptones do not scale well, both from an aspect of cost, but also potentially in terms of availability of substrate. For truly novel processes and organisms, there may be little help in the existing literature and it is important to remember that the process will take the cells far from their 'natural' conditions, and they may therefore come up against unexpected limitations. For example, for marine microorganisms it is tempting to base growth media on the composition of seawater, since the organisms have evolved in this type of environment. For some minerals, however, while they may be in effectively limitless supply in the natural environment, high biomass concentrations can result in depletion and limitation. In contrast, for other ions, the concentration present in the sea may represent a significant excess over the biological need, even at high densities.

Thought also needs to be given to downstream processing during the development of biomass production. In some cases the needs of downstream processing may even dictate where production is carried out (Curtain 2000). To bring a product into its final form, it is necessary that the characteristics of the biomass produced lend themselves to harvest and, as applicable, to the steps of drying, extraction and purification. It is necessary to choose downstream processes that are available at a relevant scale, which give an acceptable yield and throughput at an acceptable cost, and which meet with the regulatory requirements for the product.

In all of this, thought must be given to the patent landscape. Novel processes provide the opportunity for protection by filing applications, but if others are in the same market, there is the risk that certain avenues are already protected. Alternative methods may need to be found, or in some cases licensing of technology may be necessary.

16.2 Addressing Your Market

There are clearly many opportunities to produce marvelous products from marine microorganisms. Nevertheless, one must not get carried away by their potential, but must face the economic reality that, without a market to sell into, and without the ability to produce these products at a price that customers will accept, this potential is meaningless. Therefore, part of the industrialization process may be to *create* a market, and will certainly include convincing backers that there is at least a potential market for the product. It is also never too early in the industrialization process to talk to potential customers and to find out under what circumstances they would be willing to buy the product, since their requirements can have a crucial impact on the development of the process.

16.2.1 Replacement Products from Marine Microorganisms

With increasing pressures on the environment, there is a growing demand from consumers for sustainable alternatives to products derived from fossil fuels or limited natural resources. Since marine microorganisms form the base of the marine food chain and produce a huge diversity of molecules, there are plenty of opportunities for replacement products to be developed that meet the sustainability and scale criteria in these markets.

With these types of product, one of the main advantages for commercialization is the existence of preexisting markets, where demand can be demonstrated, prices have been established and customers already exist. The disadvantages are, of course, that the market already exists with established competitors, and that price expectations are already set.

One might naïvely expect that the advantages of a sustainable product would command some form of price premium over the products that they are developed to replace, but while consumers may express a desire for sustainability, they are far less willing to pay for it. In reality one should not expect to be able to sell for a greater price than exists in the market place, unless the product provides some additional advantage or value over the existing solutions. Even then, for certain markets, especially those with a high regulatory burden, entrenched products can be very difficult to displace, even with a product of lower cost.

Volume is also an important aspect in these types of markets, and financial and logistical planning on how to address the demands of the market are critical parts of the development process. Customers will already have markets of their own to serve, and will have existing supply chains. They may be unwilling to change, or at least incorporate, a new supply unless minimum quantities of material can be guaranteed. This has the potential to create a circular problem, since volumes cannot be guaranteed without finance to support production, yet finance cannot be found without a guarantee of sales, and sales cannot be made without a guarantee of volume production.

Even when looking at equivalent products to replace those already on the market, obtaining regulatory approval for both the product and the production process is another aspect of the development. This approval can take a significant amount of time, and has to be factored-in when planning an approach to market. The regulatory environment will be discussed later in this chapter.

16.2.2 New Products from Marine Microorganisms

One of the most exciting aspects of working with marine microorganisms is the huge potential that they offer. Microorganisms can be isolated from the depths of ocean trenches, from smoking vents on the sea floor, or from under the ice sheets, representing whole new areas of diversity from which new products could be developed. Due to the fantastic level of untapped variety available in the ocean, it is almost as easy to find new, amazing, and useful strains of marine microorganisms at the beach or just offshore. Even fish from the local market have been used to source strains with potential interest (Ryan et al. 2010). The challenge is almost one of working out which elements to exploit, and which market to address, rather than finding something to exploit in the first place. In some cases, the potential is there to create a whole new market, and that really can be exciting.

Even if some form of activity has already been identified at laboratory scale, the choice of market is an important element to address in the very early stages of development as it defines which types of screens are made to identify bioactives, whether they need to be extracted or purified or can remain in biomass, what choices can be made on methods of production, to what degree regulatory issues will affect the process and product, and the volume requirement and price sensitivities of the market. Choice of market will also affect the types and amounts of investment that can be attracted to fund the development of the product.

16.2.2.1 Pharmaceuticals

Bioactive compounds for use in the pharmaceutical space are an attractive area for research, and many compounds with a marine origin have been, and are being tested, although the vast majority of these have been isolated from multicellular organisms (Martins et al. 2014; Mayer et al. 2010). The diversity of molecules produced by marine microorganisms means that there are bound to be many more effective compounds for the treatment of disease and disorders waiting to be discovered. Screens for biological activity are only the very first stage of a process that can last well over a decade. Potential rewards are high, since these types of compounds have a high value, but most fail before reaching the clinic due to issues with safety or efficacy. Full development costs, including all phases of clinical trials, can stretch into billions of dollars (Paul et al. 2010).

For a new molecule, bringing a product to market, including the tasks of gaining full regulatory approval, and then marketing and distribution, is a costly and lengthy process. For those reasons, many companies choose to license technology, or partner with a large pharmaceutical company for the final stages of bringing a molecule to the clinic. Even in earlier stages of development, where a company may be performing early animal tests or clinical trials, there are significant requirements for product purity and constraints on the manner in which it is produced (for example facilities may need to be registered and audited for current good manufacturing practices (cGMP)), and these issues need to be factored into development.

Although risks can be minimized by targeting particular applications and types of molecules, they can never be eliminated. Glycomar, for example, is attempting to minimize their risks with a range of polysaccharides sourced from marine microorganisms. The nature of the molecules and the choice of topical applications decrease the risk of adverse reactions (Glycomar 2014).

The regulatory requirements when producing new, alternative sources of an active molecule are not as demanding as the process for registering a new drug, since efficacy is already demonstrated, but can still be significant, as the steps involved in the production of these molecules need to be carefully controlled.

16.2.2.2 Health, Food and Nutrition

Health products (distinct from pharmaceuticals), food ingredients, and nutritional supplements are other attractive areas for the development of products from marine microorganisms. Many of these microorganisms produce molecules essential for proper nutrition, and due to the position of these microorganisms at the base of the food web, these molecules are then transferred into many of our healthier foods. These can include essential fatty acids, antioxidant pigments, vitamins, and proteins with well balanced amino acid profiles. By producing these molecules directly within the microorganisms, we are able to offer materials that can be blended into existing foodstuffs to improve their nutritional value, supplements with general benefits to health, and products with specific nutritional properties for subgroups of the population. The market for these types of products is already large with, for example, around 75 % of microalgal production being used in health food and dietary supplements (Vigani et al. 2014). The market continues to grow as consumers become more aware of the impact of their diet on their short and long-term health. Here, products from marine microorganisms have an in-built market advantage, since both 'natural' products and the sea itself have healthy connotations.

Since health products, food ingredients and nutritional supplements are destined for human consumption, there are regulatory requirements to be borne in mind. Production of biomass must be performed in a manner consistent with food use; depending on the process of manufacture and degree of downstream processing performed, for example, ingredients used in the growth of biomass may have to be food-grade themselves, and testing for toxins, heavy metals, microbial contaminants, or other substances hazardous to health may be part of the requirements for release of material. It is also necessary to obtain permission from regulatory authorities to put the product on sale. Unfortunately, there is no one method of gaining this permission and separate approvals must be gained in Europe, the US, and other territories around the world.

In Europe, “Novel Food” approval is required for products with no history of safe use in at least one country of the European Union, and obtaining this can be a lengthy process, often taking years. While no product follows exactly the same route as another, for any truly novel product, safety testing of biomass and the final product form in animals is likely to be necessary, as is a demonstration of process reproducibility, product stability and descriptions, and limitations of intended use. For some specific uses such as food additives, including colorants, antioxidants, and stabilizers, further regulatory approval is required. Any claims to health benefits also need to be supported by scientific evidence and approved by the European authorities.

In the US, where products are permitted to be sold once generally recognized as safe (GRAS), safety is self-affirmed rather than approved and thus the process of getting a product on market is generally faster than that seen for Europe. Again, a dossier supporting the safety of a product for a certain purpose, often including evidence from animal studies, has to be produced and assessed by a scientific panel. The dossier, along with the conclusions of the panel are then presented to the US Food and Drug Administration (FDA). The authority never gives approval, but does have the ability to negatively respond to the application, indicating that it does not believe the dossier provides sufficient evidence of safety. In contrast to GRAS-designated materials, food additives, which include food colors, must be actively approved by the agency before they can be incorporated into foods. Other territories around the world require different approvals: Australia and New Zealand are regulated by Food Standards Australia and New Zealand (FSANZ), Japan by the Food Safety Commission (FSC), India by the Food Safety and Standards Authority of India (FSSAI), and China by the China Food and Drug Administration (CFDA), to name but a few.

16.2.2.3 Cosmetics

There is a longstanding use of marine products in cosmetics, with many materials derived from seaweeds having been employed for decades. These materials are exploited in a number of ways, with some being used to affect the basic physical properties of cosmetics, such as the viscosity of creams, others being used for their effects on moisture retention, and yet others being used for more active purposes with anti-inflammatory or antioxidant activity (Fitton et al. 2007).

Marine microorganisms can produce the same types of molecules, and many more besides, and are a rich resource for exploitation in this area. Cosmetics with marine algal extracts listed as ingredients are now common, polysaccharides and extracts from marine bacteria are also in use, and a number of companies are now making entries into the market with purified ingredients derived from marine microorganisms with specific benefits. Martins et al. (2014) provide a good history of the molecules and extracts on the market today, along with an explanation of their modes of action where these are understood.

Having an expanded range of molecules with useful properties is obviously of benefit to the cosmetics industry, but importantly for an industry that is, by its nature, very image conscious, these also have the potential to be produced in controlled conditions, and in a sustainable manner. This is therefore an extremely attractive industry for future developments from marine microorganisms, with both a history and huge potential.

As with foodstuffs, cosmetics and their ingredients are highly regulated. In most countries, manufacturers have responsibility for ensuring the safety of materials before they are put on the market, and their satisfaction of regulatory requirements. Responsibility for upholding the regulations then generally lies with an authority such as the FDA in the US, the Japanese Ministry of Health, Labor and Welfare, and various national bodies across Europe (although there is common Europe-wide safety legislation for cosmetics). Whilst many of the regulatory requirements are at least similar between territories, one critical difference lies in the treatment of active ingredients. In territories such as Europe, many of these may be incorporated into cosmetics with no further regulatory burden, whereas in the US these may be treated as drugs, and require further approval before they may be used (Kingham and Beirne 2011).

16.2.2.4 Feed

Animal nutrition is yet another area where marine microorganisms are being introduced as sources of ingredients or additives. The health and wellbeing of companion animals are, at least in the US and Europe, taken almost as seriously as human health and wellbeing. Many of the lessons learnt about the benefits of ingredients of marine origin, in particular antioxidants and omega-3 fatty acids, are being applied to pet foods.

Marine microorganisms have been used for decades in aquaculture, specifically in the hatcheries where the larval stages of mollusks, crustaceans and fish are reared either on the microorganisms themselves, or on prey reared on the microorganisms. Production is usually small-scale and local to the hatcheries, but is a vital part of the exploitation of these resources. Pressure on natural resources has, for many years now, also been having an impact on the feed for juvenile and adult organisms. Where once fishmeal and fish oil were readily available and cheap, they are now scarce and expensive, and so other, vegetable sources of protein and fat have been substituted into the diet (FAO 2014). Unfortunately, for many species used in aquaculture, vegetable sources may not provide all the materials required; for example, salmonid fish have a dependence on polyunsaturated fatty acids (PUFAs) in their diet and some crustaceans and mollusks require a source of cholesterol. For the provision of sterols, plant sources may in fact be counterproductive since the phytosterols can have an anti-nutritive effect. Other nutrients, while not absolutely necessary for growth, may have a positive effect on the quality of the flesh of the animals and these too may be difficult to source from vegetables. Pigments such as the carotenoid astaxanthin fall into this group.

In the natural environment, the diets of the species used in aquaculture come either directly or indirectly from marine microorganisms, which form the base of the food chain. Species can thus be found and exploited to provide an alternative to both animal and vegetable dietary components, and there is growing interest from the aquaculture industry in these types of solutions, particularly where such species can be demonstrated to have further positive effects on health or productivity.

Changes in the patterns of land usage and agriculture demand that new sources of feed for farmed animals also be investigated. Marine microorganisms can be excellent sources of protein and energy, in the form of lipids or fibers, and thus have plenty of potential for exploitation in this area. If grown in controlled culture, their consistency may also be advantageous, since food formulation is extremely sensitive to content and having a reliable source of consistent quality would be most helpful.

The major challenges for aquaculture, and even more so for animal feed, are price and required volume. In many cases soy is a dominant source of nutrition, and many of the disadvantages of this material have already been worked around (for example, higher heat may be used to destroy anti-nutritive factors). Since soy is readily available, to provide a replacement, marine microorganisms need to either be cheaper or demonstrate a distinct advantage.

The volumes required to address these markets are also staggering, and this provides further challenges; *Spirulina*, for example, could perhaps be considered as a candidate to replace some of the fishmeal used today in aquaculture, being a rich source of protein, minerals, and vitamins with the added bonus of providing pigments and antioxidants. Potential production volumes are, however, just too low to have more than a tiny impact on annual demand. Even Earthrise, the largest producer of *Spirulina* in the world only has an annual production of around 500 T dry weight (DIC 2015), whereas annual production of (and demand for) fishmeal is measured in millions of tonnes (FAO 2014). For farmed animal feed the potential volumes are equivalent or larger.

Regulation of feed materials follows much the same pattern as food in Europe as well as in the US. Feed ingredients, those that make up the bulk of the feed, are subject to the least stringent control, while feed additives, which are added in smaller quantities to improve the quality of the feed or animal performance or health, are generally subject to approval, and medicated feeds are subject to even greater control (Smedley 2011).

16.3 Products in the Real World

16.3.1 Polyunsaturated Fatty Acids

One of the best series of examples of products from marine microorganisms entering, and to some extent creating, a market can be found with the PUFAs. Here the questions of differentiation in the market, regulatory requirements, production volume, and barriers to entry for new products may all be illustrated.

PUFAs, especially the long chain omega-3s docosahexaenoic acid (DHA) and eicosapentaenoic acid (EPA), are an essential part of the human diet, since the body is very poor at synthesizing these molecules, even from similar fatty acids of shorter chain length. A higher level of consumption of these PUFAs is associated with improved eye and brain health (Horrocks and Yeo 1999), and lower risk of mortality from cardiovascular disease (Yokoyama et al. 2007). There are also indications that insufficient intake may affect learning and mental health (Deacon et al. 2015).

Terrestrial animals and vegetables are poor sources of EPA and DHA, but marine sourced ingredients are much better sources, in large part due to the production of EPA and DHA by marine microorganisms, and their concentration up through the food chain. Oily fish such as sardines, anchovies, and salmon are particularly good sources. Since the original sources of these omega-3s are marine microorganisms, one might expect there to be many opportunities for exploitation. Commercial reality is different, with only one company who has been a real success, several others who are bringing variants of the same process to market, and a host of others who have tried and failed.

16.3.1.1 The Market

The main issue facing any new entrant into the PUFA market is the existence of an existing source: fish oil. Not so long ago, fish oil was effectively a waste product and was used as a cheap source of energy for animal feed, or even as a raw material in the production of paint. The rise of aquaculture, where fish oil is often a required ingredient in feed, and later an increased awareness of the health benefits of omega-3s, have led to increased demand and pressure on the supply of these oils, the production of which is now declining due to pressures on fish stocks (FAO 2014). Due to the increased demand and decreased supply, the price of fish oil has risen markedly, especially over the last 5 years, but is generally still lower than the price of oils produced from marine microorganisms. Unless the price of fish oil continues to increase markedly, or new production technologies bring the price of oils down considerably, marine microbial PUFA products will not compete on price alone, and thus need to find other differentiating factors.

At the end of the last century, the American company Martek managed to do just that, and built a billion dollar company on the basis of DHA production processes using marine microorganisms. Their success was built on several factors:

- Martek managed to create a new market for their product, one in which the differences between the microbial product and fish oil derived products were distinct advantages, and one in which customers were willing to pay for those advantages,
- the market established was in an area where there was a large degree of regulatory oversight so that the safety associated with a controlled production process was of great benefit and allowed the product to become the de facto standard, which was hard to displace,
- being the ones to create both the process and the market, Martek was able to establish a powerful intellectual property portfolio to protect their process, and were able to establish very strong commercial relationships with their customers who, in effect, were reliant on Martek for supply.

The key to creation of this market was identifying that infant formula made from cow milk differs from human breast milk in that it contains less DHA. DHA is important for brain development, and thus an argument was made that infant formula should be supplemented with this fatty acid. This conclusion has been born out in studies of infant development comparing formula with and without DHA (Koletzko et al. 2008). In this particular case, however, the presence of EPA is undesirable, since it can be antagonistic to the effects of arachidonic acid, which is important for the growth of infants (Carlson et al. 1993). EPA is, in any case, found in lower levels in human breast milk than the other PUFAs, and there is a desire to mimic nature. Thus, a source of DHA was needed that was also low in EPA.

One area where single-celled oils differ markedly from fish oils is in their composition; whilst different from species to species and season to season, fish oils all tend to be complex mixtures of fatty acids, with EPA and DHA mixed in various proportions, usually with EPA as the dominant of the two. Single cell oils, in contrast, tend to be much simpler, with either EPA or DHA dominant, and the other either absent or present in much smaller quantities. Martek developed a high efficiency, and scalable heterotrophic process to produce an oil from *Crypthecodinium cohnii* that contained DHA and virtually no other PUFAs (FSANZ 2003), allowing them to position the single cell oils as the ideal solution for supply into the infant formula market.

Regulatory requirements for safety are much higher in infant formula than for normal food products and ingredients, and the security offered by a controlled fermentation process for production of the oils was a further advantage in promoting the microbial oil over alternative sources.

Having secured their position with this first product, Martek also developed DHA-rich oil from another group of microorganisms, the Thraustochytrids. This oil differs slightly in that it contains a non-negligible amount of the omega-6 docosapentaenoic acid, DPA (Martek 2004). The product is cheaper to produce as the organisms grow faster and make more oil. While this new oil has been used to expand into new markets including food, feed and supplements, the original oil remains dominant in the infant formula market, despite its higher cost. This is because any changes to the formulation required expensive studies into the effect of the newer oil on infants, with unknown outcome given the presence of DPA omega-6; a risk no one was willing to take until recently. Regulatory requirements, which aided in the creation of the opportunity for the microbial oil, thus also protected it from new entrants.

For EPA, supply is still dominated by fish oil in all markets. It is true that no one has yet identified microorganisms that can produce EPA-rich oils with the same economics as DHA-rich oils, but perhaps more importantly, no one has yet managed to identify and capitalize on an opportunity where microbial oils present a significant advantage over fish oils for reasons other than economics. In recent years companies such as Aurora Algae and Photonz may have come close, but they appear not to have convinced the markets of their strategies and have now fallen silent.

16.3.1.2 Future Prospects for PUFAs from Marine Microorganisms

One area where oils from marine microorganisms triumph over fish oils is in the area of sustainability. Demands on natural sources of fish oils are increasing, and climate change brings risks to even the supply we have. There is definitely an appetite in the market for alternative, secure, and sustainable sources of these types of molecules, and it is up to the industry to provide them at a price point acceptable to the user.

16.3.2 Carotenoid Pigments

The carotenoid pigments provide another good example of how materials derived from marine microorganisms have been able to differentiate themselves in an existing market, and are an illustration of how the biology of the organism and the process technology must come together in an economic manner in order to have a viable product.

16.3.2.1 Artificial Synthesis Versus Natural Sourcing

Several carotenoids, including those with the largest markets, astaxanthin and beta-carotene, can be chemically synthesized. These molecules have isomeric forms and in their use as pigments and generalized antioxidants this is not necessarily a problem since all isomers share the same color and have at least some antioxidant activity. For this reason, the synthetic forms dominate the market with over 90 % of volumes sold (Sandman 2015). The types of isomers found in biological sources is, however, often different from those in the synthetic product, and this may have implications for other functions of the molecule. For instance, beta-carotene in synthetic form is predominantly the *trans* isomer, but natural sources generally contain a significant amount of *cis* isomer. The two isomers may be metabolized to different compounds, although the body also appears be able to transform *trans* to *cis* (Parker 1996). In contrast, astaxanthin in synthetic form is a mixture of three isomers, while microbial sources tend to produce only a single isomer; the two main microbial sources of astaxanthin, *Haematococcus,* and *Xanthophyllomyces*, each produce a different single isomer, but neither form is further metabolized in the body (Liaaen-Jensen 2004).

For both of these pigments, while arguments may be made for one particular isomeric form conferring benefits over another, it is really a consumer preference for natural sources that has been the driver in establishing a market for these pigments from microorganisms. Nonetheless, even given this preference, there is only a limited premium on price available for a natural source, so achieving suitable economics of production has also been critical in establishing and growing the market.

16.3.2.2 Production of Beta-Carotene by *Dunaliella*

Much has been written about the choice of the halotolerant microalga *Dunaliella salina* as a producer of beta-carotene, and the optimization of production of biomass (Borowitzka 1990), but it is worth reiterating the major points that have an impact on its commercialization.

- The microalga can produce extremely high levels of beta-carotene, up to 14 % of biomass in ideal conditions (Borowitzka et al. 1984).
- The microalga is an obligate phototrophic organism meaning that both growth and pigment production are light-dependent.
- Conditions for maximal growth are not the same as for maximal pigment production with stresses such as nitrogen limitation, high salinity and high irradiance having positive effects on pigment production, but negative effects on biomass productivity.
- The organism can survive high salinities where numbers of predators such as protozoa and amoeba are low, and where other *Dunaliella* species, which produce lower amounts of carotenoids, cannot compete.

The need to make high amounts of pigment in the biomass constrains the process to very low cell densities due to the requirements for light penetration and growth stresses, and thus high volumes of culture are required. In this way, the economics of the process are dictated by the availability of high intensity (sun) light, large areas of land, high volumes of brine and saltwater, but most importantly, an efficient harvesting system (both physically and economically).

It is interesting to note that, for the largest and most successful producer of natural beta-carotene, Betatene (now part of BASF), it was the last of these factors that dictated the location of production; an important reminder for those focused on the microbiological side, that biomass production is only the first step of many, and that all steps need to work for a successful commercial enterprise. In Betatene's case, the choice of site in Whyalla, South Australia, was made based on the availability of a cheap source of nitrogen gas, which provided them with the ability to concentrate biomass efficiently and without risk of oxidation of product (Curtain 2000).

16.3.2.3 Future Products

Marine microorganisms are a rich source of many different carotenoid molecules, some available from terrestrial plants and others unique to the marine world. Most have some antioxidant properties, but there is ongoing and growing work into other beneficial effects that these types of molecules may have, for example lutein and zeaxanthin for eye health (Krinsky et al. 2003), fucoxanthin in management of weight and blood glucose (Maeda et al. 2008), or anti-inflammatory effects from alloxanthin and diatoxanthin (Konishi et al. 2008). Nonmicrobial sources of some of these molecules can already be found on the market, but added demand should ensure that there is a place for new products and processes to be developed.

16.4 Conclusion

Bringing new products from marine microorganisms to the market requires success in a series of areas: a market must exist or be created, economic means of production must be developed without infringing others' patents, and then scaled to a volume that serves the market, safety and efficacy must be ensured, regulators and financiers must be satisfied, and the expertise to manage these tasks needs to be grown. While these requirements present challenges, they are all surmountable, and a growing awareness of the richness available to us from the microbial inhabitants of the oceans suggests that there should be many more success stories in the future.

Acknowledgments The research leading to these results has received funding from the European Union Seventh Framework Programme (FP7/2007–2013) under grant agreement No. 311975. This publication reflects the views only of the author, and the European Union cannot be held responsible for any use which may be made of the information contained therein.

References

Abreu AP, Fernandes B, Vicente AA, Teixeira J, Dragone G (2012) Mixotrophic cultivation of *Chlorella vulgaris* using industrial dairy waste as organic carbon source. Bioresour Technol 118:61–66

Barclay W, Apt K, Dong XD (2013) Commercial production of microalgae via fermentation. In: Richmond A, Hu Q (eds) Handbook of microalgal culture: applied phycology and biotechnology, 2nd edn, Wiley, Oxford, pp 134–145. doi:10.1002/9781118567166.ch9

Bhargav S, Panda BP, Ali M, Javed S (2008) Solid-state fermentation: an overview. Chem Biochem Eng Q 22(1):49–70

Borowitzka MA (1990) The mass culture of *Dunaliella salina*. In: Regional workshop on the culture and utilization of seaweeds, Cebu City (Philippines), 27–31 Aug 1990. http://www.fao.org/docrep/field/003/ab728e/ab728e06.htm. Accessed 1 Sept 2015

Borowitzka LJ, Borowitzka MA, Moulton T (1984) Mass culture of *Dunaliella*: from laboratory to pilot plant. Hydrobiologia 116(117):115–121

Calleja P (2012) Method for culturing mixotrophic single-cell algae in the presence of a discontinuous provision of light in the form of flashes. Patent Cooperation Treaty application WO2012/035262, USA

Carlson SE, Werkman SH, Peeples JM et al (1993) Arachidonic acid status correlates with first year growth in preterm infants. Proc Natl Acad Sci 90(3):1073–1077

Curtain C (2000) Plant biotechnology—the growth of Australia's algal β-carotene industry. Australas Biotechnol 10(3):19–23

Deacon G, Kettle C, Hayes D et al (2015) Omega 3 polyunsaturated fatty acids and the treatment of depression. Crit Rev Food Sci Nutr. doi:10.1080/10408398.2013.876959

DIC (2015) Press release: DIC strengthens its position as the global leader for natural blue food coloring. Available at http://earthrise.com/dic-strengthens-its-position-as-the-global-leader-for-natural-blue-food-coloring/. Accessed 1 Sept 2015

EU (2015) Applications under regulation (EC) No 258/97 of the European Parliament and of the Council. http://ec.europa.eu/food/safety/docs/novel-food_applications-status_en.pdf. Accessed 1 Sept 2015

FAO (2014) The state of world fisheries and aquaculture 2014. FAO, Rome

Fitton JH, Irhimeh M, Falk N (2007) Macroalgal fucoidan extracts: a new opportunity for marine cosmetics. Cosmet toilet 122(8):55

FSANZ (2003) DHASCO and ARASCO oils as sources of long-chain polyunsaturated fatty acids in infant formula, a safety assessment. Technical report series no. 22, Food Standards Australia New Zealand

Glycomar (2014) Oligosaccharides in drug discovery. Available at http://www.glycomar.com/documents/Oligosaccharidesindrugdiscovery_2014_000.pdf. Accessed 1 Sept 2015

Heussner AH, Mazija L, Fastner J, Dietrich DR (2012) Toxin content and cytotoxicity of algal dietary supplements. Toxicol Appl Pharmacol 265:263–271

Horrocks LA, Yeo YK (1999) Health benefits of docosahexaenoic acid (DHA). Pharmacol Res 40 (3):211–225

Jenson GS, Ginsberg DI, Drapeau C (2001) Blue-green algae as an immuno-enhancer and biomodulator. J Nutraceuticals Nutr 3:24–30

Jiang Y, Xie P, Chen J, Liang G (2008) Detection of the hepatotoxic microcystins in 36 kinds of cyanobacteria Spirulina food products in China. Food Addit Contam Part A Chem Anal Control Expo Risk Assess 25(7):885–894

Kaplan D, Richmond A, Dubinsky Z, Aaronson S (1986) Algal nutrition. In: Richmond A (ed) Handbook for microalgal mass culture. CRC Press, Boca Raton, Fl, USA, pp 147–198

Kingham R, Beirne LE (2011) Cosmetics regulation in the United States and the European Union: different pathways to the same result. Update Food Drug Law Regul Educa 2011:37–40

Koletzko B, Lien E, Agostoni C et al (2008) The roles of long-chain polyunsaturated fatty acids in pregnancy, lactation and infancy: review of current knowledge and consensus recommendations. J Perinat Med 36(1):5–14

Konishi I, Hosokawa M, Sashima T et al (2008) Suppressive effects of alloxanthin and diatoxanthin from *Halocynthia roretzi* on LPS-induced expression of pro-inflammatory genes in RAW264.7 cells. J Oleo Sci 57(3):181–189

Krinsky NI, Landrum JT, Bone RA (2003) Biologic mechanisms of the protective role of lutein and zeaxanthin in the eye. Annu Rev Nutr 23(1):171–201

Liaaen-Jensen S (2004) Basic carotenoid chemistry. Oxid Stress Dis 13:1–30

Maeda H, Tsukui T, Sashima T et al (2008) Seaweed carotenoid, fucoxanthin, as a multi-functional nutrient. Asia Pac J Clini Nutr 17:196–199

Martek (2004) DHASCO-S product specifications. Martek Biosciences Corporation, Columbia
Martins A, Vieira H, Gaspar H, Santos S (2014) Marketed marine natural products in the pharmaceutical and cosmeceutical industries: tips for success. Mar Drugs 12(2):1066–1101
Mayer AM, Glaser KB, Cuevas C et al (2010) The odyssey of marine pharmaceuticals: a current pipeline perspective. Trends Pharmacol Sci 31(6):255–265
McBride RC, Lopez S, Meenach C et al (2014) Contamination management in low cost open algae ponds for biofuels production. Ind Biotechnol 10(3):221–227
Molinski TF, Dalisay DS, Lievens SL et al (2009) Drug development from marine natural products. Nat Rev Drug Discov 8(1):69–85
Parker RS (1996) Absorption, metabolism, and transport of carotenoids. FASEB J 10(5):542–551
Paul SM, Mytelka DS, Dunwiddie CT et al (2010) How to improve R&D productivity: the pharmaceutical industry's grand challenge. Nat Rev Drug Discov 9(3):203–214. doi:10.1038/nrd3078
Ryan J, Farr H, Visnovsky S, Vyssotski M, Visnovsky G (2010) A rapid method for the isolation of eicosapentaenoic acid-producing marine bacteria. J Microbiol Methods 82(1):49–53
Sandman G (2015) Carotenoids of biotechnological importance. Adv Biochem Eng Biotechnol 148:449–467
Smedley KO (2011) Comparison of approval process and risk-assessment procedures for feed ingredients. Available at http://www.afia.org/rc_files/205/ifif_feedingredientapproval-comparisonreport_2011.pdf. Accessed 1 Sept 2015
Spolaore P, Joannis-Cassan C, Duran E, Isambert A (2006) Commercial applications of microalgae. J Biosci Bioeng 101:87–96
Vigani M, Parisi C, Cerezo ER (eds) (2014) Microalgae-based products for the food and feed sector: an outlook for Europe. Publications Office of the European Union, Luxembourg
Yokoyama M, Origasa H, Matsuzaki M et al (2007) Effects of eicosapentaenoic acid on major coronary events in hypercholesterolaemic patients (JELIS): a randomised open-label, blinded endpoint analysis. Lancet 369(9567):1090–1098
Zittelli GC, Biondi N, Rodolfi L, Tredici MR (2013) Photobioreactors for mass production of microalgae. In: Richmond A, Hu Q (eds) Handbook of microalgal culture: applied phycology and biotechnology, 2nd edn, Wiley, Oxford, pp 225–266. doi:10.1002/9781118567166.ch13

Chapter 17
Marine Genetic Resources and the Access and Benefit-Sharing Legal Framework

Laura E. Lallier, Arianna Broggiato, Dominic Muyldermans and Thomas Vanagt

Abstract The legal landscape regulating the access to and utilization of genetic resources has changed with the entry into force of the Nagoya Protocol in 2014, and the adoption of the related EU Regulation on user compliance in 2014. Moreover, many countries are now adopting laws that regulate access to their genetic resources. This has clear implications for scientists working on genetic resources, including those doing taxonomic and biotechnology research on marine microorganisms. The first part of this chapter informs the scientific community on the Access and Benefit-Sharing (ABS) legal framework, including a focus on their application to marine genetic resources, for which the United Nations Convention on the Law of the Sea (1982) is also relevant. The difference between (domestic) access legislation to genetic resources, and the compliance mechanisms, such as the EU Regulation 511/2014 is explained in detail. A more practical description of ABS related obligations is then presented in a step-by-step approach, which can serve as a basic guideline for scientists.

Acronyms

ABNJ	Areas beyond national jurisdiction
ABS	Access and benefit-sharing
CBD	Convention on biological diversity
CNA	Competent national authority
EEZ	Exclusive economic zone
EU	European Union
GR	Genetic resources

L.E. Lallier · A. Broggiato · T. Vanagt (✉)
eCOAST Marine Research, Esplanadestraat 1, 8400 Ostend, Belgium
e-mail: thomas.vanagt@ecoast.be

L.E. Lallier
Department of European, Public and International Law, Faculty of Law, University of Ghent, Universiteitstraat 4, 9000 Ghent, Belgium

D. Muyldermans · T. Vanagt
ABS-int, Technologiepark 3, 9052 Zwijnaarde, Belgium

L.J. Stal and M.S. Cretoiu (eds.), *The Marine Microbiome*,
DOI 10.1007/978-3-319-33000-6_17

IRCC Internationally recognized certificate of compliance
LOSC Law of the sea convention
MAT Mutually agreed terms
MTA Material transfer agreement
NP Nagoya protocol
PIC Prior informed consent

17.1 Introduction

When dealing with marine microbial research or with marine bioprospecting in general one may encounter a number of success stories, such as pharmaceutical applications of red seaweeds in Carragelose® by Marinomed, or the commercialization in cosmetics of Resilience® by Estée Lauder, based on a Caribbean gorgonian (Martins et al. 2014). However, with the recent development in international law on biodiversity conservation and utilization, the chances of seeing these success stories marred by non-compliance with Access and Benefit-Sharing (ABS) requirements of the country of origin of the genetic resources (GRs) are at stake. To date, no such case exists for marine resources, but recent examples with restrictions on the use of the Castor plant, through a ruling by the National Green Tribunal in India, and controversy about the potential use of Stevia (Meienberg et al. 2015) show that legislation on ABS can have serious implications for Research and Development involving GRs. Indeed, there are many legal aspects to be considered when conducting marine microbial research, depending on which phase of the biodiscovery pipeline one is at. This chapter is positioned at the early stages of the biodiscovery pipeline, when the question of access to the targeted GR is posed, whether the samples to be further studied are collected in situ or are sourced ex situ, i.e., from a biorepository. In this regard, particular attention should be given to the Convention on Biological Diversity (CBD) and its Nagoya Protocol on Access to Genetic Resources and the Fair and Equitable Sharing of Benefits Arising from their Utilization (Nagoya Protocol), as well as the United Nations Convention on the Law of the Sea (LOSC) considering the marine dimension of this book.

17.1.1 Convention on Biological Diversity

The CBD entered into force in 1993, and addresses biodiversity through three different pillars: biodiversity conservation, sustainable use of biodiversity components, and the fair and equitable sharing of the benefits arising out of the utilization of GRs (art. 1 CBD). While the latter objective is often seen simply as a means of monetary return to biodiversity-rich countries, it was originally designed to balance costs and benefits of biodiversity conservation between developed and developing states. In fact, non-monetary benefits represent the majority of the actual returns to the providing country, since only a small proportion of biodiscovery projects actually results in commercially profitable products (Cragg et al. 2012a, b). The CBD explicitly recognizes the sovereign right of States to exploit their own resources in accordance with their own environmental policies. Thus, in the context of marine GRs, it is up to the coastal State from which the GR originates to choose whether or not to regulate access to it and under which terms.

17.1.2 Nagoya Protocol

This additional agreement to the CBD was adopted in 2010 and entered into force in 2014. The intent of the Nagoya Protocol is to ensure that researchers ('users') respect ABS rules of the State providing the GRs. More precisely, the Nagoya Protocol puts obligations both on the user and the provider: while the user of GRs must abide by the law of the providing country, the latter must ensure certainty and clarity of the requirements and measures for granting access to GRs. The main obligations provided by the Nagoya Protocol can be divided in three main pillars:

- *Access obligations*, if enacted, must be clear, transparent, fair and provide rules and procedures for prior informed consent (PIC) and mutually agreed terms (MAT);
- *Benefit-sharing obligations* must be subject to MAT between provider and user, include research and development, subsequent applications, and commercialization, and may be monetary and/or non-monetary;
- *Compliance obligations* must be ensured through the adoption by user countries of legislative and regulatory measures, the respect of contractual obligations contained in MAT, and the monitoring of the utilization of GRs by users, including by the means of checkpoints. This is particularly the object of the EU Regulation 511/2014.

17.1.3 European Union Regulation on Compliance

The adoption of the EU Regulation 511/2014 on compliance measures for users from the Nagoya Protocol on Access to Genetic Resources and the Fair and Equitable Sharing of Benefits Arising from their Utilization in the Union (EU ABS Regulation) is an effort to ensure that all EU Member States achieve the objectives of the Nagoya Protocol with regards to compliance in a uniform manner. Its provisions range from the direct obligations of the user to apply and declare due diligence to the obligations of EU Member States with regards to monitoring and checking user compliance, and include a Register of collections for ex situ access to GRs.

17.1.4 Law of the Sea Convention

Because the sovereignty of a State can extend to the sea, the CBD and the Nagoya Protocol also apply in the maritime jurisdiction of a coastal State that has adopted ABS measures. However, at sea, this ABS framework overlaps with the one of the LOSC addressing marine scientific research in general. The LOSC was adopted in 1982 and entered into force in 1994. It can be broadly described as a "global and general framework setting the boundaries of States' jurisdiction and regulating the activities taking place there, including marine scientific research" (Lallier et al. 2014).

This chapter introduces and further explains the fundamentals of the ABS regime and adequate permits to conduct marine microbial research lawfully. After presenting the basic principles of the entire framework (see: Sect. 17.2), it provides a more practical, phased approach to comply with ABS requirements (see: Sect. 17.3).

17.2 Basic Principles of Access and Benefit-Sharing of Marine Genetic Resources

This section introduces the ABS regime applicable to marine microbial research. This includes the framework derived from the CBD and its Nagoya Protocol, as well as the EU ABS Regulation (see: Sect. 17.2.1). It also takes into consideration the specific provisions of the LOSC regulating marine scientific research (see: Sect. 17.2.2).

17.2.1 Key Elements of Access and Benefit-Sharing

In this section, some explanations are provided on the principal elements found in the CBD and the Nagoya Protocol, as well as the EU ABS Regulation. In particular, it addresses the specific definition of terms (see: Sect. 17.2.1.1), the declination of ABS obligations (see: Sect. 17.2.1.2), and the EU ABS Regulation (see: Sect. 17.2.1.3).

17.2.1.1 Use of Terms and Definitions

As legal terminology is a language of its own, this part provides more clarity on the meaning of terms used in the framework for ABS of marine GRs, which does not necessarily correspond to the scientific world's lexicon. In international law, it is common practice to find a provision relating to the definition of terms used throughout the text of a treaty, for clarity and interpretation purposes. The object

(GRs), their use (research and development on the genetic and/or biochemical composition of genetic resources, including through the application of biotechnology) and the players (provider and user) are the main components of the regime hereby introduced.

Genetic Resources

In 1992, the CBD already defined GRs in rather broad terms. The Nagoya Protocol and the EU ABS regulation kept the CBD definition[1]: "Genetic resources" means "genetic material of actual or potential value". According to the same provision of the CBD, "genetic material" includes "any material of plant, animal, microbial, or other origin containing functional units of heredity".

Utilization

It is with the Nagoya Protocol that the notion of utilization of GRs was introduced in the ABS regime, to complete and broaden the definition of biotechnology provided by the CBD. Article 2 of the Nagoya Protocol reads

> (c) "Utilization of genetic resources" means to conduct *research and development* on the *genetic and/or biochemical composition* of genetic resources, including through the application of *biotechnology* [...];
> (d) "Biotechnology" as defined in Article 2 of the [CBD] means any *technological application* that uses *biological systems, living organisms, or derivatives* thereof, to make or modify *products or processes for specific use*;
> (e) "Derivative" means a naturally occurring *biochemical compound* resulting from the *genetic expression or metabolism* of biological or genetic resources, even if it does not contain functional units of heredity.

Provider and User

In the context of ABS for GRs, the two main players involved are the provider and the user. The provider is to be understood as a *country* that is a party to the Nagoya Protocol, and which is "the country of *origin* of such resources or a Party that *has acquired* the genetic resources in accordance with the CBD".[2] The user of GRs was not precisely defined until the EU ABS Regulation defined it as[3] "a natural or legal person that *utilizes* genetic resources or traditional knowledge associated with genetic resources."

The relationship between a provider and user is very much based on negotiations in good faith to ensure the equitable sharing of benefits arising of the utilization of GRs. Such sharing shall be upon MAT. This clearly means that although a providing country is perfectly entitled to regulate access to its GRs and require benefit-sharing arrangements, it cannot impose such arrangements to the user in an arbitrary manner and the user remains free to refuse an arrangement and source GRs elsewhere. Figure 17.1 shows the process, from targeting the GRs to their possible uses, may it be commercial or not. In between, providers and users' relationship is tagged with PIC and MAT. Those are two steps that need to be taken by scientists

---

[1]Article 2 CBD.

[2]Article 5(1) Nagoya Protocol.

[3]Article 3(4) EU ABS Regulation.

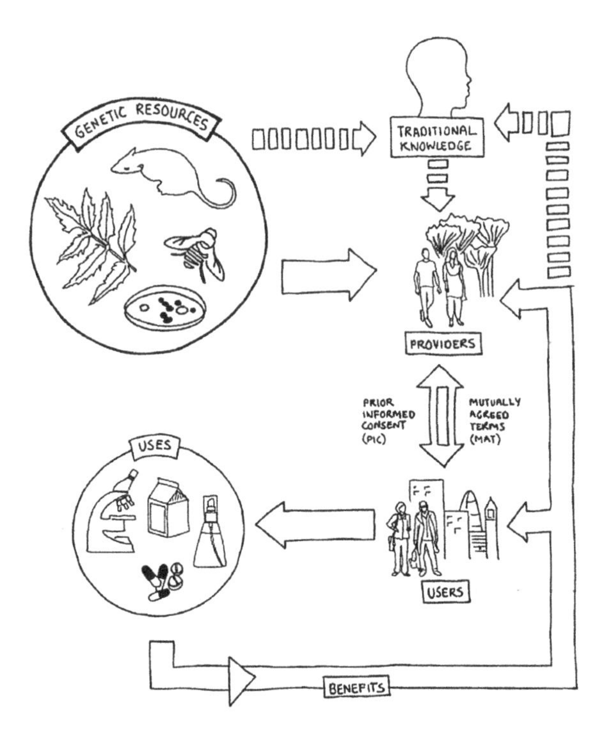

Fig. 17.1 Access and benefit-sharing framework in the Nagoya Protocol. *Source* Introduction to access and benefit-sharing, factsheet produced and published by the CBD secretariat (2011), p. 4. Available at: www.cbd.int/abs/infokit/revised/web/all-files-en.pdf (Accessed 26 October 2015)

when they wish to become "users" of GRs in the meaning of the ABS regime. Those elements are further explained in Sect. 17.2.1.2. Please note that it is a country's sovereign right to *not* regulate access to its GRs, in which case obtaining PIC and MAT are not necessary, and the use of GRs from that country are free and unrestricted.

17.2.1.2 Access and Benefit-Sharing Obligations

The core of ABS obligations lies in two distinct but intertwined elements: the user must seek the PIC of the provider, and reach agreement on MAT with regards to benefit-sharing.

Prior Informed Consent

According to the Nagoya Protocol,[4] "access to genetic resources for their utilization *shall be subject to the prior informed consent* of the Party providing such resources". In fact, PIC may be seen as the formality that crystallizes the agreement to provide access by means of fair and non-arbitrary rules and procedures to consider an application for accessing GRs. PIC should be issued in a written decision by the competent national authority (CNA) of the providing country, which can

[4]Article 6 Nagoya Protocol.

take the form of a permit or its equivalent. If this permit is registered in the CBD database called the ABS Clearing House, it will constitute an internationally recognized certificate of compliance (IRCC).[5]

- *Competent national authorities* (*CNA*): the Nagoya Protocol requires that a State designate a CNA that will be responsible for granting PIC.[6] They also have to designate a national focal point (NFP) to help applicants (future users) finding their ways around the specifics of a provider's ABS legal and administrative organization.
- *Internationally recognized certificate of compliance* (*IRCC*): The issuance of an IRCC is an important tool to facilitate compliance, as it will serve later on, in case of dispute, as evidence that the user accessed the GRs in compliance with the provider's PIC and MAT rules and procedures. In a compliance perspective, it is also a tool that EU member States will use to monitor the utilization of GRs by EU users. To this end, EU users are required to keep strict records of all the relevant documentation, including the IRCC, for 20 years after the end of the period of utilization.[7] Record tracks also include all the information relating to the content of MAT as described below.

Mutually Agreed Terms

The MAT will set up the terms and conditions for utilization of GRs and benefit sharing related thereto, and needs to be agreed upon by the provider and the user. The three elements presented below are the most common in MAT, although this list is non-exhaustive.

- *Benefit-sharing arrangements*: the benefit-sharing conditions represent without doubt the key element of the entire ABS framework. However, benefits are not necessarily monetary. In fact, in many cases the benefits will be non-monetary, e.g.,: sharing of results, participation of nationals in product development, contributions to education and training in the providing country, technology and capacity transfer.[8] In fact, such practice is not new and has already been implemented through partnerships and contracts for years (Cragg et al. 2012a, b).
- *Subsequent third-party transfer*: As there are many more players than just one provider and one user in the biodiscovery pipeline (Laird and Wynberg 2012), the transfer of the GRs and/or the research results associated to a "third-party" who was not involved in the PIC and MAT process is a possibility that should be anticipated in MAT. Some providing countries will require the user to come back to its CNA for consent prior to any transfer. Either way, the EU ABS

[5]Article 14 and 17 Nagoya Protocol.

[6]Article 13 Nagoya Protocol.

[7]Article 4 EU ABS Regulation.

[8]A non-exhaustive list of possible monetary and non-monetary benefits is provided in the annex of the Nagoya Protocol.

regulation imposes rather strict obligations on the user: any material transfer should be accompanied by the IRCC, the MAT, and where not applicable all the relevant information (including source of the GR, date and place of access, permits, associated rights, and obligations).

- *Change of intent*: In many cases, the GR is first accessed for basic research purposes and commercial use will not have been agreed upon yet (nor any (monetary) benefit-sharing arrangements related thereto). Nevertheless, there is a chance that any associated discovery might lead to applied research and eventually the development of a commercial product, and raise commercial interests. It has been recommended, therefore, to leave room in MAT for the negotiation of new terms further down the line, e.g., in a "commercial development agreement" (Cragg et al. 2012a, b).

17.2.1.3 Compliance Under the EU ABS Regulation

The compliance provision of the Nagoya Protocol requires that countries parties take "appropriate, effective and proportionate measures" in order to ensure the compliance of their users with the ABS rules in the providing country they are working with.[9] While *effectiveness* implies deterring measures and sanctions for non-compliance, *appropriate* and *proportionate* can be understood as taking into consideration the different interests and contexts at stake in a given situation, and to avoid too much administrative burden (Greiber et al. 2012). In implementing this provision, the EU therefore puts an obligation of due diligence on the users.

Due Diligence Obligation

The EU ABS Regulation states that "users shall exercise *due diligence* to ascertain that genetic resources [...] which they utilize have been accessed in accordance with applicable ABS legislation or regulatory requirements".[10] This obligation of due diligence means that the onus is on the user, when utilizing GRs, to make sure that the applicable ABS legislation of the providing country is respected and that the GRs are acquired with the appropriate PIC and MAT, if this is required by the providing country. This will have an impact on research funding, for instance, as applications for EU or national research grants will need to be accompanied by a declaration of due diligence as evidence that all the ABS obligations have been fulfilled and the relevant permit obtained.[11] The EU ABS Regulation does not apply to GRs accessed from pre-existing collections

[9]Article 15 Nagoya Protocol.

[10]Article 4 EU ABS Regulation.

[11]Article 7 EU ABS Regulation.

(i.e., material obtained by the collections before the entry into force of the Nagoya Protocol—12 October 2014).

The Register of Collections

A collection, in the meaning of the EU ABS Regulation,[12] is "a set of collected samples of GRs *and related information* that is accumulated and stored, whether held by public and private entities". For a collection to be registered, it must prove that the acquisition, storage, and transfer of the samples are complying with relevant ABS legislation, and that samples are accompanied with all the relevant information and documents (MAT, IRCC, and others). Users who obtain GRs from a collection which is included in *the Register* are considered to have exercised due diligence as regards the seeking of the relevant information, without the need to seek further information than the one provided by the collection. Of course, the user will have to stick to the intended use as specified in the PIC and MAT that comes with the GRs.

Best Practice

The EU ABS Regulation also encourages the development of best practices.[13] Mechanisms, tools or procedures designed by associations of users or other interested parties, which enable users to comply with their obligations, can be recognized as best practice by the EU Commission. While such best practice for marine GRs has yet to be developed and recognized, this chapter proposes a practical approach to the regime that has been described in this section through a step-by-step guidance to user compliance (see: Sect. 17.3).

17.2.2 Specificities to Consider in the Marine Environment

When collecting the GRs that will be the object of the research project on site, there are two distinct regimes to take into consideration. More precisely, when the research project is based on GRs sampled at sea, the regimes set by the LOSC and the CBD overlap, yet are also complementary. Indeed, the LOSC addresses marine scientific research in general, and refers to the act of sampling. The CBD refers to the utilization of the sampled GRs. However, when microbial research is based on material obtained from a pre-existing collection, thus not needing sampling expedition, only the CBD and the instruments thereby derived apply.

17.2.2.1 Marine Scientific Research: Relative Freedom

The LOSC contains many provisions that address marine scientific research. Depending on the specific maritime area where it will occur, the conduct of scientific research can either be a freedom or a right, and can be subject to regulations

[12]Articles 3(9), 4(7) and 5(3) EU ABS Regulation.

[13]Article 8 EU ABS Regulation.

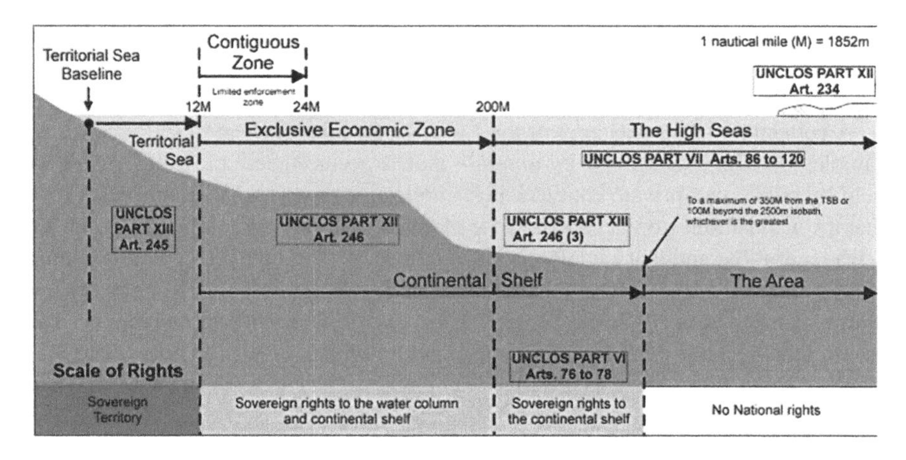

Fig. 17.2 Maritime boundaries in the law of the sea convention. *Source* Alan Evans, National Oceanography Centre, UK. At: http://www.unclosuk.org/group.html (3 Nov. 2015)

or not. The maritime spaces are delimited in the LOSC as is shown on Fig. 17.2. Up to the Exclusive Economic Zone's (EEZ) 200 nautical miles from the coastline for the water column, and up to 350 nautical miles for the continental shelf,[14] the maritime space is under the jurisdiction of the coastal State. Within these limits, all third States—including their researchers—have the right to conduct marine scientific research provided that they respect the requirements of the coastal State.[15] Beyond the 12 nautical miles of the territorial sea, though, the coastal State does not have full sovereignty but rather exclusive rights relating to specific activities only. In particular, coastal States have the right to regulate and authorize marine scientific research as they wish, as long as it respects the basic rights of third States. The LOSC gives further details on the minimum requirements to organize research campaigns in the waters of the coastal States.[16] For instance, notification should occur at least six months prior to a scientific project, and participation of national observers to the campaign is a prerogative of the coastal State. The authorization and conditions are often granted in the form of a research permit, the authority responsible being different in each case depending on the administrative organization of the coastal State. In Chile, for example, the Navy holds that competence (Supreme Decree 1975). However, under normal circumstances, the coastal State

[14]The distances mentioned are a maximum allowed by the LOSC. Depending on each coastal States maritime features and/or specific claims, the extent of the maritime zones under their jurisdiction may vary.

[15]Article 238 LOSC.

[16]Articles 245–257 LOSC.

must grant its consent to scientific activities, since the objective of the LOSC is clearly to promote marine scientific research rather than restraining it (Salpin 2013).

Beyond the limits of the EEZ and/or the continental shelf, marine scientific research is subject to different principles. The international area of the deep seabed is subject to a specific regime set out in Part XI of the LOSC, which grants to the soil, subsoil thereof and the mineral resources the status of common heritage of mankind, and places related activities under the control and administration of the International Seabed Authority. Pursuant to Part XI, marine scientific research in the international seabed is encouraged and promoted, but can only occur under the auspices of the Authority and in accordance with this specific regime. However, in theory, bioprospecting and marine microbial research as such are excluded from this regime and fall under the umbrella of the freedom of the high seas.[17] The freedoms of the high seas[18] are proclaimed by the LOSC in Part VII and include the freedom of scientific research. But while the LOSC was still protecting such freedoms, this last decade has seen policy processes occurring at the United Nations level that tend toward framing and regulating scientific research in the high seas, reducing the extent of that secular tradition (see: Sect. 17.2.2.2).

17.2.2.2 Marine Genetic Resources in Areas Beyond National Jurisdiction

In 2006, the Ad Hoc Open-ended Informal Working Group to study issues relating to the conservation and sustainable use of marine biological diversity beyond areas of national jurisdiction (BBNJ Working Group) met for the first time (Report 2006). Two years before that, the United Nations General Assembly established this consultative body (Resolution 59/24), with the following mandate:

> (a) To survey the past and present activities of the United Nations and other relevant international organizations with regard to the conservation and sustainable use of marine biological diversity beyond areas of national jurisdiction;
> (b) To examine the scientific, technical, economic, legal, environmental, socio-economic and other aspects of these issues;

[17]Indeed, the regime of Part XI was purposely designed for activities related to mineral resources and seabed mining. Thus, bioprospecting is excluded from the regime, and the regulatory mandate of the International Seabed Authority does not include activities other than those relating to minerals. However, the Authority must promote, collect, and disseminate the results of scientific research, including in order to better understand the environment and prepare impact assessments of the activities under its jurisdiction. Adding to that the complexity of organizing research campaigns in such deep and remote locations, chances are that a bioprospecting project collaborate with mining exploration campaigns to sample, and thus be dealing directly or indirectly with the Authority.

[18]Highs seas are determined as the surface and water column beyond national jurisdiction (art. 86 LOSC).

(c) To identify key issues and questions where more detailed background studies would facilitate consideration by States of these issues;
(d) To indicate, where appropriate, possible options and approaches to promote international cooperation and coordination for the conservation and sustainable use of marine biological diversity beyond areas of national jurisdiction;

Over time, the status of marine GRs and bioprospecting in areas beyond national jurisdiction (ABNJ) was vividly discussed in the BBNJ Working Group, not without divergences of opinions as to whether it should remain a freedom of the high seas or should be regulated (Rayfuse and Warner 2008). Let us recall that the Nagoya Protocol was negotiated in a parallel forum at around the same time, setting up a rather strict regime on access to and use of marine biodiversity within national jurisdiction as opposed to the greater freedom to operate in ABNJs.

In 2014, the General Assembly requested the BBNJ working Group to make recommendations on the scope, parameters, and feasibility of an international instrument under the LOSC (Resolution 69/245). In the recommendations thereby issued by the BBNJ Working Group in February 2015 (Letter 2015), the United Nations General Assembly decided to launch the development of an international legally binding instrument under the LOSC on the conservation and sustainable use of marine biological diversity of ABNJ. This new binding instrument shall address, amongst others, the topic of "the conservation and sustainable use of marine biological diversity of ABNJ, in particular, together and as a whole, marine GRs, including questions on the sharing of benefits" (Resolution 69/292).

In other words, the LOSC will be completed by an implementing tool with the binding force of a treaty that will regulate access to and use of GRs in ABNJ, in a similar way to the Nagoya Protocol being a supplementary agreement to the CBD.[19] While marine scientific research on GRs is currently relatively free in ABNJ, it is therefore important to keep in mind that it will be subject to specific ABS requirements in a near future.[20]

17.3 Practical Guide: A Step-by-Step Approach to Compliance

As already seen in Sect. 17.2 of this chapter, marine scientific research activities need to be organized in full respect of international and national obligations related to ABS and the LOSC.

This section of the chapter illustrates the steps to be followed in order to guarantee that legal compliance and certainty is achieved: this is beneficial for the

[19]The scope of the CBD being limited to areas under the jurisdiction of States, neither the CBD or the Nagoya Protocol can address the matter of access to and use of GRs in ABNJ.

[20]A first draft text of the international legally binding instrument is expected by the end of 2017 (A/RES/69/292 par. 1(a)).

research community, the provider countries and also the possible private investors. Scientists will need to take the following steps in order to comply with ABS requirements set out in the CBD, the Nagoya Protocol and the EU ABS Regulation, *and* other requirements according to the LOSC. Please note that additional national or regional laws might exist that require a scientist to get additional permits, for instance to sample in national parks.

First of all, when a scientist is planning to undertake a research project on GRs (notwithstanding if the source of the GR is the country of registration of the research institution where he/she works), the first necessary step is to refer the matter and raise awareness of the legal issue to the legal representative of the research institution and to the legal department or the technology transfer office of the institution (if any exists[21]). The only person who is entitled to sign legal documentation on behalf of a research institution is the legal representative of the institution, or any other member who has been invested of such competence; therefore it is of fundamental importance that scientists do not sign the legal documentation described in this section, unless they are entitled to do so. Receiving professional legal and technical advice and support is the best option in case the research institution does not have in-house the competences to deal with ABS.

17.3.1 Access and Benefit-Sharing Compliance

This section illustrates generally the steps to follow in order to identify what kind of permits are necessary in relation to ABS in the country where a sampling is planned, and how to obtain them. Figure 17.3 sums up the relationship between the provider and the user. Let us recall that the user needs to obtain the PIC of the providing country if required, and both provider and user need to agree on MAT that includes benefit-sharing provisions (see: Sect. 17.2.1).

In case of research undertaken within the EU and implying the utilization of GRs, *in the drafting phase of a project*, and before applying for grants, scientists already need to foresee what kind of ABS issues they might be facing during the project, and how to handle them in compliance with national and international obligations. As a matter of fact, this is required by the EU ABS Regulation through

[21]If no legal department nor technology transfer office exist it is recommended to look for an external legal consultant or consultancy private company with outstanding experiences in providing legal support on ABS. ABS-int (http://www.abs-int.eu/en/home) for example is a multi-disciplinary team that consists of professionals with different backgrounds, including science, law, and regulatory with experience in providing advice to national and international companies and institutions, including in the fields of sustainable resources management, biodiscovery, biosafety regulations, stewardship, environmental law, and intellectual property law. GeoMedia is also a company providing such kind of support (http://www.geo-media.de/consulting.html?&L=1).

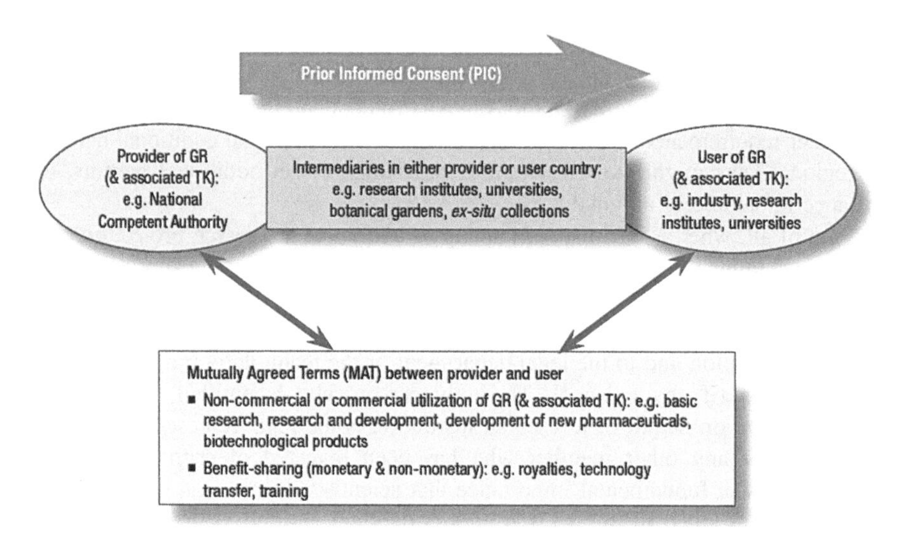

Fig. 17.3 Process and content of mutually agreed terms. *Source* CBD secretariat, frequently asked questions on access and benefit-sharing, at: www.cbd.int/doc/programmes/abs/factsheets/abs-factsheet-faqs-en.pdf

the due diligence system which also included an obligation to submit due diligence declarations at certain checkpoints. The first checkpoint is at the stage of external funding. At this checkpoint, the users have to submit to the CNA of their country a declaration of due diligence (see: Sect. 17.2.1.3). Certain funding programs of the EU, such as for example Horizon 2020, already include an ABS self-evaluation within the ethical self-assessment part of the proposal.

Therefore, scientists need to undertake a preliminary analysis in order to demonstrate that they will be able to exercise due diligence, which means being sure that the GRs they will use are or will be accessed in accordance with applicable ABS rules. Once they made sure that all required legal permits are in place prior to sourcing the desired GRs, they will then need to undertake the steps described further below. Two different scenarios can occur and will require taking different steps, depending on whether the GR are sourced in situ directly from sampling the sea, or are sourced ex situ from existing collections.

The second checkpoint is at the point of commercialization, when a due diligence declaration has to be submitted to the CNA of the country where the utilization was done. If a GR leaves the EU before the R&D process has been completed, even if a final product has not yet been developed, a similar due diligence declaration has to be made.

Note that under the EU ABS regulation, relevant information has to be stored for a period of 20 years after the end of the utilization.

17.3.1.1 Access to In Situ Marine Genetic Resources

As a step 0, scientists need to check in which maritime zone the sampling is planned:

- It can be in waters under national jurisdiction (case A: internal waters—territorial seas—exclusive economic zones);
- In ABNJ (case B: the high seas and the deep seabed beyond the continental shelf);
- Or in the Antarctic Treaty area (case C: the Antarctic).

Case A*: *sampling in internal waters, territorial seas or exclusive economic zones

If sampling is planned to take place in internal waters, territorial seas and/or exclusive economic zones, regardless of whether in a foreign country or the country of the research institution, the following steps need to be undertaken:

Step 1: First of all, users need to verify if the country from where they want to source their GRs is a party to the CBD and the Nagoya Protocol.[22] If this is the case, as soon as possible, notify the primary NFP for CBD, and the NFP and CNA of the Nagoya Protocol of the country where sampling is planned[23] (Provider State) and ask if any additional steps need to be taken.[24] It is often useful, and sometimes even required by ABS national legislation, to undertake such research project with a local partner who might know about the national procedure and might help in negotiating MAT (see intermediaries in Fig. 17.3).

In the particular context of marine GRs, the CNA is sometimes different from the one competent for terrestrial GRs. For instance, South Africa's CNA is the Department of Environmental Affairs, but when it comes to marine matters, the competence falls to the Oceans and Coasts branch as an exception to the rule (South Africa's guidelines 2012).

Step 2: Ask for advice on the specific requirements to be fulfilled prior to sampling activities according to the provider country's legislation on ABS.[25] Basic information on national ABS legislations and measures can be found on the website of the ABS Clearing House or can be obtained from the NFP.[26]

Step 3: If the Provider State has ABS legislation in place requiring PIC and MAT, contact the provider country's competent authority for ABS and start

[22]List of parties can be found here: https://www.cbd.int/information/parties.shtml.

[23]If this country is party to the Convention on Biological Diversity and to the Nagoya Protocol.

[24]Contact details of the CBD NFP of the Provider State can be found at www.cbd.int/information/nfp.shtml and more detailed information on national ABS legislations and procedures can be found in this website the country is part to the CBD only www.cbd.int/abs/measures/ or in the website of the ABS clearing House https://absch.cbd.int/countries if the country is also party to the NP.

[25]Such requirements may include a simple notification of the sampling and/or an ABS agreement.

[26]ABS Clearing House Mechanism: https://absch.cbd.int.

negotiating an ABS agreement.[27] Legal support will be needed for such negotiation.[28]

Step 4: Access and use the material only in accordance with the requirements set out in the PIC and/or MAT. In case the permit is inserted in the ABS Clearing House by the provider country, it will become an IRCC.

Step 5: Transfer the sampled material and/or the associated knowledge (which includes data and scientific results) to a third party only in accordance with the requirements set out in the PIC and/or MAT[29] and upon the signature of a Material Transfer Agreement (MTA). The MTA sets out the conditions on use (if any) of the sampled material and/or associated knowledge by the third party, in accordance with the original PIC and/or MAT agreed upon with the provider country. Together with the MTA, users need to transfer to subsequent users the IRCC and the information on the MAT. In case the IRCC does not exist, the following information has to be passed on:

- date and place of access to GR,
- description of the GR,
- the source,
- the presence or absence of rights and obligations related to ABS,
- the access permit and the MAT.

This information needs to be kept for 20 years after the end of the utilization. Finally, if this information is not complete or if there is uncertainty, the user needs to obtain an access permit or discontinue utilization.

Step 6: If a user wants to use the material and/or associated knowledge for other purposes than the ones agreed upon in the PIC and/or MAT (change of intent), then the user needs to go back to the relevant authorities of the Provider State and renegotiate a PIC and/or MAT, as should be specified in the MAT.

---

[27]One template ABS agreement targeting exactly marine microorganisms has been developed by the EU funded project Micro B3 (https://www.microb3.eu): Model Agreement on Access to Marine Microorganisms and Benefit Sharing (Micro B3 ABS Model Agreements). Its text, included commentaries is available here https://www.microb3.eu/work-packages/wp8.

The Micro B3 ABS Model Agreement can be adapted to different needs: to public domain, hybrid, and full commercial use at the point of access. It has been used as basis for ABS negotiations during the Ocean Sampling Day Campaign (an https://www.microb3.eu/osd) and it is now available as an example of best practices on the website of the CBD Secretariat. It has been subsequently endorsed by the EU funded project PharmaSea (http://www.pharma-sea.eu).

[28]The ABS agreement will not be needed in the drafting phase of a project, but it is advisable to start taking contact with the competent authorities already in the drafting phase.

[29]Users need to be absolutely sure that they have the right to transfer the material and/or associated knowledge before sending the material and data to a third party. If the permit/agreement is silent about this, it does not mean that you have the right to transfer. In the latter case, a clarification with the Provider State might be necessary.

Case B*: *sampling in areas beyond National Jurisdictions

There is neither notification nor permits required so far. As highlighted in Sect. 17.2.1 of this chapter a process is undergoing at the United Nations level to discuss an Implementing Agreement to the LOSC addressing marine biodiversity in ABNJ: this will have implications for accessing and/or utilizing GRs coming from ABNJ in the future. It is advisable to keep a record of the sampling provenance for a period of 20 years, in order to be able to demonstrate the source, in case this is questioned.

Case C*: *sampling in the Antarctic Treaty area

Any activity undertaken in the Antarctic Treaty area (south of 60° latitude) is subject to prior notification. Moreover, the national legislation of the country undertaking the research might require from the researchers to apply for and to obtain a permit.[30]

17.3.1.2 Access to Ex Situ Marine Genetic Resources

The users have to exercise due diligence according to the EU ABS Regulation (see: Sect. 17.2.1.3), which means that they have to seek either the IRCC related to the materials they want to access in the collection (if it exists), or all the relevant ABS information related to the material (see step 5 above). When this information is insufficient or uncertainties exist about the legality of access, the users have to obtain an access permit from the provider country or discontinue utilization. Of course, this is only applying to materials deposited in the collections after the entry into force of the EU ABS Regulation. The EU ABS Regulation provides for the possibility for collections to apply for being registered (either as a whole or in part) in the Register of Collections established by the European Commission. This will have the advantage for the users to assert that, when accessing materials in a registered collection, they have exercised due diligence. However, given the immense efforts collections will have to undertake in order to adjust their management system to the new rules, this is going to be challenging, both for collections and users.

17.3.2 Law of the Sea Compliance

According to the LOSC, and as explained in Sect. 17.2.1.1, a research permit is needed from the coastal state in order to undertake Marine Scientific Research in the territorial seas and exclusive economic zones of that coastal state.

[30]In case of entering and sampling in a "Special Protected Area" a special permit is needed: basic information can be found here http://www.ats.aq/e/ep_protected.htm.

If scientists are planning to sample in

(a) Their own State's national waters (meaning the research institution's country), they need to contact the authority which is competent to release such a marine research permit (it is usually the Ministry of Environment, the Ministry of Research or the Ministry of Transports) and provide full details on the research project.
(b) In a foreign country's waters, they need to contact the Embassy of the coastal State at least six months in advance of the expedition, and provide full details on the research project. The Embassy should assist in obtaining the necessary consent and permits from the competent authority for maritime matters.

17.4 Conclusions

This chapter introduced the legal framework for ABS associated to the utilization of marine GRs. These are requirements to be taken into consideration at the early stages of a research project involving GRs, including marine microbes. However, it does not cover all the legal issues that may arise along the marine biodiscovery pipeline. Compliance with ABS requirements is crucial at the time of accessing, collecting, or acquiring the GRs that will be the object of a given project, and there is a need for continuous follow-up through the monitoring mechanisms put in place by countries, e.g. pursuant to the EU ABS Regulation.

The purpose of this chapter is to inform scientists working with GRs on the new regulatory framework brought by the CBD and its Nagoya Protocol, as well as the EU ABS regulation on compliance, while raising awareness on the potential overlap with permit requirements due to sampling GRs at sea, where the law of the sea guides coastal states' legislation on marine scientific research. Compliance with all ABS requirements ensures that the value of interesting discoveries at the non-commercial research level is safeguarded for further development and commercialization, by utilizing only GRs acquired with the necessary documentation, which in turn provides the legal certainty sought by subsequent parties further down the value chain. It is equally important to scientists to comply with ABS requirements to avoid being accused of committing biopiracy.

ABS is not only important to avoid legal and ethical issues while conducting research. The ultimate goal of this legal framework is to encourage the conservation of biological diversity, through the promotion of research and development activities on GRs and the fair and equitable sharing of the benefits arising out of their utilization. As mentioned in this chapter, a great majority of such benefits are non-monetary and are meant to enhance capacity-building, training, and the sharing of knowledge with the developing world that is, in many cases, where the most biodiversity-rich countries are.

Acknowledgments The research leading to these results has received funding from the European Union Seventh Framework Programme (FP7/2007–2013) under grant agreement no 311975. This publication reflects the views only of the authors, and the European Union cannot be held responsible for any use which may be made of the information contained therein.

References

CBD (1992) Convention on biological diversity, Rio de Janeiro, 5 Jan 1992, 1760 UNTS 79. Available via https://www.cbd.int/doc/legal/cbd-en.pdf. Accessed 14 Oct 2015

Cragg GM, Katz F, Newman DJ, Rosenthal J (2012a) The impact of the United Nations Convention on biological diversity on natural products research. Nat Prod Rep 29:1407–1423

Cragg GM, Katz F, Newman DJ, Rosenthal J (2012b) Legal and ethical issues involving marine biodiscovery and development. In: Fattorusso E, Gerwick WH, Taglialatela-Scafati O (eds) Handbook of marine natural products. Springer, New York, p 1314

EU ABS Regulation (2014) Regulation (EU) No 511/2014 of the European Parliament and of the Council of 16 Apr 2014 on compliance measures for users from the Nagoya Protocol on Access to Genetic Resources and the Fair and Equitable Sharing of Benefits Arising from their Utilization, OJ L 150, 20 May 2014, pp 59–71

Greiber T, Moreno SP, Åhrén M, Carrasco JN, Kamau EC, Medaglia JC, Oliva MJ, Perron-Welch F, Ali N, Williams C (2012) An explanatory guide to the Nagoya Protocol on access and benefit-sharing. IUCN, Gland, Switzerland, pp 161–162. Available at https://cmsdata.iucn.org/downloads/an_explanatory_guide_to_the_nagoya_protocol.pdf. Accessed 16 Oct 2015

Laird S, Wynberg R (2012) Biosciences at a crossroads: implementing the Nagoya Protocol on access and benefit-sharing in a time of scientific, technological and industrial change. Secretariat of the CBD, Policy Brief. Available at https://www.cbd.int/abs/doc/protocol/factsheets/policy/policy-brief-01-en.pdf. Accessed 14 Oct 2015

Lallier LE, McMeel O, Greiber T, Vanagt T, Dobson AD, Jaspars M (2014) Access to and use of marine genetic resources: understanding the legal framework. Nat Prod Rep 31(5):612–616

Letter dated 13 Feb 2015 from the Co-Chairs of the Ad Hoc Open-ended Informal Working Group to the President of the General Assembly, A/69/780 (13 Feb 2015). Available at http://undocs.org/A/69/780. Accessed 14 Oct 2015

LOSC (1982) United Nations Convention on the Law of the Sea, Montego Bay, 10 Dec 1982. Available at: http://www.un.org/Depts/los/convention_agreements/texts/unclos/unclos_e.pdf. Accessed 14 Oct 2015

Martins A, Vieira H, Gaspar H, Santos S (2014) Marketed marine natural products in the pharmaceutical and cosmeceutical industries: tips for success. Mar Drugs 12(2):1066–1101

Meienberg F, Sommer L, Lebrecht T, Lovera M, Gonzalez S, Luig B, von Bremen V, Steiner K, Glauser M, Kienle U (2015) The Bitter Sweet Taste of Stevia—Commercialisation of Stevia-derived sweeteners by violating the rights of indigenous peoples, misleading marketing and controversial SynBio production. Berne Declaration, CEIRAD, Misereor, Pro Stevia Switzerland, SUNU, University of Hohenheim (publishers), pp 32

Nagoya Protocol (2010) Nagoya Protocol on access to genetic resources and the fair and equitable sharing of benefits arising from their utilization to the convention on biological diversity, Nagoya, 29 Oct 2010, UNEP/CBD/COP/DEC/X/1. https://www.cbd.int/abs/doc/protocol/nagoya-protocol-en.pdf. Accessed 14 Oct 2015

Rayfuse R, Warner R (2008) Securing a sustainable future for the oceans beyond national jurisdiction: the legal basis for an integrated cross-sectoral regime for high seas governance for the 21st century. Int J Mar Coast Law 23(3):399–421

Report of the Ad Hoc Open-ended Informal Working Group to study issues relating to the conservation and sustainable use of marine biological diversity beyond areas of national jurisdiction, A/61/65 (9 Mar 2006). Available at http://undocs.org/A/61/65. Accessed 14 Oct 2015

Salpin C (2013) The law of the sea: a before and an after nagoya? In: Morgera E, Buck M, Tsioumami E (eds) The 2010 nagoya protocol on access and benefit-sharing in perspective, implications for international law and implementation challenges. Martinus Nijhoff Publishers, Leiden, p 149

South Africa's Bioprospecting, Access and Benefit-Sharing Regulatory Framework: Guidelines for Providers, Users and Regulators (2012). Department of Environmental Affairs, South Africa, p 32. Available at: https://www.environment.gov.za/sites/default/files/legislations/bioprospecting_regulatory_framework_guideline.pdf. Accessed 16 Oct 2015

Supreme Decree no 711, 22 Aug 1975, Chile. Available at http://www.shoa.cl/tramites/decreto711_ing.pdf. Accessed 14 Oct 2015

UNEP Report of the Meeting of the Group of Legal and Technical Experts on Concepts, Terms, Working Definitions and Sectoral Approaches, UNEP/CBD/ABSWG/7/2 (12 Dec 2008). Available at https://www.cbd.int/doc/meetings/abs/absgtle-01/official/absgtle-01-abswg-07-02-en.pdf. Accessed 14 Oct 2015

United Nations General Assembly resolution 59/24, Oceans and the law of the sea, A/RES/59/24 (17 Nov 2004). Available at http://undocs.org/A/RES/59/24. Accessed 14 Oct 2015

United Nations General Assembly resolution 69/245, Oceans and the law of the sea, A/RES/59/24 (29 Dec 2014). Available at http://undocs.org/A/RES/69/245. Accessed 14 Oct 2015

United Nations General Assembly resolution 69/292, Development of an international legally-binding instrument under the United Nations Convention on the Law of the Sea on the conservation and sustainable use of marine biological diversity of areas beyond national jurisdiction, A/RES/69/292 (19 June 2015). Available at http://undocs.org/A/RES/69/292. Accessed 14 Oct 2015

Chapter 18
Outreach of the Unseen Majority

Marieke Reuver, Georgia Bayliss-Brown, Tanja Calis, Pamela Cardillo, Cliona Ní Cheallachaín and Niamh Dornan

Abstract Traditionally, scientists communicate results from their research projects by writing scientific articles published by scientific journals and nowadays this is still the preferred way for the majority of scientists to communicate their results. However, scientific interaction outside the traditional peer-reviewed journal space is becoming more important to academic communication in recent years, due to a number of important motives. Research projects have the potential to have great environmental, social, and economic benefits; but, in practice, only well communicated research tends to have an effect on policy, industry, or society in general. Within the marine domain, communicating research results is particularly important. It has been identified that the inability to transfer research results into goods and services is affecting knowledge intensive sectors, such as the marine sector. The need for a more focused knowledge transfer (KT) approach came from the understanding that potential benefits can only be realized when research results are actually accepted, adopted, and exploited by its relevant end users. As the concept of KT becomes popularized; its practice more widespread; and its impact noted, the importance of clarifying the definitions of terminology associated with outreach activity becomes increasingly important. Two main current developments in marine outreach are Ocean Literacy (OL) and KT. Increasing OL is a highly effective and promising means by which to change people's behavior towards our seas and ocean in a positive and constructive way. KT is increasingly recognized as a necessity, and a trend is forming where it is becoming a condition of funding for researchers to demonstrate the wider relevance of their research, and communicate beyond traditional academic publications. Well-considered KT is one of the most cost-effective methods for gaining a measurable return on research investment. It will also ensure that innovation and ideas are used for the creation of new products and services; improvements to the environment and society; and, changes in policy. When performing outreach, dissemination, and KT activities, one needs to be conscious of the fact that it is not always feasible, or indeed prudent, to openly and freely share the products of knowledge creation. IP generators need to carefully

M. Reuver (✉) · G. Bayliss-Brown · T. Calis · P. Cardillo · C.N. Cheallachaín · N. Dornan
AquaTT, 8989, Dublin 2, Ireland
e-mail: marieke@aquatt.ie

L.J. Stal and M.S. Cretoiu (eds.), *The Marine Microbiome*,
DOI 10.1007/978-3-319-33000-6_18

consider whether or not they might want to protect their creation before sharing it, as disclosure in any format can adversely affect any subsequent Intellectual Property Rights (IPR) and can render the IP not patentable. Unfortunately, there are still a plethora of barriers, and even disincentives, stalling the transfer of research and science outreach. Barriers to KT are identified along the entire length of the research life cycle and include communication challenges. In Europe, policy demands that publicly funded research contributes more directly to society, the economy and the environment; hence, incentivizing the focus on communication. Outreach and KT have the potential and task to inspire critical thinking, inform public policy, foster a faster knowledge exchange, and prevent duplication of research efforts. Improving outreach and KT is an important challenge for the successful development of marine biotechnology in Europe.

18.1 Introduction

Traditionally, scientists communicate results from their science projects by writing scientific articles published by scientific journals. Science is public, not private knowledge, and so each scientist's goal is to disseminate his or her work and to make a significant contribution to the public domain of science. Centuries ago, scientists shared their observations and results by writing letters, and even anagrams, to their colleagues that were hand copied and passed from person to person (Sarton 1936). Established in 1665, the world's longest-running scientific journal, Philosophical Transactions of the Royal Society, believed that science could only move forward through a transparent and open exchange of ideas backed by experimental evidence and so contributed to the increasing acceptance of publication of papers in modern academic journals (Royal Society 2015).

Nowadays, publication through peer-reviewed scientific journals is still the preferred way for the majority of scientists to communicate their results. In an era of heightened competition for scarce research positions and funding, the mantra of modern academia—"publish or perish"—continues to intensify (Fischer et al. 2012). Scientists are under pressure to produce as many publications as possible in peer-reviewed "high-impact" journals to raise their profile among peers and influence their discipline. However, scientific interaction outside the traditional peer-reviewed journal space is becoming more important to academic communication in recent years, in particular through open access and online publishing leading to an improved communication and availability of research results.

Moreover, communication of scientific results to users beyond the scientific community has increased in significance, focusing on a scientist's reach beyond his or her field and capturing societal impact. Almost twenty years ago, Jane Lubchenco codified the idea of a "new social contract for science" (Lubchenco 1998). She asserted that society expects two outcomes from its investment of public funds in science: "the production of the best possible science and the production of something useful." Lubchenco challenged scientists to not only make their research relevant to today's most pressing problems, but also to embrace their responsibility to share their findings (Smith et al. 2013). It can even be argued that engagement with the public beyond the scientific community is a responsibility and moral imperative for researchers (Marincola 2003).

A major rationale for scientists to engage in science outreach is to build a broad base of public support for science funding. When scientists do not convince the public that their science matters, funding for their particular field of science will likely vanish, as will science as a whole. Other important motives are: the need to engage young people to do science; to inspire the next generation of scientists; to promote scientific literacy; to change behaviors; and contribute to economic growth.

More recently, when launching the EU 2020 Strategy, the European Commissioner for Competition released a policy statement highlighting Europe's determination to link science and research and innovation to market (European Commission 2010); thus, laying the foundation for a knowledge-based economy.

To achieve sustainable economic growth within the marine environment, citizens must be engaged, responsive, and sympathetic to political action. This can be accomplished when citizens understand the influence of seas and oceans on their lives; and, the contrasting impact of their behavior on the marine ecosystem. A prerequisite to achieving this is by using outreach to promote the inherent value of marine resources, as an "ecosystem service".

In Europe, there has been an increased focus on outreach, dissemination, and knowledge transfer (KT) activities at a policy level, evidenced by increased policy documentation referring to the importance of and funding directed towards transferring research results (European Commission 2007a, b, 2009). One form of outreach regularly utilized by the scientific research community is conferences. The EurOCEAN conferences are major marine science-policy conferences, hosted in Europe. They provide a forum for the marine and maritime research community, and wider stakeholders, to interface with European and Member State policymakers and strategic planners. At the EurOCEAN conferences, delegates consider, discuss, and respond to new marine science and technology developments, challenges and opportunities. Acting as a catalyst for the marine and maritime research community to respond to and impact on European science-policy developments, these conferences have provided inputs to a number of strategic science-policy developments in Europe, including in relation to marine outreach. These developments are evidenced in the adoption of the Integrated Maritime Policy for Europe (2007), its environmental pillar the Marine Strategy Framework Directive (2008) and the European Strategy for Marine and Maritime Research (2008).

From these developments, it can be seen that marine outreach has grown to become a more important focus area in recent years

- The Galway Declaration (EurOCEAN 2004) noted that the marine science community must improve its communication skills in explaining the contribution that its work can make to economic and social development.
- The Aberdeen Declaration (EurOCEAN 2007) called for an integrated "European Marine and Maritime Science, Research, Technology and Innovation

Strategy" that must enable cooperation between all stakeholders to enhance KT, of which outreach is an element.

- The Ostend Declaration (EurOCEAN 2010) stated that the European marine and maritime research community recognizes that the seas and ocean are one of the grand challenges for the twenty-first century and so acknowledges the need to translate the related messages to all sectors of society.
- The Rome Declaration (EurOCEAN 2014) shows the increasing significance of outreach as the first of its four goals states, "Valuing the ocean—Promoting a wider awareness and understanding of the importance of the seas and ocean in the everyday lives of European citizens".
- The Galway Statement on Atlantic Ocean Cooperation (2013) specifically identified Ocean Literacy (OL), a marine outreach concept, as one of five key areas for cooperation in policy dialog between top marine scientists from the EU, US and Canada; "we further intend to promote our citizens' understanding of the value of the Atlantic by promotion ocean literacy".

This book chapter on "Outreach of the unseen majority" aims to outline the different developments in outreach, dissemination, and KT in recent years, the importance of outreach, discussing best practice methodologies and challenges and opportunities for the future. The ultimate message is to reinforce public awareness of our shared maritime heritage, the value of investing in marine science research, and the importance of teaching our younger generation to care for our seas.

18.2 Why Is Outreach Important?

> Knowledge creation and its dissemination are two sides of the same coin—knowledge does not impact on society if it is unable to disseminate (Bartling and Friesike 2014).

There is no value in doing research unless others know about it, by communicating it one way or another. Researchers often underestimate the importance of communicating their research and results outside the scientific community. However, unless research is communicated to and understood by the people who need to use it, even the best quality research is read merely by a limited cohort of other scientists. Research projects have the potential to have great environmental, social, and economic benefits; but, in practice, only well communicated research tends to have an effect on policy, industry, or society in general. Moreover, throughout history it can be seen that breakthroughs in knowledge creation have gone hand-in-hand with breakthroughs in its dissemination (Bartling and Friesike 2014).

Within the marine domain, communicating research results is particularly important. With the global population expected to reach nine billion people by 2050, we are increasingly looking at the oceans to provide us with sufficient food and energy to support this growing populace. To ensure that the world's marine space is used safely and sustainably, public knowledge of the ocean and its role in our lives is crucial. Society needs to appreciate the seas and ocean, by becoming 'Ocean Literate', to ensure we minimize anthropogenic impacts on its delicate and complex ecosystem. Many people have little awareness of the need to protect vital ocean resources and the importance of the marine environment, including the impact it has on their daily lives and human well-being; its role in global change; its rich natural and cultural heritage; and its importance to the maritime economy (EurOCEAN 2014). It is essential that the marine community is better able to transfer its knowledge to industry, policy, and society at large as protecting the ocean is a "grand challenge" for humankind. By achieving a transformation in appreciation and understanding of the ocean's role across society as a whole, we can create better conditions for investment and sustainable blue growth (EurOCEAN 2014).

Within the European marine biotechnology sector, the European Science Foundation Marine Board has identified that there is an urgent need to improve information exchange among those who are actively involved in European marine biotechnology (European Science Foundation 2010). The report states that an effective dissemination of novel marine biotechnology research discoveries can greatly improve Europe's capacity to generate new commercial opportunities and would decrease the apparent gap, which currently exists between researchers and high-tech companies. The report recommends that mechanisms need to be developed to mobilize and facilitate the efficient pooling of knowledge, data, and research capacities distributed throughout Europe.

Internationally, the Organisation for Economic Co-operation and Development (OECD) similarly identified that "there may be a need, as evidenced by the emergence of marine biotechnology in national bioeconomy strategies, for a communication strategy around marine biotechnology, perhaps with a focus on environmental issues and sustainability more broadly". They claimed that such a strategy would be able to highlight the importance of marine biotechnology and how its application would affect socio-economically important activities such as food and biofuel production. The OECD also stated that communication among stakeholders, at national and international levels, and consultation with civil society is required to stimulate the development of marine biotechnology and optimize diffusion of its proceeding innovations in the marketplace. They identified that "focused and effective international dialog will be needed to address hurdles such as development of indicators, R&D infrastructure and sustainable development of marine resources" (OECD 2013).

It has been identified that the inability to transfer research results into goods and services is affecting knowledge intensive sectors, such as the marine sector (European Commission 2013a). Europe 2020, the European Union's growth strategy, states that Europe's future jobs and economic growth will increasingly have to come from innovation in business models, products, and services (European Commission 2014c). With this in mind, communication about European research projects should aim to demonstrate the ways in which research and innovation is contributing to a European 'Innovation Union' and account for public spending by providing tangible proof that scientific evidence and collaborative research adds value (European Commission 2011). Activities to disseminate information, exploit research and innovation results, and carry out communication activities are an important and integral part in the European Commission's current Research and Development Programme, Horizon 2020.

Evidently, Europe's funding agencies are progressively embracing a KT model for communication and outreach. The need for a more focused KT approach came from the understanding that potential benefits can only be realized when research results are actually accepted, adopted, and exploited by its relevant end users. Consequently, it is increasingly the case that for a research project to be successful (both initially funded and impactful), an effective communication strategy is needed to transfer findings to the relevant end users. There has never been a greater need for researchers to think about this knowledge flow; processes for achieving successful outreach and transfer; and, to clearly distinguish between dissemination and KT.

18.3 Terminology to Facilitate Understanding

The language we use influences the way we think (Pinker 2007).

The essence of outreach is the application of language to share concepts, views, and knowledge. Ironically, there is discord in the way that terminology relating to outreach is used by communication professionals, marine scientists, and funding agencies, as the following subsection provides evidence of. The term "**outreach**" is used to describe the provision of a service to a population who might not otherwise have access to this service (adapted from Oxford Dictionary 2015). In this chapter, we consider outreach to be any activity that involves "reaching out" beyond the field of research.

The differences between outreach, communication, dissemination, and KT are subtle to the untrained eye. The four terms are often used interchangeably causing a problem for stakeholders, specifically those responsible for administering research funding. If a funding agency specifies a need for, and expects, impact to be driven by KT as the prescribed method, they will not be satisfied when the knowledge is distributed by dissemination activities alone. The project partner, as well as the funding agency, would be frustrated at this outcome, and resources would have been used inefficiently. A clear terminology of these terms is required for mitigatory purposes. The terminology used in this chapter originates from the work undertaken by AquaTT, a renowned European KT organization who have had an outreach and KT role in numerous European research projects.

"**Communication**" is considered to be the overarching term, which covers dissemination, outreach and KT. It is the act of imparting or exchanging information, ideas or feelings by speaking, writing, or using some other medium. "**Dissemination**" is a one-way form of communication, spreading knowledge widely, often to a non-specific audience. Dissemination is commonly used to promote activity and raise awareness of research projects' aims and objectives, using a range of media such as leaflets, websites, and events. "**Knowledge transfer**" describes a two-way process through which a knowledge output moves from a knowledge source to a targeted potential user who then applies that knowledge; where a "**knowledge output**" is a unit of knowledge or learning generated by or through research activity[1] (AquaTT 2012). The reason that KT is described as a two-way process is because its core philosophy is to frame communication activities around target users' needs. Whilst dissemination activities determine an audience and develop materials suitable to that group of people, effective KT requires that a specific target user is profiled and bespoke materials are developed in a medium that is framed for and specific to that user's behaviors, needs, and interests.

Within the academic world, there is a different understanding and interpretation in the meaning of KT. There is undoubted agreement that KT seeks to organize, create, capture, or distribute knowledge and ensure its availability for future users. This mentality has also been referred to as "**knowledge management**" since the 1990s (Nonaka and Takeuchi 1995). In organizational theory, KT is performed in response to the practical problem of losing knowledge, often tacit knowledge, when staff leaves an organization (Argote et al. 2000). In marine science, KT is more selfless in its manner with its primary purpose being to increase the likelihood that research evidence will be applied by policy, industry, science, and society (Mitton et al. 2007). It is performed to spread any knowledge (explicit and tacit) more widely to those that may benefit from or exploit the knowledge, via commercial and non-commercial settings (Thérin 2013); thus, increasing its overall impact. "**Impact**", in this instance, is the demonstrable contribution that excellent research makes to society and the economy and can be described in a number of ways (ESRC 2015; Mitton et al. 2007; RCUK 2011)

- Instrumental impact—leads to action and directly influences and shapes legislating, policy, practice, or service provision
- Conceptual impact—contributes to a change in awareness of understanding of policy and other issues, such as increased OL[2] in citizens, or reframing debates
- Symbolic impact—legitimizes and strengthens existing positions and policies
- Capacity building impact—through technical and skill development.

Within the world of communication professionals, a whole lexicon surrounds KT and is often overlapping in its meaning and commonly disputed, and has not yet been entirely overcome. Whilst KT is defined in this chapter as a two-way process through which a knowledge output moves from a knowledge source to a targeted potential user, Lavis et al. (2003) imply that the direction of KT is not specified, and

[1]Definition developed by AquaTT in the context of Knowledge Management in the MarineTT project.

[2]An Ocean Literate person understands the essential principles and fundamental concepts about the ocean; can communicate about the ocean in a meaningful way; and, is able to make informed and responsible decisions regarding the ocean and its resources (Ocean Literacy Network 2013).

could include just one-way communication as well. They argue that there is increasing evidence to show that successful uptake of knowledge requires a genuine interaction or "**knowledge exchange**" between researchers and stakeholders, and these interactions can be driven from either party. "**Knowledge translation**" appears to be interchangeable with this definition, yet is often used within the health sector as an umbrella term for all of the activities involved in carrying research from the laboratory to practical use.

The term most closely related to KT, and arguably considered a subset of KT, is "**technology transfer**" (Thérin 2013). It is the process of transferring technologies, skills, technical knowledge, facilities, and methodologies among technology developers and end users to ensure that technological developments are accessible to a wide range of users. In some countries, Technology Transfer Offices are a popular vehicle for commercializing university intellectual property (IP) and are the technological version of a "**knowledge broker**" (Siegel et al. 2007). In a research world that favors grant acquisition and academic publication, a role for knowledge brokers has developed. This role provides knowledge synthesis and sits at the interface between the worlds of science, industry, policy, and society, facilitates engagement and action (Lomas 2007). Knowledge brokers are seen as the human force behind KT (Ward et al. 2009).

As the concept of KT becomes popularized, its practice more widespread and its impact noted, the importance of clarifying the definitions of terminology associated with outreach activity becomes increasingly important. The terms provided in this section are merely the tip of the iceberg and further peripheral terms will appear with increased familiarity within this discipline.

18.4 Current Developments in Marine Outreach

As can be seen from the previous sections, there are different terms and concepts used for the area of outreach. Here we focus on two main current developments in marine outreach, OL and KT.

18.4.1 Ocean Literacy

> We know more about the surface of the Moon and about Mars than we do about [the deep sea floor], despite the fact that we have yet to extract a gram of food, a breath of oxygen or a drop of water from those bodies. Paul Snelgrove, Oceanographer

OL is an understanding of the ocean's influence on you—and your influence on the ocean. It is a relatively new concept, which is still evolving. Effective marine outreach efforts ideally result in increased OL. Increasing OL is a highly effective and promising means by which to change people's behavior towards our seas and ocean in a positive and constructive way.

18.4.2 The Origins of the Ocean Literacy Concept

The development of the OL concept began about two decades ago in the United States of America (USA). Members of the ocean sciences and ocean education communities were concerned that the education standards in the USA at the time (1996 National Science Education Standards and most state standards) provided little guidance for teaching about the ocean and atmosphere. Consequently, the teaching of ocean sciences was largely ignored in most kindergarten to 12th grade (K-12) classrooms (Cava et al. 2005; Schoedinger et al. 2010).

Ocean educators and scientists began to realize that without a coherent framework of concepts and messages, these topics would not become part of mainstream teaching and learning about science, which could lead to a lack of awareness of and indifference towards ocean issues. It became clear that the first step in the campaign for OL should be reaching a consensus on what was important for people to learn about the ocean. It was also recognized that agreement was needed on the fundamental issues of the need for OL; the definition of OL; identification of key ocean concepts for inclusion in K-12 curricula; and alignment of ocean concepts to the National Science Education Standards.

In 2004, key American OL stakeholders came together and reached a definition of OL and a draft set of principles (Schoedinger et al. 2010), which were then further developed through a community-wide consensus-building process. The resulting documents, OL: The Essential Principles of Ocean Sciences for Learners of All Ages and OL Scope and Sequence for Grades K-12, outline the knowledge required to be considered Ocean Literate. These two documents were designed to be a practical resource for educators and policymakers in the USA (NOAA 2013). They have also been used as the basis for the development and adoption of the OL concept in Europe.

As a result of the US OL campaign, ocean concepts were incorporated into A Framework for K-12 Science Education (National Academy of Sciences 2012) and the Next Generation Science Standards (Achieve, Inc. 2013) in the USA (NOAA 2013), which can be seen as a major milestone in marine outreach.

18.4.3 Definition of Ocean Literacy (NOAA 2013)

The definition of OL as agreed upon in the USA is that OL is an understanding of the ocean's influence on you—and your influence on the ocean.

An Ocean Literate person:

- Understands the essential principles and fundamental concepts about the ocean
- Can communicate about the ocean in a meaningful way
- Is able to make informed and responsible decisions regarding the ocean and its resources.

18.4.4 Ocean Literacy in Europe

In Europe, OL is an even more recent concept in marine outreach, and it has been building upon its American origins. Many European nations champion global conservation issues, and the key role the EU plays in international ocean affairs, understanding European citizens' awareness, concerns, and priorities is of global importance.

Portugal was one of the first countries in Europe to adapt and adopt the OL concept, with the "Conhecer o Oceano" ("Knowing the Ocean") campaign lead by Ciencia Viva (www.cienciaviva.pt/oceano/home) in 2011 (NOAA 2013).

At a European policy level, the Galway Statement on Atlantic Ocean Cooperation (2013) specifically identified OL as one of five key areas for cooperation in policy dialog between top marine scientists from the EU, US, and Canada (Domegan et al. 2015). This research alliance between the EU, Canada, and the USA says "we further intend to promote our citizens' understanding of the value of the Atlantic by promotion ocean literacy" (Galway Statement 2013).

In 2014, the EU again reinforced the need for OL in Europe by defining a specific call topic in their €80 billion Research and Innovation Funding Programme, Horizon 2020, addressing 'Ocean literacy—Engaging with society'. The original topic description stated: "We will not achieve a sustainable exploitation of marine resources and a good environmental status of our seas and ocean unless citizens understand the influence of seas and ocean on their lives and how their behavior can have an impact on marine ecosystems. This is a pre-requisite to develop the ecosystem based approach for marine activities and promote the understanding/protection of marine ecosystem services" (European Union 2013). This resulted in the funding of two projects, allocating a substantial European research budget to increasing OL through marine outreach.

One of the projects funded under the Horizon 2020 OL topic is Sea Change. Sea Change aims to establish a fundamental "Sea Change" in the way European citizens view their relationship with the sea, by empowering them, as Ocean Literate citizens, to take direct and sustainable action towards a healthy ocean and seas, healthy communities and ultimately a healthy planet.

While in the US OL framework focuses on formal K-12 education, this was considered to be too narrow and difficult a focus for Europe, where curricula vary widely between and sometimes even within countries. Instead, Sea Change has three broader focus areas: the general public, formal educators, and policy makers. The Sea Change partnership, along with its expert International Advisory Group, have adapted the definition of an Ocean Literate person developed by in the US slightly for a European perspective, defining an Ocean Literate person as someone who

- Understands the importance of the ocean to humankind
- Can communicate about the ocean in a meaningful way
- Is able to make informed and responsible decisions regarding the ocean and its resources.

18.4.5 Co-creating Ocean Literacy

While education and traditional advertising can be effective in raising awareness, many studies (Hastings and Domegan 2014) indicate that behavior change rarely occurs due to simply informing, but through thought-out initiatives which take place at the community level and focus on removing barriers to an activity while simultaneously enhancing the activity's benefits (Domegan et al. 2015). The Sea Change project implements behavioral and social change methodologies in order to effectively increase OL and spread a 'Sea Change' movement across Europe. A key feature of the Sea Change methodology is the cocreation of OL principles and protocols. These guidelines emphasize who to engage, what to work on together and how change happens. They are to be implemented by expert "change agents" who collaborate to add value, act interdependently, share knowledge, and build trust in innovative, scaled-out ideas and solutions to social challenges (Domegan et al. 2015).

Sea Change developed a methodology for cocreating OL, which was created by combining global best practice and original research on OL, behavioral change, and social innovation (Domegan et al. 2015). The methodology aims to create ownership for a new and innovative way of OL thinking and behaving. It explains the best practices for embedding the five OL Co-Creation Principles and nine OL Co-Creation Protocols within OL activities, which together, are designed to form the foundation of successful OL interventions (Domegan et al. 2015).

The following five OL Co-Creation Principles are the necessary ingredients to bring about an Ocean Literate population (Domegan et al. 2015)

- The Change Principle: This directs you to value cocreation in order to close value-action gaps.
- The Client Principle: This directs you to "really know" your target groups.
- The Competitive Principle: This directs you to pay attention to other choices and alternatives.
- The Collective Principle: This directs you to ensure you do not separate yourself from your environment.
- The Creative Principle: This directs you to seem imaginative and innovative solutions.

The following nine protocols outline a basic guide on how to effectively implement OL activities (Domegan et al. 2015):

- The Values Protocol: Identify and contain harmful "me" values, co-create "Our" values for OL and your OL activity.
- Situation and Scoping Analysis Protocol: Scope out what is going on with your surrounding environment.
- Boundary and Stakeholder Analysis Protocol: Identify the stakeholders who can affect or be affected by your OL activity.
- The Competitive Analysis Protocol: Identify the competition and whether you should compete or collaborate with them.

- The Research Protocol: Research guides your planning process and informs all of the other protocols.
- The Theory Protocol: Learn from other people's work.
- The Segmentation and Targeting Protocol: Identify your population, segment your population into segments, and choose a segment to target.
- The Intervention Mix Protocol (6 P's): Co-design the 6 P's (product, price, place, promotion, partnerships, and policy) in order to design an offering that appeals to your target group.
- The Impact Protocol: Reflection and feedback loops are essential ingredients for impact.

Although this chapter provides an overview of OL at this point in time, as mentioned previously, it is a new concept that is continually evolving. Through a process of collaboration and building upon experiences, OL methodologies will continue to be optimized in the future.

The inherent link between science and society forms the foundation for the need for interdisciplinary involvement in OL. As outlined earlier in this chapter, outreach activities are within the interest of researchers in terms of self-preservation of research. An Ocean Literate society will understand the importance of marine research for protecting the marine environment and supporting sustainable blue growth, and will therefore support sustained investment in marine research and other related activities. Therefore, marine researchers would benefit from putting the development of an Ocean Literate society as the central aim of their outreach efforts.

18.4.6 From Scientific Publications to Knowledge Transfer

Research not communicated is research not done (Glover 2014).

As defined in the terminology section, KT and outreach are both types of communication. In the marine sector, outreach is defined as the provision of information to someone who would otherwise not receive it. KT is performing outreach but contains an additional methodological aspect; tailoring any subsequent outreach to user needs.

In 1945, the United States Office of Scientific Research and Development identified that there must be a continual interaction between a scientist in the laboratory and wider society for the effective development of scientific advances, and speedy conversion from scientific discovery to practical applications. In a seminal report published on their behalf, Bush (1945) states that "advances in science when put to practical use mean more jobs, higher wages, shorter hours, more abundant crops [and] more leisure for recreation…but to achieve these objectives…the flow of new scientific knowledge must be both continuous and substantial". Increasingly, nowadays funding agencies are recognizing this as a necessity and forming a trend where it is becoming a condition of funding for researchers to demonstrate the wider relevance of their research, and communicate beyond traditional academic publications.

The issue of economic return from investment into scientific research has been on the European science-policy agenda for decades; but, today, there is a pronounced need for the evaluation of the social, cultural, and ecological impact of scientific research also (European Commission 2009). This new approach promotes a culture of KT and rewards researchers delivering on all three institutional missions: teaching; research, and transfer knowledge. The Europe 2020 Flagship Initiative, Innovation Union, identifies the "need to get more innovation out of our research" and calls for "cooperation between the worlds of science and the world of business [to] be enhanced, obstacles removed and incentives put in place" (European Commission 2011). The Innovation Union provides a number of short-, medium- and long-term commitments that will deliver long-term economic and social benefits, including a commitment to develop a common approach to the "dissemination, transfer and use of research results". Well-considered KT is one of the most cost-effective methods for gaining a measurable return on research investment. It will also ensure that innovation and ideas are used for the creation of new products and services, improvements to the environment and society, and changes in policy.

In addition to making commitments to promote a common approach in the translation of research to impact, the European Commission (2008) recommends that

- KT between universities and industry is made a permanent political and operational priority for all public research funding bodies within a Member State, at both national and regional level.
- Sufficient resources and incentives are available to public research organizations and their staff to engage in KT activities.

In 2014, an Independent Expert Group on Boosting Open Innovation and KT in the European Union recommended that KT, along with open innovation, should be put in the spotlight (European Commission 2014a), boosting open innovation and KT in the European Union. They stated that KT can be seen as a major tool for open innovation (OI).

Three well-known models of KT are described in the literature (Reardon et al. 2006)

- Producer Push model—where researchers are responsible for transferring and facilitating the uptake of research knowledge, i.e., to push knowledge towards audiences they identify as needing to know.
- User Pull model—where the user is responsible for identifying and using research knowledge, i.e., to pull knowledge from sources they identify as producing research useful to their own decision-making/situation.
- Exchange model—where researchers and end users are jointly responsible for the uptake of research knowledge.

The discipline is moving towards the Exchange model, stimulating careers in knowledge brokerage, but there are inherent difficulties in the transfer of knowledge from research (Lavis et al. 2003; Mitton et al. 2007)—see challenges and opportunities also. Researchers must be encouraged and supported to engage in KT to transmit good ideas, research results and skills between universities and other research organizations, industry, policy makers, and/or the wider community (European Commission 2007a, b).

18.4.7 Knowledge Transfer in the European Marine Domain

KT can consist of a range of activities that aim to capture knowledge, skills, and competence from a source, those who generate them, and transmit them to an end user, those who will apply them. KT can include commercial and non-commercial activities; such as research collaborations, consultancy, licensing, spin-off creation, researcher mobility, and publication (European Commission 2007a, b).

Horizon 2020, the European Commission's largest research and innovation funding program to date (2014–2020) is dedicated to facilitating "breakthroughs, discoveries and world-firsts by taking great ideas from the lab to the market" (European Commission 2015a). Activities to disseminate information and exploit research and innovation results as well as carry out communication activities will be an important and integral part of Horizon 2020. The European Commission will thus implement information and communication actions for Horizon 2020, which will include communication measures concerning supported projects and results (European Commission 2013b). An example of how KT is carried out in a European marine research project can be taken from the European Commission-funded MaCuMBA project (MaCuMBA 2015).

The MaCuMBA project applies a "knowledge management methodology" based on an approach originally developed by the Irish knowledge management organization AquaTT[3] in the EC-funded FP7 MarineTT project. AquaTT was the coordinator of the pioneering MarineTT project which was recognized as an 'exemplar' project in the ex post evaluation of FP7 to the EC. MarineTT was a European flagship project addressing the need for more effective KT in European publicly funded research projects. One of its outcomes was the development of a robust knowledge management and transfer methodology, focusing on knowledge outputs from research projects. This approach, and its subsequent iterations, has been consistently applied to several subsequent FP7 and Horizon 2020 projects, in particular to marine-related research projects, including AQUAEXCEL, AQUAEXCEL 2020, Aquainnova, COEXIST, COLUMBUS, COMMON SENSE, ECsafeSEAFOOD, MG4U, ParaFishControl, and STAGES. Most recently and most notable for its focus on KT is COLUMBUS,[4] a European Blue Growth

[3]www.aquatt.ie.

[4]www.columbusproject.eu.

Knowledge Transfer flagship project which started only very recently (2015). COLUMBUS will ensure that applicable knowledge generated through EC-funded science and technology research, including projects like MaCuMBA, can be transferred effectively to advance the governance of the marine and maritime sectors while improving competitiveness of European companies and unlocking the potential of the oceans to create future jobs and economic "blue" growth in Europe (AquaTT 2015).

The MaCuMBA knowledge management methodology ensures that the transfer of the collected knowledge is strategic, coordinated, and effective. Furthermore, the methodology focuses on knowledge outputs, rather than generalized information of the project, where a "knowledge output" is a unit of knowledge that has been generated in a scientific project.[5] It is not limited to de novo or pioneering discoveries but may also include new methodologies, processes, adaptations, insights, alternative applications of prior know-how, and knowledge (AquaTT 2012).

The methodology consists of three distinct phases

- Identification and collection of knowledge output generated by the project.
- Analysis and validation; and,
- Development of a KT plan incorporating impact measurement metrics.

These phases are progressive with each segment informing the next, building towards the development of a targeted, strategic and timely KT plan. Before devising a KT plan, it is essential to have an in-depth and critical understanding of each knowledge output. A "knowledge output table" has been developed by AquaTT, MaCuMBA's knowledge management partner, to facilitate the knowledge collection process. This table, contributed to by MaCuMBA's research partners on a regular basis, contains fields describing valuable information about each of its knowledge outputs, such as the potential end user and application. In addition to several focused KT activities such as targeted industry workshops, the collected knowledge outputs are made publicly available through the Marine Knowledge Gate.[6] This platform can be used to feed into both User Pull and Exchange models of KT. As MaCuMBA is dedicated to maximizing KT efforts, its focus is on taking knowledge to potential target users via a Producer Push model.

To initiate the Producer Push model, the methodology includes an analysis phase. In this phase, the key actors and sectors that might be involved in applying a knowledge output are profiled. This information is then used to frame and guide more impactful KT activities. The intention of this exercise is to identify the knowledge output's pathway to impact or "knowledge output pathway". A knowledge output pathway is a term coined by the COLUMBUS project and is used to represent the series of steps (and actors) that connect a knowledge output with its end user, who will apply the knowledge output and result in an eventual

[5]Definition developed by AquaTT in the context of Knowledge Management in the MarineTT project.

[6]www.kg.eurocean.org.

impact. Eventual impacts can vary widely depending on the knowledge and end user. Some examples might include but are not exclusive to

- The development of Blue Economy[7]—commercializing a product or service; improving existing business performance; creating new markets for an existing product or service; establishing of a new businesses or a strategic collaboration between businesses to market a new service; or, attracting inward investment, i.e., by finding new ways to exploit environmental resources and services optimally.
- Sustainable Blue Growth—applying knowledge to inform policy and regulation; improving environmental monitoring programs; or, enabling the development of ecosystem services.
- A Blue Society—enhancing public health and well-being; saving public sector money; or, enabling a resilient society (e.g., protecting vulnerable people, places and infrastructure; providing a secure supply of food, energy, water).

As knowledge output pathways can be lengthy, particularly if the intended application is to influence policy, it may not be possible for a research project to ensure transfer to the end user and reach eventual impacts of all its knowledge outputs. Rather, the focus is on identifying a target user within the knowledge output pathway who will facilitate the knowledge output being carried further along the pathway to reach the end user. To ensure that the knowledge output is carried along the pathway, KT activities must be successful. An essential part of the analysis phase, therefore, is to profile the target user and gain an in-depth understanding of their motivations, technical level, role and responsibility as well as their preferred source and method of receiving information.

These first two phases (knowledge output collection and analysis) are time-consuming and require comprehensive investigation and contextualization of the knowledge itself, as well as its potential application. The invested resource in these two phases, however, is essential as this evidence is required for the final phase, namely carrying out KT and impact measurement. The knowledge management methodology developed by AquaTT recognizes that a thorough level of detail is required to develop a tailor-made, bespoke, and informed KT plan.

Scientists have knowledge, but typically limited authority to change behavior. Decision-makers have power, but may lack the in-depth understanding and scientific evidence required to support and to defend their decisions. Our knowledge-based society is progressively becoming more dependent on scientific and technological breakthroughs. Carlos Moedas, the EC Commissioner for Research, Science and Innovation, stated: "global knowledge, to solve global challenges, is a web we weave together" (European Commission 2015b) and this is important for science to enable advances in policy, industry, and society. The ambition, however, to turn Europe into a knowledge intensive and innovative society is still affected by the European Paradox (European Commission 1995).

---

[7]A definition of the Blue Economy is found in European Commission (2014a, b).

This term explains the simultaneous circumstance of Europe enjoying a high level and quality of scientific production (peer-reviewed publications) while maintaining concern that research is not being converted into successful wealth-generating innovations, new businesses, and societal impact. Implementing efficient KT protocols can limit any loss of knowledge and can be used to overcome these challenges and drive Europe into a new era of value creation from publicly funded research.

18.5 Intellectual Property Issues in Marine Outreach

When performing outreach, dissemination, and KT activities, one needs to be conscious of the fact that it is not always feasible, or indeed prudent, to openly and freely share the products of knowledge creation. IP is an almost inevitable by-product of engaging in research activities. IP refers to "creations of the mind: inventions; literary and artistic works; and symbols, names and images used in commerce. IP is divided into two categories: Industrial Property includes patents for inventions, trademarks, industrial designs and geographical indications" (World Intellectual Property Organization 2015).

IP generators need to carefully consider whether or not they might want to protect their creation before sharing it, as disclosure in any format can adversely affect any subsequent Intellectual Property Rights (IPR) and can render the IP not patentable. In fact, in many instances open sharing of IP is expressly prohibited by funding contracts. These contracts can often mandate that certain results should be kept secret until a decision has been reached regarding their protection, ownership, commercially viability, and future exploitation.

IPR allow the creators of the IP to benefit from their own work or investment in its development. IPR protecting the exclusivity of IP can be achieved by a number of means: patents, trademarks, designs, copyrights, or geographical indications. These protection mechanisms prevent unauthorized exploitation of IP; in return, creators can potentially profit from their intellectual investment. It is important to note that these IPR mechanisms can stimulate knowledge sharing. The creation of IP, and its subsequent protective rights (IPR), can create new channels through which knowledge can be transferred from its origin to end users. These end users may then apply or develop the IP with a view to commercializing it, i.e., through the patent system or an IP license agreement. For example, patents often contain detailed unique technical information and the patent system acts as a repository for state-of-the-art information on applied technology (European Patent Office 2007).

In the context of publicly funded research, it is important to ask whether the knowledge produced in universities and other public research institutions can be transferred for use in industry and other entities such that it yields benefit for the societies which have funded it. In an area such as health, an area where marine biotechnology can make important contributions, the development of new medicines is driven by societal needs. A culture of secrecy where "hoarding" results is allowed is antithetic when society has funded the initial discovery. Swift dissemination of data and research outputs, therefore, is now a mandatory clause in the contracts of research projects funded by the European Commission (Article 29: European Commission 2015c). In addition, there is also an obligation to protect generated knowledge, which is seen to have high potential to be commercially exploitable (Article 27: European Commission 2015c). Remaining honorable to these clauses simultaneously can be a delicate balancing act.

Some parties claim that, particularly in the field of biotechnology, drugs and medical science, an over emphasis on patenting university research could restrict access to public knowledge, and at worst jeopardize the culture of open science (Montobbio 2009). It is difficult to ignore, however, the need to justify public spending on research at the individual, institutional and European level; and pursuing patents and commercialization of knowledge can provide this defense. Indeed, the European Commission (2007a) has clearly stated that "compared to North America, the average university in Europe generates far fewer inventions and patents. This is largely due to a less systematic and professional management of knowledge and intellectual property by European universities". Consequently, it is pertinent to determine which knowledge should be made publicly accessible swiftly and which should remain private pending possible commercial interest and application. Taking MaCuMBA, a European research project where its majority of funding comes from public sources, as an example, it has found itself traversing these gray areas as its consortium consists of both research and industry partners. Positively, through working closely together, this collaboration provides an immediate route to commercialization for high value research outputs and facilitates a timely transfer to market. Indeed, the utility of involving industrial partners in publicly funded research projects has long been recognized and the obvious advantages such partners confer—such as their expertise, manufacturing capability, and market access—cannot be disputed.

MaCuMBA, like many of its EU-funded project counterparts, has an Exploitation Committee who oversees the development and transfer of the research outputs. This Exploitation Committee is primarily tasked with investigating the possibility for commercial exploitation of research results, supplying additional expertise and advice regarding possible IPR routes and appropriate exploitation channels. Specifically, they are able to recommend strategic routes for knowledge to pass from research creators to society. This transfer occurs via various intermediate entities and exploitation partners; each of whom are well-positioned to amplify and add value to the research outputs.

In summary, the prevailing view in Europe is that IP is integral to ensuring the stimulation of investment in innovation and supported by a robust and effectively enforced IP infrastructure will avoid commercial-scale IPR infringements that result in economic harm. The European Commission's aim is to ensure that this infrastructure allows European knowledge creators to reap the appropriate returns from welfare-enhancing innovation for EU citizens (European Commission 2015d).

18.6 Challenges, Opportunities, and Recommendations

If you always do what you've always done, you'll always get what've you've always got (Henry Ford 1863–1947).

There are a plethora of barriers stalling the transfer of research and science outreach, even disincentives (Lavis et al. 2003; Mitton et al. 2007; Smith et al. 2013). These barriers to transfer can be observed at both individual and organizational levels (Mitton et al. 2007) and exist for all actors across all stages of the funding lifecycle, including the structures around funding programs.

At an individual level, researchers are often unaware of the value of their knowledge to potential users outside the science community; how to transfer it to those users effectively, as well as how to develop the correct relationships to support effective transfer so that it can be applied. When it comes to science outreach, researchers cite not only a lack of time and funding, but also the lack of knowledge and training as an impediment (Smith et al. 2013). Whilst actively undertaking research, they are often expected to teach, publish highly cited articles regularly, develop funding proposals, and manage projects. As can be seen in the terminology section, another challenge is the different interpretations of terminology related to outreach and, hence, difference expectations of actors involved. Researchers must be encouraged and supported to engage in KT and outreach to transmit their ideas, research results and skills between universities and other research organizations, industry, policy makers, and/or the wider community (European Commission 2007a, b).

At an organizational level, cultural bias against engagement with outreach and KT afflicts some universities, departments, and disciplines. More so, organizations can often fail to reward the efforts made to transfer knowledge and have been known to actively discourage it, as non-academic dissemination is not recognized in traditional evaluation measures, i.e., citations and impact factors (Smith et al. 2013). The identified challenges experienced within organizations, in relation to achieving efficient KT in Europe, include less systematic and professional management of knowledge and intellectual property by European universities compared to American universities; cultural differences between the business and science communities; lack of incentives; legal barriers; and fragmented markets for knowledge and technology (European Commission 2007a). Overall, the authors believe a culture change among the scientific community is needed to move away from traditional metric for measuring the success and impact of science.

In the marine domain, the European Science Foundation (2010) noted that marine research centers in academic institutions are largely focused on fundamental research and that there is insufficient interaction between these scientists and potential end users in the private sector or in governmental decision-making bodies. The obstacles they identified, that were blocking effective transfer of knowledge from science-to-industry, were insufficient support and motivation, legal issues, and cultural disparity (European Science Foundation 2010). The OECD (2013) identified a main challenge being the timing of the engagement between researchers and industry. Engagement with industry is often regarded as incidental to basic research and development (R&D) or at most a post-research, downstream activity. This can leave R&D results stranded and unable to reach the anticipated market for technical or feasibility reasons, or the inexistence of an appropriate ready-formed market (OECD 2013).

The MarineTT project, "European Marine Research KT and Uptake of Results", was formed in 2010 to respond to the need for more efficient KT in Europe, and was specific to the marine field. MarineTT was a Framework Programme 7 Support Action that piloted new methodologies and tools for capturing, analyzing, and transferring knowledge from past and in-progress European Commission (EC) projects. The overall aim was to develop improved systems that can measurably demonstrate value creation from research investments. One of its findings was that there currently exist challenges concerning stakeholder access and uptake of relevant knowledge and innovation. Although the outcomes were based on European funding mechanisms, the results are relevant to any research system and can be used to inform best practice for future KT and outreach activities.

Barriers to KT were identified along the entire length of the research life cycle and included communication challenges. A key conclusion was that innovation would be enhanced if there was an open dialog between researchers and end users throughout the lifetime of a research project; thus, there is a need to identify, engage, and communicate with end users early on in a project's lifetime. This concept has since been confirmed by the OECD (2013) who stated that effective earlier links between academic researchers and industry could mobilize knowledge and foster innovation. Building trust and understanding between researchers and end users is paramount to effective KT. The main challenge identified was researchers' lack of understanding on how to carry out KT (Table 18.1).

Recommendations for resolving the most critical barriers were made and these included making training in KT mandatory for all funded partners and involving professional science communication in every European research project (MarineTT 2012).

Fortunately, political will and corresponding communications on the need to achieve greater impact from publicly funded research is growing. Realizing the importance of KT for innovation, the Commission developed an innovation strategy where one of its ten key areas for action was to improve KT between public research institutions and third parties, including industry and civil society organizations (European Commission 2006). The EC are also recognizing that in order for research institutions to develop relationships with industry (to maximize the impact of their research results), specialist staff is required to manage the knowledge resources (European Commission 2007c).

In Europe, policy demands that publicly funded research contributes more directly to society, the economy and the environment; hence, incentivizing the focus on communication. Furthermore, each EC-funded research project is expected

Table 18.1 Top ranked barriers to knowledge transfer and innovation from research in Europe (MarineTT 2012)

Barriers identified
• Lack of understanding on how to carry out knowledge transfer
• Lack of investment in knowledge transfer and uptake
• Lack of incentives for knowledge generators to transfer knowledge
• Lack of transparency and accessibility to publicly funded research
• Ineffective knowledge transfer strategies resulting in low impact from research
• Publicly funded research agendas do not always address the needs of end users
• The system of working in closed research consortia and not collaborating/sharing externally can limit innovation
• Lack of flexibility in the research implementation phase which restricts consortia from adapting or responding to interim results
• Failure to engage in systematic analysis of research knowledge outputs that are essential to identifying potential end user(s), applications of the knowledge and understanding of realistic timelines for innovation
• The gap between the worlds of science and end user groupings (industry, policy and wider society)
• End users do not always have the capacity or motivation to take up results and use them
• The established scientific research infrastructure and culture is not designed for rapid and responsive innovation

to carry out meaningful and effective outreach and KT activities. Within the framework of the EC funding programs, larger budgets have become available for European projects that specifically addressing KT, including the COLUMBUS project. COLUMBUS represents the most substantial investment by the European Commission into KT activities to date. The project's overarching objective is to ensure that applicable knowledge generated through EC-funded science and technology research can be transferred effectively to advance the governance of the marine and maritime sectors while contributing to Blue Growth—improving competitiveness of European companies and unlocking the potential of the oceans to create future jobs and economic growth in Europe (European Commission 2014b).

To do this, COLUMBUS has set up a "knowledge fellowship", a network of "knowledge transfer fellows" whose role is to build capacity in their organizations and sectors in KT. The COLUMBUS approach aligns with the philosophy of community capacity building in the future blue economy and is in-line with the policy request to create new "Knowledge Innovation Communities" (European Institute of Innovation and Technology 2015).

This knowledge management methodology used by COLUMBUS has recently attracted the attention of the EC.[8] The EC recently announced their plans to establish an information platform on marine research across the whole Horizon 2020 program (European Commission 2014b). The EC foresees that this platform becomes "a gateway into insights emerging from research projects that can accelerate the uptake of new ideas by industry [to] help ensure that public research funding pays off through innovation by business" (European Commission 2014b). COLUMBUS assisted the EC in piloting this initiative by providing them with more than 400 knowledge outputs from over 30 Oceans of Tomorrow projects (European Commission 2014d). This platform is considered an important step forward in reaching real impact that makes visible change in the society through marine outreach (European Commission 2014b).

Another important development is the launch of the Marine Strategy Framework Directive (MSFD) Competence Centre (MCC) at the EC's Joint Research Centre. The MCC is intended to help EU countries achieve 'Good Environmental Status' of their marine waters by 2020. Knowledge brokerage is an integral aspect to the planned work of MCC with the key objective being to "extract MSFD-relevant knowledge from past projects and channel it into the MSFD implementation and adaptation process" (Joint Research Centre 2014).

To conclude, outreach and KT have the potential and task to inspire critical thinking, inform public policy, foster a faster knowledge exchange and prevent duplication of research efforts. Improving outreach and KT is an important challenge for the successful development of marine biotechnology in Europe.

Acknowledgments The research leading to these results has received funding from the European Union Seventh Framework Programme (FP7/2007–2013) under grant agreement No. 311975 (MaCuMBA). This publication has also been informed and enriched by two more research projects funded under the European Union Horizon 2020 Framework Programme under grant agreement No. 652690 (COLUMBUS) and No. 652644 (Sea Change). This publication reflects the views only of the author, and the European Union cannot be held responsible for any use which may be made of the information contained therein.

References

AquaTT (2012) Overview of the MarineTT initiative [online]. 21 February, RAC Coordination Meeting, Brussels. Available from: http://documents.mx/documents/unlocking-marine-knowledge-overview-of-the-marinett-initiative-dr-gill-marmelstein-aquatt-scientific-project-officer-rac-coordination-meeting-brussels.html. Accessed Dec 2015

AquaTT (2015) Knowledge demand prioritised by EC Flagship Blue Growth project COLUMBUS —July 2015 [html]. Available from: http://www.columbusproject.eu/index.php?option=com_content&view=category&layout=blog&id=239&Itemid=1604. Accessed Dec 2015

[8]COLUMBUS builds upon the knowledge management methodology first devised in MarineTT as outlined in the 'current developments in marine outreach' section of this chapter.

Argote L, Ingram P, Levine JM, Moreland RL (2000) Knowledge transfer in organizations: learning from the experience of others. Organ Behav Human Decis Processes 82(1):1–8
Bartling S, Friesike S (2014) Towards another scientific revolution. In: Bartling S, Friesike S (eds) Opening science: the evolving guide on how the internet is changing research, collaboration and scholarly publishing. Springer, Cham, pp 3–15
Bush V (1945) Science—the endless frontier: a report to the President on a programme for post-war scientific research. National Science Foundation [html], Washington, DC. Available from: https://archive.org/details/scienceendlessfr00unit. Accessed December 2015
Cava F, Schoedinger S, Strang C Tuddenham P (2005) Science content and standards for ocean literacy: a report on ocean literacy. Available from: http://www.coexploration.org/oceanliteracy/documents/OLit2004-05_Final_Report.pdf
Domegan C, Devaney M, McHugh P, Hastings G, Piwowarczyk J (2015) Ocean literacy sea change guiding principles manual. EU Sea Change Project
ESRC (2015) What is impact? [html]. Available from: http://www.esrc.ac.uk/research/evaluation-and-impact/what-is-impact/. Accessed Dec 2015
EurOCEAN (2004) The galway declaration. http://www.euroceanconferences.eu/eurocean-2004
EurOCEAN (2007) The Aberdeen declaration. http://www.euroceanconferences.eu/eurocean-2007
EurOCEAN (2010) The Ostend declaration. http://www.euroceanconferences.eu/eurocean-2010
EurOCEAN (2014) The Rome declaration. http://www.euroceanconferences.eu/eurocean-2014
European Commission (1995) Green paper on innovation [pdf]. Available at: http://europa.eu/documents/comm/green_papers/pdf/com95_688_en.pdf. Accessed Dec 2015
European Commission (2006) Communication from the commission to the council, the European Parliament, the European economic and social committee and the committee of the regions—putting knowledge into practice: a broad-based innovation strategy for the EU [COM (2006) 0502] [html]. Available from: http://ec.europa.eu/invest-in-research/pdf/download_en/knowledge_transfe_07.pdf. Accessed Dec 2015
European Commission (2007a) Improving knowledge transfer between research institutions and industry across Europe [pdf]. Available from: http://ec.europa.eu/invest-in-research/pdf/download_en/knowledge_transfe_07.pdf. Accessed Dec 2015
European Commission (2007b) Knowledge transfer between research institutions and industry—frequently asked questions (MEMO/07/127) [pdf]. Available from: http://europa.eu/rapid/press-release_MEMO-07-127_en.htm. Accessed Dec 2015
European Commission (2007c) Communication from the commission to the council, the European Parliament, the European economic and social committee and the committee of the regions—improving knowledge transfer between research institutions and industry across Europe: embracing open innovation—Implementing the Lisbon agenda [COM (2007) 182] [html]. Available from: http://europa.eu/rapid/press-release_MEMO-07-127_en.htm. Accessed Dec 2015
European Commission (2008) Recommendations on the management of intellectual property in knowledge transfer activities and code of practice for universities and other public research organisations [pdf]. Available from: http://ec.europa.eu/invest-in-research/pdf/download_en/ip_recommendation.pdf. Accessed Dec 2015
European Commission (2009) Metrics for knowledge transfer from public research organisations in Europe: report from the European Commission's expert group on knowledge transfer metrics [pdf]. Available from: http://ec.europa.eu/invest-in-research/pdf/download_en/knowledge_transfer_web.pdf. Accessed Dec 2015
European Commission (2010) Competition, state aid and subsidies in the European Union. Speech by European Commissioner for Competition, Joaquin Almunia [html]. http://europa.eu/rapid/press-release_SPEECH-10-29_en.htm?locale=en
European Commission (2011) Europe 2020 flagship initiative innovation union: SEC(2010) 1161 [pdf]. Available from: https://ec.europa.eu/research/innovation-union/pdf/innovation-union-communication-brochure_en.pdf. Accessed Dec2015

European Commission (2013a) Communication from the commission, annual growth survey 2014 COM (2013) 800 [pdf]. Available from: http://www.eesc.europa.eu/resources/docs/annual-grouth-survey-2014_en.doc. Accessed Dec 2015

European Commission (2013b) HORIZON 2020 work programme 2014–2015 17. Communication, dissemination and exploitation. Available at: http://ec.europa.eu/research/participants/data/ref/h2020/wp/2014_2015/main/h2020-wp1415-comm-diss_v1.0_en.pdf. Accessed Dec 2015

European Commission (2014a) Communication from the commission to the European Parliament, the Council, the European economic and social committee and the committee of the regions: innovation in the blue economy: realising the potential of our seas and oceans for jobs and growth [html]. Available from: http://eur-lex.europa.eu/legal-content/EN/TXT/?uri=celex%3A52014DC0254R(01). Accessed Dec 2015

European Commission (2014b) Innovation in the blue economy [html]. Available at: https://ec.europa.eu/dgs/maritimeaffairs_fisheries/magazine/en/policy/innovation-blue-economy. Accessed Dec 2015

European Commission (2014c) The European Union explained: Europe 2020: Europe's growth strategy [pdf]. Available from: http://bookshop.europa.eu/en/europe-2020-pbNA0414862/?CatalogCategoryID=sciep2OwkgkAAAE.xjhtLxJz. Accessed Dec 2015

European Commission (2014d) The oceans of tomorrow projects (2010–2013) [pdf]. Available from: http://ec.europa.eu/research/bioeconomy/pdf/ocean-of-tomorrow-2014_en.pdf. Accessed Dec 2015

European Commission (2015a) HORIZON 2020. The EU framework programme for research and innovation website "What is Horizon 2020" [html]. Available at: https://ec.europa.eu/programmes/horizon2020/en/what-horizon-2020. Accessed Dec 2015

European Commission (2015b) Keynote speech of Carlos Moedas—Commissioner for research, science and innovation, 4th Dec 2015 [online]. Lund Revisited—next steps in tackling societal challenges, Lund, Sweden. Available from: https://ec.europa.eu/commission/2014-2019/moedas/announcements/european-research-and-innovation-global-challenges_en. Accessed Dec 2015

European Commission (2015c) H2020 general model grant agreement—multi version 2.1 [pdf]. Available at: http://ec.europa.eu/research/participants/data/ref/h2020/mga/gga/h2020-mga-gga-multi_en.pdf. Accessed Dec 2015

European Commission (2015d) Enforcement of intellectual property rights [html]. Available at: http://ec.europa.eu/growth/industry/intellectual-property/enforcement/index_en.htm. Accessed Dec 2015

European Institute of Innovation and Technology (2015) Knowledge and innovation communities (KICs) [html]. Available from: http://eit.europa.eu/activities/innovation-communities. Accessed Dec 2015

European Patent Office (2007) Why researchers should care about patents [pdf]. Available at: https://www.uab.cat/Document/165/238/Patents_for_researchers,0.pdf. Accessed Dec 2015

European Science Foundation (2010) Marine biotechnology: a new vision and strategy for Europe. ESF Marine Board Position Paper 15, 2010 [pdf]. Available from: http://www.esf.org/fileadmin/Public_documents/Publications/marine_biotechnology_01.pdf. Accessed Dec 2015

European Union (2013) H2020-BG-2014-2015: ocean literacy—engaging with society. Available from: https://ec.europa.eu/research/participants/portal/desktop/en/opportunities/h2020/topics/531-bg-13-2014.html. Accessed: 15 Dec 2015

Fischer J, Ritchie EG, Hanspach J (2012) Academia's obsession with quantity. Trends Ecol Evol 27:473–474. doi:10.1016/j.tree.2012.05.010

Galway Statement on Atlantic Ocean Cooperation (2013) https://ec.europa.eu/research/iscp/pdf/galway_statement_atlantic_ocean_cooperation.pdf. Accessed Dec 2015

Glover A (2014) Research not communicated is research not done [online] 17 March, Science Centre World Summit, Mechelen. Available from: http://www.scws2014.org/wp-content/uploads/2014/08/Science-and-decision-making.pdf. Accessed Dec 2015

Hastings G, Domegan C (2014) Social marketing from tunes to symphonies, 2nd edn. Routledge, London

Joint Research Centre (2014) Sharing knowledge to protect our marine environment [html]. Available from: https://ec.europa.eu/jrc/en/news/jrc-launches-marine-competence-centre. Accessed Dec 2015

Lavis J, Ross S, McLeod C, Gildiner A (2003) Measuring the impact of health research. J Health Serv Res Policy 8:165–170

Lomas J (2007) The in-between world of knowledge brokering. Br Med J 334(7585):129–132

Lubchenco J (1998) Entering the century of the environment: a new social contract for science. Science 279:491–497. doi:10.1126/science.279.5350.491

MaCuMBA (2015) MaCuMBA project website [html]. Available at http://macumbaproject.eu/. Accessed Dec 2015

Marincola E (2003) Research advocacy: why every scientist should participate. PLoS Biol 1(3): e71. doi:10.1371/journal.pbio.0000071

MarineTT (2012) Deliverable 4.4. Handbook on Marine TT: European marine research knowledge transfer and uptake of results [pdf]. Available from: http://marinett.eu/images/MARINETT/materials/D%204.4%20Handbooks.pdf. Accessed Dec 2015

Mitton C, Adair CE, McKenzie E, Patten SB, Perry BW (2007) Knowledge transfer and exchange: review and synthesis of the literature. Millbank Q 85(4):729–768

Montobbio F (2009) Intellectual property rights and knowledge transfer from public research to industry in the US and Europe: which lessons for innovation systems in developing countries? In: World Intellectual Property Organization (2009) The Economics of Intellectual Property [pdf]. Available at: http://www.wipo.int/export/sites/www/ip-development/en/economics/pdf/wo_1012_e_ch_6.pdf. Accessed Dec 2015

National Oceanic and Atmospheric Administration (NOAA) (2013) Ocean literacy: the essential principles and fundamental concepts of ocean sciences for learners of all ages version 2, a brochure resulting from the 2-week on-line workshop on ocean literacy through science standards. Available from: http://www.coexploration.org/oceanliteracy/documents/OceanLitChart.pdf. Accessed: 15 Dec 2015

Nonaka I, Takeuchi H (1995) The knowledge-creating company. Oxford University Press, New York

Ocean Literacy Network (2013) Ocean literacy: the essential principles of ocean sciences for learners of all ages, Version 2. [pdf]. Available from: http://www.coexploration.org/oceanliteracy/documents/OceanLitChart.pdf. Accessed Dec 2015

OECD (2013) Marine biotechnology: enabling solutions for ocean productivity and sustainability [pdf]. Available from: http://www.marinebiotech.eu/sites/marinebiotech.eu/files/public/library/MBT%20publications/2013%20OECD.pdf. Accessed Dec 2015

Pinker S (2007) The stuff of thought: language as a window into human nature. Viking Penguin, New York

RCUK (2011) RCUK impact requirements: frequently asked questions [pdf]. Available from: http://www.rcuk.ac.uk/RCUK-prod/assets/documents/impacts/RCUKImpactFAQ.pdf. Accessed Dec 2015

Reardon R, Lavis J, Gibson J (2006) From research to practice: a knowledge transfer planning guide [pdf]. Available at: file://C:/Users/georgia/Downloads/kte_planning_guide_2006.pdf. Accessed Dec 2015

Royal Society (2015) Publishing the philosophical transactions: the economic, social and cultural history of a learned journal, 1665–2015 [html]. https://arts.st-andrews.ac.uk/philosophicaltransactions/

Sarton G (1936) Notes on the history of anagrammatism. Isis A J Hist Sci 26:132–138

Schoedinger S, Uyen Tran L, Whitley L (2010) From the principles to the scope and sequence: a brief history of the ocean literacy campaign. *National Marine Educators Association Special Report* #3, p 3. Available from: http://www.coexploration.org/oceanliteracy/NMEA_Report_3/NMEA_2010.pdf. Accessed: 15 Dec 2015

Siegel DS, Veugelera R, Wright M (2007) Technology transfer offices and commercialization of university intellectual property: performance and policy implications. Oxford Rev Econ Policy 23(4):640–660

Smith B, Baron N, English C, Galindo H, Goldman E, McLeod K et al (2013) COMPASS: navigating the rules of scientific engagement. PLoS Biol 11(4):e1001552. doi:10.1371/journal.pbio.1001552

Thérin F (2013) Handbook of research on techno-entrepreneurship, second edition: how technology and entrepreneurship are shaping the development of industries and companies. Edward Elgar Publishing, Cheltenham

Ward V, House A, Hamer S (2009) Knowledge brokering: the missing link in the evidence to action chain? Evid Policy 5(3):267–279

World Intellectual Property Organisation (2015) What is intellectual property? WIPO Publication No. 450(E) [pdf]. Available from: http://www.wipo.int/edocs/pubdocs/en/intproperty/450/wipo_pub_450.pdf. Accessed Dec 2015

Lightning Source UK Ltd.
Milton Keynes UK
UKOW06n1556050716

277738UK00002B/48/P

9 783319 329987